PROTEIN SCIENCE AND ENGINEERING

SMALL STRESS PROTEINS AND HUMAN DISEASES

PROTEIN SCIENCE AND ENGINEERING

Additional books in this series can be found on Nova's website
under the Series tab.

Additional E-books in this series can be found on Nova's website
under the E-book tab.

PROTEIN SCIENCE AND ENGINEERING

SMALL STRESS PROTEINS AND HUMAN DISEASES

STÉPHANIE SIMON

AND

ANDRÉ-PATRICK ARRIGO

EDITORS

Nova Science Publishers, Inc.

New York

For permission to use material from this book please contact us:
Telephone 631-231-7269; Fax 631-231-8175
Web Site: http://www.novapublishers.com

LIBRARY OF CONGRESS CATALOGING-IN-PUBLICATION DATA

Small stress proteins & human diseases /Editor, Stéphanie Simon & André-Patrick Arrigo.
492 p. ; ill. (some col.) cm.
Includes index.
ISBN 978-1-61470-636-6 (softcover)
1. Heat shock proteins --Physiological effect. I. Simon, Stephanie and Arrigo, André-Patrick.
(OCoLC)ocn619195515
572.6

2011286582

Published by Nova Science Publishers, Inc. † New York

CONTENTS

PREFACE

This book is concerned with a group of proteins, denoted small Heat shock (or stress) proteins (sHsps), that play major roles in human diseases. sHsps, which are denoted so because of their small molecular masses (15 to 30 kDa), belong to the family of Heat shock proteins (Hsps) originally characterized in cells exposed to heat shock. Nowdays, Hsps are also denoted stress proteins consequently of the large spectrum of conditions or agents that induce or stimulate their expression in virtually all living organisms, including procaryots. It is now well established that Hsps expression in response to stress results of cellular alterations in protein folding and membrane integrity. In addition, Hsps, which encompass a large number of related proteins, are also constitutively expressed in several types of cells, particularly those in pathological conditions. The family of human small stress proteins contains ten (or eleven, see table below) members sharing a C-terminal structural domain, refered as the α-crystallin domain, first detected in the mammalian lens protein α-crystallin. Up until recently, these proteins were designated as HspXX, where XX corresponded to the first digits of the apparent molecular weights with the exception of αA- and αB-crystallin which were called by their own name. A new nomenclature, presented below, has been proposed to better characterize human sHsps: the HspB1 to HspB10 terminology. HspB1 refers to Hsp27, one of the first sHsps to be characterized. Only HspB1, HspB4, HspB5, HspB8 plus the less conserved Hsp16.2 polypeptide share an ATP-independent chaperone « holdase » activity that favors their interaction, through dynamic and complex modulations of their hetero- and homo-oligomerization/phosphorylation profiles, with misfolded polypeptides. By doing so, sHsps attenuate the irreversible and deleterious aggregation of stress-altered polypeptides. The chaperone activity of sHsps is therefore different from that of the ATP-dependent « foldase » chaperones (Hsp70, Hsp90 and Hsp60) aimed at refolding stress-induced misfolded polypeptides. However, the holdase and foldase chaperones are part of a coordinated protein refolding network that provide cells with a better resistance to numerous different types of aggressions, such as those that can be encountered in several human diseases.

The aim of this book is therefore to bring together chapters written by top scientific leaders to give a comprehensive view of the fast emerging contrasting roles played by sHsps in human diseases as diverses as neurodegenerations, myopathies, cardiomyopathies, cataracts and cancers. For example, sHsps attenuate stress- or pathologically-induced protein aggregation and can therefore be beneficial to counteract pathologies induced by aggregation-prone proteins (i.e Alzheimer, Parkinson, Huntington). On the other hand, the protective activity of sHsps, particularly against apoptosis, can be deleterious for cancer patients as it can protect cells that should normally be eliminated.

After an introductory chapter providing an overview of these proteins, the book is divided in four main sections each of which deals with one important aspect of sHsps. The first section contains chapters that deals with the beneficial protecting effects mediated by sHsps in diseases related to protein conformation, cytoskeleton, redox state and autophagy as well as their role in protecting neuronal and cardiac cells. The second section of this book deals with the pathological state and amyloid structures induced by mutations in sHsps genes. Pathologies such as cataract, myopathies, cardiopathies and neuropathies are described. In contrast to the preceding sections, the third one contains chapters describing the deleterious effect of the protective activity of sHsps against apoptosis; a phenomenon which leads to tumorigenesis and tumor expansion. The newly described but distantly related Hsp16.2 polypeptide expressed in tumor cells fits well in this section. Finally, the use of sHsp as prognosis cancer markers is discussed. The last section of the book describes different therapeutic approaches aimed at either inhibiting sHsps expression or modulating their chaperone activity by specific peptides. The last chapter summarizes these approaches and points to the complex protective role of sHsps: beneficial when it protects cells that should not die (i.e neurons, cardiomyocytes) and deleterious by protecting cells that should die, such as cancer cells. Hence, future therapeutic drugs will have to be carefully designed to overcome these problems.

Overall, it is hoped that this book will provide a comprehensive review of the many features the small stress proteins and their dual and contrasting role play in human pathologies. We are most grateful to all authors who contributed so efficiently to this book and to the Editorial board of Nova Sciences for its support and strong interest.

Stéphanie Simon and André-Patrick Arrigo
October 2009

ABBREVIATIONS

New name	*Other (previous) Name*
HspB1	Hsp27, Hsp28
HspB2	MKPB
HspB3	-
HspB4	αA-crystallin
HspB5	αB-crystallin
HspB6	Hsp20
HspB7	cvHsp
HspB8	H11, Hsp22
HspB9	-
HspB10	ODF1

In: Small Stress Proteins and Human Diseases
Editors: Stéphanie Simon et al.
ISBN: 978-1-61470-636-6

Introduction to Small Heat Shock Proteins

Chantal van de Schootbrugge and Wilbert C. Boelens*
Department of Biomolecular Chemistry 271, Nijmegen Center for Molecular Life Sciences, Radboud University, P.O. Box 9101, 6500 HB Nijmegen, The Netherlands

Abstract

Small heat shock proteins (sHSPs) are present in all life forms on earth, apart from some pathogenic bacteria [1,2]. Unicellular organisms usually contain one or two sHSPs, but higher eukaryotes can contain up to 20 different sHSPs, each with a unique or complementary function. sHSPs form a diverse family of which all members are characterized by the presence of a conserved α-crystallin domain of ~90 residues [3]. The human sHSP family contains ten authentic members, termed HSPB1-10 [4,5] and a related member called HSP16.2 (Table 1).

The human sHSPs vary in monomeric size from 16-28 kDa and are organized in homo- or hetero-oligomeric structures. These can vary dramatically in complex size, containing from 2 to up to about 50 subunits. Under physiological conditions the levels of expression vary between tissues from nil to more than 1% of the total cellular protein content [6]. Most sHSPs are ubiquitously expressed, although some of them are expressed only in highly specific tissues, such as lens or testis.

Originally, the protein family was thought to function only as molecular chaperones. They can inhibit protein aggregation by binding unfolding proteins. Nowadays, it has become clear that sHSPs are implicated in many other essential functions in the cell as well, like signal transduction, protein degradation, apoptosis and stabilizing the cellular structure under stress situations. Therefore, malfunctioning of sHSPs, leading to aggregate formation with target proteins or misplaced anti-apoptotic activity, may cause a broad spectrum of diseases, like neuropathy, myopathy, cataract and cancer (Table 1). In this chapter several general aspects will be discussed, such as the evolutionary relationship and structural and functional characteristics of the different human sHSPs.

* Corresponding Author: Wilbert C. Boelens
Radboud University
Nijmegen Center for Molecular Life Sciences
Department of Biomolecular Chemistry 271,
P.O. Box 9101, 6500 HB Nijmegen,
The Netherlands
Email: W.Boelens@ncmls.ru.nl

1. Evolutionary Aspects

Back in 1894 α-crystallin, now known to be composed of αA- and αB-crystallin, was discovered as a structural protein in the mammalian eye lens [7]. Differentiated lens fiber cells have the unique property, like in erythrocytes, of lacking protein turnover due to the absence of cellular organelles, such as nuclei and mitochondria. Yet, these cells and their proteins must survive for life. To maintain transparency of the lens, aggregation of the ageing lens proteins must be avoided. The role of αA- and αB-crystallin, now also called HSPB4 and HSPB5, was stepwise unravelled in this process. Around 1973, the sequence of bovine αA- and αB-crystallin was determined by direct protein sequencing [8,9]. Later, in 1982, it was realized that HSPB4 and HSPB5 belong to the family of sHSPs based on conspicuous sequence similarities with four *Drosophila melanogaster* sHSPs [10]. Ten years later, by virtue of their ability to prevent the aggregation of unfolding proteins, HSPB4 and HSPB5 were recognized as molecular chaperones [11]. Only, their activity is generally considered as 'chaperone-like', because of their inability to refold proteins.

In the following decades, one by one the other human sHSPs were discovered that share conspicuous sequence identity with HSPB4 and HSPB5. At present, there are ten HSPB family members known, called HSPB1-10. These possess the α-crystallin domain as the defining feature of their evolutionary relationship (Figure 1 position 113 - 201). Human HSP16.2 resembles the sHSPs in monomeric size, heat inducibility and chaperone-like activity [12]. Only, this protein contains a F5/8 type C domain, which is characteristic for the discoidin (DS) domain family [13]. Moreover, it has a marginal sequence similarity with the human sHSPs (Figure 1) and lacks the distinctive fold of the α-crystallin domain [14]. Considering that tertiary structures of diverging proteins are much better conserved in evolution than their primary structures, the inclusion of HSP16.2 in the sHSP family seems not legitimate.

Outside the α-crystallin domain, sHSPs have a low sequence similarity, although a few recurring motifs can be recognized. Noticeable is the "I-P-I/V" motif, which is localized in the C-terminal extension of a number of sHSPs (Figure 1, positions 218 - 220). This motif is thought to play both a structural and functional role in those proteins [15]. In the N-terminal region several sHSPs display a SRLFDQxFG-like motif (Fig 1, positions 51 - 59); this motif has been pinpointed as structurally and functionally important as well [16]. Deleting this part affect the higher order assembly of the subunits and structural stability as well as the chaperone-like activity.

The genes encoding sHSPB1-10 are scattered over the different chromosomes, apart from HSPB2 and HSPB5, which form a head-to-head gene pair on chromosome 11 [17] (Table 1).The chromosomal dispersal together with the sequence divergence, which amounts to 45 to 85%, suggests that the duplications responsible for these multiple genes occurred before the earliest divergence of extant vertebrates. Indeed, in lower vertebrates orthologs of all mammalian sHSPs have been identified, except for HSPB10 [18,19]. The highly deviating and specialized function of HSPB10 as a structural sperm tail protein might be the reason why no orthologs have yet been recognized in amphibians or fish.

Table 1. Characteristics of sHSPs and involvement in disease[1]

Formal name[2]	Alternative name	pI	Chromosomal location[3]	Molecular Mass (kD)	Length (a.a.)	Tissue distribution	Diseases
HSPB1	Hsp27	6.4	7q11.23	22.3	205	Ubiquitous	Neuropathy[4] Ischemia/reperfusion[5,6,7,8] Cancer[9,10]
HSPB2	MKBP	4.8	11q22-q23	20.2	182	Heart and muscle	Myopathy[11, 12, 13] Ischemia/reperfusion[14]
HSPB3	HSPL27	5.9	5q11.2	17.0	150	Heart and muscle	
HSPB4	CRYAA	6.2	21q22.3	19.9	173	Eye lens	Cataract[15]
HSPB5	CRYAB	7.4	11q22.3-q23.1	20.2	175	Ubiquitous	Neuropathy[16] Myopathy[17,18,19] Ischemia/reperfusion[20,21] Cancer[22] Cataract[23,24]
HSPB6	Hsp20	6.4	19q13.12	17.1	160	Ubiquitous	Neuropathy[25] Ischemia/reperfusion[26,27]
HSPB7	cvHSP	6.5	1p36.23-p34.3	18.6	170	Heart and muscle	
HSPB8	H11 and HSP22	4.7	12q24.23	21.6	196	Ubiquitous	Neuropathy[28, 29] Cancer[30,31] Ischemia[32]
HSPB9	CT51	9.0	17q21.2	17.5	159	Testis	Cancer[33]
HSPB10	ODF1	-	8q22.3	28.4	250	Testis	
HSP16.2	HSPCO34, PP25	-	1p32.1-p33	16.3	144	Placenta	Cancer[34]

[1]Based on [4;155] [2]Formal names according to the guidelines of the HUGO Gene Nomenclature Committee ((http://www.genenames.org/); [3]Chromosomal locations are from the Swissprot database; [4] [156], [5] [157], [6] [158], [7] [159], [8] [160], [9] [161], [10] [146], [11] [162], [12] [163], [13] [164], [14] [165], [15] [140], [16] [166], [17] [162], [18] [121], [19] [163], [20] [37], [21] [167], [22] [147], [23] [168], [24] [169], [25] [170], [26] [171], [27] [172], [28] [173], [29] [174], [30] [175], [31] [148], [32] [176], [33] [149], [34] [150].

Based on a sequence alignment of human, mouse and chicken sHSPs and HSP16.2, a maximum likelihood tree has been constructed, which facilitates the comparison of the evolutionary relationships between these proteins (Figure 2). It confirms that HSPB4-6 are each others closest relatives, as are HSPB2 and B3. HSPB9 and B10 are clearly the earliest diverging and most deviating sHSPs. HSP16.2 is placed outside the sHSPs, as should be expected for a non-homologous protein.

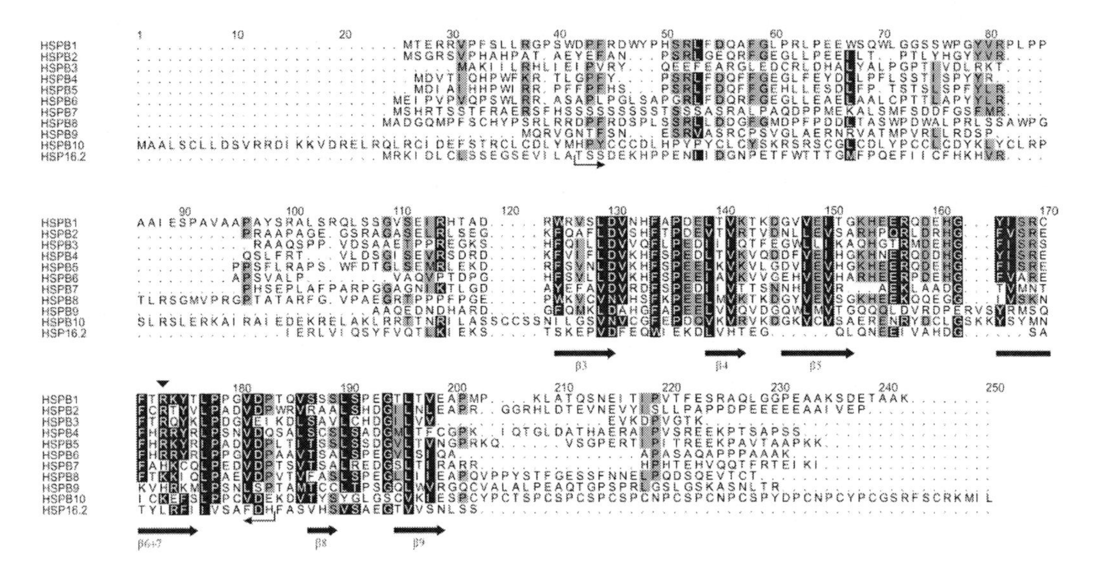

Figure 1. Alignment of the human sHSPs and HSP16.2. ClustalW v1.83 [177] was used to align the sequences at default settings. The C-terminal part (residues 200-250) was manually edited using GeneDoc [178]. Similarity groups used for shading are DN, EQ, ST, KR, FYW and LIVM. Residues in black are conserved in 8-11 sequences and those in grey in 6-7. The β-strands are indicated that were identified by solid state NMR in HSPB5 [24]. The arrowhead indicates the structurally and functionally important Arg in the α-crystallin domain. The bracketed arrows underneath the HSPB16.2 sequence mark the boundaries of the f5/8 domain found in this sequence [179]. This f5/8 domain does not coincide with the α-crystallin domain, which is located between residues 113 and 200.The N-terminal SRLFDQFFG motif is at positions 51 to 59, the C-terminal IXI/V motif at positions 218 to 220.

2. Structure of Small Stress Proteins

One of the most notable features of sHSPs is their organization as large oligomeric structures (Figure 3). Human sHSPs, such as HSPB1, HSPB4, HSPB5 and HSPB6, exhibit a remarkable polydispersity in their oligomeric states. Through interactions with themselves or with other sHSPs, homo- and hetero-oligomers can be formed containing up to 50 subunits [11,20]. The quaternary structures are dynamic, which is reflected by rapid subunit exchange under native and stress conditions [21,22]. The subunit exchange is markedly enhanced at high temperature, which may be a key factor in preventing protein aggregation during heat denaturation.

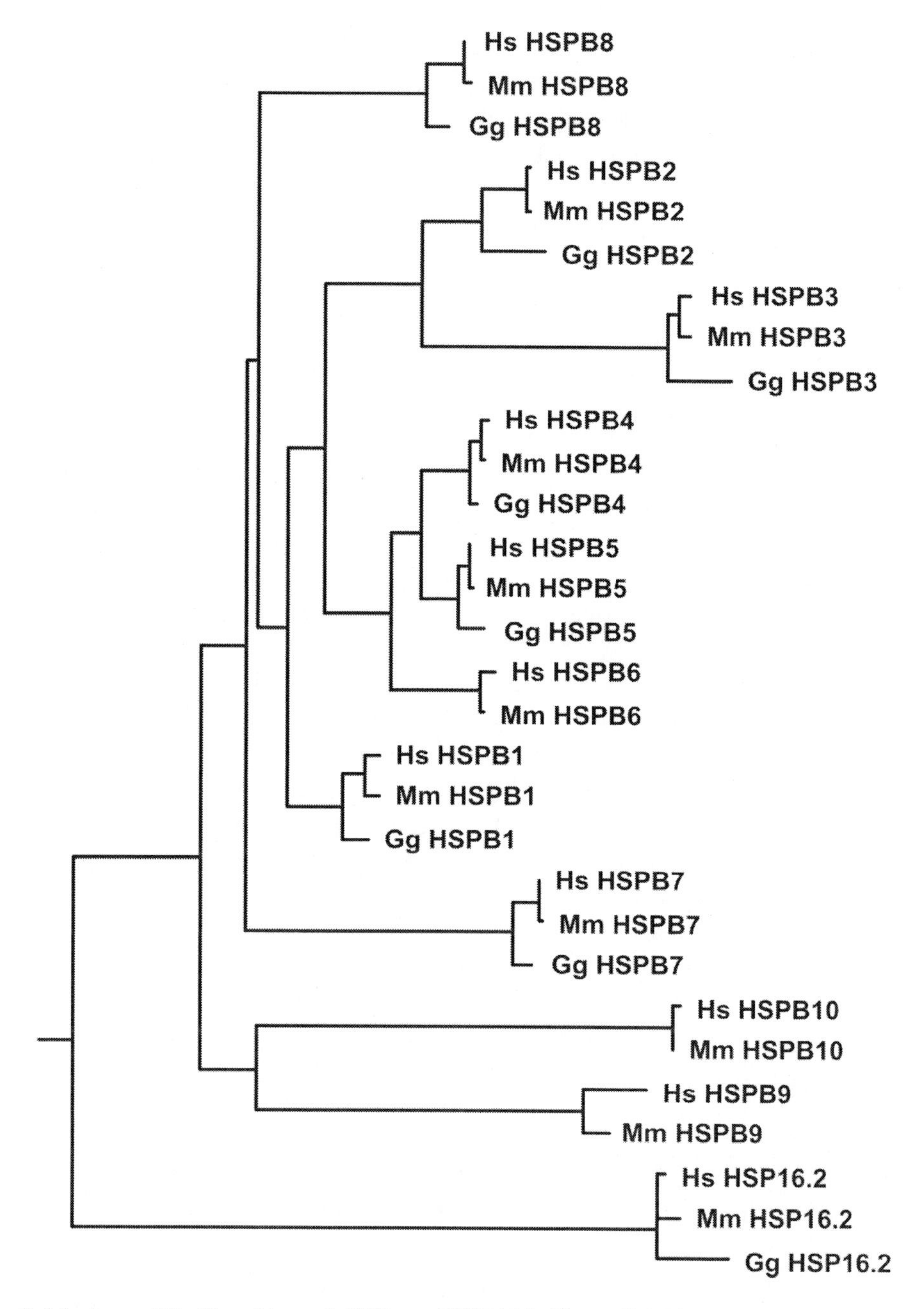

Figure 2. Maximum Likelihood tree of sHSPs and HSP16.2. The online PhyML program (http://atgc.lirmm.fr/phyml/ [180]) was used to calculate the Maximum Likelihood tree of the human (Hs), mouse (Mm) and chicken (Gg) proteins, using 500 bootstrap sets, the WAG model of substitution, estimated proportion of invariable sites, estimated gamma correction parameter and a BIONJ tree as starting point. The alignment used for this calculation was made with ClustalW v1.83 at default settings.

The α-crystallin domain plays a central role in sHSP oligomerisation across the evolutionary spectrum. Structurally important residues in the amino acid sequences of the α-crystallin domain are well conserved, and the predicted secondary structure shows a clear conservation of β-strand structures (Figure 1, [1,19]). Most of these β-strands have been confirmed to be present in HSPB4 by site-directed spin labelling [23] and in HSPB5 by NMR studies [24]. Due to their polydisperse oligomeric nature it has not been possible to obtain crystal structures of the human sHSPs. However, from other organisms three sHSPs with well-defined oligomeric structures have been analyzed by X-ray crystallography: HSP16.5 from *Methanoccocus jannaschii* [25], HSP16.9 from wheat, *Triticum aestivum* [26], and Tsp36 from the tapeworm, *Taenia saginata* [27], the latter containing two α-crystallin domains per subunit.

The four resulting X-ray structures of α-crystallin domains are very similar and form a compact β-sheet sandwich. The β-sandwich is composed of two layers of anti-parallel sheets, constructed from four to five β-strands. In HSP16.5 and HSP16.9 one β-strand (β6) comes from an inter-domain loop of the adjacent monomer and is an important part of the interface between the subunits in a dimer. This strand is not present in Tsp36 [27]. An important outcome of these crystallization studies is the pinpointing of a highly conserved arginine, residue 120 in HSPB5 (arrowhead in Figure 1). Mutations of this residue in human sHSPs are associated with cataract, myopathy and neuropathy (see part II). In the crystal structures this arginine residue forms intermolecular polar interactions that are localized at positions affecting either higher-order assembly or substrate binding. The α-crystallin domain is surrounded by a hydrophobic N-terminal region and a rather short and polar C-terminal extension. The last 10-18 residues of the C-terminal extension are highly flexible as determined by NMR studies [28]. This extension acts as a solubilizer to counteract the hydrophobicity associated with the target protein sequestration. Many of these C-terminal extensions contain a short β-strand formed by the conserved I-P-I/V or I/V-x-I/V motif. This strand is able to make contact with a hydrophobic patch in the α-crystallin domain of a neighbouring subunit, which is critical for the oligomer formation [26] and is important for the functioning of sHSPs [29,30]. Also, the N-terminal region plays an important role in the stabilization of the assembly. This likely occurs via hydrophobic contacts between the N-terminal regions of the subunits that are buried inside the oligomer [26].

Beside the three crystal structures, the quaternary structures of four other sHSPs have been determined by cryo-EM (Figure 3). The oligomeric arrangements of the seven structures show great variability. Tsp36 forms a dimeric structure with an open space in the middle [27]. Wheat Hsp16.9 assembles into a doughnut-shaped structure formed by a dodecameric double disk, of which each disc is organized as a trimer of dimers. Acr1 from *Mycobacterium tuberculosis* is a dodecameric assembly with a hollow, ball-like structure formed from a stable tetrahedral arrangement [31]. Hsp16.5 from *Methanoccocus jannaschii* [25], HSP20.2 from *Archaeoglobus fulgidus* [32] and Hsp26 from *Saccharomyces cerevisiae* [33] form hollow ball-like structures as well, which all contain 24 subunits. Combining all the structural information, the predominant oligomeric structure of sHSPs appears to be a hollow structure from which the C-termini protrude freely from the surface [34,35].

Haley et al. have analyzed the quaternary structure of human HSPB5 using cryo-electron microscopy, which due to the heterogeneity of the complex had a low resolution [36]. They show that this protein also forms a hollow structure. It displays a great variation, both outside and within the protein shell, suggesting that the subunits do not pack in a specific manner.

One way by which HSPB5 can modulate its quaternary structure is by assembling from either monomeric or dimeric building blocks [37]. The assembly from these different substructures affects the exposure of hydrophobic surface and could thus be a way to provide a molecular mechanism for regulating its binding activity. This may enable the assembly to bind up to one unfolded target protein per monomer [38].

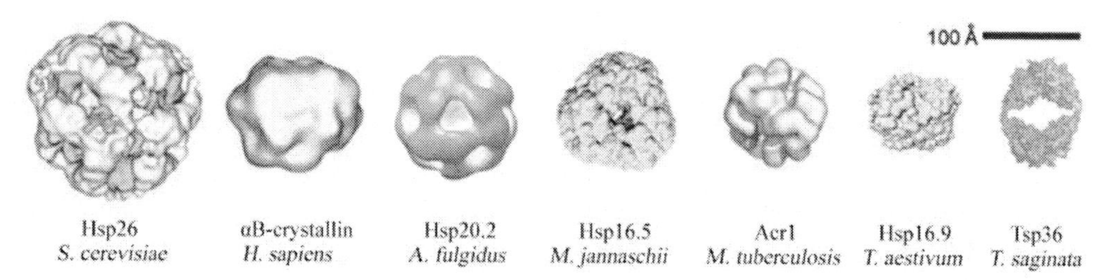

[1] Figure based on [35].

Figure 3. Quaternary structures of sHSPs, determined by either cryo-EM or X-ray crystallography[1]. Hsp26 from *S. cerevisiae* (24 subunits, cryo-EM)[35], αB-crystallin from *H. sapiens* (16 subunits, cryo-EM)[36], Hsp20.2 from *A. fulgidus* (24 subunits, EM)[32], Hsp16.5 from *M. jannaschii* (24 subunits, crystal)[25], Acr1 from *M. tuberculosis* (12 subunits, cryo-EM)[31], Hsp16.9 from *T. aestivum* (12 subunits, crystal)[26] , Tsp36 from *T. saginata* (4 subunits, crystal)[27].

3. Expression of sHSPs

The expression of the sHSPs changes during development, differs in spatial distribution and may vary depending on the local stress conditions. Under non-stress conditions, HSPB1, HSPB5, HSPB6 and HSPB8 are ubiquitously expressed in most tissues, with a high expression in cardiac and skeletal muscle [39-46]. HSPB4 is abundantly expressed together with HSPB5 in the eye lens [47,48], but is almost undetectable in other tissues [40,41,44,49,50]. HSPB2 and HSPB3 expression is restricted to heart and skeletal muscle cells [40,51,52], while HSPB7 is exclusively expressed in cardiac tissues, where it is a very abundant protein [40, 53]. HSPB9 and HSPB10 are both testis specific. Interestingly, the expression patterns of the head-to-head located genes for HSPB2 and HSPB5 differ totally, indicating that within the intergenic sequence beside shared, promoter-preferred cis-control elements are present as well [54,55].

Several, but not all, members of the sHSP family exhibit a pattern of inducible expression in response to heat or other physiological stimuli. Heat shock is a commonly applied form of stress to determine whether the transcription of the target gene is promoted by the heat shock transcription factor 1 (HSF1). So far it has been shown that HSPB1, and to a lesser extent B5 and B8, are heat shock inducible, while no induction has been observed for HSPB2, B3, B4 and B6 [56]. Hypoxia, during which cells or tissues are deprived of oxygen, causes the activation of the hypoxia-inducible transcription factor HIF-1. In newborn piglets, hypoxia effected a tissue dependent induction of sHSPs. HSPB5 was strongly induced in heart, stomach and intestine, while HSPB1 and HSP6 were markedly increased in brain tissues [57,58]. Many other types of stress, such as osmotic or arsenite stress [59], hypoxia/reoxygenation [60], oxidative stress [61], exposure to heavy metals [62], and enforced exercise [63,64] can increase sHSP expression as well .

Pathological conditions are also stressful for cells. Therefore, it is not surprising that enhanced expression of sHSPs is associated with certain diseases, especially neurodegenerative disorders and cancer. In brains of Alzheimer's patients, an increased expression of HSPB1 and B5 has been observed in the activated glial cells around the pathological lesions [65]. In transgenic mouse models of familial amyotrophic lateral sclerosis (fALS) and Parkinson's disease a similar effect on HSPB1 and HSPB5 expression has been detected [66]. These observations indicate that sHSPs may have a specific role in protecting activated glial cells in neurodegenerative diseases. High levels of constitutive HSPB1 and HSPB5 expression have been detected in several cancers, particularly those of carcinoma origin [67]. Both proteins may increase the stress resistance of cancer cells, for instance by influencing the apoptotic process [68]. In this way, the tumorigenic potential of those cells can be affected.

4. Post-Translational Modifications of sHSPs

An important property of some if not most sHSPs concerns their ability to become phosphorylated. This allows to control their functioning by several signal transduction pathways [69-71]. Most sHSPs have phosphorylatable serine residues in their N-terminal region. HSPB1 is phosphorylated at S15, S78 and S82 by MAPKAPK-2 [72]. Phosphorylation of HSPB5 occurs at S19 by an unknown protein kinase, at S45 by p44/p42 MAPK and at S59 by MAPKAPK-2 [73]. HSPB6 is phosphorylated at S16 by cyclic nucleotide-dependent protein kinases [74]. cAMP-dependent protein kinase phosphorylates HSPB8 at S24 and S57 [75], while cyclin-dependent kinase 5, together with its activator p35, phosphorylates HSPB10 at S193 [76]. Most is known about the effect of phosphorylation on HSPB1, HSPB5 and HSPB6. In response to a wide variety of stimuli HSPB1 and HSPB5 show rapid phosphorylation that modulates their activities. Stimuli can be as diverse as those mediated by growth factors, differentiating agents, tumor necrosis factor, oxidative stress or heat shock [67]. Non-phosphorylated HSPB1 and B5 form large oligomeric structures and upon phosphorylation the equilibrium shifts toward smaller oligomeric structures [77,78]. Large non-phosphorylated HSPB1 oligomers have a greater potential to protect the cell through their ability to display chaperone-like activity. In contrast, small oligomers may act at the level of F-actin polymerization/depolymerization [79]. In cells under stress conditions, HSPB1 stabilizes the actin cytoskeleton organization [80].

The triple phosphorylation-mimicking mutant of HSPB5, of which serines 19, 45 and 59 are replaced by aspartic acids, has a higher exchange rate of subunits between oligomers. Moreover, mimicking phosphorylation enhanced chaperone activity in preventing the aggregation of most target proteins studied [81]. The cellular distribution of HSPB5 can change as well. Phosphorylation of serines 45 and 59 leads to translocation into the cell nuclei. HSPB5 is localized here in so-called "speckles", which are nuclear bodies that function as storage compartments for splicing factors [82]. HSPB1 can enter cell nuclei as well, but phosphorylation alone is not sufficient for effective nuclear translocation [83,84]. In response to stress, HSPB5 is also known to interact with the three major cytoskeletal components, i.e. microtubules, intermediate filaments and microfilaments [85]. Phosphorylation is probably important for these interactions, since HSPB5 phosphorylated at S59 colocalizes with the cytoskeleton components, such as vinculin in focal adhesion and desmin in aggregates [86].

Phosphorylation of HSPB6 plays a role in the relaxation of smooth muscle and prevents vasospasm of human artery smooth muscle [74,87]. The effect of the phosphorylation of HSPB6 might be controlled by the phospho-binding protein 14-3-3, which specifically interacts with phosphorylated HSPB6. Beside phosphorylation other modifications of sHSPs occur as well. Deamidation is an aging-related process, resulting in the replacement of an amide group by a carboxylic group. This introduces a negative charge, which has been shown to affect the oligomeric structure and chaperone-like activity of HSPB4 and HSPB5 [88]. Other well known modifications are truncation [89,90], oxidation [91], glycation [92-95] and racemization/isomerization [96], which all can effect the functioning of sHSPs. Most of these modifications have been detected in the sHSPs isolated from eye lenses, which can accumulate more types and higher levels of modification due to their very low turnover.

5. Interactions with Unfolding Proteins

During unfolding of a protein hydrophobic residues from the interior will be exposed and are responsible for aggregation, which may lead to age-related diseases. Progress in understanding the mechanism of chaperone function of sHSPs has been achieved mainly by a biochemical approach using purified sHSPs and model substrate proteins. The activity of a sHSP is measured by its ability to prevent thermally or chemically induced aggregation of a substrate. Aggregation is a late step in the unfolding pathway. By interacting with partially unfolded proteins, sHSPs inhibit the aggregation [26]. Both the N-terminal and C-terminal regions have been postulated to be involved in the chaperone function of sHSPs. This is based on cross-linking studies with a hydrophobic probe or small substrates [35]. Once formed, complexes between substrates and sHSPs are very stable. Electron microscope images have revealed large complexes, of which the morphology depends on the identities of the substrate proteins and the respective sHSPs [97]. In addition to these large complexes, still larger assemblies of substrate proteins and sHSPs have been observed when an excess of non-native proteins is present. During heat shock, HSPB1 is involved in the trapping of eIF4G in insoluble heat shock granules, which affect the mRNA translation [98]. HSPB1 can also translocate into the nucleus upon heat shock, where it forms granules that co-localize with interchromatin granule clusters [83]. Those granular accumulations may reflect a situation of failed refolding, where the substrate is stored for subsequent degradation. Under pathological conditions large accumulations of sHSPs with protein aggregates have been observed. Well known examples are β-amyloid in Alzheimer's disease [65], Rosenthal fibers in Alexander disease [99], Lewy bodies in Parkinson's disease [100] and inclusion bodies in myopathies [101]. In these situations, the incorporation of sHSPs into protein aggregates is likely beneficial for reactivation or degradation of substrates.

Despite a critical role in conferring stress tolerance to cells [35], the involvement of sHSP in cellular protein refolding is less clear. The binding capacity can reach one substrate protein per sHSP subunit, indicating that sHSPs can be a major contributor to the chaperoning capacity of a cell. sHSP chaperone activity does not require the input of ATP energy, and release of the bound substrate with subsequent refolding can be accomplished in vitro through interaction with HSP70 [102]. Studies in E. coli and S. cerevisiae revealed the involvement of a cooperative cellular chaperone network connected to the sHSP-substrate complexes that involves HSP70-HSP40 and members of the HSP100 family [103-106] (Figure 4).

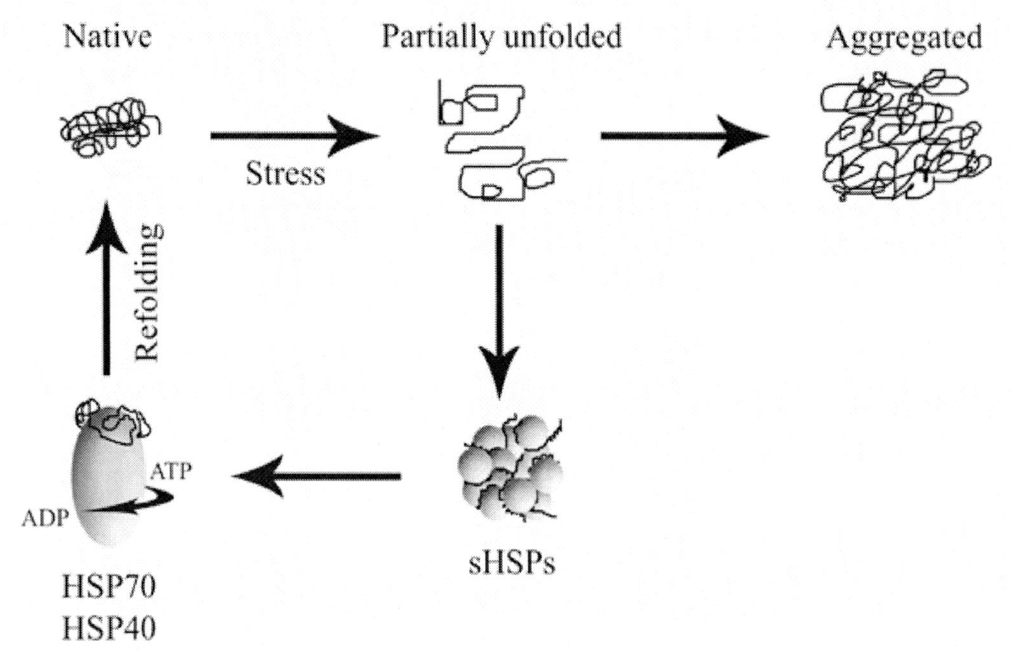

[1] Figure is based on [181].

Figure 4. Model for the chaperone functioning of sHSPs under stress conditions.[1] Unfolding of native proteins can lead to (insoluble) aggregations. Partially unfolded proteins can be bound by sHSPs, keeping them in a soluble state, which is accessible for HSP70 and HSP40. These ATP-dependent chaperones can refold the protein to its native state.

6. Interactions with Partner Proteins

Interaction between proteins is a key element in building structures and functional protein complexes. These interactions are often transient, and are essential for all activities within the biological processes. SHSPs can interact with each other to form mixed oligomeric complexes and with unrelated partners to fulfil specific functions. Here we focus on four important functional interactions, involving hetero-oligomer formation, cytoskeletal organization, apoptosis and protein degradation.

6.1. Interactions amongst sHSPs

In muscle cells seven sHSPs are abundantly expressed. In these cells, hetero-oligomers of sHSPs may form and fulfil specific functions. A number of studies describe such mixed sHSP complexes, e.g. complexes containing HSPB5/HSPB6 [107], HSPB1/HSPB6 [108], HSPB1/HSPB5 [109]. Using two-hybrid, Förster resonance energy transfer (FRET), immunoprecipitation and crosslinking analyses several more interactions between the seven muscle-type sHSPs have been detected (see Figure 5). Although several sHSPs can interact with each other in vitro, not all possible interactions may actually occur in cells. For instance, HSPB2 may interact with HSPB3 HSPB6 and HSPB8, but it was found that HSPB2 is predominantly associated with the less abundant HSPB3 in muscle extract [52]. Not much is known about the occurrence, composition and dynamics of hetero-oligomeric sHSP complexes in vivo, and whether particular compositions are preferred above others. This is especially due to the very rapid subunit exchange, which makes it difficult to quantify subunit ratios. Furthermore, the affinity between subunits is variable and can be regulated in a way which will affect the size of the oligomers [52]. Well known factors that influence the interaction between different sHSPs are phosphorylation [110] and temperature [108,111]. Changing the composition and size of hetero-oligomers may alter the accessibility of particular sites within the complex and in this way will influence the functioning of sHSPs in the cellular environment.

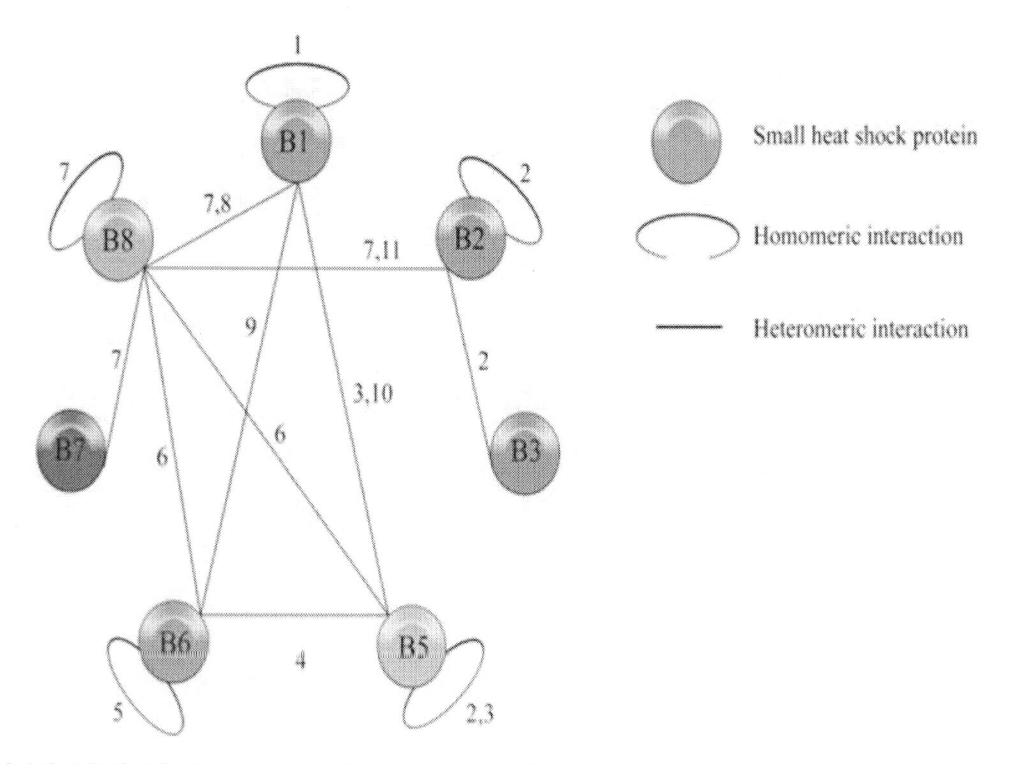

Figure 5. Interactions between the seven sHSPs expressed in muscle cells. Several complexes can be formed through interactions of sHSPs with themselves or with other sHSPs.

6.2. Interactions with the Cytoskeleton

sHSPs are proposed to modulate the structure of the cytoskeleton, and this activity is linked to their protective function. Some members of the sHSP family can interact with actin filaments, intermediate filaments and microtubules [112-115]. HspB1, B5 and B6 are able to associate with actin [114]. The interaction appears to be vital for the stability of the actin network and is important for cardiac and skeletal muscle development [116]. HSPB6 also interacts with actinin, which is an actin-crosslinking protein [117]. HSPB5 associates with intermediate filaments, especially type III filaments such as vimentin [118], desmin [119] and glial fibrillary acidic protein GFAP [120]. Here, it regulates assembly and assists intermediate filaments in recovery from stress by preventing untimely filament–filament interactions that would otherwise lead to aggregation. The missense mutation R120G in HSPB5, which reduces the chaperone-like activity [121], leads to accumulations of desmin-positive sarcomeric inclusions [122]. HSPB5 also plays a critical role in Alexander disease to prevent aggregation of GFAP and the formation of Rosenthal fibers [123]. In the absence of stress, HSPB5 may interact directly with tubulin and microtubule-associated proteins to promote microtubule assembly and under stress may protect against microtubule depolymerization [124-126]. The function of these and other interactions between sHSPs and cytoskeletal proteins are described in chapter 1-2.

6.3. Interaction with Apoptotic Proteins

Apoptosis is a controlled way of cell death and several stimuli can trigger this event. There are two pathways leading to apoptosis: the intrinsic (mitochondrial) pathway and the extrinsic (death receptor) pathway. Increased expression of sHSPs during the stress response correlates with better survival, which might be an effect of the inhibition of the apoptotic process. Several sHSPs play a role in modulating apoptosis by interacting with components of the apoptotic pathways (see chapter 1-3). HSPB1 functions as an anti-apoptotic protein by interfering in both the pathways. In the mitochondrial pathway, HSPB1 interacts with cytochrome C [127] and inhibits caspase 3 activation [128]. In the death receptor pathway, phosphorylated HSPB1 interacts with Daxx, thereby inhibiting TGFβ stimulation and ultimately apoptosis [129]. HSPB4 and HSPB5 interact with the anti-apoptotic factors Bax and Bcl-xs, inhibiting the intrinsic pathway [130]. In vitro, HSPB5 is capable of binding procaspase 3 cleavage products [131] and this way might be involved in the inhibition of the intrinsic apoptosis pathway [132]. HSPB6 protects cardiomyocytes against apoptosis, and phosphorylation at Ser16 enhances its cardioprotection [133]. Strikingly, HSPB8 has a kind of dual-function; it can function anti- as well as pro-apoptotic depending on cell type and expression level [134]. Its chaperone-like activity and the interaction with and regulation of certain protein kinases are likely important for the influence on the apoptotic process [134,135]. The functionally related HSP16.2, is also an anti-apoptotic protein [12].

From all the reported observations it is clear that most members of the sHSP family may help in cell survival in several different ways. However, the significance and physiological importance of each interaction in the regulation of the apoptotic process is still not well understood and needs further investigations.

6.4. Interaction with the Degradation Machinery

Three sHSPs have been implicated in the ubiquitin-proteasome pathway, a process in which ubiquitin ligases mark substrate proteins for degradation by the proteasome. This is executed by covalent addition of multiple ubiquitin molecules. HSPB5 can interact with F-Box protein X4 (FBX4), which forms an E3 ubiquitin ligase, together with Skp1 and Cullin1 [136]. HSPB5 can both in vitro and in vivo bind to one of the fourteen subunits of the 20 S proteasome, possibly guiding the substrate to the proteasome [137]. One of the targets of this FBX4/HSPB5 E3 ligase is phosphorylated cyclin D1 [138]. This protein accounts for the G1 to S phase transition in the cell cycle and breakdown leads to diminished cell growth. In this way HSPB5 interferes with cell growth.

HSPB1 can interfere in the cell cycle as well. It enhances the ubiquitination and proteasomal degradation of p27 (kip1), which is a cyclin-dependent kinase inhibitor (CDKi). To perform this function, HSPB1 is present in small oligomers and phosphorylation of HSPB1 is not required. When this inhibitor is broken down, the cylin/CDK complexes will be able to function again, leading to cell cycle progression [139].

HSPB10 can interact with outer dense fibers interacting protein 1 (OIP1), which is a candidate E3 ubiquitin protein ligase. Both are localized in the sperm tail. Cyclin-dependent kinase 5 (Cdk5)/p35, of which the activity is associated with spermiogenesis, phosphorylates HSPB10 and thereby increases the affinity of the interaction and possibly leads to the degradation of HSPB10 [76].

7. Involvement of sHSPs in Diseases

As molecular chaperones, the sHSPs protect protein structure and activity, thereby counteracting the deleterious effects of disease processes. The expression of various sHSPs is therefore upregulated under different pathological conditions, as already mentioned. sHSPs guard against ischemic and reperfusion injury due to heart attack and stroke (see chapter 1-5), prevent cataract in the mammalian eye lens (see chapter 2-2), and may protect against aggregation and/or toxicity of α-synuclein in Parkinson's disease, huntingtin in Huntington's disease or amyloid in Alzheimer's disease (see chapters 1-1, 1-4 and 1-7). The exact role of sHSPs in these diseases might be linked to their chaperone activity in providing a first line of defense against misfolded proteins, and in modulating the aberrant protein interactions that trigger pathogenic cascades.

On the other hand, mutated sHSPs may contribute to cell malfunction. For instance, mutants of HSPB1, HSPB4, HSPB5 and HSPB8 are implicated in diseases such as myofibrillar myopathy, cardiomyopathy, neuropathies and cataract (detailed in part 2 of the book). These pathologies result from the misfolding and progressive aggregation of the mutated sHSP, although the molecular mechanism is likely to be different for each type of disease [67,115]. The cataract-causing HSPB4 mutants R49C and R116C have been shown to have an enhanced affinity for substrates, thereby reducing the refolding toward fully native proteins [140]. The implied molecular basis is a gain of function that leads to and increased binding of damaged proteins and subsequent precipitation of the saturated HSPB4 complexes in the developing lens of affected individuals. On the other hand the K141E mutation in

HSPB8, causing neuromuscular disorders, destabilizes the structure and decreases the chaperone-like activity [141,142]. The mutation-induced destabilization makes denaturation of HSPB8 more probable and in this way might stimulate aggregate formation in cells [141]. Moreover, the myofibrillar myopathy associated R120G mutation in HSPB5 has been found to modify the properties of HSPB5, such as oligomerization and in vitro chaperone-like activity [121], and to increase its affinity to desmin [143]. This mutation in HSPB5 results in misfolding and progressive aggregation, and subsequent association with desmin filaments to form HSPB5/desmin/amyloid positive aggresomes [144].

Several members of the HSPB family are implicated in cancer development, of which HSPB1, HSPB5 and HSPB8 are the best studied (see Part III). High levels of constitutive expression of these sHSPs have been detected in several cancer cells [67]. A possible mechanism by which HSPB1 may increase the tumorigenic potential of cells is to resist the apoptotic process [145]. This protein may also play a role in metastasis formation, since higher expression levels of HSPB1 have been observed in metastatic as compared to non-metastatic tissues [146]. HSPB5 can acts as an oncoprotein in breast cancer and its expression is correlated with a poor clinical outcome [147]. High expression induces neoplastic changes and invasive properties in human mammary epithelial cells, which are inhibited by HSPB5 phosphorylation. Phosphorylation affects oligomerization and anti-apoptotic activities of HSPB5, and shows the importance of post-translational modification on the functioning of this sHSP. Remarkably, HSPB5 does not increase the tumorigenic potential under all circumstances. In fact, the protein can reduce cancer progression by forming, together with FBX4, an SCF E3 ligase able to ubiquitinate and subsequently degrade cylcin D1 [138]. Cyclin D1 is a well known oncoprotein, and decrease in HSPB5 would mean an increase in cyclin D1, driving tumor formation. Indeed, in many tumor samples HSPB5 actually is decreased [138]. HSPB8 might have a dual role in tumor development as well. In several tumor tissues and cell lines (melanoma, prostate cancer, sarcoma) HSPB8 expression is reduced, and forced expression increased the percentage of apoptotic cells, while in estrogen-receptor positive breast cancer the expression was increased and associated with cancer progression that blocked apoptosis and promoted oncogenic transformation. These observations indicate that depending on the cell type and extent of expression HSPB8 possesses both pro- and anti-tumorigenic potential [148].

Beside the three well studied HSPBs, other family members are associated with cancer development as well. HSPB9, normally only present in the testis, is a cancer-testis antigen since it is expressed by certain lung and germ cell tumours [149]. Cancer-testis antigens include a series of proteins typically synthesized in primitive germ cells and as a result of malignant transformation those proteins reappear in tumors. The prognostic value of the presence of HSPB9 is unknown. HSP16.2 expression correlates directly with the grade of brain tumors, and its level increases with the increase of cell anaplasia [150]. On the contrary, the expression of HSPB6 inversely correlates with tumor stage of hepatocellular carcinoma tissues, suggesting that HSPB6 plays a role against tumor progression.

The expression of sHSPs is also involved in another aspect of cancer biology. sHSP expression is associated with cellular resistance to cytostatic anticancer drugs [151,152]. Some of these drugs, in particular cisplatin [153], vincristine and colchicine [125] even enhance HSPB1 and HSPB5 expression, leading to even higher resistance to these drugs. High expression of sHSPs may impair the efficiency of the clinical treatments using chemotherapeutic agents.

As the understanding of the role of sHSPs in human diseases is advancing, opportunities for disease prevention and treatment become more apparent. Several kinds of strategies can be followed in which sHSPs may be used as targets to affect their pathological roles. Forced expression of HSPB1 and HSPB8 by oral administration of the anti-ulcer drug geranylgeranylacetone have been shown to reduce amyloid oligomer formation in transgenic mice overexpressing HSPB5 R120G in the heart. This could be a new strategy for treating patients with amyloid-based neurodegenerative diseases [154]. Other strategies might be used as well. For instance, anti-sense or nucleotide-based therapies could be developed with the aim to inhibit pathological sHSP expression. Also, drugs that stimulate the degradation of putative sHSP client proteins, or drugs that maintain sHSPs in the form of large oligomers to stimulate the chaperone properties or mask specific mutations in these chaperones can be considered. Furthermore, therapeutic exploitation of sHSPs can be performed by using sHSP expression as a biomarker to help to select and improve treatments. Hence, it is likely that in the near future drugs will be available for clinical trials to test whether compounds that modulate sHSP functions are active towards the deleterious effects induced by these diseases.

ACKNOWLEDGMENTS

We want to give many thanks to Guido Kappé for the generation of figures 1 and 2 and Wilfried de Jong for critical reading the manuscript.

REFERENCES

[1] de Jong WW, Caspers GJ, Leunissen JA. Genealogy of the alpha-crystallin-small heat-shock protein superfamily. *Int J Biol Macromol.* 1998; 22(3-4):151-162.

[2] Kappe G, Leunissen JA, de Jong WW. Evolution and diversity of prokaryotic small heat shock proteins. *Prog Mol Subcell Biol.* 2002; 28:1-17.

[3] de Jong WW, Leunissen JA, Voorter CE. Evolution of the alpha-crystallin/small heat-shock protein family. *Mol Biol Evol.* 1993; 10(1):103-126.

[4] Kappe G, Franck E, Verschuure P, Boelens WC, Leunissen JA, de Jong WW. The human genome encodes 10 alpha-crystallin-related small heat shock proteins: HspB1-10. *Cell Stress Chaperones,* 2003; 8(1):53-61.

[5] Fontaine JM, Rest JS, Welsh MJ, Benndorf R. The sperm outer dense fiber protein is the 10th member of the superfamily of mammalian small stress proteins. *Cell Stress & Chaperones,* 2003; 8(1):62-69.

[6] Kato K, Shinohara H, Kurobe N, Goto S, Inaguma Y, Ohshima K. Immunoreactive alpha A crystallin in rat non-lenticular tissues detected with a sensitive immunoassay method. *Biochim Biophys Acta,* 1991; 1080(2):173-180.

[7] Mörner CT. Untersuchung der Proteinsubstanzen in den sichtbrechenden Medien de Auges. *Hoppe-Seyler's Z Physiol.Chemie.* 18, 16. 1894.

[8] van der Ouderaa FJ, de Jong WW, Groenendijk GW, Bloemendal H. Sequence of the first 68 residues of the alpha B2 chain of bovine alpha-crystallin. *Biochem Biophys Res Commun.* 1974; 57(1):112-119.

[9] van der Ouderaa FJ, de Jong WW, Bloemendal H. The amino-acid sequence of the alphaA2 chain of bovine alpha-crystallin. *Eur J Biochem.* 1973; 39(1):207-222.

[10] Ingolia TD, Craig EA. Four small Drosophila heat shock proteins are related to each other and to mammalian alpha-crystallin. *Proc Natl Acad Sci U S A,* 1982; 79(7):2360-2364.

[11] Horwitz J. Alpha-crystallin can function as a molecular chaperone. *Proc Natl Acad Sci U S A*, 1992; 89(21):10449-10453.

[12] Bellyei S, Szigeti A, Pozsgai E, Boronkai A, Gomori E, Hocsak E et al. Preventing apoptotic cell death by a novel small heat shock protein. *Eur J Cell Biol.* 2007; 86(3):161-171.

[13] Gough J, Karplus K, Hughey R, Chothia C. Assignment of homology to genome sequences using a library of hidden Markov models that represent all proteins of known structure. *J Mol Biol.* 2001; 313(4):903-919.

[14] Ramelot TA, Raman S, Kuzin AP, Xiao R, Ma LC, Acton TB et al. Improving NMR protein structure quality by Rosetta refinement: a molecular replacement study. *Proteins*, 2009; 75(1):147-167.

[15] Pasta SY, Raman B, Ramakrishna T, Rao C. The IXI/V motif in the C-terminal extension of alpha-crystallins: alternative interactions and oligomeric assemblies. *Mol Vis.* 2004; 10:655-662.

[16] Pasta SY, Raman B, Ramakrishna T, Rao C. Role of the conserved SRLFDQFFG region of alpha-crystallin, a small heat shock protein. Effect on oligomeric size, subunit exchange, and chaperone-like activity. *J Biol Chem.* 2003; 278(51):51159-51166.

[17] Iwaki A, Nagano T, Nakagawa M, Iwaki T, Fukumaki Y. Identification and characterization of the gene encoding a new member of the alpha-crystallin/small hsp family, closely linked to the alphaB-crystallin gene in a head-to-head manner. *Genomics,* 1997; 45(2):386-394.

[18] Elicker KS, Hutson LD. Genome-wide analysis and expression profiling of the small heat shock proteins in zebrafish. *Gene,* 2007; 403(1-2):60-69.

[19] Franck E, Madsen O, van Rheede T, Ricard G, Huynen MA, de Jong WW. Evolutionary diversity of vertebrate small heat shock proteins. *J Mol Evol.* 2004; 59(6):792-805.

[20] Gusev NB, Bogatcheva NV, Marston SB. Structure and properties of small heat shock proteins (sHsp) and their interaction with cytoskeleton proteins. *Biochemistry,* (Mosc) 2002; 67(5):511-519.

[21] van den Oetelaar PJ, van Someren PF, Thomson JA, Siezen RJ, Hoenders HJ. A dynamic quaternary structure of bovine alpha-crystallin as indicated from intermolecular exchange of subunits. *Biochemistry,* 1990; 29(14):3488-3493.

[22] Bova MP, Ding LL, Horwitz J, Fung BK. Subunit exchange of alphaA-crystallin. *J Biol Chem.* 1997; 272(47):29511-29517.

[23] Koteiche HA, Mchaourab HS. Folding pattern of the alpha-crystallin domain in alphaA-crystallin determined by site-directed spin labeling. *J Mol Biol.* 1999; 294(2):561-577.

[24] Jehle S, van Rossum B, Stout JR, Noguchi SM, Falber K, Rehbein K et al. alphaB-crystallin: a hybrid solid-state/solution-state NMR investigation reveals structural aspects of the heterogeneous oligomer. *J Mol Biol.* 2009; 385(5):1481-1497.

[25] Kim KK, Kim R, Kim SH. Crystal structure of a small heat-shock protein. *Nature,* 1998; 394(6693):595-599.

[26] van Montfort RL, Basha E, Friedrich KL, Slingsby C, Vierling E. Crystal structure and assembly of a eukaryotic small heat shock protein. *Nat Struct Biol.* 2001; 8(12):1025-1030.

[27] Stamler R, Kappe G, Boelens W, Slingsby C. Wrapping the alpha-crystallin domain fold in a chaperone assembly. *J Mol Biol.* 2005; 353(1):68-79.

[28] Morris AM, Treweek TM, Aquilina JA, Carver JA, Walker MJ. Glutamic acid residues in the C-terminal extension of small heat shock protein 25 are critical for structural and functional integrity. *FEBS J.* 2008; 275(23):5885-5898.

[29] Li Y, Schmitz KR, Salerno JC, Koretz JF. The role of the conserved COOH-terminal triad in alphaA-crystallin aggregation and functionality. *Mol Vis.* 2007; 13:1758-1768.

[30] Saji H, Iizuka R, Yoshida T, Abe T, Kidokoro S, Ishii N et al. Role of the IXI/V motif in oligomer assembly and function of StHsp14.0, a small heat shock protein from the acidothermophilic archaeon, Sulfolobus tokodaii strain 7. *Proteins,* 2008; 71(2):771-782.

[31] Kennaway CK, Benesch JL, Gohlke U, Wang L, Robinson CV, Orlova EV et al. Dodecameric structure of the small heat shock protein Acr1 from Mycobacterium tuberculosis. *J Biol Chem.* 2005; 280(39):33419-33425.

[32] Haslbeck M, Kastenmuller A, Buchner J, Weinkauf S, Braun N. Structural dynamics of archaeal small heat shock proteins. *J Mol Biol.* 2008; 378(2):362-374.

[33] White HE, Orlova EV, Chen S, Wang L, Ignatiou A, Gowen B et al. Multiple distinct assemblies reveal conformational flexibility in the small heat shock protein Hsp26. *Structure*, 2006; 14(7):1197-1204.

[34] Carver JA, Aquilina JA, Truscott RJ, Ralston GB. Identification by 1H NMR spectroscopy of flexible C-terminal extensions in bovine lens alpha-crystallin. *FEBS Lett.* 1992; 311(2):143-149.

[35] Haslbeck M, Franzmann T, Weinfurtner D, Buchner J. Some like it hot: the structure and function of small heat-shock proteins. *Nat Struct Mol Biol.* 2005; 12(10):842-846.

[36] Haley DA, Bova MP, Huang QL, Mchaourab HS, Stewart PL. Small heat-shock protein structures reveal a continuum from symmetric to variable assemblies. *J Mol Biol.* 2000; 298(2):261-272.

[37] Benesch JL, Ayoub M, Robinson CV, Aquilina JA. Small heat shock protein activity is regulated by variable oligomeric substructure. *J Biol Chem.* 2008; 283(42):28513-28517.

[38] Wang K, Spector A. The chaperone activity of bovine alpha crystallin. Interaction with other lens crystallins in native and denatured states. *J Biol Chem.* 1994; 269(18):13601-13608.

[39] Dillmann WH. Small heat shock proteins and protection against injury. *Ann N Y Acad Sci.* 1999; 874:66-68.

[40] Verschuure P, Tatard C, Boelens WC, Grongnet JF, David JC. Expression of small heat shock proteins HspB2, HspB8, Hsp20 and cvHsp in different tissues of the perinatal developing pig. *Eur J Cell Biol.* 2003; 82(10):523-530.

[41] Tallot P, Grongnet JF, David JC. Dual perinatal and developmental expression of the small heat shock proteins [FC12]aB-crystallin and Hsp27 in different tissues of the developing piglet. *Biol Neonate.* 2003; 83(4):281-288.

[42] van de Klundert FA, Gijsen ML, van den IJssel PR, Snoeckx LH, de Jong WW. alpha B-crystallin and hsp25 in neonatal cardiac cells--differences in cellular localization under stress conditions. *Eur J Cell Biol.* 1998; 75(1):38-45.

[43] Simkhovich BZ, Marjoram P, Poizat C, Kedes L, Kloner RA. Brief episode of ischemia activates protective genetic program in rat heart: a gene chip study. *Cardiovasc Res.* 2003; 59(2):450-459.

[44] Lutsch G, Vetter R, Offhauss U, Wieske M, Grone HJ, Klemenz R et al. Abundance and location of the small heat shock proteins HSP25 and alphaB-crystallin in rat and human heart. *Circulation*, 1997; 96(10):3466-3476.

[45] Ciocca DR, Oesterreich S, Chamness GC, McGuire WL, Fuqua SA. Biological and clinical implications of heat shock protein 27,000 (Hsp27): a review. *J Natl Cancer Inst.* 1993; 85(19):1558-1570.

[46] Efthymiou CA, Mocanu MM, de Belleroche J, Wells DJ, Latchmann DS, Yellon DM. Heat shock protein 27 protects the heart against myocardial infarction. *Basic Res Cardiol.* 2004; 99(6):392-394.

[47] Srinivasan AN, Nagineni CN, Bhat SP. alpha A-crystallin is expressed in non-ocular tissues. *J Biol Chem.* 1992; 267(32):23337-23341.

[48] Egwuagu CE, Chepelinsky AB. Extralenticular expression of the alpha A-crystallin promoter/gamma interferon transgene. *Exp Eye Res.* 1997; 64(3):491-495.

[49] Dubin RA, Wawrousek EF, Piatigorsky J. Expression of the murine alpha B-crystallin gene is not restricted to the lens. *Mol Cell Biol.* 1989; 9(3):1083-1091.

[50] Bhat SP, Nagineni CN. alpha B subunit of lens-specific protein alpha-crystallin is present in other ocular and non-ocular tissues. *Biochem Biophys Res Commun.* 1989; 158(1):319-325.

[51] Lam WY, Wing Tsui SK, Law PT, Luk SC, Fung KP, Lee CY et al. Isolation and characterization of a human heart cDNA encoding a new member of the small heat shock protein family--HSPL27. *Biochim Biophys Acta,* 1996; 1314(1-2):120-124.

[52] Sugiyama Y, Suzuki A, Kishikawa M, Akutsu R, Hirose T, Waye MM et al. Muscle develops a specific form of small heat shock protein complex composed of MKBP/HSPB2 and HSPB3 during myogenic differentiation. *J Biol Chem.* 2000; 275(2):1095-1104.

[53] Golenhofen N, Perng MD, Quinlan RA, Drenckhahn D. Comparison of the small heat shock proteins alphaB-crystallin, MKBP, HSP25, HSP20, and cvHSP in heart and skeletal muscle. *Histochem Cell Biol.* 2004; 122(5):415-425.

[54] Swamynathan SK, Piatigorsky J. Regulation of the mouse alphaB-crystallin and MKBP/HspB2 promoter activities by shared and gene specific intergenic elements: the importance of context dependency. *Int J Dev Biol.* 2007; 51(8):689-700.

[55] Doerwald L, van Rheede T, Dirks RP, Madsen O, Rexwinkel R, van Genesen ST et al. Sequence and functional conservation of the intergenic region between the head-to-head genes encoding the small heat shock proteins alphaB-crystallin and HspB2 in the mammalian lineage. *J Mol Evol.* 2004; 59(5):674-686.

[56] Taylor RP, Benjamin IJ. Small heat shock proteins: a new classification scheme in mammals. *J Mol Cell Cardiol.* 2005; 38(3):433-444.

[57] Louapre P, Grongnet JF, Tanguay RM, David JC. Effects of hypoxia on stress proteins in the piglet heart at birth. *Cell Stress Chaperones*, 2005; 10(1):17-23.

[58] Nefti O, Grongnet JF, David JC. Overexpression of alphaB crystallin in the gastrointestinal tract of the newborn piglet after hypoxia. *Shock*, 2005; 24(5):455-461.

[59] Golenhofen N, Ness W, Wawrousek EF, Drenckhahn D. Expression and induction of the stress protein alpha-B-crystallin in vascular endothelial cells. *Histochem Cell Biol.* 2002; 117(3):203-209.

[60] Yu AL, Fuchshofer R, Birke M, Priglinger SG, Eibl KH, Kampik A et al. Hypoxia/reoxygenation and TGF-beta increase alphaB-crystallin expression in human optic nerve head astrocytes. *Exp Eye Res*. 2007; 84(4):694-706.

[61] Omar R, Pappolla M. Oxygen free radicals as inducers of heat shock protein synthesis in cultured human neuroblastoma cells: relevance to neurodegenerative disease. *Eur Arch Psychiatry Clin Neurosci.* 1993; 242(5):262-267.

[62] Hawse JR, Cumming JR, Oppermann B, Sheets NL, Reddy VN, Kantorow M. Activation of metallothioneins and alpha-crystallin/sHSPs in human lens epithelial cells by specific metals and the metal content of aging clear human lenses. Invest Ophthalmol Vis Sci 2003; 44(2):672-679.

[63] Boluyt MO, Brevick JL, Rogers DS, Randall MJ, Scalia AF, Li ZB. Changes in the rat heart proteome induced by exercise training: Increased abundance of heat shock protein hsp20. *Proteomics,* 2006; 6(10):3154-3169.

[64] Vissing K, Bayer ML, Overgaard K, Schjerling P, Raastad T. Heat shock protein translocation and expression response is attenuated in response to repeated eccentric exercise. *Acta Physiol.* (Oxf) 2008.

[65] Wilhelmus MM, Otte-Holler I, Wesseling P, de Waal RM, Boelens WC, Verbeek MM. Specific association of small heat shock proteins with the pathological hallmarks of Alzheimer's disease brains. *Neuropathol Appl Neurobiol.* 2006; 32(2):119-130.

[66] Wang J, Martin E, Gonzales V, Borchelt DR, Lee MK. Differential regulation of small heat shock proteins in transgenic mouse models of neurodegenerative diseases. *Neurobiol Aging*, 2008; 29(4):586-597.

[67] Arrigo AP, Simon S, Gibert B, Kretz-Remy C, Nivon M, Czekalla A et al. Hsp27 (HspB1) and alphaB-crystallin (HspB5) as therapeutic targets. *FEBS Lett.* 2007; 581(19):3665-3674.

[68] Hishiya A, Takayama S. Molecular chaperones as regulators of cell death. *Oncogene.* 2008; 27(50):6489-6506.

[69] Aggeli IK, Beis I, Gaitanaki C. Oxidative stress and calpain inhibition induce alpha B-crystallin phosphorylation via p38-MAPK and calcium signalling pathways in H9c2 cells. *Cell Signal,* 2008; 20(7):1292-1302.

[70] Golenhofen N, Ness W, Koob R, Htun P, Schaper W, Drenckhahn D. Ischemia-induced phosphorylation and translocation of stress protein alpha B-crystallin to Z lines of myocardium. *Am J Physiol.* 1998; 274(5 Pt 2):H1457-H1464.

[71] Ito H, Okamoto K, Nakayama H, Isobe T, Kato K. Phosphorylation of alphaB-crystallin in response to various types of stress. *J Biol Chem.* 1997; 272(47):29934-29941.

[72] Landry J, Lambert H, Zhou M, Lavoie JN, Hickey E, Weber LA et al. Human HSP27 is phosphorylated at serines 78 and 82 by heat shock and mitogen-activated kinases that recognize the same amino acid motif as S6 kinase II. *J Biol Chem.* 1992; 267(2):794-803.

[73] Kato K, Ito H, Kamei K, Inaguma Y, Iwamoto I, Saga S. Phosphorylation of alphaB-crystallin in mitotic cells and identification of enzymatic activities responsible for phosphorylation. *J Biol Chem.* 1998; 273(43):28346-28354.

[74] Beall A, Bagwell D, Woodrum D, Stoming TA, Kato K, Suzuki A et al. The small heat shock-related protein, HSP20, is phosphorylated on serine 16 during cyclic nucleotide-dependent relaxation. *J Biol Chem.* 1999; 274(16):11344-11351.

[75] Shemetov AA, Seit-Nebi AS, Bukach OV, Gusev NB. Phosphorylation by cyclic AMP-dependent protein kinase inhibits chaperone-like activity of human HSP22 in vitro. *Biochemistry-Moscow,* 2008; 73(2):200-208.

[76] Rosales JL, Sarker K, Ho N, Broniewska M, Wong P, Cheng M et al. ODF1 phosphorylation by Cdk5/p35 enhances ODF1-OIP1 interaction. *Cell Physiol Biochem.* 2007; 20(5):311-318.

[77] Kato K, Hasegawa K, Goto S, Inaguma Y. Dissociation as a result of phosphorylation of an aggregated form of the small stress protein, hsp27. *J Biol Chem.* 1994; 269(15):11274-11278.

[78] Ito H, Kamei K, Iwamoto I, Inaguma Y, Nohara D, Kato K. Phosphorylation-induced change of the oligomerization state of alpha B-crystallin. *J Biol Chem.* 2001; 276(7):5346-5352.

[79] Benndorf R, Hayess K, Ryazantsev S, Wieske M, Behlke J, Lutsch G. Phosphorylation and supramolecular organization of murine small heat shock protein HSP25 abolish its actin polymerization-inhibiting activity. *J Biol Chem.* 1994; 269(32):20780-20784.

[80] Lavoie JN, Gingras-Breton G, Tanguay RM, Landry J. Induction of Chinese hamster HSP27 gene expression in mouse cells confers resistance to heat shock. HSP27 stabilization of the microfilament organization. *J Biol Chem.* 1993; 268(5):3420-3429.

[81] Ahmad MF, Raman B, Ramakrishna T, Rao C. Effect of phosphorylation on alpha B-crystallin: differences in stability, subunit exchange and chaperone activity of homo and mixed oligomers of alpha B-crystallin and its phosphorylation-mimicking mutant. *J Mol Biol.* 2008; 375(4):1040-1051.

[82] den Engelsman J, Bennink EJ, Doerwald L, Onnekink C, Wunderink L, Andley UP et al. Mimicking phosphorylation of the small heat-shock protein alphaB-crystallin recruits the F-box protein FBX4 to nuclear SC35 speckles. *Eur J Biochem.* 2004; 271(21):4195-4203.

[83] Bryantsev AL, Kurchashova SY, Golyshev SA, Polyakov VY, Wunderink HF, Kanon B et al. Regulation of stress-induced intracellular sorting and chaperone function of Hsp27 (HspB1) in mammalian cells. *Biochem J.* 2007; 407(3):407-417.

[84] Bryantsev AL, Chechenova MB, Shelden EA. Recruitment of phosphorylated small heat shock protein Hsp27 to nuclear speckles without stress. *Exp Cell Res.* 2007; 313(1):195-209.

[85] Maddala R, Rao VP. alpha-Crystallin localizes to the leading edges of migrating lens epithelial cells. *Exp Cell Res.* 2005; 306(1):203-215.

[86] Launay N, Goudeau B, Kato K, Vicart P, Lilienbaum A. Cell signaling pathways to alphaB-crystallin following stresses of the cytoskeleton. *Exp Cell Res.* 2006; 312(18):3570-3584.

[87] Flynn CR, Smoke CC, Furnish E, Komalavilas P, Thresher J, Yi Z et al. Phosphorylation and activation of a transducible recombinant form of human HSP20 in Escherichia coli. *Protein Expr Purif.* 2007; 52(1):50-58.

[88] Gupta R, Srivastava OP. Deamidation affects structural and functional properties of human alphaA-crystallin and its oligomerization with alphaB-crystallin. *J Biol Chem.* 2004; 279(43):44258-44269.
[89] Kamei A, Iwase H, Masuda K. Cleavage of amino acid residue(s) from the N-terminal region of alpha A- and alpha B-crystallins in human crystalline lens during aging. *Biochem Biophys Res Commun.* 1997; 231(2):373-378.
[90] Takeuchi N, Ouchida A, Kamei A. C-terminal truncation of alpha-crystallin in hereditary cataractous rat lens. *Biol Pharm Bull.* 2004; 27(3):308-314.
[91] Chen SJ, Sun TX, Akhtar NJ, Liang JJ. Oxidation of human lens recombinant alphaA-crystallin and cysteine-deficient mutants. *J Mol Biol.* 2001; 305(4):969-976.
[92] Blakytny R, Carver JA, Harding JJ, Kilby GW, Sheil MM. A spectroscopic study of glycated bovine alpha-crystallin: investigation of flexibility of the C-terminal extension, chaperone activity and evidence for diglycation. *Biochim Biophys Acta,* 1997; 1343(2):299-315.
[93] Nagaraj RH, Oya-Ito T, Padayatti PS, Kumar R, Mehta S, West K et al. Enhancement of chaperone function of alpha-crystallin by methylglyoxal modification. *Biochemistry,* 2003; 42(36):10746-10755.
[94] Satish KM, Mrudula T, Mitra N, Bhanuprakash RG. Enhanced degradation and decreased stability of eye lens alpha-crystallin upon methylglyoxal modification. *Exp Eye Res.* 2004; 79(4):577-583.
[95] Seidler NW, Yeargans GS, Morgan TG. Carnosine disaggregates glycated alpha-crystallin: an in vitro study. *Arch Biochem Biophys.* 2004; 427(1):110-115.
[96] Fujii N, Matsumoto S, Hiroki K, Takemoto L. Inversion and isomerization of Asp-58 residue in human alphaA-crystallin from normal aged lenses and cataractous lenses. *Biochim Biophys Acta,* 2001; 1549(2):179-187.
[97] Stromer T, Ehrnsperger M, Gaestel M, Buchner J. Analysis of the interaction of small heat shock proteins with unfolding proteins. *J Biol Chem.* 2003; 278(20):18015-18021.
[98] Cuesta R, Laroia G, Schneider RJ. Chaperone hsp27 inhibits translation during heat shock by binding eIF4G and facilitating dissociation of cap-initiation complexes. Genes Dev 2000; 14(12):1460-1470.
[99] Quinlan RA, Brenner M, Goldman JE, Messing A. GFAP and its role in Alexander disease. Exp Cell Res 2007; 313(10):2077-2087.
[100] Outeiro TF, Klucken J, Strathearn KE, Liu F, Nguyen P, Rochet JC et al. Small heat shock proteins protect against alpha-synuclein-induced toxicity and aggregation. *Biochem Biophys Res Commun.* 2006; 351(3):631-638.
[101] Sharma MC, Goebel HH. Protein aggregate myopathies. *Neurol India,* 2005; 53(3):273-279.
[102] Lee GJ, Vierling E. A small heat shock protein cooperates with heat shock protein 70 systems to reactivate a heat-denatured protein. *Plant Physiol.* 2000; 122(1):189-198.
[103] Matuszewska M, Kuczynska-Wisnik D, Laskowska E, Liberek K. The small heat shock protein IbpA of Escherichia coli cooperates with IbpB in stabilization of thermally aggregated proteins in a disaggregation competent state. *J Biol Chem.* 2005; 280(13):12292-12298.
[104] Mogk A, Deuerling E, Vorderwulbecke S, Vierling E, Bukau B. Small heat shock proteins, ClpB and the DnaK system form a functional triade in reversing protein aggregation. *Mol Microbiol.* 2003; 50(2):585-595.

[105] Haslbeck M, Miess A, Stromer T, Walter S, Buchner J. Disassembling protein aggregates in the yeast cytosol. The cooperation of Hsp26 with Ssa1 and Hsp104. *J Biol Chem.* 2005; 280(25):23861-23868.

[106] Cashikar AG, Duennwald M, Lindquist SL. A chaperone pathway in protein disaggregation. Hsp26 alters the nature of protein aggregates to facilitate reactivation by Hsp104. *J Biol Chem.* 2005; 280(25):23869-23875.

[107] Kato K, Goto S, Inaguma Y, Hasegawa K, Morishita R, Asano T. Purification and characterization of a 20-kDa protein that is highly homologous to alpha B crystallin. *J Biol Chem.* 1994; 269(21):15302-15309.

[108] Bukach OV, Glukhova AE, Seit-Nebi AS, Gusev NB. Heterooligomeric complexes formed by human small heat shock proteins HspB1 (Hsp27) and HspB6 (Hsp20). *Biochim Biophys Acta,* 2008.

[109] Zantema A, Verlaan-De Vries M, Maasdam D, Bol S, van der EA. Heat shock protein 27 and alpha B-crystallin can form a complex, which dissociates by heat shock. *J Biol Chem.* 1992; 267(18):12936-12941.

[110] Sun X, Welsh MJ, Benndorf R. Conformational changes resulting from pseudophosphorylation of mammalian small heat shock proteins--a two-hybrid study. *Cell Stress Chaperones,* 2006; 11(1):61-70.

[111] Lelj-Garolla B, Mauk AG. Self-association and chaperone activity of Hsp27 are thermally activated. *J Biol Chem.* 2006; 281(12):8169-8174.

[112] Wang X, Klevitsky R, Huang W, Glasford J, Li F, Robbins J. AlphaB-crystallin modulates protein aggregation of abnormal desmin. *Circ Res.* 2003; 93(10):998-1005.

[113] Panasenko OO, Kim MV, Marston SB, Gusev NB. Interaction of the small heat shock protein with molecular mass 25 kDa (hsp25) with actin. *Eur J Biochem.* 2003; 270(5):892-901.

[114] Mounier N, Arrigo AP. Actin cytoskeleton and small heat shock proteins: how do they interact? *Cell Stress Chaperones,* 2002; 7(2):167-176.

[115] Sun Y, MacRae TH. Small heat shock proteins: molecular structure and chaperone function. *Cell Mol Life Sci.* 2005; 62(21):2460-2476.

[116] Brown DD, Christine KS, Showell C, Conlon FL. Small heat shock protein Hsp27 is required for proper heart tube formation. *Genesis*, 2007; 45(11):667-678.

[117] Tessier DJ, Komalavilas P, Panitch A, Joshi L, Brophy CM. The small heat shock protein (HSP) 20 is dynamically associated with the actin cross-linking protein actinin. *J Surg Res.* 2003; 111(1):152-157.

[118] Perng MD, Cairns L, van den IJ, Prescott A, Hutcheson AM, Quinlan RA. Intermediate filament interactions can be altered by HSP27 and alphaB-crystallin. *J Cell Sci.* 1999; 112 (Pt 13):2099-2112.

[119] Bennardini F, Wrzosek A, Chiesi M. Alpha B-crystallin in cardiac tissue. Association with actin and desmin filaments. *Circ Res.* 1992; 71(2):288-294.

[120] Nicholl ID, Quinlan RA. Chaperone activity of alpha-crystallins modulates intermediate filament assembly. *EMBO J.* 1994; 13(4):945-953.

[121] Bova MP, Yaron O, Huang Q, Ding L, Haley DA, Stewart PL et al. Mutation R120G in alphaB-crystallin, which is linked to a desmin-related myopathy, results in an irregular structure and defective chaperone-like function. *Proc Natl Acad Sci U S A*, 1999; 96(11):6137-6142.

[122] Vicart P, Caron A, Guicheney P, Li Z, Prevost MC, Faure A et al. A missense mutation in the alphaB-crystallin chaperone gene causes a desmin-related myopathy. *Nat Genet.* 1998; 20(1):92-95.

[123] Hagemann TL, Boelens WC, Wawrousek EF, Messing A. Suppression of GFAP toxicity by {alpha}B-crystallin in mouse models of Alexander disease. *Hum Mol Genet.* 2009.

[124] Arai H, Atomi Y. Chaperone activity of alpha B-crystallin suppresses tubulin aggregation through complex formation. *Cell Struct Funct.* 1997; 22(5):539-544.

[125] Kato K, Ito H, Inaguma Y, Okamoto K, Saga S. Synthesis and accumulation of alphaB crystallin in C6 glioma cells is induced by agents that promote the disassembly of microtubules. *J Biol Chem.* 1996; 271(43):26989-26994.

[126] Xi JH, Bai F, McGaha R, Andley UP. Alpha-crystallin expression affects microtubule assembly and prevents their aggregation. *FASEB J.* 2006; 20(7):846-857.

[127] Bruey JM, Ducasse C, Bonniaud P, Ravagnan L, Susin SA, Diaz-Latoud C et al. Hsp27 negatively regulates cell death by interacting with cytochrome c. *Nat Cell Biol.* 2000; 2(9):645-652.

[128] Pandey P, Farber R, Nakazawa A, Kumar S, Bharti A, Nalin C et al. Hsp27 functions as a negative regulator of cytochrome c-dependent activation of procaspase-3. *Oncogene.* 2000; 19(16):1975-1981.

[129] Charette SJ, Lavoie JN, Lambert H, Landry J. Inhibition of Daxx-mediated apoptosis by heat shock protein 27. *Mol Cell Biol.* 2000; 20(20):7602-7612.

[130] Mao YW, Liu JP, Xiang H, Li DW. Human alphaA- and alphaB-crystallins bind to Bax and Bcl-X(S) to sequester their translocation during staurosporine-induced apoptosis. *Cell Death Differ.* 2004; 11(5):512-526.

[131] Kamradt MC, Chen F, Cryns VL. The small heat shock protein alpha B-crystallin negatively regulates cytochrome c- and caspase-8-dependent activation of caspase-3 by inhibiting its autoproteolytic maturation. *J Biol Chem.* 2001; 276(19):16059-16063.

[132] Stegh AH, Kesari S, Mahoney JE, Jenq HT, Forloney KL, Protopopov A et al. Bcl2L12-mediated inhibition of effector caspase-3 and caspase-7 via distinct mechanisms in glioblastoma. *Proc Natl Acad Sci U S A,* 2008; 105(31):10703-10708.

[133] Nicolaou P, Knoll R, Haghighi K, Fan GC, Dorn GW, Hasenfub G et al. Human mutation in the anti-apoptotic heat shock protein 20 abrogates its cardioprotective effects. *J Biol Chem.* 2008; 283(48):33465-33471.

[134] Shemetov AA, Seit-Nebi AS, Gusev NB. Structure, properties, and functions of the human small heat-shock protein HSP22 (HspB8, H11, E2IG1): a critical review. *J Neurosci Res.* 2008; 86(2):264-269.

[135] Sui X, Li D, Qiu H, Gaussin V, Depre C. Activation of the bone morphogenetic protein receptor by H11kinase/Hsp22 promotes cardiac cell growth and survival. *Circ Res.* 2009; 104(7):887-895.

[136] den Engelsman J, Keijsers V, de Jong WW, Boelens WC. The small heat-shock protein alpha B-crystallin promotes FBX4-dependent ubiquitination. *J Biol Chem.* 2003; 278(7):4699-4704.

[137] Boelens WC, Croes Y, de Jong WW. Interaction between alphaB-crystallin and the human 20S proteasomal subunit C8/alpha7. *Biochim Biophys Acta,* 2001; 1544(1-2):311-319.

[138] Lin DI, Barbash O, Kumar KG, Weber JD, Harper JW, Klein-Szanto AJ et al. Phosphorylation-dependent ubiquitination of cyclin D1 by the SCF(FBX4-alphaB crystallin) complex. *Mol Cell,* 2006; 24(3):355-366.

[139] Parcellier A, Brunet M, Schmitt E, Col E, Didelot C, Hammann A et al. HSP27 favors ubiquitination and proteasomal degradation of p27Kip1 and helps S-phase re-entry in stressed cells. *FASEB J.* 2006; 20(8):1179-1181.

[140] Koteiche HA, McHaourab HS. Mechanism of a hereditary cataract phenotype. Mutations in alphaA-crystallin activate substrate binding. *J Biol Chem.* 2006; 281(20):14273-14279.

[141] Irobi J, Van Impe K, Seeman P, Jordanova A, Dierick I, Verpoorten N et al. Hot-spot residue in small heat-shock protein 22 causes distal motor neuropathy. *Nature Genetics,* 2004; 36(6):597-601.

[142] Kim MV, Kasakov AS, Seit-Nebi AS, Marston SB, Gusev NB. Structure and properties of K141E mutant of small heat shock protein HSP22 (HspB8, H11) that is expressed in human neuromuscular disorders. *Archives of Biochemistry and Biophysics,* 2006; 454(1):32-41.

[143] Perng MD, Wen SF, van den IJ, Prescott AR, Quinlan RA. Desmin aggregate formation by R120G alphaB-crystallin is caused by altered filament interactions and is dependent upon network status in cells. *Mol Biol Cell,* 2004; 15(5):2335-2346.

[144] Sanbe A, Osinska H, Villa C, Gulick J, Klevitsky R, Glabe CG et al. Reversal of amyloid-induced heart disease in desmin-related cardiomyopathy. *Proc Natl Acad Sci U S A,* 2005; 102(38):13592-13597.

[145] Garrido C, Fromentin A, Bonnotte B, Favre N, Moutet M, Arrigo AP et al. Heat shock protein 27 enhances the tumorigenicity of immunogenic rat colon carcinoma cell clones. *Cancer Res.* 1998; 58(23):5495-5499.

[146] Xu L, Chen S, Bergan RC. MAPKAPK2 and HSP27 are downstream effectors of p38 MAP kinase-mediated matrix metalloproteinase type 2 activation and cell invasion in human prostate cancer. *Oncogene*. 2006; 25(21):2987-2998.

[147] Moyano JV, Evans JR, Chen F, Lu M, Werner ME, Yehiely F et al. AlphaB-crystallin is a novel oncoprotein that predicts poor clinical outcome in breast cancer. *J Clin Invest.* 2006; 116(1):261-270.

[148] Sun X, Fontaine JM, Bartl I, Behnam B, Welsh MJ, Benndorf R. Induction of Hsp22 (HspB8) by estrogen and the metalloestrogen cadmium in estrogen receptor-positive breast cancer cells. *Cell Stress & Chaperones,* 2007; 12(4):307-319.

[149] de Wit NJ, Verschuure P, Kappe G, King SM, de Jong WW, van Muijen GN et al. Testis-specific human small heat shock protein HSPB9 is a cancer/testis antigen, and potentially interacts with the dynein subunit TCTEL1. *Eur J Cell Biol.* 2004; 83(7):337-345.

[150] Pozsgai E, Gomori E, Szigeti A, Boronkai A, Gallyas F, Jr., Sumegi B et al. Correlation between the progressive cytoplasmic expression of a novel small heat shock protein (Hsp16.2) and malignancy in brain tumors. *BMC Cancer*, 2007; 7:233.

[151] Rocchi P, So A, Kojima S, Signaevsky M, Beraldi E, Fazli L et al. Heat shock protein 27 increases after androgen ablation and plays a cytoprotective role in hormone-refractory prostate cancer. *Cancer Res.* 2004; 64(18):6595-6602.

[152] Oesterreich S, Weng CN, Qiu M, Hilsenbeck SG, Osborne CK, Fuqua SA. The small heat shock protein hsp27 is correlated with growth and drug resistance in human breast cancer cell lines. *Cancer Res.* 1993; 53(19):4443-4448.

[153] Oesterreich S, Schunck H, Benndorf R, Bielka H. Cisplatin induces the small heat shock protein hsp25 and thermotolerance in Ehrlich ascites tumor cells. *Biochem Biophys Res Commun.* 1991; 180(1):243-248.

[154] Sanbe A, Daicho T, Mizutani R, Endo T, Miyauchi N, Yamauchi J et al. Protective effect of geranylgeranylacetone via enhancement of HSPB8 induction in desmin-related cardiomyopathy. *PLoS ONE,* 2009; 4(4):e5351.

[155] Hu Z, Chen L, Zhang J, Li T, Tang J, Xu N et al. Structure, function, property, and role in neurologic diseases and other diseases of the sHsp22. *J Neurosci Res.* 2007; 85(10):2071-2079.

[156] Ackerley S, James PA, Kalli A, French S, Davies KE, Talbot K. A mutation in the small heat-shock protein HSPB1 leading to distal hereditary motor neuronopathy disrupts neurofilament assembly and the axonal transport of specific cellular cargoes. *Hum Mol Genet.* 2006; 15(2):347-354.

[157] An JJ, Lee YP, Kim SY, Lee SH, Lee MJ, Jeong MS et al. Transduced human PEP-1-heat shock protein 27 efficiently protects against brain ischemic insult. *FEBS J.* 2008.

[158] Sakamoto K, Urushidani T, Nagao T. Translocation of HSP27 to sarcomere induced by ischemic preconditioning in isolated rat hearts. *Biochem Biophys Res Commun.* 2000; 269(1):137-142.

[159] Bluhm WF, Martin JL, Mestril R, Dillmann WH. Specific heat shock proteins protect microtubules during simulated ischemia in cardiac myocytes. *Am J Physiol.* 1998; 275(6 Pt 2):H2243-H2249.

[160] Schober A, Burger-Kentischer A, Muller E, Beck FX. Effect of ischemia on localization of heat shock protein 25 in kidney. *Kidney Int Suppl.* 1998; 67:S174-S176.

[161] Uozaki H, Horiuchi H, Ishida T, Iijima T, Imamura T, Machinami R. Overexpression of resistance-related proteins (metallothioneins, glutathione-S-transferase pi, heat shock protein 27, and lung resistance-related protein) in osteosarcoma. Relationship with poor prognosis. *Cancer,* 1997; 79(12):2336-2344.

[162] Benjamin IJ, Guo Y, Srinivasan S, Boudina S, Taylor RP, Rajasekaran NS et al. CRYAB and HSPB2 deficiency alters cardiac metabolism and paradoxically confers protection against myocardial ischemia in aging mice. *Am J Physiol Heart Circ Physiol.* 2007; 293(5):H3201-H3209.

[163] Morrison LE, Whittaker RJ, Klepper RE, Wawrousek EF, Glembotski CC. Roles for alphaB-crystallin and HSPB2 in protecting the myocardium from ischemia-reperfusion-induced damage in a KO mouse model. *Am J Physiol Heart Circ Physiol.* 2004; 286(3):H847-H855.

[164] Suzuki A, Sugiyama Y, Hayashi Y, Nyu-i N, Yoshida M, Nonaka I et al. MKBP, a novel member of the small heat shock protein family, binds and activates the myotonic dystrophy protein kinase. *J Cell Biol.* 1998; 140(5):1113-1124.

[165] Yoshida K, Aki T, Harada K, Shama KM, Kamoda Y, Suzuki A et al. Translocation of HSP27 and MKBP in ischemic heart. *Cell Struct Funct.* 1999; 24(4):181-185.

[166] Agius MA, Kirvan CA, Schafer AL, Gudipati E, Zhu S. High prevalence of anti-alpha-crystallin antibodies in multiple sclerosis: correlation with severity and activity of disease. *Acta Neurol Scand.* 1999; 100(3):139-147.

[167] Bullard B, Ferguson C, Minajeva A, Leake MC, Gautel M, Labeit D et al. Association of the chaperone alphaB-crystallin with titin in heart muscle. *J Biol Chem.* 2004; 279(9):7917-7924.

[168] Berry V, Francis P, Reddy MA, Collyer D, Vithana E, MacKay I et al. Alpha-B crystallin gene (CRYAB) mutation causes dominant congenital posterior polar cataract in humans. *Am J Hum Genet.* 2001; 69(5):1141-1145.

[169] Graw J. Genetics of crystallins: Cataract and beyond. Exp Eye Res 2008.

[170] Lee S, Carson K, Rice-Ficht A, Good T. Hsp20, a novel alpha-crystallin, prevents Abeta fibril formation and toxicity. *Protein Sci.* 2005; 14(3):593-601.

[171] Islamovic E, Duncan A, Bers DM, Gerthoffer WT, Mestril R. Importance of small heat shock protein 20 (hsp20) C-terminal extension in cardioprotection. *J Mol Cell Cardiol.* 2007; 42(4):862-869.

[172] Zhu YH, Ma TM, Wang X. Gene transfer of heat-shock protein 20 protects against ischemia/reperfusion injury in rat hearts. *Acta Pharmacol Sin.* 2005; 26(10):1193-1200.

[173] Houlden H, Laura M, Wavrant-De Vrieze F, Blake J, Wood N, Reilly MM. Mutations in the HSP27 (HSPB1) gene cause dominant, recessive, and sporadic distal HMN/CMT type 2. *Neurology,* 2008; 71(21):1660-1668.

[174] Wilhelmus MMM, Boelens WC, Otte-Holler I, Kamps B, Kusters B, Maat-Schieman MLC et al. Small heat shock protein HspB8: its distribution in Alzheimer's disease brains and its inhibition of amyloid-beta protein aggregation and cerebrovascular amyloid-beta toxicity. *Acta Neuropathologica*, 2006; 111(2):139-149.

[175] Yu YX, Heller A, Liehr T, Smith CC, Aurelian L. Expression analysis and chromosome location of a novel gene (H11) associated with the growth of human melanoma cells. *Int J Oncol.* 2001; 18(5):905-911.

[176] Depre C, Wang L, Sui X, Qiu H, Hong C, Hedhli N et al. H11 kinase prevents myocardial infarction by preemptive preconditioning of the heart. *Circ Res.* 2006; 98(2):280-288.

[177] Thompson JD, Higgins DG, Gibson TJ. CLUSTAL W: improving the sensitivity of progressive multiple sequence alignment through sequence weighting, position-specific gap penalties and weight matrix choice. *Nucleic Acids Res*. 1994; 22(22):4673-4680.

[178] Nicholas KB. GeneDoc: Analysis and Visualization of Genetic Variation. Nicholas H.B., Deerfield DW, editors. *EMBNEW.NEWS,* 4(14): 1997.

[179] Baumgartner S, Hofmann K, Chiquet-Ehrismann R, Bucher P. The discoidin domain family revisited: new members from prokaryotes and a homology-based fold prediction. *Protein Sci.* 1998; 7(7):1626-1631.

[180] Guindon S, Gascuel O. A simple, fast, and accurate algorithm to estimate large phylogenies by maximum likelihood. *Syst Biol.* 2003; 52(5):696-704.

[181] Haslbeck M. sHsps and their role in the chaperone network. *Cell Mol Life Sci.* 2002; 59(10):1649-1657.

[182] Fu L, Liang JJ. Detection of protein-protein interactions among lens crystallins in a mammalian two-hybrid system assay. *J Biol Chem.* 2002; 277(6):4255-4260.

[183] Pipkin W, Johnson JA, Creazzo TL, Burch J, Komalavilas P, Brophy C. Localization, macromolecular associations, and function of the small heat shock-related protein HSP20 in rat heart. *Circulation*, 2003; 107(3):469-476.

[184] Fontaine JM, Sun X, Benndorf R, Welsh MJ. Interactions of HSP22 (HSPB8) with HSP20, alphaB-crystallin, and HSPB3. *Biochem Biophys Res Commun.* 2005; 337(3):1006-1011.
[185] Sun X, Fontaine JM, Rest JS, Shelden EA, Welsh MJ, Benndorf R. Interaction of human HSP22 (HSPB8) with other small heat shock proteins. *J Biol Chem* 2004; 279(4):2394-2402.
[186] Kato K, Shinohara H, Goto S, Inaguma Y, Morishita R, Asano T. Copurification of small heat shock protein with alpha B crystallin from human skeletal muscle. *J Biol Chem.* 1992; 267(11):7718-7725.
[187] Benndorf R, Sun X, Gilmont RR, Biederman KJ, Molloy MP, Goodmurphy CW et al. HSP22, a new member of the small heat shock protein superfamily, interacts with mimic of phosphorylated HSP27 ((3D)HSP27). *J Biol Chem.* 2001; 276(29):26753-26761.

Part 1: Protective Functions of Small Heat Shock Proteins, Roles in Conformational, Inflammation and Redox State Related Diseases

In: Small Stress Proteins and Human Diseases
Editors: Stéphanie Simon et al.
ISBN: 978-1-61470-636-6

Chapter 1.1

CYTOPROTECTIVE FUNCTIONS OF SMALL STRESS PROTEINS IN PROTEIN CONFORMATIONAL DISEASES

Serena Carra and Harm H. Kampinga*
Department of Cell Biology, Section of Radiation and Stress Cell Biology, University Medical Center Groningen, University of Groningen, Groningen, The Netherlands

ABSTRACT

The mammalian small heat shock protein family comprises 10 members (HspB1-10) some of which have been implicated indirectly or directly in several neurodegenerative and neuromuscular disorders. Upregulation of some HspB members has been found in brain amyloidosis, where they are often trapped within inclusion bodies, and mutations of several HspB proteins have been associated with muscular and neurological disorders. These two findings strongly suggest an important role for the HspB proteins in the maintaining of neuronal and muscular cells viability. How these molecular chaperones can alleviate neurodegeneration and how their mutation can result in toxicity is still largely unknown. This review will summarize the current knowledge about HspB protein function and implication in neurodegenerative diseases. From this, we suggest that, besides the two most characterized biochemical properties of HspB members (chaperone activity and cytosketelal stabilization), also non-canonical pathways seem relevant to the neuroprotective actions of some HspB members.

* Corresponding Author: Harm H. Kampinga
University Medical Center Groningen
Department of Cell Biology
Section of Radiation and Stress Cell Biology
The Netherlands
Email: h.h.kampinga@med.umcg.nl .

1. Age-Related Protein Conformational Disorders and Small Heat Shock Proteins

Accumulation of aggregated proteins in the form of fibrils and filamentous aggregates, referred to as amyloids, is a pathological hallmark of many age-related protein conformational neurodegenerative disorders, such as Alzheimer, Parkinson, polyglutamine and Creutzfeldt-Jacob diseases [1-3]. The nature of the amyloidogenic protein that forms fibrils defines the type of the disease. Thus, accumulation of synuclein is characteristic of synucleopathies (e.g. Parkinson disease, dementia with Lewy's bodies and Multiple System Atrophy) [4]; aggregation of the microtubule-associated protein tau is typical of tauopathies (e.g. Alzheimer disease, Pick disease, Cortical Basal Degeneration and Progressive Supranuclear Palsy) [5]; aggregation of proteins containing polyglutamine extension is characteristic of the polyglutamine disorders (e.g. Huntington disease, several forms of Spinocerebellar Ataxias, Dentatorubral pallidoluysian atrophy and the Kennedy disease) [1]. In all cases, the disease-causing mutant proteins accumulate within the brain as fibrillar aggregates that are visible by light microscopy, consist of compact β–sheet structures, are highly insoluble and resistant to proteolytic degradation [3]. Such large intracellular or extracellular structures originate from a multi-step nucleation-dependent process in which early stage detergent-soluble pre-fibrillar aggregates are gradually converted into detergent-insoluble fibrillar structures that are associated with a rearranged intermediate filament network. They are heterogeneous in composition and contain, beside the mutated aggregating proteins, several key intracellular elements, including molecular chaperones, component of the proteasome system, transcription factors and/or elements of the vesicular transport machinery [6, 7].

For a long time, these heterogeneous proteinaceous inclusions have been considered as toxic species, which, by entrapping cellular players, can indirectly impair key processes such as proteasome-mediated protein degradation, gene expression and vesicular trafficking [8, 9]. However, the causal relationship between inclusions and disease progression has not yet been demonstrated. In fact, recent findings show that the presence of inclusions does not always correlate with cell death and cytotoxicity [10-14]. These findings have lead to the current hypothesis that the microaggregated and diffusing intermediate species would be toxic, whereas, the macromolecular larger aggregates (inclusions), visible in the postmortem brain, may play a protective function [15, 16]. Evidence from *in vitro* studies showed that the formation of these macromolecular aggregates, also referred to as inclusion bodies, is a well-orchestrated process that requires a functional axonal transport. The inclusions bodies are often found at the microtubule organizing centre area (MTOC), where they originate from the fusion of the toxic microaggregated species, which are retrogradely transported to the MTOC. Here at the MTOC, the microaggregated species are compacted together and are encaged by intermediate filament elements in so-called aggresomes. The encaging decreases the chances that the microaggregated species could interact with and impair the function of key cellular elements [17-20]. Moreover, transporting the intermediate toxic species to the MTOC allows confining them in a region of the cells enriched in autophagic vacuoles and lysosomes, which can easily engulf and digest them. Indeed, a growing body of evidence suggests a major role for autophagy in the degradation of protein aggregates associated with conformational diseases [21-28]. Thus, in order to decrease the cytotoxicity mediated by the intermediate mutated species the cells rely on a functional axonal transport and on the autophagic proteolytic process, which are intimately connected. In fact, fusion of the autophagic vacuoles with the lysosomes, a step that ensure the digestion of their inner content, is impaired by microtubules destabilization [29].

2. Molecular Chaperones as Cellular Defense Mechanism Against Aggregating Substrates

An initial step in the cellular defense mechanisms against aggregating proteins is represented by the molecular chaperones, which include the heat shock protein (Hsp) Hsp70/Hsp40 and the small Hsp (HspB) families [30]. Under physiological conditions, molecular chaperones recognize and bind to unfolded proteins and assist in their folding (e.g. assembly of nascent polypeptide chains) or, when refolding is not possible, target them for degradation [31]. In the presence of mutated aggregating substrates, molecular chaperones can bind them, thus preventing or slowing down their aggregation and can participate in their proteasome-assisted or autophagy-mediated degradation (e.g. chaperone-mediated autophagy) [32, 33]. The neuroprotective role of molecular chaperones in protein conformational disorders has been extensively studied [34]. This is particularly true for members of the Hsp70 machine that showed protective effects in both cellular and sometimes also in animal models of various protein folding diseases [35, 36]. In contrast to Hsp70, the efficacy of the HspB members against mutated aggregate-prone substrates has been studied less intensively and mainly in cellular models [37-40]. Here, we will focus on the various lines of evidence that point to possible canonical and non-canonical roles of the HspB members in neurodegeneration.

3. Histological and Genetic Evidence for a Role of HspB Members in Protein Misfolding-Related Neurodegenerative Diseases

The first suggestion that HspB proteins may exert a protective role in protein conformational neurodegenerative disorders comes from immunohistochemical studies on post-mortem brain. Here it was found that HspB1, HspB5 and HspB8 are up-regulated both in glial cells and neurons in areas characterized by reactive gliosis [38, 39, 41-51]. Reactive gliosis is a reaction to brain injury and is characterized by morphological and functional changes of astrocytes aimed at protecting the surrounding neuronal population from toxic insults. Following brain damage, astrocytes undergo hypertrophy and become phagocytic to digest cellular debris [52]. Interestingly, the highest expression levels of HspB proteins are often observed in reactive astrocytes, which provide essential activities that preserve neuronal function. Although the significance of HspB up-regulation in glial cells (mainly astrocytes) is not yet completely understood, it may be part of the adaptive change and response to neuronal damage.

The second evidence that suggests a neuroprotective role for HspB comes from recent findings that mutations in several members of the HspB family (HspB1, HspB4, HspB5 and

HspB8) were found to be associated with neurological and muscular disorders [53-63]. It is thus likely that perturbing the HspB function may have detrimental effects for specific cell types. In particular, these mutations can lead to a gain of toxic function and/or to the loss of the chaperone activity, which may both contribute to the development of these diseases (for detailed information see also part 3 of the book).

Third, several HspB members can attenuate the aggregation of mutated toxic proteins *in vitro* and after overexpression in mammalian cells to protect against their mediated cytotoxicity. Furthermore, recently, some members of the *Drosophila melanogaster* small heat shock protein family have been shown to protect against mutated polyglutamine-induced neurodegeneration *in vivo* [64].

All together these results suggest that HspB proteins may exert important functions for the maintenance and viability of neuronal and muscular cells and that their up-regulation may increase the cellular defense against misfolded aggregating proteins. However, the precise mechanism of action of the HspB proteins still remains elusive. Interestingly, some members of the HspB family play an important role in cytoskeleton stabilization (actin-based microfilaments and intermediate filaments), whereas a role in the modulation of autophagy is emerging for HspB8 [65-72]. As previously described, both axonal transport and autophagy are closely connected and are critical for aggresome formation and clearance [17, 29]. The relevance of HspB role in cytoskeleton stabilization and autophagy modulation, as well as the relevance of other specialized HspB functions (e.g. anti-oxidant function), in their neuroprotective role will be discussed in more detail below.

4. Mammalian HspB Neuroprotective Function May Be, at Least in Part, Independent on their "Chaperone-Like" Activity

The human small Hsp family, also called HspB, comprises ten low molecular weight (15-40 kDa) members called, according to the new nomenclature, HspB1-HspB10 (see Table 1) [73]. HspB protein expression differs according to the type of tissue; some members are quite ubiquitously expressed, while others are found only in specific tissues (see Table 1); moreover, the expression of several HspB members can be induced by heat and/or other stresses [74]. Human HspB, like small Hsp from other organisms (e.g. Archea, bacteria, yeast, plant), are characterized by the presence in their C-terminus of a highly conserved sequence of 80-100 amino acids called the α-crystallin domain [30]. The α-crystallin domain is responsible for intra- and inter-molecular interactions leading to the formation of homo- and hetero-dimers [75]. Dimers are considered as the basic unit of the small Hsps; they can associate into higher molecular weight structures (both homo- and hetero-oligomers can exist). The oligomerization of small Hsps is a highly dynamic process, which is believed to regulate their function, and (in part) depends on their (de)phosphorylation status as demonstrated using pseudo-phosphorylation and non-phosphorylatable mutants [76]. Under physiological conditions most small Hsps mainly exist as oligomers, which are considered to be a reservoir of the cellular chaperone power. Upon stressful conditions, the oligomers would dissociate into dimers, which represent the active units that can bind unfolded/misfolded substrates, thus avoiding their irreversible aggregation [66, 77].

To date, two biochemical functions have been attributed to HspB proteins: the chaperone-like activity (first characterized for the Archea, yeast and plant small Hsps) and the ability to stabilize the cytoskeleton.

Like for the small Hsps from other organisms, the chaperone-like activity of human HspB has been initially studied using *in vitro* systems. The first studies were focussing on the few members discovered in the past, namely HspB1 (Hsp27), HspB4 (αA-crystallin) and HspB5 (αB-crystallin) [78, 79]. In these assays, recombinant HspB proteins were incubated with either pure unfolded substrates or with cellular extracts containing both exogenous unfolded substrates (e.g. luciferase) and all the other members of the molecular chaperone superfamily. This allowed directly testing of the ability of these HspB members to act as "pure" molecular chaperones. Using such cell-free assays, it has been shown that HspBs can bind to unfolded substrates and prevent their irreversible aggregation. In general, substrate dissociation and its renaturation is not supported by HspB members alone and depends on the Hsp70 ATP-dependent molecular chaperones [80]. In contrast to the Hsp70s, HspB binding to unfolded substrates is ATP-independent and is tightly correlated to their dynamic nature (dimers oligomerization/dissociation) [30, 81]. Thus, for example, the dissociation of HspB1 oligomers is essential for binding non-native substrates [82]. Several of the observations done in these cell-free systems were confirmed using cell lines, thus underlying the importance of such studies in the understanding of HspB structure and chaperone-like properties [30]. However, in such cell-free systems, HspBs interaction with other intracellular elements, which may be required in order to exert their protective functions, may be absent. In addition, post-translational modifications of HspBs, which can e.g. affect their oligomerization dynamics, may also be altered and, eventually contribute to modify their mode of action in the cellular versus the cell-free context.

The second, and best studied, function of the mammalian HspB is the ability to interact with and/or modulate the structure, stability and dynamics of the cytoskeleton (see chapter 1.2) [70]. In contrast to the chaperone activity, the effects of mammalian HspB on the structure and dynamics of the cytoskeleton have been mostly studied in cell culture. In particular, the role of HspB1, HspB4, HspB5 and HspB6 in the stabilization of microfilaments and intermediate filaments is well established [65-70, 83, 84]. In all cases, like for the chaperone activity, the effect of HspB on cytoskeletal elements polymerization/depolymerization and stability depends on their phosphorylation state. The dissociation of the HspB oligomers is essential for their interaction with and/or modulation of the structure and dynamics of the cytoskeleton [65, 66, 85]. This is in line with the *in vitro* data showing that the HspB dimers, as well as the oligomers that are able to rapidly change their oligomerization state, mainly exert chaperone activity [86].

The role of some HspB members in cytoskeleton stabilization (actin based microfilaments and intermediate filaments) could be crucial for axonal transport, aggresome formation and autophagy that help clearing protein aggregates [17, 29]. Indeed, a role for HspB8 in the modulation of autophagy is emerging [71, 72]. It may thus be argued that in the presence of misfolded aggregate-prone proteins, the HspBs recognize and bind them, facilitating either directly (through binding; chaperone function; e.g. HspB1, HspB4, HspB5) or indirectly (through the stabilization of the cytoskeleton; e.g. HspB1, HspB4, HspB5, HspB6) their transport towards the MTOC, where they are confined and can be easily digested within the autophagic vacuoles. This would be sufficient to protect the cells.

However, the data obtained using both cellular and animal models of protein conformational disorders suggest a picture that is much more complex and that goes beyond these two characterized functions of HspBs. In contrast, some data suggest that, depending on the type of misfolded substrate, the HspB chaperone-like function does not seem to play a role towards unfolded protein *per se*; instead, other specialized functions like the ability to modulate the oxidative-stress response (HspB1) or protein synthesis (HspB8) seem to be involved.

Table 1. The mammalian small Hsp (HspB) family: nomenclature and expression

HspB name	previous alternative names	tissue expression	function
HspB1	Hsp27, Hsp25 (mouse) not related to *D. Melanogaster* Hsp27	muscles (smooth, skeletal and cardiac), brain, colon, kidney, liver, lens, lung, stomach, testis	cytoskeleton stabilization; anti-apoptotic; anti-oxidant
HspB2	MKBP	skeletal and cardiac muscles	activation of DMPK, maintaining myofibrillar integrity
HspB3	HspL27	muscles (smooth, skeletal and cardiac)	maintaining myofibrillar integrity
HspB4	αA-crystallin	lens	maintaining the proper refractive index in the lens
HspB5	αB-crystallin	muscles (smooth, skeletal and cardiac), brain, colon, kidney, liver, lens, lung, stomach, testis	cytoskeleton stabilization; maintaining the proper refractive index in the lens
HspB6	Hsp20, p20	muscles (smooth, skeletal and cardiac), brain, colon, kidney, liver, lung, stomach	smooth muscle relaxation; cardioprotection
HspB7	cvHsp	skeletal and cardiac muscles	maintaining myofibrillar integrity
HspB8	Hsp22, H11, E2IG1, not related to *D. Melanogaster* Hsp22	muscles (smooth, skeletal and cardiac), brain, colon, kidney, lung, placenta, skin, stomach, testis	anti-aggregation; protein synthesis inhibitor; autophagy stimulator
HspB9	none	testis	cancer/testis antigen
HspB10	ODFP, ODF1	testis	cytoskeleton stabilization

In this review, we will further concentrate on three members of the HspB family, namely HspB1, HspB5 and HspB8, for which examples of their neuroprotective role are provided in the literature. For all these proteins, both canonical chaperone-like and non-canonical non-chaperone like activities, according to the kind of disease and misfolded substrate, are described.

4.1. HspB1

HspB1 is one of the most well studied members of the HspB family, highly expressed under physiological conditions in several tissues [87]. The neuroprotective effects of HspB1 have been experimentally demonstrated using both cell cultures and animal models of acute diseases (e.g. ischaemia/reperfusion, nerve injury and kainate-induced seizures) [88-90]. In these models, the ability of HspB1 to decrease neurodegeneration and cell death was supposed to be due to a complex mechanism of action, which includes its canonical chaperone-like function, but also its role in the oxidative stress response, its anti-apoptotic function and its role in the cytoskeleton stabilization [91-94]. How these events are interconnected is, however, not yet completely understood.

Interestingly, HspB1 seems to also display protective functions in chronic, late-onset protein conformation neurodegenerative disorders, where its expression levels have been found up-regulated, both in glial and neuronal cells [41, 44, 46, 48, 51]. Among the disease models that have been extensively studied and show a protective function of HspB1 are the models of Huntington's Disease, Amyotrophic Lateral Sclerosis and tauopathies. Even if some discrepancies exist between the data obtained using cellular versus animal models of these diseases, a general assumption can be proposed. Whereas the protective function of HspB1 towards mutated huntingtin and SOD1 seems more likely due to HspB1 role in the modulation of the oxidative stress signalling pathway, its ability to protect against hyperphosphorylated tau would be mainly ascribed to its chaperone-like activity. HspB1 implication in these three disorders and its putative mechanism of action will be discussed in detail.

4.1.1. HspB1 and Huntington Disease

Huntington disease is a neurological disorder due to mutation of the gene encoding for the protein huntingtin. The mutation causes an abnormal expansion of the CAG repeat, which results in the abnormal length of the polyglutamine tract of the huntingtin proteins, causing its instability, aggregation and gain of toxicity [1].

Overexpression of HspB1 has shown some beneficial effects in cellular models of Huntington disease (HD). Here, HspB1 was not able to decrease mutated huntingtin aggregation, nor was found to interact or colocalize with the aggregates [40]. Moreover, in this cellular model, the large HspB1 oligomers, and not the low-molecular weight phosphorylated oligomers (that show chaperone-like activity *in vitro*), were able to protect the cells against polyglutamine-mediated oxidative stress and cell death [94, 95]. Combined, these data suggest a mechanism of action unrelated to a direct chaperone action on the mutant protein. One possibility could be the suggested effects of HspB1 on the cellular redox state [92, 94, 96]. Indeed, in the cell models HspB1 reduced the mutant huntingtin-related production of reactive oxygen species (ROS) [40], a characteristic feature that is also seen in

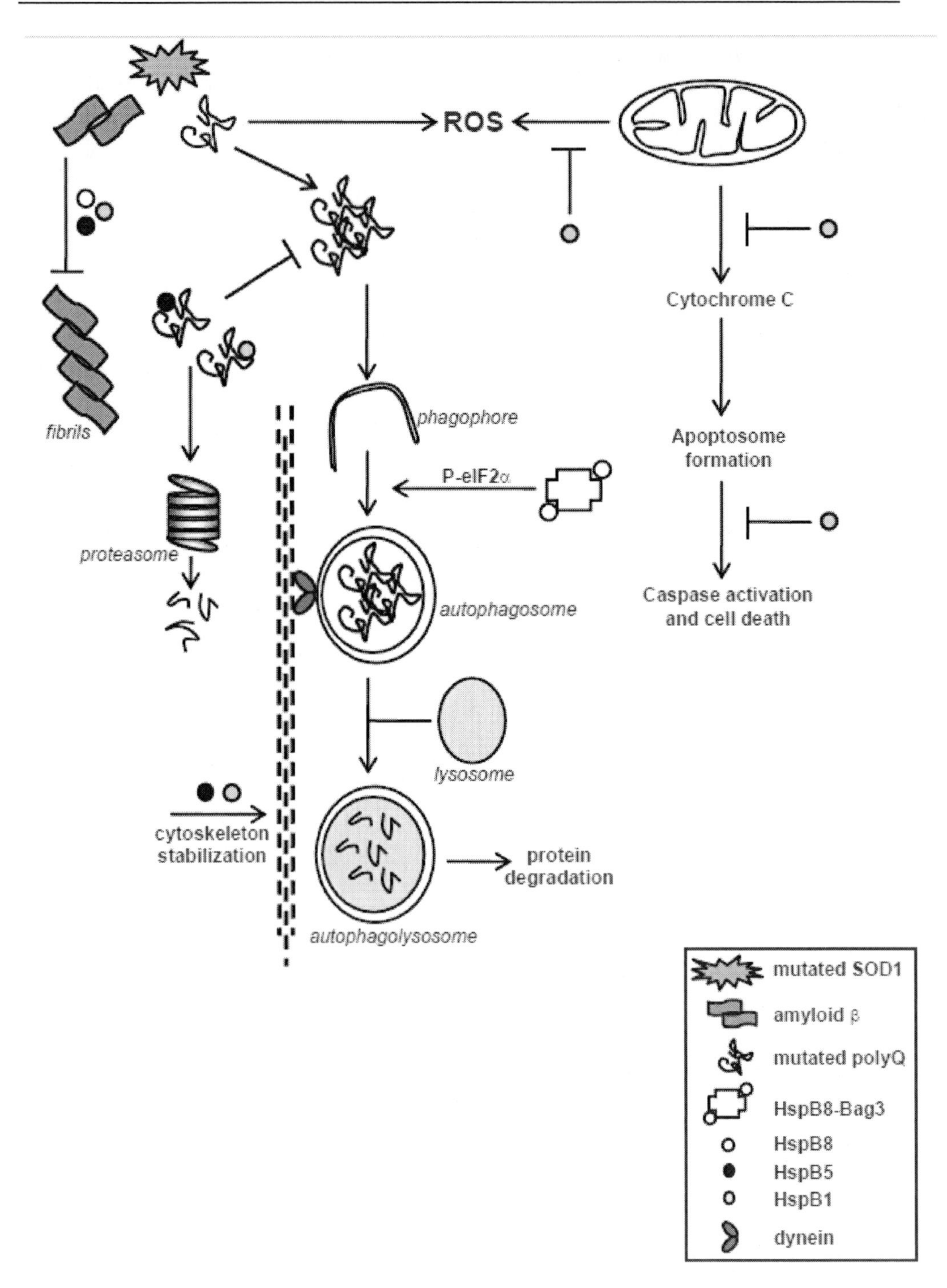

Figure 1. Molecular chaperones facilitate protein folding, prevent protein aggregation and exert neuroprotective functions. The protective activity of the HspB proteins can be due to the direct binding to the mutated protein and/or can result from the modulation of specific stress pathways (e.g. inhibition of ROS production and of caspase activation by HspB1), from the stimulation of the autophagy pathway (e.g. HspB8-Bag3 complex) or from the stabilization of the cytoskeleton (e.g. HspB1, HspB5), whose integrity is essential for dynein-mediated vesicle and autophagosome trafficking.

both human and mouse HD models [97-101]. Yet, in the mouse R6/2 HD animal model, HspB1 overexpression did not show any protection [102]. However, in this mouse model, HspB1 was not able to form high molecular weight oligomers, which were held responsible for the reduction of ROS levels in cell culture models of HD [40, 94]. In a rat HD model, however, HspB1 did show a protective effect against mutated huntingtin mediated cytotoxicity, which, like in cell models, was not associated with reduced aggregation [103]. Although no data are available neither on the oligomeric status of HspB1 nor on ROS levels, one may speculate that the protective function of HspB1 towards mutated polyglutamine toxicity is mainly due to its ability to modulate the oxidative stress response. How exactly HspB1 decreases the accumulation of ROS is still unknown.

4.1.2. HspB1 and Amyotrophic Lateral Sclerosis

ALS is an adult-onset neurodegenerative disorder characterized by the death of motor neurons in the cortex, brainstem and spinal cord, which include both sporadic and familiar cases. In 20% of the familial cases of ALS, the mutations in superoxide dismutase-1 (Cu,Zn SOD/SOD1), are causative for the disease through a toxic gain of function [104, 105]. Several factors have been implicated in the development of ALS and in the toxicity of mutated SOD1. These include: 1) the aggregation of mutated SOD1 and the sequestration of important intracellular elements, like molecular chaperones, ubiquitin [106-108]; 2) the production and accumulation of harmful superoxide radicals [109]; 3) mitochondrial dysfunction, which impairs the energy production, affecting particularly motor neurons, characterized by high energy demands [110]; 4) alteration of the neurofilament network structure and stability [111, 112].

HspB1 overexpression was found to protect cultured cells against mutated SOD1-mediated toxicity and HspB1 was found to be up-regulated in the spinal cord of mutant SOD1(G93A) mice [113, 114]. Moreover, in double transgenic mice expressing mutated SOD1(G93A) and human HspB1, the small heat shock protein could provide a significant improvement in motor strength and increased survival of spinal motor neurons, although this effect was only significant in the early and not late stage of the disease, and could not be reproduced by another independent group [115, 116].

In analogy with the R6/2 Huntington disease mouse model, discrepancies between the protective effects of HspB1 in cell models versus mouse models exist. However, part of these discrepancies could be explained by the observation that in both the R6/2 Huntington disease and the SOD1(G93A) mouse models, a failure in HspB1 oligomerization *in vivo* occurs. This change in HspB1 oligomerization dynamics has been suggested as one of the causes for the inability of HspB1 to protect against neurodegeneration *in vivo*. Further studies are needed to better understand how changes in HspB1 oligomerization can influence its neuroprotective function.

4.1-3. HspB1 and Tauopathies

Another group of diseases in which the modulation of HspB1 expression seems to have beneficial effects is represented by the tauopathies that are a class of neurodegenerative disorders characterized by the aggregation of tau in neurofibrillar tangles. Tauopathies include Alzheimer and Pick diseases, Progressive supranuclear palsy and Corticobasal degeneration. The aggregation of tau, which is a microtubule-associated protein, occurs in the absence of any mutation, but is due to its hyperphosphorylation. Hyperphosphorylated tau becomes partially insoluble and accumulates in form of amorphous aggregates in the cell body and axons of neurons, thus initiating a cascade of events leading to microtubule destabilization, axonal transport disruption and, finally, cell death. In fact, increasing tau phosphorylation (by either knocking out Pin1 or inhibiting the phosphatase PP2A activity) resulted in increased tau-mediated toxicity [117-119]. In contrast to the HD and ALS models, the experimental data available so far suggest for the tauopathies a canonical chaperone-like mechanism of action of HspB1. Using Alzheimer Disease post-mortem brain tissue and a cellular model of tauopathies it has been shown, by both immunohistochemical and immunoprecipitation techniques, that HspB1 selectively interacts with the pathological hyperphosphorylated form of tau. Moreover, HspB1 was able to facilitate the proteasome-mediated degradation of hyperphosphorylated tau and to decrease the global phosphorylation of tau likely by interacting with it and modifying its conformation [120]. Although the tauopathies may represent a class of disorders in which HspB1 protection may be due to its canonical chaperone-like function, HspB1-mediated cytoskeleton stabilization may also be involved.

4.2. HspB5

HspB5 (αB-crystallin) has been initially characterized as a lens protein, where it interacts with HspB4 (αA-crystallin). HspB5, however, is expressed in many other tissues, including heart, skeletal muscles, brain, kidney, placenta and lungs [87]. HspB5 has been implicated, either directly or indirectly in several neurological and/or muscular disorders. Mutations in HspB5 have been directly associated with congenital cataract and desmin-related myopathy [59, 60, 63, 121], whereas abnormal expression and/or localization of HspB5 to specific brain areas have been documented in Alzheimer, Huntington, Pick, Diffuse Lewy body and Parkinson diseases [45-47, 49-51, 122].

To date, the neuroprotective role of HspB5 has been studied using cellular and/or animal models of the following disorders: congenital cataract, Parkinson disease, Alexander disease and Multiple Sclerosis. As observed for HspB1, depending on the type of disease and of the misfolded substrate, HspB5 may exert protective effects which depend essentially on its chaperone-like function (Congenital cataract, Parkinson and Alexander diseases) or which rely on the modulation of complex stress response pathways (Multiple Sclerosis).

4.2.1. HspB5 and Congenital Cataract

In the eyes, HspB5, together with the HspB4 and the other crystallin proteins (beta and gamma), is responsible for transparency. Thus, it is not surprising that mutations in the crystallin proteins (HspB4, HspB5, beta and gamma crystallins) have been associated with the development of cataract, which is a condition characterized by the progressive opacification of the eye and obstruction of the passage of light [123]. One of the best studied functions of HspB5 and HspB4 in the eyes is their chaperone activity, which protects crystallins themselves, as well as other key structural and functional components of the lens, from chemical- or heat-induced denaturation and aggregation, thus maintaining the transparency of the lens. Intriguingly, the mutations of HspB5 and HspB4 cause protein instability and aggregation, thus compromising their chaperone-like function [60, 124]. Besides, gain of dominant negative properties, which inhibits the chaperone activity of other wild type crystallins, have been also reported [60, 125].

More details on the functions of HspB4 and HspB5 are given in chapters 1-3 and 2-2.

4.2.2. HspB5 and Parkinson

Parkinson disease is a common neurodegenerative disorder, with a complex pathogenesis, characterized by the dysfunctional regulation/mutation of many different genes, including the genes encoding for α–synuclein, parkin, PTEN induced putative kinase 1 (PINK1), DJ-1, Leucine Rich Repeat Kinase 2 (LRRK2) and ATP13A2 [126]. Parkinson disease is characterized by the presence of Lewy bodies which are α–synuclein filamentous structures [127, 128]. Interestingly, HspB5 is also found in Lewy bodies. The precise significance of HspB5 localization to Lewy bodies is not fully understood, but it is suggested that both the up-regulation of HspB5 and its localization to the pathological inclusions may be aimed at decreasing the aggregation rate and may represent an adaptive response to chronic stress [129]. This is further suggested by the evidence that, *in vitro*, HspB5 interacts with and modulates the fibrillization of both wild-type and mutated α–synuclein (A30P and A53T), thus favouring the formation of amorphous aggregates which can be more easily digested [129].

4.2.3. HspB5 and Alexander Disease

Alexander disease is yet another example of neurological disorder in which modulation of HspB5 shows protective effects, which are likely due to its chaperone function. Alexander disease is a neurodegenerative genetic disorder caused by mutations of the gene encoding for the glial fibrillary acidic protein (GFAP), the major astrocytic intermediate filament protein [42]. Alexander disease is characterized by the progressive loss of the myelin sheath and by the aggregation of mutated GFAP, which accumulates forming astrocytic inclusion bodies called Rosenthal fibers [130]. Together with GFAP, HspB5 is one of the main components of Rosenthal fibers [41, 42]. Using mouse astrocytes as cellular system, it has been demonstrated that overexpression of HspB5 inhibits GFAP aggregation, facilitating its relocalization into the filament network [131]. This suggests that the chaperone-like function of HspB5 may participate in its protective role against GFAP inclusion formation. The GFAP mutants also induce a stress response resulting in the activation of the stress kinase Jnk and, consequently, in induction of apoptosis [131, 132]. Increased activity of Jnk in the presence of mutated GFAP has been confirmed in samples from Alexander disease patients [133] and HspB5 can attenuate the activation of the MAP kinase signal transduction pathway [134]. The reduced activation of the Jnk stress kinase pathway may, however, just be an indirect consequence of the HspB5-mediated decrease of mutated GFAP aggregation and stabilization of the intermediate filaments. In such a scenario, HspB5 chaperone activity would represent the first step required to inhibit the toxic cascade of events leading to protein aggregation and to the stimulation of the Jnk stress kinase pathway [131]. However, it is not yet clear whether the chaperone-like and anti-apoptotic activities of HspB5 are mutually linked.

4.2.4. HspB5 and Multiple Sclerosis

Multiple Sclerosis, also known as disseminated sclerosis or encephalomyelitis disseminata, is an autoimmune disease characterized by inflammation and progressive demyelination, due to the loss of oligodendrocytes [135]. Curiously, HspB5 is expressed at elevated levels in oligodendrocytes and astrocytes around the affected regions in Multiple Sclerosis [136]. Initially, a pathogenic role for HspB5 has been proposed: HspB5 has been for long considered as a candidate auto-antigen involved in the inflammatory response and in the progression of the disease [136]. However, recently it has been shown that HspB5 plays a protective function in autoimmune demyelination by inhibiting the inflammatory response by suppression of the NFkB pathway [137]. Which biochemical function of HspB5 causes this anti-inflammatory response remains to be elucidated.

4.3. HspB8

HspB8 (Hsp22/H11/E2IG1) has been cloned only recently [138] and its functional properties only start to be characterized. However, several findings suggest a potential implication for HspB8 in neurodegenerative disorders, in particular in peripheral neuropathies (two mutations in HspB8 have been identified as the cause of the disease, see chapter 2-5), Alzheimer and polyglutamine disorders.

4.3.1. HspB8 and Alzheimer Disease

Alzheimer disease is the most common form of dementia, characterized by the presence in the brain of abnormally folded amyloid-β deposits. These deposits found outside and around neurons, are insoluble and are often referred to as plaques (extracellular) or tangles (intracellular).

In analogy with HspB1 and HspB5, HspB8 has been found associated with senile plaques in Alzheimer disease brains, where it may play a role in the protective stress response to prevent the aggregation of the amyloid-β protein [38, 39]. Such a protective effect has been further suggested by experimental evidences using cell-free systems and mammalian cells. In cell-free systems, HspB8 can directly interact with amyloid-β, thus decreasing its aggregation through a canonical chaperone-like mechanism [39]. Consistently, in cells its transient transfection decreased both amyloid-β aggregation and mediated cell death [39]. Interestingly, similar results were obtained using HspB1 and HspB5, thus suggesting that several members of the HspB family, with different structural properties, share chaperone-like activities towards the amyloid-β substrate [38]. A role for HspB8 in the stimulation of the macroautophagy pathway has been recently shown [71] (see also paragraph below). This raises the possibility that, in cells, the ability of HspB8 to decrease the toxicity mediated by amyloid-β may not only depends on its chaperone-like activity, but it may also be due to the HspB8-mediated stimulation of the macroautophagy machinery, which would facilitate amyloid-β degradation.

4.3.2. HspB8 and Polyglutamine Disorders

HspB8 also shows "chaperone-like" activity in cells towards other aggregate-prone substrates, including mutated polyglutamine proteins [37]. Although no data are yet available about HspB8 expression levels and distribution in human brains affected by polyglutamine disorders, HspB8 overexpression in mammalian cells resulted in decreased mutated polyglutamine proteins aggregation (both mutated huntingtin and androgen receptor, responsible for Huntington and Kennedy diseases, respectively), thus extending its chaperone-like effect to different and structurally unrelated misfolded substrates [37]. However, from the experimental data available, it seems unlikely that the protective effects of HspB8 towards mutated polyglutamine proteins are due to its chaperone function. To date no experimental evidence is available for a direct interaction of HspB8 with mutated polyglutamine proteins nor in cell-free systems nor in cells. In fact, recent data suggest a novel, non-canonical mechanism of action for HspB8 [71]. Unlike many other HspB members, which form high molecular weight oligomers, HspB8 mainly forms monomers or dimers both *in vitro* and when (over)expressed in cells [71, 139]. In cells, HspB8 shows no major interaction with other endogenous HspB proteins but rather exists in a stoichiometric and stable complex with the Hsc70/Hsp70 co-chaperone Bag3 [71]. The interaction of HspB8 with Bag3 is important not only for HspB8 stability, but also for its function. Transient transfection of both HspB8 and Bag3 decreased mutated polyglutamine aggregation through a mechanism that requires protein synthesis inhibition and autophagy stimulation, via the induction of phosphorylation of the translation initiation factor eIF2α [140]. Interestingly, a direct link between eIF2α phosphorylation and autophagy activation has been demonstrated and both mechanisms help the cells to protect against the accumulation of unfolded/misfolded proteins [141]. This has been further supported by the recent finding that the pharmacological drug rapamycin, which simultaneously inhibits protein synthesis and stimulates autophagy, exerts neuroprotective functions in cellular and animal models of polyglutamine disorders [22, 24, 142]. The novelty and the non-canonical aspect of this function of HspB8-Bag3 was further strengthened by the observation that it is totally independent on Hsc70/Hsp70, as a deletion mutant of Bag3 lacking the BAG domain and thus unable to bind to Hsc70/Hsp70, is as efficient as full length Bag3 in decreasing mutated huntingtin total levels and aggregation [72, 140]. All together, these observations point to a non-canonical mode of action for the HspB8-Bag3 chaperone complex that seems not to involve direct substrate-chaperone interaction.

Interestingly, the eIF2α-signalling pathway is activated upon stressful conditions, including treatments with heavy metals, heat shock, and pathogen infection. All these stresses result in the accumulation of misfolded proteins and in the formation of aggresome-like structures (like observed in many protein conformational disorders). Curiously, all these stresses also induce HspB8 and Bag3 expression [143-145]. It is thus tempting to speculate that HspB8 and Bag3 may be part of a stress signalling pathway activated by the cells to protect against the accumulation of toxic aggregating species, with a mechanism that seems to extend beyond their role as "chaperones". Yet the precise mechanism of action of HspB8-Bag3 and how they induce the eIF2α signalling pathway remains to be elucidated.

Conclusion

The human small heat shock protein family comprises ten members. Several members of the family are up-regulated in protein conformational disorders or their mutations are directly associated with neurological and neuromuscular disorders. Both these findings strongly suggest an important role for HspB proteins in the modulation of neuronal and muscular cell viability. To date, how HspB proteins protect against misfolded protein-mediated toxicity is not completely understood. Do HspBs only recognize and bind to the misfolded substrates? Do they hold these misfolded substrates for the Hsp70 chaperone machinery, which would then further process them? Or do HspBs also participate, non-canonically, in the refolding or degradation of clients? Moreover, can the HspB proteins directly modulate the stress signaling pathways aimed at decreasing the misfolded substrate concentration and mediated-toxicity? All together the data available in the literature strongly suggest that the HspB proteins probably do not only hold the aggregate-prone substrate, but also participate in a more active manner in the modulation of its processing (e.g. HspB8 and stimulation of macroautophagy). Moreover, HspB proteins may help the cells in the (in)activation of crucial stress response signaling pathways aimed at neuroprotection (e.g. inhibition of ROS production by HspB1, inhibition of the inflammatory response by HspB5, inhibition of protein synthesis and autophagy stimulation by HspB8). All these functions likely participate in HspB pro-survival functions. The future challenge will be to understand whether and how HspB chaperone activity and their ability to modulate the stress signaling pathways are interconnected.

REFERENCES

[1] Cummings CJ, Zoghbi HY. Trinucleotide repeats: mechanisms and pathophysiology. *Annu Rev Genomics Hum Genet.* 2000;1:281-328.

[2] Bucciantini M, Giannoni E, Chiti F, Baroni F, Formigli L, Zurdo J, et al. Inherent toxicity of aggregates implies a common mechanism for protein misfolding diseases. *Nature,* 2002 Apr 4;416(6880):507-11.

[3] Skovronsky DM, Lee VM, Trojanowski JQ. Neurodegenerative diseases: new concepts of pathogenesis and their therapeutic implications. *Annu Rev Pathol.* 2006;1:151-70.

[4] Bennett MC. The role of alpha-synuclein in neurodegenerative diseases. *Pharmacol Ther.* 2005 Mar;105(3):311-31.

[5] Lee VM, Goedert M, Trojanowski JQ. Neurodegenerative tauopathies. *Annu Rev Neurosci.* 2001;24:1121-59.

[6] Ross CA, Pickart CM. The ubiquitin-proteasome pathway in Parkinson's disease and other neurodegenerative diseases. *Trends Cell Biol.* 2004 Dec;14(12):703-11.

[7] Waelter S, Boeddrich A, Lurz R, Scherzinger E, Lueder G, Lehrach H, et al. Accumulation of mutant huntingtin fragments in aggresome-like inclusion bodies as a result of insufficient protein degradation. *Mol Biol Cell,* 2001 May;12(5):1393-407.

[8] Bence NF, Sampat RM, Kopito RR. Impairment of the ubiquitin-proteasome system by protein aggregation. *Science,* 2001 May 25;292(5521):1552-5.

[9] Sugars KL, Rubinsztein DC. Transcriptional abnormalities in Huntington disease. *Trends Genet.* 2003 May;19(5):233-8.

[10] Cummings CJ, Reinstein E, Sun Y, Antalffy B, Jiang Y, Ciechanover A, et al. Mutation of the E6-AP ubiquitin ligase reduces nuclear inclusion frequency while accelerating polyglutamine-induced pathology in SCA1 mice. *Neuron.* 1999 Dec;24(4):879-92.

[11] Saudou F, Finkbeiner S, Devys D, Greenberg ME. Huntingtin acts in the nucleus to induce apoptosis but death does not correlate with the formation of intranuclear inclusions. *Cell,* 1998 Oct 2;95(1):55-66.

[12] Klement IA, Skinner PJ, Kaytor MD, Yi H, Hersch SM, Clark HB, et al. Ataxin-1 nuclear localization and aggregation: role in polyglutamine-induced disease in SCA1 transgenic mice. *Cell,* 1998 Oct 2;95(1):41-53.
[13] Taylor JP, Tanaka F, Robitschek J, Sandoval CM, Taye A, Markovic-Plese S, et al. Aggresomes protect cells by enhancing the degradation of toxic polyglutamine-containing protein. *Hum Mol Genet.* 2003 Apr 1;12(7):749-57.
[14] Arrasate M, Mitra S, Schweitzer ES, Segal MR, Finkbeiner S. Inclusion body formation reduces levels of mutant huntingtin and the risk of neuronal death. *Nature,* 2004 Oct 14;431(7010):805-10.
[15] Schaffar G, Breuer P, Boteva R, Behrends C, Tzvetkov N, Strippel N, et al. Cellular toxicity of polyglutamine expansion proteins: mechanism of transcription factor deactivation. *Mol Cell.* 2004 Jul 2;15(1):95-105.
[16] Bennett EJ, Bence NF, Jayakumar R, Kopito RR. Global impairment of the ubiquitin-proteasome system by nuclear or cytoplasmic protein aggregates precedes inclusion body formation. *Mol Cell,* 2005 Feb 4;17(3):351-65.
[17] Kopito RR. Aggresomes, inclusion bodies and protein aggregation. *Trends Cell Biol.* 2000 Dec;10(12):524-30.
[18] Johnston JA, Illing ME, Kopito RR. Cytoplasmic dynein/dynactin mediates the assembly of aggresomes. *Cell Motil Cytoskeleton,* 2002 Sep;53(1):26-38.
[19] Johnston JA, Ward CL, Kopito RR. Aggresomes: a cellular response to misfolded proteins. *J Cell Biol.* 1998 Dec 28;143(7):1883-98.
[20] Garcia-Mata R, Bebok Z, Sorscher EJ, Sztul ES. Characterization and dynamics of aggresome formation by a cytosolic GFP-chimera. *J Cell Biol.* 1999 Sep 20;146(6):1239-54.
[21] Ravikumar B, Acevedo-Arozena A, Imarisio S, Berger Z, Vacher C, O'Kane CJ, et al. Dynein mutations impair autophagic clearance of aggregate-prone proteins. *Nat Genet.* 2005 Jul;37(7):771-6.
[22] Ravikumar B, Duden R, Rubinsztein DC. Aggregate-prone proteins with polyglutamine and polyalanine expansions are degraded by autophagy. *Hum Mol Genet.* 2002 May 1;11(9):1107-17.
[23] Ravikumar B, Rubinsztein DC. Role of autophagy in the clearance of mutant huntingtin: a step towards therapy? *Mol Aspects Med.* 2006 Oct-Dec;27(5-6):520-7.
[24] Ravikumar B, Vacher C, Berger Z, Davies JE, Luo S, Oroz LG, et al. Inhibition of mTOR induces autophagy and reduces toxicity of polyglutamine expansions in fly and mouse models of Huntington disease. *Nat Genet.* 2004 Jun;36(6):585-95.
[25] Bjorkoy G, Lamark T, Brech A, Outzen H, Perander M, Overvatn A, et al. p62/SQSTM1 forms protein aggregates degraded by autophagy and has a protective effect on huntingtin-induced cell death. *J Cell Biol.* 2005 Nov 21;171(4):603-14.
[26] Bjorkoy G, Lamark T, Johansen T. p62/SQSTM1: a missing link between protein aggregates and the autophagy machinery. *Autophagy,* 2006 Apr-Jun;2(2):138-9.
[27] Iwata A, Christianson JC, Bucci M, Ellerby LM, Nukina N, Forno LS, et al. Increased susceptibility of cytoplasmic over nuclear polyglutamine aggregates to autophagic degradation. *Proc Natl Acad Sci U S A.* 2005 Sep 13;102(37):13135-40.
[28] Iwata A, Riley BE, Johnston JA, Kopito RR. HDAC6 and microtubules are required for autophagic degradation of aggregated huntingtin. *J Biol Chem.* 2005 Sep 28.

[29] Webb JL, Ravikumar B, Rubinsztein DC. Microtubule disruption inhibits autophagosome-lysosome fusion: implications for studying the roles of aggresomes in polyglutamine diseases. *Int J Biochem Cell Biol.* 2004 Dec;36(12):2541-50.

[30] Vos MJ, Hageman J, Carra S, Kampinga HH. Structural and functional diversities between members of the human HSPB, HSPH, HSPA, and DNAJ chaperone families. *Biochemistry,* 2008 Jul 8;47(27):7001-11.

[31] Hartl FU, Hayer-Hartl M. Molecular chaperones in the cytosol: from nascent chain to folded protein. *Science*, 2002 Mar 8;295(5561):1852-8.

[32] Goldberg AL. Protein degradation and protection against misfolded or damaged proteins. *Nature*, 2003 Dec 18;426(6968):895-9.

[33] Cuervo AM, Stefanis L, Fredenburg R, Lansbury PT, Sulzer D. Impaired degradation of mutant alpha-synuclein by chaperone-mediated autophagy. *Science,* 2004 Aug 27;305(5688):1292-5.

[34] Barral JM, Broadley SA, Schaffar G, Hartl FU. Roles of molecular chaperones in protein misfolding diseases. *Semin Cell Dev Biol.* 2004 Feb;15(1):17-29.

[35] Auluck PK, Chan HY, Trojanowski JQ, Lee VM, Bonini NM. Chaperone suppression of alpha-synuclein toxicity in a Drosophila model for Parkinson's disease. *Science,* 2002 Feb 1;295(5556):865-8.

[36] Warrick JM, Chan HY, Gray-Board GL, Chai Y, Paulson HL, Bonini NM. Suppression of polyglutamine-mediated neurodegeneration in Drosophila by the molecular chaperone HSP70. *Nat Genet.* 1999 Dec;23(4):425-8.

[37] Carra S, Sivilotti M, Chavez Zobel AT, Lambert H, Landry J. HspB8, a small heat shock protein mutated in human neuromuscular disorders, has in vivo chaperone activity in cultured cells. *Hum Mol Genet.* 2005 Jun 15;14(12):1659-69.

[38] Wilhelmus MM, Boelens WC, Otte-Holler I, Kamps B, de Waal RM, Verbeek MM. Small heat shock proteins inhibit amyloid-beta protein aggregation and cerebrovascular amyloid-beta protein toxicity. *Brain Res.* 2006 May 17;1089(1):67-78.

[39] Wilhelmus MM, Boelens WC, Otte-Holler I, Kamps B, Kusters B, Maat-Schieman ML, et al. Small heat shock protein HspB8: its distribution in Alzheimer's disease brains and its inhibition of amyloid-beta protein aggregation and cerebrovascular amyloid-beta toxicity. *Acta Neuropathol.* (Berl). 2006 Feb;111(2):139-49.

[40] Wyttenbach A, Sauvageot O, Carmichael J, Diaz-Latoud C, Arrigo AP, Rubinsztein DC. Heat shock protein 27 prevents cellular polyglutamine toxicity and suppresses the increase of reactive oxygen species caused by huntingtin. Hum Mol Genet. 2002 May 1;11(9):1137-51.

[41] Iwaki T, Iwaki A, Tateishi J, Sakaki Y, Goldman JE. Alpha B-crystallin and 27-kd heat shock protein are regulated by stress conditions in the central nervous system and accumulate in Rosenthal fibers. *Am J Pathol.* 1993 Aug;143(2):487-95.

[42] Iwaki T, Kume-Iwaki A, Liem RK, Goldman JE. Alpha B-crystallin is expressed in non-lenticular tissues and accumulates in Alexander's disease brain. *Cell,* 1989 Apr 7;57(1):71-8.

[43] Iwaki T, Wisniewski T, Iwaki A, Corbin E, Tomokane N, Tateishi J, et al. Accumulation of alpha B-crystallin in central nervous system glia and neurons in pathologic conditions. *Am J Pathol.* 1992 Feb;140(2):345-56.

[44] Renkawek K, Bosman GJ, de Jong WW. Expression of small heat-shock protein hsp 27 in reactive gliosis in Alzheimer disease and other types of dementia. *Acta Neuropathol.* (Berl). 1994;87(5):511-9.

[45] Renkawek K, de Jong WW, Merck KB, Frenken CW, van Workum FP, Bosman GJ. alpha B-crystallin is present in reactive glia in Creutzfeldt-Jakob disease. *Acta Neuropathol.* (Berl). 1992;83(3):324-7.

[46] Renkawek K, Stege GJ, Bosman GJ. Dementia, gliosis and expression of the small heat shock proteins hsp27 and alpha B-crystallin in Parkinson's disease. *Neuroreport*, 1999 Aug 2;10(11):2273-6.

[47] Renkawek K, Voorter CE, Bosman GJ, van Workum FP, de Jong WW. Expression of alpha B-crystallin in Alzheimer's disease. *Acta Neuropathol.* (Berl). 1994;87(2):155-60.

[48] Aquino DA, Capello E, Weisstein J, Sanders V, Lopez C, Tourtellotte WW, et al. Multiple sclerosis: altered expression of 70- and 27-kDa heat shock proteins in lesions and myelin. *J Neuropathol Exp Neurol.* 1997 Jun;56(6):664-72.

[49] Lowe J, Errington DR, Lennox G, Pike I, Spendlove I, Landon M, et al. Ballooned neurons in several neurodegenerative diseases and stroke contain alpha B crystallin. *Neuropathol Appl Neurobiol.* 1992 Aug;18(4):341-50.

[50] Lowe J, McDermott H, Pike I, Spendlove I, Landon M, Mayer RJ. alpha B crystallin expression in non-lenticular tissues and selective presence in ubiquitinated inclusion bodies in human disease. *J Pathol.* 1992 Jan;166(1):61-8.

[51] Wilhelmus MM, Otte-Holler I, Wesseling P, de Waal RM, Boelens WC, Verbeek MM. Specific association of small heat shock proteins with the pathological hallmarks of Alzheimer's disease brains. *Neuropathol Appl Neurobiol.* 2006 Apr;32(2):119-30.

[52] Mucke L, Eddleston M. Astrocytes in infectious and immune-mediated diseases of the central nervous system. *Faseb J.* 1993 Oct;7(13):1226-32.

[53] Evgrafov OV, Mersiyanova I, Irobi J, Van Den Bosch L, Dierick I, Leung CL, et al. Mutant small heat-shock protein 27 causes axonal Charcot-Marie-Tooth disease and distal hereditary motor neuropathy. *Nat Genet.* 2004 Jun;36(6):602-6.

[54] Houlden H, Laura M, Wavrant-De Vrieze F, Blake J, Wood N, Reilly MM. Mutations in the HSP27 (HSPB1) gene cause dominant, recessive, and sporadic distal HMN/CMT type 2. *Neurology*, 2008 Nov 18;71(21):1660-8.

[55] Irobi J, Van Impe K, Seeman P, Jordanova A, Dierick I, Verpoorten N, et al. Hot-spot residue in small heat-shock protein 22 causes distal motor neuropathy. *Nat Genet.* 2004 Jun;36(6):597-601.

[56] Tang BS, Zhao GH, Luo W, Xia K, Cai F, Pan Q, et al. Small heat-shock protein 22 mutated in autosomal dominant Charcot-Marie-Tooth disease type 2L. *Hum Genet.* 2005 Feb;116(3):222-4.

[57] Graw J, Klopp N, Illig T, Preising MN, Lorenz B. Congenital cataract and macular hypoplasia in humans associated with a de novo mutation in CRYAA and compound heterozygous mutations in P. *Graefes Arch Clin Exp Ophthalmol.* 2006 Feb 2:1-8.

[58] Litt M, Kramer P, LaMorticella DM, Murphey W, Lovrien EW, Weleber RG. Autosomal dominant congenital cataract associated with a missense mutation in the human alpha crystallin gene CRYAA. *Hum Mol Genet.* 1998 Mar;7(3):471-4.

[59] Liu M, Ke T, Wang Z, Yang Q, Chang W, Jiang F, et al. Identification of a CRYAB mutation associated with autosomal dominant posterior polar cataract in a Chinese family. *Invest Ophthalmol Vis Sci.* 2006 Aug;47(8):3461-6.

[60] Liu Y, Zhang X, Luo L, Wu M, Zeng R, Cheng G, et al. A Novel {alpha}B-Crystallin Mutation Associated with Autosomal Dominant Congenital Lamellar Cataract. *Invest Ophthalmol Vis Sci.* 2006 Mar;47(3):1069-75.

[61] Mackay DS, Andley UP, Shiels A. Cell death triggered by a novel mutation in the alphaA-crystallin gene underlies autosomal dominant cataract linked to chromosome 21q. Eur *J Hum Genet.* 2003 Oct;11(10):784-93.

[62] Mackay DS, Boskovska OB, Knopf HL, Lampi KJ, Shiels A. A nonsense mutation in CRYBB1 associated with autosomal dominant cataract linked to human chromosome 22q. *Am J Hum Genet.* 2002 Nov;71(5):1216-21.

[63] Vicart P, Caron A, Guicheney P, Li Z, Prevost MC, Faure A, et al. A missense mutation in the alphaB-crystallin chaperone gene causes a desmin-related myopathy. *Nat Genet.* 1998 Sep;20(1):92-5.

[64] Bilen J, Bonini NM. Genome-wide screen for modifiers of ataxin-3 neurodegeneration in Drosophila. *PLoS genetics.* 2007 Oct;3(10):1950-64.

[65] Lavoie JN, Gingras-Breton G, Tanguay RM, Landry J. Induction of Chinese hamster HSP27 gene expression in mouse cells confers resistance to heat shock. HSP27 stabilization of the microfilament organization. *J Biol Chem.* 1993 Feb 15;268(5):3420-9.

[66] Lavoie JN, Lambert H, Hickey E, Weber LA, Landry J. Modulation of cellular thermoresistance and actin filament stability accompanies phosphorylation-induced changes in the oligomeric structure of heat shock protein 27. *Mol Cell Biol.* 1995 Jan;15(1):505-16.

[67] Nicholl ID, Quinlan RA. Chaperone activity of alpha-crystallins modulates intermediate filament assembly. *Embo J.* 1994 Feb 15;13(4):945-53.

[68] Perng MD, Cairns L, van den IP, Prescott A, Hutcheson AM, Quinlan RA. Intermediate filament interactions can be altered by HSP27 and alphaB-crystallin. *J Cell Sci.* 1999 Jul;112 (Pt 13):2099-112.

[69] Dreiza CM, Brophy CM, Komalavilas P, Furnish EJ, Joshi L, Pallero MA, et al. Transducible heat shock protein 20 (HSP20) phosphopeptide alters cytoskeletal dynamics. *Faseb J.* 2005 Feb;19(2):261-3.

[70] Liang P, MacRae TH. Molecular chaperones and the cytoskeleton. *J Cell Sci.* 1997 Jul;110 (Pt 13):1431-40.

[71] Carra S, Seguin SJ, Lambert H, Landry J. HspB8 chaperone activity toward poly(Q)-containing proteins depends on its association with Bag3, a stimulator of macroautophagy. *J Biol Chem.* 2008 Jan 18;283(3):1437-44.

[72] Carra S, Seguin SJ, Landry J. HspB8 and Bag3: a new chaperone complex targeting misfolded proteins to macroautophagy. *Autophagy,* 2008 Mar-Apr;4(2):237-9.

[73] Kappe G, Franck E, Verschuure P, Boelens WC, Leunissen JA, de Jong WW. The human genome encodes 10 alpha-crystallin-related small heat shock proteins: HspB1-10. *Cell Stress Chaperones,* 2003 Spring;8(1):53-61.

[74] Arrigo AP. [Heat shock proteins as molecular chaperones]. *Med Sci.* (Paris). 2005 Jun-Jul;21(6-7):619-25.

[75] Kim KK, Kim R, Kim SH. Crystal structure of a small heat-shock protein. *Nature,* 1998 Aug 6;394(6693):595-9.

[76] Gaestel M. sHsp-phosphorylation: enzymes, signaling pathways and functional implications. *Prog Mol Subcell Biol.* 2002;28:151-69.

[77] Morrison LE, Hoover HE, Thuerauf DJ, Glembotski CC. Mimicking phosphorylation of alphaB-crystallin on serine-59 is necessary and sufficient to provide maximal protection of cardiac myocytes from apoptosis. *Circ Res*. 2003 Feb 7;92(2):203-11.

[78] Jakob U, Gaestel M, Engel K, Buchner J. Small heat shock proteins are molecular chaperones. *J Biol Chem*. 1993 Jan 25;268(3):1517-20.

[79] Horwitz J. Alpha-crystallin can function as a molecular chaperone. *Proc Natl Acad Sci U S A*. 1992 Nov 1;89(21):10449-53.

[80] Lee GJ, Vierling E. A small heat shock protein cooperates with heat shock protein 70 systems to reactivate a heat-denatured protein. *Plant Physiol*. 2000 Jan;122(1):189-98.

[81] Giese KC, Vierling E. Changes in oligomerization are essential for the chaperone activity of a small heat shock protein in vivo and in vitro. *J Biol Chem*. 2002 Nov 29;277(48):46310-8.

[82] Shashidharamurthy R, Koteiche HA, Dong J, McHaourab HS. Mechanism of chaperone function in small heat shock proteins: dissociation of the HSP27 oligomer is required for recognition and binding of destabilized T4 lysozyme. *J Biol Chem*. 2005 Feb 18;280(7):5281-9.

[83] Iwaki T, Iwaki A, Tateishi J, Goldman JE. Sense and antisense modification of glial alpha B-crystallin production results in alterations of stress fiber formation and thermoresistance. *J Cell Biol*. 1994 Jun;125(6):1385-93.

[84] Lee JS, Zhang MH, Yun EK, Geum D, Kim K, Kim TH, et al. Heat shock protein 27 interacts with vimentin and prevents insolubilization of vimentin subunits induced by cadmium. *Exp Mol Med*. 2005 Oct 31;37(5):427-35.

[85] During RL, Gibson BG, Li W, Bishai EA, Sidhu GS, Landry J, et al. Anthrax lethal toxin paralyzes actin-based motility by blocking Hsp27 phosphorylation. *Embo J*. 2007 May 2;26(9):2240-50.

[86] Giese KC, Vierling E. Mutants in a small heat shock protein that affect the oligomeric state. Analysis and allele-specific suppression. *J Biol Chem*. 2004 Jul 30;279(31):32674-83.

[87] Klemenz R, Andres AC, Frohli E, Schafer R, Aoyama A. Expression of the murine small heat shock proteins hsp 25 and alpha B crystallin in the absence of stress. *J Cell Biol*. 1993 Feb;120(3):639-45.

[88] Akbar MT, Lundberg AM, Liu K, Vidyadaran S, Wells KE, Dolatshad H, et al. The neuroprotective effects of heat shock protein 27 overexpression in transgenic animals against kainate-induced seizures and hippocampal cell death. *J Biol Chem*. 2003 May 30;278(22):19956-65.

[89] Efthymiou CA, Mocanu MM, de Belleroche J, Wells DJ, Latchmann DS, Yellon DM. Heat shock protein 27 protects the heart against myocardial infarction. *Basic research in cardiology*. 2004 Nov;99(6):392-4.

[90] Sharp P, Krishnan M, Pullar O, Navarrete R, Wells D, de Belleroche J. Heat shock protein 27 rescues motor neurons following nerve injury and preserves muscle function. *Exp Neurol*. 2006 Apr;198(2):511-8.

[91] Bruey JM, Paul C, Fromentin A, Hilpert S, Arrigo AP, Solary E, et al. Differential regulation of HSP27 oligomerization in tumor cells grown in vitro and in vivo. *Oncogene*. 2000 Oct 5;19(42):4855-63.

[92] Mehlen P, Kretz-Remy C, Preville X, Arrigo AP. Human hsp27, Drosophila hsp27 and human alphaB-crystallin expression-mediated increase in glutathione is essential for the

protective activity of these proteins against TNFalpha-induced cell death. *Embo J.* 1996 Jun 3;15(11):2695-706.

[93] Garrido C, Bruey JM, Fromentin A, Hammann A, Arrigo AP, Solary E. HSP27 inhibits cytochrome c-dependent activation of procaspase-9. *Faseb J.* 1999 Nov;13(14):2061-70.

[94] Mehlen P, Hickey E, Weber LA, Arrigo AP. Large unphosphorylated aggregates as the active form of hsp27 which controls intracellular reactive oxygen species and glutathione levels and generates a protection against TNFalpha in NIH-3T3-ras cells. *Biochem Biophys Res Commun.* 1997 Dec 8;241(1):187-92.

[95] Wyttenbach A, Carmichael J, Swartz J, Furlong RA, Narain Y, Rankin J, et al. Effects of heat shock, heat shock protein 40 (HDJ-2), and proteasome inhibition on protein aggregation in cellular models of Huntington's disease. *Proc Natl Acad Sci U S A.* 2000 Mar 14;97(6):2898-903.

[96] Mehlen P, Preville X, Chareyron P, Briolay J, Klemenz R, Arrigo AP. Constitutive expression of human hsp27, Drosophila hsp27, or human alpha B-crystallin confers resistance to TNF- and oxidative stress-induced cytotoxicity in stably transfected murine L929 fibroblasts. *J Immunol.* 1995 Jan 1;154(1):363-74.

[97] Sian J, Dexter DT, Lees AJ, Daniel S, Agid Y, Javoy-Agid F, et al. Alterations in glutathione levels in Parkinson's disease and other neurodegenerative disorders affecting basal ganglia. *Ann Neurol.* 1994 Sep;36(3):348-55.

[98] Browne SE, Bowling AC, MacGarvey U, Baik MJ, Berger SC, Muqit MM, et al. Oxidative damage and metabolic dysfunction in Huntington's disease: selective vulnerability of the basal ganglia. *Ann Neurol.* 1997 May;41(5):646-53.

[99] Perez-Severiano F, Santamaria A, Pedraza-Chaverri J, Medina-Campos ON, Rios C, Segovia J. Increased formation of reactive oxygen species, but no changes in glutathione peroxidase activity, in striata of mice transgenic for the Huntington's disease mutation. *Neurochem Res.* 2004 Apr;29(4):729-33.

[100] Bogdanov MB, Andreassen OA, Dedeoglu A, Ferrante RJ, Beal MF. Increased oxidative damage to DNA in a transgenic mouse model of Huntington's disease. *J Neurochem.* 2001 Dec;79(6):1246-9.

[101] Tabrizi SJ, Workman J, Hart PE, Mangiarini L, Mahal A, Bates G, et al. Mitochondrial dysfunction and free radical damage in the Huntington R6/2 transgenic mouse. *Ann Neurol.* 2000 Jan;47(1):80-6.

[102] Zourlidou A, Gidalevitz T, Kristiansen M, Landles C, Woodman B, Wells DJ, et al. Hsp27 overexpression in the R6/2 mouse model of Huntington's disease: chronic neurodegeneration does not induce Hsp27 activation. *Hum Mol Genet.* 2007 May 1;16(9):1078-90.

[103] Perrin V, Regulier E, Abbas-Terki T, Hassig R, Brouillet E, Aebischer P, et al. Neuroprotection by Hsp104 and Hsp27 in lentiviral-based rat models of Huntington's disease. *Mol Ther.* 2007 May;15(5):903-11.

[104] Rosen DR, Siddique T, Patterson D, Figlewicz DA, Sapp P, Hentati A, et al. Mutations in Cu/Zn superoxide dismutase gene are associated with familial amyotrophic lateral sclerosis. *Nature,* 1993 Mar 4;362(6415):59-62.

[105] Cleveland DW, Rothstein JD. From Charcot to Lou Gehrig: deciphering selective motor neuron death in ALS. *Nat Rev Neurosci.* 2001 Nov;2(11):806-19.

[106] Bruijn LI, Houseweart MK, Kato S, Anderson KL, Anderson SD, Ohama E, et al. Aggregation and motor neuron toxicity of an ALS-linked SOD1 mutant independent from wild-type SOD1. *Science*, 1998 Sep 18;281(5384):1851-4.

[107] Mather K, Martin JE, Swash M, Vowles G, Brown A, Leigh PN. Histochemical and immunocytochemical study of ubiquitinated neuronal inclusions in amyotrophic lateral sclerosis. *Neuropathol Appl Neurobiol.* 1993 Apr;19(2):141-5.

[108] Bruijn LI, Becher MW, Lee MK, Anderson KL, Jenkins NA, Copeland NG, et al. ALS-linked SOD1 mutant G85R mediates damage to astrocytes and promotes rapidly progressive disease with SOD1-containing inclusions. *Neuron.* 1997 Feb;18(2):327-38.

[109] Pasinelli P, Houseweart MK, Brown RH, Jr., Cleveland DW. Caspase-1 and -3 are sequentially activated in motor neuron death in Cu,Zn superoxide dismutase-mediated familial amyotrophic lateral sclerosis. *Proc Natl Acad Sci U S A.* 2000 Dec 5;97(25):13901-6.

[110] Wong PC, Pardo CA, Borchelt DR, Lee MK, Copeland NG, Jenkins NA, et al. An adverse property of a familial ALS-linked SOD1 mutation causes motor neuron disease characterized by vacuolar degeneration of mitochondria. *Neuron.* 1995 Jun;14(6):1105-16.

[111] Cote F, Collard JF, Julien JP. Progressive neuronopathy in transgenic mice expressing the human neurofilament heavy gene: a mouse model of amyotrophic lateral sclerosis. *Cell,* 1993 Apr 9;73(1):35-46.

[112] [112] Kong J, Xu Z. Overexpression of neurofilament subunit NF-L and NF-H extends survival of a mouse model for amyotrophic lateral sclerosis. Neurosci Lett. 2000 Mar 3;281(1):72-4.

[113] Vleminckx V, Van Damme P, Goffin K, Delye H, Van Den Bosch L, Robberecht W. Upregulation of HSP27 in a transgenic model of ALS. *J Neuropathol Exp Neurol.* 2002 Nov;61(11):968-74.

[114] Patel YJ, Payne Smith MD, de Belleroche J, Latchman DS. Hsp27 and Hsp70 administered in combination have a potent protective effect against FALS-associated SOD1-mutant-induced cell death in mammalian neuronal cells. *Brain Res Mol Brain Res.* 2005 Apr 4;134(2):256-74.

[115] Krishnan J, Vannuvel K, Andries M, Waelkens E, Robberecht W, Van Den Bosch L. Over-expression of Hsp27 does not influence disease in the mutant SOD1(G93A) mouse model of amyotrophic lateral sclerosis. *J Neurochem.* 2008 Sep;106(5):2170-83.

[116] Sharp PS, Akbar MT, Bouri S, Senda A, Joshi K, Chen HJ, et al. Protective effects of heat shock protein 27 in a model of ALS occur in the early stages of disease progression. *Neurobiol Dis.* 2008 Apr;30(1):42-55.

[117] Liou YC, Sun A, Ryo A, Zhou XZ, Yu ZX, Huang HK, et al. Role of the prolyl isomerase Pin1 in protecting against age-dependent neurodegeneration. *Nature,* 2003 Jul 31;424(6948):556-61.

[118] Sun L, Liu SY, Zhou XW, Wang XC, Liu R, Wang Q, et al. Inhibition of protein phosphatase 2A- and protein phosphatase 1-induced tau hyperphosphorylation and impairment of spatial memory retention in rats. *Neuroscience,* 2003;118(4):1175-82.

[119] Gong CX, Lidsky T, Wegiel J, Grundke-Iqbal I, Iqbal K. Metabolically active rat brain slices as a model to study the regulation of protein phosphorylation in mammalian brain. *Brain Res Brain Res Protoc.* 2001 Feb;6(3):134-40.

[120] Shimura H, Miura-Shimura Y, Kosik KS. Binding of tau to heat shock protein 27 leads to decreased concentration of hyperphosphorylated tau and enhanced cell survival. *J Biol Chem.* 2004 Apr 23;279(17):17957-62.

[121] Berry V, Francis P, Reddy MA, Collyer D, Vithana E, MacKay I, et al. Alpha-B crystallin gene (CRYAB) mutation causes dominant congenital posterior polar cataract in humans. *Am J Hum Genet.* 2001 Nov;69(5):1141-5.

[122] Shinohara H, Inaguma Y, Goto S, Inagaki T, Kato K. Alpha B crystallin and HSP28 are enhanced in the cerebral cortex of patients with Alzheimer's disease. J Neurol Sci. 1993 Nov;119(2):203-8.

[123] Hejtmancik JF. Congenital cataracts and their molecular genetics. *Semin Cell Dev Biol.* 2008 Apr;19(2):134-49.

[124] Cobb BA, Petrash JM. Structural and functional changes in the alpha A-crystallin R116C mutant in hereditary cataracts. *Biochemistry,* 2000 Dec 26;39(51):15791-8.

[125] Li H, Li C, Lu Q, Su T, Ke T, Li DW, et al. Cataract mutation P20S of alphaB-crystallin impairs chaperone activity of alphaA-crystallin and induces apoptosis of human lens epithelial cells. Biochim Biophys *Acta,* 2008 May;1782(5):303-9.

[126] Yang YX, Wood NW, Latchman DS. Molecular basis of Parkinson's disease. Neuroreport. 2009 Jan 28;20(2):150-6.

[127] Spillantini MG, Schmidt ML, Lee VM, Trojanowski JQ, Jakes R, Goedert M. Alpha-synuclein in Lewy bodies. *Nature,* 1997 Aug 28;388(6645):839-40.

[128] Lowe J, Landon M, Pike I, Spendlove I, McDermott H, Mayer RJ. Dementia with beta-amyloid deposition: involvement of alpha B-crystallin supports two main diseases. *Lancet,* 1990 Aug 25;336(8713):515-6.

[129] Rekas A, Adda CG, Andrew Aquilina J, Barnham KJ, Sunde M, Galatis D, et al. Interaction of the molecular chaperone alphaB-crystallin with alpha-synuclein: effects on amyloid fibril formation and chaperone activity. J Mol Biol. 2004 Jul 23;340(5):1167-83.

[130] Brenner M, Johnson AB, Boespflug-Tanguy O, Rodriguez D, Goldman JE, Messing A. Mutations in GFAP, encoding glial fibrillary acidic protein, are associated with Alexander disease. *Nat Genet.* 2001 Jan;27(1):117-20.

[131] Quinlan RA, Brenner M, Goldman JE, Messing A. GFAP and its role in Alexander disease. *Exp Cell Res.* 2007 Jun 10;313(10):2077-87.

[132] Xia Z, Dickens M, Raingeaud J, Davis RJ, Greenberg ME. Opposing effects of ERK and JNK-p38 MAP kinases on apoptosis. *Science,* 1995 Nov 24;270(5240):1326-31.

[133] Tang G, Xu Z, Goldman JE. Synergistic effects of the SAPK/JNK and the proteasome pathway on glial fibrillary acidic protein (GFAP) accumulation in Alexander disease. *J Biol Chem.* 2006 Dec 15;281(50):38634-43.

[134] Webster KA. Serine phosphorylation and suppression of apoptosis by the small heat shock protein alphaB-crystallin. *Circ Res.* 2003 Feb 7;92(2):130-2.

[135] Compston A, Coles A. Multiple sclerosis. Lancet. 2002 Apr 6;359(9313):1221-31.

[136] van Noort JM, van Sechel AC, Bajramovic JJ, el Ouagmiri M, Polman CH, Lassmann H, et al. The small heat-shock protein alpha B-crystallin as candidate autoantigen in multiple sclerosis. *Nature,* 1995 Jun 29;375(6534):798-801.

[137] Ousman SS, Tomooka BH, van Noort JM, Wawrousek EF, O'Connor KC, Hafler DA, et al. Protective and therapeutic role for alphaB-crystallin in autoimmune demyelination. *Nature,* 2007 Jul 26;448(7152):474-9.

[138] Benndorf R, Sun X, Gilmont RR, Biederman KJ, Molloy MP, Goodmurphy CW, et al. HSP22, a new member of the small heat shock protein superfamily, interacts with mimic of phosphorylated HSP27 ((3D)HSP27). J Biol Chem. 2001 Jul 20;276(29):26753-61.

[139] Chowdary TK, Raman B, Ramakrishna T, Rao CM. Mammalian Hsp22 is a heat-inducible small heat-shock protein with chaperone-like activity. Biochem J. 2004 Jul 15;381(Pt 2):379-87.

[140] Carra S, Brunsting JF, Lambert H, Landry J, Kampinga HH. HspB8 Participates in Protein Quality Control by a Non-chaperone-like Mechanism That Requires eIF2{alpha} Phosphorylation. J Biol Chem. 2009 Feb 27;284(9):5523-32.

[141] Talloczy Z, Jiang W, Virgin HWt, Leib DA, Scheuner D, Kaufman RJ, et al. Regulation of starvation- and virus-induced autophagy by the eIF2alpha kinase signaling pathway. *Proc Natl Acad Sci U S A.* 2002 Jan 8;99(1):190-5.

[142] King MA, Hands S, Hafiz F, Mizushima N, Tolkovsky AM, Wyttenbach A. Rapamycin inhibits polyglutamine aggregation independently of autophagy by reducing protein synthesis. *Mol Pharmacol.* 2008 Apr;73(4):1052-63.

[143] Sun X, Fontaine JM, Bartl I, Behnam B, Welsh MJ, Benndorf R. Induction of Hsp22 (HspB8) by estrogen and the metalloestrogen cadmium in estrogen receptor-positive breast cancer cells. *Cell Stress Chaperones,* 2007 Winter;12(4):307-19.

[144] Rosati A, Leone A, Del Valle L, Amini S, Khalili K, Turco MC. Evidence for BAG3 modulation of HIV-1 gene transcription. *J Cell Physiol.* 2007 Mar;210(3):676-83.

[145] Pagliuca MG, Lerose R, Cigliano S, Leone A. Regulation by heavy metals and temperature of the human BAG-3 gene, a modulator of Hsp70 activity. *FEBS Lett.* 2003 Apr 24;541(1-3):11-5.

In: Small Stress Proteins and Human Diseases
Editors: Stéphanie Simon et al.
ISBN: 978-1-61470-636-6

Chapter 1.2

FUNCTIONAL SYMBIOSIS BETWEEN THE INTERMEDIATE FILAMENT CYTOSKELETON AND SMALL HEAT SHOCK PROTEINS

Andrew Landsbury[1], Ming Der Perng[2], Ehmke Pohl[3,4], and Roy A. Quinlan[1,4]*

[1]School of Biological and Biomedical Sciences, Durham University, South Road, Durham DH1 3LE, UK

[2]Institute of Molecular Medicine, College of Life Science, National Tsing Hua University, 30013 Hsinchu, Taiwan

[3]Department of Chemistry, Durham University, South Road, Durham DH1 3LE, UK

[4]Biophysical Sciences Institute, Durham University, South Road, Durham DH1 3LE, UK

ABSTRACT

The complexity of the cytoskeleton and its regulatory mechanisms has increased as multicellular organisms have evolved. The small heat shock proteins (sHSPs) are key to maintaining the important cytoskeletal proteins actin, myosin, titin, tubulin and most critically, the intermediate filament (IF) proteins. Mutations in either sHSPs or their interacting IF proteins cause similar diseases in humans, confirming the key functional relationship between these two very distinct protein families. This demonstrates how sHSPs and IFs depend upon each other for correct function and begs the question whether this should be recognised as such by naming the sHSP-IF complex as the *SIF complex*. In the cell, the sHSPs will chaperone the cytoskeleton and specifically the IF network. We

* Roy A. Quinlan
School of Biological and Biomedical Sciences,
Durham University,
South Raod,
Durham DH1 3LE,
UK
Tel: (+44)191-334-1331
Fax (+44)191-334-1201
Email: r.a.quinlan@dur.ac.uk

propose this sHSP-IF interaction is symbiotic, with the chaperones becoming more efficient once bound to the surface of a filament. We also take this opportunity to highlight a second key concept for sHSPs. In addition to chaperoning individual client proteins, the sHSPs have an additional function that has been tacitly assumed, but which needs be overtly recognised. This is to chaperone the protein complexes and polymeric structures formed by the individual client proteins. Such a function will prevent unintentional polymer or complex interactions that could lead ultimately to protein aggregation. In the case of the cytoskeleton, the presence of sHSPs prevents filament entanglement and their subsequent aggregation. The sHSP-IF complex is a *prima facie* example of this specific chaperone role.

1. The Evolutionary Context – Bacterial Cytoskeleton, Stress and the Origins of the *SIF* Complex

It is well known that bacteria have equivalents to all the major eukaryotic filamentous systems [1] and that chaperones are involved in their initial folding and subsequent activity [2]. However, it is only just being appreciated that the different cytoskeletal systems in bacteria are all dependant upon each other and link with stress responses and stress proteins, as seen in eukaryotes. Charbon et al have recently shown that the proposed bacterial IF equivalent protein crescentin requires MreB, the actin-equivalent protein, in order to function correctly [3]. It has also been shown that assembly of the MreB division complex requires FtsZ, the tubulin-equivalent protein [4] and that formation of the FtsZ ring structure requires a small coiled protein, ZapB [5], another putative bacterial IF protein [6]. This illustrates the important interplay between the different bacterial cytoskeletal elements, as also seen in mammalian cells [7-9].

Bacteria, like eukaryotic cells, can respond to stimuli by changing their cytoskeleton [10], highlighting another response which is conserved from bacteria to man. In actinomycetes, environmental and nutrient stresses induce aerial hyphal growth [11], which is dependent upon FiLP, a structural protein proposed to be IF related with homologues in many other bacterial species [6]. This shows how bacterial coiled coil proteins with IF-like features are required for shape changes and suggests that environmental stresses can be the trigger for such events. It is worth noting that syncoilin, a mammalian IF protein, is assemble incompetent on its own, has a non-conventional coiled-coil arrangement [12] and is associated with membrane complexes [13]. This illustrates that flexibility must be exercised in identifying bacterial IF candidate proteins [6]. Another IF trait in mammalian cells is increased expression to combat stress and this is another selection criterion [9, 14] that highlights an evolutionary conserved and important cell response.

In mammalian cells, the cytoskeleton helps determine the location, distribution and function of stress protein complexes [7-9]. In bacteria, stress proteins also have been shown to associate with the cytoskeleton [15]. In gram-negative bacteria, the plasma membrane located phage stress protein (PSP) complex depends upon the actin-related protein MreB for location and mobility. The PSP complex is induced in response to stresses that disturb the proton motive force in cells. This highlights another functional interaction between the cytoskeleton and stress proteins in bacteria. Further examples of this will likely emerge as the bacterial filament field progresses and functional parallels with eukaryotic systems become apparent.

Table 1. Mutations in shsps and if proteins linked to human disease

A. sHSPs Genes and Mutations with Links to Human Disease

sHSP	Gene Symbol (synonyms)	Chromosonal Location (marker)	Mutation Details (MIM)	Phenotypes	Reference
HSP27	HSPB1 (CM2F; DKFZp586P1322; HMN2B; HS.76067; Hsp25; HSP27; HSP28; SRP27)	7q11.23 (D7S672-D7S806 see [78] for details; STS-N23152; D7S2435, D7S2490 or D7S2518, D7S2470, D7S2421, D7S669, D7S2204 see [158] for details)	P39L G84R L99M (recessive) R127W S135F S135M R136W R140G K141Q T151I P182L P182S (c.-217T>C) (602195)	Neuronopathy, distal type IIB Charcot Marie Tooth Disease type IIF amyotrophic lateral sclerosis – genetic modifier?	[159] [159, 160] [159] [78, 158, 161] [78] [159] [78] [78] [78, 162] [78] [78, 163] [164] [165]
HSPB2/MKBP	HSPB2 (Hs.78846; LCH11CR1K; MGC133245; MKBP)	11q22-q23 (D11S3407)	(602179)		
HSPB3/HSPL27	HSPB3 (HSPL27)	5q11.2 (D5S1355 and D5S1885)	(604624)		

Table 1. Continued.

sHSP	Gene Symbol (synonyms)	Chromosonal Location (marker)	Mutation Details (MIM)	Phenotypes	Reference
aA-crystallin HSPB4	CRYAA (HSPB4, CRYA1)	21q22.3(D21S1912)	W9X:RECESSIVE R12C R21L R21W R49C R54C G98R R116C R116H (123580)	Cataract Cataract with microcornea Cataract Cataract with micro- ophthalmia Cataract Cataract with microcornea Cataract Cataract sometimes associated with micro- ophthalmia and microcornea Cataract	[166] [167, 168] [168, 169] [169] [167] [168] [168, 170, 171] [172] [173-175] [176] [177]
αB-crystallin HSPB5	CRYAB (CRYA2; CTPP2; HSPB5)	11q22.3--q23.1 (D11S1986, D11S908)	P20S R56W recessive) R120G D140N del450A del464CT Q151X R157H A171T (123590)	Cataract Cataract Desmin related myopathy and Cataract Cataract Cataract Myofibrillar myopathy Myofibrillar myopathy Dilated cardiomyopathy Cataract	[178] [179] [27] [180] [181] [66] [66] [121] [168]
HSP20	HSPB6 (AA387366; Gm479; Hsp20; MGC107687)	19q13.13 (D19S615E)	P2OL (610695)	Diminished cardioprotection – genetic modifier to cardiomyopathy	[182]
CVHSP	HSPB7 (CVHSP)	1p36.23-p34.3 (HSPB7_990)	(610692)		
HSP22	HSPB8	12q24 (D12S349 and PLA2GiB see [77])	K141E K141N (608014)	Charcot Marie Tooth Disease type IIF	[77, 183]

Table 1. Continued.

sHSP	Gene Symbol (synonyms)	Chromosonal Location (marker)	Mutation Details (MIM)	Phenotypes	Reference
HSPB9	HSPB9 (CT51; FLJ27437)	17q21.2 (D17S2018)	(608014)		
ODF1	HSP10 (HSPB10; MGC129928; MGC129929; ODF; ODF2;ODF27; ODFP; ODFPG; ODFPGA; ODFPGB; RT7; SODF)	8q22 (D8S1049)	182878		[184]

B. Mutations in Cytoskeletal Proteins Linked with sHSPs

sHSP	Gene Symbol (synonyms)	Chromosonal Location (marker)	Mutation Details (MIM)	Phenotypes	Reference
Vimentin	VIM (FLJ36605)	10p13	E151K	Cataract	[185]
Filensin	BFSP1 (CP115; CP94; LIFL-H)	20p12.1-p11.23(D20S860)	Deletion encompassing exon 6,c.736-1384_c.957-66 del – frameshift and loss of exon 6; RECESSIVE (603307)	Cataract	[186]
CP49	BFSP2 (CP47; CP49; LIFL-L; MGC142078; MGC142080; PHAKOSIN)	3q21-25 (D3S1292, D3S2322, D3S1541)	E223del; R287W R339H (603212)	Cataract and myopia Cataract Cataract	[187-189] [190] [191]

Table 1. Continued.

sHSP	Gene Symbol (synonyms)	Chromosonal Location (marker)	Mutation Details (MIM)	Phenotypes	Reference
NFL	NEFL(CMT1F; CMT2E; FLJ53642; NF-L; NF68; NFL)	8p21	P8R	Charcot Marie Tooth Disease type IIE	[192-194]
			P8R	Charcot Marie Tooth Disease type IF	[195]
			(13-BP,DUP/INS, NT61)E21X	Charcot Marie Tooth Disease type IIE	[196]
			P22S		[197,198]
			L94P		
			E140X		[194]
			(162280)	Charcot Marie Tooth Disease type IF	[199]
Desmin	DES (CMD1I; CSM1; CSM2; FLJ12025; FLJ39719; FLJ41013; FLJ41793)	2q35	S2I	Cardioskeletal myopathy	[200]
			S13F	Cardioskeletal myopathy	[201, 202]
			R16C	Restrictive cardiomyopathy	[203]
			S46F	Cardioskeletal myopathy	[200]
			S46Y	Cardioskeletal myopathy	[200]
			E108K	Cardiomyopathy	[204, 205]
			delR173-E179	Cardioskeletal myopathy	[206]
			A213V	Cardioskeletal myopathy	[205]
			delD214-E245	Cardioskeletal myopathy	[203, 205,
			K240del	Skeletal myopathy	207]
			E245D	Skeletal myopathy	[208]
			S298L	Cardiomyopathy	[209]
			D312N	Cardiomyopathy	[204]
			A337P	Cardiomyopathy	[204]
			L338R	Cardiomyopathy	[69, 210,
			N342D	Cardiomyopathy	211]
			L345P	Cardioskeletal myopathy	[211]
			R350P	Scapuloperoneal syndrome type Kaeser and myopathies	[210, 212] [213, 214]
			R350W	Cardiomyopathy	[215, 216]
			R355P	Cardioskeletal myopathy	[204]
			A357P	Cardiomyopathy	[217]
					[218]

Table 1. Continued.

sHSP	Gene Symbol (synonyms)	Chromosonal Location (marker)	Mutation Details (MIM)	Phenotypes	Reference
			delE359-S361	Skeletal myopathy	[219]
			A360P	Cardiomyopathy	[69, 210]
			delN366	Cardioskeletal myopathy	[219]
			I367F	Cardioskeletal myopathy	[220]
			L370P	Cardiomyopathy	[221]
			L385P	Cardioskeletal myopathy	[218]
			Q389P	Cardioskeletal myopathy	[222]
			L392P	Cardioskeletal myopathy	[220]
			N393I	Cardioskeletal myopathy	[69, 210,
			D399Y	Cardioskeletal myopathy	211]
			E401K	Cardioskeletal myopathy	[211]
			R406W	Cardioskeletal myopathy	[211]
			E413K	Restrictive cardiomyopathy	[203, 210,
			P419S	Cardioskeletal myopathy	223-225]
			K449T	Cardioskeletal myopathy	[226]
			I451M	Cardioskeletal myopathy	[204, 220]
			T453I	Restrictive cardiomyopathy	[200]
			R454W	Cardiomyopathy	[210, 227,
			V459I	Restrictive cardiomyopathy	228]
			V469M	EmeryDreyfuss	[203]
				musculardystrophy (associated	[229]
				withLMNA mutation)	[204]
			(125660)		[230]

A significant difference between bacterial and mammalian cytoskeletons is that the bacterial cytoskeleton appears largely membrane associated. Whilst this feature is consistent with eukaryotic IFs such as nuclear lamins [16], cytoplasmic vimentin [17], lenticular Bfsp1 and 2 [18, 19] and synemin [20] as well as specialist membrane IF proteins like syncoilin [13], the mammalian cytoskeleton largely comprises filaments that traverse the cell cytoplasm between many different membrane compartments [21-23]. This occupation of the cytoplasmic space requires additional mechanisms to retain the individuality of the cytoskeletal filaments as well as assist their integration into a transcellular cytoskeletal network. The sHSP chaperones can provide such activities and in this chapter we will review these functions with respect to the cytoskeleton.

2. The Cytoskeleton – The Key to Cell Mechanics and Resisting Stress

The cytoskeleton forms a dynamic transcellular complex that integrates both internal and external signalling pathways to the mechanical properties of the cell [21-24]. It comprises an interconnected network of IFs, microfilaments (MFs) and microtubules (MTs) that allows cells to become integrated within tissues and adopt different functions [22, 24], which is essential to the emergence of multicellular organisms. The cytoskeleton comprises micron-long filaments with subunits that can easily be rearranged to adjust their surface geometries to alter their association with other subcellular structures and associated proteins.

Accessory (associated) proteins are required to organise cytoskeletal filaments and to prevent their entanglement due to their length and also the repeated charged and hydrophobic patches on their filament surface, as illustrated by actin [25]. The presence of actin binding proteins modulates the mechanical properties of the actin filaments and their degree of cross-linking. The sHSPs are one of many accessory proteins that help to stabilise and to organise the cytoskeleton and protect it against stressful conditions eg [9]. Evidence for this comes from those human diseases caused by mutations in sHSPs that are diagnosed by histopathological aggregates comprising IFs (see Table 1). Indeed, the R120G mutation in αB-crystallin (HSPB5) promotes filament-filament interactions [26] and their accumulation into aggregates [27].

3. Intermediate Filaments are *Bona Fide* Stress Proteins and Symbiotic Partners in the *SIF* Complex

A surprising point to emerge from early studies on the heat shock response of cells was the identity of so called "prompt" HSPs [28]. These proteins are produced from existing mRNA within the first 30-45 minutes after heat shock and most are associated with the cytoskeleton and nucleoskeleton fractions of the cell [28]. This group of prompt HSPs by definition includes vimentin, a major cytoplasmic IF protein, lamin B, a key lamina component [29] and emerin, a lamina associated protein [30]. These data indicate that the cytoplasmic and nuclear IF cytoskeletons are involved in the earliest responses of the cell to stress. This is supported by changes in the IF cytoskeleton which occur during chemical and physical stresses as well as disease [14]. We therefore suggest that IF proteins are also stress proteins.

IF involvement in the stress response is also supported by its association with HSPs (eg HSP70 and HSP90) and particularly sHSPs [7]. In a range of different tissue culture cells, two prominent sHSPs, HSP27 (HSPB1) and αB-crystallin, associate with the cytoplasmic IFs in unstressed cells [7, 31] and both proteins are also found in the nucleus in a "speckle" compartment containing splicing factors such as SC35 [32]. Recently, αB-crystallin and HSP27 have been found to concentrate at intranuclear foci with lamins A/C in the muscle cell line C2C12 after heat shock [33]. Both protein chaperones shift to the cytoskeletal fraction of cells [34] and tissues [35, 36] after stress or as a result of disease [37-39] and remain associated with the cytoplasmic IFs [7, 40], but some αB-crystallin [41, 42] and HSP27 [43] shift to the nuclear compartment into stress granules and nuclear speckles, a site where lamins and sHSPs can potentially interact [33]. These observations suggest that both cytoplasmic and nuclear IFs assist in the cell stress response through their association with key chaperone proteins, such as αB-crystallin and HSP27.

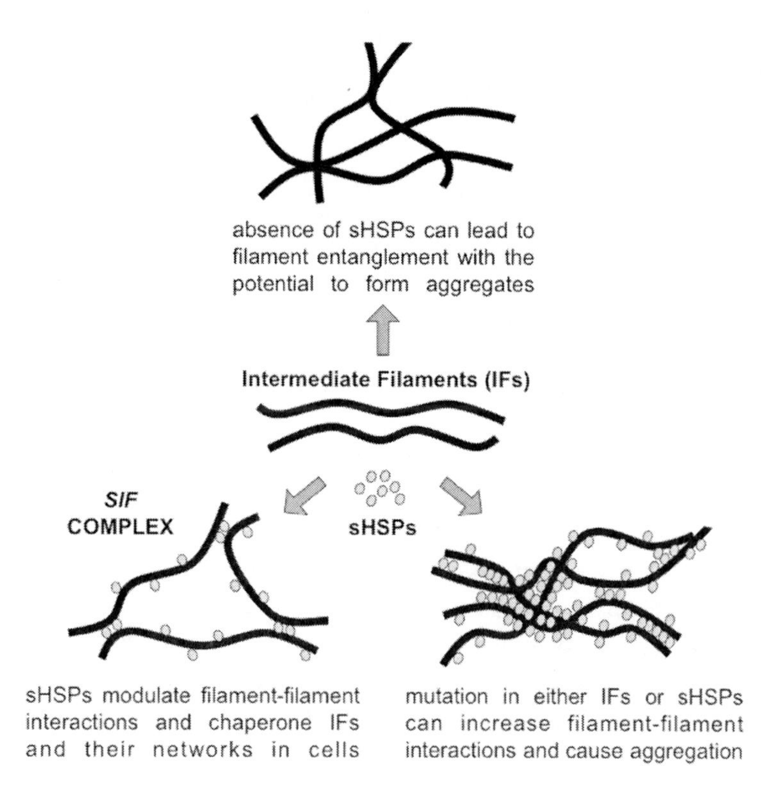

Figure 1. Summary of sHSP interactions with IFs. sHSPs chaperone IFs and their networks in cells. The absence of sHSP can lead to IF entanglement and mutations in sHSPs or IFs can alter filament-filament interactions with the potential to form aggregates.

The diseases caused by sHSP mutations and the characteristic histopathological IF aggregates that result (Table 1), demonstrate the potential importance of sHSPs to IF function. sHSPs maintain the individuality of IFs within their cytoskeletal networks, preventing unintended filament-filament interactions that could lead to IF aggregation (Figure 1). In other words, the sHSPs actually chaperone individual IFs in the cytoskeletal network. However, as well as chaperoning IFs, sHSPs also bind to filament surfaces [7, 44], which we suggest will increase their chaperone efficiency, since protein chaperones are more productive when bound to a surface than when free in solution [45, 46]. This demonstrates the proposed symbiosis between IFs and sHSPs that we wish to highlight in this review. The ability of sHSPs to chaperone IFs as well as individual client proteins is a key point to understand.

4. sHSP Interactions with IFs

Initial studies on the lens cytoskeleton led to the discovery that α-crystallins interact with vimentin [44]. Subsequently it has been shown that both αB-crystallin and HSP27 have temperature-dependent interactions with IFs [7, 40, 47]. In vitro assembly studies have also shown that αB-crystallin can inhibit IF assembly and in the case of GFAP, inhibition is independent of phosphorylation [7, 44]. The inhibitory effect of αB-crystallin upon vimentin and GFAP assembly provides a direct regulatory mechanism. Co-immunoprecipitation has demonstrated the associations of sHSPs with soluble IF subunits including vimentin and GFAP [7, 44]. Association of sHSP with IF networks has also been visualised in a range of cell lines with different IF compositions using immunofluorescence microscopy [7, 31].

We have proposed and provided data to support the hypothesis that the main functions of the sHSP-IF interaction is; to maintain the individuality of IFs [7]; to modulate inter-filament interactions in their networks [26] and to stabilise assembly intermediates [48]. The most compelling evidence to confirm the importance of the sHSP-IF interaction comes from the description of a mutation in αB-crystallin (R120G), which causes IF aggregation and a phenotype that mimics desmin related myopathies [27]. This mutation changes the dissociation constant (K_d=1.14μM vs 2.37μM), causing the mutant to bind more strongly to desmin filaments, promoting more filament-filament interactions leading to IF aggregation [26]. The ability of αB-crystallin to chaperone individual IFs has been compromised despite the fact that filament binding has increased [26, 48]. As shown in Figure 1, the R120G site in αB-crystallin occupies one of the desmin-selective binding sites, spanning the region containing this residue. Others have suggested that the degree of protein aggregation is key to the ensuing cardiomyopathy [49] and in a mouse model of the R120G αB-crystallin induced cardiomyopathy, evidence was presented to show that autophagy was increased in response to the presence of the mutant protein aggregates. When the autophagy pathway was further compromised, disease severity increased [50]. Similar results were found for GFAP-based Alexander disease [51].

In Alexander disease and cardiomyopathy, we suggest that autophagy is a consequence of and not the cause of disease due to sHSP or IF mutations (Figure 3). Both R120G αB-crystallin [49] and mutant desmin [52] are reported to impair the proteasome system as do many IF mutants [53-55], because IF proteins are normally turned over by the proteasome [56]. R120G αB-crystallin itself also forms protein amyloid [57, 58], which also inhibits the proteasome [59] and induces autophagy [50]. Inhibition of the proteasome will induce the accumulation of αB-crystallin and HSP27 into aggresomes [60] and cause IF aggregation

[61]. Recent evidence even suggests a direct correlation between rate of protein turnover by the proteasome and the level of IF protein [62] suggesting that this link between IFs, sHSPs and the proteasome machinery is indeed critical to cell function. In fact αB-crystallin stimulates FBX4 dependent protein degradation by the proteasome [63]. Whatever might be the downstream consequences of mutations in sHSPs or IF proteins, these will exacerbate the primary problem, which is IF aggregation and the loss of *SIF* complex function. We suggest that it is the loss of the sHSP chaperone activity toward IFs that leads to their initial inappropriate associations, entanglement and subsequent aggregation (Figure 1).

Analysis of the myopathy causing mutation Q151X in αB-crystallin [64] shows that although loss of the C-terminal extension reduces solubility and prevents oligomerisation of αB-crystallin, it actually increases desmin binding and chaperone activity [64]. We interpret this as an increase in the number of sites available for client protein binding due to the loss of sHSP oligomerisation (see Figure 2 and [65]). The protein has a strong tendency to aggregate as the loss of the C-terminal tail domain destabilizes αB-crystallin [64], but still the clinical feature of this mutant includes desmin-containing aggregates [66] rather than those that contain exclusively αB-crystallin. These data support the view that there is significant overlap between inter-subunit binding sites and those that bind clients such as IF proteins (Figure 2). It also highlights that despite the instability of both Q151X [64] and R120G αB-crystallin [67, 68], both have similar muscle pathologies [27, 66] with desmin aggregates as a key histological feature rather than αB-crystallin-exclusive aggregates. The direct physical association of αB-crystallin oligomers with IFs has been well documented [7, 26, 44, 64] and these observations further support the proposed symbiotic relationship between sHSPs and IFs.

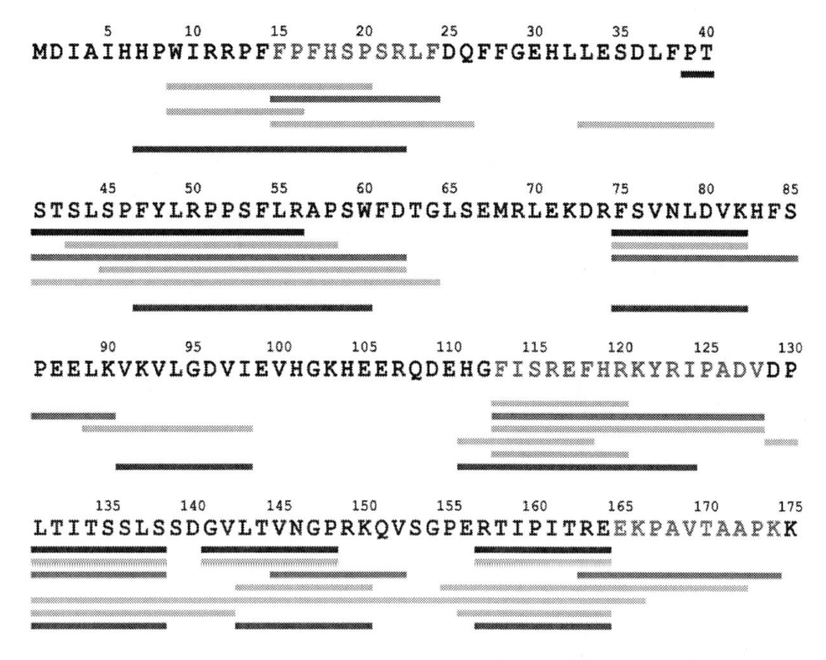

Figure 2. Sequences in αB-crystallin involved in oligomerisation and binding to selected client proteins, such as desmin, GFAP, actin, microtubles and growth factors. The sequence for αB-crystallin is shown and red residues identify desmin binding regions, which are not involved in oligomer assembly. Regions being used to form αB-crystallin oligomers will be inaccessible to other client

proteins. The R120G mutation in human αB-crystallin, which lies within a desmin interaction region, causes abnormal interactions with desmin filaments leading to cataracts and desmin-related myopathy [26]. Ser-59, which lies within an actin interaction region, has been shown to modulate interactions with actin filaments where phosphorylation causes increased interaction [120]. Sequences in αB-crystallin, which bind both filament proteins and client proteins have shown preferential interaction with filaments relative to other clients at 23°C and preferential interactions with unfolding clients at 45°C relative to the filaments [154]. This suggests that under normal physiological conditions αB-crystallin interacts with and stabilises filaments, while under conditions of stress αB-crystallin dissociates from the filaments and used the same domains to interact with and stabilise unfolding clients.

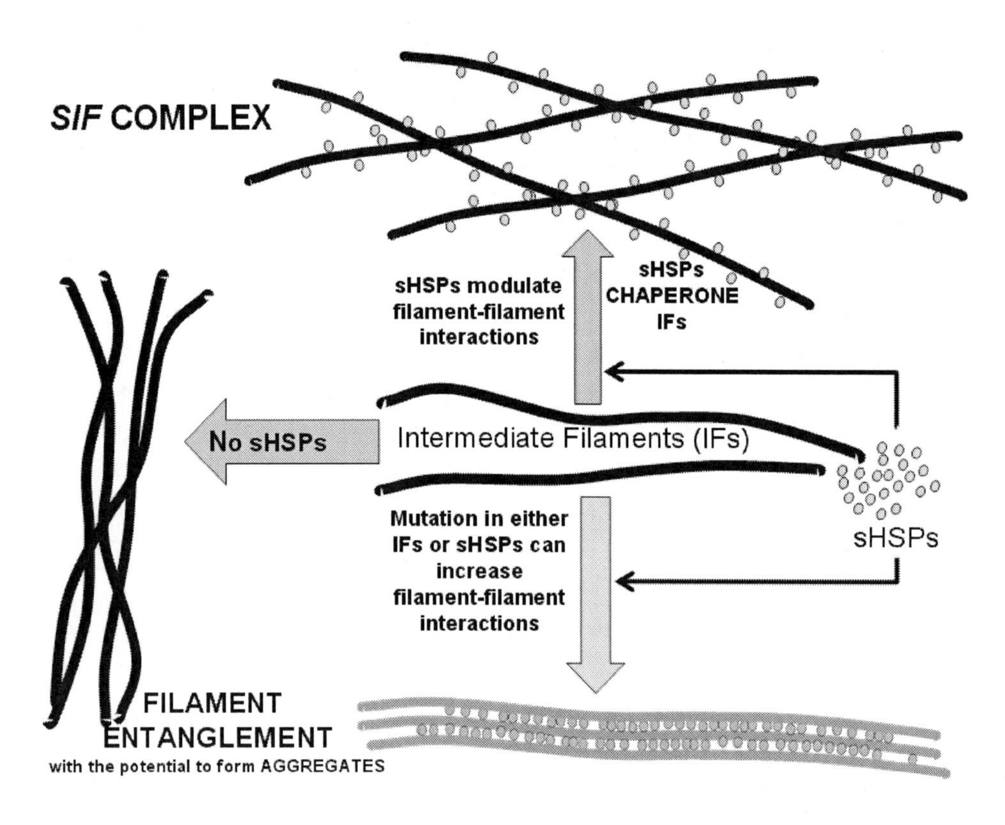

Figure 3. Consequences of mutations in *SIF* complex components for IF assemblies. Mutations in either IF proteins or their associated sHSPs lead to IF aggregation. Consequences of this include activation of stress kinases, proteasome inhibition, oxido-reductive stress and the induction of autophagy. Once the balance has shifted in favour of these processes, then the system becomes self-perpetuating increasing the stress-load experienced by the cell.

5. Lessons From IF Based Diseases

Table 1 gives a summary of all the currently known mutations involving sHSPs and the diseases they cause. Mutations in IF proteins that cause similar diseases have also been identified and are also listed. The clear phenocopy produced by mutations in both desmin [69] and αB-crystallin [27] provides the strongest evidence of the functional link between IFs and sHSPs. A large number of mutations have now been reported for desmin and are spread throughout the protein sequence and cause a range of different skeletal and cardiac muscle phenotypes, but they all cause IF aggregation [70]. Moreover the desmin mutations change the assembly [71] and the mechanical properties of the filaments [72] and the R120G αB-crystallin mutation induces IF aggregation, which also compromises the mechanical properties of the filament network in the cardiomycyte [24, 70].

There are three main disease types caused by sHSP and IF mutations; myopathy, cataract and neuropathy. The common histopathological phenotype characteristic of the myopathies [73, 74] and neuropathies [75-78] are protein aggregates containing sHSPs and IFs. In the case of cataract [19], the composition of the lens opacities have not yet been analysed in detail and so the presence of both the lenticular IFs and sHSPs in the same aggregates has yet to be demonstrated (see part 3 of the book).

The most dramatic demonstration of the potential for sHSPs as a treatment for IF-based diseases comes from the prevention of morbidity by the overexpression of αB-crystallin in a mouse model of a neuropathy, Alexander disease [79]. Alexander disease (OMIM #203450) is a rare neurological disorder caused by mutations in GFAP [80]. The disease results from the accumulation of mutant GFAP into characteristic aggregates called Rosenthal fibres that sequester αB-crystallin and HSP27 [37, 38, 81], activate stress kinases [8, 82] and the stress response [82]. It has been shown that the overexpression of even wild type GFAP leads to IF aggregation [83], which compromises the ubiquitin–proteosome system [82] and causes accumulation of αB-crystallin, as reported in an Alexander disease mouse model [84] and in samples from Alexander disease brain tissue [37, 81]. Mutations in GFAP have been shown to alter filament organization [8, 81] and solubility [85]. Mutant GFAPs can also change the association with other cellular proteins as well as αB-crystallin such as plectin, a cytoskeletal linker protein [86]. Rosenthal fibers contain αB-crystallin [37] and αB-crystallin can regulate GFAP filament interactions in vitro [7]. The overexpression of αB-crystallin is capable of disaggregating wild type GFAP accumulations in tissue culture cells [87] and in the ultimate test, overexpression of αB-crystallin prevented GFAP-induced morbidity in a mouse model of Alexander disease [79]. These data suggest that the association of αB-crystallin with GFAP filaments is essential for normal astrocyte function and the formation of Rosenthal fibres during Alexander disease not only perturbs this interaction, but also the function of both GFAP IFs and sHSPs in astrocytes.

In the case of desmin related myopathies (desminopathies) no such direct inhibition of desmin IF aggregation by the overexpression of αB-crystallin has been attempted in an animal model, although it was successful in a tissue culture model [88]. Inducing the overexpression of HSPB8 by geranylgeranylacetone [89] can reduce the effects of R120G αB-crystallin induced cardiomyopathy and the realization that it is αB-crystallin oligomers [90] rather than the amyloid aggregates in the animal model that are responsible for the developing cardiomyopathy only confirms our assertion that it is the effects upon the IF cytoskeleton that is the important primary lesion from which the pathology develops.

6. Interaction of sHSPs with Other Client Proteins of the Cytoskeleton

6.1. Actin Cytoskeleton

Actin was the first component of the cytoskeleton to be identified as a client protein for sHSPs [91]. Both actin and sHSPs are highly dynamic polymers and their interactions depend on the phosphorylation status of the sHSPs [92]. HSP25 (HSPB1) purified from turkey gizzard was first identified as an inhibitor of actin polymerization (IAP), which caps the barbed end of the actin filament, thus preventing monomer addition [91, 93]. Inhibition of actin polymerization in vitro appears to require nonphosphorylated monomers, whereas phosphorylated monomers and non-phosphorylated multimers seem to affect actin cytoskeleton dynamics [94]. Subsequent studies show that HSP25/HSP27 and particularly their pseudo-phosphorylated variants also interact with actin and stablise MFs against stress conditions, thus preventing the formation of insoluble aggregates [95-98]. The other sHSP αB-crystallin has also been shown to stabilize actin filaments and regulate actin assembly dynamics in a phosphorylation-dependent manner both in vitro [99, 100] and in cultured cells [101]. Reducing αB-crystallin gene expression using RNAi technology leads to disruption of the actin microfilament network, further supporting an important role of αB-crystallin in the maintenance of microfilament integrity and ultimately cellular survival [102]. In addition to HSP27/25 and αB-crystallin, other sHSPs such as αA-crystallin [103] and HSP20 [104] also interact with actin, although for HSP20 (HSPB6), this interaction might be indirect [105]. Recently, Gosh et al. reported the identification of several actin interactive sequences in αB-crystallin that could potentially promote actin assembly and also inhibit filament disassembly and aggregation in vitro [65]. Whatever the precise binding sites might be, a direct actin-sHSP interaction probably also occurs in vivo and may participate in the regulation of the actin filament assembly.

6.2. Myosin Interactions – Preserving ATPase Activity

Myosin is the molecular motor that drives actomyosin-based motility, including muscle shortening and cytokinesis, by ATP-dependent conformational changes [106]. Interactions between αB-crystallin and myosin have been shown to maintain the enzymatic activity of the motor and prevent its aggregation during stress. Under heat shock conditions in vitro (43°C), myosin unfolding causes loss of enzymatic activity followed by formation of aggregates [107]. However, at heat-shock temperatures in the presence of αB-crystallin, myosin retains most of its Ca-ATPase and Mg-ATPase activity and is protected from thermal aggregation [108]. αB-crystallin may thus be critical for maintaining myosin integrity in vivo and sustaining muscle function during heat shock conditions. It has been found that αB-crystallin is more highly expressed in the skeletal muscles of athletes compared to non-athletes where vigorous exercise results in heat-shock conditions [109]. In this situation as with the other cytoskeletal proteins except IFs there is the demonstrated chaperone activity for the protein rather than the polymer (*cf* Figure 1).

6.3. The Third Filament System of Muscle – Titin

Titin is a large filamentous protein involved in the assembly and function of vertebrate striated muscles. Titin regulates sarcomere assembly and force transmission at the Z-line, maintains resting tension in the I-band region [110, 111] and also acts as an anchoring protein for the other sarcomere/Z-disc proteins [110]. Titin contains a molecular spring region which develops force [112] and determines the passive stiffness and thus the filling characteristics of the heart [113]. This molecular spring region contains Ig domains and a heart-specific N2B region, which modulates the elasticity of the titin filament [114].

Titin is an important binding partner of αB-crystallin and HSP27 in cardiac muscle [115, 116]. Antibody labeling in sectioned fibers has shown that αB-crystallin binds to the N2B region of cardiac titin, although weak binding to titin Ig domains has also been detected [116, 117]. Under stress conditions, aB-crystallin moves from the cytosol to the I-band of skeletal muscle [116, 118] and binds to myofibrils via the titin N2B region which lowers the probability of Ig domain unfolding and maintains the elasticity of the N2B spring region [117, 119]. αB-crystallin binding is phosphorylation activated and mediated by p38-MAPKAPK2 [120].

αB-crystallin mutants R157H and R120G, which have been identified in patients with cardiomyopathy, have decreased chaperone activity for the N2B and Ig domains of titin which become more vulnerable to unfolding [117, 121]. Mutations in titin are also associated with cardiomyopathy [122] and it has been shown that the Gln4053ter mutation in the titin N2B domain reduces binding to αB-crystallin [121]. This suggests that the interaction between αB-crystallin and titin is important for normal heart function.

6.4. Cadherin – Kidney Specific (Ksp) – Interactions

αB-crystallin binds to kidney specific cadherin [123] via its N-terminal domain of αB-crystallin. Although no function of Ksp-cadherin has been directly demonstrated, it is believed to act as a cell–cell adhesion molecule to maintain kidney tissue integrity and may also play a role in cell signaling, as described for other cadherins [124]. αB-crystallin may act as a linker between Ksp-cadherin and the cytoskeleton. Co-localization of F-actin with both Ksp-cadherin and αB-crystallin has been observed in collecting duct cells and Ksp-cadherin and αB-crystallin had been shown to co-sediment with actin from kidney lysates [123].

6.5. Microtubules

sHSPs are not only important in regulating stability and assembly of actin-based MFs, their interactions with tubulin and MTs have also been documented. Fujita et al. reports that αB-crystallin interacts with MTs through MT-associated proteins (MAPs), and this interaction gives MTs resistance to disassembly both in vitro and in unstressed cells [125]. Further studies revealed that the highly conserved αB-crystallin domain is essential to prevent denatured tubulin from aggregation [126]. In a parallel study, HSP27 has also been shown to interact with tubulin and microtubules in HeLa cells [127]. Under stress conditions, sHSPs can protect microtubules against depolymerization [128-131], thus maintaining MT integrity

[132]. Protein pin array technology identifies several tubulin interactive sequences on the surface of human αB-crystallin, providing direct experimental evidence that αB-crystallin can selectively stabilise tubulin as a client protein and collectively modulate MT assembly/disassembly through a dynamic mechanism of sHSP subunit exchange [133].

7. Filament Binding Sites

As shown in Figure 2, there is significant overlap between the binding sites predicted for crystallin oligomerisation and those for client proteins. However, some sites are specific to a restricted subset of client proteins and in the case of desmin these have been highlighted on a model of αB-crystallin (Figure 4) based on the crystal structure of HSP16.9. These sites also overlap with other filament forming proteins (Figure 2) including actin [65] and tubulin [126, 133] and amyloidogenic proteins like α-synuclein [134]. The very recent crystal structure of the crystallin domain for HSP20 and αB-crystallin will refine this prediction of potential oligomerisation [135], but higher-order assembly is driven in part by sequences in the C-terminal extension. Whilst the homology approach presents an obvious drawback to representing αB-crystallin oligomers on the HSP16.9 oligomer, the main message is that the desmin sites are not apparently readily accessible to bind a desmin filament. Electron micrographs and biochemical data show that the αB-crystallin oligomers most certainly bind to IFs [7, 26, 44]. The fact that the αB-crystallin oligomer is highly dynamic by the exchange of oligomer subunits [136, 137] such as dimers or even monomers in conjunction with phosphorylation [135, 138] offers potential mechanisms. In addition, the fact that such subunits will bind to desmin and other IFs is supported by our studies on the myopathy causing mutant Q151X αB-crystallin [64].

These data highlight an interesting structural conundrum concerning sHSP function with respect to client protein binding and chaperone activity. This conundrum has also been recognised for pea Hsp18.1 [139]. A photoactivatable cross-linker was incorporated at multiple positions throughout Hsp18.1 to tag directly the interaction sites between this sHSP and two client proteins, firefly luciferase and malate dehydrogenase under substrate denaturing conditions [139]. These clients bound the β7 region within the conserved α-crystallin domain as well as multiple residues in the N-terminal region. These binding sites in the Hsp18.1 dodecamer model are not accessible to client proteins and would only be fully exposed in the sHSP dimer [139]. However, it has been demonstrated that at substrate denaturing temperatures, the equilibrium between the sHSP oligomeric and dimeric forms is shifted toward the dimer [140, 141]. sHSP subunit exchange also increases with elevated temperatures, which would further facilitate substrate binding to a suboligomeric sHSP species [142, 143].

The fact that substrate binding occurs throughout the sHSP (see Figure 2) indicates that there is not a discrete substrate-binding surface as seen for other protein chaperones such as Hsp90, Hsp70, and GroEL [144, 145]. Previous studies have suggested that there are both high and low affinity substrate binding sites on sHSPs [146, 147] and different substrates have been shown to bind different sHSP regions [139, 148]. Hydrogen-deuterium exchange studies of the sHSP-substrate complex have shown that the N-terminal region remains unstructured even when bound to a substrate [149]. These data suggest that the disordered N-terminal region of the sHSP [141, 150] may present variable binding site conformations, which allows efficient binding to a wide range of unfolding protein.

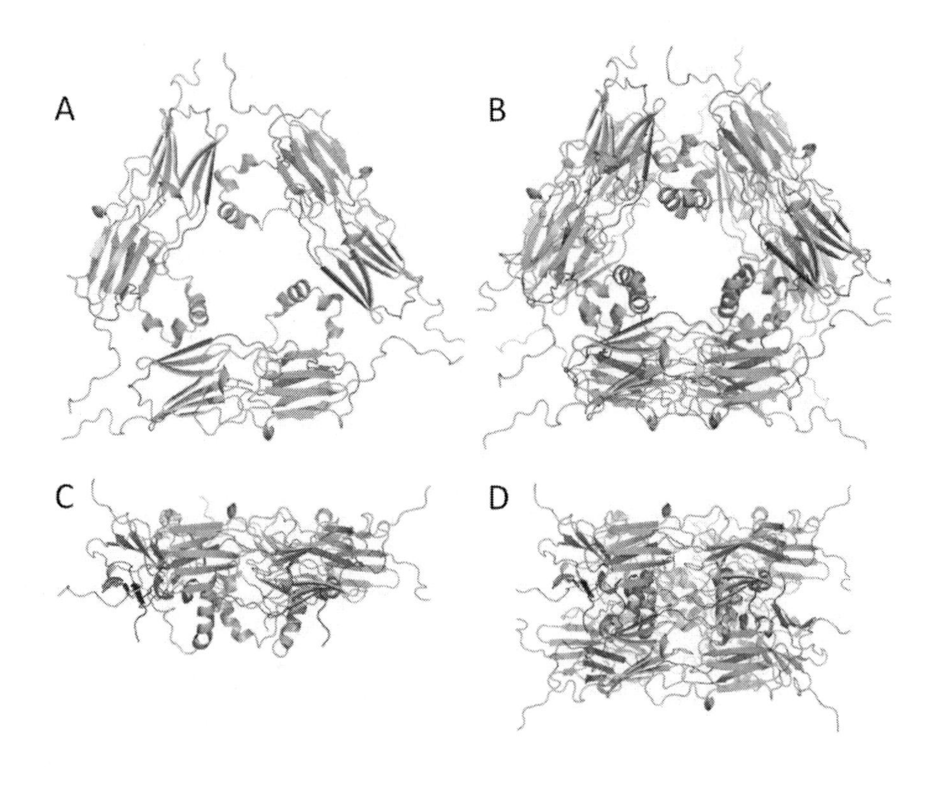

Figure 4. Localisation of sequences in HSP16.5 equivalent to the desmin binding sequences in αB-crystallin. The homology model was built using the program MODELLER [155] and based on the crystal structure of the small heat shock protein HSP16.9 from wheat [141]. HSP16.9 shares 27% sequence identity with αB-crystallin and forms a dodecameric structure with four independent chains in the asymmetric unit, only two of which contain the N-terminal domain. The chain including the N-terminus was used to model the monomer of aB-crystallin, which was then superimposed by least squares methods using COOT [156, 157] onto each monomer from the HSP16.9 dodecamer to create dodecameric and hexameric models for αB-crystallin. The N-terminal domain in one monomer of each αB-crystallin dimer was removed in order to avoid steric clash. Ribbon representation of a hexameric (A, C) and dodecameric (B, D) arrangment of an αB-crystallin model. A top down view (A, B) and side view (C, D) is show for both arrangments. The desmin binding regions are colored in red (residues 113-128) and blue (residues 15-24) respectively.

The solution to this conundrum affecting sHSP function will be evident once we understand fully the properties of αB-crystallin itself as we know that the quaternary structure of αB-crystallin is highly variable [151, 152] and by implication essential to its function.

Conclusion

In this chapter, we have highlighted the potential importance of the IF to sHSP function and vice versa. We have encapsulated this idea by proposing the *SIF* complex to highlight the symbiotic relationship between IFs and sHSPs. We have reviewed the evidence supporting loss of *SIF* complex function as the major cause for the inherited diseases caused by mutations in sHSPs and IFs, a problem that is exacerbated by for example the onset of autophagy [50], proteasome inhibition [59] and concomitant oxido-reductive stress [153]. This recognises that IFs are a key part of the stress response along with the sHSPs and that both enhance the functional spectrum of each other. We have drawn attention to the ability of sHSPs to chaperone individual IFs by inhibiting their inate tendency to become entangled by virtue of their filamentous nature. Therefore sHSPs can chaperone the polymers as well as the individual subunits and proteins of cytoskeletal filaments. This exposes a new function for sHSPs when associated with the cytoskeleton, which translates into the modulation of the mechanical properties of the cytoskeleton and therefore the cellular response(s) to their environment. We have summarized these concepts in Figure 5. The ability to "chaperone" the cytoskeleton was demonstrated first for IFs and fits our proposal to include IFs as *bona fide* stress proteins along with protein chaperones.

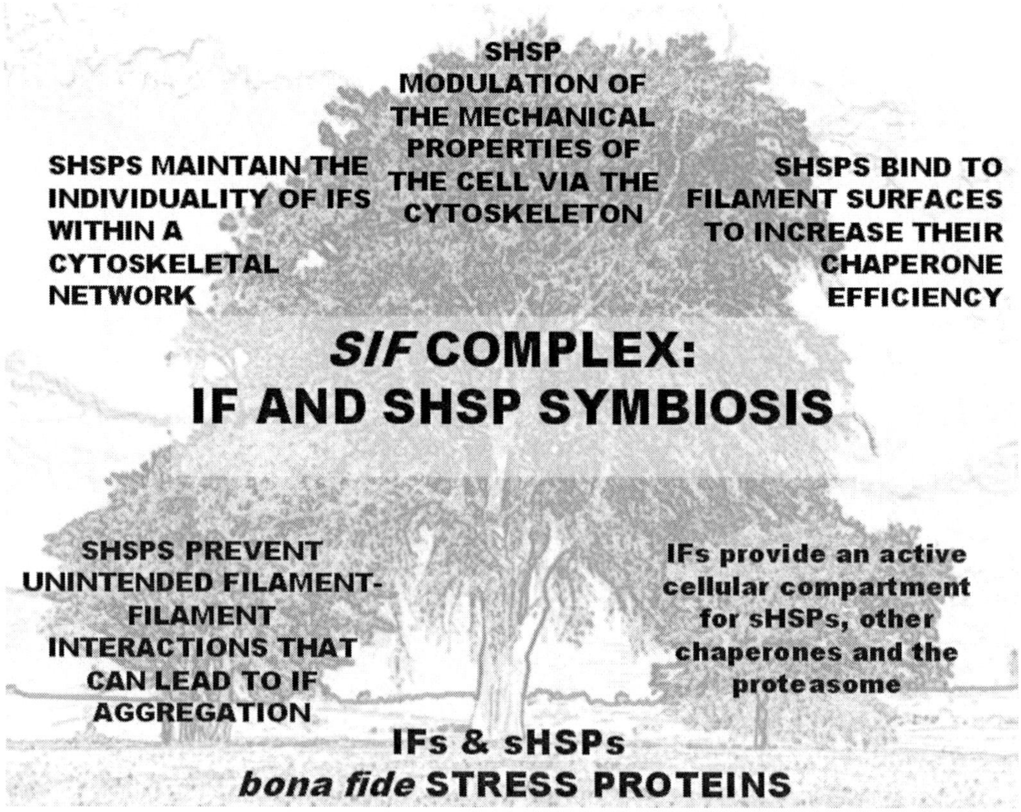

Figure 5. Summary of the concepts proposed in this article concerning the *SIF* complex.

Acknowledgments

RAQ is a Research Fellow supported by the Leverhulme Trust. AL is a BBSRC CASE student supported by the BBSRC and Farfield Scientific. September 2009.

References

[1] Erickson HP. Evolution of the cytoskeleton. Bioessays. 2007 Jul;29(7):668-77.

[2] Weaver VM, Carson CE, Walker PR, Chaly N, Lach B, Raymond Y, et al. Degradation of nuclear matrix and DNA cleavage in apoptotic thymocytes. *J Cell Sci.* 1996;109(Pt 1):45-56.

[3] Charbon G, Cabeen MT, Jacobs-Wagner C. Bacterial intermediate filaments: in vivo assembly, organization, and dynamics of crescentin. *Genes Dev.* 2009 May 1;23(9):1131-44.

[4] Vats P, Shih YL, Rothfield L. Assembly of the MreB-associated cytoskeletal ring of Escherichia coli. Mol Microbiol. 2009 Apr;72(1):170-82.

[5] Ebersbach G, Galli E, Moller-Jensen J, Lowe J, Gerdes K. Novel coiled-coil cell division factor ZapB stimulates Z ring assembly and cell division. *Mol Microbiol.* 2008 May;68(3):720-35.

[6] Bagchi S, Tomenius H, Belova LM, Ausmees N. Intermediate filament-like proteins in bacteria and a cytoskeletal function in Streptomyces. *Mol Microbiol.* 2008 Nov;70(4):1037-50.

[7] Perng MD, Cairns L, van den IP, Prescott A, Hutcheson AM, Quinlan RA. Intermediate filament interactions can be altered by HSP27 and alphaB-crystallin. *J Cell Sci.* 1999 Jul;112 (Pt 13):2099-112.

[8] Perng MD, Wen SF, Gibbon T, Middeldorp J, Sluijs JA, Hol EM, et al. GFAP Filaments Can Tolerate the Incorporation of Assembly-compromised GFAP-{delta}, but with Consequences for Filament Organization and {alpha}B-crystallin Association. *Mol Biol Cell*, 2008 Aug 13.

[9] Quinlan R. Cytoskeletal competence requires protein chaperones. *Prog Mol Subcell Biol.* 2002;28:219-33.

[10] Cabeen MT, Charbon G, Vollmer W, Born P, Ausmees N, Weibel DB, et al. Bacterial cell curvature through mechanical control of cell growth. *EMBO J.* 2009 May 6;28(9):1208-19.

[11] Claessen D, de Jong W, Dijkhuizen L, Wosten HA. Regulation of Streptomyces development: reach for the sky! *Trends Microbiol.* 2006 Jul;14(7):313-9.

[12] Kemp MW, Edwards B, Burgess M, Clarke WT, Nicholson G, Parry DA, et al. Syncoilin isoform organization and differential expression in murine striated muscle. *J Struct Biol.* 2009 Mar;165(3):196-203.

[13] Newey SE, Howman EV, Ponting CP, Benson MA, Nawrotzki R, Loh NY, et al. Syncoilin, a novel member of the intermediate filament superfamily that interacts with alpha-dystrobrevin in skeletal muscle. *J Biol Chem.* 2001 Mar 2;276(9):6645-55.

[14] Pekny M, Lane EB. Intermediate filaments and stress. *Exp Cell Res.* 2007 Jun 10;313(10):2244-54.

[15] Engl C, Jovanovic G, Lloyd LJ, Murray H, Spitaler M, Ying L, et al. In vivo localizations of membrane stress controllers PspA and PspG in Escherichia coli. *Mol Microbiol.* 2009 Jun 26.

[16] Goldberg MW, Fiserova J, Huttenlauch I, Stick R. A new model for nuclear lamina organization. *Biochem Soc Trans.* 2008 Dec;36(Pt 6):1339-43.

[17] Franke WW, Hergt M, Grund C. Rearrangement of the vimentin cytoskeleton during adipose conversion: formation of an intermediate filament cage around lipid globules. *Cell,* 1987;49(1):131-41.

[18] Sandilands A, Prescott AR, Carter JM, Hutcheson AM, Quinlan RA, Richards J, et al. Vimentin and CP49 / Filensin form distinct networks in the lens which are independantly modulated during lens fibre cell differentiation. *J Cell Sci.* 1995;108:1397-406.

[19] Song S, Landsbury A, Dahm R, Liu Y, Zhang Q, Quinlan RA. Functions of the intermediate filament cytoskeleton in the eye lens. *J Clin Invest.* 2009 Jul;119(7):1837-48.

[20] Bellin RM, Huiatt TW, Critchley DR, Robson RM. Synemin may function to directly link muscle cell intermediate filaments to both myofibrillar Z-lines and costameres. *J Biol Chem.* 2001 Aug 24;276(34):32330-7.

[21] Janmey PA, Winer JP, Murray ME, Wen Q. The hard life of soft cells. *Cell Motil Cytoskeleton,* 2009 May 28.

[22] Kim S, Coulombe PA. Intermediate filament scaffolds fulfill mechanical, organizational, and signaling functions in the cytoplasm. Genes Dev. 2007 Jul 1;21(13):1581-97.

[23] Vogel V, Sheetz MP. Cell fate regulation by coupling mechanical cycles to biochemical signaling pathways. *Curr Opin Cell Biol.* 2009 Feb;21(1):38-46.

[24] Herrmann H, Strelkov SV, Burkhard P, Aebi U. Intermediate filaments: primary determinants of cell architecture and plasticity. J Clin Invest. 2009 Jul;119(7):1772-83.

[25] Dichtl MA, Sackmann E. Microrheometry of semiflexible actin networks through enforced single-filament reptation: frictional coupling and heterogeneities in entangled networks. *Proc Natl Acad Sci U S A.* 2002 May 14;99(10):6533-8.

[26] Perng MD, Wen SF, van den IP, Prescott AR, Quinlan RA. Desmin aggregate formation by R120G alphaB-crystallin is caused by altered filament interactions and is dependent upon network status in cells. *Mol Biol Cell,.* 2004 May;15(5):2335-46.

[27] Vicart P, Caron A, Guicheney P, Li Z, Prevost MC, Faure A, et al. A missense mutation in the alphaB-crystallin chaperone gene causes a desmin-related myopathy. *Nat Genet.* 1998;20(1):92-5.

[28] Reiter T, Penman S. "Prompt" heat shock proteins: translationally regulated synthesis of new proteins associated with the nuclear matrix-intermediate filaments as an early response to heat shock. *Proc Natl Acad Sci U S A.* 1983 Aug;80(15):4737-41.

[29] Dynlacht JR, Story MD, Zhu WG, Danner J. Lamin B is a prompt heat shock protein. *J Cell Physiol.* 1999 Jan;178(1):28-34.

[30] Haddad N, Paulin-Levasseur M. Effects of heat shock on the distribution and expression levels of nuclear proteins in HeLa S3 cells. *J Cell Biochem.* 2008 Dec 15;105(6):1485-500.

[31] Wisniewski T, Goldman JE. alpha B-crystallin is associated with intermediate filaments in astrocytoma cells. *Neurochemical Research,* 1998;23(3):385-92.

[32] van den IJssel P, Wheelock R, Prescott A, Russell P, Quinlan RA. Nuclear speckle localisation of the small heat shock protein alpha B-crystallin and its inhibition by the R120G cardiomyopathy-linked mutation. *Exp Cell Res.* 2003 Jul 15;287(2):249-61.

[33] Adhikari AS, Sridhar Rao K, Rangaraj N, Parnaik VK, Mohan Rao C. Heat stress-induced localization of small heat shock proteins in mouse myoblasts: intranuclear

lamin A/C speckles as target for alphaB-crystallin and Hsp25. *Exp Cell Res*. 2004 Oct 1;299(2):393-403.

[34] Kato K, Goto S, Hasegawa K, Shinohara H, Inaguma Y. Responses to heat shock of alpha B crystallin and HSP28 in U373 MG human glioma cells. *Biochim Biophys Acta*, 1993;1175(3):257-62.

[35] Chiesi M, Longoni S, Limbruno U. Cardiac alpha-crystallin. III. Involvement during heart ischemia. Mol Cell Biochem. 1990;97(2):129-36.

[36] Golenhofen N, Htun P, Ness W, Koob R, Schaper W, Drenckhahn D. Binding of the stress protein alpha B-crystallin to cardiac myofibrils correlates with the degree of myocardial damage during ischemia/reperfusion in vivo. *J Mol Cell Cardiol.* 1999; 31(3):569-80.

[37] Iwaki T, Kume-Iwaki A, Liem RKH, Goldman JE. AB-crystallin is expressed in non-lenticular tissues and accumulates in Alexander's disease brain. Cell. 1989;57:71-8.

[38] Iwaki T, Iwaki A, Tateishi J, Sakaki Y, Goldman JE. Alpha-b-crystallin and 27-kd heat-shock protein are regulated by stress conditions in the central-nervous-system and accumulate in rosenthal fibers. *Am J Path*. 1993; 143(2):487-95.

[39] Lowe J, McDermott H, Pike I, Spendlove I, Landon M, Mayer RJ. Alpha B crystallin expression in non-lenticular tissues and selective presence in ubiquitinated inclusion bodies in human disease. *J Pathol*. 1992 Jan; 166(1):61-8.

[40] Djabali K, deNechaud B, Landon F, Portier MM. alpha B-crystallin interacts with intermediate filaments in response to stress. J Cell Sci. 1997;110(Pt21):2759-69.

[41] Klemenz R, Frohli E, Steiger RH, Schafer R, Aoyama A. Alpha B-crystallin is a small heat shock protein. *Proc Natl Acad Sci U S A*. 1991; 88(9):3652-6.

[42] Wiesmann KE, Coop A, Goode D, Hepburne-Scott HW, Crabbe MJ. Effect of mutations of murine lens alphaB crystallin on transfected neural cell viability and cellular translocation in response to stress. *FEBS Lett*. 1998 Oct 30;438(1-2):25-31.

[43] Bryantsev AL, Kurchashova SY, Golyshev SA, Polyakov VY, Wunderink HF, Kanon B, et al. Regulation of stress-induced intracellular sorting and chaperone function of Hsp27 (HspB1) in mammalian cells. *Biochem J*. 2007 Nov 1;407(3):407-17.

[44] Nicholl ID, Quinlan RA. Chaperone activity of a-crystallins modulates intermediate filament assembly. *EMBO J.* 1994;13:945-53.

[45] Altamirano MM, Garcia C, Possani LD, Fersht AR. Oxidative refolding chromatography: folding of the scorpion toxin Cn5 [see comments]. *Nat Biotechnol.* 1999;17(2):187-91.

[46] Altamirano MM, Golbik R, Zahn R, Buckle AM, Fersht AR. Refolding chromatography with immobilized mini-chaperones. Proc Natl Acad Sci U S A. 1997;94(8):3576-8.

[47] Muchowski PJ, Valdez MM, Clark JI. aB-crystallin selectively targets intermediate filament proteins during thermal stress. Invest Ophthalmol Vis Sci. 1999;40:951-8.

[48] Perng MD, Muchowski PJ, van den IJssel P, Wu GJS, Clark JI, Quinlan RA. The cardiomyopathy and lens cataract mutation in aB-crystallin compromises secondary, tertiary and quaternary protein structure and reduces in vitro chaperone activity. J Biol Chem. 1999; 274:33235-43.

[49] Liu J, Tang M, Mestril R, Wang X. Aberrant protein aggregation is essential for a mutant desmin to impair the proteolytic function of the ubiquitin-proteasome system in cardiomyocytes. J Mol Cell Cardiol. 2006 Apr; 40(4):451-4.

[50] Tannous P, Zhu H, Johnstone JL, Shelton JM, Rajasekaran NS, Benjamin IJ, et al. Autophagy is an adaptive response in desmin-related cardiomyopathy. *Proc Natl Acad Sci U S A*. 2008 Jul 15; 105(28):9745-50.

[51] Tang G, Yue Z, Talloczy Z, Goldman JE. Adaptive autophagy in Alexander disease-affected astrocytes. *Autophagy*, 2008 Sep-Oct;4(5):701-3.

[52] Liu J, Chen Q, Huang W, Horak KM, Zheng H, Mestril R, et al. Impairment of the ubiquitin-proteasome system in desminopathy mouse hearts. *FASEB J*. 2006 Feb; 20(2):362-4.

[53] Cho W, Messing A. Properties of astrocytes cultured from GFAP over-expressing and GFAP mutant mice. *Exp Cell Res*. 2009 Apr 15; 315(7):1260-72.

[54] Yoneda K, Furukawa T, Zheng YJ, Momoi T, Izawa I, Inagaki M, et al. An autocrine/paracrine loop linking keratin 14 aggregates to tumor necrosis factor alpha-mediated cytotoxicity in a keratinocyte model of epidermolysis bullosa simplex. *J Biol Chem*. 2004 Feb 20; 279(8):7296-303.

[55] Harada M, Strnad P, Toivola DM, Omary MB. Autophagy modulates keratin-containing inclusion formation and apoptosis in cell culture in a context-dependent fashion. *Exp Cell Res*. 2008 May 1; 314(8):1753-64.

[56] Ku NO, Omary MB. Keratins turn over by ubiquitination in a phosphorylation-modulated fashion. *J Cell Biol*. 2000 May 1; 149(3):547-52.

[57] Sanbe A, Osinska H, Saffitz JE, Glabe CG, Kayed R, Maloyan A, et al. Desmin-related cardiomyopathy in transgenic mice: A cardiac amyloidosis. *Proc Natl Acad Sci U S A*. 2004 Jul 6; 101(27):10132-6.

[58] Meehan S, Knowles TP, Baldwin AJ, Smith JF, Squires AM, Clements P, et al. Characterisation of amyloid fibril formation by small heat-shock chaperone proteins human alphaA-, alphaB- and R120G alphaB-crystallins. *J Mol Biol*. 2007 Sep 14; 372(2):470-84.

[59] Chen Q, Liu JB, Horak KM, Zheng H, Kumarapeli AR, Li J, et al. Intrasarcoplasmic amyloidosis impairs proteolytic function of proteasomes in cardiomyocytes by compromising substrate uptake. *Circ Res*. 2005 Nov 11;97(10):1018-26.

[60] Ito H, Kamei K, Iwamoto I, Inaguma Y, Garcia-Mata R, Sztul E, et al. Inhibition of Proteasomes Induces Accumulation, Phosphorylation, and Recruitment of HSP27 and alphaB-Crystallin to Aggresomes. *J Biochem*. (Tokyo). 2002 Apr; 131(4):593-603.

[61] Bardag-Gorce F, Vu J, Nan L, Riley N, Li J, French SW. Proteasome inhibition induces cytokeratin accumulation in vivo. *Exp Mol Pathol*. 2004 Apr; 76(2):83-9.

[62] Nitahara-Kasahara Y, Fukasawa M, Shinkai-Ouchi F, Sato S, Suzuki T, Murakami K, et al. Cellular vimentin content regulates the protein level of hepatitis C virus core protein and the hepatitis C virus production in cultured cells. *Virology*, 2009 Jan 20; 383(2):319-27.

[63] den Engelsman J, Bennink EJ, Doerwald L, Onnekink C, Wunderink L, Andley UP, et al. Mimicking phosphorylation of the small heat-shock protein alphaB-crystallin recruits the F-box protein FBX4 to nuclear SC35 speckles. *Eur J Biochem*. 2004 Nov; 271(21):4195-203.

[64] Hayes VH, Devlin G, Quinlan RA. Truncation of alphaB-crystallin by the myopathy-causing Q151X mutation significantly destabilizes the protein leading to aggregate formation in transfected cells. *J Biol Chem*. 2008 Apr 18;283(16):10500-12.

[65] Ghosh JG, Houck SA, Clark JI. Interactive sequences in the stress protein and molecular chaperone human alphaB crystallin recognize and modulate the assembly of filaments. *Int J Biochem Cell Biol.* 2007; 39(10):1804-15.

[66] Selcen D, Engel AG. Myofibrillar myopathy caused by novel dominant negative alpha B-crystallin mutations. *Ann Neurol.* 2003 Dec; 54(6):804-10.

[67] Meehan S, Knowles TP, Baldwin AJ, Smith JF, Squires AM, Clements P, et al. Characterisation of Amyloid Fibril Formation by Small Heat-shock Chaperone Proteins Human alphaA-, alphaB- and R120G alphaB-Crystallins. J Mol Biol. 2007 Jun 29.

[68] Treweek TM, Rekas A, Lindner RA, Walker MJ, Aquilina JA, Robinson CV, et al. R120G alphaB-crystallin promotes the unfolding of reduced alpha-lactalbumin and is inherently unstable. *Febs J.* 2005 Feb; 272(3):711-24.

[69] Goldfarb LG, Park KY, Cervenakova L, Gorokhova S, Lee HS, Vasconcelos O, et al. Missense mutations in desmin associated with familial cardiac and skeletal myopathy. *Nat Genet.* 1998; 19(4):402-3.

[70] Goldfarb LG, Dalakas MC. Tragedy in a heartbeat: malfunctioning desmin causes skeletal and cardiac muscle disease. J Clin Invest. 2009 Jul;119(7):1806-13.

[71] Bar H, Mucke N, Kostareva A, Sjoberg G, Aebi U, Herrmann H. Severe muscle disease-causing desmin mutations interfere with in vitro filament assembly at distinct stages. *Proc Natl Acad Sci U S A.* 2005 Oct 18;102(42):15099-104.

[72] Kreplak L, Bar H. Severe myopathy mutations modify the nanomechanics of desmin intermediate filaments. *J Mol Biol.* 2009 Jan 30;385(4):1043-51.

[73] Goldfarb LG, Olive M, Vicart P, Goebel HH. Intermediate filament diseases: desminopathy. *Adv Exp Med Biol.* 2008; 642:131-64.

[74] Schroder R, Schoser B. Myofibrillar myopathies: a clinical and myopathological guide. *Brain Pathol.* 2009 Jul; 19(3):483-92.

[75] Quinlan RA, Brenner M, Goldman JE, Messing A. GFAP and its role in Alexander disease. *Exp Cell Res.* 2007 Jun 10;313(10):2077-87.

[76] Li R, Johnson AB, Salomons G, Goldman JE, Naidu S, Quinlan R, et al. Glial fibrillary acidic protein mutations in infantile, juvenile, and adult forms of Alexander disease. *Ann Neurol.* 2005 Mar;57(3):310-26.

[77] Irobi J, Van Impe K, Seeman P, Jordanova A, Dierick I, Verpoorten N, et al. Hot-spot residue in small heat-shock protein 22 causes distal motor neuropathy. *Nat Genet.* 2004 Jun; 36(6):597-601.

[78] Evgrafov OV, Mersiyanova I, Irobi J, Van Den Bosch L, Dierick I, Leung CL, et al. Mutant small heat-shock protein 27 causes axonal Charcot-Marie-Tooth disease and distal hereditary motor neuropathy. *Nat Genet.* 2004 Jun; 36(6):602-6.

[79] Hagemann TL, Boelens WC, Wawrousek EF, Messing A. Suppression of GFAP toxicity by alphaB-crystallin in mouse models of Alexander disease. *Hum Mol Genet.* 2009 Apr 1; 18(7):1190-9.

[80] Brenner M, Johnson AB, Boespflug-Tanguay O, Rodriguez D, Goldman JE. Mutations in GFAP, encoding glial fibrillary acidic protein are associated with Alexander Disease. Nat Gen. 2001;27:117-20.

[81] Der Perng M, Su M, Wen SF, Li R, Gibbon T, Prescott AR, et al. The Alexander disease-causing glial fibrillary acidic protein mutant, R416W, accumulates into Rosenthal fibers by a pathway that involves filament aggregation and the association of alpha B-crystallin and HSP27. Am J Hum Genet. 2006 Aug;79(2):197-213.

[82] Tang G, Xu Z, Goldman JE. Synergistic effects of the SAPK/JNK and the proteasome pathway on glial fibrillary acidic protein (GFAP) accumulation in Alexander disease. *J Biol Chem.* 2006 Dec 15; 281(50):38634-43.

[83] Messing A, Head MW, Galles K, Galbreath EJ, Goldman JE, Brenner M. Fatal encephalopathy with astrocyte inclusions in GFAP transgenic mice. *Am J Path.* 1998; 152(2):391-8.

[84] Bensch KG, Fleming JE, Lohmann W. The role of ascorbic acid in senile cataract. *Proc Natl Acad Sci U S A.* 1985;82(21):7193-6.

[85] Hsiao VC, Tian R, Long H, Der Perng M, Brenner M, Quinlan RA, et al. Alexander-disease mutation of GFAP causes filament disorganization and decreased solubility of GFAP. *J Cell Sci.* 2005 May 1;118(Pt 9):2057-65.

[86] Tian R, Gregor M, Wiche G, Goldman JE. Plectin regulates the organization of glial fibrillary acidic protein in Alexander disease. Am J Pathol. 2006 Mar;168(3):888-97.

[87] Koyama Y, Goldman JE. Formation of GFAP cytoplasmic inclusions in astrocytes and their disaggregation by alphaB-crystallin [In Process Citation]. *Am J Pathol.* 1999; 154(5):1563-72.

[88] Wang X, Klevitsky R, Huang W, Glasford J, Li F, Robbins J. AlphaB-crystallin modulates protein aggregation of abnormal desmin. *Circ Res.* 2003 Nov 14;93(10):998-1005.

[89] Sanbe A, Daicho T, Mizutani R, Endo T, Miyauchi N, Yamauchi J, et al. Protective effect of geranylgeranylacetone via enhancement of HSPB8 induction in desmin-related cardiomyopathy. *PLoS One,* 2009;4(4):e5351.

[90] Sanbe A, Osinska H, Villa C, Gulick J, Klevitsky R, Glabe CG, et al. Reversal of amyloid-induced heart disease in desmin-related cardiomyopathy. *Proc Natl Acad Sci U S A.* 2005 Sep 20; 102(38):13592-7.

[91] Miron T, Wilchek M, Geiger B. Characterization of an inhibitor of actin polymerization in vinculin- rich fraction of turkey gizzard smooth-muscle. *Eur J Biochem.* 1988; 178(2):543-53.

[92] Mounier N, Arrigo AP. Actin cytoskeleton and small heat shock proteins: how do they interact? *Cell Stress Chaperones*, 2002 Apr;7(2):167-76.

[93] Miron T, Vancompernolle K, Vandekerckhove J, Wilchek M, Geiger B. A 25-kd inhibitor of actin polymerization is a low-molecular mass heat-shock protein. *J Cell Biol.* 1991; 114(2):255-61.

[94] Benndorf R, Hayess K, Ryazantsev S, Wieske M, Behlke J, Lutsch G. Phosphorylation and supramolecular organization of murine small heat shock protein HSP25 abolish its actin polymerization-inhibiting activity. *J Biol Chem.* 1994; 269(32):20780-4.

[95] Guay J, Lambert H, GingrasBreton G, Lavoie JN, Huot J, Landry J. Regulation of actin filament dynamics by p38 map kinase-mediated phosphorylation of heat shock protein 27. *J Cell Sci.* 1997; 110(Pt3):357-68.

[96] Panasenko OO, Kim MV, Marston SB, Gusev NB. Interaction of the small heat shock protein with molecular mass 25 kDa (hsp25) with actin. *Eur J Biochem.* 2003 Mar; 270(5):892-901.

[97] Pivovarova AV, Chebotareva NA, Chernik IS, Gusev NB, Levitsky DI. Small heat shock protein Hsp27 prevents heat-induced aggregation of F-actin by forming soluble complexes with denatured actin. *FEBS J.* 2007 Nov; 274(22):5937-48.

[98] Schafer C, Clapp P, Welsh MJ, Benndorf R, Williams JA. HSP27 expression regulates CCK-induced changes of the actin cytoskeleton in CHO-CCK-A cells. *Am J Physiol.* 1999 Dec; 277(6 Pt 1):C1032-43.

[99] Wang K, Spector A. a-crystallin stabilises actin filaments and prevents cytochalasin-induced depolymerisation in a phosphorylation-dependent manner. *Eur J Biochem.* 1996; 242:56-66.

[100] Wieske M, Benndorf R, Behlke J, Dolling R, Grelle G, Bielka H, et al. Defined sequence segments of the small heat shock proteins HSP25 and alphaB-crystallin inhibit actin polymerization. *Eur J Biochem*. 2001 Apr; 268(7):2083-90.

[101] Singh BN, Rao KS, Ramakrishna T, Rangaraj N, Rao Ch M. Association of alphaB-crystallin, a small heat shock protein, with actin: role in modulating actin filament dynamics in vivo. *J Mol Biol.* 2007 Feb 23; 366(3):756-67.

[102] Iwaki T, Iwaki A, Tateishi J, Goldman JE. Sense and antisense modification of glial alpha B-crystallin production results in alterations of stress fiber formation and thermoresistance. *J Cell Biol.* 1994; 125(6):1385-93.

[103] Gopalakrishnan S, Takemoto L. Binding of actin to lens alpha crystallins. *Curr Eye Res.* 1992;11(9):929-33.

[104] Brophy CM, Lamb S, Graham A. The small heat shock-related protein-20 is an actin-associated protein. *J Vasc Surg*. 1999; 29(2):326-33.

[105] Bukach OV, Marston SB, Gusev NB. Small heat shock protein with apparent molecular mass 20 kDa (Hsp20, HspB6) is not a genuine actin-binding protein. *J Muscle Res Cell Motil.* 2005;26(4-5):175-81.

[106] Geeves MA, Fedorov R, Manstein DJ. Molecular mechanism of actomyosin-based motility. *Cell Mol Life Sci.* 2005 Jul; 62(13):1462-77.

[107] Nozais M, Bechet JJ, Houadjeto M. Inactivation, subunit dissociation, aggregation, and unfolding of myosin subfragment 1 during guanidine denaturation. *Biochemistry,* 1992 Feb 4; 31(4):1210-5.

[108] Melkani GC, Cammarato A, Bernstein SI. AlphaB-crystallin maintains skeletal muscle myosin enzymatic activity and prevents its aggregation under heat-shock stress. *J Mol Biol.* 2006 May 5; 358(3):635-45.

[109] Yoshioka M, Tanaka H, Shono N, Snyder EE, Shindo M, St-Amand J. Serial analysis of gene expression in the skeletal muscle of endurance athletes compared to sedentary men. *FASEB J.* 2003 Oct; 17(13):1812-9.

[110] Tskhovrebova L, Trinick J. Titin: properties and family relationships. *Nat Rev Mol Cell Biol.* 2003 Sep; 4(9):679-89.

[111] Granzier HL, Labeit S. The giant protein titin: a major player in myocardial mechanics, signaling, and disease. Circ Res. 2004 Feb 20;94(3):284-95.

[112] Linke WA, Granzier H. A spring tale: new facts on titin elasticity. *Biophys J.* 1998 Dec; 75(6):2613-4.

[113] Wu Y, Cazorla O, Labeit D, Labeit S, Granzier H. Changes in titin and collagen underlie diastolic stiffness diversity of cardiac muscle. *J Mol Cell Cardiol.* 2000 Dec; 32(12):2151-62.

[114] Helmes M, Trombitas K, Centner T, Kellermayer M, Labeit S, Linke WA, et al. Mechanically driven contour-length adjustment in rat cardiac titin's unique N2B sequence: titin is an adjustable spring. *Circ Res.* 1999 Jun 11; 84(11):1339-52.

[115] Golenhofen N, Arbeiter A, Koob R, Drenckhahn D. Ischemia-induced association of the stress protein alpha B-crystallin with I-band portion of cardiac titin. *J Mol Cell Cardiol.* 2002 Mar;34(3):309-19.
[116] Bullard B, Ferguson C, Minajeva A, Leake MC, Gautel M, Labeit D, et al. Association of the chaperone alphaB-crystallin with titin in heart muscle. *J Biol Chem.* 2004 Feb 27; 279(9):7917-24.
[117] Zhu Y, Bogomolovas J, Labeit S, Granzier H. Single molecule force spectroscopy of the cardiac titin N2B element: effects of the molecular chaperone alphaB-crystallin with disease-causing mutations. *J Biol Chem.* 2009 May 15; 284(20):13914-23.
[118] Golenhofen N, Perng MD, Quinlan RA, Drenckhahn D. Comparison of the small heat shock proteins alphaB-crystallin, MKBP, HSP25, HSP20, and cvHSP in heart and skeletal muscle. *Histochem Cell Biol.* 2004 Nov; 122(5):415-25.
[119] Kumarapeli AR, Wang X. Genetic modification of the heart: chaperones and the cytoskeleton. *J Mol Cell Cardiol.* 2004 Dec;37(6):1097-109.
[120] Launay N, Goudeau B, Kato K, Vicart P, Lilienbaum A. Cell signaling pathways to alphaB-crystallin following stresses of the cytoskeleton. *Exp Cell Res.* 2006 Nov 1;312(18):3570-84.
[121] Inagaki N, Hayashi T, Arimura T, Koga Y, Takahashi M, Shibata H, et al. Alpha B-crystallin mutation in dilated cardiomyopathy. *Biochem Biophys Res Commun.* 2006 Apr 7; 342(2):379-86.
[122] Gerull B, Gramlich M, Atherton J, McNabb M, Trombitas K, Sasse-Klaassen S, et al. Mutations of TTN, encoding the giant muscle filament titin, cause familial dilated cardiomyopathy. *Nat Genet.* 2002 Feb;30(2):201-4.
[123] Thedieck C, Kalbacher H, Kratzer U, Lammers R, Stevanovic S, Klein G. alpha B-crystallin is a cytoplasmic interaction partner of the kidney-specific cadherin-16. *J Mol Biol.* 2008 Apr 18; 378(1):145-53.
[124] McLachlan RW, Yap AS. Not so simple: the complexity of phosphotyrosine signaling at cadherin adhesive contacts. *J Mol Med.* 2007 Jun;85(6):545-54.
[125] Fujita Y, Ohto E, Katayama E, Atomi Y. alphaB-Crystallin-coated MAP microtubule resists nocodazole and calcium-induced disassembly. *J Cell Sci.* 2004 Apr 1; 117(Pt 9):1719-26.
[126] Ohto-Fujita E, Fujita Y, Atomi Y. Analysis of the alphaB-crystallin domain responsible for inhibiting tubulin aggregation. Cell Stress Chaperones. 2007 *Summer*; 12(2):163-71.
[127] Hino M, Kurogi K, Okubo MA, Murata-Hori M, Hosoya H. Small heat shock protein 27 (HSP27) associates with tubulin/microtubules in HeLa cells. *Biochem Biophys Res Commun.* 2000 Apr 29; 271(1):164-9.
[128] Arai H, Atomi Y. Chaperone activity of alpha B-crystallin suppresses tubulin aggregation through complex formation. *Cell Structure and Function*, 1997; 22(5):539-44.
[129] Kato K, Ito H, Inaguma Y, Okamoto K, Saga S. Synthesis and accumulation of alpha-b-crystallin in c6 glioma-cells is induced by agents that promote the disassembly of microtubules. *Journal of Biological Chemistry*, 1996;271(43):26989-94.
[130] Day RM, Gupta JS, MacRae TH. A small heat shock/alpha-crystallin protein from encysted Artemia embryos suppresses tubulin denaturation. Cell Stress Chaperones. 2003 *Summer*; 8(2):183-93.
[131] Xi JH, Bai F, McGaha R, Andley UP. Alpha-crystallin expression affects microtubule assembly and prevents their aggregation. *FASEB J.* 2006 May;20(7):846-57.

[132] Bluhm WF, Martin JL, Mestril R, Dillmann WH. Specific heat shock proteins protect microtubules during simulated ischemia in cardiac myocytes. *Am J Physiol.* 1998;275(6 Pt 2):H2243-9.

[133] Ghosh JG, Houck SA, Clark JI. Interactive domains in the molecular chaperone human alphaB crystallin modulate microtubule assembly and disassembly. *PLoS One,* 2007;2(6):e498.

[134] Ghosh JG, Houck SA, Clark JI. Interactive sequences in the molecular chaperone, human alphaB crystallin modulate the fibrillation of amyloidogenic proteins. *Int J Biochem Cell Biol.* 2008;40(5):954-67.

[135] Bagneris C, Bateman OA, Naylor CE, Cronin N, Boelens WC, Keep NH, et al. Crystal Structures of alpha-Crystallin Domain Dimers of alphaB-Crystallin and Hsp20. *J Mol Biol.* 2009 Jul 30.

[136] Bova MP, Ding LL, Horwitz J, Fung BK. Subunit exchange of alphaA-crystallin. *J Biol Chem.* 1997;272(47):29511-7.

[137] Shashidharamurthy R, Koteiche HA, Dong J, McHaourab HS. Mechanism of chaperone function in small heat shock proteins: dissociation of the HSP27 oligomer is required for recognition and binding of destabilized T4 lysozyme. *J Biol Chem.* 2005 Feb 18;280(7):5281-9.

[138] Aquilina JA, Benesch JL, Ding LL, Yaron O, Horwitz J, Robinson CV. Phosphorylation of alphaB-crystallin alters chaperone function through loss of dimeric substructure. *J Biol Chem.* 2004 Jul 2;279(27):28675-80.

[139] Jaya N, Garcia V, Vierling E. Substrate binding site flexibility of the small heat shock protein molecular chaperones. *Proc Natl Acad Sci U S A.* 2009 Aug 26.

[140] van Montfort R, Slingsby C, Vierling E. Structure and function of the small heat shock protein/alpha-crystallin family of molecular chaperones. *Adv Protein Chem.* 2001;59:105-56.

[141] van Montfort RL, Basha E, Friedrich KL, Slingsby C, Vierling E. Crystal structure and assembly of a eukaryotic small heat shock protein. *Nat Struct Biol.* 2001 Dec;8(12):1025-30.

[142] Liu L, Ghosh JG, Clark JI, Jiang S. Studies of alphaB crystallin subunit dynamics by surface plasmon resonance. *Anal Biochem.* 2006 Mar 15;350(2):186-95.

[143] Friedrich KL, Giese KC, Buan NR, Vierling E. Interactions between small heat shock protein subunits and substrate in small heat shock protein-substrate complexes. *J Biol Chem.* 2004 Jan 9;279(2):1080-9.

[144] Mayer MP, Bukau B. Hsp70 chaperones: cellular functions and molecular mechanism. *Cell Mol Life Sci.* 2005 Mar;62(6):670-84.

[145] Horwich AL, Fenton WA, Chapman E, Farr GW. Two families of chaperonin: physiology and mechanism. *Annu Rev Cell Dev Biol.* 2007;23:115-45.

[146] Sathish HA, Stein RA, Yang G, McHaourab HS. Mechanism of chaperone function in small heat-shock proteins. Fluorescence studies of the conformations of T4 lysozyme bound to alphaB-crystallin. *J Biol Chem.* 2003 Nov 7;278(45):44214-21.

[147] Ahrman E, Lambert W, Aquilina JA, Robinson CV, Emanuelsson CS. Chemical cross-linking of the chloroplast localized small heat-shock protein, Hsp21, and the model substrate citrate synthase. *Protein Sci.* 2007 Jul;16(7):1464-78.

[148] Ghosh JG, Shenoy AK, Jr., Clark JI. Interactions between important regulatory proteins and human alphaB crystallin. *Biochemistry,* 2007 May 29;46(21):6308-17.

[149] Cheng G, Basha E, Wysocki VH, Vierling E. Insights into small heat shock protein and substrate structure during chaperone action derived from hydrogen/deuterium exchange and mass spectrometry. *J Biol Chem.* 2008 Sep 26;283(39):26634-42.

[150] Kim KK, Kim R, Kim SH. Crystal structure of a small heat-shock protein. *Nature,* 1998;394(6693):595-9.

[151] Haley DA, Horwitz J, Stewart PL. The small heat-shock protein, alphaB-crystallin, has a variable quaternary structure. *J Mol Biol.* 1998;277(1):27-35.

[152] Aquilina JA, Benesch JL, Bateman OA, Slingsby C, Robinson CV. Polydispersity of a mammalian chaperone: mass spectrometry reveals the population of oligomers in alphaB-crystallin. *Proc Natl Acad Sci U S A.* 2003 Sep 16;100(19):10611-6.

[153] Rajasekaran NS, Connell P, Christians ES, Yan LJ, Taylor RP, Orosz A, et al. Human alpha B-crystallin mutation causes oxido-reductive stress and protein aggregation cardiomyopathy in mice. *Cell,* 2007 Aug 10;130(3):427-39.

[154] Ghosh JG, Estrada MR, Clark JI. Interactive domains for chaperone activity in the small heat shock protein, human alphaB crystallin. *Biochemistry,* 2005 Nov 15;44(45):14854-69.

[155] Eswar N, Eramian D, Webb B, Shen MY, Sali A. Protein structure modeling with MODELLER. Methods Mol Biol. 2008;426:145-59.

[156] Emsley P, Cowtan K. Coot: model-building tools for molecular graphics. *Acta Crystallogr D Biol Crystallogr.* 2004 Dec;60(Pt 12 Pt 1):2126-32.

[157] Krissinel E, Henrick K. Secondary-structure matching (SSM), a new tool for fast protein structure alignment in three dimensions. *Acta Crystallogr D Biol Crystallogr.* 2004 Dec;60(Pt 12 Pt 1):2256-68.

[158] Dierick I, Baets J, Irobi J, Jacobs A, De Vriendt E, Deconinck T, et al. Relative contribution of mutations in genes for autosomal dominant distal hereditary motor neuropathies: a genotype-phenotype correlation study. *Brain,* 2008 May;131(Pt 5):1217-27.

[159] Houlden H, Laura M, Wavrant-De Vrieze F, Blake J, Wood N, Reilly MM. Mutations in the HSP27 (HSPB1) gene cause dominant, recessive, and sporadic distal HMN/CMT type 2. *Neurology,* 2008 Nov 18;71(21):1660-8.

[160] James PA, Rankin J, Talbot K. Asymmetrical late onset motor neuropathy associated with a novel mutation in the small heat shock protein HSPB1 (HSP27). *J Neurol Neurosurg Psychiatry,* 2008 Apr;79(4):461-3.

[161] Tang B, Liu X, Zhao G, Luo W, Xia K, Pan Q, et al. Mutation analysis of the small heat shock protein 27 gene in chinese patients with Charcot-Marie-Tooth disease. *Arch Neurol.* 2005 Aug;62(8):1201-7.

[162] Ikeda Y, Abe A, Ishida C, Takahashi K, Hayasaka K, Yamada M. A clinical phenotype of distal hereditary motor neuronopathy type II with a novel HSPB1 mutation. *J Neurol Sci.* 2009 Feb 15;277(1-2):9-12.

[163] Ackerley S, James PA, Kalli A, French S, Davies KE, Talbot K. A mutation in the small heat-shock protein HSPB1 leading to distal hereditary motor neuronopathy disrupts neurofilament assembly and the axonal transport of specific cellular cargoes. *Hum Mol Genet.* 2006 Jan 15;15(2):347-54.

[164] Kijima K, Numakura C, Goto T, Takahashi T, Otagiri T, Umetsu K, et al. Small heat shock protein 27 mutation in a Japanese patient with distal hereditary motor neuropathy. *J Hum Genet.* 2005;50(9):473-6.

[165] Dierick I, Irobi J, Janssens S, Theuns J, Lemmens R, Jacobs A, et al. Genetic variant in the HSPB1 promoter region impairs the HSP27 stress response. *Hum Mutat.* 2007 Aug;28(8):830.
[166] Pras E, Frydman M, Levy-Nissenbaum E, Bakhan T, Raz J, Assia EI, et al. A nonsense mutation (W9X) in CRYAA causes autosomal recessive cataract in an inbred Jewish Persian family. *Invest Ophthalmol Vis Sci.* 2000 Oct;41(11):3511-5.
[167] Hansen L, Yao W, Eiberg H, Kjaer KW, Baggesen K, Hejtmancik JF, et al. Genetic heterogeneity in microcornea-cataract: five novel mutations in CRYAA, CRYGD, and GJA8. *Invest Ophthalmol Vis Sci.* 2007 Sep;48(9):3937-44.
[168] Devi RR, Yao W, Vijayalakshmi P, Sergeev YV, Sundaresan P, Hejtmancik JF. Crystallin gene mutations in Indian families with inherited pediatric cataract. *Mol Vis.* 2008;14:1157-70.
[169] Graw J, Klopp N, Illig T, Preising MN, Lorenz B. Congenital cataract and macular hypoplasia in humans associated with a de novo mutation in CRYAA and compound heterozygous mutations in P. *Graefes Arch Clin Exp Ophthalmol.* 2006 Aug;244(8):912-9.
[170] Mackay DS, Andley UP, Shiels A. Cell death triggered by a novel mutation in the alphaA-crystallin gene underlies autosomal dominant cataract linked to chromosome 21q. *Eur J Hum Genet.* 2003 Oct;11(10):784-93.
[171] Khan AO, Aldahmesh MA, Meyer B. Recessive congenital total cataract with microcornea and heterozygote carrier signs caused by a novel missense CRYAA mutation (R54C). *Am J Ophthalmol.* 2007 Dec;144(6):949-52.
[172] Santhiya ST, Soker T, Klopp N, Illig T, Prakash MV, Selvaraj B, et al. Identification of a novel, putative cataract-causing allele in CRYAA (G98R) in an Indian family. Mol Vis. 2006;12:768-73.
[173] Litt M, Kramer P, LaMorticella DM, Murphey W, Lovrien EW, Weleber RG. Autosomal dominant congenital cataract associated with a missense mutation in the human alpha crystallin gene CRYAA. *Hum Mol Genet.* 1998;7(3):471-4.
[174] Beby F, Commeaux C, Bozon M, Denis P, Edery P, Morle L. New phenotype associated with an Arg116Cys mutation in the CRYAA gene: nuclear cataract, iris coloboma, and microphthalmia. *Arch Ophthalmol.* 2007 Feb;125(2):213-6.
[175] Vanita V, Singh JR, Hejtmancik JF, Nuernberg P, Hennies HC, Singh D, et al. A novel fan-shaped cataract-microcornea syndrome caused by a mutation of CRYAA in an Indian family. *Mol Vis.* 2006;12:518-22.
[176] Richter L, Flodman P, Barria von-Bischhoffshausen F, Burch D, Brown S, Nguyen L, et al. Clinical variability of autosomal dominant cataract, microcornea and corneal opacity and novel mutation in the alpha A crystallin gene (CRYAA). Am J Med Genet A. 2008 Apr 1;146(7):833-42.
[177] Gu F, Luo W, Li X, Wang Z, Lu S, Zhang M, et al. A novel mutation in AlphaA-crystallin (CRYAA) caused autosomal dominant congenital cataract in a large Chinese family. *Hum Mutat.* 2008 May;29(5):769.
[178] Liu M, Ke T, Wang Z, Yang Q, Chang W, Jiang F, et al. Identification of a CRYAB mutation associated with autosomal dominant posterior polar cataract in a Chinese family. *Invest Ophthalmol Vis Sci.* 2006 Aug;47(8):3461-6.

[179] Safieh LA, Khan AO, Alkuraya FS. Identification of a novel CRYAB mutation associated with autosomal recessive juvenile cataract in a Saudi family. *Mol Vis.* 2009;15:980-4.

[180] Liu Y, Zhang X, Luo L, Wu M, Zeng R, Cheng G, et al. A novel alphaB-crystallin mutation associated with autosomal dominant congenital lamellar cataract. *Invest Ophthalmol Vis Sci.* 2006 Mar;47(3):1069-75.

[181] Berry V, Francis P, Reddy MA, Collyer D, Vithana E, MacKay I, et al. Alpha-B crystallin gene (CRYAB) mutation causes dominant congenital posterior polar cataract in humans. *Am J Hum Genet.* 2001 Nov;69(5):1141-5.

[182] Nicolaou P, Knoll R, Haghighi K, Fan GC, Dorn GW, 2nd, Hasenfub G, et al. Human mutation in the anti-apoptotic heat shock protein 20 abrogates its cardioprotective effects. *J Biol Chem.* 2008 Nov 28;283(48):33465-71.

[183] Tang BS, Zhao GH, Luo W, Xia K, Cai F, Pan Q, et al. Small heat-shock protein 22 mutated in autosomal dominant Charcot-Marie-Tooth disease type 2L. *Hum Genet.* 2005 Feb;116(3):222-4.

[184] Fontaine JM, Rest JS, Welsh MJ, Benndorf R. The sperm outer dense fiber protein is the 10th member of the superfamily of mammalian small stress proteins. *Cell Stress Chaperones,* 2003 Spring;8(1):62-9.

[185] Müller M, Bhattacharya SS, Moore T, Prescott Q, Wedig T, Herrmann H, et al. Dominant cataract formation in association with a vimentin assembly disrupting mutation. *Hum Mol Genet.* 2009 Mar 15;18(6):1052-7.

[186] Ramachandran RD, Perumalsamy V, Hejtmancik JF. Autosomal recessive juvenile onset cataract associated with mutation in BFSP1. *Hum Genet.* 2007 May;121(3-4):475-82.

[187] Jakobs PM, Hess JF, FitzGerald PG, Kramer P, Weleber RG, Litt M. Autosomal-Dominant Congenital Cataract Associated with a Deletion Mutation in the Human Beaded Filament Protein Gene BFSP2. *Am J Hum Genet.* 2000;66(4):1432-6.

[188] Zhang Q, Guo X, Xiao X, Yi J, Jia X, Hejtmancik JF. Clinical description and genome wide linkage study of Y-sutural cataract and myopia in a Chinese family. *Mol Vis*. 2004 Nov 17;10:890-900.

[189] Cui X, Gao L, Jin Y, Zhang Y, Bai J, Feng G, et al. The E233del mutation in BFSP2 causes a progressive autosomal dominant congenital cataract in a Chinese family. *Mol Vis*. 2007;13:2023-9.

[190] Conley YP, Erturk D, Keverline A, Mah TS, Keravala A, Barnes LR, et al. A Juvenile-Onset, Progressive Cataract Locus on Chromosome 3q21-q22 Is Associated with a Missense Mutation in the Beaded Filament Structural Protein-2. *Am J Hum Genet.* 2000;66(4):1426-31.

[191] Ma X, Li FF, Wang SZ, Gao C, Zhang M, Zhu SQ. A new mutation in BFSP2 (G1091A) causes autosomal dominant congenital lamellar cataracts. *Mol Vis.* 2008; 14:1906-11.

[192] De Jonghe P, Mersivanova I, Nelis E, Del Favero J, Martin JJ, Van Broeckhoven C, et al. Further evidence that neurofilament light chain gene mutations can cause Charcot-Marie-Tooth disease type 2E. *Ann Neurol*. 2001 Feb;49(2):245-9.

[193] Perez-Olle R, Jones ST, Liem RK. Phenotypic analysis of neurofilament light gene mutations linked to Charcot-Marie-Tooth disease in cell culture models. *Hum Mol Genet.* 2004 Oct 1; 13(19):2207-20.

[194] Miltenberger-Miltenyi G, Janecke AR, Wanschitz JV, Timmerman V, Windpassinger C, Auer-Grumbach M, et al. Clinical and electrophysiological features in Charcot-Marie-Tooth disease with mutations in the NEFL gene. *Arch Neurol.* 2007 Jul;64(7):966-70.

[195] Jordanova A, De Jonghe P, Boerkoel CF, Takashima H, De Vriendt E, Ceuterick C, et al. Mutations in the neurofilament light chain gene (NEFL) cause early onset severe Charcot-Marie-Tooth disease. *Brain,* 2003 Mar; 126(Pt 3):590-7.

[196] Leung CL, Nagan N, Graham TH, Liem RK. A novel duplication/insertion mutation of NEFL in a patient with Charcot-Marie-Tooth disease. *Am J Med Genet A.* 2006 May 1; 140(9):1021-5.

[197] Georgiou DM, Zidar J, Korosec M, Middleton LT, Kyriakides T, Christodoulou K. A novel NF-L mutation Pro22Ser is associated with CMT2 in a large Slovenian family. *Neurogenetics,* 2002 Oct;4(2):93-6.

[198] Fabrizi GM, Cavallaro T, Angiari C, Bertolasi L, Cabrini I, Ferrarini M, et al. Giant axon and neurofilament accumulation in Charcot-Marie-Tooth disease type 2E. *Neurology*, 2004 Apr 27; 62(8):1429-31.

[199] Abe A, Numakura C, Saito K, Koide H, Oka N, Honma A, et al. Neurofilament light chain polypeptide gene mutations in Charcot-Marie-Tooth disease: nonsense mutation probably causes a recessive phenotype. *J Hum Genet.* 2009; 54(2):94-7.

[200] Selcen D, Ohno K, Engel AG. Myofibrillar myopathy: clinical, morphological and genetic studies in 63 patients. *Brain*, 2004 Feb; 127(Pt 2):439-51.

[201] Bergman JE, Veenstra-Knol HE, van Essen AJ, van Ravenswaaij CM, den Dunnen WF, van den Wijngaard A, et al. Two related Dutch families with a clinically variable presentation of cardioskeletal myopathy caused by a novel S13F mutation in the desmin gene. *Eur J Med Genet.* 2007 Sep-Oct;50(5):355-66.

[202] Pica EC, Kathirvel P, Pramono ZA, Lai PS, Yee WC. Characterization of a novel S13F desmin mutation associated with desmin myopathy and heart block in a Chinese family. *Neuromuscul Disord.* 2008 Feb; 18(2):178-82.

[203] Arbustini E, Pasotti M, Pilotto A, Pellegrini C, Grasso M, Previtali S, et al. Desmin accumulation restrictive cardiomyopathy and atrioventricular block associated with desmin gene defects. *Eur J Heart Fail.* 2006 Aug; 8(5):477-83.

[204] Taylor MR, Slavov D, Ku L, Di Lenarda A, Sinagra G, Carniel E, et al. Prevalence of desmin mutations in dilated cardiomyopathy. *Circulation,* 2007 Mar 13; 115(10):1244-51.

[205] Kostareva A, Gudkova A, Sjoberg G, Kiselev I, Moiseeva O, Karelkina E, et al. Desmin mutations in a St. Petersburg cohort of cardiomyopathies. Acta Myol. 2006 Dec; 25(3):109-15.

[206] Munoz-Marmol AM, Strasser G, Isamat M, Coulombe PA, Yang Y, Roca X, et al. A dysfunctional desmin mutation in a patient with severe generalized myopathy [In Process Citation]. *Proc Natl Acad Sci U S A.* 1998;95(19):11312-7.

[207] Park KY, Dalakas MC, Goebel HH, Ferrans VJ, Semino-Mora C, Litvak S, et al. Desmin splice variants causing cardiac and skeletal myopathy. *J Med Genet.* 2000 Nov; 37(11):851-7.

[208] Schroder R, Vrabie A, Goebel HH. Primary desminopathies. *J Cell Mol Med.* 2007 May-Jun; 11(3):416-26.

[209] Vrabie A, Goldfarb LG, Shatunov A, Nagele A, Fritz P, Kaczmarek I, et al. The enlarging spectrum of desminopathies: new morphological findings, eastward geographic spread, novel exon 3 desmin mutation. *Acta Neuropathol.* 2005 Apr; 109(4):411-7.

[210] Dalakas MC, Park KY, Semino-Mora C, Lee HS, Sivakumar K, Goldfarb LG. Desmin myopathy, a skeletal myopathy with cardiomyopathy caused by mutations in the desmin gene. *N Engl J Med.* 2000 Mar 16;342(11):770-80.

[211] Goudeau B, Rodrigues-Lima F, Fischer D, Casteras-Simon M, Sambuughin N, de Visser M, et al. Variable pathogenic potentials of mutations located in the desmin alpha-helical domain. *Hum Mutat.* 2006 Sep;27(9):906-13.

[212] Schroder R, Goudeau B, Simon MC, Fischer D, Eggermann T, Clemen CS, et al. On noxious desmin: functional effects of a novel heterozygous desmin insertion mutation on the extrasarcomeric desmin cytoskeleton and mitochondria. *Hum Mol Genet.* 2003 Mar 15; 12(6):657-69.

[213] Carlsson L, Fischer C, Sjoberg G, Robson RM, Sejersen T, Thornell LE. Cytoskeletal derangements in hereditary myopathy with a desmin L345P mutation. Acta Neuropathol. 2002 Nov;104(5):493-504.

[214] Sjoberg G, Saavedra-Matiz CA, Rosen DR, Wijsman EM, Borg K, Horowitz SH, et al. A missense mutation in the desmin rod domain is associated with autosomal dominant distal myopathy, and exerts a dominant negative effect on filament formation. *Hum Mol Genet.* 1999 Nov; 8(12):2191-8.

[215] Walter MC, Reilich P, Huebner A, Fischer D, Schroder R, Vorgerd M, et al. Scapuloperoneal syndrome type Kaeser and a wide phenotypic spectrum of adult-onset, dominant myopathies are associated with the desmin mutation R350P. *Brain,* 2007 Jun;130(Pt 6):1485-96.

[216] Bar H, Fischer D, Goudeau B, Kley RA, Clemen CS, Vicart P, et al. Pathogenic effects of a novel heterozygous R350P desmin mutation on the assembly of desmin intermediate filaments in vivo and in vitro. *Hum Mol Genet.* 2005 May 15; 14(10):1251-60.

[217] Fidzianska A, Kotowicz J, Sadowska M, Goudeau B, Walczak E, Vicart P, et al. A novel desmin R355P mutation causes cardiac and skeletal myopathy. *Neuromuscul Disord.* 2005 Aug; 15(8):525-31.

[218] Dagvadorj A, Goudeau B, Hilton-Jones D, Blancato JK, Shatunov A, Simon-Casteras M, et al. Respiratory insufficiency in desminopathy patients caused by introduction of proline residues in desmin c-terminal alpha-helical segment. *Muscle Nerve*, 2003 Jun; 27(6):669-75.

[219] Kaminska A, Strelkov SV, Goudeau B, Olive M, Dagvadorj A, Fidzianska A, et al. Small deletions disturb desmin architecture leading to breakdown of muscle cells and development of skeletal or cardioskeletal myopathy. *Hum Genet.* 2004 Feb;114(3):306-13.

[220] Olive M, Armstrong J, Miralles F, Pou A, Fardeau M, Gonzalez L, et al. Phenotypic patterns of desminopathy associated with three novel mutations in the desmin gene. *Neuromuscul Disord.* 2007 Jun;17(6):443-50.

[221] Sugawara M, Kato K, Komatsu M, Wada C, Kawamura K, Shindo PS, et al. A novel de novo mutation in the desmin gene causes desmin myopathy with toxic aggregates. *Neurology*, 2000 Oct 10;55(7):986-90.

[222] Goudeau B, Dagvadorj A, Rodrigues-Lima F, Nedellec P, Casteras-Simon M, Perret E, et al. Structural and functional analysis of a new desmin variant causing desmin-related myopathy. *Hum Mutat.* 2001 Nov; 18(5):388-96.

[223] Park KY, Dalakas MC, Semino-Mora C, Lee HS, Litvak S, Takeda K, et al. Sporadic cardiac and skeletal myopathy caused by a de novo desmin mutation. *Clin Genet.* 2000 Jun; 57(6):423-9.

[224] Olive M, Goldfarb L, Moreno D, Laforet E, Dagvadorj A, Sambuughin N, et al. Desmin-related myopathy: clinical, electrophysiological, radiological, neuropathological and genetic studies. *J Neurol Sci.* 2004 Apr 15; 219(1-2):125-37.

[225] Dagvadorj A, Olive M, Urtizberea JA, Halle M, Shatunov A, Bonnemann C, et al. A series of West European patients with severe cardiac and skeletal myopathy associated with a de novo R406W mutation in desmin. J Neurol. 2004 Feb; 251(2):143-9.

[226] Pruszczyk P, Kostera-Pruszczyk A, Shatunov A, Goudeau B, Draminska A, Takeda K, et al. Restrictive cardiomyopathy with atrioventricular conduction block resulting from a desmin mutation. *Int J Cardiol.* 2007 Apr 25;117(2):244-53.

[227] Li D, Tapscoft T, Gonzalez O, Burch PE, Quinones MA, Zoghbi WA, et al. Desmin mutation responsible for idiopathic dilated cardiomyopathy. *Circulation,* 1999;100(5):461-4.

[228] Tesson F, Sylvius N, Pilotto A, Dubosq-Bidot L, Peuchmaurd M, Bouchier C, et al. Epidemiology of desmin and cardiac actin gene mutations in a European population of dilated cardiomyopathy. *Eur Heart J.* 2000 Nov; 21(22):1872-6.

[229] Bar H, Goudeau B, Walde S, Casteras-Simon M, Mucke N, Shatunov A, et al. Conspicuous involvement of desmin tail mutations in diverse cardiac and skeletal myopathies. *Hum Mutat.* 2007 Apr; 28(4):374-86.

[230] Muntoni F, Bonne G, Goldfarb LG, Mercuri E, Piercy RJ, Burke M, et al. Disease severity in dominant Emery Dreifuss is increased by mutations in both emerin and desmin proteins. *Brain,* 2006 May; 129(Pt 5):1260-8.

In: Small Stress Proteins and Human Diseases
Editors: Stéphanie Simon et al.
ISBN: 978-1-61470-636-6

Chapter 1.3

THE TWO LENS STRUCTURE PROTEINS, αA-CRYSTALLIN (HSPB4) AND αB-CRYSTALLIN (HSPB5), CONTROL APOPTOSIS THROUGH REGULATION OF MULTIPLE SIGNALING TRANSDUCTION PATHWAYS

David W. Li[1, 2, 3*]***, Lili Gong***[1]***, Mi Deng***[1]***, Jinping Liu***[1]***, Mugen Liu***[4]***, and Ying-Wei Mao***[5 6]

[1]Department of Biochemistry & Molecular Biology, College of Medicine, University of Nebraska Medical Center, 985870 Nebraska Medical Center, Omaha, NE 68198-5870
[2]Key Laboratory of Protein Chemistry & Developmental Biology of the Ministry of Education, College of Life Sciences, Hunan Normal University, Changsha, P.R. China
[3]Department of Ophthalmology & Visual Sciences, College of Medicine, University of Nebraska Medical Center, 985870 Nebraska Medical Center, Omaha, NE 68198-5870
[4]Key Laboratory of Molecular Biophysics of the Ministry of Education, College of Life Science and Technology and Center for Human Genome Research, Huazhong University of Science and Technology, Wuhan, P.R. China
[5]Howard Hughes Medical Institute
[6]Picower Institute for Learning and Memory, Department of Brain and Cognitive Sciences, Massachusetts Institute of Technology, Cambridge, MA 02139, U.S.

---

[*] Corresponding Author David Li
University of Nebraska Medical Center
College of Medicine
Department of Biochemistry and Molecular Biology
Department of Ophthalmology and Visual Science
985870 Nebraska Medical Center
Omaha, Nebraska 68198-5870, USA
dwli1688@hotmail.com

ABSTRACT

α-Crystallins are major lens structural proteins, belonging to the family of the small heat shock proteins (HSPs). α-Crystallins consist of two polypeptides, αA and αB that share 55% amino acid sequence identity. The two 20-kDa subunits form soluble aggregates with an average molecular mass of 600-800 kDa and can be isolated from lens fiber cells as a heteroaggregate containing αA- and αB-peptides in a ratio of 3 to 1. αA-crystallin (HspB4) is predominantly expressed in the ocular lens with small amounts present in spleen and thymus. In contrast, αB-crystallin (HspB5) is expressed mainly in the lens but also expressed outside of the lens in a number of tissues such as skeletal and cardiac muscle and to lesser extent in skin, brain, and kidney. Besides their structural role, α-crystallins are important chaperones in the ocular lens, act as autokinases, and also play important roles in suppressing both developmental and stress-induced apoptosis to guard the lens differentiation and prevent pathogenesis in the lens as well as several other tissues. In this chapter, we will focus on the current knowledge regarding the molecular mechanisms by which αA- and αB-crystallins suppress apoptosis in both lens and non-lens tissues.

INTRODUCTION

αCrystallins are major lens proteins which have structural roles in maintaining lens transparency [1]. Various mutations identified in α-crystallins have been shown to cause congenital cataracts and also other human diseases (Table 1) [2-11]. α-Crystallins consist of two isoforms: αA and αB, which share 55% amino acid sequence identity [12]. While αA-crystallin is largely expressed in the lens with a small amount in the spleen and thymus [13], αB-crystallin has been found in the lens and heart with relatively abundant level, and also in several other tissues including spinal chord, skin, muscle, brain and kidney in relatively small amount [14-16]. The two α-crystallins share a homologous C-terminal domain of about 80 amino acid residues with numerous other proteins and this feature characterizes them into a special group named small heat shock protein family [17-18]. Besides their structural roles in the lens, studies in the past two decades have revealed that α-crystallins have extraordinary non-structural functions.

The first non-structural role of α-crystallins was demonstrated by Horwitz in 1992 [19]. Using a simple light scattering assay at 360 nm to measure interactions between α-crystallins and other proteins, he demonstrated that the purified bovine α-crystallins can prevent heat-induced protein aggregation of numerous substrates that include alcohol dehydrogenase, β_L-crystallin and total lens soluble proteins. Since his pioneering study, the chaperone function of α-crystallins has been extensively studied by many laboratories [20-26] and several chapters in this book address this subject.

Soon after the discovery of the chaperone function, Kantorow and Piatigorsky [27] showed that both αA- and αB-crystallins can act as autokinases to phosphorylate themselves. Autophosphorylation of α-crystallins may have important effects in their functions, which wait to be further elucidated.

Ubiquitin/proteasome-mediated protein degradation plays essential roles in many cellular processes [28-29]. αB-Crystallin is found playing a critical role in regulating protein degradation. It binds to certain components of the 20S proteasome both in vitro and in vivo [30-34]. For example, αB-crystallin binds to an F-box containing protein, FBX-4, a component of the ubiquitin-protein isopeptide ligase SCF (SKP1/CUL1/F-box), translocates FBX-4 to the detergent-insoluble faction and stimulates the ubiquitin/proteasome pathway, which is required for cyclinD1 degradation [31].

Another important function of α-crystallins is their ability to regulate cell death in different systems through various mechanisms. This chapter will summarize the current knowledge on the anti-apoptotic function of α-crystallins with the emphasis on the mechanistic aspects. Before we discuss the anti-apoptotic role of α-crystallins and the related molecular mechanisms, we will briefly describe the biology of the ocular lens and the role of apoptosis in lens development and pathology (involvement of sHsps in cataracts is also treated in chapter 2-2).

Table 1. Mutations in the two α-crystallins causing cataracts and other human diseases

Crystallin	Mutations	Diseases	References
αA	W9X	Cataract	2
	R49C	Cataract	3
	R116C	Cataract	4
	R116H	Cataract	5
	G98R	Cataract	6
αB	R120G	Desmin-Related Myopathy	7
	450delA[a]	Cataract	8
	Q151X	Myofibrillar myopathy	9
	464delCTX[b]	Myofibrillar myopathy	9
	D140N	Cataract	10
	P20S	Cataract	11

[a]450delA: nucleotide deletion at position 450 that resulted in a framshift in codon 150 and produced an aberrant protein consisting 184 residues.
[b]464CTX, 2bp nucleotide deletion at position 464 (464delCT) that generates eight missense codons (RAHHSHHP) followed by a stop codon.

1. The Biology of the Ocular Lens

The lens of the vertebrate eye is a nonvascularized, noninnervated transparent organ. It contains only a single layer of epithelial cells covering its anterior surface. The remainder of the lens is principally composed of enucleated and non-organelle fiber cells [35-36]. The single layer of epithelial cells is primarily responsible for maintenance of the homeostasis of the entire lens and detoxification of noxious components in the ocular lens environment throughout life [37]. During embryonic development, the ectoderm cells destined to become the ocular lens are determined through multiple steps, initiated in the gastrula and early neurula, and completed when the cells make contact with the optic vesicle [38-39]. These ectoderm cells located in the forebrain region are further induced by the optic vesicle and become thickened into lens placodes [38-39]. The lens placode cells invaginate into a lens vesicle and finally they become separated from the neighboring ectoderm and form a complete lens vesicle [35-36, 38-39]. In the developing human lens, it takes about 8 days to form a complete lens vesicle [40]. Lens differentiation begins when the posterior cells of the lens vesicle become elongated, filling the lumen of the lens vesicle and moving towards the anterior epithelium. During mouse embryonic development, α-crystallin genes are turned on at very early stage and αB-crystallin is detected in the lens placode at E9.5, and αA-crystallin detected in the lens cup at E10 and E10.5 [41]. The elongated cells form the primary lens fiber cells through an extensive differentiation process, which culminates by denucleation, a process called pycnosis [42-43]. At the same time, the anterior lens epithelial cells divide, migrate, and differentiate into secondary lens fiber cells in the equatorial region [35-36, 38-39]. Both growth and differentiation of the ocular lens are regulated by different growth factors [44]. The lens continues to grow throughout life, building up layers of new fiber cells from the equatorial region [45]. During lens development and morphogenesis, controlled cell death through apoptosis is an important aspect, which is described in the next section.

2. Apoptosis and Cataractogenesis in the Ocular Lens

Apoptosis is programmed cellular suicide and differs from necrotic cell death [46-47]. Necrotic cell death is a pathological form of cell death resulting from acute cellular injury, which is typified by rapid cell swelling and lysis, leading to the release of cytoplasmic materials, which often trigger an inflammatory response [47]. In contrast, apoptotic cell death is characterized by controlled autodigestion of the dying cell [46-47]. The apoptotic cells often break into small fragments called apoptotic bodies, which are phagocytosed and digested by macrophages or neighboring cells. The genetic control of apoptotic death was initially explored in *C. elegans*. Horvitz and his associates have now elucidated the genetic program and discovered that more than a dozen of regulators mediate the apoptotic death of 129 cells in *C. elegans* [48]. In mammals, apoptosis is controlled by several different families of genes including those coding for members of death-domain-containing protein family, the Bcl-2 family, and the caspase family. The regulators of these families are assembled into two major death pathways: the extrinsic death pathway and intrinsic death pathway [49-51].

Apoptosis is a fundamental feature of animal development and occurs in many different tissues [49-50]. Apoptosis is also closely associated with diseases, which are derived from either increased apoptosis (such as AIDS) or inhibited apoptosis (such as cancer) [52].

Apoptosis plays an important role in lens development. In the developing rat lens, Gluckmann [53] documented cell death in two prominent regions: the lens vesicle ectoderm connecting to the determined corneal ectoderm and the lens stalk region. About two-decade later, Silver and Hughes reported in more details the cell death in the lens and other compartments of the rat eye [54]. Using Hoechst staining, Ishizaki et al. [55] confirmed that apoptotic cells were frequently observed in rat embryonic lens and also in postnatal rat lenses. These studies have recently been further confirmed during development of murine and human lenses [56-58].

Cataract, referring to any opacification in the ocular lens, is a disease that accounts for 22% of all cases of blindness and thus causes an important health problem. Cataract can develop at any time in life due to inherited mutations (congenital cataracts such as α-crystallin gene mutations described in table 1) or non-inherited conditions (non-congenital cataracts) [59].

Several line of evidence points to the conclusion that stress-induced apoptosis of lens epithelial cells leads to non-congenital cataractogenesis in animal and human. First, during development, apoptosis induced by gene mutations, expression of exogenous genes, stress factors, or mechanical damage all causes microphthalmia or cataractogenesis (see Ref. 60 for a detail summary). One of the most striking examples came from the study of CREB-2 knockout mice [61]. Normal lens development occurs between EDs 12.5 and 14.5 in CREB-2(-/-) mice. These mice displayed normal formation of the early lens vesicle, normal elongation of posterior primary fiber cells, and normal formation of the anterior lens epithelial cells. However, the anterior lens epithelial cells and their direct descendants at the equatorial poles of the lens underwent massive and synchronous p53-dependent apoptosis between EDs14.5 and 16.5. The complete loss of these cells accounts directly for the embryonic lens degeneration seen in these animals [61]. Mutation of the FoxE3 gene also induced abnormal lens epithelial cell apoptosis in the anterior epithelium of the lens vesicle and led to absence of secondary lens fibers and formation of a dysplastic, cataractous lens [62]. Introduction of a number of exogenous genes into lenses also causes lens cell apoptosis followed by cataractogenesis [63-70]. Suppression of the lens stalk cell apoptosis at the lens vesicle stage by hyaluronic acid led to faulty separation of the lens vesicle, resulting in abnormal lens formation [56]. Second, we previously demonstrated that in the in vitro rat lens organ culture, stress factors such as hydrogen peroxide, calcimycin, and UVB all induce lens epithelial cell apoptosis followed by development of lens opacification [71-73]. In vitro organ culture study by Menko and her associates demonstrated that inhibition of Src kinase pathway led to cataract formation which was initiated by apoptosis of lens epithelial cells in the treated lens [74-75]. Third, in the *in vivo* animal model studies with rat and rabbit, several laboratories have demonstrated that UV irradiation [76-78], diabetic condition [79-80], selenite [81], and N-methyl-N-nitrosourea [82] all induced early apoptosis of lens epithelial cells followed by cataractogenesis. In a recent report, Wolf and his associates demonstrated that death of lens epithelial cells is directly related to aging-related cataractogenesis [83]. Fourth, in human cataractous lenses, we have previously observed that varying levels of apoptotic cells were present in the capsular epithelia from cataractous patients [71]. These results have been confirmed by numerous laboratories [84-87] but in contrast to one report [88]. In a recent study [87], Charakidas A et al. reported that TUNEL staining was observed in 25 (92.6%) specimens of cataractous lenses, whereas cells undergoing apoptosis were identified only in 2 (8%) of the epithelia from non-cataractous lenses. This group concluded that the accumulation of small-scale epithelial cell losses during lifetime may induce alterations in lens fiber formation and homeostasis and result in loss of lens transparency. Apoptosis is also closely associated with human polar cataracts [89-90]. Finally, recent studies have revealed that ER stresses cause cataractogenesis. In exploring the mechanism, it was found that apoptosis induced by ER stress plays an important role in ER stress-induced cataractgenesis [91].

The ocular lens is immersed in the aqueous humor and constantly faces the challenges from various stress conditions. To avoid stress-induced death of the single layer of lens epithelial cells through apoptosis and thus cataractogenesis, the lens must carry the strong defense system to guard the apoptotic process. Two of the most important guardians in the ocular lens are the lens structural proteins, αA- and αB-crystallins. We now turn to the discussion of the anti-apoptotic function of these two crystallins.

3. Discovery of the Anti-apoptotic Role of α-Crystallins

The protective role of α-crystallin was initially explored in NIH 3T3 fibroblast by Klemenz group who found that ectopic expression of αB-crystallin cDNA or induced expression of αB-crystallin by the synthetic glucocorticoid hormone dexamethasone renders NIH 3T3 cells thermoresistant [92]. Around the same time, Arrigo and his associates used a murine fibrosarcoma cell line L929 to express various members of the small heat shock proteins including *Drosophila* hsp27, human hsp27, and human αB-crystallin and then tested their functions in conferring the transfected stable lines of L929 clones the ability to resist on different stress conditions such as heat, oxidative stress, tumor necrosis factor and kinase inhibitor treatments [93-94]. During these studies, they elegantly demonstrated that αB-crystallin, like hsp27, displays strong ability to resist apoptosis induced by Fas/Fas ligand, TNF-α and staurosporine [95]. This pioneering work was soon followed by numerous laboratories and the anti-apoptotic role of α-crystallins in development and stress-induced pathology has been vigorously explored in different systems [96-124].

4. α-Crystallins Prevents Developmental and Stress-Induced Apoptosis

The direct evidence that α-crystallins prevent developmental apoptosis came from the study by Moorozov and Waurousek [114]. These authors demonstrated that knockout of the αA- and αB-crystallins caused elevated apoptosis of the cortical lens fiber cells, leading to microphthalmia and cataractogenesis. Ousman et al. [118] reported that the αB(-/-) astrocytes showed more cleaved caspase-3 and more TUNEL staining, and thus increased apoptosis. In addition, αB-crystallin is an important regulator of tubular morphogenesis and survival of endothelial cell during tumor angiogenesis [119]. Tumor grown in αB(-/-) mice are significantly less vascularized than wild-type tumors and displayed increased areas of apoptosis/necrosis [119].

In the in vitro cultured epithelial cells derived from human and rabbit lenses, we have shown that the two α-crystallins can prevent apoptosis induced by staurosporine (a general kinase inhibitor), okadaic acid and calyculin A (the protein serine/threonine phosphatases-1 and 2A inhibitors) or oxidative stress [101-102, 108]. Andley et al. [97] demonstrated that the αA(-/-) lens epithelial cells are 40 times more sensitive to UVA damage than the wild type mouse lens epithelial cells. Introduction of the exogenous αA-crystallin into cultured lens epithelial cells enhanced the resistance of the transfected cells against UVA-induced apoptosis. The same group also compared the relative antiapoptotic abilities of the αA- and αB-crystallins under treatment of UVA irradiation, TNFα and staurosporine, and concluded that αA-crystallin is a stronger antiapoptotic regulator than αB-crystallin does [98]. However, when the two α-crystallins were overexpressed in *E. coli* and tested for their ability against cold or other stress conditions, an opposite conclusion was obtained [99]. We also tested the possible differential anti-apoptotic abilities of the two α-crystallins against stress-induced apoptosis in three different cell lines: the human lens epithelial cell line (HLE) which contains virtually undetectable endogenous crystallins; the human retina pigment epithelial cells (ARPE-19) and the rat embryonic myocardium cell line (H9c2), both of which expresses endogenous αB-crystallin [108]. The expression of the endogenous αB-crystallin in ARPE-19 and H9c2 cells makes them much more resistant to staurosporine-induced apoptosis in the absence of exogenous α-crystallin expression. Regardless absence or presence of the endogenous α-crystallin, expression of HαA and HαB in these cells provides additional protections. Under treatment of 100 nM staurosporine, 10 μM etoposide or 400 mM sorbitol (generating osmotic stress), HαA and HαB consistently show similar protections against a given stress condition in HLE, ARPE-19 or H9c2 cells. The inconsistence of the results from different laboratories may reflect the difference of the physiological conditions of the transfected cells used for the comparative analysis. On the other hand, αA and αB-crystallins, as we will discuss below, can regulate different signaling pathways, and thus provide the possibility of differential antiapoptotic abilities depending upon the physiological environment of the tested cells.

As mentioned above, α-crystallins are expressed in retinal pigment epithelial cells [108]. Recent studies have shown that the protection conferred by α-crystallins in this tissue is very important for retinal functions [117, 120].

Outside the ocular tissues, Martin et al. [96] first reported that αB-crystallin can protect cardiac myocytes from ischemic injury soon after Arrigo's pioneering study in L929 cells [95]. This in vitro work was soon followed by the in vivo studies carried out by Das's group who showed that overexpression of αB-crystallin confers simultaneous protection against cardiomyocyte apoptosis and necrosis during myocardial ischemia and reperfusion [100].

The dominant mutant, R116C in αA-crystallin and R120G in αB-crystallin displayed diminished antiapoptotic functions as tested in both lens epithelial cells [105,108] and cardiomyocytes [112]. The possible explanation for this phenomenon is discussed later is this chapter.

5. α-Crystallins Suppress Apoptosis through Regulation of Members of the Caspase Family

To explore the possible mechanism by which α-crystallins prevent apoptosis, we initially made an expression construct in which the mouse αB crystallin cDNA was inserted into the green fluorescence protein expression vector, pEGFPC3 and then this construct, pEGFP-mαB and the corresponding vector were introduced into rabbit lens epithelial cells, N/N1003A. When the two types of cells were treated with okadaic acid (inhibitor of protein phosphatase-1 and –2A) or DMSO (mock), it was found that mouse αB-crystallin protects the transfected N/N1003A cells from okadaic acid-induced apoptosis. To determine the possible mechanism by which mouse αB-crystallin inhibits apoptosis, we examined the caspase-3 activity in both types of cells after okadaic acid treatment. It was found that caspase-3 activity was up-regulated by okadaic acid about 5-fold in non-transfected and vector-transfected N/N1003A cells. In contrast, only 1.5-fold up-regulation of caspase-3 activity was observed in mouse αB-crystallin-transfected N/N1003A cells after okadaic acid treatment. Thus, the protection against okadaic acid-induced apoptosis by mouse αB-crystallin is associated with its ability to suppress caspase-3 activation. This is the initial discovery that αB-crystallin regulates caspase-3 activity [101]. To explore how αB-crystallin suppresses caspase-3 activation, we then conducted immunoprecipitation-linked Western blot analysis [102]. Our results demonstrated that mouse αB-crystallin can interact with procaspase-3 and partially processed procaspase-3 to prevent stress-induced apoptosis [102].

Our discovery, that αB-crystallin interacts with procaspase-3 and partially processed caspase-3 to prevent caspase-3 activation, has been confirmed in several other types of cells. DePinho and his associates [122-123] recently found that the Bcl-2 like 12 gene (*bcl2L12*) is universally overexpressed in the primary glioblastoma multiforme (GBM) and functions to block post-mitochondrial apoptosis signaling by neutralizing effector caspase-3 and caspase-7 maturation. They found that while Bcl2L12 can prevent caspase-7 maturation through direct interaction between Bcl2L12 and procaspase-7, neutralization of caspase-3 maturation by Bcl2L12 occurs through an indirect mechanism in which Bcl2L12 induces transcriptional upregulation of the small heat shock protein, αB-crystallin, which binds to procaspase-3 and its cleavage intermediates *in vitro* and *in vivo*. In rat cardiomyocyte H9c2 cells, Aggeli et al. [121] reported that in responding to oxidative stress insult, αB-crystallin became phosphorylated by p38-MAPK/MSK1 in the presence of intracellular free calcium and the phosphorylated αB-crystallin interacted with procaspase-3 to block its cleavage and subsequent activation.

Around the same time when we used co-immunoprecipitation to reveal the interaction between αB-crystallin and procaspase/partially processed procaspase-3, Cryns' group [103] demonstrated that αB-crystallin interacted with the cleavage intermediate of the procaspase-3 using in vitro interaction-linked cleavage assays.

The importance of the interactions between αB-crystallin and the procaspase-3/partially processed caspase-3 is recently demonstrated *in vivo* by Waurousek's group [114], who found that knockout of both α-crystallins leads to upregulation of caspase-3 and caspase-6 activities in the fiber cell zone of mouse lens where secondary lens fiber cell disintegration occurs, suggesting that the elevated levels of caspase-3 and caspase-6 activities are probably causing the observed fiber cell disintegration, and the eventual cataractogenesis. The same group also demonstrated that αA-crystallin directly interacts with caspase-6 through co-immunoprecipitation-linked western blot analysis [114].

Using surface plasmon resonance (SPR) analysis, we have recently found that both αA- and αB-crystallins can interact with caspase-3 (Gong et al., in preparation). We found that caspase-3 exhibited a stronger binding to αB-crystallin (K_D=155 nM) than that to αA-crystallin (K_D=240 nM). These results are somewhat different from a recent report [125] where a protein pin array was used to explore the interactions between αB-crystallin and other proteins. The oligo peptide with a few amino acids from αB-crystallin had little affinity to caspase-3. The plausible explanation for the difference is that the oligos in protein pin array assay chops αB-crystallin into small fragments with a few amino acids and thus break down the binding domain or binding site in αB-crystallin response for the interaction with procaspase-3/caspase-3.

6. α-Crystallins Prevent Apoptosis through Control of Members of the Bcl-2 Family

To explore whether αA- and αB-crystallins are able to prevent apoptosis at the mitochondrial level, upstream of caspase-3 activation, we cloned human αA- and αB-crystallin cDNAs from human eye lenses and generated the in frame constructs expressing fusion proteins of human αA- and αB-crystallins (HαA and HαB) with green fluorescence protein (GFP). These expression constructs (pEGFP-HαA and pEGFP-HαB) and the vector (pEGFP-neo) were transfected into the human lens epithelial cells and the stable clones expressing only GFP from the vector (pEGFP-HLE), GFP-HαA (pEGFP-HαA-HLE) or GFP-HαB (pEGFP-HαB-HLE) were establish in the presence of G418 selection.

Using these stable cell lines, we demonstrated that both αA- and αB-crystallins can prevent staurosporine-induced apoptosis. Staurosporine, a potent apoptosis inducer demonstrated in a broad spectrum of cells activates apoptosis through the mitochondrial death pathway [126-130]. The pro-apoptotic members of the Bcl-2 family play a critical role in staurosporine-induced apoptosis [129-130]. Cells lacking both Bax and Bak are known to be completely resistant to multiple apoptotic stimuli including staurosporine [131-132]. Analysis of cell death in Bax- or Bak-deficient cell lines indicates that both regulators are necessary for staurosporine-induced apoptosis [133-134]. Consistent with these studies, we found that in human lens epithelial cells staurosporine regulates both Bak and Bax at different ways.

Staurosporine was found to upregulate Bak expression up to 5-fold in vector-transfected human lens epithelial cells, and the upregulated Bak was accumulated into mitochondria of the vector-transfected cells. In the GFP-HαA- and GFP-HαB-transfected cells, however, Bak upregulation was different. First, expression of HαA and HαB slightly enhanced Bak expression in unknown mechanism. On the other hand, expression of HαA- and HαB prevented staurosporine-induced additional Bak upregulation [108]. Since cells lacking both Bax and Bak are completely resistant to staurosporine [132-133], inhibition of additional Bak upregulation by HαA and HαB, to some degree, contributed to their antiapoptotic abilities against staurosporine-induced apoptosis. Bak upregulation is also observed in human colonic adenoma AA/C1 cells during butyrate-induced apoptosis [135], and in human stomach epithelial cells during bacterium–induced apoptosis [136].

Staurosporine did not induce Bax upregulation in the vector or α-crystallin-transfected cells. This kinase inhibitor, however, promoted Bax translocation from cytosol into mitochondria in vector-transfected cells. The death signal-induced Bax conformation change followed by its insertion into mitochondrial membrane is an important step for Bax to promote apoptosis [137-140]. Thus, staurosporine-induced translocation of Bax turned on the death program in human lens epithelial cells. This process was largely abrogated in cells expressing either GFP-HαA or GFP-HαB. Both αA- and αB-crystallins can directly bind to Bax as demonstrated by GST pulldown assay and co-immunoprecipitation [108].

In addition, our results also demonstrated the regulation of another proapoptotic member of the Bcl-2 family by the two alpha-crystallins. Staurosporine stimulated translocation of Bcl-X_S from cytosol into mitochondria in vector-transfected pEGFP-HLE cells. Again, HαA and HαB directly bind to Bcl-X_S and prevent its translocation into mitochondria [108].

The two major mutants, R116C in HαA and R120G in HαB, cause cataractogenesis in the lens [4, 7]. Consistent with these observations, the chaperone-like ability of R116C and R120G was found substantially decreased [141-142]. These mutants apparently displayed much weaker antiapoptotic ability [105, 108, 112]. The attenuated antiapoptotic abilities were derived from their weak interactions with Bax and Bcl-X_S and thus decreased abilities to sequester translocation of Bax and Bcl-X_S from cytosol into mitochondria induced by staurosporine treatment [108].

In vector-transfected cells, once the pro-apoptotic members, Bax and Bcl-X_S, were mobilized into mitochondria, the integrity of mitochondria was interrupted as evidenced by the release of cytochrome c, followed by activation of executioner caspase, caspase-3 and degradation of PARP. These downstream events induced by staurosporine in vector-transfected human lens epithelial cells were similar to those reported in other cell lines [127-129]. In either HαA- or HαB-transfected cells, the staurosporine-induced downstream events were largely turned off [108]. Thus, these results demonstrate that the two α-crystallins can regulate Bcl-2 family members to prevent stress-induced apoptosis.

Besides the two α-crystallins, Hsp60 has also been shown to directly bind to Bax to halt its pro-apoptotic role [143].

7. α-Crystallin Activates AKT Pathway to Resist on Stress-Induced Apoptosis

c-Akt is the cellular homolog of the transforming oncogene of the AKT retrovirus [144-147]. Simultaneous with the identification of c-Akt, the protein kinase B and a kinase related to A- and C-Protein kinases were cloned [148-149]. c-Akt1, PKB and RAC-PK were found to be encoded by the same gene (herein referred to as Akt). The Akt family contains two other members: c-Akt2 and c-Akt3 [150-153]. Akt family proteins contain a central kinase domain with specificity for serine or threonine residues in substrate proteins. In addition, the amino terminus of Akt includes a pleckstrin homology (PH) domain, which mediates lipid-protein and/or protein-protein interactions [154-156]. Phosphorylation of Akt at Thr-450 is sometimes necessary to prime it for activation [157]. Growth factor stimulation of PI3K leads to increased intracellular level of PIP3. Binding of PIP3 to the Akt PH domain results in Akt translocation from the cytoplasm to the plasma membrane. In addition, the interaction of the PH domain with the lipid products of PI3K causes a conformational change in Akt, allowing its activation through phosphorylation at Thr-308 and Ser-473 by PDK-1 and other kinases [158-162].

Akt signaling pathway components are highly expressed in the ocular tissues [Lan Zhang et al. in preparation]. Our recent results also demonstrated that at both mRNA and protein levels, Akt-1 is a major surviving kinase in all the four ocular tissues examined. Compared with Akt-1, Akt-2 is much reduced in all the tissues examined and Akt-3 is only expressed in the retina [Zhang et al. in preparation].

AKT signaling pathway plays an important role in lens differentiation and survival [163-168]. In our previous study of the UVA-induced apoptosis, we found that irradiation of both vector- and human αA-transfected HLE cells with a dose of 150 KJ/m^2 induced activation of both ERK1 and ERK2 [109]. In contrast, UVA-induced activation of ERK1/2 was suppressed in human αB-crystallin-transfected cells [109]. Thus, UVA-induced apoptosis was associated with activation of ERK1/2 and human αB but not αA suppresses UVA-induced activation of ERK1/2 and apoptosis of HLE. Since HαA was unable to suppress UVA-induced activation of the RAF/MEK/ERK pathway, one possibility would be that human αA may be involved in regulation of other pathways. The possible activation of Akt kinase in the three types of stably transfected cells was examined. As expected, UVA-irradiation with a dose of 150 KJ/m^2 induced exclusive activation of Akt-1 in human αA-transfected cells but not observed in the vector- and human αB-transfected cells. In addition, UVA irradiation led to some down-regulation of Akt-1 protein in the vector- and human αB-transfected cells but visible up-regulation in the human αA-transfected cells [109]. Thus, under UVA irradiation, human αA upregulates and activates Akt to counteract UVA-induced apoptosis. To further confirm that activation of AKT is important for human αA-crystallin to prevent UVA-induced apoptosis, we inhibited AKT-1 activation. When αA-crystallin-regulated AKT-1 activation by UVA was blocked, the protection against UVA-induced apoptosis by αA-crystallin was significantly attenuated [109].

To explore how αA-crystallin may activate Akt pathway, we examined its possible interaction with Akt-1. First, we performed the surface plasmon resonance (SPR) analysis between αA, or αB and Akt-1 using BIACORE 3000 with research grade CM5 sensor chips and purified Akt-1, αA and αB crystallins [Gong et al., submitted]. αA- and αB-crystallins were covalently coupled to individual flow cell surfaces of a carboxymethyldextran-coated BIAcore sensor chip CM5 using amine-coupling chemistry. To determine the kinetics of the interactions between Akt-1 and αA-/αB-crystallins, Akt-1 with five different concentrations (50, 100, 250, 500, 1000 nM) was passed over the surfaces of both αA- and αB-crystallin chips at 30 μl/min. A buffer injection was used as a negative control. The association rates, dissociation rates, and affinity constants were calculated with BIA evaluation 4.1 software (Biacore AB, Uppsala, Sweden). Our results suggest that Akt-1 only binds to αA-crystallin (Kd=5.6 μM) but not αB-crystallin [Gong et al., submitted].

To further explore the interaction between αA and Atk-1, co-immunoprecipitation was conducted and the obtained results showed that αA and Akt-1 form *in vivo* interacting complex in mouse lens [Gong et al., submitted]. Immunocytochemistry revealed that Akt1 and αA-crystallin were co-localized in both epithelium and fiber cells of mouse eye lens [Gong et al., submitted].

Activation of Akt pathway by αA-crystallin was further confirmed by in vivo studies [Gong et al., submitted]. An examination of the phosphorylation status of Akt 1 at Ser 473 in normal, αA-/-, and αA-/αB- DKO mice revealed that the normal Akt-1 activity detected in wild type mouse was dropped more than 50% in both αA-/-, and αA-/αB- DKO mice. These results clearly indicate that αA-crystallin displays strong regulation on Akt-1 activity.

While the details regarding how αA-crystallin activates Akt-1 remains to be worked out, the regulation of Akt activation by Hsp27 provides important hint. Hsp 27 is found in the same signaling complex with p38 MAPK, MAPK-activated protein kinase 2 (MAPKAPK-2), and Akt [161, 169-171]. Moreover, in polymorphonuclear leukocyte, Hsp27 activates Akt by scaffolding MAPKAPK-2 to Akt signal complex [171]. Deletion of the Hsp27 binding site in Akt (amino acids 117-128) results in loss of its interactions with Hsp27 and MAPKAPK-2 but not with Hsp90 as demonstrated by co-immunoprecipitation and GST pulldown studies [171]. On the other hand, phosphorylation of Hsp27 at Ser-82 by Akt induces dissociation of Hsp27 from Akt1 [171]. In addition, Hsp 90 can directly interact with Akt and help it to stay in its active status [172].

8. αB-Crystallin Suppresses ERK Pathway to Inhibit Stress-Induced Apoptosis

The mitogen-activated protein kinases (MAPKs) are actively involved in mediation of stress-induced apoptosis [173-174]. In the ocular lens, all of the three types of MAP kinases are expressed [175]. These kinases play important role in lens development and pathogenesis [176-186]. During stress-induced apoptosis of lens epithelial cells, activation of ERK1/2 is necessary for calcimycin-induced apoptosis [110] and for UVA-induced apoptosis [109]. This property distinguishes lens epithelial cells from most other tissue cells and also explains why the ocular lens never develops natural tumor [110]. Considering that the ERK pathway is normally involved in promotion of surviving in most other tissues and that oncogenic RAS persistently activates the ERK1 and ERK2 pathways, which contributes to the increased proliferative rate of tumor cells [187], lack of natural tumor in the ocular lens may partially be due to the fact that in the lens epithelial cells (the only cells that have complete sets of cellular organelles including nuclei in the lens organ), the RAF/MEK/ERK pathway mediates stress-induced apoptosis. Upon DNA damages induced by various stress factors, activation of the RAF/MEK/ERK pathway, instead of promoting survival of the cells with genetic lesions, signals apoptosis of the damaged cells. In this way, the cells bearing the genetic damages are removed in a timely manner and thus tumor development is avoided [110].

αB-crystallin can prevent both calcimycin- and UVA-induced apoptosis [109-110]. To understand the possible mechanism, we tested whether αB-crystallin could abrogate calcimycin-induced activation of ERK1/2. Western blot analysis revealed that ERK1/2 activity became downregulated during calcimycin treatment in pEGFP-mαB-N/N1003A cells [110]. Similarly, the activity of the p38 kinase became steady downregulated during calcimycin treatment. αB-crystallin also prevented the calcimycin-induced transient activation of JNK2 [110]. These results indicate that the mechanism for αB-crystallin to prevent calcimycin-induced apoptosis is derived from its suppression of calcimycin-induced activation of ERK1/2. The fact that in αB-crystallin expression cells, calcimycin-induced activation of ERK1/2, p38 and JNK2 was all inhibited indicates that αB-crystallin may act on the initial steps of RAS/RAF/MEK/ERK signaling pathway. For this reason, we investigated the effect of αB-crystallin on Ras activation.

Both pEGFP-N/N1003A and pEGFP-mαB-N/N1003A cells were labelled with ^{32}P-orthophosphate and then treated with DMSO and 5 μM calcimycin. Cell extracts were prepared from labeled and treated cells, and the RAS protein in each sample was immunoprecipitated. After that, the levels of GTP-bound RAS and GDP-bound RAS were determined with thin-layer chromatography. While DMSO treatment of both vector and αB-crystallin expression cells only induced background levels of RAS activation [110], treatment of the vector-transfected cells with 5 μM calcimycin induced substantially activation of RAS. However, such activation of RAS was not observed in the αB-crystallin expression cells. Thus, αB-crystallin blocks Ras activation to suppress activation of the downstream MAP kinases after treatment by calcimycin [110]. To explore how αB-crystallin may block the activation of RAS, we examined the upstream regulators of RAS. The RAS small GTPases are normally activated by the quanine nucleotide exchange factors (GEFs), which induce the dissociation of GDP to allow association of GTP [188]. Upon response to growth factor stimulation, two common RAS GEFs, son-of-sevenless 1 (Sos 1) and 2 (Sos 2), are activated to transfer from cytosol to the plasma membrane and there they activate RAS [189]. However, under calcium treatment, RAS is activated by other factors, the RAS-guanine-nucleotide-releasing factor 1 (RAS-GRF1) and 2 (RAS-GRF2) [188]. Because RAS-GRF1 is mainly expressed in the nerve system and RAS-GRF2 is widely expressed in many tissues [190-191], we therefore examined the activation of RAS by RAS-GRF2 after calcimycin treatment. After calcimycin treatment, the 135 kd RAS-GRF2 was transferred from cytosol to plasma membrane. In the αB-crystallin expression cells, however, the translocation of the RAS-GRF2 from cytosol to plasma membrane was largely inhibited [110]. These results and also the cross-linking experiments demonstrated that in N/N1003A cells, calcimycin activated RAS mainly through RAS-GRF2 and αB-crystallin negatively regulates activation of RAS through suppression of RAS-GRF2 translocation [110].

A recent study reported that the total and phospho-ERK1/2 and p38 were much enhanced in the astrocytes of the αB(-/-) mice than in those from normal mice with the same genetic background [119]. These results are consistent with our observation that αB-crystallin negatively regulates RAS/RAF/MEK/ERK signaling pathway. On the other hand, in breast cancer cells, Cryns et al [192] reported that overexpression of αB-crystallin constitutively activated the MEK/EKR signaling pathway and transformed the immortalized human mammary epithelial cells. The mechanism underlining this function of αB-crystallin remains to be elucidated. Thus, depending upon the types of cells, αB-crystallin may have differential effects on the MAPK pathway.

9. α Crystallins Regulate DNA Damaging Pathway to Control Stress-Induced Apoptosis

In eukaryotes, DNA damage induces rapid response of the damaging signaling transduction pathway. The major components of this pathway include the upstream kinases, ATM and ATR, the intermediate kinases CHK1 and CHK2, and the G1 checkpoint substrates, p53 and Mdm2; S phase and G2/M checkpoint substrates [193-194]. UV or ionizing radiation (IR) induces activation of ATM through autophosphorylation, and the activated ATM then regulate the downstream G1 checkpoint substrates p53, Mdm2 and CHK2 [195]. Recent studies have revealed that α-crystallins can regulated the DNA damaging pathway components. Andley and her associates first reported that the primary cultures of the lens epithelial cells derived from αB-/- mice display genomic instability and undergo hyperproliferation. This property is at least partially related to the upregulated expression of the functionally impaired p53 [196]. Xiao and his associates [197] demonstrated that in mouse myogenic C2C12 cells, hydrogen peroxide induced apoptosis of the treated cells. During hydrogen peroxide treatment, p53 in the C2C12 cells was found translocated from cytoplasm into mitochondria. Overexpression of αB-crystallin in the C2C12 cells prevented hydrogen peroxide-induced apoptosis. The mechanism mediating this protection was largely related to the interaction between αB-crystallin and p53, which blocked p53 translocation from cytosol into mitochondria. We recently found that UVA irradiation induced hyperphosphorylation of p53 at several residues in human lens epithelial cells. Both αA- and αB-crystallins distinctly attenuated p53 hyperphosphorylation at these sites (Liu et al., in preparation). Furthermore, the two α-crystallins seemed to interfere with the activation of several p53 upstream kinases (Liu et al., in preparation).

10. Modulation of the Anti-Apoptotic Ability of αB-Crystallin by Phosphorylation

Early studies have shown that α-crystallins are phosphoproteins [198]. Phosphorylation of αB-crystallin at Ser-59 is important for its function against apoptosis in cardiac myocytes. Hoover et al. [99] demonstrated that MKK6-activated p38 MAP kinase can phosphorylate αB-crystallin at Ser-59, which seems to provide maximal protection of cardiac myocytes from apoptosis [99, 199]. A recent study [200] demonstrated that phosphorylation of αB-crystallin occurred after its translocation from cytosol into mitochondria. Once αB-crystallin is phosphorylated at Ser-59 in mitochondria, it stabilizes mitochondrial membrane potential and thus inhibits apoptosis.

Although it has not be examined weather phosphorylation of αB-crystallin at Ser-19 has a role in modulating its antiapoptotic ability, analysis of the P20S mutant αB-crystallin provides some hint. The immunofluorescence confocal microscopy revealed that wild-type αB-crystallin was localized mainly in the cytoplasm and surrounded the cell nucleus, whereas mutant P20S αB-crystallin was present in both the cytoplasm and nucleus. To investigate whether mutation P20S of αB-crystallin acquires any detrimental function to cellular homeostasis, we examined the cytotoxic effect of mutation P20S. HLE cells with stable expression of wild-type αB-crystallin or mutant P20S αB-crystallin were stained using Hoechst 33342 in order to study the changes in nuclear morphology. The stable expression of P20S αB-crystallin increased chromatin condensation and fragmentation compared with cells with stable expression of wild-type αB-crystallin or cells transfected with empty vector only [124]. These results demonstrate that the P20S mutation increases apoptosis of human lens epithelial cells. Thus, phosphorylation modulates the antiapoptotic function of α-crystallins.

Conclusion

Both *in vitro* and *in vivo* studies have shown that α-crystallins act as strong anti-apoptotic regulators. In lens and retina of the eye, cardiac myocytes of the heart, and neuron and glial cells of the brain, α-crystallins are important guardians preventing apoptosis triggered by stress factors and pathological conditions and thus protect these tissues from running into various diseases such as cataract, retina degeneration, heart failure, and various brain diseases. α-crystallins prevent stress-induced apoptosis through regulating members of the Bcl-2 and caspase families. In addition, αA-crystallin can help to activate the Akt survival pathway to suppress apoptosis, while αB-crystallin can intervened the Ras/Raf/MEK/ERK signaling pathway to inhibit apoptosis, displaying differential anti-apoptotic mechanisms dependent upon the stress conditions implicated. The two α-crystallins can also modulate the DNA damaging pathway to halt apoptosis. Together, these results demonstrate that the two α-crystallins can regulate multiple signaling transduction pathways to regulate apoptosis (Figure 1). It is expected that in the near future, the detail mechanisms regarding how α-crystallins regulate these different signaling transduction pathways to halt apoptosis will be elucidated.

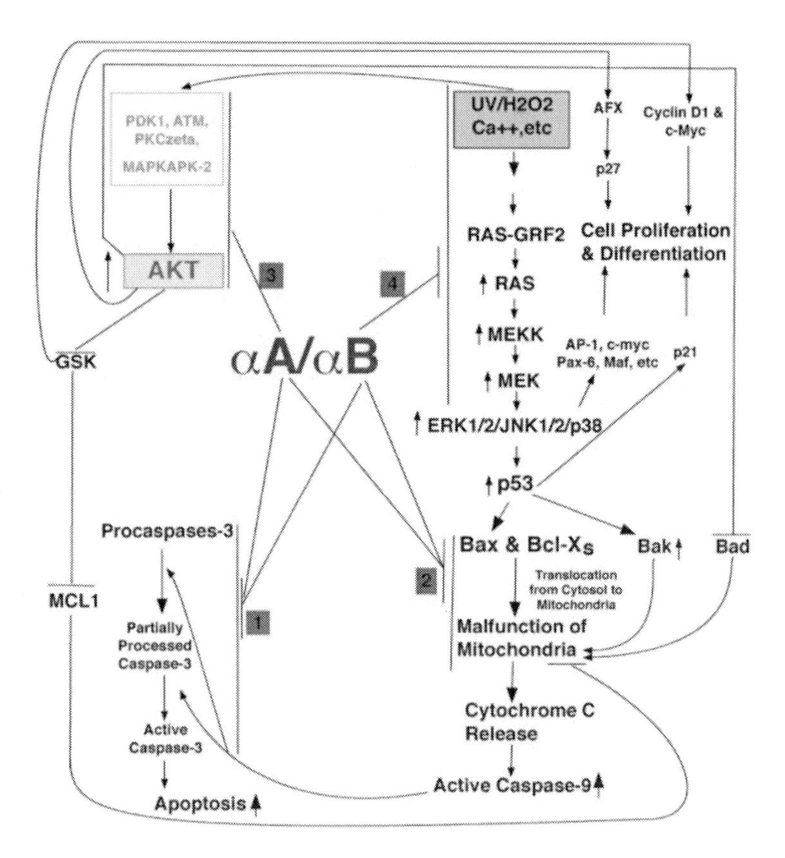

Figure 1. α-Crystallins Regulate Multiple Signaling Transduction Pathways to Prevent Developmental or Stress-Induced Apoptosis. αA- and αB-crystallins can directly bind to the precursors of caspase-3 to prevent its activation so that abnormal developmental apoptosis is suppressed and normal lens differentiation is ensured (Pathway #1). Both α-crystallins may directly bind to Bax and Bcl-X_S to repress their translocation into mitochondria, thus suppressing stress-induced apoptosis (Pathway #2). αA-crystallin may bind to Akt to promote its activation to counteract the stress-induced apoptosis (Pathway #3). αB-crystallin may negatively regulate activation of the ERK signaling pathway to prevent stress-induced apoptosis (Pathway #4). Finally, αA- and αB-crystallins may negatively regulate the DNA damaging pathway to halt stress-induced apoptosis (Pathway #5, not highlighted).

Acknowledgments

This work is supported in part by the National Institute of Health Grants 1 R01EY15765 and 1R01EY018380 (DWL), the University of Nebraska Medical Center startup funds (DWL), the Changjiang Scholar Team Award Funds from Education Ministry of China (DWL), National Science Foundation Funds of China (ML), National Key Laboratory Funds of China (DWL), the Lotus Scholar Program Funds from Hunan Province Government (DWL) and the Special Honorary Professorship Funds from Hunan Normal University (DWL). We apologize to those colleagues whose valuable publications could not be cited here due to space limitation.

References

[1] Bloemendal, H. (1982). Lens proteins. *CRC Crit. Rev. Biochem.* 12:1-38.

[2] Pras E, Frydman M, Levy-Nissenbaum E, Bakhan T, Raz J, Assia EI et al. (2000). A nonsense mutation (W9X) in CRYAA causes autosomal recessive cataract in an inbred Jewish Persian family. *Invest Ophthalmol Vis Sci* 41: 3511–3515.

[3] Mackay DS, Andley UP, Shiels A. (2003). Cell death triggered by a novel mutation in the alphaA-crystallin gene underlies autosomal dominant cataract linked to chromosome 21q. *Eur J Hum Genet* 11: 784–793.

[4] Litt M, Kramer P, LaMorticella DM, Murphey W, Lovrien EW, Weleber RG. (1998). Autosomal dominant congenital cataract associated with a missense mutation in the human alpha crystallin gene CRYAA. *Hum Mol Genet 7*: 471–474.

[5] Gu F, Luo W, Li X, Wang Z, Lu S, Zhang M et al. (2008). A novel mutation in AlphaA-crystallin (CRYAA) caused autosomal dominant congenital cataract in a large Chinese family. *Hum Mutat* 29: 769.

[6] Santhiya ST, Soker T, Klopp N, Illig T, Prakash MV, Selvaraj B et al. (2006). Identification of a novel, putative cataract-causing allele in CRYAA (G98R) in an Indian family. *Mol Vis* 12: 768–773.

[7] Vicart P, Caron A, Guicheney P, Li Z, Prevost MC, Faure A et al. (1998). A missense mutation in the alph αB-crystallin chaperone gene causes a desmin-related myopathy. *Nat Genet* 20: 92–95.

[8] Berry V, Francis P, Reddy MA, Collyer D, Vithana E, MacKay I et al. (2001). Alpha-B crystallin gene (CRYαB) mutation causes dominant congenital posterior polar cataract in humans. *Am J Hum Genet* 69: 1141–1145.

[9] Selcen D, Engel AG. (2003). Myofibrillar myopathy caused by novel dominant negative alpha B-crystallin mutations. *Ann Neurol* 54: 804–810.

[10] Liu M, Ke T, Wang Z, Yang Q, Chang W, Jiang F et al. (2006a). Identification of a CRY αB mutation associated with autosomal dominant posterior polar cataract in a Chinese family. *Invest Ophthalmol Vis Sci* 47: 3461–3466.

[11] Liu Y, Zhang X, Luo L, Wu M, Zeng R, Cheng G et al. (2006b). A novel alphaB-crystallin mutation associated with autosomal dominant congenital lamellar cataract. *Invest Ophthalmol Vis Sci* 47: 1069–1075.

[12] Andley, UA. (2007). Crystallins in the eye: Function and pathology. *Progress in Retinal and Eye Research* 26: 78–98.

[13] Kato K, Shinohara H, Kurobe N, Goto S, Inaguma Y, Ohshima K. (1991). Immunoreactive alpha A crystallin in rat non-lenticular tissues detected with a sensitive immunoassay method. *Biochim Biophys Acta.* 1080: 173-180.

[14] Bhat SP, Nagineni CN. (1989). αB subunit of lens specific protein α-crystallin is present in other ocular and non-ocular tissues. *Biochem. Biophys. Res. Commun.* 158: 319-325.

[15] Iwaki T, Kume-Iwaki A, Liem RKH, Goldman JE. (1989). αB-crystallin is expressed in non lenticular tissues and accumulates in Alexander's disease brain. *Cell.* 57: 71-78.

[16] Dubin RA, Wawrousek EF, Piatigorsky J. (1989). Expression of the murine αB-crystallin gene is not restricted to lens. *Mol. Cell. Biol.* 9: 1083-1091.

[17] Ingolia TD, Craig EA. (1982). Four small Drosophila heat shock proteins are related to each other and to mammalian a-crystallin. *Proc. Natl. Acad. Sci. USA.* 79, 2360-2364.

[18] Klemenz R, Frohli E, Steiger RH, Schafer R, and Aoyama A (1991). Alpha B-crystallin is a small heat shock protein. . *Proc. Natl. Acad. Sci. USA.* 88:3652-3656.

[19] Horwitz J. (1992). ãCrystallin can function as a molecular chaperone. *Proc. Natl. Acad. Sci. USA.* 89:10449-10453.

[20] Rao PV, Horwitz J, Zigler JS Jr. (1993). Alpha-crystallin, a molecular chaperone, forms a stable complex with carbonic anhydrase upon heat denaturation. *Biochem Biophys Res Commun.* 190:786-793.

[21] Kelley MJ, David LL, Iwasaki N, Wright J, Shearer TR. (1993). alpha-Crystallin chaperone activity is reduced by calpain II in vitro and in selenite cataract. *J Biol Chem.* 268:18844-18849.

[22] Boyle D, Takemoto L. (1994). Characterization of the alpha-gamma and alpha-beta complex: evidence for an *in vivo* functional role of alpha-crystallin as a molecular chaperone. *Exp Eye Res.* 58:9-15.

[23] Nicholl ID, Quinlan RA. (1994). Chaperone activity of alpha-crystallins modulates intermediate filament assembly. *EMBO J.* 13:945-53.

[24] Clark JI, Huang QL. (1996). Modulation of the chaperone-like activity of bovine alpha-crystallin. *Proc Natl Acad Sci U S A.* 93:15185-15189.

[25] Sun TX, Das BK, Liang JJ. (1997). Conformational and functional differences between recombinant human lens alphaA- and alphaB-crystallin. *J Biol Chem.* 272:6220-6225.

[26] Reddy GB, Das KP, Petrash JM, Surewicz WK. (2000). Temperature-dependent chaperone activity and structural properties of human alphaA- and alphaB-crystallins. *J Biol Chem.* 275:4565-4570.

[27] Kantorow M and Piatigorsky J (1994). Alpha-crystallin/small heat shock protein has autokinase activity. *Proc. Natl. Acad. Sci. USA* 91: 3112-3116.

[28] Arrigo AP, Tanaka K, Goldberg AL, Welch WJ. (1988). Identity of the 19S 'prosome' particle with the large multifunctional protease complex of mammalian cells (the proteasome). *Nature.* 331(6152), 192-4.

[29] Finley D, Ciechanover A, Varshavsky A. (2004). Ubiquitin as a central cellular regulator. *Cell.* S116, S29–S32.

[30] den Engelsman J, Keijsers V, de Jong WW, Boelens WC. (2003). The small heat-shock protein alpha B-crystallin promotes FBX4-dependent ubiquitination. *J Biol Chem.* 278(7), 4699-704.

[31] Lin DI, Barbash O, Kumar KG, Weber JD, Harper JW, Klein-Szanto AJ, Rustgi A, Fuchs SY, Diehl JA. (2006). Phosphorylation-dependent ubiquitination of cyclin D1 by the SCF(FBX4-alph αB crystallin) complex. *Mol Cell.* 24(3), 355-66.

[32] Barbash O, Lin DI, Diehl JA. (2007). SCF Fbx4/alphaB-crystallin cyclin D1 ubiquitin ligase: a license to destroy. *Cell Div.* 2, 2.

[33] Lin DI, Lessie MD, Gladden AB, Bassing CH, Wagner KU, Diehl JA. (2008). Disruption of cyclin D1 nuclear export and proteolysis accelerates mammary carcinogenesis. *Oncogene.* 27(9):1231-42.

[34] Barbash O, Diehl JA. (2008). SCF(Fbx4/alphaB-crystallin) E3 ligase: when one is not enough. *Cell Cycle.* 7(19):2983-6.

[35] McAvoy, J.W. (1980). Induction of the eye lens. *Differentiation* 17:137-149.

[36] Piatigorsky J. (1981). Lens differentiation in vertebrates. A review of cellular and molecular features. *Differentiation.* 19:134-153.

[37] Spector A. (1995). Oxidative stress-induced cataract: mechanism of action. *FASEB J.* 9:1173-1182.

[38] Grainger RM. (1992). Embryonic lens induction: shedding light on vertebrate tissue determination. *Trends Genet.* 8:349-355.

[39] Wride, M.A. (1996). Cellular and molecular features of lens differentiation: a review of recent advances. *Differentiation.* 61:77-93.

[40] O'Rahilly, R. The early development of the eye in staged human embryos. *Contrib. Embryo. Carneg. Inst.* 38:1-42.

[41] Robinson, M.L., Overbeek, P.A., (1996). Differential expression of alpha A- and alpha B-crystallin during murine ocular development. *Invest. Ophthalmol. Vis. Sci.* 37:2276–2284.

[42] Kuw αBara, T., and Imaizumi, M. (1974). Denucleation process of the lens. *Invest. Ophthalmol. Vis. Sci.* 13:973-981.

[43] Bassnett S, Mataic D. (1997). Chromatin degradation in differentiating fiber cells of the eye lens. *J. Cell Biol.* 137:37-49.

[44] Lang RA. (1999). Which factors stimulate lens fiber cell differentiation in vivo? *Invest Ophthalmol Vis Sci.* 40:3075-3078.

[45] Harding JJ, Crabble JC. (1984). The lens: development, proteins, met αBolism and cataract. In "*The Eye*". (ed. Davison, H.). pp 207-492. Academic press, New York.

[46] Kerr, J.F.R., Wyllie, A.H., and Currrie, A.R. (1972). Apoptosis: a basic biological phenomenon with wide-ranging implications in tissue kinetics. *Br. J. Cancer.* 26:239-257.

[47] Arends, M. J. and Wyllie, A. H. (1991). Apoptosis: mechanisms and roles in pathology. *Int. Rev. Exp. Path.* 32:223-254.

[48] Horvitz, H.R., Shaham, S. and Hengartner, M.O. (1994) The genetics of programmed cell death in the nematode *Caenorh αBditis elegans*. *Cold Spring Harbor Symp Quant Biol* 59:377-385.

[49] Vaux, D.L., Korsmeyer, S.J. (1999). Cell death in development. *Cell.* 96(2):245-254.

[50] Gross, A., McDonnell, J.M., Korsmeyer, S.J. (1999). BCL-2 family members and the mitochondria in apoptosis. *Genes Dev.* 13:1899-1911.

[51] Danial NN, Korsmeyer SJ (2006). Cell Death: Critical Control Points. *Cell*. 116:205–219.

[52] Thompson, C. B. (1995). Apoptosis in the pathogenesis and treatment of disease. *Science*. 267:1456-1462.

[53] Glucksmann, A. (1951). Cell deaths in normal vertebrate ontogeny. *Biol. Rev*. 26:59-86.

[54] Silver J, AFW Hughes (1973). The role of cell death during morphogenesis of the mammalian eye. *J. Morph*. 140:159-170.

[55] Ishizaki, Y., Voyvodic, J.T., Burne, J.F., and Raff, M.C. (1993). Control of lens epithelial cell survival. *J. Cell Biol*. 121:899-908.

[56] Ozeki H, Ogura Y, Hir αBayashi Y, Shimada S. (2001). Suppression of lens stalk cell apoptosis by hyaluronic acid leads to faulty separation of the lens vesicle. *Exp Eye Res*. 72:63-70.

[57] Bozanic D, Tafra R, Saraga-B αBic M.(2003). Role of apoptosis and mitosis during human eye development. *Eur J Cell Biol*. 82:421-429.

[58] Mohamed YH, Amemiya T. (2003). Apoptosis and lens vesicle development. *Eur J Ophthalmol*. 13(1):1-10.

[59] Harding, J.J. and Cr αBbe, J.C. (1984) The lens: development, proteins, metabolism and cataract. In: Davison, H. (ed.): The Eye. Academic press, New York

[60] Yan, Q., Liu, J-P and Li, D. W-C. (2006) Apoptosis in lens development and pathology. *Differentiation*. 74, 195-211.

[61] Hettmann T, Barton K, Leiden JM. (2000). Microphthalmia due to p53-mediated apoptosis of anterior lens epithelial cells in mice lacking the CREB-2 transcription factor. *Dev. Biol*. 222:110-123.

[62] Blixt A, Mahlapuu M, Aitola M, Pelto-Huikko M, EneRback S, Carlsson P. (2000). A forkhead gene, FoxE3, is essential for lens epithelial proliferation and closure of the lens vesicle. *Genes Dev*. 14:245-254.

[63] Fromm L, Shawlot W, Gunning K, Butel JS, OveRbeek PA. (1994). The retinoblastoma protein-binding region of simian virus 40 large T antigen alters cell cycle regulation in lenses of transgenic mice. *Mol Cell Biol*. 14:6743-6754.

[64] Pan H, Griep AE.(1995). Temporally distinct patterns of p53-dependent and p53-independent apoptosis during mouse lens development. *Genes Dev*. 9:2157-2169.

[65] Morgenbesser SD, Schreiber-Agus N, Bidder M, Mahon KA, OveRbeek PA, Horner J, DePinho RA. (1995). Contrasting roles for c-Myc and L-Myc in the regulation of cellular growth and differentiation in vivo. *EMBO J*. 14:743-756.

[66] Nakamura T, Pichel JG, Williams-Simons L, Westphal H. (1995). An apoptotic defect in lens differentiation caused by human p53 is rescued by a mutant allele. *Proc Natl Acad Sci U S A*. 92:6142-6146.

[67] Robinson ML, MacMillan-Crow LA, Thompson JA, OveRbeek PA. (1996). Expression of a truncated FGF receptor results in defective lens development in transgenic mice. *Development*. 121:3959-3967.

[68] Gomez Lahoz E, Liegeois NJ, Zhang P, Engelman JA, Horner J, Silverman A, Burde R, Roussel MF, Sherr CJ, Elledge SJ, DePinho RA. (1997). Cyclin D- and E-dependent kinases and the p57(KIP2) inhibitor: cooperative interactions in vivo. *Mol Cell Biol*. 19:353-363.

[69] McCaffrey J, Yamasaki L, Dyson NJ, Harlow E, Griep AE. (1999). Disruption of retinoblastoma protein family function by human papillomavirus type 16 E7

oncoprotein inhibits lens development in part through E2F-1. *Mol Cell Biol.* 19:6458-68.

[70] de Iongh RU, Lovicu FJ, OveRbeek PA, Schneider MD, Joya J, Hardeman ED, McAvoy JW. (2001). Requirement for TGFbeta receptor signaling during terminal lens fiber differentiation. *Development.*128:3995-4010.

[71] Li WC, Kuszak JR, Dunn K, Wang RR, Ma W, Wang GM, Spector A, et al. (1995). Lens epithelial cell apoptosis appears to be a common cellular basis for non-congenital cataract development in humans and animals. *J Cell Biol.* 130:169-181.

[72] Li W-C, Kuszak JR, Wang G-M, Wu Z-Q, Spector A. (1995). Calcimycin-induced lens epithelial cell apoptosis contributes to cataract formation. *Exp. Eye Res.* 61:89-96.

[73] Li W-C, Spector A. (1996). Lens epithelial cell apoptosis is an early event in the development of UVB-induced cataract. *Free Radic Biol Med.* 20:301-311.

[74] Menko S and Zhou J (2004). Src kinase signaling formation of cataract through a mechanism involving caspase activation, cadherin cleavage and the induction of apoptosis. *Invest Ophthalmol Vis Sci.* 45 (supplement): Abstract #3507.

[75] Zhou J, Leonard M, Van Bockstaele E, Menko AS. (2007).Mechanism of Src kinase induction of cortical cataract following exposure to stress: destablization of cell-cell junctions. *Mol Vis.* 13:1298-310.

[76] Michael R, Vrensen GF, van Marle J, Gan L, SodeRberg PG. (1998). Apoptosis in the rat lens after *in vivo* threshold dose ultraviolet irradiation. *Invest Ophthalmol Vis Sci.* 39:2681-2687.

[77] Michael, R., Vrensen, G.F., van Marle, J., Lofgren, S., Soderberg, P.G. (2000). Repair in the rat lens after threshold ultraviolet radiation injury. *Invest Ophthalmol Vis Sci.* 41(1):204-212.

[78] Ayala M, Strid H, Jacobsson U, Söderberg PG. (2007). p53 expression and apoptosis in the lens after ultraviolet radiation exposure. *Invest Ophthalmol Vis Sci.*, 48(9):4187-4191.

[79] Takamura Y, Kubo E, Tsuzuki S, Akagi Y. (2003). Apoptotic cell death in the lens epithelium of rat sugar cataract. *Exp Eye Res*. 77:51-57.

[80] Murata M, Ohta N, Sakurai S, Alam S, Tsai J, Kador PF, Sato S. (2001). The role of aldose reductase in sugar cataract formation: aldose reductase plays a key role in lens epithelial cell death (apoptosis). *Chem Biol Interact.* 130-132(1-3):617-25.

[81] Tamada Y, Fukiage C, Nakamura Y, Azuma M, Kim YH, Shearer TR. (2000). Evidence for apoptosis in the selenite rat model of cataract. *Biochem Biophys Res Commun*. 275:300-306.

[82] Yoshizawa K, Oishi Y, Nambu H, Yamamoto D, Yang J, Senzaki H, Miki H, Tsubura A. (2000). Cataractogenesis in neonatal Sprague-Dawley rats by N-methyl-N-nitrosourea. *Toxicol Pathol.* 28:555-564.

[83] Wolf N, et al. 2005. Age-related cataract progression in five mouse models for anti-oxidant protection or hormonal influence. *Exp Eye Res*. 81:276-85

[84] Nishi O, Nishi, K. (1998). Apoptosis in lens epithelial cells of human cataracts. *Journal of Eye (In Japanese).* 15:1309-1313.

[85] Takamura, Y., Sugimoto, Y., Kubo, E., Takahashi, Y. and Akagi,Y. (2001) Immunohistochemical study of apoptosis of lens epithelial cells in human and di αBetic rat cataracts. *Jpn J Ophthalmol* 45:559–563.

[86] Okamura, N., Ito, Y., Shibata, M.A., Ikeda, T. and Otsuki, Y. (2002) Fas-mediated apoptosis in human lens epithelial cells of cataracts associated with diabetic retinopathy. *Med Electron Microsc.* 35:234–241.

[87] Charakidas A. Kalogeraki A, Tsilimbaris M, Koukoulomatis P, Brouzas D, Delides G. (2005). Lens epithelial apoptosis and cell proliferation in human age-related cortical cataract. *Eur J Ophthalmol.* 15(2):213-20.

[88] Harocopos GJ, Alvares KM, Kolker AE, Beebe DC. (1998). Human age-related cataract and lens epithelial cell death. *Invest Ophthalmol Vis Sci.* 39:2696-2706.

[89] Mihara E, Miyata H, Nagata M, Ohama E., (2000). Lens epithelial cell damage and apoptosis in atopic cataract-histopathologcal and immunohistochemical studies. *Jpn J Ophthalmol.* 44:695-696.

[90] Lee, J.H., Wan, L., Song, J., Kang, J.J., Chung, W., Lee, E.H. and Kim, E.K. (2002) TGF-b-induced apoptosis and reduction of Bcl-2 in human lens epithelial cells in vitro. *Curr Eye Res* 25:147-153.

[91] Mulhern ML, Madson CJ, Danford A, Ikesugi K, Kador PF, Shinohara T. (2006). The unfolded protein response in lens epithelial cells from galactosemic rat lenses. *Invest Ophthalmol Vis Sci.* 47(9):3951-3959.

[92] Aoyama A, Frohli E, Schafer R, Klemenz R. (1993). Alpha B-crystallin expression in mouse NIH 3T3 fibroblasts: glucocorticoid responsiveness and involvement in thermal protection. *Mol Cell Biol.* 13:1824-1835.

[93] Mehlen P, Preville X, Chareyron P, Briolay J, Klemenz R, Arrigo AP.(1995). Constitutive expression of human hsp27, Drosophila hsp27, or human alpha B-crystallin confers resistance to TNF- and oxidative stress-induced cytotoxicity in st αBly transfected murine L929 fibroblasts. *J Immunol.* 154:363-374.

[94] Mehlen P, Kretz-Remy C, Preville X and Arrigo A-P (1996) Human hsp27, Drosophila hsp27 and human alph αB-crystallin expression-mediated increase in glutathione is essential for the protective activity of these proteins against TNFalpha-induced cell death. *EMBO J.* 15:2695-2706.

[95] Mehlen P, Schulze-Osthoff K, Arrigo AP. (1996). Small stress proteins as novel regulators of apoptosis. Heat shock protein 27 blocks Fas/APO-1- and staurosporine-induced cell death. *J Biol Chem.* 271(28):16510-4.

[96] Martin JL, Mestril R, Hilal-Dandan R, Brunton LL, Dillmann WH. (1997). Small heat shock proteins and protection against ischemic injury in cardiac myocytes. *Circulation.* 96(12):4343.

[97] Andley UP, Song Z, Wawrousek EF, Bassnett S. (1998). The molecular chaperone alphaA-crystallin enhances lens epithelial cell growth and resistance to UVA stress. *J Biol Chem.* 273:31252-61.

[98] Andley UP, Song Z, Wawrousek EF, Fleming TP, Bassnett S. (2000). Differential protective activity of alpha A- and alpha B-crystallin in lens epithelial cells. *J Biol Chem.* 275:36823-31.

[99] Hoover HE, Thuerauf DJ, Martindale JJ, Glembotski CC. (2000). alpha B-crystallin gene induction and phosphorylation by MKK6-activated p38. A potential role for alpha B-crystallin as a target of the p38 branch of the cardiac stress response. *J Biol Chem.* 275: 23825-33.

[100] Ray PS, Martin JL, Swanson EA, Otani H, Dillmann WH, Das DK. (2001). Transgene overexpression of alph αB crystallin confers simultaneous protection against

cardiomyocyte apoptosis and necrosis during myocardial ischemia and reperfusion. *FASEB J.* 15:393-402.

[101] Li DW-C, Xiang H, Mao Y-W, Wang J, Fass U, Zhang X-Y, and Xu C. (2001). Caspase-3 is actively involved in okadaic acid-induced lens epithelial cell apoptosis. *Exp. Cell Res.* 266:279-291.

[102] Mao Y-W, Xiang H, Wang J, Korsmeyer S, Reddan J, Li D W-C. (2001). Human Bcl-2 gene attenuates the ability of rabbit lens epithelial cells against H_2O_2-induced apoptosis through downregulation the alpha B crystallin gene. *J. Biol. Chem.* 27: 43435-43445.

[103] Kamradt MC, Chen F, Cryns VL. (2001). The small heat shock protein alpha B-crystallin negatively regulates cytochrome c- and caspase-8-dependent activation of caspase-3 by inhibiting its autoproteolytic maturation. *J Biol Chem.* 276:16059-63.

[104] Kamradt MC, Chen F, Sam S, Cryns VL. (2002). The small heat shock protein alpha B-crystallin negatively regulates apoptosis during myogenic differentiation by inhibiting caspase-3 activation. *J Biol Chem.* 277:38731-38736.

[105] Andley UP, Patel HC, Xi JH. (2002). The R116C mutation in alpha A-crystallin diminishes its protective ability against stress-induced lens epithelial cell apoptosis. *J Biol Chem.* 277:10178-10186.

[106] Alge CS, Priglinger SG, Neubauer AS, Kampik A, Zillig M, Bloemendal H, Welge-Lussen U. (2002). Retinal pigment epithelium is protected against apoptosis by alpha B-crystallin. *Invest Ophthalmol Vis Sci.* 43(11):3575-82.

[107] Morrison LE, Hoover HE, Thuerauf DJ, Glembotski CC. (2003). Mimicking phosphorylation of alph αB-crystallin on serine-59 is necessary and sufficient to provide maximal protection of cardiac myocytes from apoptosis. *Circ Res.* 92(2):203-11.

[108] Mao, Y-W., Liu, J-P., Xiang, H., and Li, D. W-C. (2004). Human αA- and αB-crystallins bind to Bax and Bcl-XS to sequester their translocation during staurosporine-induced apoptosis. *Cell Death and Differ.* 11:512-526.

[109] Liu, J-P., R. Schlosser, W-Y. Ma, Z. Dong, H. Feng, L. Liu, X.-Q. Huang, Y. Liu, and D.W-C. Li. (2004). Human αA- and αB-crystallins prevent UVA-induced apoptosis through regulation of PKCα, RAF/MEK/ERK and AKT signaling pathways. *Exp. Eye Res.* 79:393-403.

[110] Li, D. W-C., J-P., Liu, Mao, Y-W., H. Xiang, J. Wang, Z. Dong and J. Reed (2005). Calcium-activated RAF/MEK/ERK signaling pathway mediates p53-dependent apoptosis and is abrogated by alpha B-crystallin through inhibition of RAS activation. *Mol. Biol. Cell.* 16:4437-4453.

[111] Liu B, Bhat M, Nagaraj RH. (2004). Alpha B-crystallin inhibits glucose-induced apoptosis in vascular endothelial cells. *Biochem Biophys Res Commun.* 321(1):254-8.

[112] Maloyan A, Sanbe A, Osinska H, Westfall M, Robinson D, Imahashi K, Murphy E, Robbins J. (2005). Mitochondrial dysfunction and apoptosis underlie the pathogenic process in alpha-B-crystallin desmin-related cardiomyopathy. *Circulation.* 112:3451-61.

[113] Kamradt MC, Lu M, Werner ME, Kwan T, Chen F, Strohecker A, Oshita S, Wilkinson JC, Yu C, Oliver PG, Duckett CS, Buchsbaum DJ, LoBuglio AF, Jordan VC, Cryns VL. The small heat shock protein alpha B-crystallin is a novel inhibitor of TRAIL-

induced apoptosis that suppresses the activation of caspase-3. *J Biol Chem.* 280:11059-66.

[114] Moorozov, V., Wawrousek, E.F. (2006). Caspase-dependent secondary lens fiber cell disintegration in αA/αB-crystallin double knockout mice. *Development.* 133:813-821.

[115] Ikeda R, Yoshida K, Ushiyama M, Yamaguchi T, Iwashita K, Futagawa T, Shibayama Y, Oiso S, Takeda Y, Kariyazono H, Furukawa T, Nakamura K, Akiyama S, Inoue I, Yamada K. (2006). The small heat shock protein alpha B-crystallin inhibits differentiation-induced caspase 3 activation and myogenic differentiation. *Biol Pharm Bull.* 29(9):1815-9.

[116] Kadono T, Zhang XQ, Srinivasan S, Ishida H, Barry WH, Benjamin IJ. (2006). CRY AB and HSPB2 deficiency increases myocyte mitochondrial permeability transition and mitochondrial calcium uptake. *J Mol Cell Cardiol.* 40:783-9.

[117] Yaung J, Jin M, Barron E, Spee C, Wawrousek EF, Kannan R, Hinton DR. (2007). alpha-Crystallin distribution in retinal pigment epithelium and effect of gene knockouts on sensitivity to oxidative stress. *Mol Vis.* 13:566-77.

[118] Ousman S, Beren H. Tomooka, Johannes M. van Noort, Eric F. Wawrousek, Kevin C. O'Connor, David A. Hafler, Raymond A. Sobel, William H. Robinson & Lawrence Steinman. (2007). Protective and therapeutic role for αB-crystallin in autoimmune demyelination. *Nature.* 448:474-479.

[119] Dimberg A, Rylova S, Dieterich LC, Olsson AK, Schiller P, Wikner C, Bohman S, Botling J, Lukinius A, Wawrousek EF, Claesson-Welsh L. (2008). Alpha B-crystallin promotes tumor angiogenesis by increasing vascular survival during tube morphogenesis. *Blood.* 111(4):2015-23.

[120] Whiston EA, Sugi N, Kamradt MC, Sack C, Heimer SR, Engelbert M, Wawrousek EF, Gilmore MS, Ksander BR, Gregory MS. (2008). Alpha B-crystallin protects retinal tissue during Staphylococcus aureus-induced endophthalmitis. *Infect Immun.* 76(4):1781-90.

[121] Aggeli IK, Beis I, Gaitanaki C. (2008). Oxidative stress and calpain inhibition induce alpha B crystallin phosphorylation via p38-MAPK and calcium signalling pathways in H9c2 cells. *Cell Signal.* 20 (7):1292-302. Erratum in: *Cell Signal.* 2008. 20(9):1695.

[122] Stegh AH, Kesari S, Mahoney JE, Jenq HT, Forloney KL, Protopopov A, Louis DN, Chin L, DePinho RA.(2008). Bcl2L12-mediated inhibition of effector caspase-3 and caspase-7 via distinct mechanisms in glioblastoma. *Proc Natl Acad Sci U S A.* 105(31):10703-8.

[123] Stegh AH, Chin L, Louis DN, DePinho RA. (2008). What drives intense apoptosis resistance and propensity for necrosis in glioblastoma? A role for Bcl2L12 as a multifunctional cell death regulator. *Cell Cycle.* 7(18):2833-9.

[124] Li H, Li C, Lu Q, Su T, Ke T, Li DW-C, Yuan M, Liu J, Ren X, Zhang Z, Zeng S, K Q. Wang Q and M Liu (2008). Cataract mutation P20S of αB-crystalline impairs chaperone activity of αA-crystalline and induces apoptosis of human lens epithelial cells. *Biophy. Biochem. Acta-Human Diseases.* 1782:303-309.

[125] Ghosh JG, Shenoy AK Jr, Clark JI. (2007). Interactions between important regulatory proteins and human alpha B crystallin. *Biochemistry.* 46(21):6308-17.

[126] Vander Heiden MG, Chandel NS, Williamson EK, Schumacker PT and Thompson CB (1997) Bcl-xL regulates the membrane potential and volume homeostasis of mitochondria. *Cell* 91:627-637.
[127] Bossy-Wetzel E, Newmeyer DD and Green DR (1998) Mitochondrial cytochrome c release in apoptosis occurs upstream of DEVD-specific caspase activation and independently of mitochondrial transmembrane depolarization. *EMBO J.* 17:37-49.
[128] Deshmukh M and Johnson EM Jr (2000) Staurosporine-induced neuronal death: multiple mechanisms and methodological implications. *Cell Death. Differ.* 7:250-261.
[129] Murphy KM, Ranganathan V, Farnsworth ML, Kavallaris M and Lock RB (2000) Bcl-2 inhibits Bax translocation from cytosol to mitochondria during drug-induced apoptosis of human tumor cells. *Cell Death Differ.* 7:102-111.
[130] Kashkar H, Kronke M and Jurgensmeier JM (2002) Defective Bax activation in Hodgkin B-cell lines confers resistance to staurosporine-induced apoptosis. *Cell Death Differ.* 9:750-757.
[131] Wei MC, Zong WX, Cheng EH, Lindsten T, Panoutsakopoulou V, Ross AJ, Roth KA, MacGregor GR, Thompson CB and Korsmeyer SJ (2001) Proapoptotic BAX and BAK: a requisite gateway to mitochondrial dysfunction and death. *Science* 292:727-730.
[132] Cheng EH, Wei MC, Weiler S, Flavell RA, Mak TW, Lindsten T and Korsmeyer SJ (2001) BCL-2, BCL-X(L) sequester BH3 domain-only molecules preventing BAX- and BAK-mediated mitochondrial apoptosis. *Mol. Cell.* 8:705-11.
[133] Wang GQ, Gastman BR, Wieckowski E, Goldstein LA, Gambotto A, Kim TH, Fang B, Rabinovitz A, Yin XM and R αBinowich H (2001) A role for mitochondrial Bak in apoptotic response to anticancer drugs. *J. Biol. Chem.* 276:34307-34317.
[134] Theodorakis P, Lomonosova E and Chinnadurai G (2002) Critical requirement of BAX for manifestation of apoptosis induced by multiple stimuli in human epithelial cancer cells. *Cancer Res.* 62:3373-3376.
[135] Hague A, Diaz GD, Hicks DJ, Krajewski S, Reed JC and Paraskeva C (1997) Bcl-2 and Bak may play a pivotal role in sodium butyrate-induced apoptosis in colonic epithelial cells; however overexpression of bcl-2 does not protect against bak-mediated apoptosis. *Int. J. Cancer.* 72:898-905.
[136] Chen G, Sordillo EM, Ramey WG, Reidy J, Holt PR, Krajewski S, Reed JC, Blaser MJ and Moss SF (1997) Apoptosis in gastric epithelial cells is induced by Helicobacter pylori and accompanied by increased expression of BAK. *Biochem. Biophys. Res. Commun.* 239:626-32.
[137] Goping IS, Gross A, Lavoie JN, Nguyen M, Jemmerson R, Roth K, Korsmeyer SJ and Shore GC (1998) Regulated Targeting of BAX to Mitochondria. *J. Cell Biol.* 143:207-215.
[138] Nechushtan A, Smith CL, Hsu YT and Youle RJ (1999) Conformation of the Bax C-terminus regulates subcellular location and cell death. *EMBO J.* 18:2330-2341.
[139] Wang K, Gross A, Waksman G and Korsmeyer SJ (1998) Mutagenesis of the BH3 domain of BAX identifies residues critical for dimerization and killing. *Mol. Cell. Biol.* 18:6083–6089.
[140] Suzuki M, Youle RJ and Tjandra N (2000) Structure of Bax: coregulation of dimer formation and intracellular localization. *Cell* 103:645–654.

[141] Shroff NP, Cherian-Shaw M, Bera S, Abraham EC. (2000). Mutation of R116C results in highly oligomerized alpha A-crystallin with modified structure and defective chaperone-like function. *Biochemistry*. 39(6):1420-1426.

[142] Bova MP, Yaron O, Huang Q, Ding L, Haley DA, Stewart PL, Horwitz J. (1999). Mutation R120G in alphaB-crystallin, which is linked to a desmin-related myopathy, results in an irregular structure and defective chaperone-like function. *Proc Natl Acad Sci U S A*. 96(11):6137-42.

[143] Kirchhoff SR, Gupta S, Knowlton AA. (2002). Cytosolic heat shock protein 60, apoptosis and myocardial injury. *Circulation*. 105:2899-2904.

[144] Staal, S.P. (1987). Molecular cloning of the *akt* oncogene and its human homologs *Akt1* and *Akt2*: Amplification of *Akt1* in a primary human gastric adenocarcinoma. *Proc. Natl. Acad. Sci.* 84:5034-5037.

[145] Staal, S.P., J.W. Hartley, and W.P. Rowe. (1977). Isolation of transforming murine leukemia viruses from mice with a high incidence of spontaneous lymphoma. *Proc. Natl. Acad. Sci.* 74:3065-3067.

[146] Bellacosa, A., J.R. Testa, S.P. Staal, and P.N. Tsichlis. (1991). A retroviral oncogene, *akt*, encoding a serine threonine kinase containing an SH2-like region. *Science* 254:274-277.

[147] Bellacosa, A., T.F. Franke, M.E. Gonzalez-Portal, K. Datta, T. Taguichi, J. Gardner, J.Q. Cheng, J.R. Testa, and P.N. Tsichlis. (1993). Structure, expression and chromosomal mapping of *c-akt*: Relationship to *v-akt* and its implications. *Oncogene* 8:745-754.

[148] Coffer, P.J. and J.R. Woodgett. (1991). Molecular cloning and characterization of a novel putative protein-serine kinase related to the cAMP-dependent and protein kinase C families. *Eur. J. Biochem.* 201:475-481.

[149] Jones, P.F., T. Jakubowicz, F.J. Pitossi, F. Maurer, and B.A. Hemmings. (1991). Molecular cloning and identification of a serine/threonine protein kinase of the second messenger subfamily. *Proc. Natl. Acad. Sci.* 88:4171-4175.

[150] Cheng, J.Q., B. Ruggeri, W.M. Klein, G. Sonoda, D.A. Altomare, D.K. Watson, and J.R. Testa. 1996. Amplification of AKT2 in human pancreatic cancer cells and inhibition of AKT2 expression and tunorigeneicity by antisense RNA. *Proc. Natl. Acad. Sci.* 93:3636-3641.

[151] Konishi, H., S. Kuroda, M. Tanaka, H. Matsuzaki, Y. Ono, K. Kameyama, T. Haga, and U. Kikkawa. 1995. Molecular cloning and characterization of a new member of the RAC protein kinase family: Association of the pleckstrin homology domain of three types of RAC protein kinase with protein kinase C subspecies and beta gamma subunits of G proteins. *Biochem. Biophys. Res. Comm.* 216:526-534.

[152] Brodbeck, D., P. Cron, and B.A. Hemmings. 1999. A human protein kinase B with regulatory phosphorylation sites in the acitvation loop and in the C-terminal hydrophobic domain. *J. Biol. Chem.* 274:9133-9136.

[153] Nakatani, K., H. Sakaue, D.A. Thompson, R.J. Weigel, and R.A. Roth. 1999. Identification of a human Akt3 (protein kinase B) which contains the regulatiory serine phosphorylation. *Biochem. Biophys. Res. Comm.* 257:906-910.

[154] Mayer, B.J., R. Ren, K.L. Clark, and D. Baltimore. 1993. A putative modular domain present in diverse signalling proteins. *Cell* 73:629-630.

[155] Musacchio, A., T. Gibson, P. Rice, J. Thompson, and M. Saraste. 1993. The PH domain: A common piece in the structural pathwork of signaling proteins. *Trends Biochem. Sci.* 18:343-348.

[156] Datta, K., T.F. Franke, T.O. Chan, A. Makris, S.-I. Yang, D.R. Kaplan, D.K. Morrison, E.A. Golemis, and P.N. Tsichlis. 1995. AH/PH domain-mediated interaction between Akt molecules and its potential role in Akt regulation. *Mol. Cell. Biol.* 15:2304-2310.

[157] Shao Z, Bhattacharya K, Hsich E, Park L, Walters B, Germann U, Wang Y-M, Kyriakis J, Mohanlal R, Kuida K, Namchuk M, Salituro F, Yao Y-M, Hou W-M, Chen X, Aronovitz M, Tsichlis PN, Bhattacharya S, Force T, Kilter H. C-Jun N-terminal kinases mediate reaction of Akt and cardiomyocyte survival after hypoxic injury in vitro and in vivo. *Circ Res.* 98:111–118.

[158] Datta SR, Brunet A, Greenberg ME. (1999). Cellular survival: a play in three Akts. *Genes Dev.*13:2905-2927.

[159] Vivanco I, Sawyers CL. (2002). The phosphatidylinositol 3-Kinase AKT pathway in human cancer. *Nat Rev Cancer*. 2:489-501.

[160] Michael P. Scheid and James R.Woodgett (2001). PKB/AKT: functional insight from genetic models. *Nat. Rev. Mol. Cell. Biol.* 2:760-768

[161] Zheng C, Lin Z, Zhao ZJ, Yang Y, Niu H, Shen X. (2006). MAPK-activated protein kinase-2 (MK2)-mediated formation and phosphorylation-regulated dissociation of the signal complex consisting of p38, MK2, Akt, and Hsp27. *J Biol Chem.* 281(48):37215-26.

[162] Viniegra JG, Martinez N, Modirassari P, Losa JH, Parada Cobo C, Lobo VJ, Luquero CI, Alvarez-Vallina L, Ramon y Cajal S, Rojas JM, Sanchez-Prieto R. (2005). Full activation of PKB/Akt in response to insulin or ionizing radiation is mediated through ATM. *J Biol Chem.* 280(6):4029-36.

[163] Chandrasekher G, Sailaja D. (2003). Differential activation of phosphatidylinositol 3 kinase signaling during proliferation and differentiation of lens epithelial cells. *Invest Ophthalmol Vis Sci.* 44(10):4400-11.

[164] Chandrasekher G, Sailaja D. (2004) Phosphatidylinositol 3-kinase (PI-3K)/Akt but not PI-3K/p70 S6 kinase signaling mediates IGF-1-promoted lens epithelial cell survival. *Invest Ophthalmol Vis Sci.* 45(10):3577-88.

[165] Gotoh N, Ito M, Yamamoto S, Yoshino I, Song N, Wang Y, Lax I, Schlessinger J, Shibuya M, Lang RA. (2004). Tyrosine phosphorylation sites on FRS2alpha responsible for Shp2 recruitment are critical for induction of lens and retina. Proc *Natl Acad Sci U S A.* 101(49):17144-9.

[166] James C, Collison DJ, Duncan G. (2005) Characterization and functional activity of thrombin receptors in the human lens.*Invest Ophthalmol Vis Sci.* 46(3):925-32.

[167] Iyengar L, Patkunanathan B, Lynch OT, McAvoy JW, Rasko JE, Lovicu FJ. Aqueous humour- and growth factor-induced lens cell proliferation is dependent on MAPK/ERK1/2 and Akt/PI3-K signalling. *Exp Eye Res.* 83(3):667-78.

[168] Weber GF, Menko AS. (2006). Phosphatidylinositol 3-kinase is necessary for lens fiber cell differentiation and survival. *Invest Ophthalmol Vis Sci.* 47(10):4490-9.

[169] Madhavi J. Rane, Patricia Y. Coxon, Dave W. Powell, Rose Webster, Jon B. Klein, William Pierce, Peipei Ping, and Kenneth R. McLeish (2001). p38 Kinase-dependent MAPKAPK-2 Activation Functions as 3-Phosphoinositide-dependent Kinase-2 for Akt in Human Neutrophils. *J. Biol. Chem.* 276:3517-3523.

[170] Madhavi J. Rane, Yong Pan, Saur αBh Singh, David W. Powell, Rui Wu, Timothy Cummins, Qingdan Chen, Kenneth R. McLeish, and Jon B. Klein. (2003). Heat Shock Protein 27 Controls Apoptosis by Regulating Akt Activation. *J. Biol. Chem.* 278: 27828-27836.

[171] Rui Wu, Hina Kausar, Paul Johnson, Diego E. Montoya-Durango, Michael Merchant, and Madhavi J. Rane. (2007). Hsp27 Regulates Akt Activation and Polymorphonuclear Leukocyte Apoptosis by Scaffolding MK2 to Akt Signal Complex. *J. Biol. Chem.* 282: 21598-21608.

[172] Saori Sato, Naoya Fujita, and Takashi Tsuruo (2000). Modulation of Akt kinase activity by binding to Hsp90. *Proc. Natl. Acad. Sci. USA*,. 97:10832-10837.

[173] Xia Z, Dickens M, Raingeaud J, Davis RJ, Greenberg ME.(1995). Opposing effects of ERK and JNK-p38 MAP kinases on apoptosis. *Science*. 270:1326-1331.

[174] Ballif BA, Blenis J. (2001). Molecular mechanisms mediating mammalian mitogen-activated protein kinase (MAPK) kinase (MEK)-MAPK cell survival signals. *Cell Growth Differ*. 12:397-408.

[175] Li, D. W-C., Jin-Ping Liu, J. Wang, Mao, Y-W., and Li-Hui Hou. (2003). Protein expression patterns of the signaling molecules for MAPK pathways in human, bovine and rat lenses. *Invest. Ophthalmol. Vis. Sci.* 44:5277-5286.

[176] Gong X, Wang X, Han J, Niesman I, Huang Q, Horwitz J. (2001). Development of cataractous macrophthalmia in mice expressing an active MEK1 in the lens. *Invest Ophthalmol Vis Sci.* 42:539-548.

[177] Zatechka SD Jr, Lou MF. (2002). Studies of the mitogen-activated protein kinases and phosphatidylinositol-3 kinase in the lens. 1. The mitogenic and stress responses. *Exp Eye Res.* 74(6):703-17.

[178] Zatechka SD Jr, Lou MF. (2002). Studies of the mitogen-activated protein kinases and phosphatidylinositol-3 kinase in the lens. 2. The intercommunications. *Exp Eye Res.* 75(2):177-92.

[179] Govindarajan V, Overbeek PA. (2001). Secreted FGFR3, but not FGFR1, inhibits lens fiber differentiation. *Development.* 128:1617-27.

[180] Lovicu FJ, McAvoy JW. (2001). FGF-induced lens cell proliferation and differentiation is dependent on MAPK (ERK1/2) signalling. *Development.* 128:5075-5084.

[181] Le AC, Musil LS. (2001). A novel role for FGF and extracellular signal-regulated kinase in gap junction-mediated intercellular communication in the lens. *J Cell Biol.* 154:197-216.

[182] Le AC, Musil LS. (2001). FGF signaling in chick lens development. *Dev Biol.* 233:394-411.

[183] Seth RK, Haque MS, Zelenka PS. (2001). Regulation of c-fos induction in lens epithelial cells by 12(S)HETE-dependent activation of PKC. *Invest Ophthalmol Vis Sci.* 42(13):3239-3246.

[184] Zhou J, Fariss RN, Zelenka PS. (2003). Synergy of epidermal growth factor and 12(s)-hydroxyeicosatetraenoate on protein kinase C activation in lens epithelial cells. *J Biol Chem.* 278:5388-98.

[185] Benkhelifa S, Provot S, N αBais E, Eychene A, Calothy G, Felder-Schmittbuhl MP. (2001). Phosphorylation of MafA is essential for its transcriptional and biological properties. *Mol Cell Biol.* 21:4441-4452.

[186] Ochi H, Ogino H, Kageyama Y, Yasuda K. (2003). The st αBility of the lens-specific Maf protein is regulated by fibroblast growth factor (FGF)/ERK signaling in lens fiber differentiation. *J Biol Chem*. 278:537-44.

[187] Johnson GL, Lapadat R. (2002). Mitogen-activated protein kinase pathways mediated by ERK, JNK, and p38 protein kinases. *Science.* 298:1911-1912.

[188] Cullen, P.J. and Lockyer, P.J. (2002). Integration of calcium and Ras signalling. *Nat. Rev. Mol. Cell Biol. 3*:339–348.

[189] Campbell SL, Khosravi-Far R, Rossman KL, Clark GJ, Der CJ. 1998. Increasing complexity of Ras signaling. *Oncogene.* 17:1395-413.

[190] Farnsworth, C.L., Freshney, N.W., Rosen, L.B., Ghosh, A., Greenberg, M.E. and Feig, L.A. 1995. Calcium activation of Ras mediated by neuronal exchange factor Ras-GRF. *Nature 376*:524–527.

[191] Fam, N.P., Fan, W.T., Wang, Z.X., Zhang, L.J., Chen, H. and Moran, M.F. 1997. Cloning and characterization of Ras-GRF2, a novel guanine nucleotide exchange factor for Ras. *Mol. Cell. Biol. 17*:1396–1406.

[192] Moyano JV, Evans JR, Chen F, Lu M, Werner ME, Yehiely F, Diaz LK, Turbin D, Karaca G, Wiley E, Nielsen TO, Perou CM, Cryns VL. (2006). αB-crystallin is a novel oncoprotein that predicts poor clinical out come in breast cancer. *J. Clin. Invest*. 116, 261-270.

[193] Hartwell, L. H. & Kastan, M. B. (1994). Cell cycle control and cancer. *Science* 266: 1821–1828.

[194] Kastan MB, Lim D-S. (2000). The many substrates and functions of ATM. *Nature Rev. Mol. Cell Biol.* 1:179–186.

[195] Bakkenist CJ, Kastan MB (2003). DNA damage activates ATM through intermolecular autophosphorylation and dimmer dissociation *Nature.* 421:499-506.

[196] Bai F, Xi JH, Wawrousek EF, Fleming TP, and Andley UP. (2003). Hyperproliferation and p53 status of lens epithelial cells derived from αB-crystallin knockout mice. *J. Biol. Chem.* 278:36876-36886.

[197] Liu S, Li J, Tao Y, Xiao X. (2007). Small heat shock protein αB-crystallin binds to p53 to sequester its translocation to mitochondria during hydrogen peroxide-induced apoptosis. *Biochem. Biophy. Res. Commun.* 354:109–114.

[198] Spector, A., Chiesa, R., Sredy, J. & Garner, W. (1985). cAMP-dependent phosphorylation of bovine lens alpha-crystallin. *Proc. Natl. Acad. Sci. USA* 82:4712-4716.

[199] Morrison LE, Hoover HE, Thuerauf DJ, Glembotski CC. (2003). Mimicking phosphorylation of alphaB-crystallin on serine-59 is necessary and sufficient to provide maximal protection of cardiac myocytes from apoptosis. *Circ Res*. 92(2):203-11.

[200] Whittaker R, Glassy MS, Gude N, Sussman MA, Gottlieb RA, Glembotski CC. (2009). Kinetics of the Translocation and Phosphorylation of αB-crystallin in mouse heart mitochondria during ex vivo ischemia. *Am J Physiol Heart Circ Physiol.* 286(5):H1633-1642.

In: Small Stress Proteins and Human Diseases
Editors: Stéphanie Simon et al.
ISBN: 978-1-61470-636-6

Chapter 1.4

SMALL STRESS PROTEINS IN THE CENTRAL NERVOUS SYSTEM: FROM NEURODEGENERATION TO NEUROPROTECTION

Andreas Wyttenbach*, Sarah Hands and Vincent O'Connor
Southampton Neuroscience Group, School of Biological Sciences, University of Southampton, Basset Crescent East, Southampton SO16 7PX, UK

ABSTRACT

We summarize data on the physiological and stress-induced expression of the small HSPs HspB1, HspB5, HspB6, HspB8, heme-oxygenase 1 (Hsp32) and metallothioneins MT1, -2 and 3 in the central nervous system (CNS). We contrast the stress induction of these proteins during chronic neurodegeneration due to protein misfolding in age-related diseases with their role during acute brain injury (traumatic brain injury and stroke). This evidence is discussed with respect to potential brain region and cell-type specific responses and vulnerability to stress. Although the basal expression of all these proteins is well documented in neurons and astrocytes, their physiological expression *in vivo* in oligodendro- and microglia is less clear. During acute stress some of these proteins (typically HspB1 and HspB5) are quickly upregulated in the neuronal compartment to likely interfere with cell death cascades. This is followed by a more sustained increase in expression in astroglial and microglial cells of sHSPs, HO-1 and MTs. A clear functional role in neuroprotection against ischemia has been shown only for HspB1 using genetic knockout approaches, HspB1 inducing drugs and viral delivery in rodent models paving the way to potential therapeutic treatments during stroke in humans. Such mechanistic studies have not been performed in models of protein misfolding diseases (proteinopathies such as Alzheimer's-Parkinson's-Polyglutamine-and Prion diseases), but

* Corresponding Author: Andreas Wyttenbach
University of Southampton
School of Biological Sciences
Southampton Neuroscience Group
Basset Crescent East,
Southampton SO16 7PX, UK
A.Wyttenbach@soton.ac.uk

a key similarity is the increased expression of small stress proteins in glial cells, especially associated with an astrocytic stress response. This likely affords cytoprotection through a sustained defense against oxidative stress and modulation of neuroinflammation. The apparent redistribution of many small stress proteins to extracellular protein deposits, and in some cases to intracellular protein inclusion bodies points to a further modulatory function during protein aggregation and/or degradation that could be linked to the neuroprotective activities of these proteins in both neurons and glial cells. Given the chronic nature of proteinopathies and the complex interdependence between protein aggregation, imbalance of the cellular redox-homeostasis and inflammatory and other signalling events, it will be difficult to detail the neuroprotective role(s) of each stress protein *in vivo*. Nevertheless the induction and manipulation of multiple stress proteins in different cell types and cell compartments has to be considered for the development of effective neuroprotective strategies in the CNS.

1. Introduction

The central nervous system (CNS) may experience acute (eg traumatic brain injury, ischemia) or chronic stress (e.g. in protein misfolding diseases called "proteinopathies") during CNS development and in the adult brain. In both situations cells in the CNS may induce a regulated stress response that appears to differ substantially depending on the cell type, designed to protect and/or repair damaged molecules and cellular structures. Experimental models of acute forms of brain injury (ischemia or several forms of excitotoxic lesions) have determined a relevant functional role for small stress proteins in neuroprotection. However, their involvement during chronic neurodegeneration associated with proteinopathies such as Alzheimer's (AD)- Parkinson's (PD)- Huntington's (HD) and prion disease is less clear. In view of the central importance of these proteinopathies in the context of an ageing population, small stress proteins are coming under increasing scrutiny.

As research into this subject intensifies, it has also become clear that basic knowledge of such proteins are often missing. Therefore, we summarize here the current knowledge on the physiological expression and distribution of small stress proteins in the rodent and human brain and whether there are significant differences across brain regions. This point is important as it may inform the debate about why neuropathology during chronic neurodegeneration in proteinopathies is often, at least initially, restricted to specific brain regions. Given the selective vulnerability of certain cell types in all chronic CNS diseases and during acute forms of brain injury, another relevant consideration is whether expression and/or induction of small stress proteins are fundamentally different in various cells types and how this impacts on neuroprotection. Finally, in the light of the evidence that cell-cell interactions play key roles in modulating disease processes in the CNS, we will discuss whether small stress proteins are likely to be involved in these acute and chronic disease processes.

We will consider two groups of small stress proteins for which there are exemplars for a role in neuronal function and neuroprotective mechanisms in the CNS: 1) the group of small heat shock proteins (sHSPs) induced by various forms of stressors and also considered in chapters 1-1 and 2-5 in a neurological context. We have recently shown that under physiological conditions there are five out of 10 annotated HspB members expressed in the rodent brain from which four (HspB1, HspB5, HspB6 and HspB8) can be detected at the protein level [1] 2) there is evidence that the small, stress-inducible metallothionein (MTs) and heme oxygenase (HO-1) (Hsp32) proteins, that are regulated by oxidative stress and heavy metals, are expressed in various neurodegenerative conditions and play key roles in neuroprotection. Hence we will also consider MTs and HOs. These two groups of proteins are regulators of basic cellular functions such as the cellular redox-status, bioenergetics, inflammation, cytoskeleton dynamics and they also participate in many signaling cascades (e.g. apoptosis, see chapters 1-3 and 3-4). Because most of these functions have been addressed in other chapters in detail, we will not discuss how these proteins regulate such processes, but give examples of where we believe specific functions of stress proteins are particularly relevant for maintaining neuronal homeostasis.

2. From Brain Regions to Cell-Cell Interactions

The mammalian CNS is subdivided into many regions and interlinked circuits. This requires highly specialized cell types such as neurons dealing with a high energy demand and information processing, immuno-reactive microglial cells, astrocytes that interact with synapses and blood vessels regulating key aspects of brain bioenergetics and the myelin producing oligodendroglia. Hence it is not surprising that during stressful conditions a multifaceted response occurs. Depending on the quality and magnitude of a stress stimulus, whole brain regions and/or specific cell types compensate via e.g. gene expression changes that ultimately modulate the behavioural response of the whole organism in order to adjust to changed environmental and/or physiological conditions.

Such cellular stress if not appropriately engaged can lead to brain-region and cell type specific degeneration resulting in neuronal loss and is particularly evident in conditions of age-related chronic neurodegeneration. For example AD is a disorder that affects the brain in a complicated fashion: structural brain abnormalities (atrophy) are consistently revealed in the hippocampus, entorhinal cortex and corpus callosum while subcortical and infratentorial brain structures such as the basal ganglia and cerebellum are relatively preserved (see refs in [2]). Neuropathological studies in AD indicate specific loss of layer III and V large pyramidal neurons in the association cortex. In HD, an autosomal dominant genetic disorder caused by a CAG/polyglutamine (polyQ) expansion mutation leading to misfolding and aggregation of the ubiquitously expressed huntingtin (Htt) protein, it is the basal ganglia (caudate and putamen) that are severely affected with an early loss of medium-spiny neurons and a relative sparing of interneurons [3]. In the spinocerebellar ataxias (SCA's) that also belong to the polyQ diseases, it is the cerebellum that is the target of pathology where GABAergic Purkinje neurons (eg in SCA-1) located in the cerebellar cortex are most severely affected [4]. In PD, the second most common neurodegenerative disease with most cases arising sporadically like in AD (ca 90%), cell loss is most prominent in the substantia nigra pars compacta (SNc), a subcortical nucleus belonging to the basal ganglia ([5] and refs therein). In prion-and related disorders neuronal demise in the CNS is more widespread: in post-mortem brains of people who suffered from Creutzfeldt-Jacob's disease (CJD) severe neuronal loss is found in the cerebellum and thalamus, but also in the cerebral cortex [6]. In variant CJD (vCJD) neuronal death in the cerebral cortex is most severe in the primary visual cortex, but also found in the thalamus and midbrain [6] with subsequent neuronal loss in other areas (eg basal ganglia) in cases with severe spongiform change. Many of these findings are based on analysis of a human brain at end-stage pathology. In contrast, experimental animal models of prion disease show neuronal death in more specific brain regions preceded by loss of synapses [7]. Alterations at and loss of synapses appear to be associated with all the above-mentioned proteinopathies and could be one of the early signs driving cognitive decline and neuropathology. In AD it has indeed been suggested that synaptic changes best mimic the region specific brain pathology [8].

Postmitotic cells such as neurons seem to be particularly vulnerable to stress (e.g. protein misfolding, oxidative stress) and show an attenuated response regarding the induction of stress proteins such as the sHSPs. Nevertheless neurons are known to up-regulate sHSPs *in vivo* under conditions of acute brain injury and chronic stress. Such up-regulation is more widely known for astrocytes and microglia. In contrast, the role of small stress proteins in oligodendroglia are less clear although *in vitro* evidence shown that oligodendrocytes are able to mount a stress response and induce several members of the HSP family [9]. This broad cellular response highlights the need to distinguish between cell-autonomous neuroprotective mechanisms or intercellular mechanisms driven by small stress proteins. This is complicated by the fact that all neural cells are highly compartmentalized. While understanding this complexity is challenging enough during acute brain injury, it is even more difficult to integrate the complexity over the time periods during which age-related neurodegeneration occurs (years to decades).

While there is extensive evidence of sHSPs as intracellular regulators of signaling cascades, their recently proposed roles as extracellular mediators, mainly in inflammatory processes [10], adds another layer of complexity. This is highly relevant given that sHSPs could play important roles in neuroprotection via receptor-mediated signaling, like neurotrophins. The knowledge on both the intra-and potential extracellular physiological expression of sHSPs, MTs and HOs in the human brain is presently scarce and the evidence on brain region and cell-type specific expression is even less complete. Often their expression is studied under conditions of neurodegeneration. Below we summarize such studies and also highlight what is known about expression under physiological conditions.

3. Changes in Expression and Distribution of Small Stress Proteins During Proteinopathies and Neuroprotection

3.1. Small Stress Proteins and Neurodegeneration

The acronym "small stress proteins" can be misleading to the naive reader reinforcing the view that these proteins and the widespread cellular functions they engaged in are used predominantly in a stressed or pathological situation. It is clear that much of the cell's normal physiology generates associated activity, which would otherwise be deleterious (or appear stressful) unless balanced by associated ameliorating pathways. This appears to be the case for the HSPs in the context of the CNS as they are found expressed constitutively and appear to perform key functions in normal physiology [11]. This is most aptly demonstrated by considering the deleterious effects of the growing numbers of mutations found in HspB1, HspB5 and HspB8 that appear to selectively perturb motor neuron function [12]. This is a well studied paradigm of sHSP function and is informative as it details disruptive mutations that prevent constitutively expressed proteins function that are presumably contributing to normal function. Secondly, these mutations highlight the role of cell type specificity in so much as motor neurons are not the only neurons that express the sHSPs but appear to selectively show a degenerative phenotype. This is a cameo for the broader significance of neuron specific expression/degeneration that largely remains a puzzle in many important diseases. It would thus appear that sHSPs may be able to inform on the basis of selective cell or brain region dysfunction that lies at the centre of many major neuropathologies. Finally, the selective loss of the motor neurons is apparent although the sHSPs that underlie the peripheral motor neuron pathology are surrounded by a supporting cast of non-neuronal cells that also express mutant protein. This highlights a potential role for these cell types in ordinarily protecting against inappropriate function leading to neurodegeneration. The latter point builds on the one above and highlights that understanding sHSPs will inform on the relative importance of the non-neuronal cells in mediating or imposing a modulatory exacerbation or protection during the disease processes that ultimately lead to nerve degeneration.

We will focus on four major diseases that appear primarily driven by protein misfolding. The key features of these diseases with respect to the major misfolded proteins responsible, the region in which there is primary dysfunction and indications for the major neurons that are lost, are listed in Table 1. Our aim will be to highlight the major sHSPs and other small stress proteins, summarize how their expression changes in response to disease and where possible discuss the likely consequence of this regulation in the context of what is known about the function of these proteins.

Table 1. List of the proteinopathies considered in this chapter. The major proteins that through mutation or activation of misfolding pathways accumulate as aggregated proteins in extra- or intracellular compartments are listed. Where tractable, the brain region classically associated with disease is also given. These diseases are often associated with the selective loss of a neuronal subtype and other cellular changes.

Disease	Misfolded protein	Brain region mostly affected	Neuronal loss	Other associated cellular changes
Alzheimer's Disease	ABeta tau	Cortex	Cholinergic neurons	Astrogliosis Microgliosis
Prion Disease	Prion (PrPsc)	Various	Interneurons	Astrogliosis Microgliosis
Huntington's Disease	Huntingtin	Cortex, Striatum	Medium spiny neurons	Gliosis not well studied
Parkinson's Disease	α-synuclein	Substantia Nigra, Putamen	Dopaminergic neurons	Astrogliosis Microgliosis

3.2. Alzheimer's Disease

The protein misfolding insults that cause AD are two fold. Firstly, the inappropriate processing of the membrane bound APP leads to a sequential proteolytic cleavage and the production of two major peptides, Abeta 1-40 and Abeta 1-42. By virtue of the parent protein's membrane topology, these peptides are released into the extracellular milieu. These processed peptides undergo sequential oligomerization and aggregation pathways that ultimately lead to the deposition of the protein aggregates outside the cell. These aggregates lead to the neuropathological feature that is routinely used to score AD brains, namely the amyloid plaques. The biophysical and biochemical features of these plaques include birefringes, staining with congo red and the association of ubiquitin and other proteins that are recruited into the plaques. Although the neurotoxicity of plaques was presumed from their association with neurodegeneration it is now clear that the roles played by Abeta and its amyloidogenic deposits are more complicated. Firstly and perhaps most damning is the observation that cognitive ability, the clinical score of AD, does not correlate well with the deposition of amyloid that posthumously diagnoses the disease [13]. Secondly, there is a wealth of evidence that smaller oligomers of Abeta peptides that precede overt deposition may be as, or more important determinants of disease [14]. Finally conventional processing of the parent protein suggests that Abeta release by proteolytic activity would lead to extracellular deposition. However there is increasing evidence for the presence and pathological consequence of Abeta in the intracellular compartment in neurons [15] that allows these agents to more directly engage the sHSPs and other stress proteins.

The second misfolding determinant of AD was defined by neuropathological investigation of diseased brains that highlight the accumulation of intracellular neurofibrillary tangles (NFTs). NFTs are now known to arise through the misfolding and modified phosphorylation of the major tubulin binding protein tau. This involves accumulation of intracellular aggregates of tau that disrupt a number of microtubule dependent processes that otherwise support neuronal viability. There is a growing realization that resolving the cause-effect relationships of the cognitive dysfunction in AD through the tau or Abeta pathways maybe unsatisfying. Accordingly there is an increasing compliance to integrating how the two pathways may interact with each other to give the disease that is clinically recognized [16]. This maturing view reaffirms that complex cellular and integrative signaling interactions underpin disease and this is certainly reflected in the limited observations that pertain to the sHSPs in the context of AD.

3.2.1. HspB1 (Hsp27)

Despite its relatively defined role in the ontogeny of motor neuron disease the role of HspB1 in other proteinopathies is less clear. In AD it shows a selective expression, that may include an up-regulation in the glial cells in which we know it is otherwise constitutively expressed [12]. This predominant expression in non-neuronal cells appears consistent with other reports showing that where there is expression in the CNS, it is at a relatively low level and restricted to non neuronal cells [1]. These observations are consistent with description of the other major sHSPs and suggest that it plays a modulatory role in AD derived from a reactive response of supporting CNS cells. Of note is the limited evidence that suggests that HspB1 immunoreactivity associates with extracellular Abeta plaques [17] and when paired with the expression pattern supports the idea that non-neuronal cells may liberate sHSPs. One can only speculate that this will have chaperone activity or provide other neuromodulatory activities associated with sHSPs. Neuroprotective activities of increased expression are evident particularly in models of acute stress (see below). These observations of increased neuroprotection normally involve perturbation of intact systems encompassing both neuronal and non-neuronal cells. Detailed investigations of the source of the neuroprotective activity are lacking and it is possible that neural AD associated up-regulation of HspB1 in non-neural cells has an important neuroprotective role (see discussion for HspB5 below).

3.2.2. HspB5 (αB-crystallin, αBC)

Early studies highlighted the association of HspB5 with dementia brains, clearly extending its role beyond a lens chaperone [18]. Further studies in AD tissue showed an association with areas of extracellular deposit. In agreement with our own investigations in the mouse (unpublished data) the predominant expression of HspB5 in human control tissue was in both astrocytes and oligodendrial cells [17]. This does not rule out a low or cell type restricted neuronal expression of HspB5 in CNS neurons. In AD tissue it appears that HspB5 expression is similar to control tissue and dominated by an appearance in astrocytes and microglia that occur around the plaque deposits [17]. Nevertheless, studies that have quantified levels of HspB5 in control and AD tissue suggest there is an up-regulation in protein expression and this is again best reconciled by it being driven via a reactive response in glia [19]. Although this illustrates that the glial cells support a reactive response, it remains unresolved as to how HspB5 might interact with the neurodegenerative process. A more recent documentation provides evidence that some of the HSPs that are seen in AD are extracellularly accumulated within the plaques [20]. There is *in vitro* evidence to support an association of HspB5 and Abeta although its significance for chaperoning of the aggregates or intermediates of the misfolding pathway is unclear [21]. There is some suggestion of a preferential association of the small oligomers that precede aggregate deposition, but the *in vivo* significance of this remains unclear.

The induction and potential extracellular release of HSP may have significant indirect consequences [22]. In other neurodegenerative models the accumulation of extracellular HSP has been ascribed a role in the regulation of an inflammatory response. Indeed the action of sHSPs, likely acting on non-neuronal cells through cell surface Toll receptors, is one of down regulation of the inflammatory response associated with neurodegeneration [23]. Thus, the non–neuronal up-regulation and potential liberation of HSPs, including HspB5, may act indirectly to modulate the inflammatory response. The role of inflammatory cascades is under intense investigation and the role of this biology in ameliorating neurodegeneration or promoting the removal of Abeta plaques highlights routes by which the non-neuronal potential of HspB5 and other sHSPs may be utilized both endogenously and therapeutically [23].

In contrast to Abeta, the intracellular misfolding associated with tau might make a more obvious way in which sHSPs interact with the misfolded protein. However, there remains some controversy as to whether HspB5 interacts with neuronal tau with various studies finding no or limited evidence for an association of HspB5 with the intra-neuronal aggregates [24]. In contrast, and in keeping with the predominant expression of HspB5 in non-neuronal cells, there appears to be an association with HspB5 and tau aggregates that are sometimes found in diseases with tauopathy [25]. The consequence for the tau misfolding in glia is likely to be deleterious but how this impacts on neurodegeneration is not clear.

3.2.3. HspB6 (Hsp20) and HspB8 (Hsp22)

Few studies have investigated expression of HspB6 in the human CNS. In a study by Wilhelmus et al the authors observed HspB6 immunoreactivity in astrocytes of the white matter and occasionally grey matter in the normal brain [20]. In AD tissue HspB6 expression was reported not to be changed as measured by western blotting, but clear evidence was provided that HspB6 may distribute into senile plaques (SPs), both in the cortex and hippocampus with elevated HspB6 reactivity also found in astrocytes surrounding these sites. Interestingly, neurofibrillary tangles (NFTs) were not immunopositive for HspB6 [20]. This finding is similar for HspB1 and HspB5 (see above) that are also found in SPs, but not in NFTs [20].

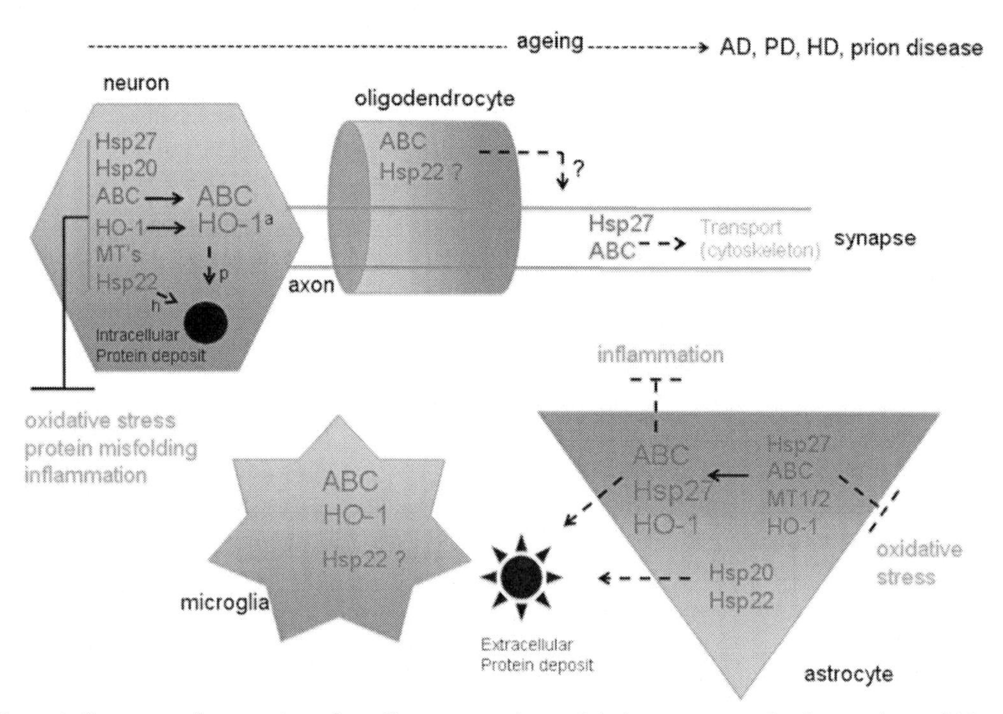

Figure 1. Summary of expression of small stress proteins and their neuroprotective interactions within- and between different cell types in the CNS during chronic neurodegeneration associated with protein misfolding and aggregation. The larger size of letter indicating small stress proteins indicates an increase in expression. Only changes that have been confirmed by several studies (see text) have been included in this diagram. If a change has only been observed in a particular disease, this has been indicated by a small letter a: AD, p: PD, h: HD. Many changes of small stress proteins under acute conditions of brain injury are similar (but see text for details).

Expression of HspB8 was examined in a different study by Wilhelmus and colleagues. HspB8 appeared to have a similar distribution as HspB6 in that mainly astrocytes in both grey and white matter and cerebrovascular cells were immunoreactive [26]. Oligodendrocytes and notably microglia appeared HspB8 positive, but the authors identified these cells by morphological criteria only. As for the other sHSPs, HspB8 was found in SPs, but not in NFTs.

Hence overall, it appears that the sHSPs in AD are induced in some neuronal populations, but predominantly up-regulated in astrocytes and found in extracellular Abeta plaques. It is likely that the neuroprotective actions include the regulation of protein aggregation, inflammatory response and protection against oxidative stress, both inside and outside cells (see Figure 1).

3.3. sHSPs in Prion Disease

Prion disease is associated with the misfolding of a ubiquitously expressed glycolipid anchored cell surface protein. The physiological role of this protein is not well understood with various observations suggesting a function in cell adhesion, cell surface re-cycling and by virtue of a debated copper–binding as a regulator of oxidative stress [27]. The widely expressed protein which is denoted cellular prion (PrPc) is liable to undergo protein misfolding to generate pathological PrPsc. This misfolding is triggered by one of three modifications encompassing an infectious route, genetic mutation in the PrPc or sporadic cases with no defined trigger. In an infectious context prion disease is triggered by intestinal intake of diseased tissue in which misfolded prion from the donor enters the host via a subsequent peripheral neurotrophic route and penetrates the recipient's CNS. This allows the donor PrPsc to interact with the host's endogenous PrPc and promotes the catalytic conversion of the recipient protein to PrPsc. The infectious route is clinically characterized in sheep Scrapie disease, cow Bovine Spongiform Encephalitis (BSE) and human or variant Creuzfeldt-Jacob (CJD) and Kuru disease in humans. The diseases that arise from this misfolding cause a selective neuronal degeneration in the CNS despite the proteins wide cellular expression. The ensuing neurodegeneration is associated with brain vacuolation (spongiosis), astrogliosis, synaptic degeneration and neuronal loss [28]. Whether the disease is triggered by infection or genetic mutations, the disease will lead to extracelullar deposition of misfolded PrPsc. The biophysical features of the extracellular misfolded protein are defined by its ability to resist full digestion by proteinase K. In a way that parallels the major plaques seen in AD, but this experimental definition may be of limited value in defining pathogenesis as lower molecular weight aggregation states may be the more pathogenic form of the protein. Indeed such intermediates appear likely to underlie prion disease in which pathogenesis occurs in the complete absence of classic proteinase K resistant deposits [29].

The major members of the large HSP family are implicated in prion disease particularly with respect to the role they played in the misfolding pathways that propagate infectious protein that underlies disease [30]. This is reinforced at a functional level by the observation that disease progression is modulated by Hsf-1 that regulates the expression of the stress induced HSPs [31]. However, Hsf-1's ability to regulate members of the neuronal sHSPs opens up to their potential in prion diseases. Further, in view of the supporting role played by the sHSPs in modulating large HSP or other stress pathways it is surprising that there is very limited consideration of the role played by the sHSPs in prion disease. As indicated above, investigation of the sHSPs in mice brain suggest that the sub-classes that are expressed seem to be predominately expressed in non-neuronal cells. In view of the astrocytosis that is the defining feature of the major forms of prion disease this might predict that HSPB1, HSPB5, HSPB6 and HSPB8 will be upregulated in neuropathological prion disease tissue. However, besides a few studies that show an increase in HspB5 and HspB1 expression close to PrPsc plaques in astrocytes and microglia [31], no systematic study on the expression of these proteins in human or animal clinical disease has been performed.

Our own unpublished investigation in an animal model of prion disease triggered by directly inoculating PrPsc into the brain (ME7 strain) support the notion that disease progression is associated with a significant increase in three of the 10 HspBs that are normally expressed in the mouse CNS [1]. Further, we screened the other sHSPs that were not detected in the undiseased brain and confirmed that they were not induced by the misfolding stress associated with prion disease (unpublished data). These observations have been supported at the transcript level by a recent transcriptomic analysis in five distinct mouse models of prion disease listing HSPB1, HSPB6 and HSPB8 among a cohort of 233 genes that make up a core transcriptional response common to prion disease [32]. Although a detailed cellular expression of the genes was beyond the scope of this systematic study it is interesting that the supplementary information reports that HSPB6 expression is neuronal based, shown by a follow up screen of 120 genes using *in situ* hybridization. This opens up potential understanding of a gene whose clear presence in the brain as shown by western blotting and transcript analysis had not been defined at a cellular level [1] (also see the Allen Brain Atlas on-line).

Interestingly, HSPB5 is not among the reported genes that satisfy criteria for inclusion within the assembled prion disease gene networks [32]. This would appear at odds with our own more limited analysis that indicates a strong up-regulation of both HSPB5 transcript and protein in the ME7 model of prion disease [33]. Although preliminary, this study of sHSPs in prion disease highlights their regulated expression affording them the opportunity to use their glial and potentially the neuronal expression to engage in chaperone- and other activities that may modulate disease progression. The significance of the changes associated with prion disease are awaiting investigation of the role played by astrocytes or the direct mechanistic testing of the key features of disease progression in models deficient in one or several of the sHSPs.

3.4. Parkinson's Disease

PD is delineated into those with a genetic or idiopathic trigger. In either case disease is associated with the degeneration of the substantia nigra and the nigrostriatal system. As these mid brain structures are central to direct and indirect pathways that underlies motor control, this explains the movement disorders that characterize PD [34]. Nevertheless, there are other symptoms associated with this disease including depression and there is clear evidence for neuropathological changes beyond the nigra striatal system including the amygdala and the limbic system. This extra nigra striatal pathology might provide an explanation for the less often described anxiety and depressive symptoms of this disorder.

The cardinal, but not definitive, neuropathological features of PD are Lewy bodies (LBs), proteinaceous intra-neuronal inclusions which are immunopositive for the neuronal protein α-synuclein and ubiquitin. α-synuclein is normally a highly unstructured relatively soluble or membrane associated protein and LBs are made of beta sheet enriched structures. This indicates that the disease is a protein misfolding disorder, an argument supported by the decoration of these aggregates with ubiquitin [35]. Mutation in α-synuclein that lead to changes in the proteins primary sequence or cause the wild-type protein's over expression promote the formation of these aggregates and the associated pathology seen in disease. In addition to the cluster of mutations in the human α-synuclein (SNCA) there are a number of other genetic changes associated with PD including Parkin, Leucine Repeat R Kinase, PINK-1 and DJ-1 [5].

Whether the stress associated with accumulated misfolded protein uniquely triggers Parkinsonism is not clear. In particular the susceptibility of the dopaminergic pathways to direct oxidative stress is well recognized. This is the basis of chemical toxicity by MPP+ and other mitochondrial dysfunctions that generate an ensuing oxidative stress that leads to degeneration. Indeed the supporting role played by mitochondria in dopamine synthetic and degradative pathways might contribute to the selective vulnerability of these cells that is a key feature of PD pathology. This is further reinforced by the utilization of free radical generating iron based Fenton reactions in the biosynthetic pathways of dopamine synthesis that generates stress.

The recent identification of several gene loci whose mutation is associated with PD can be well reconciled with a contribution from protein misfolding and mitochondrial stress induced misfolding pathways. In particular LRRK, whose mutations are associated with LBs suggests that kinase cascades can contribute to a downstream protein homeostasis which if dysfunctional leads to the initiation of misfolded protein and the accumulation of aggregates [5]. Similarly the emergence of mutations in Parkin, a ubiquitin-E3 ligase, can be linked to abnormal protein homeostasis causing disease. However, this simplistic view of convergent mechanism acting in disease is less easily reconciled with the observation that patients with the Parkin mutations appear not to display LBs. In contrast to the above, PINK-1 and DJ-1 mutations are associated with PD, but these proteins directly regulate mitochondrial function and couple this to the sensing of oxidative stress [36]. This would suggest that the oxidative dysfunction associated with PD may operate independent of the protein misfolding pathways. Overall, the above highlights a complexity of underlying causes of PD, but supports the notion that there is a pivotal dependence on protein misfolding and/or oxidative stressors, both of which might be modulated by the known neuroprotective activities that are ascribed to the sHSPs (see chapter 1-1).

3.4.1. HspB1 (HSP27)

Neuropathological investigation of sHSP expression in PD provides both transcriptional and immunocytochemical evidence of increased expression. In the case of gene array studies evidence supports the induction of HspB1 transcripts [37]. Quantitative PCR and *in situ* hybridization confirm this induction and the latter indicates that induced transcript is abundant around the region of disintegrating neurons, likely associated with glial cells. This report is consistent with additional studies that identify increased expression of HspB1 protein in tissue of PD patients and its association with reactive glial cells in the regions of selective degeneration [38]. Further, in dementia with LBs (DLB) a 2-3-fold up-regulation of both HspB1 mRNA and protein over control cases has been reported [39]. However, uncertainty remains around the conclusion that the HSP response is restricted to reactive astrocytes. Indeed, reports argue against a reactive astrocytosis in PD and suggest that the gliosis is related to reactive microglia [40]. Interestingly, reactive astrocytosis seem to be selectively observed in advanced disease or PD with dementia. It is worth noting that studies which directly compare the gliosis and associated sHSPs staining in AD and PD note a more robust response in AD [38]. Thus part of the discrepancy reported in these studies may arise due to the stage of disease progression, the etiology of disease in each cohort or the relative sensitivity of the methodologies used, upon which comparisons are based.

3.4.2. HspB5 (αBC)

The induction of HspB5 seems to mirror that of HspB1 although the increase in expression that is reported for PD is less dramatic compared to examples of neurodegeneration associated with extracellular protein deposition [38]. Whether this reflects the more modest gliosis and/or less well-defined cellular composition (astrocytes/microglia) which makes the reactive response in PD remains unclear. In the case of HspB5 there is an additional controversy as studies of PD tissues indicate that HspB5 is highly expressed in neuronal cells. This study shows staining for HspB5 in neurons of the cortex and amygdala and indicates that immunopositive cells are devoid of LBs or dysmorphic neurites [41] hinting at a potential neuroprotective role for intraneuronal HspB5. This is interesting as *in vitro* studies have shown a chaperoning function of HspB5 towards α-synuclein [42]. However, these *in vitro* observations do not preclude a neuroprotective role for HspB5 in antioxidant pathways that appear central to PD (see above).

Overall the studies on HspB5 favor a predominant role in non-neuronal cells and hint at a potential glio-degenerative contribution in PD. These observations are reinforced when considering some reports that have investigated HspB5 (and HspB1) in the various animal models of PD. In particular, it was shown that the relative level of sHSP induction varies when comparing the effects of distinct mutations in the mouse models of distinct proteinopathies [43]. Thus driving similar levels of different mutant proteins appear to result in distinct mutant specific consequences. In particular, it is only the transgenic PD models that give rise to degeneration and motor deficits that drive an increased HspB5 and HspB1 induction. These disease associated increases in HspB5 (and HspB1) are restricted to the associated astrocytosis, thus the PD animal models also support a non-neuronal focus for sHSPs. There is no available data on HspB6 and HspB8 expression in PD and the currently published PD mouse models.

3.5. Polyglutamine (polyQ) Diseases and sHSPs

Another group of 10 disorders associated with intracellular misfolding and aggregation are the polyQ diseases. With the exception of spinobulbar muscular atrophy (SBMA), which is X-linked, the polyQ diseases are autosomal dominantly inherited and all lead to death [44]. Onset of the human disease is generally in mid-life with a debilitating disease progression lasting 10-20 years. The length of the polyQ expansion in the relevant and different proteins in each disease is inversely correlated with the age-of onset of disease and disease severity (a longer polyQ stretch leads to earlier onset). When the polyQ stretch is above ca. 40 glutamines the polyQ containing proteins misfolds and form intracellular inclusion bodies (IBs), mainly within neurons [44]. The location of polyQ aggregation is similar to PD in the cytoplasm, but can also occur in the nucleus and axons depending on the particular disease. A key feature of polyQ disorders is the relative brain region specific pathology and loss of neurons [4], similar to other proteinopathies. For example in HD, it is mainly neurons in the striatum (caudate and putamen) that are most severely affected during disease with later involvement of the cortex and other brain regions. For a recent review on polyQ protein functions and the variety of neuronal dysfunctions caused in the CNS the reader is referred to Hands et al. [44].

While many studies examined the expression and role of the Hsp40/70 family members in regulating polyQ pathology in cell- and animal models, no studies have rigorously investigated the expression levels of HSPs or sHSPs in the human brain in polyQ disorders (reviewed in [45, 46]). However, one report showed that transformed lymhoblastoid cells from patients with spinocerebellar ataxia type 7 (SCA-7) exhibited a decreased expression of HspB1 protein [47]. This finding is consistent with a reduced expression of HspB1 found in other cell models of polyQ disorders including SCA-3 [48] and HD (our own unpublished observation), but the relevance of these findings to human disease is questionable. Indeed in the study of Chang and colleagues it appeared that HspB1 expression is increased in the SCA-3 post-mortem brain [48]. The most rigorous study on HSPs in a polyQ mouse model so far has been performed by Hay and colleagues in which it was shown that the R6/2 mouse (expressing an N-terminal fragment of Htt) exhibited a progressive loss of several HSPs (Hsp40, Hsp70, SGTs), but not HspB1 over time [49]. Expression of HspB6 and HspB8 was not examined, but in an earlier study the expression of HspB5 was shown to be reduced in the R6/2 transgenic brain [50]. Expression of HspB5 and HspB1 were unchanged in yet another mouse model of HD and of dentatorubral and pallidoluysian atrophy (DRPLA) [51]. In an elegant study of a SCA-17 mouse, a disease that is caused by a polyQ expansion in the TATA-box binding protein (TBP), Friedman and colleagues found that HspB1 was significantly down-regulated [52]. The interaction of polyQ expanded TBP with the general transcription factor IIB (TFIIB) was enhanced and TFIIB occupancy of the HspB1 promoter was decreased. Whether HspB1 expression is also lost in the human SCA-17 CNS is unknown. Overall the current evidence suggests that sHSPs are mostly unchanged or expression is lost in rodent models.

Despite some studies showing that sHSPs are cytoprotective in cell models of HD via counteracting free radicals (eg HspB1 [53]) or reducing polyQ aggregation itself by perhaps accelerating the turnover of polyQ aggregation-prone proteins (e.g. HspB8, see chapter 1-1) a systematic examination of changes in expression levels in mouse models and in the human brain are missing, but underway in our laboratories. It appears that sHSPs could play roles *in vivo* by modulating intracellular polyQ misfolding itself, counteracting the redox- imbalance reported in HD and stabilizing axonal transport abnormalities via their known regulatory effects on the (neuronal) cytoskeleton (see chapter 1-2). It is less clear whether sHSPs have a role to play in inflammatory processes that have so far not been well elucidated in the polyQ disorders.

3.6. Other Small Stress Inducible Proteins: MTs and HOs

3.6.1. Expression of MTs Under Basal and Disease Conditions

Metallothioneins are a family of low-molecular-weight (6-7 kDa), cysteine-rich (25-30%) proteins which bind heavy metals including zinc and copper (Duncan et al 2006). Metallothioneins can exert cell-protective effects both through both their ability to sequester heavy metals and their anti-oxidant properties resulting from their multiple sulfhydryl groups. Four MT isoforms have been identified in the mammalian genome (MT1-4). Under stress conditions, the metal-responsive transcription factor (MTF-1) activates expression of metallothioneins, with metal-response elements (MREs) in their promoter regions [54]. MT 1 and 2 are expressed in nearly all tissues, including the CNS, MT3 is mainly expressed in the CNS whereas MT4 is mainly expressed in keratinizing epithelia (for a review see [55]). There are variations in reports on the cellular localization of MTs within the CNS, as expression levels depend on several factors including age and stress. However, it appears that MT1 and 2 are expressed mainly in glial cells and MT3 is expressed in neurons.

Abnormalities in metallothionein expression have been reported during multiple neurodegenerative disorders and ageing [56-68] consistent with the dysregulation of metals (including copper and zinc) and oxidative damage observed in these disorders including AD, PD, HD and prion diseases (for a review see [69]).

3.6.2. MTs and Alzheimer's Disease

In AD MT1 and 2 are upregulated in astrocytes and their expression is increased in cells surrounding amyloid plaques, consistent with the astrocytosis and microgliosis known to occur in this region [56, 57]. Changes in the expression levels of MT3 during AD, however, remain controversial. Some studies show that MT3 is down-regulated in the brains of AD patients, whereas others show that MT3 remains unaltered [56, 70], which may reflect alterations in age or stage of disease of the brains investigated in these studies. MT3 has been shown to be protective against Abeta toxicity in a cell model via a metal swap between the zinc of the metallothionein and the copper of the Abeta [71].

3.6.3. MTs Expression in Prion Diseases

Metallothionein 1/2 has been shown to be upregulated in scrapie-infected hamster brain, transgenic mice infected with brain homogenate from BSE infected cattle and in cattle with BSE, particularly in astrocytes surrounding PrP deposition [66, 67, 72]. In CJD patients with a relatively long disease progression, however, MT 1, 2 and 3 were all significantly reduced in astrocytes, particularly surrounding extracellular plaques, although Western blotting revealed MT1 and 2 levels were up-regulated in CJD patients with a short disease duration [68]. In order to investigate the role of MTs in prion disease, MT1/2 knockout mice were infected with scrapie [73]. Whilst the incubation period and symptoms were unaltered in these knockout mice, small increases were seen in protein deposition and in astroglial and microglial activation in the hippocampus. The relatively small changes may be due to a compensatory response by MT3, not investigated in this study. Hence despite changes in expression of MTs in prion disease further experiments must demonstrate a neuroprotective role for MT's.

3.6.4. MTs, Parkinson's Disease and Neuroprotection

6-Hydroxydopamine (6-OHDA) and 1-methyl-4-phenylpyridinium (MPP+), both of which are toxic to dopaminergic neurons and induce Parkinsonism, have been shown to reduce the levels of MTs selectively in the rodent striatum, whereas addition of dopamine upregulated MT synthesis. However, no change in MT1 was seen when comparing substantia nigra and putamen from PD patients and control subjects. This is consistent with the lack of reactive astrocytosis seen in these patients assuming that MT induction by astrocytes participates in the protective response towards neuronal loss that is apparent in these CNS regions. MT knockout mice are more susceptible to 1-methyl-4-phenyl-1,2,3,6-

tetrahydropyridine (MPTP)-induced Parkinsonism [74] and MT overexpression protects against 3-morpholinosydnonimine (SIN-1) mediated nitrative stress and toxicity in neuronal cells and mouse models [75]. Furthermore, a MT knockout mouse is more sensitive to SIN-1 toxicity [76]. MT3 has also been shown to prevent ROS production in a dopaminergic cell line treated with 6-OHDA via induction of heme oxygenase-1 [77]. These findings suggest that MTs are playing a neuroprotective role in models of PD. One study investigated the role of MTs in a genetic mouse model of PD: loss of MT1/2 lead to lethargic behavior, reduced body size, mitochondrial function and lifespan [74] implicating MT1/2 in synuclein pathology.

3.6.5. MT expression in Polyglutamine Diseases is Unclear

An examination of gene expression data in HD cell models and human brain tissue shows that MT1 and 2 mRNAs are up-regulated due to polyQ expanded Htt (or its N-terminal fragments), but further studies have not been undertaken as of yet [64, 65].

3.6.6. Implications for MTs during Neuroprotection

As outlined above, MTs are induced in conditions of protein misfolding stress and in acute models of PD and this occurs mainly in glial cells (astrocytes). Hence it is reasonable to suggest that MT induction serves a neuroprotective role. This notion is also supported by studies using MT ko mice during acute brain injury (see below). During these stress conditions astrocytes known to be "activated" (normally determined by increased GFAP expression and changes in morphology), do not appear to undergo cell death, like neurons. Therefore, MT up-regulation may protect astrocytes, especially by dealing with redox-and metal related stresses. Another way of neuroprotection, and similar to sHSPs, is that astrocytes secrete MTs in response to neuronal injury leading to neuronal regeneration [78].

However, it is likely that MT induction in neurons also plays a role during neuroprotection. It has indeed been demonstrated that addition of Zn (II) to cultured neurons causes an up-regulation of metallothioneins and this enhances the survival of dopaminergic neurons [79]. Furthermore, MTs could play a role via their involvement in neurite plasticity. MT2 causes an increase in initial neurite outgrowth in cultured cortical, hippocampal and cerebellar rat neurons [80, 81]. In contrast MT3 has a neurite outgrowth inhibitory effect and was therefore initially given the name Growth Inhibitory Factor (GIF) [82]. The mechanisms of induction during chronic disease are expected to be complex and the regulation of several factors such as protein aggregation, oxidative stress and inflammation could drive neuropathology. Indeed, MT1/2 are known to also be upregulated during neuroinflammation [56]. But given the importance of metals in AD, PD, HD and prion diseases and the abnormalitites in MT expression during these diseases, MTs appear suitable candidates for therapeutic intervention. For example, it has been suggested that metals accelerate Abeta aggregation, localize to extracellular Abeta plaques that are centres of redox-reactions that involve metal- (eg copper) dependent processes (Fenton reaction) and MTs are found to also co-localise to these sites, similar to sHSPs (see above) [83]. Hence MTs may play a role in Abeta-related pathology.

3.6.7. Expression and Stress Induction of HOs

Heme oxygenases (HO-1 and HO-2), involved in the degradation of heme to biliverdin/bilirubin, free iron and carbon monoxide in the endoplasmic reticulum, have been strongly implicated in neurodegenerative diseases (see [84] for a review). In the normal brain, HO-2 is constitutively and ubiquitously expressed throughout the brain whereas HO-1 (HSP32) expression occurs in distinct populations of neurons in the cerebellum, thalamus, hypothalamus, brain stem, hippocampal dentate gyrus and cerebral cortex and in some glial cells. HO-1 is under the control of metal response elements (MREs, as in the case of MTs), as well as heat shock elements (HSE). Therefore HO-1 can be upregulated in both neurons and glia by a variety of stressors including oxidants, heavy metals and inflammatory stimuli [84, 85].

3.6.8. Role of HO-1 in Alzheimer's Disease

In the brains of AD patients, HO-1 is induced in neurons and astrocytes, particularly in the hippocampus, and also localizes to plaques and tangles [86, 87]. HO-1 protein was also shown to be upregulated in neurons in transgenic mice overexpressing APP and surrounded the amyloid plaques [88]. Transient transfection of the HO-1 gene (Hmox1) caused proteasomal degradation of tau in neuroblastoma cells, indicating that the up-regulation of HO-1 may also be an attempt by the cell to upregulate the proteasomal system to remove aggregation-prone tau [89]. A yeast two hybrid screen and co-immunoprecipitation identified an inhibitory interaction of APP with HO, an interaction that may take place in the endoplasmic reticulum [90]. APP with mutations associated with familial AD inhibited HO activity more than wild-type APP in a cell line [90].

3.6.9. HO-1 Expression in Prion Diseases

Exposure to an amyloidogenic peptide of human prion protein caused an up-regulation of HO-1 mRNA in astroglial cell cultures, but not in cortical or hippocampal neuronal cultures [91]. HO-1 mRNA expression is increased in the cortex, striatum, hippocampus, cerebellum and brainstem of scrapie infected mice [92, 93]. HO-1 protein expression was also shown to increase in the extracellular matrix and neuropil of the cortex of scrapie infected mice, whilst remaining mainly absent from GFAP-immunoreactive astrocytes and neurons [94]. However, HO-1 protein levels were induced in GFAP-immunoreactive astrocytes in the hippocampus of scrapie-infected mice [93]. HO-1 expression was only marginally increased in human brains affected by vCJD and Gerstmann-Straussler-Scheinker (GSS) and associated with prion deposits, but HO-1 levels were unchanged in brains with sCJD [95].

3.6.10. HO-1 in Parkinson's Disease

HO-1 is over-expressed in astrocytes of the substantia nigra and is associated with LBs in dopaminergic neurons [96]. Overexpression of HO-1 has been shown to protect against MPP+ -induced toxicity [97]. As for tau, HO-1 up-regulation has also been shown to induce proteasomal degradation of α-synuclein. This is likely to occur via activity of HO-1 directly on the proteaseome rather than by an interaction between HO-1 and α-synuclein, as inhibiting downstream effects of HO-1 catabolites, prevented the effect of α -synuclein clearance [98].

3.6.11. HO-1 and Polyglutamine Diseases

There are no published studies in which HO-1 expression has been examined in polyQ disorders and we did not find any significant changes in HO-1 mRNA by data mining the published gene changes that have been systematically performed in the polyQ disorders.

3.6.12. HOs and Neuroprotection

HO-1 has been shown to be protective against oxidative stress in neurons and as discussed above has been shown to be protective against MPP+-induced toxicity [97] and in traumatic and ischemic brain injury (see below). Given the trend throughout studies of the various neurodegenerative disorders for HO-1 to be mainly up-regulated in astrocytes, rather than neurons, it has been suggested that the reduced ability of neurons to induce HO-1 and therefore their limited ability to respond to oxidative stress could partly account for their selective loss in these disorders [84, 99]. This would be consistent with the finding that the number of human cortical and hippocampal neurons and glial cells that are immunoreactive for HO-1 were shown to increase with age [100]. It has been suggested that the increased HO-1 staining dopaminergic neurons in the substantia nigra in elderly subjects indicate that during normal ageing these cells are subjected to increased oxidative stress compared to other neurons [84]. However, a rigorous and systematic analysis of HO-1 expression during human ageing has not been performed.

It should also be noted that HO-1 has been implicated in exacerbation of neurotoxicity, possibly via one or more of the biologically active substances produced from the breakdown of heme, i.e. iron, CO and bilirubin. For example suppression of HO-1 expression by metalloporphyrin has been shown to be neuroprotective in some models of brain injury [101, 102] and elevated HO-1 can promote dopaminergic cell toxicity induced by polychlorinated biphenyls [103]. The effect of HO-1 is therefore likely to depend on the extent and length of toxic insult and the level and duration of induction of HO-1.

4. Neuroprotection by Small Stress Proteins During Acute Forms of Brain Damage

4.1. Traumatic Brain Injury and Stroke

We consider two major forms of acute brain damage: traumatic brain injury (TBI) and several forms of stroke (e.g. ischemia). TBI is a form of acquired brain injury that occurs when a sudden trauma (e.g. an object) causes damage to the CNS either by hitting or piercing the skull. TBI can be associated with a cerebral hemorrhage (bleeding), but intracranial hemorrhage can also be caused by a stroke. Stroke is a rapidly developing loss of brain function caused by a disturbance of blood supply (and hence oxygen and glucose availability) due to a hemorrhage or the lack of blood supply due to ischemia. Stroke is the second most common cause of death worldwide and the leading cause of adult disability [104]. Ischemic stroke accounts for ca. 80% of all strokes from a thrombotic or embolic occlusion of a major cerebral artery, often the middle cerebral artery (MCA) [105]. During ischemia a complex sequence of spatio-temporal pathophysiological events occurs within minutes and the brain injury may last hours or even days. The major events that occur after cerebral vessel occlusion and subsequent reperfusion are a failure of energy metabolism with acidosis and the generation of free radicals, membrane depolarization and glutamate release with activation of excitotoxic cascades resulting in apoptosis and necrosis [105]. The characteristics of acute brain injuries depend on the severity and duration of the cerebral blood flow reduction and in animal models of acute ischemic stroke blood flow is crucially reduced in the central region of the brain (infarct core with necrotic death) and in a graded fashion centrifugally from the core (ischemic penumbra with cells dying by necrosis and apoptosis). Given the cascade of events occurring during acute brain injury small stress proteins could interfere at various steps and provide neuroprotection as they are known to interfere with several (if not all) of these events. Not surprisingly, there is ample evidence that both MTs/HO-1 and sHSPs are induced during acute brain injuries and several reports have already elucidated how stress proteins mediate their neuroprotective functions using genetic approaches.

4.2. Expression and Neuroprotective Role of sHSPs in TBI and Stroke

Increased synthesis of HSPs after transient ischemia has been observed in the rat brain early on [106], but a more detailed analysis of specific sHSPs localized to certain brain cells only occurred over the last decade. Increased expression of HspB5 has been localized to ballooned neurons at the edges of areas with cerebral infarction (CI) and it was suggested that HspB5, not normally expressed in neurons under basal conditions, could be involved in regeneration [18]. Interestingly, in this study HspB5 was also found in ballooned neurons of cases of AD and undefined dementia indicating that selected neurons up-regulate HspB5 also in conditions of chronic neurodegeneration (see above). A cytoprotective effect by HspB5 in neurons subjected to ischemic stress was also proposed in the study of Minami and colleagues [107]. Interestingly, these authors found an elevated HspB5 immunoreactivity in morphologically normal, convex-and ballooned neurons after CI in the human brain that significantly increased over time. In order to study the relationship between induction of HspB5 and cell type specificity over time in an animal model [108] induced MCA occlusion (MCAO) in the rat brain. Both HspB5 transcripts and protein were transiently expressed only 4 hours after MCAO/reperfusion in pyramidal neurons in the peri-infarct region of the ischemic hemisphere. However, 2 days after MCAO a significant induction of αBC appeared in reactive astrocytes in the penumbra that was sustained for several more days [108]. Hence it appears that during ischemia both neurons and astrocytes may be involved in the neuroprotective response although the action of elevated HspB5 could be different in the two cell types.

More studies have examined the role of HspB1 during stroke and TBI. Endogenous induction of HspB1 has been observed in cells surviving ischemic insults [109] and also in models of ischemic preconditioning [110] paving the way to understand the protective role of HspB1 during stroke. While both Hsp70 and HspB1 appeared to be up-regulated in neurons in the rat brain, a significant and sustained HspB1 expression was associated with an ongoing astrogliosis, similar to what has been shown for HspB5 [110, 111]. HspB1 was also suggested to be involved in tolerance to oxygen-glucose deprivation (OGD) and NMDA toxicity in organotypic hippocampal cultures [112]. While a clear astrocytic upregulation of HspB1 was observed after NMDA injection in the immature rat brain [113], TBI in the form of an aspiration lesion in the rat cortex resulted in an induction of HspB1 in astrocytes with a transient increase also observed in microglial cells [114].

Apart from one study showing that after focal ischemia HspB6 appears to be induced in apical dendrites of CA1 pyramidal neurons [115], no further reports on HspB6 or HspB8 have been published.

Given these findings one would predict that HSP inducing drugs or viral delivery of HSP *in vivo* are valid neuroprotective strategies. Indeed, in an important study by Lu and colleagues geldanamycin (a benzoquinone ansamycin known to induce HSPs) was injected in the lateral cerebral ventricle 24 hours before MCAO [116]. Immunocytochemical analysis showed that while Hsp70 was up-regulated in neurons an increase in HspB1 expression was observed in glia and arteries in cortex, hippocampus and other brain regions leading to a more than 50% decrease in infarct volume and significant reduction in TUNEL positive cells. This study suggested that an elevation of glial HspB1 expression is involved in neuroprotection. A facilitation of the HSP response by HspB1 was also proposed by 17-beta-estradiol during ischemia [117]. Further evidence for neuroprotection by HspB1 was provided by the study of Badin et al. who showed that viral delivery of this protein (but not Hsp70) into the post-ischemic brain afforded significant protection [118]. Hence it may be important to sustain increased levels of HspB1 not only to suppress neuronal death, but also for ongoing repair by astroglia while other HSPs (eg Hsp70) could act on the more immediate stress (e.g. protein denaturation) during stroke. This view is also supported by injury studies in the cerebellum where HspB1 expression was immediately expressed and sustained up to 20 days post injury, whilst Hsp72 was not (reviewed in [119]).

Transgenic approaches also show that HspB1 protects against excitotoxicity and ischemic brain injury. Belleroche and colleagues found a reduced lesion size and less seizures and neuronal death in the brain of HspB1overexpressing mice after kainate injection compared to wild-type controls [120]. In an impressive study by Stetler et al. it was shown that both transgenic and viral overexpression of HspB1 conferred long-lasting tissue preservation and neurobehavioural recovery measured by infarct volume, sensorimotor function and cognitive tasks up to 3 weeks following focal cerebral ischemia [121]. Importantly, these authors demonstrated that HspB1 acted on neuronal ASK1 activity blocking neuronal death instead of interfering with the known downstream cell death execution cascades. Yet a different approach was chosen in a study by An and colleagues injecting PEP- HspB1 fusion proteins intraperitoneally into gerbils which prevented neuronal death and lipid peroxidation in the CA1 region of the hippocampus in response to transient focal ischemia [122].

Together these studies demonstrate that elevation of HspB1 under ischemic conditions affords neuroprotection. Protection by HspB1 of neuronal cells occurs by blocking known steps in death cascades (also see chapter 3.4), but HspB1 is also involved in neuroprotection through blocking increased oxidative stress and support for the neuronal cytoskeleton. The functional role of the glial response for neuroprotection that is mainly astrocytic is less clear. Given the neuroprotective roles of these cells in oxidative stress, the production of neurotrophic factors and inflammatory mediators and the demonstration that HspB1 can modulate these processes, it is tempting to speculate that HspB1 also provides neuroprotection in this way (see chapter 1-2). More generally, sHSPs could play such roles intra-and extracellularly and hence the neuroprotective signaling and cell-cell communication between neurons and astrocytes will be complex and a challenge to disentangle. Given that HspB5 is mainly expressed in the white matter in at least the rodent brain (oligodendrocytes) under physiological conditions [1] it is possible that these cells participate in the protection of neuronal cells via axonal support or perhaps oligodendrocytes need a high level of protection due to their enhanced sensitivity to stress. No study has addressed this question yet. Finally, because both HspB5 and HspB1 have been shown to be axonally synthesized in addition to other cytoskeletal elements [123], it is likely that the neuronal up-regulation of these proteins observed in many studies is a neuroprotective response to injury that disrupts axonal transport processes.

4.3. Induction and Neuroprotection by MTs and HO-1 during TBI and Stroke

MT1/2 proteins were shown to be induced in astrocytes at the edges of infarcts in the ischemic human brain [124]. More recent studies showed an induction of MT mRNA during cerebral ischemia in mice (Schneider et al., 2004, others). The protection against cerebral ischemia by several chemical compounds including resveratrol [125], erythropoietin (EPO) [126] and cilostazol (a arterial vasodilator and platelet aggregation inhibitor) have also been linked to the induction of MT1/2 mainly in the penumbral areas. In the study of Wakida and colleagues the neuroprotective effect of EPO against permanent focal cerebral ischemia was significantly reduced in MT1/2 knockout mice demonstrating a functional role for MT1/2 in neuroprotection.

HO-1 (Hsp32) expression is mainly induced in microglia, and astrocytes, but also found in neurons and associated with ischemia and TBI in the rodent and human brain [127, 128]. As outlined above, HO-1 metabolises heme to carbon monoxide (CO), iron and biliverdin and its by product bilirubin, both powerful antioxidants. Hence neuronal overexpression of HO-1 protected against MCAO in mice [129] demonstrating a functional role for HO-1 in protection of the CNS. Curcumin upregulates NFR-2 and HO-1 expression (mRNA and protein) and protects against focal ischemia in the rat CNS [130]. Furthermore, Saleem and colleagues showed that in a model of ischemic reperfusion brain injury the neuroprotective action of EGb761, a standardized Ginkgo biloba extract known to induce HO-1, is lost in HO-1 knockout mice [131].

Given these studies, it appears that HO-1 is mainly induced in astroglial cells during acute brain injury, similar to conditions of chronic neurodegeneration (see above). A more prominent HO-1 glial response is also observed during kainate -mediated excitotoxiciy in the cerebellum [132] where Bergman glia but not Purkinje cells induce HO-1. This demonstrates again to a cell-type specific stress response where cell-cell communication may be important. This concept has been experimentally demonstrated in an elegant study by Imuta and colleagues: these authors found that in HO-1 was induced in the ischemic rat brain in astrocytes that surrounded dying TUNEL-positive neurons. In an *in vitro* paradigm for ischemia they exposed cultured astrocytes to normobaric hypoxia which triggered a quick (4h) induction of HO-1 mRNA and protein. Upon reoxygenation an increase of CO in the medium in turn decreased the hypoxia-mediated neuronal death of cerebral neurons suggesting a potential mechanism for a protective paracellular pathway regulated by HO-1 [133].

CONCLUSION

The potential neuroprotective functions of small stress proteins is manifold and ranges from the modulation of protein misfolding, aggregation and degradation, the redox homeostasis, the inflammatory response, the remodeling of the cytoskeleton via interactions with actin and microtubules to interfering with various signaling cascades (e.g. neuronal death). The protective actions may occur inside and/or outside cells and some neuroprotective mechanisms are expected to be specific to different cell types in the CNS. While the impact of stress proteins in neurons appears directed towards death signaling cascades and protection of the axonal and perhaps the synaptic compartment, glial cells, especially astrocytes, upregulate small stress proteins in a sustained manner and as a consequence likely regulate redox-and inflammatory reactions in a cell- and non-cell autonomous fashion. In the future more studies should be directed towards understanding the physiological roles of sHSPs and MTs and in examining changes in expression and cell-specificity of all small stress proteins considered in this chapter both in the rodent and human brain during chronic neurodegeneration and ageing.

ACKNOWLEDGMENTS

AW is grateful for funding by the Medical Research Council (UK), the BBSRC, Gerald Kerkut Trust, HighQ Foundation, Wessex Medical Research Trust and the University of Southampton.

REFERENCES

[1] Quraishe S, Asuni A, Boelens WC, O'Connor V and Wyttenbach A. Expression of the small heat shock protein family in the mouse CNS: differential anatomical and biochemical compartmentalization. *Neuroscience,* 2008; 153: 483-491.

[2] Hampel H, Teipel SJ, Alexander GE, Pogarell O, Rapoport SI and Möller H. In vivo imaging of region and cell type specific neocortical neurodegeneration in Alzheimer's disease. Perspectives of MRI derived corpus callosum measurement for mapping disease progression and effects of therapy. Evidence from studies with MRI, EEG and PET. *J Neural Transm.* 2002; 109: 837-855.

[3] Cowan CM and Raymond LA. Selective neuronal degeneration in Huntington's disease. *Curr Top Dev Biol.* 2006; 75: 25-71.

[4] Orr HT and Zoghbi HY. Trinucleotide Repeat Disorders. *Annual Review of Neuroscience,* 2007; 30: 575 621.

[5] Gasser T. Mendelian forms of Parkinson's disease. *Biochim Biophys Acta,* 2009; 1792: 587-596.

[6] Liberski PP and Ironside JW. An outline of the neuropathology of transmissible spongiform encephalopathies (prion diseases). *Folia Neuropathol.* 2004; 42 Suppl B: 39-58

[7] Cunningham C, Deacon R, Wells H, Boche D, Waters S, Diniz CP, Scott H, Rawlins JN and Perry VH. Synaptic changes characterize early behavioural signs in the ME7 model of murine prion disease. *Eur J Neurosci.* 2003; 17: 2147-2155.

[8] Thind K and Sabbagh MN. Pathological correlates of cognitive decline in Alzheimer's disease. *Panminerva Med.* 2007; 49: 191-195.

[9] Goldbaum O and Richter-Landsberg C. Stress proteins in oligodendrocytes: differential effects of heat shock and oxidative stress. *J Neurochem.* 2001; 78: 1233-1242.

[10] van Noort JM. Stress proteins in CNS inflammation. J Pathol. 2008; 214: 267-275.

[11] Haslbeck M, Franzmann T, Weinfurtner D and Buchner J. Some like it hot: the structure and function of small heat-shock proteins. *Nat Struct Mol Biol.* 2005; 12: 842-846.

[12] Der Perng M and Quinlan R. Neuronal diseases: small heat shock proteins calm your nerves. *Curr Biol.* 2004; 14: R625-626.

[13] Aizenstein H, Nebes R, Saxton J, Price J, Mathis C, Tsopelas N, Ziolko S, James J, Snitz B, Houck P, Bi W, Cohen A, Lopresti B, DeKosky S, Halligan E and Klunk W. Frequent amyloid deposition without significant cognitive impairment among the elderly. *Arch Neurol.* 2008; 65: 1509-1517.

[14] Shankar G, Li S, Mehta T, Garcia-Munoz A, Shepardson N, Smith I, Brett F, Farrell M, Rowan M, Lemere C, Regan C, Walsh D, Sabatini B and Selkoe D. Amyloid-beta protein dimers isolated directly from Alzheimer's brains impair synaptic plasticity and memory. *Nat Med.* 2008; 14: 837-842.

[15] Ohyagi Y. Intracellular Amyloid β-Protein As a Therapeutic *Target for Treating Alzheimer's Disease Current Alzheimer Research,* 2008; 5: 555-561

[16] Adalbert, Gilley J and Coleman M. Abeta, tau and ApoE4 in Alzheimer's disease: the axonal connection. *Trends Mol Med.* 2007; 13: 135-142.

[17] Shinohara H, Inaguma Y, Goto S, Inagaki T and K. K. Alpha B crystallin and HSP28 are enhanced in the cerebral cortex of patients with Alzheimer's disease. *J Neurol Sci.* 1993; 119: 203-208

[18] Lowe J, McDermott H, Pike I, Spendlove I, Landon M and Mayer RJ. alpha B crystallin expression in non-lenticular tissues and selective presence in ubiquitinated inclusion bodies in human disease. *J Pathol.* 1992; 166

[19] Renkawek K, Voorter CE, Bosman GJ, van Workum FP and de Jong WW. Expression of alpha B-crystallin in Alzheimer's disease. *Acta Neuropathol.* (Berl). 1994; 87: 155

[20] Wilhelmus MM, Otte-Höller I, Wesseling P, de Waal RM, Boelens WC and MM V. Specific association of small heat shock proteins with the pathological hallmarks of Alzheimer's disease brains. *Neuropathol Appl Neurobiol.* 2006; 32: 119-130

[21] Raman B, Ban T, Sakai M, Pasta SY, Ramakrishna T, Naiki H, Goto Y and ChM. R. AlphaB-crystallin, a small heat-shock protein, prevents the amyloid fibril growth of an amyloid beta-peptide and beta2-microglobulin. *Biochem J.* 2005; 392

[22] Sherman M and G. M. Heat shock proteins in cancer. *Ann N Y Acad Sci.* 2007; 1113: 192-201.

[23] Ousman SS, Tomooka BH, van Noort JM, Wawrousek EF, O'Connor KC, Hafler DA, Sobel RA, Robinson WH and L. S. Protective and therapeutic role for alphaB-crystallin in autoimmune demyelination. *Nature,.* 2007; 448: 474-479

[24] Iwaki T, Wisniewski T, Iwaki A, Corbin E, Tomokane N, Tateishi J and Goldman JE. Accumulation of alpha B-crystallin in central nervous system glia and neurons in pathologic conditions. *Am J Pathol.* 1992; 140: 345-356
[25] Dabir DV, Trojanowski JQ, Richter-Landsberg C, Lee VM and Forman MS. Expression of the small heat-shock protein alphaB-crystallin in tauopathies with glial pathology. *Am.J.Pathol.* 2004; 164: 155
[26] Wilhelmus MM, Boelens WC, Otte-Holler I, Kamps B, de Waal RM and Verbeek MM. Small heat shock proteins inhibit amyloid-beta protein aggregation and cerebrovascular amyloid-beta protein toxicity. *Brain Res.* 2006; 1089: 67
[27] Le Pichon CE, Valley MT, Polymenidou M, Chesler AT, Sagdullaev BT, Aguzzi A and Firestein S. Olfactory behavior and physiology are disrupted in prion protein knockout mice. *Nature Neruoscience,* 2009; 12: 60-69
[28] Ironside JW. Pathology of variant Creutzfeldt-Jakob disease. *Arch Virol Suppl.* 2000; 16
[29] Silveira JR, Raymond GJ, Hughson AG, Race RE, Sim VL, Hayes SF and Caughey B. The most infectious prion protein particles. *Nature*, 2005; 437: 257-261
[30] Tatzelt J, Voellmy R and Welch W. Abnormalities in stress proteins in prion diseases. *Cell Mol Neurobiol.* 1998; 18: 721-729
[31] Renkawek K, De Jong WW, Merck KB, Frenken CW, Van Workum FP and Bosman GJ. alpha B-crystallin is present in reactive glia in Creutzfeldt-Jakob disease. *Acta Neuropathol.* 1992; 83: 324-327.
[32] Hwang D, Lee IY, Yoo H, Gehlenborg N, Cho JH, Petritis B, Baxter D, Pitstick R, Young R, Spicer D, Price ND, Hohmann JG, Dearmond SJ, Carlson GA and Hood LE. A systems approach to prion disease. *Mol Syst Biol.* 2009; 5: 252.
[33] O'Connor V, Quraishe S, Asuni A, Lunnon K, Baxter M, Brown J, Perry VH and Wyttenbach A. A co-ordinated and selective small heat shock protein response is associated with the ME7 model of prion disease. Neuroprion Conference. 2008; Madrid.
[34] Robinson PA. Protein stability and aggregation in Parkinson's disease. *Biochem J.* 2008; 413: 1-13.
[35] Spillantini MG, Schmidt ML, Lee VM, Trojanowski JQ, Jakes R and Goedert M. Alpha-synuclein in Lewy bodies. *Nature,* 1997; 388: 839-840
[36] Henchcliffe C and Beal MF. Mitochondrial biology and oxidative stress in Parkinson disease pathogenesis. *Nat Clin Pract Neurol.* 2008; 4: 600-609
[37] Zhang Y, James M, Middleton FA and Davis RL. Transcriptional analysis of multiple brain regions in Parkinson's disease supports the involvement of specific protein processing, energy metabolism, and signaling pathways, and suggests novel disease mechanisms. *Am J Med Genet B Neuropsychiatr Genet.* 2005; 137B: 5-16.
[38] Renkawek K, Stege GJ and Bosman GJ. Dementia, gliosis and expression of the small heat shock proteins hsp27 and alpha B-crystallin in Parkinson's disease. *Neuroreport.* 1999; 10: 2273-2276
[39] Outeiro TF, Klucken J, Strathearn KE, Liu F, Nguyen P, Rochet J-C, Hyman BT and McLean PJ. Small heat shock proteins protect against [alpha]-synuclein-induced toxicity and aggregation. *Biochemical and Biophysical Research Communications,* 2006; 351: 631-638

[40] Mirza B, Hadberg H, Thomsen P and Moos T. The absence of reactive astrocytosis is indicative of a unique inflammatory process in Parkinson's disease. *Neuroscience*, 2000; 95: 425-432

[41] Braak H, Del Tredici K, Sandmann-Kiel D, Rüb U and Schultz C. Nerve cells expressing heat-shock proteins in Parkinson's disease. *Acta Neuropathol.* (Berl). 2001; 102: 449-454

[42] Rekas A, Adda CG, Andrew Aquilina J, Barnham KJ, Sunde M, Galatis D, Williamson NA, Masters CL, Anders RF, Robinson CV, Cappai R and Carver JA. Interaction of the molecular chaperone alphaB-crystallin with alpha-synuclein: effects on amyloid fibril formation and chaperone activity. J Mol Biol. 2004; 340: 1167-1183

[43] Wang J, Martin E, Gonzales V, Borchelt DR and Lee MK. Differential regulation of small heat shock proteins in transgenic mouse models of neurodegenerative diseases. *Neurobiology of Ageing,* 2008; 29: 586-597.

[44] 44Hands S, Sinadinos C and Wyttenbach A. Polyglutamine gene function and dysfunction in the ageing brain. Biochimica et Biophysica Acta (BBA) - *Gene Regulatory Mechanisms,* 2008; 1779: 507-521

[45] Wyttenbach A. Role of Heat Shock Proteins During Polyglutamine Neurodegeneration: Mechanisms and Hypothesis. Journal of Molecular Neuroscience. 2004; 23: 69-96.

[46] Wyttenbach A, Arrigo AP and Richter-Landsberg C. (2006) Role of Heat Shock Proteins during Neurodegeneration in Alzheimer's, Parkinson's and Huntington's Disease. In Heat Shock Proteins in Neural Cells, Eurekah.com.

[47] Tsai HF, Lin SJ, Li C and Hsieh M. Decreased expression of Hsp27 and Hsp70 in transformed lymphoblastoid cells from patients with spinocerebellar ataxia type 7. *Biochem Biophys Res Commun.* 2005; 334: 1279-1286.

[48] Chang WH, Cemal CK, Hsu YH, Kuo CL, Nukina N, Chang MH, Hu HT, Li C and Hsieh M. Dynamic expression of Hsp27 in the presence of mutant ataxin-3. *Biochem Biophys Res Commun.* 2005; 336: 258-267

[49] Hay DG, Sathasivam K, Tobaben S, Stahl B, Marber M, Mestril R, Mahal A, Smith DL, Woodman B and Bates GP. Progressive decrease in chaperone protein levels in a mouse model of Huntington's disease and induction of stress proteins as a therapeutic approach. *Hum.Mol.Genet.* 2004; 13: 1389

[50] Zabel C, Chamrad DC, Priller J, Woodman B, Meyer HE, Bates GP and Klose J. Alterations in the mouse and human proteome caused by Huntington's disease. *Mol.Cell Proteomics,* 2002; 1: 366

[51] Wang J, Martin E, Gonzales V, Borchelt DR and Lee MK. Differential regulation of small heat shock proteins in transgenic mouse models of neurodegenerative diseases. *Neurobiol Aging,* 2008; 29: 586-597

[52] Friedman MJ, Shah AG, Fang ZH, Ward EG, Warren ST, Li S and Li XJ. Polyglutamine domain modulates the TBP-TFIIB interaction: implications for its normal function and neurodegeneration. *Nat Neurosci.* 2007; 10: 1519-1528.

[53] Wyttenbach A, Sauvageot O, Carmichael J, Diaz-Latoud C, Arrigo AP and Rubinsztein DC. Heat shock protein 27 prevents cellular polyglutamine toxicity and suppresses the increase of reactive oxygen species caused by huntingtin. *Human Molecular Genetics*, 2002; 11: 1137-1151

[54] Laity JH and Andrews GK. Understanding the mechanisms of zinc-sensing by metal-response element binding transcription factor-1 (MTF-1). *Archives of Biochemistry and Biophysics*, 2007; 463: 201-210.

[55] Hidalgo J, Aschner M, Zatta P and Vasák M. Roles of the metallothionein family of proteins in the central nervous system. *Brain Research Bulletin,* 2001; 55: 133-145

[56] Hidalgo J, Penkowa M, Espejo C, Martinez-Caceres EM, Carrasco J, Quintana A, Molinero A, Florit S, Giralt M and Ortega-Aznar A. Expression of Metallothionein-I, -II, and -III in Alzheimer Disease and Animal Models of Neuroinflammation. *Experimental Biology and Medicine,* 2006; 231: 1450-1458.

[57] Zambenedetti P, Giordano R and Zatta P. Metallothioneins are highly expressed in astrocytes and microcapillaries in Alzheimer's disease. *Journal of Chemical Neuroanatomy,* 1998; 15: 21-26.

[58] Miyazaki I, Asanuma M, Higashi Y, Sogawa CA, Tanaka K-i and Ogawa N. Age-related changes in expression of metallothionein-III in rat brain. *Neuroscience Research,* 2002; 43: 323-333

[59] Yu WH, Lukiw WJ, Bergeron C, Niznik HB and Fraser PE. Metallothionein III is reduced in Alzheimer's disease. *Brain Research,* 2001; 894: 37-45

[60] Colangelo V, Schurr J, Ball MJ, Pelaez RP, Bazan NG and Lukiw WJ. Gene expression profiling of 12633 genes in Alzheimer hippocampal CA1: Transcription and neurotrophic factor down-regulation and up-regulation of apoptotic and pro-inflammatory signaling. *Journal Of Neuroscience Research,* 2002; 70: 462-473

[61] Tsuji S, Kobayashi H, Uchida Y, Ihara Y and Miyatake T. Molecular cloning of human growth inhibitory factor cDNA and its down-regulation in Alzheimer's disease. *EMBO J.* 1992; 11: 4843–4850

[62] Shiraga H, Pfeiffer RF and Ebadi M. The effects of 6-hydroxydopamine and oxidative stress on the level of brain metallothionein. *Neurochem Int.* 1993; 23: 561-566

[63] Rojas P, Hidalgo J, Ebadi M and Rios C. Changes of Metallothionein I + II Proteins in the Brain after 1-Methyl-4-Phenylpyridinium Administration in Mice. *Progress in Neuro-Psychopharmacology and Biological Psychiatry,* 2000; 24: 143-154

[64] Hodges A, Strand AD, Aragaki AK, Kuhn A, Sengstag T, Hughes G, Elliston LA, Hartog C, Goldstein DR, Thu D, Hollingsworth ZR, Collin F, Synek B, Holmans PA, Young AB, Wexler NS, Delorenzi M, Kooperberg C, Augood SJ, Faull RLM, Olson JM, Jones L and Luthi-Carter R. Regional and cellular gene expression changes in human Huntington's disease brain. *Hum. Mol. Genet.* 2006; 15: 965-977.

[65] Wyttenbach A, Swartz J, Kita H, Thykjaer T, Carmichael J, Bradley J, Brown R, Maxwell M, Schapira A, Orntoft TF, Kato K and Rubinsztein DC. Polyglutamine expansions cause decreased CRE-mediated transcription and early gene expression changes prior to cell death in an inducible cell model of Huntington's disease. *Human Molecular Genetics*, 2001; 10: 1829-1845.

[66] Duguid JR, Bohmont CW, Liu NG and Tourtellotte WW. Changes in brain gene expression shared by scrapie and Alzheimer disease. *Proc Natl Acad Sci U S A.* 1989; 86: 7260–7264

[67] Hanlon J, Monks E, Hughes C, Weavers E and Rogers M. Metallothionein in Bovine Spongiform Encephalopathy. *Journal of Comparative Pathology,* 2002; 127: 280-289

[68] Kawashima T, Doh-ura K, Torisu M, Uchida Y, Furuta A and Iwaki T. Differential Expression of Metallothioneins in Human Prion Diseases. *Dement Geriatr Cogn Disord.* 2000; 11: 251-262.
[69] Halliwell B. Oxidative stress and neurodegeneration: where are we now? *Journal of Neurochemistry,* 2006; 97: 1634-1658
[70] Erickson JC, Sewell AK, Jensen LT, Winge DR and Palmiter RD. Enhanced neurotrophic activity in Alzheimer's disease cortex is not associated with down-regulation of metallothionein-III (GIF). *Brain Research*, 1994; 649: 297-304
[71] Meloni G, Sonois V, Delaine T, Guilloreau L, Gillet A, Teissie J, Faller P and Vasak M. Metal swap between Zn7-metallothionein-3 and amyloid-[beta]-Cu protects against amyloid-[beta] toxicity. *Nat Chem Biol.* 2008; 4: 366-372
[72] Tortosa R, Vidal E, Costa C, Alamillo E, Torres JM, Ferrer I and Pumarola M. Stress response in the central nervous system of a transgenic mouse model of bovine spongiform encephalopathy. *The Veterinary Journal,* 2008; 178: 126-129.
[73] Vidal E, Tortosa R, Márquez M, Serafin A, Hidalgo J and Pumarola M. Infection of metallothionein 1 + 2 knockout mice with Rocky Mountain Laboratory scrapie. *Brain Research,* 2008; 1196: 140-150
[74] Ebadi M, Brown-Borg H, El Refaey H, Singh BB, Garrett S, Shavali S and Sharma SK. Metallothionein-mediated neuroprotection in genetically engineered mouse models of Parkinson's disease. *Molecular Brain Research,* 2005; 134: 67-75
[75] Sharma SK and Ebadi M. Metallothionein Attenuates 3-Morpholinosydnonimine (SIN-1)-Induced Oxidative Stress in Dopaminergic Neurons. *Antioxidants & Redox Signaling*, 2003; 5: 251-264
[76] Ebadi M and Sharma S. Metallothioneins 1 and 2 Attenuate Peroxynitrite-Induced Oxidative Stress in Parkinson Disease. *Experimental Biology and Medicine*, 2006; 231: 1576-1583
[77] Hwang YP, Kim HG, Han EH and Jeong HG. Metallothionein-III protects against 6-hydroxydopamine-induced oxidative stress by increasing expression of heme oxygenase-1 in a PI3K and ERK/Nrf2-dependent manner. *Toxicology and Applied Pharmacology,* 2008; 231: 318-327
[78] Chung RS, Penkowa M, Dittmann J, King CE, Bartlett C, Asmussen JW, Hidalgo J, Carrasco J, Leung YKJ, Walker AK, Fung SJ, Dunlop SA, Fitzgerald M, Beazley LD, Chuah MI, Vickers JC and West AK. Redefining the Role of Metallothionein within the Injured Brain: Extracellular metallothioneins play an important role in the astroctyte-neuron response to injury. *J. Biol. Chem.* 2008; 283: 15349-15358
[79] Gauthier MA, Eibl JK, Crispo JA and Ross GM. Covalent arylation of metallothionein by oxidized dopamine products: a possible mechanism for zinc-mediated enhancement of dopaminergic neuron survival. *Neurotox Res.* 2008; 14: 317-328
[80] Chung RS, Vickers JC, Chuah MI and West AK. Metallothionein-IIA Promotes Initial Neurite Elongation and Postinjury Reactive Neurite Growth and Facilitates Healing after Focal Cortical Brain Injury. *J. Neurosci.* 2003; 23: 3336-3342
[81] Køhler LB, Berezin V, Bock E and Penkowa M. The role of metallothionein II in neuronal differentiation and survival. *Brain Research*, 2003; 992: 128-136
[82] Uchida Y and Ihara Y. The N-terminal Portion of Growth Inhibitory Factor Is Sufficient for Biological Activity. *J. Biol. Chem.* 1995; 270: 3365-3369

[83] Bush A and Tanzi RE. Therapeutics for Alzheimer's disease based on the metal hypothesis. *Neurotherapeutics*, 2008; 5: 421-432
[84] Hyman M. Schipper WS, Hillel Zukor, Jacob R. Hascalovici, David Zeligman,. Heme Oxygenase-1 and Neurodegeneration: Expanding Frontiers of Engagement. *Journal of Neurochemistry*,. 2009; 9999
[85] Dennery PA. Regulation and role of heme oxygenase in oxidative injury. *Curr Top Cell Regul.* 2000; 36: 181-199
[86] Smith MA, Kutty RK, Richey PL, Yan SD, Stern D, Chader GJ, Wiggert B, Petersen RB and Perry G. Heme oxygenase-1 is associated with the neurofibrillary pathology of Alzheimer's disease. *Am J Pathol.* 1994; 145: 42-47
[87] Schipper HM, Cissé S and Stopa EG. Expression of heme oxygenase-1 in the senescent and Alzheimer-diseased brain. Ann Neurol. 1995; 37: 758-768
[88] Smith M, Keisuke Hirai, Karen Hsiao, Miguel A. Pappolla, Peggy L. R. Harris, Sandra L. Siedlak, Massimo Tabaton and George Perry. Amyloid-beta Deposition in Alzheimer Transgenic Mice Is Associated with Oxidative Stress. Journal of Neurochemistry. 1998; 70: 2212-2215
[89] Song W, Patel A, Dong H, Paudel HK and Schipper HM. Heme oxygenase-1 promotes proteosomal degradation of tau and α-synuclein in human neuroblastoma cells. Alz Demen. 2008; 4: T410
[90] Takahashi M, Doré S, Ferris CD, Tomita T, Sawa A, Wolosker H, Borchelt DR, Iwatsubo T, Kim S-H, Thinakaran G, Sisodia SS and Snyder SH. Amyloid Precursor Proteins Inhibit Heme Oxygenase Activity and Augment Neurotoxicity in Alzheimer's Disease. *Neuron.* 2000; 28: 461-473.
[91] Rizzardini M, Chiesa R, Angeretti N, Lucca E, Salmona M and Forloni G. Prion Protein Fragment 106-126 Differentially Induces Heme Oxygenase-1 mRNA in Cultured Neurons and Astroglial Cells. *Journal of Neurochemistry,* 1997; 68: 715-720
[92] Yun S-W, Gerlach M, Riederer P and Klein MA. Oxidative stress in the brain at early preclinical stages of mouse scrapie. *Experimental Neurology,* 2006; 201: 90-98
[93] Choi Y-G, Kim J-I, Lee H-P, Jin J-K, Choi E-K, Carp RI and Kim Y-S. Induction of heme oxygenase-1 in the brains of scrapie-infected mice. *Neuroscience Letters,* 2000; 289: 173-176.
[94] Guentchev M, Voigtländer T, Haberler C, Groschup MH and Budka H. Evidence for Oxidative Stress in Experimental Prion Disease. *Neurobiology Of Disease,* 2000; 7: 270-273.
[95] Petersen R, Siedlak S, Lee H-g, Kim Y-S, Nunomura A, Tagliavini F, Ghetti B, Cras P, Moreira P, Castellani R, Guentchev M, Budka H, Ironside J, Gambetti P, Smith M and Perry G. Redox metals and oxidative abnormalities in human prion diseases. *Acta Neuropathologica*, 2005; 110: 232-238.
[96] Schipper HM, Liberman A and Stopa EG. Neural Heme Oxygenase-1 Expression in Idiopathic Parkinson's Disease. *Experimental Neurology*, 1998; 150: 60-68
[97] Hung S-Y, Liou H-C, Kang K-H, Wu R-M, Wen C-C and Fu W-M. Overexpression of Heme Oxygenase-1 Protects Dopaminergic Neurons against 1-Methyl-4-Phenylpyridinium-Induced Neurotoxicity. *Mol Pharmacol.* 2008; 74: 1564-1575.
[98] Song W, Patel A, Qureshi HY, Han D, Schipper HM and Paudel HK. The Parkinson disease-associated A30P mutation stabilizes alpha-synuclein against proteasomal

degradation triggered by heme oxygenase-1 over-expression in human neuroblastoma cells. Journal of Neurochemistry. 2009; May 13 Epub ahead of print.

[99] Dwyer BE, Nishimura RN and Lu S-Y. Differential expression of heme oxygenase-1 in cultured cortical neurons and astrocytes determined by the aid of a new heme oxygenase antibody. Response to oxidative stress. *Molecular Brain Research*, 1995; 30: 37-47.

[100] Hirose W, Ikematsu K and Tsuda R. Age-associated increases in heme oxygenase-1 and ferritin immunoreactivity in the autopsied brain. *Legal Medicine,* 2003; 5: S360-S366.

[101] Kadoya C, Domino EF, Yang G-Y, Stern JD and Betz AL. Preischemic But Not Postischemic Zinc Protoporphyrin Treatment Reduces Infarct Size and Edema Accumulation After Temporary Focal Cerebral Ischemia in Rats. *Stroke,* 1995; 26: 1035-1038.

[102] Koeppen AH, Dickson AC and Smith J. Heme oxygenase in experimental intracerebral hemorrhage: the benefit of tin-mesoporphyrin. *J Neuropathol Exp Neurol.* 2004; 63: 587-597

[103] Lee DW, Gelein RM and Opanashuk LA. Heme-Oxygenase-1 Promotes Polychlorinated Biphenyl Mixture Aroclor 1254-Induced Oxidative Stress and Dopaminergic Cell Injury. *Toxicol. Sci.* 2006; 90: 159-167.

[104] Murray CJ and Lopez A. Mortality by cause for eight regions of the world: Global Burden of Disease Study. *LANCET*, 1997; 349: 1269-1276.

[105] Durukan A and Tatlisumak T. Acute ischemic stroke: overview of major experimental rodent models, pathophysiology, and therapy of focal cerebral ischemia. *Pharmacol Biochem Behav*. 2007; 87: 179-197.

[106] Dienel GA, Kiessling M, Jacewicz M and Pulsinelli WA. Synthesis of heat shock proteins in rat brain cortex after transient ischemia. *J Cereb Blood Flow Metab.* 1986; 6: 505-510.

[107] Minami M, Mizutani T, Kawanishi R, Suzuki Y and Mori H. Neuronal expression of alphaB crystallin in cerebral infarction. *Acta Neuropathol.* (Berl). 2003; 105: 549-554

[108] Piao CS, Kim SW, Kim JB and Lee JK. Co-induction of alphaB-crystallin and MAPKAPK-2 in astrocytes in the penumbra after transient focal cerebral ischemia. *Exp Brain Res.* 2005; 163: 421-429.

[109] Kato H, Kogure K, Liu XH, Araki T, Kato K and Y. I. Immunohistochemical localization of the low molecular weight stress protein HSP27 following focal cerebral ischemia in the rat. *Brain Res*. 1995; 679

[110] Currie RW, Ellison JA, White RF, Feuerstein GZ, Wang X and Barone FC. Benign focal ischemic preconditioning induces neuronal Hsp70 and prolonged astrogliosis with expression of Hsp27. *Brain Res.* 2000; 863: 169-168

[111] Villapol S, Acarin L, Faiz M, Castellano B and Gonzalez B. Survivin and heat shock protein 25/27 colocalize with cleaved caspase-3 in surviving reactive astrocytes following excitotoxicity to the immature brain. *Neuroscience,* 2008; 153: 108-119

[112] Valentim LM, Rodnight R, Geyer AB, Horn AP, Tavares A, Cimarosti H, Netto CA and Salbego CG. Changes in heat shock protein 27 phosphorylation and immunocontent in response to preconditioning to oxygen and glucose deprivation in organotypic hippocampal cultures. *Neuroscience,* 2003; 118: 379-378.

[113] Acarin L, Paris J, González B and Castellano B. Glial expression of small heat shock proteins following an excitotoxic lesion in the immature rat brain. *Glia.* 2002; 38: 1-14

[114] Sanz O, Acarin L, González B and Castellano B. Expression of 27 kDa heat shock protein (Hsp27) in immature rat brain after a cortical aspiration lesion. *Glia.* 2001; 36: 259-270.

[115] Niwa M, Hara A, Taguchi A, Aoki H, Kozawa O and Mori H. Spatiotemporal expression of Hsp20 and its phosphorylation in hippocampal CA1 pyramidal neurons after transient forebrain ischemia. *Neurol Res.* 2008.

[116] Lu A, Ran R, Parmentier-Batteur S, Nee A and Sharp FR. Geldanamycin induces heat shock proteins in brain and protects against focal cerebral ischemia. *J Neurochem.* 2002; 81: 355-364.

[117] Lu A, Ran RQ, Clark J, Reilly M, Nee A and Sharp FR. 17-beta-estradiol induces heat shock proteins in brain arteries and potentiates ischemic heat shock protein induction in glia and neurons. *J Cereb Blood Flow Metab.* 2002; 22: 183-195.

[118] Badin RA, Modo M, Cheetham M, Thomas DL, Gadian DG, Latchman DS and Lythgoe MF. Protective effect of post-ischaemic viral delivery of heat shock proteins in vivo. *J Cereb Blood Flow Metab.* 2009; 29: 254-263.

[119] Reynolds LP and Allen GV. A review of heat shock protein induction following cerebellar injury. *Cerebellum,* 2003; 2: 171-177

[120] Akbar MT, Lundberg AM, Liu K, Vidyadaran S, Wells KE, Dolatshad H, Wynn S, Wells DJ, Latchman DS and de Belleroche J. The neuroprotective effects of heat shock protein 27 overexpression in transgenic animals against kainate-induced seizures and hippocampal cell death. *J Biol Chem.* 2003; 278: 19956-19965.

[121] Stetler RA, Cao G, Gao Y, Zhang F, Wang S, Weng Z, Vosler P, Zhang L, Signore A, Graham SH and Chen J. Hsp27 protects against ischemic brain injury via attenuation of a novel stress-response cascade upstream of mitochondrial cell death signaling. *J Neurosci.* 2008; 28: 13038-13055.

[122] An JJ, Lee YP, Kim SY, Lee SH, Lee MJ, Jeong MS, Kim DW, Jang SH, Yoo KY, Won MH, Kang TC, Kwon OS, Cho SW, Lee KS, Park J, Eum WS and Choi SY. Transduced human PEP-1-heat shock protein 27 efficiently protects against brain ischemic insult. *FEBS Journal,* 2008; 275: 1296-1308.

[123] Willis D, Li KW, Zheng JQ, Chang JH, Smit A, Kelly T, Merianda TT, Sylvester J, van Minnen J and Twiss JL. Differential transport and local translation of cytoskeletal, injury-response, and neurodegeneration protein mRNAs in axons. *J Neurosci.* 2005; 25: 778-791

[124] Neal JW, Singhrao SK, Jasani B and Newman GR. Immunocytochemically detectable metallothionein is expressed by astrocytes in the ischaemic human brain. *Neuropathol Appl Neurobiol.* 1996; 22: 243-247.

[125] Yousuf S, Atif F, Ahmad M, Hoda N, Ishrat T, Khan B and Islam F. Resveratrol exerts its neuroprotective effect by modulating mitochondrial dysfunctions and associated cell death during cerebral ischemia. *Brain Res.* 2009; 1250: 242-253

[126] Wakida K, Shimazawa M, Hozumi I, Satoh M, Nagase H, Inuzuka T and Hara H. Neuroprotective effect of erythropoietin, and role of metallothionein-1 and -2, in permanent focal cerebral ischemia. *Neuroscience,* 2007; 148: 105-114

[127] Fukuda K, Panter SS, Sharp FR and Noble LJ. Induction of heme oxygenase-1 (HO-1) after traumatic brain injury in the rat. *Neurosci Lett.* 1995; 199: 127-130

[128] Beschorner R, Adjodah D, Schwab JM, Mittelbronn M, Pedal I, Mattern R, Schluesener HJ and Meyermann R. Long-term expression of heme oxygenase-1 (HO-1, HSP-32)

following focal cerebral infarctions and traumatic brain injury in humans. *Acta Neuropathol.* 2000; 100: 377-384

[129] Panahian N, Yoshiura M and Maines MD. Overexpression of heme oxygenase-1 is neuroprotective in a model of permanent middle cerebral artery occlusion in transgenic mice. *J Neurochem.* 1999; 72: 1187-1203

[130] Yang C, Zhang X, Fan H and Liu Y. Curcumin upregulates transcription factor Nrf2, HO-1 expression and protects rat brains against focal ischemia. Brain Res. 2009; May 13 Epub ahead of print

[131] Saleem S, Zhuang H, Biswal S, Christen Y and Doré S. Ginkgo biloba extract neuroprotective action is dependent on heme oxygenase 1 in ischemic reperfusion brain injury. *Stroke*, 2008; 39: 3389-3396

[132] Nakaso K, Kitayama M, Fukuda H, Kimura K, Yanagawa T, Ishii T, Nakashima K and Yamada K. Oxidative stress-related proteins A170 and heme oxygenase-1 are differently induced in the rat cerebellum under kainate-mediated excitotoxicity. *Neurosci Lett.* 2000; 282: 57-60.

[133] Imuta N, Hori O, Kitao Y, Tabata Y, Yoshimoto T, Matsuyama T and Ogawa S. Hypoxia-mediated induction of heme oxygenase type I and carbon monoxide release from astrocytes protects nearby cerebral neurons from hypoxia-mediated apoptosis. *Antioxid Redox Signal,* 2007; 9: 543-552.

In: Small Stress Proteins and Human Diseases
Editors: Stéphanie Simon et al.
ISBN: 978-1-61470-636-6

Chapter 1.5.

SMALL STRESS PROTEINS IN PROTECTING THE HEART

Guo-Chang Fan [*] ***and Evangelia G. Kranias***
Department of Pharmacology and Cell Biophysics, University of Cincinnati College of Medicine, Cincinnati, OH 45267, U.S.

ABSTRACT

Mammalian small heat shock proteins (sHsps), including at least ten members of HspB1-B10, are constitutively expressed during development and are believed to be important in cellular processes that include cell morphogenesis, inflammatory response, cell proliferation, mitogenesis, platelet function, and cell differentiation. They make up a class of molecular chaperones with molecular masses of <43 kDa, contain a highly conserved immunoglobulin-like structural fold called the α-crystallin core domain, and are upregulated in response to cellular stress. Recently, increased evidence has shown the potential importance of sHsps in the regulation of key components of signaling pathways under stress and nonstress conditions, as well as in response to stress associated with myocardial disease. sHsps have been demonstrated to affect the development of cardiac disease and over-expression of sHsps results in cardioprotection against physiological or patho/physiological stresses. This chapter will focus on reviewing the protective effects of sHsp in the heart and their underlying mechanisms.

---

[*] Corresponding Author: Guo-Chang Fan
University of Cincinnati
College of Medicine
Department of Pharmacology and Cell Biophysics
Cincinnati, OH 45267-0575. USA
Phone: (513) 558-2340.
Fax: (513) 558- 2269.
fangg@ucmail.uc.edu

INTRODUCTION

A multitude of physiologic stresses and diseases result in protein damage and misfolded protein structure, leading to cellular injury and dysfunction. The cell has evolved a number of protective measures to promote survival during these periods of environmental stress and disease. One of the most highly conserved mechanisms of cellular protection involves the expression of a polypeptide family known as heat shock or stress proteins (Hsps).

Stress proteins belong to multigene families that range in molecular size from 10 to 105 kDa and are found in all major cellular compartments. They are identified by their molecular weights (e.g., Hsp110, Hsp90, Hsp70, and Hsp27) [1, 2]. The small heat shock protein (sHsp) family, with molecular weight ranging from 12 to 43 kDa, is characterized by a highly conserved C-terminal domain of about 90 amino acids, the "α-crystallin domain", and a variable N-terminal domain, which appears to be responsible for the significant variations of the sizes of the oligomers [3-7]. The human sHsp family is comprised of at least 10 known members [8]. Some of them are ubiquitously expressed such as HspB1, HspB5, HspB6, and HspB8; the others are expressed in a tissue-restricted manner like HspB2, HspB3, HspB4, HspB7, HspB9, and HspB10 [9]. It is well-documented that sHsps promote cell survival in otherwise lethal conditions.

Various mechanisms have been proposed to account for this cytoprotective effect [10]. The protective mechanism relies partly on the molecular chaperone function of sHsps, which prevents the aggregation of nascent and stress-accumulated misfolded proteins [11-15]. On the other hand, sHsps interact directly with various components of the tightly regulated programmed cell death machinery, upstream and downstream of the mitochondrial events [16-20]. Therefore, sHsps have been suggested to affect the development of cardiac disease and over-expression of sHsps may result in cardioprotection against physiological or patho/physiological stresses (see also chapter 1-6). This chapter will focus on reviewing the protective effects of sHsp in the heart and their underlying mechanisms.

1. CARDIOPROTECTIVE EFFECTS OF HSPB1 (HSP25/27)

Among the best characterized members of the sHsp family, HspB1, also referred to as the Estrogen-Regulated 24K protein and Hsp27 (mouse Hsp25), is one of several sHsps expressed constitutively and ubiquitously in most cells types, including cardiac and skeletal muscle [21-24]. The expression of HspB1 can be activated in diverse cells and tissues by various stressors such as heavy metals, oxidative stress, hormones, hypoxia, and ischemia [22, 25, 26]. In turn, such maneuvers for up-regulation of HspB1 are likely to confer thermo-tolerance and cross-tolerance under stressful conditions [26, 27].

1.1. Cardiac Protective Roles of HspB1 are Associated with Its Endogenous Levels

It has been reported that HspB1 and other sHsps are differentially regulated during heart development [21, 23]. In the rat, the levels of HspB1 are much lower (≈6%) in the adult hearts (0.2 to 0.3 μg Hsp25/mg protein), compared to those in early fetal stages (3 to 4 μg Hsp25/mg protein) [24]. However, in the adult human failing and transplanted hearts, the amounts of Hsp25 were found to be 5-10-fold higher than in fetal human hearts [24].

Interestingly, in neonatal cardiac myocytes, adenoviral-mediated Hsp27 transgene expression did not result in a protective effect against ischemic injury; this is most likely due to the already high endogenous level of Hsp25 protein in neonatal cardiac myocytes, levels that markedly exceed those found in adult cardiac myocytes [28]. In line with the lower level of Hsp25 in adult cardiac myocytes, increasing Hsp27 levels through adenoviral-based transgene expression in adult cardiac myocytes resulted in increased protection against ischemic injury [29-33]. On the other hand, the importance of high Hsp25 levels in neonatal cardiac myocytes for normal cell function and integrity is further emphasized by the results from Hsp27 antisense expression studies [28]. Infection of neonatal cardiac myocytes with adenoviral vectors expressing antisense Hsp27 resulted in a marked decrease of Hsp25 levels, which led to enhanced ischemic injury [28]. Accordingly, in adult cardiomyocytes, Hsp27 antisense expression did not lower Hsp25 protein levels and ischemic injury was unchanged, again emphasizing the close correlation between Hsp25/27 protein levels and the extent of ischemic injury [28].

Recently, Lu et al. [33] using adenoviral-mediated Hsp27 gene transfer into adult rat heart revealed that 3.5-fold increase in Hsp27 level *in vivo* improved post-ischemic cardiac performance in isolated perfused hearts, subjected to no-flow global ischemia/reperfusion (I/R: 30-min/30-min); such improvement of cell contractions was associated with myofilament Ca^{2+} responsiveness but not the SR Ca transient. More importantly, Hsp27 can interact with cTnI at the COOH-terminus and cTnT at the NH_2-terminus, which may prevent protease μ-calpain from cTnI and cTnT degradation under I/R. These results confirm and extend previous findings, indicating that Hsp27 plays an important role in regulation of contractile function during I/R and provide new insight into the mechanisms mediating the protective effects of Hsp27 in myofilament regulatory proteins.

1.2. Protective Role of HspB1 in Atrial Fibrillation and Heart Failure

Atrial fibrillation (AF) is the most common cardiac arrhythmia, which has the tendency to become more persistent over time. When the arrhythmia continues, AF induces changes at the structural level, predominantly myolysis, which are associated with the progression of AF [34, 35]. Using the HL-1 cell model for AF, Brundel and colleagues [36] provided the first evidence that upregulation of Hsp by a mild heat shock and pharmacologically by the drug GGA, attenuates tachy-pacing-induced myolysis. Furthermore, transfection experiments directly demonstrated that elevated expression of Hsp27 alone was sufficient for this protection. These results were extended to human AF, where a highly significant increase of Hsp27 expression was observed only in atrial appendages of patients with paroxysmal AF. In addition, the expression levels of Hsp27 correlated inversely with the duration of the arrhythmia and with the amount of myolysis in paroxysmal and persistent AF. Finally, like in the tachy-paced cell model for AF, Hsp27 was found to be localized at the myofilaments in human atrial myocytes. Taken together, these data imply that Hsp27 upregulation protects against atrial fibrillation.

Hsp27 was also reported to be altered during heart failure in human and animal models [37-41]. Knowlton and colleagues [37] observed that Hsp27 was increased about two-fold in dilated cardiomyopathy (DCM) and ischemic cardiomyopathy (ICM) hearts, compared to normal hearts. Scheler et al. [38] compared two-dimensional electrophoresis patterns of heat shock protein Hsp27 in normal and cardiomyopathic hearts. They found that twelve spots corresponding to Hsp27 due to post-translational modifications were significantly altered in DCM samples, and ten of these were significantly changed in ICM. Furthermore, in tachycardia induced congestive heart failure (CHF) dogs, Hsp27 and its phosphorylation at both Ser-78 and Ser-82 sites were increased by ~2 fold [39]. These observations indicate that Hsp27 plays an important role in the process of heart failure.

In addition, overexpression of Hsp27 in cardiomyocytes protects against doxorubicin (DOX)-induced cardiomyopathy by reducing cardiomyocyte apoptosis, mitochondria damage and protein carbonylation [42-45]. The possible mechanisms are associated with: 1) regulation of p53 transcriptional activity [43]; 2) inhibition of ROS generation and the augmentation of Akt activation [32, 44]; and 3) activation of MAPK, leading to phosphorylation of Hsp27 and stabilization of cytoskeleton [45], as described below.

1.3. Phosphorylation of HspB1 and Its Cardioprotective Effects

The precise mechanism by which Hsp27 mediates this protective effect is not completely clear, but it has been demonstrated that Hsp27 associates with actin microfilaments and influences microfilament organization [29, 45, 46]. Also the phosphorylation status of Hsp27 may markedly influence its function [30, 47-50]. In unstressed cells, Hsp25/27 is present as oligomers that depolymerise into smaller forms on stress-induced phosphorylation [30]. There are three potential sites of phosphorylation, Ser15, Ser78 and Ser82 in human Hsp27 [50]. The kinases responsible for phosphorylation of Hsp25/Hsp27 are mitogen-activated protein kinase (MAPK) activated protein kinases 2 and 3 [47-49]. These are in turn activated by the p38-MAPK cascade. Protein kinase C (PKC) also phosphorylates Hsp25 on Ser15 and Ser86 [51, 52]. Recently human Hsp27 has been found to be phosphorylated on Ser82 by protein kinase D [54, 55]. Furthermore, phosphorylation of Hsp27 by PKC and MAPKs has been implicated in its cardioprotective effects, which are, in part, mediated through actin polymerization as cap-binding protein with the cytoskeletal thin-filament proteins [50].

The biological functions of Hsp25/27 are supposedly dependent on their degree of phosphorylation/oligomerization though this may not be true for some cytoprotective effects in heart. Recently, Dillmann and associates [56] used the serine-to-alanine mutants to explore the role of phosphorylation in the protective effect of Hsp27. In these mutants, serines 15, 78, and 82 were changed to alanine or glycine. Using these phosphorylation mutants, they obtained similar protective effects to that obtained with wild-type Hsp27. These data indicate that, for the general chaperoning effects of Hsp27, phosphorylation at the serines is not an essential requirement. In addition, they showed that a higher protective effect was observed by combining Hsp27 with αB-crystallin than from αB-crystallin or Hsp27 alone [56]. Similarly, an increased protective effect was observed by combination of αB-crystallin and Hsp70 [56]. Clearly, more data regarding the functional roles of Hsp27 and its phosphorylation in the heart will be required before a complete understanding is achieved.

2. HSPB2 AND ITS CARDIOPROTECTIVE ROLES

The HspB2 gene encodes for a 22 kDa protein arranged in a head-to-head orientation with CRYAB (αB-crystallin) [57]. Interestingly, this HspB2 is identical to MKBP, a binding protein for myotonic dystrophy protein kinase (DMPK, the protein product of the gene responsible for myotonic dystrophy) [58]. MKBP/HspB2 binds specifically to the kinase domain of DMPK, thus activating the kinase activity [58]. HspB2 also shows a chaperone-like activity that protects the kinase from heat-induced inactivation [57, 58]. Importantly, the expression of HspB2, but not other sHsps, is specifically up-regulated in the skeletal muscle of myotonic dystrophy patients as if to compensate for the reduced amount of DMPK [59]. Together with the fact that DMPK knock-out mice develop a late-onset, progressive myopathy, those findings suggest that DMPK is involved in a stress-response system in muscle cells by being a specific target of HspB2 [60].

HspB2, constitutively expressed in skeletal and cardiac muscles, shows homo-oligomeric activity and forms aggregates in muscle cytosol [57, 61]. Furthermore, it redistributes to the insoluble fraction in response to heat shock, suggesting that it is one of the stress responsive proteins found in muscle cells [62]. However, the present results also indicate that HspB2 is a unique member of the sHsp family, since it forms an independent complex distinct from the complex formed by other sHsps such as Hsp27 and αB-crystallin [57, 62]. In addition, its expression is not induced by heat treatment [9]. Based on these results, HspB2 is believed to be involved in a novel stress responsive system distinct from the known system composed of Hsp27, αB-crystallin, and Hsp20, and both systems may work independently to confer stress resistance to cardiomyocytes [62-65]. The intracellular localization of HspB2 at the Z-membranes and neuromuscular junction suggests a role for the protein in protecting the components of these muscle-specific structures [57]. More importantly, it has been observed that stretch stimulation in primary cultures of neonatal mouse cardiomyocytes could induce the expression of HspB2, suggesting that HspB2 might play a role in a system by which muscle cells respond to mechanical stress [64].

Using an ex vivo model of ischemia–reperfusion, Morrison et al. [66] recently demonstrated reduced contractile performance and increased apoptosis and necrosis in HspB2/CRYAB double knockout (DKO) mouse hearts. Although these findings are consistent with the requirements of CRYAB in cytoskeletal remodeling, this study could not distinguish whether one or both sHsps were required for either contractile performance and/or ischemic protection. Surprisingly, using an in situ model of ischemia–reperfusion, Benjamin et al.[67] reported that, following 30 min ischemia and 24 h of reperfusion, infarct size was significantly reduced in the hearts of HspB2/CRYAB DKO mice, compared with those of wild-type mice. Beyond differences in experimental approaches, the mechanisms for these paradoxical findings are presently unknown. To unmask the distinct roles of CryAB and HspB2, Pinz, et al. [68] generated a mouse model of HspB2 KO by crossing CryAB overexpressing mice with HspB2/CRYAB DKO mice, and observed that CryAB and HspB2 had distinct and nonredundant functions in the heart, with CryAB protecting diastolic performance and HspB2 required for normal systolic performance and normal cardiac energetics. Without doubt, the further studies are necessary to better understand the role of HspB2 in the heart.

3. HspB3 and HspB4/CRYAA

HspB3, named HSPL27, was recognized to have a cardiac-specific expression pattern in the literature [69, 70]. Although heat shock does not induce HspB3 expression, this protein appears to translocate to the cytoskeleton during stressful conditions [9, 62]. The mechanism of translocation is not completely understood, but may be related to its ability to interact exclusively with HspB2 [62, 70]. It has been reported that muscle develops a specific form of sHsp complex composed of MKBP/HspB2 and HspB3 during myogenic differentiation, suggesting that the sHsp oligomers comprising MKBP/HspB2 and HspB3 represents an additional system closely related to muscle function [62]. HspB3 is usually devoid of the N-terminal conserved region and C-terminal tail, and is among the most divergent of cardiac-specific sHsps [70]. The lack of inducibility, combined with a high basal level of expression, suggests that HspB3 has a significant role as a constitutive chaperone in the myocardium.

HspB4, αA-crystallin or CRYAA, is among the earliest recognized sHsps and it is abundantly expressed in the lens [71, 72]. Thus far, there have been no studies on the cardioprotective effects of HspB4.

4. HspB5/CRYAB and Its Role in Protecting the Heart

CRYAB, also named as αB-crystallin or HspB5, has been well characterized [73, 74]. αB-crystallin is most closely related to αA-crystallin with 53% sequence homology, but unlike αA-crystallin, αB-crystallin is normally expressed in the heart, skeletal muscle, and kidney [74, 75]. The CryAB gene and another sHsp (HspB2) gene reside head-to-head in chromosome 9 in mice [57]. Controlled by the shared bidirectional promoter, CryAB and HspB2 are co-expressed in mammalian hearts [57]. CryAB is the most abundant sHsp in cardiomyocytes [73-75]. The molecular chaperone properties of CryAB may entail prevention of stress-induced aggregation of denaturing proteins, as well as trapping aggregation-prone proteins in large, soluble, multimeric reservoirs [15, 73, 76].

4.1. αB-crystallin and Ischemic Stress

αB-crystallin expression can be induced in response to stress stimuli such as heat shock and oxidative stress [77, 78]. Evidence is rapidly accumulating to support the protective effects of αB-crystallin against noxious stresses like hyperthermia and environmental stress. For example, ectopic expression of αB-crystallin has been found to render NIH 3T3 cells thermo-resistant as well as glioma cells through stabilization of cytoskeletal structure [79, 80]. In ischemic rats, αB-crystallin was found to bind with the myofibrils, suggesting its cardioprotective role [81-84]. Recovery from exercise was associated with specific transient changes in the expression of αB-crystallin, which suggests that this heat shock protein may have specific roles in the remodeling process evoked by repeated bouts of contractile activity.

Recent studies point to the involvement of αB-crystallin in mediating cellular protection against myocardial ischemia reperfusion injury [85]. Ischemia was found to cause a rapid redistribution of αB-crystallin from the cytosolic pool to intercalated disks and Z lines of the myofibrils, which suggests its role in the protection of myocardial contractile apparatus against ischemia reperfusion injury [82]. Several studies showed that ischemia-induced phosphorylation of cardiac αB-crystallin is accompanied by a complete shift of this protein from the soluble to insoluble fraction, followed by its translocation to intercalated disks and myofibrillar Z lines, two cytoskeletal structures known to become destabilized during prolonged ischemia [83-84]. An increased expression of αB-crystallin through an adenoviral vector system protected the adult cardiomyocytes against injury mediated by simulated ischemia [28]. In another related study, Golenhofen et al. found that microtubular integrity was significantly preserved after ischemia in cells overexpressing αB-crystallin [83]. *Ex vivo* perfused hearts from transgenic (TG) mice that ubiquitously overexpress CryAB tolerated ischemia/reperfusion better [85]. By contrast, CryAB/HspB2 DKO mouse hearts displayed poorer functional recovery, a higher cell death rate, increased stiffness, and, hence, poor relaxation of the myocardium following ischemia/reperfusion, compared with wild-type (WT) controls [66]. In addition, CryAB was found to interact with mitochondria, which was evidenced by increases in mitochondrial permeability transition and calcium uptake in cardiomyocytes from CryAB/HspB2 DKO mice [18, 19, 81]. Taken together, these data indicate that αB-crystallin provides cardioprotection against ischemic insults via stabilization of cytoskeleton and interaction with mitochondria.

4.2. αB-crystallin and Its Role in Cardiac Hypertrophy

Interestingly, cardiac CryAB protein expression can be induced by pressure overload [76]. Moreover, recent studies indicate that CryAB/HspB2 deficiency activates the NFAT signaling and induces cardiac hypertrophic responses under no stress or minimal stress conditions, and exacerbates cardiac malfunction on pressure overload, whereas CryAB overexpression significantly attenuates pressure-overload hypertrophic responses and associated NFAT activation in mouse hearts [76]. Further experiments revealed that CryAB overexpression suppressed adrenergic stimulation–induced nuclear translocation of NFAT, leading to inhibition of the expression of a bona fide NFAT target gene and attenuation of adrenergic-induced hypertrophy in cultured cardiomyocytes [76]. CryAB overexpression failed to further suppress hypertrophic growth, when the calcineurin–NFAT pathway was blocked by CsA. These new findings demonstrate that CryAB negatively regulates pressure overload-induced cardiac hypertrophic responses through inhibition of NFAT signaling in cardiomyocytes.

4.3. αB-crystallin Phosphorylation and Its Cardiac Protective Effects

In response to various types of cellular stress, αB-crystallin is phosphorylated on Ser-19, 45, and 59 [78]. Although the signaling pathways leading to Ser-19 phosphorylation are unknown, the extracellular signal-regulated protein kinase (ERK) MAPK pathway appears to be responsible for phosphorylation of Ser-45, and the p38 MAPK pathway is responsible for phosphorylation of Ser-59 [77, 86, 87]. In the heart, oxidative stress can lead to the activation of p38 and MAPKAPK-2, the latter of which phosphorylates αB-crystallin on Ser-59 [77].

Conversely, blocking p38-mediated MAPKAPK-2 activation inhibits αB-crystallin phosphorylation on Ser-59 and enhances apoptotic myocyte death [77]. These results are consistent with a role for phosphorylation of αB-crystallin on Ser-59 in mediating the cytoprotective actions of this small heat shock protein [88]. Actually, such phosphorylation is required for αB-crystallin-mediated cytoprotection [88]. The αB-crystallin mutant (AAE), which mimics phosphorylation on Ser-59, protected cardiac myocytes against apoptosis, whereas the non-phosphorylatable αB-crystallin mutant (AAA) conferred increased susceptibility to apoptosis [88]. These data indicate that phosphorylation of αB-crystallin on Ser-59 alone is necessary and sufficient for this sHsp to confer maximal cytoprotection. The molecular mechanisms underlying the cytoprotection mediated by αB-crystallin (S59P) or increased apoptosis conferred by αB-crystallin (AAA) are unknown. However, since αB-crystallin (AAE) reduced caspase-3 activation, the antiapoptotic effects of αB-crystallin (S59P) may involve inhibition of the conversion of p24 caspase-3 to active caspase-3. Although it has been shown that αB-crystallin can bind to p24 and decrease the rate of activated caspase-3 generation [18], the role, if any, of αB-crystallin serine-59 phosphorylation has not been established.

It is most likely that αB-crystallin affects cytoprotection by interacting with many targets in addition to caspase-3. αB-crystallin monomers, which are ≈25 kDa, associate into oligomers of ≈32 subunits, reaching a mass of 645 kDa [89]. The mass of the oligomers decreases in response to cellular stresses that also increase the phosphorylation of αB-crystallin [90, 91], suggesting that the phosphorylated αB-crystallin monomers translocate to sites where they mediate cytoprotection. For example, cardiac ischemia causes αB-crystallin translocation from a diffuse localization in the cytosol, to myofibrils [83], where it binds to the I-band region of the Z-lines [82]. This translocation takes place on a time frame consistent with a possible role for phosphorylation in mediating αB-crystallin relocation [82-84]. The localization of αB-crystallin to sarcomeres may allow myofibrils to maintain their correctly folded state and optimal function, similar to the way Hsp27 stabilizes actin filaments in endothelial cells after stresses that activate p38 [83]. Although these results imply that αB-crystallin monomers would serve as effective chaperones, this concept is controversial. For example, in glioma cells, expression of αB-crystallin (DDD), where serines-19, 45, and 59 have been replaced by aspartic acid, which mimics phosphorylation at these sites, resulted in oligomers that are smaller and display reduced chaperone-like activity [91]. However, a naturally occurring form of αB-crystallin carrying an R120G mutation, which is known to cause desmin-related cardiomyopathy, forms oligomers that exceed 645 kDa; yet, this form is a poorer chaperone [92]. Thus, the relationships between αB-crystallin oligomer size, phosphorylation status, and chaperone and cytoprotective activities remain unclear.

5. HspB6/Hsp20 in Protecting the Heart

Hsp20, also referred as P20/HspB6, was originally co-purified with the small heat-shock protein Hsp27/Hsp28 from rat and human skeletal muscles [93]. Its genome contains 3 exons and 2 introns. Human, rat and mouse Hsp20 are composed of 157, 162 and 162 amino acids, respectively [94]. Although hyperthermia did not induce Hsp20 in rat tissue [93], heat pretreatment of swine carotid artery was associated with increased Hsp20 levels [95]. This protein is detected in all tissues by a sandwich-type immunoassay system, reaching a maximal level of 1.3% of total protein in skeletal, heart and smooth muscles [93], and its expression levels are altered during development [96]. Recent studies have revealed the significance of Hsp20 in relaxation of vascular muscle and inhibition of platelet aggregation [16, 97]. Of particular interest, recent work indicates that cardiac Hsp20 overexpression renders protection against stress-induced injury *in vitro* and *in vivo*, through its anti-apoptotic properties [98-104]. Therefore, it is plausible that Hsp20 provides multifaceted beneficial effects in the heart.

5.1. Hsp20 and Its Phosphorylation Under Stress Conditions

Hsp20 is normally located in the cytoplasm of cardiac myocytes. After heat stress, a subpopulation of Hsp20 migrates into the nucleus, while the other part remains in the cytoplasm. In very few cells, a faint sarcomeric association of Hsp20 is observed [105]; however, Hsp20 is prominently translocated to the myofibrils in adult rat heart and skeletal muscle under ischemic conditions [74]. A similar phenomenon with Hsp20 redistributed to stress fibers has also been observed in the rat cardiac myoblast cell line H9C2 after proteasomal inhibition [106]. Conversely, one study demonstrated that Hsp20 was predominantly in transverse bands, same as the pattern of actin staining in adult rat heart tissue and neonatal cardiomyocytes in the absence of any stimuli [107]. These data suggest that Hsp20 may protect specific elements of the cyto-skeleton in myocytes upon stress stimuli.

It has been known that Hsp20 contains 3 phosphorylation sites: Serine 16 by PKA/PKG; Serine 59, presumptively phosphorylated by PKC; and Serine 157 phosphorylated by insulin stimulation [97]. However, the serine 157 phosphorylation site does not exist in human Hsp20, and the physiological role of Serine 59 phosphorylation is not clear. So far, most results about the role of Hsp20 phosphorylation have been based on Serine 16 phosphorylation. More recently, it has been shown that sustained β-adrenergic signaling is associated with expression and phosphorylation of cardiac Hsp20, and the phosphorylation site has been identified as Ser16 [94]. Endurance training increases the phosphorylation of Hsp20 at serine 16, which improves cardiac function and protects against heart disease [108, 109]. In a congestive heart failure (CHF) model, both the expression of phosphorylated and unphosphorylated forms of Hsp20 were observed to be significantly increased compared with the normal group [110]. Together, these results support the therapeutic effects of Hsp20 and its phosphorylation in cardiac disorders.

5.2. Hsp20 and Myocardial Contraction

Hsp20 has been shown to enhance myocardial contraction [94, 103, 107]. Isolated adult cardiomyocytes transiently permeabilized with phosphopeptide analogues of Hsp20 displayed an increase in the rate of shortening [107], which was associated with an increase in the rate of lengthening and a more rapid decay of the calcium transient, suggesting that the phosphopeptide analogue of Hsp20 facilitates the increased rate by stimulating a more rapid uptake of Ca2+ by the sarcoplasmic reticulum. Chu et al. [94] used adenoviral-mediated full-length Hsp20 transfer into adult cardiomyoctes, and found that increased Hsp20 expression was associated with significant increases in contractility and Ca^{2+} transient peak. Furthermore, in a transgenic mouse model with cardiac specific-overexpression of Hsp20 (10-fold), heart function was significantly enhanced, and all Hsp20 transgenic (TG) mice were healthy without any apparent morphological/pathological abnormalities, compared with wild-types [101, 102].

Currently, the mechanisms underlying the enhancement of myocardial contraction by Hsp20 are not clear. In 1998, Damer et al [111] used microcystin-affinity chromatography to purify type 1 phosphatase (PP1)-binding proteins from the myofibrillar fraction of rabbit skeletal muscle, and surprisingly found that human Hsp20 complexed with the catalytic subunit of PP-1 (PP-1C), suggesting that Hsp20 possibly regulates the activity of PP1. Actually, we have observed that PP1 activity was significantly decreased in Hsp20 transgenic hearts [104]. Increases in PP1 activity have been reported in end stage human heart failure, whereas inactivation of PP1 greatly enhanced cardiac function [112]. Thus, Hsp20 may inhibit PP1 activity, leading to increased myocardial contraction.

5.3. Hsp20 and beta-Agonist-Induced Cardiac Remodeling

Chronic stimulation of the beta-adrenergic neurohormonal axis contributes to the progression of heart failure and mortality in animal models and human patients. In cardiomyocytes, activation of the beta-adrenergic pathway has been shown to result in transiently increased expression of a cardiac small heat-shock protein Hsp20 [100]. To gain further insights into the role of Hsp20 and its cAMP-dependent phosphorylation, Fan et al.[100] over-expressed this protein and its constitutively phosphorylated (S16D) or nonphosphorylated (S16A) mutants in adult rat cardiomyocytes. Hsp20 protected cardiomyocytes from apoptosis triggered by activation of the cAMP-PKA pathway, as indicated by decreases in the number of pyknotic nuclei, terminal deoxynucleotidyltransferase-mediated dUTP nick-end labeling, and DNA laddering, which were associated with inhibition of caspase-3 activity. These protective effects were further increased by the constitutively phosphorylated Hsp20 mutant (S16D), which conferred full protection from apoptosis. In contrast, the nonphosphorylatable mutant (S16A) exhibited no antiapoptotic properties. These findings suggest that Hsp20 and its phosphorylation at Ser16 may provide cardioprotection against β-agonist-induced apoptosis [100].

The cardioprotective effects of Hsp20 were further confirmed *in vivo* using a cardiac-specific over-expression mouse model [102]. Hsp20 transgenic hearts (10-fold over-expression) are resistant to chronic beta-agonist-induced cardiac remodeling, associated with attenuation of the cardiac hypertrophic response, markedly reduced interstitial fibrosis, and decreased apoptosis. Contractility was also preserved in hearts with increased Hsp20 levels. The possible mechanisms underlying these beneficial effects involved attenuation of the ASK1-JNK/p38 (apoptosis signal-regulating kinase 1/c-Jun NH(2)-terminal kinase/p38) signaling cascade triggered by isoproterenol. Parallel *in vitro* experiments supported the inhibitory role of Hsp20 on enforced ASK1-JNK/p38 activation in both H9c2 cells and adult rat cardiomyocytes. Immunostaining studies further demonstrated that Hsp20 colocalizes with ASK1 in cardiomyocytes. Taken together, these findings indicate that increased expression of

Hsp20 attenuates the induction of remodeling, dysfunction, and apoptosis in response to sustained beta-adrenergic stimulation.

5.4. Hsp20 and Ischemia/Reperfusion Injury

Furthermore, both Hsp20-transgenic hearts and Ad.Hsp20-infected hearts, subjected to global no-flow ischemia/reperfusion using the Langendorff preparation, exhibited improved recovery of contractile performance over the whole reperfusion period [98, 101]. This improvement was accompanied by a 2-fold decrease in lactate dehydrogenase released from transgenic hearts. The extent of infarction and apoptotic cell death was also significantly decreased, which was associated with increased protein ratio of Bcl-2/Bax and reduced Caspase-3 activity in transgenic hearts. Moreover, *in vivo* experiments of 30 min myocardial ischemia, via coronary artery occlusion, followed by 24 h reperfusion, showed that the infarct region-to-risk region ratio was significantly reduced in Hsp20-hearts, compared to wild-type hearts. Interestingly, the levels of Hsp20 phosphorylation were significantly increased in post-ischemic/reperfused transgenic hearts, suggesting that Hsp20 phosphorylation may play an important role in cardioprotection against ischemic injury. Actually, a recently identified C59T substitution in the human Hsp20 gene was associated with its diminished phosphorylation at Ser16, compared with WT-Hsp20, and complete abrogation of the Hsp20 cardioprotective effects against ischemia/reperfusion [113].

5.5. Hsp20 and Doxorubicin-Induced Cardiomyopathy

Doxorubicin (DOX), an anthracycline antibiotic, is a highly effective chemotherapeutic agent used in the treatment of solid and hematopoietic tumors; however, a major limiting factor for the clinical use of DOX is its cumulative, irreversible cardiac toxicity [114]. In fact, multiple intravenous DOX treatments over a period of several months result in the development of cardiomyopathy and congestive heart failure in humans [115]. Several studies have shown that some Hsps can act as negative regulators of DOX-triggered apoptotic and necrotic cell death, such as Hsp10 and Hsp60, which have been found to modulate DOX-induced mitochondrial apoptosis signaling in neonatal cardiomyocytes [116]. Furthermore, cardiac-specific over-expression of Hsp20 *in vivo* significantly ameliorated acute DOX-triggered cardiomyocyte apoptosis and animal mortality [104]. Hsp20 transgenic mice also showed improved cardiac function and prolonged survival after chronic administration of DOX. The mechanisms underlying these beneficial effects were associated with preserved Akt phosphorylation/activity and attenuation of DOX-induced oxidative stress. Co-immunoprecipitation studies revealed an interaction between Hsp20 and phosphorylated Akt. Accordingly, BAD phosphorylation was preserved, and cleaved caspase-3 was decreased in DOX-treated Hsp20 transgenic hearts, consistent with the antiapoptotic effects of Hsp20. Parallel *ex vivo* experiments showed that either infection with a dominant-negative Akt adenovirus or preincubation of cardiomyocytes with the phosphatidylinositol 3-kinase inhibitors significantly attenuated the protective effects of Hsp20. Taken together, these findings indicate that over-expression of Hsp20 attenuates DOX-triggered cardiomyopathy, and these beneficial effects appear to be dependent on Akt activation [104].

6. HspB7/cvHsp and Its Cardioprotective Effects

HspB7, also referred to as cvHsp, was identified using expressed sequence tag (EST) database searches [117]. This 25-kDa protein is highly expressed in adult and fetal heart as well as in skeletal muscle, while a fainter expression is evidenced in the aorta [118-120]. In contrast, cvHsp expression was virtually undetectable in other organs tested, including brain, digestive tract, liver, lung, adrenal, thyroid, spleen, thymus, gonads, and placenta [117]. This restricted tissue distribution confirmed by immunohistochemical staining in the myocardium of pigs, rats and mice[117, 119]. SAGE analysis shows that HspB7 is among the most expressed sHsps in cardiac tissue, which suggests that it may have a key role in cardiac metabolism [117]. Western blot analyses in human heart homogenates revealed the presence of two specific bands at 23 and 25 kDa for cvHsp [117]. The major posttranscriptional modifications of the cvHsp proteins are phosphorylations notably by cyclic AMP-dependent protein kinase and cyclic GMP-dependent protein kinase, mitogen-activated protein kinase-activated protein kinase-2, protein kinase C, and p44/42 mitogen-activated protein kinase. Putative consensus phosphorylation functional motifs were found in the cvHsp protein (175-amino acid isoform) sequence: five protein kinase C phosphorylation sites (Ser-2, Thr-8, Thr-71, Thr-153, and Thr-168) and two casein kinase II phosphorylation sites (Thr-82 and Ser-97). However, it remains to be established whether these two 23- and 25-kDa bands correspond to the 170- and 175-amino acid residue isoforms and/or to different phosphorylation states of cvHsp [117].

While the functions of HspB7 are presently unknown, recent studies have implicated its role in oligomerization with other sHsps and cytoskeletal proteins such as α-filamin in the myocardium [118, 121]. In addition, the gene localization of human cvHsp in 1p34-36 comprises that for several dystrophies and myopathies, including cardiomyopathy, suggesting that cvHsp could be a candidate gene for these diseases [9]. Future studies regarding the significance of cvHsp under stress and disease will be beneficial in our better understanding of the function of this protein.

7. HspB8/Hsp22 and Its Cardioprotective Effects

HspB8 is also identified as H11 kinase and Hsp22 [122-124]. While some groups considered Hsp22 to be a protein kinase homologous to the large subunit of Herpex simplex virus type 2 ribonucleotide reductase (ICP10) which has a protein kinase activity, the results of alignment between human Hsp22 (accession AAF65562 [GenBank]) and ICP10 (accession 1813262A) revealed only weak similarity (11% identical, 13% similar; or 14% identical, 17% similar using the C-terminal part of ICP10) [122]. However, phylogenetic analysis showed that Hsp22 is a true member of the sHsp superfamily, most closely related to Hsp27 and interacting with phosphorylated Hsp27 [122].

Intriguingly, in adult mammals, HspB8/H11/Hsp22 kinase gene expression is most abundant in the heart and skeletal muscle [125-127]. An increased level of Hsp22 was detected in a chronic canine model of cardiac hypertrophy, suggesting that it might participate in mechanisms of cardiac cell growth [125]. Indeed, over-expression of Hsp22 in a transgenic (TG) mouse leads to a 30% increase in myocyte cross-sectional area compared to wild-type

(WT) mice [126]. However, this TG mouse shows a pattern of myocardial hypertrophy with normal cardiac function. In particular, over-expression of Hsp22 does not induce a transition into heart failure, in contrast to the more frequently used models of cardiac overload by aortic constriction [128]. Of interest, Hsp22 interacts and co-localizes with the perinuclear proteasome [125]. Adeno-mediated over-expression of Hsp22 in isolated cardiac myocytes increased both cell growth and proteasome activity, and both were prevented upon inhibition of the proteasome [125]. Similarly, stimulation of cardiac cell growth by pro-hypertrophic stimuli increased Hsp22 expression and proteasome activity [125]. Taken together, these data demonstrate a regulation of the proteasome during cardiac hypertrophy mediated by Hsp22.

The levels of Hsp22 are also observed to be increased in a swine model of transient ischemia, and in patients with prolonged ischemia, along with an array of genes promoting cell survival [129, 130], suggesting that Hsp22 could participate in the prevention of irreversible ischemic damage. Recent studies by Depre et al. [129] have demonstrated that transgenic mice overexpressing Hsp22 significantly protected hearts from ischemia–reperfusion injury. Importantly, increased expression of HspB8 confers a cardioprotection that is equivalent to ischemic preconditioning.

The potential mechanisms underlying the cardioprotective role of Hsp22 may include: 1) activation of the serine/threonine kinase Akt/PKB, which can prevent cell death through an inhibition by phosphorylation of proapoptotic effectors, including glycogen synthase kinase-3ß (GSK-3ß), caspase-9, Bad, and the transcription factor forkhead [126, 129]; and 2) enhancement of glucose metabolism in the heart in vivo [127]. In addition to activating signaling pathways, Hsp22 promotes the subcellular translocation and crosstalk of intracellular messengers [125]. The interaction of Hsp22 with Akt was found in both cytosolic and nuclear fractions but largely predominated in the latter. Interaction of Hsp22 and AMPK was detected exclusively in the nuclear fraction. As a result, the TG mice were characterized by an increased interaction between Akt and AMPK specifically in the nucleus. Interestingly, a slight signal for Hsp22 was showed in the mitochondrial fraction of the TG mouse heart [125], which might be of importance because Hsp22 is associated with the mitochondria in *Drosophila* [131] and therefore might prevent the activation of the intrinsic pathway of apoptosis.

8. HspB9 and HspB10/ODF1

HspB9 and ODF1, the 9th and 10th members, were the last to be identified and are perhaps the final members of the sHsp family to be discovered by genome database survey [132, 133]. Both proteins are expressed in the testis where they have been previously identified and characterized to play important roles in sperm flagellar integrity [134-136]. They might share some functions with other better-studied sHsps (eg, Hsp27, αB-Cry) that are known to be involved in the organization of cytoskeletal elements such as microfilaments and intermediate filaments and in the function of muscles [137]. However, currently it is not clear if HspB9 and HspB10 are expressed in cardiac tissue.

Conclusion

We have reviewed herein the multifaceted roles of sHsps in cardioprotection and their underlying mechanisms (Table 1). Genetic and functional redundancy among sHsps triggers a common question: Why are there so many sHsps expressed in the cardiomyocyte? Actually, each sHsp is unique and appears to collaborate with others to defend against stress conditions. For example, CryAB and HspB2 proteins play nonredundant roles in the heart, CryAB in structural remodeling and HspB2 in maintaining energetic balance. Data from studies using knockout models of sHsps, combined with clinical observations of mutations in various sHsps, suggest important requirements for each sHsp in maintaining cellular homeostasis. Increased evidence indicates that the progression of disease ensues, when specific functions associated with the sHsp are absent. Future lines of investigation using cardiac conditional knock-out animal models and models of human disease related to the sHsp's mutations may accelerate our understanding of these proteins' specific roles in cardioprotection.

Table 1. Cardioprotection of sHsps and Their Underlying Mechanisms

Name (Alternative)	Cardioprotection	Mechanisms	References
HspB1 Hsp27 (human rat) Hsp25 (mouse)	Ischemic injury Doxorubicin Atrial fibrillation	ROS↓, Akt activity↑, MAPK ↑, Cytoskeleton (↔)	[28-33, 36-50, 56]
HspB2 MKBP	Ischemic injury	Cardiac energetics?	[66-68]
HspB3 HSPL27	Unknown		
HspB4 αA-crystallin	Unknown		
HspB5 αB-crystallin	Ischemic injury Hypertrophy Heart failure	Cytoskeleton (↔), NFAT signaling↓	[28, 76, 81-85, 88]
HspB6 Hsp20	Ischemic injury Doxorubicin Hypertrophy	ROS↓, Akt activity↑, ASK1↓, PP1 activity↓ Bcl-2/Bax ↑, Cytoskeleton (↔)	[98-104]
HspB7 cvHSP	Cardiomyopathy?	Cardiac metabolism?	[117, 118]
HspB8 Hsp22, H11K	Ischemic injury Hypertrophy	Akt activity↑, Glucose metabolism ↑, AMPK↑	[125-130]
HspB9	Unknown		
HspB10 ODF1	Unknown		

REFERENCES

[1] Williams RS, Benjamin IJ. Stress proteins and cardiovascular disease. *Mol Biol Med.* 1991, 8(2):197-206.

[2] Benjamin IJ, McMillan DR. Stress (heat shock) proteins: molecular chaperones in cardiovascular biology and disease. *Circ Res.* 1998, 83(2):117-32.

[3] Sun Y, MacRae TH. The small heat shock proteins and their role in human disease. *FEBS J.* 2005, 272(11):2613-27.

[4] De Jong WW, Caspers GJ, Leunissen JA. Genealogy of the alpha-crystallin-small heat-shock protein superfamily, *Int. J. Biol. Macromol.* 1998, 22: 151–162.

[5] Horwitz J. Alpha-crystallin can function as a molecular chaperone, *Proc. Natl. Acad. Sci. USA.* 1992, 89: 10449–10453.

[6] Mornon JP, Halaby D, Malfois M, Durand P, Callebaut I , Tardieu A. Alpha-crystallin C-terminal domain: on the track of an Ig fold, *Int. J. Biol. Macromol.* 1998, 22: 219–227.

[7] Narberhaus N. Alpha-crystallin-type heat shock proteins: socializing minichaperones in the context of a multichaperone network, *Microbiol. Mol. Biol. Rev.* 2002, 66: 64–93.

[8] Kappe G, Franck E, Verschuure P, Boelens WC, Leunissen JA, de Jong WW. The human genome encodes 10 alpha-crystallin-related small heat shock proteins: HspB1–10. *Cell Stress Chaperones.* 2003, 8: 53–61.

[9] Taylor RP, Benjamin IJ: Small heat shock proteins: a new classification scheme in mammals. *J Mol Cell Cardiol.* 2005, 38:433-444.

[10] Arrigo AP. In search of the molecular mechanism by which small stress proteins counteract apoptosis during cellular differentiation,*J. Cell. Biochem.* 2005, 94:241–246.

[11] Ehrnsperger M, Graber S, Gaestel M, Buchner J. Binding of non-native protein to Hsp25 during heat shock creates a reservoir of folding intermediates for reactivation, *EMBO J.* 1997, 16: 221–229.

[12] Lee GJ, Roseman AM, Saibil HR, Vierling E. A small heat shock protein stably binds heat-denatured model substrates and can maintain a substrate in a folding-competent state, *EMBO J.* 1997, 3: 659–671.

[13] Franzmann TM, Wuhr M, Richter K, Walter S, Buchner J. The activation mechanism of Hsp26 does not require dissociation of the oligomer, *J. Mol. Biol.* 2005, 350:1083–1093.

[14] Haslbeck M, Franzmann T, Weinfurtner D, Buchner J. Some like it hot: the structure and function of small heat-shock proteins, *Nat. Struct. Mol. Biol.* 2005, 12:842–846.

[15] Kumarapeli AR, Wang X. Genetic modification of the heart: chaperones and the cytoskeleton. *J Mol Cell Cardiol.* 2004, 37(6):1097-109.

[16] Fan GC, Chu G, Kranias EG. Hsp20 and its cardioprotection. *Trends Cardiovasc Med.* 2005, 15:138-41.

[17] Nakagawa M, Tsujimoto N, Nakagawa H, Iwaki T, Fukumaki Y, Iwaki A. Association of HSPB2, a member of the small heat shock protein family, with mitochondria. *Exp Cell Res.* 2001, 271(1):161-8.

[18] Kamradt MC, Chen F, Cryns VL. The small heat shock protein alpha B-crystallin negatively regulates cytochrome c- and caspase-8-dependent activation of caspase-3 by inhibiting its autoproteolytic maturation. *J Biol Chem.* 200, 276(19):16059-63.

[19] Kadono T, Zhang XQ, Srinivasan S, Ishida H, Barry WH, Benjamin IJ. CRYAB and HSPB2 deficiency increases myocyte mitochondrial permeability transition and mitochondrial calcium uptake. *J Mol Cell Cardiol.* 2006, 40(6):783-9.

[20] Chernik IS, Seit-Nebi AS, Marston SB, Gusev NB. Small heat shock protein Hsp20 (HspB6) as a partner of 14-3-3gamma. *Mol Cell Biochem.* 2007, 295(1-2):9-17.

[21] Gernold M, Knauf U, Gaestel M, Stahl J, Kloetzel P-M. Development and tissue-specific distribution of mouse small heat shock protein HSP25. *Dev Genet.* 1993, 14:103-111.

[22] Ciocca DR, Oesterreich S, Chamness GC, McGuire WL, Fuqua SA. Biological and clinical implications of heat shock protein 27,000 (Hsp27): a review. *J. Natl. Cancer Inst.* 1993, 85: 1558–1570.

[23] Tallot P, Grongnet JF, David JC. Dual perinatal and developmental expression of the small heat shock proteins alphaB-crystallin and Hsp27 in different tissues of the developing piglet. *Biol. Neonate* 2003, 83: 281–288.

[24] Lutsch G, Vetter R, Offhauss U, Wieske M, Grone HJ, Klemenz R et al. Abundance and location of the small heat shock proteins HSP25 and alphaB-crystallin in rat and human heart. *Circulation* 1997, 96:3466–3476.

[25] Van de Klundert FA, Gijsen ML, van den IPR, Snoeckx LH, de Jong WW. Alpha B-crystallin and hsp25 in neonatal cardiac cells—differences in cellular localization under stress conditions. *Eur. J. Cell Biol.* 1998, 75:38–45.

[26] Hoch B, Lutsch G, Schlegel WP, Stahl J, Wallukat G, Bartel S.et al. HSP25 in isolated perfused rat hearts: localization and response to hyperthermia. *Mol. Cell. Biochem.* 1996, 160: 231–239.

[27] Dillmann WH. Small heat shock proteins and protection against injury. *Ann N Y Acad Sci.* 1999, 874:66-8.

[28] Martin JL, Mestril R, Hilal-Dandan R, Brunton LL, Dillmann WH. Small heat shock proteins and protection against ischemic injury in cardiac myocytes. *Circulation.* 1997, 96(12):4343-8.

[29] Bluhm WF, Martin JL, Mestril R, Dillmann WH. Specific heat shock proteins protect microtubules during simulated ischemia in cardiac myocytes. *Am J Physiol.* 1998, 275(6 Pt 2):H2243-9.

[30] Martin JL, Hickey E, Weber LA, Dillmann WH, Mestril R. Influence of phosphorylation and oligomerization on the protective role of the small heat shock protein 27 in rat adult cardiomyocytes. *Gene Expr.* 1999, 7(4-6):349-55.

[31] Vander Heide RS. Increased expression of HSP27 protects canine myocytes from simulated ischemia-reperfusion injury. *Am J Physiol Heart Circ Physiol.* 2002, 282:H935-41.

[32] Liu L, Zhang XJ, Jiang SR, Ding ZN, Ding GX, Huang J, Cheng YL. Heat shock protein 27 regulates oxidative stress-induced apoptosis in cardiomyocytes: mechanisms via reactive oxygen species generation and Akt activation. *Chin Med J (Engl).* 2007, 120(24):2271-7.

[33] Lu XY, Chen L, Cai XL, Yang HT. Overexpression of heat shock protein 27 protects against ischaemia/reperfusion-induced cardiac dysfunction via stabilization of troponin I and T. *Cardiovasc Res.* 2008, 79(3):500-8.

[34] Brundel BJ, Ke L, Dijkhuis AJ, Qi X, Shiroshita-Takeshita A, Nattel S, Henning RH, Kampinga HH. Heat shock proteins as molecular targets for intervention in atrial fibrillation. *Cardiovasc Res.* 2008, 78(3):422-8.

[35] Brundel BJ, Shiroshita-Takeshita A, Qi X, Yeh YH, Chartier D, van Gelder IC, Henning RH, Kampinga HH, Nattel S. Induction of heat shock response protects the heart against atrial fibrillation. *Circ Res.* 2006,99(12):1394-402

[36] Brundel BJ, Henning RH, Ke L, van Gelder IC, Crijns HJ, Kampinga HH. Heat shock protein upregulation protects against pacing-induced myolysis in HL-1 atrial myocytes and in human atrial fibrillation.*J Mol Cell Cardiol.* 2006, 41(3):555-62.

[37] Knowlton AA, Kapadia S, Torre-Amione G, Durand JB, Bies R, Young J, Mann DL. Differential expression of heat shock proteins in normal and failing human hearts. *J Mol Cell Cardiol.* 1998, 30(4):811-8.

[38] Scheler C, Li XP, Salnikow J, Dunn MJ, Jungblut PR. Comparison of two-dimensional electrophoresis patterns of heat shock protein Hsp27 species in normal and cardiomyopathic hearts. *Electrophoresis.* 1999, 20(18):3623-8.

[39] Tanonaka K, Yoshida H, Toga W, Furuhama K, Takeo S. Myocardial heat shock proteins during the development of heart failure. *Biochem Biophys Res Commun.* 2001, 283(2):520-5.

[40] Dohke T, Wada A, Isono T, Fujii M, Yamamoto T, Tsutamoto T, Horie M. Proteomic analysis reveals significant alternations of cardiac small heat shock protein expression in congestive heart failure. *J Card Fail.* 2006, 12(1):77-84.

[41] Wellner M, Dechend R, Park JK, Shagdarsuren E, Al-Saadi N, Kirsch T, Gratze P, Schneider W, Meiners S, Fiebeler A, Haller H, Luft FC, Muller DN. Cardiac gene expression profile in rats with terminal heart failure and cachexia. *Physiol Genomics.* 2005, 20(3):256-67.

[42] Liu L, Zhang X, Qian B, Min X, Gao X, Li C, Cheng Y, Huang J. Over-expression of heat shock protein 27 attenuates doxorubicin-induced cardiac dysfunction in mice. *Eur J Heart Fail.* 2007, 9(8):762-9.

[43] Venkatakrishnan CD, Dunsmore K, Wong H, Roy S, Sen CK, Wani A, Zweier JL, Ilangovan G. HSP27 regulates p53 transcriptional activity in doxorubicin-treated fibroblasts and cardiac H9c2 cells: p21 upregulation and G2/M phase cell cycle arrest. *Am J Physiol Heart Circ Physiol.* 2008, 294(4):H1736-44.

[44] Turakhia S, Venkatakrishnan CD, Dunsmore K, Wong H, Kuppusamy P, Zweier JL, Ilangovan G. Doxorubicin-induced cardiotoxicity: direct correlation of cardiac fibroblast and H9c2 cell survival and aconitase activity with heat shock protein 27. *Am J Physiol Heart Circ Physiol.* 2007, 293(5):H3111-21.

[45] Venkatakrishnan CD, Tewari AK, Moldovan L, Cardounel AJ, Zweier JL, Kuppusamy P, Ilangovan G. Heat shock protects cardiac cells from doxorubicin-induced toxicity by activating p38 MAPK and phosphorylation of small heat shock protein 27. *Am J Physiol Heart Circ Physiol.* 2006, 291(6):H2680-91

[46] Sakamoto K, Urushidani T, Nagao T. Translocation of HSP27 to sarcomere induced by ischemic preconditioning in isolated rat hearts. *Biochem Biophys Res Commun.* 2000, 269(1):137-42.

[47] Armstrong SC, Delacey M, Ganote CE. Phosphorylation state of hsp27 and p38 MAPK during preconditioning and protein phosphatase inhibitor protection of rabbit cardiomyocytes. *J Mol Cell Cardiol.* 1999, 31(3):555-67.

[48] Peart JN, Gross ER, Headrick JP, Gross GJ. Impaired p38 MAPK/HSP27 signaling underlies aging-related failure in opioid-mediated cardioprotection. *J Mol Cell Cardiol.* 2007, 42(5):972-80.

[49] Li G, Ali IS, Currie RW. Insulin-induced myocardial protection in isolated ischemic rat hearts requires p38 MAPK phosphorylation of Hsp27. *Am J Physiol Heart Circ Physiol.* 2008, 294(1):H74-87.

[50] Bitar KN. HSP27 phosphorylation and interaction with actin-myosin in smooth muscle contraction. *Am J Physiol Gastrointest Liver Physiol.* 2002, 282(5):G894-903.

[51] Faucher C, Capdevielle J, Canal I, Ferrara P, Mazarguil H, McGuire WL, Darbon JM. The 28-kDa protein whose phosphorylation is induced by protein kinase C activators in MCF-7 cells belongs to the family of low molecular mass heat shock proteins and is the estrogen-regulated 24-kDa protein. *J Biol Chem.* 1993,268(20):15168-73.

[52] Arnaud C, Joyeux-Faure M, Bottari S, Godin-Ribuot D, Ribuot C. New insight into the signalling pathways of heat stress-induced myocardial preconditioning: protein kinase Cepsilon translocation and heat shock protein 27 phosphorylation. *Clin Exp Pharmacol Physiol.* 2004, 31(3):129-33.

[53] Dana A, Skarli M, Papakrivopoulou J, Yellon DM. Adenosine A(1) receptor induced delayed preconditioning in rabbits: induction of p38 mitogen-activated protein kinase activation and Hsp27 phosphorylation via a tyrosine kinase- and protein kinase C-dependent mechanism. *Circ Res.* 2000, 86(9):989-97.

[54] Döppler H, Storz P, Li J, Comb MJ, Toker A. A phosphorylation state-specific antibody recognizes Hsp27, a novel substrate of protein kinase D. *J Biol Chem.* 2005, 280:15013-9.

[55] Evans IM, Britton G, Zachary IC. Vascular endothelial growth factor induces heat shock protein (HSP) 27 serine 82 phosphorylation and endothelial tubulogenesis via protein kinase D and independent of p38 kinase. *Cell Signal.* 2008, 20(7):1375-84.

[56] Hollander JM, Martin JL, Belke DD, Scott BT, Swanson E, Krishnamoorthy V, Dillmann WH. Overexpression of wild-type heat shock protein 27 and a nonphosphorylatable heat shock protein 27 mutant protects against ischemia-reperfusion injury in a transgenic mouse model. *Circulation.* 2004, 110(23):3544-52.

[57] Iwaki A, Nagano T, Nakagawa M, Iwaki T, Fukumaki Y. Identification and characterization of the gene encoding a new member of the alpha-crystallin/small hsp family, closely linked to the alphaB-crystallin gene in a head-to-head manner. Genomics. 1997, 45(2):386-94.

[58] Suzuki A, Sugiyama Y, Hayashi Y, Nyu-i N, Yoshida M, Nonaka I, Ishiura S, Arahata K, Ohno S. MKBP, a novel member of the small heat shock protein family, binds and activates the myotonic dystrophy protein kinase. *J Cell Biol.* 1998, 140(5):1113-24.

[59] Jansen, G., P.J.T.A. Groenen, D. Bachner, P.H.K. Jap, M. Coerwinkel, F. Oerlemans, W. van den Broek, B. Gohlsch, D. Pette, J.J. Plomp, *et al* . Abnormal myotonic dystrophy protein kinase levels produce only mild myopathy in mice. *Nat. Genet.* 1996, 13: 316-324.

[60] Reddy, S., D.B.J. Smith, M.M. Rich, J.M. Leferovich, P. Reilly, B.M. Davis, K. Taran, H. Rayburn, R Bronson, D. Cros, *et al* . Mice lacking the myotonic dystrophy protein kinase develop a late onset progressive myopathy. *Nat. Genet.* 1996, 13: 325-335

[61] Verschuure P, Tatard C, Boelens WC, Grongnet JF, David JC. Expression of small heat shock proteins HspB2, HspB8, Hsp20 and cvHsp in different tissues of the perinatal developing pig. *Eur. J. Cell Biol.* 2003, 82: 523–530.

[62] Sugiyama Y, Suzuki A, Kishikawa M, Akutsu R, Hirose T , Waye MM. et al. Muscle develops a specific form of small heat shock protein complex composed of MKBP/HSPB2 and HSPB3 during myogenic differentiation. *J. Biol. Chem.* 2000, 275: 1095–1104.

[63] Yoshida K, Aki T, Harada K, Shama KM, Kamoda Y, Suzuki A, Ohno S. Translocation of HSP27 and MKBP in ischemic heart. *Cell Struct Funct.* 1999, 24(4):181-5.

[64] Shama KM, Suzuki A, Harada K, Fujitani N, Kimura H, Ohno S, Yoshida K.Transient up-regulation of myotonic dystrophy protein kinase-binding protein, MKBP, and HSP27 in the neonatal myocardium. *Cell Struct Funct.* 1999, 24(1):1-4.

[65] Bukach OV, Glukhova AE, Seit-Nebi AS, Gusev NB. Heterooligomeric complexes formed by human small heat shock proteins HspB1 (Hsp27) and HspB6 (Hsp20). *Biochim Biophys Acta.* 2009; 1794(3):486-95.

[66] Morrison LE, Whittaker RJ, Klepper RE, Wawrousek EF, Glembotski CC. Roles for alphaB-crystallin and HSPB2 in protecting the myocardium from ischemia-reperfusion-induced damage in a KO mouse model. *Am J Physiol Heart Circ Physiol.* 2004; 286(3):H847-55.

[67] Benjamin IJ, Guo Y, Srinivasan S, Boudina S, Taylor RP, Rajasekaran NS, Gottlieb R, Wawrousek EF, Abel ED, Bolli R. CRYAB and HSPB2 deficiency alters cardiac metabolism and paradoxically confers protection against myocardial ischemia in aging mice. *Am J Physiol Heart Circ Physiol.* 2007, 293(5):H3201-9.

[68] Pinz I, Robbins J, Rajasekaran NS, Benjamin IJ, Ingwall JS. Unmasking different mechanical and energetic roles for the small heat shock proteins CryAB and HSPB2 using genetically modified mouse hearts. *FASEB J.* 2008, 22(1):84-92.

[69] Lam WY, Wing Tsui SK, Law PT, Luk SC, Fung KP, Lee CY, Waye MM. Isolation and characterization of a human heart cDNA encoding a new member of the small heat shock protein family--HSPL27. *Biochim Biophys Acta.* 1996, 1314(1-2):120-4.

[70] Boelens WC, Van Boekel MA, De Jong WW. HspB3, the most deviating of the six known human small heat shock proteins. *Biochim Biophys Acta.* 1998, 1388(2):513-6.

[71] Maddala R, Rao VP. alpha-Crystallin localizes to the leading edges of migrating lens epithelial cells. *Exp Cell Res.* 2005, 306(1):203-15.

[72] Yaung J, Jin M, Barron E, Spee C, Wawrousek EF, Kannan R, Hinton DR. alpha-Crystallin distribution in retinal pigment epithelium and effect of gene knockouts on sensitivity to oxidative stress. *Mol Vis.* 2007, 13:566-77.

[73] Wang X, Klevitsky R, Huang W, Glasford J, Li F, Robbins J. AlphaB-crystallin modulates protein aggregation of abnormal desmin. *Circ Res.* 2003, 93(10):998-1005.

[74] Wang X, Osinska H, Gerdes AM, Robbins J. Desmin filaments and cardiac disease: establishing causality. *J Card Fail.* 2002, 8(6 Suppl):S287-92.

[75] Iwaki T, Kume-Iwaki A, Goldman JE. Cellular distribution of αB-crystallin in non-lenticular tissues. *J Histochem Cytochem.* 1990, 38: 31–39.

[76] Kumarapeli AR, Su H, Huang W, Tang M, Zheng H, Horak KM, Li M, Wang X. Alpha B-crystallin suppresses pressure overload cardiac hypertrophy. *Circ Res.* 2008, 103(12):1473-82.

[77] Hoover HE, Thuerauf DJ, Martindale JJ, Glembotski CC. αB-Crystallin gene induction and phosphorylation by MKK6-activated p38. *J Biol Chem.* 2000, 275: 23825–23833.

[78] Ito H, Okamoto K, Nakayama H, Isobe T, Kato K. Phosphorylation of αB-crystallin in response to various types of stress. *J Biol Chem.* 1997, 272: 29934–29941.

[79] Kegel KB, Iwaki A, Iwaki T, Goldman JE. AlphaB-crystallin protects glial cells from hypertonic stress. *Am J Physiol.* 1996, 270(3 Pt 1):C903-9.

[80] Mehlen P, Kretz-Remy C, Préville X, Arrigo AP. Human hsp27, Drosophila hsp27 and human alphaB-crystallin expression-mediated increase in glutathione is essential for the protective activity of these proteins against TNFalpha-induced cell death. *EMBO J.* 1996, 15(11):2695-706.

[81] Jin JK, Whittaker R, Glassy MS, Barlow SB, Gottlieb RA, Glembotski CC. Localization of phosphorylated alphaB-crystallin to heart mitochondria during ischemia-reperfusion. *Am J Physiol Heart Circ Physiol.* 2008, 294(1):H337-44.

[82] Golenhofen N, Ness W, Koob R, Htun P, Schaper W, Drenckhahn D. Ischemia-induced phosphorylation and translocation of stress protein αB-crystallin to Z lines of myocardium. *Am J Physiol.* 1998, 274: H1457–H1464.

[83] Golenhofen N, Htun P, Ness W, Koob R, Schaper W, Drenckhahn D. Binding of the stress protein αB-crystallin to cardiac myofibrils correlates with the degree of myocardial damage during ischemia/reperfusion in vivo. *J Mol Cell Cardiol.* 1999, 31: 569–580.

[84] Golenhofen N, Arbeiter A, Koob R, Drenckhahn D. Ischemia-induced association of the stress protein αB-crystallin with I-band portion of cardiac titin. *J Mol Cell Cardiol.* 2002, 34: 309–319.

[85] Ray PS, Martin JL, Swanson EA, Otani H, Dillmann WH, Das DK. Transgene overexpression of alphaB crystallin confers simultaneous protection against cardiomyocyte apoptosis and necrosis during myocardial ischemia and reperfusion. *FASEB J.* 2001, 15(2):393-402.

[86] Kato K, Ito H, Kamei K, Inaguma Y, Iawamoto I, Saga S. Phosphorylation of αB-crystallin in mitotic cells and identification of enzymatic activities responsible for phosphorylation. *J Biol Chem.* 1998, 273: 28346–28354.

[87] Eaton P, Fuller W, Bell JR, Shattock MJ. αB Crystallin translocation and phosphorylation: signal transduction pathways and preconditioning in the isolated rat heart. *J Mol Cell Cardiol.* 2001; 33: 1659–1671.

[88] Morrison LE, Hoover HE, Thuerauf DJ, Glembotski CC. Mimicking phosphorylation of alphaB-crystallin on serine-59 is necessary and sufficient to provide maximal protection of cardiac myocytes from apoptosis. *Circ Res.* 2003, 92(2):203-11.

[89] Haley DA, Horwitz J, Stewart PL. The small heat-shock protein, αB-crystallin, has a variable quaternary structure. *J Mol Biol.* 1998; 277: 27–35.

[90] Muchowski PJ, Wu GJ, Liang JJ, Adman ET, Clark JI. Site-directed mutations within the core "α-crystallin" domain of the small heat-shock protein, human αB-crystallin, decrease molecular chaperone functions. *J Mol Biol.* 1999, 289: 397–411.

[91] Ito H, Kamei K, Iwamoto K, Inaguma Y, Nohara D, Kato K. Phosphorylation-induced change of the oligomerization state of αB-crystallin. *J Biol Chem.* 2001, 276: 5346–5352.

[92] Bova MP, Yaron O, Huang Q, Ding L, Haley DA, Steward PL, Horwitz J. Mutation R120G in αB-crystallin, which is linked to a desmin-related myopathy, results in an

irregular structure and defective chaperone-like function. *Proc Natl Acad Sci U S A.* 1999, 96: 6137–6142.

[93] Kato K, Goto S, Inaguma Y, et al.: Purification and characterization of a 20-kDa protein that is highly homologous to alpha B crystallin. *J Biol Chem.* 1994, 269:15302-15309.

[94] Chu G, Egnaczyk GF, Zhao W, Jo SH, Fan GC, Maggio JE, Xiao RP, Kranias EG. Phosphoproteome analysis of cardiomyocytes subjected to beta-adrenergic stimulation: identification and characterization of a cardiac heat shock protein p20. *Circ Res.* 2004, 94(2):184-93.

[95] O'Connor MJ, Rembold CM: Heat-induced force suppression and HSP20 phosphorylation in swine carotid media. *J Appl Physiol* 2002, 93:484-488.

[96] Inaguma Y, Hasegawa K, Kato K, Nishida Y. cDNA cloning of a 20-kDa protein (p20) highly homologous to small heat shock proteins: developmental and physiological changes in rat hindlimb muscles. *Gene.* 1996, 178:145-150.

[97] Gusev NB, Bukach OV, Marston SB. Structure, properties, and probable physiological role of small heat shock protein with molecular mass 20 kD (Hsp20, HspB6). *Biochemistry (Mosc).* 2005, 70(6):629-37.

[98] Zhu YH, Ma TM, Wang X. Gene transfer of heat-shock protein 20 protects against ischemia/reperfusion injury in rat hearts. *Acta Pharmacol Sin.* 2005, 26(10):1193-200.

[99] Zhu YH, Wang X. Overexpression of heat-shock protein 20 in rat heart myogenic cells confers protection against simulated ischemia/reperfusion injury. *Acta Pharmacol Sin.* 2005, 26(9):1076-80.

[100] Fan GC, Chu G, Mitton B, Song Q, Yuan Q, Kranias EG. Small heat-shock protein Hsp20 phosphorylation inhibits beta-agonist-induced cardiac apoptosis. *Circ Res.* 2004,94(11):1474-82

[101] Fan GC, Ren X, Qian J, Yuan Q, Nicolaou P, Wang Y, Jones WK, Chu G, Kranias EG. Novel cardioprotective role of a small heat-shock protein, Hsp20, against ischemia/reperfusion injury. *Circulation.* 2005, 111(14):1792-9.

[102] Fan GC, Yuan Q, Song G, Wang Y, Chen G, Qian J, Zhou X, Lee YJ, Ashraf M, Kranias EG. Small heat-shock protein Hsp20 attenuates beta-agonist-mediated cardiac remodeling through apoptosis signal-regulating kinase 1. *Circ Res.* 2006, 99(11):1233-42.

[103] Islamovic E, Duncan A, Bers DM, Gerthoffer WT, Mestril R. Importance of small heat shock protein 20 (hsp20) C-terminal extension in cardioprotection. *J Mol Cell Cardiol.* 2007, 42(4):862-9.

[104] Fan GC, Zhou X, Wang X, Song G, Qian J, Nicolaou P, Chen G, Ren X, Kranias EG. Heat shock protein 20 interacting with phosphorylated Akt reduces doxorubicin-triggered oxidative stress and cardiotoxicity. *Circ Res.* 2008, 103(11):1270-9.

[105] van de Klundert FA, de Jong WW. The small heat shock proteins Hsp20 and alphaB-crystallin in cultured cardiac myocytes: differences in cellular localization and solubilization after heat stress. *Eur J Cell Biol* 1999, 78:567-572.

[106] Verschuure P, Croes Y, van den IJssel PR. Translocation of small heat shock proteins to the actin cytoskeleton upon proteasomal inhibition. *J Mol Cell Cardiol* 2002, 34:117-128.

[107] Pipkin W, Johnson JA, Creazzo TL, et al. Localization, macromolecular associations, and function of the small heat shock-related protein HSP20 in rat heart. *Circulation* 2003, 107:469-476.

[108] Boluyt MO, Brevick JL, Rogers DS, Randall MJ, Scalia AF, Li ZB. Changes in the rat heart proteome induced by exercise training: Increased abundance of heat shock protein hsp20. *Proteomics*. 2006, 6(10):3154-69.

[109] Burniston JG. Adaptation of the rat cardiac proteome in response to intensity-controlled endurance exercise. *Proteomics*. 2009, 9(1):106-15.

[110] Dohke T, Wada A, Isono T, Fujii M, Yamamoto T, Tsutamoto T, Horie M. Proteomic analysis reveals significant alternations of cardiac small heat shock protein expression in congestive heart failure. *J Card Fail*. 2006, 12(1):77-84.

[111] Damer CK, Partridge J, Pearson WR, Haystead TA. Rapid identification of protein phosphatase 1-binding proteins by mixed peptide sequencing and data base searching. Characterization of a novel holoenzymic form of protein phosphatase 1. *J Biol Chem*. 1998, 273:24396-24405.

[112] Carr AN, Schmidt AG, Suzuki Y, et al. Type 1 phosphatase, a negative regulator of cardiac function. *Mol Cell Biol*. 2002, 22:4124-4135.

[113] Nicolaou P, Knöll R, Haghighi K, Fan GC, Dorn GW 2nd, Hasenfub G, Kranias EG. Human mutation in the anti-apoptotic heat shock protein 20 abrogates its cardioprotective effects. *J Biol Chem*. 2008, 283(48):33465-71.

[114] Singal PK, Iliskovic N. Doxorubicin-induced cardiomyopathy. *N Engl J Med*. 1998, 339: 900–905.

[115] Ganey PE, Carter LS, Mueller RA, Thurman RG. Doxorubicin toxicity in perfused rat heart. Decreased cell death at low oxygen tension. *Circ Res*. 1991, 68: 1610–1613

[116] Shan YX, Liu TJ, Su HF, Samsamshariat A, Mestril R, Wang PH. Hsp10 and Hsp60 modulate Bcl-2 family and mitochondria apoptosis signaling induced by doxorubicin in cardiac muscle cells. *J Mol Cell Cardiol*. 2003, 35: 1135–1143.

[117] Krief S, Faivre JF, Robert P, Le Douarin B, Brument-Larignon N, Lefrère I, Bouzyk MM, Anderson KM, Greller LD, Tobin FL, Souchet M, Bril A. Identification and characterization of cvHsp. A novel human small stress protein selectively expressed in cardiovascular and insulin-sensitive tissues. *J Biol Chem*. 1999, 274(51):36592-600.

[118] Golenhofen N, Perng MD, Quinlan RA, Drenckhahn D. Comparison of the small heat shock proteins alphaB-crystallin, MKBP, HSP25, HSP20, and cvHSP in heart and skeletal muscle. *Histochem Cell Biol*. 2004, 122(5):415-25.

[119] Verschuure P, Tatard C, Boelens WC, Grongnet JF, David JC. Expression of small heat shock proteins HspB2, HspB8, Hsp20 and cvHsp in different tissues of the perinatal developing pig. *Eur J Cell Biol*. 2003, 82(10):523-30.

[120] Doran P, Gannon J, O'Connell K, Ohlendieck K. Aging skeletal muscle shows a drastic increase in the small heat shock proteins alphaB-crystallin/HspB5 and cvHsp/HspB7. *Eur J Cell Biol*. 2007, 86(10):629-40.

[121] Sun X, Fontaine JM, Rest JS, Shelden EA, Welsh MJ, Benndorf R. Interaction of human HSP22 (HSPB8) with other small heat shock proteins. *J Biol Chem*. 2004, 279(4):2394-402.

[122] Benndorf R, Sun X, Gilmont RR, Biederman KJ, Molloy MP, Goodmurphy CW, Cheng H, Andrews PC, Welsh MJ. HSP22, a new member of the small heat shock

protein superfamily, interacts with mimic of phosphorylated HSP27 ((3D)HSP27). *J Biol Chem.* 2001, 276(29):26753-61.

[123] Kim MV, Seit-Nebi AS, Gusev NB. The problem of protein kinase activity of small heat shock protein Hsp22 (H11 or HspB8). *Biochem Biophys Res Commun.* 2004, 325(3):649-52.

[124] Shemetov AA, Seit-Nebi AS, Gusev NB. Structure, properties, and functions of the human small heat-shock protein HSP22 (HspB8, H11, E2IG1): a critical review. *J Neurosci Res.* 2008, 86(2):264-9.

[125] Hedhli N, Wang L, Wang Q, Rashed E, Tian Y, Sui X, Madura K, Depre C. Proteasome activation during cardiac hypertrophy by the chaperone H11 Kinase/Hsp22. *Cardiovasc Res.* 2008, 77(3):497-505.

[126] Depre C, Hase M, Gaussin V, Zajac A, Wang L, Hittinger L, Ghaleh B, Yu X, Kudej RK, Wagner T, Sadoshima J, Vatner SF. H11 kinase is a novel mediator of myocardial hypertrophy in vivo. *Circ Res.* 2002, 91: 1007–1014

[127] Wang L, Zajac A, Hedhli N, Depre C. Increased expression of H11 kinase stimulates glycogen synthesis in the heart. *Mol Cell Biochem.* 2004, 265: 71–78.

[128] Hase M, Depre C, Vatner S, Sadoshima J. H11 has dose-dependent and dual hypertrophic and proapoptotic functions in cardiac myocytes. *Biochem J.* 2005, 388: 475–483.

[129] Depre C, Wang L, Sui X, Qiu H, Hong C, Hedhli N, Ginion A, Shah A, Pelat M, Bertrand L, Wagner T, Gaussin V, Vatner SF. H11 kinase prevents myocardial infarction by preemptive preconditioning of the heart. *Circ Res.* 2006, 98(2):280-8.

[130] Danan IJ, Rashed ER, Depre C. Therapeutic potential of H11 kinase for the ischemic heart. *Cardiovasc Drug Rev.* 2007, 25(1):14-29.

[131] Morrow G, Samson M, Michaud S, Tanguay RM. Overexpression of the small mitochondrial Hsp22 extends Drosophila life span and increases resistance to oxidative stress. *FASEB J.* 2004, 18(3):598-9.

[132] Kappe G, Verschuure P, Philipsen RL, Staalduinen AA, Van de Boogaart P, Boelens WC, et al., Characterization of two novel human small heat shock proteins: protein kinase-related HspB8 and testis-specific HspB9. *Biochim. Biophys. Acta* 2001, 1520:1–6.

[133] Fontaine JM, Rest JS, Welsh MJ, Benndorf R. The sperm outer dense fiber protein is the 10th member of the superfamily of mammalian small stress proteins. *Cell Stress Chaperones* 2003, 8: 62–69.

[134] Van der Hoorn FA, Tarnasky HA, Nordeen SK. A new rat gene RT7 is specifically expressed during spermatogenesis. *Dev. Biol.* 1990, 142: 147–154.

[135] Higgy NA, Pastoor T, Renz C, Tarnasky HA and Van der Hoorn FA. Testis-specific RT7 protein localizes to the sperm tail and associates with itself. *Biol. Reprod.* 1994, 50:1357–1366.

[136] Gastmann O, Burfeind P, Gunther E, Hameister H, Szpirer C, Hoyer-Fender S. Sequence, expression, and chromosomal assignment of a human sperm outer dense fiber gene. *Mol. Reprod.* Dev. 1993, 36: 407–418.

[137] Yamboliev IA, Hedges JC, Mutnick JL, Adam LP, Gerthoffer WT. Evidence for modulation of smooth muscle force by the p38 MAP kinase/HSP27 pathway. *Am J Physiol Heart Circ Physiol.* 2000, 278(6):H1899-907.

In: Small Stress Proteins and Human Diseases
Editors: Stéphanie Simon et al.
ISBN: 978-1-61470-636-6

Chapter 1.6

PROTEIN QUALITY CONTROL IN HEART DISEASE: SMALL HEAT SHOCK PROTEINS AND AUTOPHAGY

***Paul Tannous*[1]*, Beverly A. Rothermel*[1] *and Joseph A. Hill*[1,2*]**

Departments of [1]Internal Medicine (Cardiology), and [2]Molecular Biology, University of Texas Southwestern Medical Center, Dallas, Texas, U.S.

ABSTRACT

Disease-related stress to the myocardium, such as unremitting hypertension or myocardial injury, triggers hypertrophic growth that increases the risk of functional decompensation and malignant rhythm disturbance [1]. Hypertrophy and other forms of cardiac plasticity are associated with energy-dependent structural changes in cardiac myocytes and fibroblasts, coordinated angiogenic alterations in blood supply, adjustments in the consumption of energy-providing substrates, and, in the case of pathological cardiac hypertrophy, activation of a previously dormant "fetal gene program". Our understanding of mechanisms governing these remodeling reactions is incomplete, and deciphering them may yield insights into the pathogenesis of heart failure, a major source of morbidity and mortality worldwide.

Diverse catabolic and anabolic processes contribute to remodeling of the healthy and diseased heart. Recent studies have demonstrated the importance of two inter-related mechanisms of protein quality control, small heat shock proteins (sHsp) and autophagy. The sHsps are molecular chaperones that bind unfolded aggregate-prone proteins to facilitate proper refolding or direct them for degradation by the autophagosome. Here, we will review this topic, focusing especially on the roles of two sHsps, HspB8 and αB-crystallin (CryAB, HspB5), in cardiac health and disease.

* Corresponding Author: Joseph A. Hill
University of Texas Southwestern Medical Center,
Division of Cardiology,
Dallas, TX 75390-8573, USA
Tel: 214-648-1400
Fax: 214-648-1450
joseph.hill@UTSouthwestern.edu

1. Dynamics of Myocardial Growth

Hypertrophic growth is a primary mechanism whereby the ventricle normalizes wall stress. It is characterized by enhanced protein synthesis and an increase in the size and organization of force-generating units (sarcomeres) within individual myocytes [2, 3]. Importantly, depending on the inciting stressor, myocardial growth may or may not be associated with adverse sequelae, such as functional decompensation or serious rhythm disturbance.

The size of the heart is governed largely by myocyte growth and atrophy; myocyte proliferation is not a major contributor. That being said, in some cases, the cumulative effects of myocyte death contribute to ventricular wall thinning and associated declines in heart mass. Cell growth or shrinkage, in turn, is dictated by a delicate counterpoise between pro-growth and anti-growth mechanisms. Contributing to cellular pro-growth are activation of protein anabolic pathways and de-activation of protein catabolism. In the case of myocardial atrophy, catabolic pathways are activated and anabolic cascades switched off. And throughout these events, coordinated alterations in myocyte metabolism – with rapid changes in energy substrate preferences from fatty acids to glucose – take place. Further, cardiac myocytes, like other post-mitotic cells (e.g. neurons) have evolved a set of sophisticated mechanisms governing protein quality control [4, 5], as maintaining integrity of the rapidly changing proteome is of utmost importance as the heart grows or shrinks.

2. Small Heat Shock Proteins: Guardians of Proteome Integrity

Mechanisms governing proteome integrity involve two main components. First, the cell must maintain constant vigilance to assess protein folding and structure at all times. Likewise, there must be an equally robust mechanism for rapid and efficient clearance of damaged or non-native substrates which can be toxic to the cell. The family of sHsps is a unique component of the protein quality control pathway as these molecular chaperones play a central role in both protein surveillance and protein clearance [6, 7].

The expression of heat shock proteins is both ubiquitous and enhanced by various acute and chronic stimuli, such as heat shock, heavy metals, low molecular weight toxins, infections, and oxidative stress [8-14]. Heat shock proteins act to ensure proper protein folding, to prevent protein misfolding, and to assist in protein refolding to the correct, native state. In addition to their role in protein quality control, heat shock proteins participate in diverse signaling pathways including those controlling cell death [15].

The human genome encodes 10 sHsps which share the following structural and functional characteristics: a) an 80-100 amino acid α-crystallin domain near the C-terminus; b) a small molecular mass ranging from 17.0 to 28.4 kDa; c) the ability to form large homo or hetero-oligomers; d) increased expression in response to diverse forms of stress; and e) an ATP-independent chaperone-like activity [16]. HspB1/Hsp27, HspB5/CryAB, HspB6/Hsp20 and HspB8/Hsp22 are expressed in all tissues, whereas, sHspB2/MKBP, sHspB3, and sHspB7 are specific to heart and other muscle types. HspB4 is expressed in the lens of the eye. HspB9 and HspB10 are testis-specific (see Introduction chapter).

Expression of sHsp is highly dynamic. For example, CryAB protein levels are elevated at 2 weeks after acutely increased afterload by surgical thoracic aortic constriction (TAC) [17]. In many instances, heat shock transcription factor 1 (HSF1) is a key determinant of heat shock protein expression [18]. And Hsf1 plays a significant role in the preservation of systolic function during the development of cardiac hypertrophy under both pathologic and physiologic conditions [19, 20]. The insulin-like growth factor-1/phosphatidylinositol 3-kinase pathway, a well-established physiological cardiac hypertrophic pathway, is involved in HSF1 activation [18].

The sHsps function during polypeptide translation at the ribosome, where the vast majority of nascent proteins rely on chaperones in order to assume correct tertiary and quaternary conformations. Following translation, sHsps continually monitor the proteome for proteins which have acquired a damaged or non-native state. Upon binding non-native proteins, sHsps sequester them in a stable intermediate state, where the bound protein's fate is determined by interaction with additional components of the protein quality control system [7].

sHsp binding to a non-native conformation is an important triage point in the life of a protein. It is here that interactions with various cofactors determine whether the protein undergoes refolding or degradation. Refolding, for example, is mediated by interactions of the sHsp-bound peptide with Hsp70 and the co-chaperone Bag2, where the chaperone complex repeatedly binds and releases the non-native protein in an ATP-dependent process, facilitating refolding and a return to the native state [21, 22]. From an energetic standpoint, refolding is the preferred fate, as studies of mitochondrial protein turnover indicate that Hsp70-mediated refolding consumes less energy overall than degradation followed *de novo* generation of replacement proteins [23].

When refolding is not a viable option, the non-native protein is then targeted for degradation. The best characterized pathway of protein degradation involves the ubiquitin-proteasome system (UPS), a mechanism where individual polypeptides undergo rapid proteolysis. In this case, the sHsp-non-native protein complex interacts with Hsp70, Bag1, and the E3 ligase CHIP, with the net result being rapid shuttling towards the proteasome [24]. While the proteasome is an efficient proteolytic mechanism, its capacity and range of substrates is somewhat limited. For example, the proteasome core is a structurally narrow barrel into which substrate must fit in order to undergo proteolysis. This constraint on substrate size thus prohibits entry and proteolysis of organelles, complex oligomers, and large protein aggregates [25, 26]. In fact, complex proteinaceous material is not only a poor proteasome substrate, but some non-native proteins and mid-to-large-sized aggregates can interfere sterically with the system, "clogging it up", and causing global proteasome dysfunction [27].

As an alternative to proteasome-mediated proteolysis, proteins can be degraded via a lysosome-mediated bulk degradation process known as autophagy. In this process, large products are targeted towards the autophagosome through either direct or indirect mechanisms. Importantly, there is complex interplay between sHsp mechanisms of protein quality control and autophagy. For example, and as discussed below, HspB8 is capable of directly targeting protein substrates to the autophagosome, providing an elegant example of substrate specificity [28].

3. sHsp-Associated Proteinopathy

The importance of chaperone-mediated protein quality control is underscored by the growing list of diseases resulting from its failure. To date, mutations associated with human disease have been identified in 4 of the 10 identified sHsp [29-32]. These diseases fall within the category of protein conformational disorders or proteinopathies. In each instance, disease phenotype and clinical presentation are determined by the level of expression of the non-native sHsp and the cellular compartment in which it is expressed. For example, in the nervous system, sHsp mutations are associated with hereditary motor neuropathies, such as Charcot-Marie-Tooth disease [30]. In the heart, sHsps are associated with severe and progressive cardiomyopathy and sudden cardiac death [31] (see chapter 1-5).

Pathogenesis in sHsp-associated proteinopathies is often multifactorial. Clearly, depletion of functional, native protein can result in failure of downstream processes. Further, in many instances, pathogenesis stems, at least in part, from emergence of a toxic gain-of-function to the chaperone protein itself, resulting in widespread intracellular protein misfolding and extensive accumulation of aggregates. Furthermore, aggregates can bind to signaling components resulting in cellular toxicity and dysfunction [33]. Conversely, in some cases it is actually pre-aggregated amyloid-like substrates (pre-amyloid oligomers) that are the greater source of toxicity, with sequestration of denatured proteins in aggregates representing an adaptive cellular response [34]. In this setting, pre-aggregated substrates cause disease by binding to and disrupting transcription factors and other cytosolic signaling elements, thereby interfering with numerous cellular pathways. A number of studies have shown that heat shock proteins confer protection against cardiovascular disease by antagonizing these events. Among these beneficial effects are resistance to ischemia/reperfusion injury, including reduction in infarct size, blunting of doxorubicin-induced cardiomyopathy, and antagonism of atrial fibrillation [18, 19, 35, 36].

Intracellular aggregates of denatured protein are amassed in peri-nuclear organelles called aggresomes [37]. These structures form when the cytoskeletal machinery binds diffuse cytosolic amyloid/aggregated proteins and actively shuttles them retrograde along microtubules toward the microtubule organizing center (MTOC), the net result being a coalescence of toxic substrates in one location [37]. This mechanism is thought to be protective, as non-native proteins are sequestered and hence unable to interfere with other cellular events. While it appears that whether the presence of protein aggregates is protective or pathologic is context dependent, in no case has the presence of a non-native protein proven to be advantageous when compared to the presence of the native protein.

Extensive accumulation of pre-amyloid intermediates and intracellular aggregates, with or without the formation of aggresomes, necessitates proteolytic mechanisms to process and eliminate them. As aggregated protein is a poor proteasome substrate, in some settings actually causing proteasome dysfunction, an alternative pathway is required, *viz.* autophagy.

4. Autophagy

Autophagy is a lysosome-dependent pathway of bulk protein and organelle recycling, capable of processing complex and non-denaturable substrates, which is present in all eukaryotic cells. Three forms of autophagy have been described: chaperone-mediated autophagy, microautophagy, and macroautophagy. Chaperone-mediated autophagy is a highly specific process whereby substrates are directly targeted to the lysosome by a specific pentapeptide motif (KFERQ) and the binding of a molecular chaperone complex containing Hsp70, Hsp90, and other co-chaperones. This substrate-chaperone complex binds Lamp2, a lysosomal surface receptor, leading to substrate translocation into the lumen where it undergoes proteolysis [38]. Failure of this process is associated with human disease, including Danon's disease, an X-linked vacuolar myopathy that affects cardiac muscle and skeletal muscle and is characterized by cognitive developmental delay [39, 40]. In contrast, microautophagy is a less well characterized process in which substrates are engulfed directly by the lysosome and are degraded. It is thought that microautophagy is a constitutive form of autophagy and functions in a housekeeping-type role [41, 42].

Macroautophagy (herein termed autophagy) is the best characterized form of this process and involves multiple steps prior to lysosome-mediated proteolysis. Here, the autophagic pathway begins with formation of an autophagosome, a double-membrane structure of unknown origin which engulfs cytoplasmic contents without clearly defined substrate specificity. As the autophagosome matures it then fuses with a lysosome, forming the autolysosome, with subsequent proteolysis of engulfed materials [43]. Unlike proteasome-mediated proteolysis and its significant size constraints, the autophagic machinery is able to engulf entire organelles such as peroxisomes or mitochondria, allowing for large complexes to be broken down to basic molecular building blocks to be reutilized by the cell [43].

Accumulating data demonstrate important links between the two major pathways of protein degradation – the ubiquitin proteasome system (UPS) and autophagy – in cellular events. For example, inhibition of proteasome activity in the heart leads to accumulation of poly-ubiquitinated proteins and subsequent activation of autophagy [44, 45]. Also, the Akt-FoxO pathway links growth factor signaling to the coordinated regulation of the proteasomal and lysosomal systems. FoxO3 mediates both cardiac and skeletal muscle atrophy through its direct activation of the muscle-specific ubiquitin ligase atrogin-1/MAFbx, promoting degradation of proteins via the UPS [46]. More recent studies have shown that FoxO3 positively regulates the expression of a cadre of autophagy-related genes [47]. In other words, FoxO3 has a unique role in its regulation of both proteasomal and lysosomal systems.

A link between autophagy and proteinopathic disease has been established in a number of neurodegenerative disorders, all of which manifest extensive intracellular aggregate accumulation and eventual cellular demise [48]. In fact, many similarities exist between neurodegenerative disease and sHsp-induced proteinopathy. In each case, failure of the cell's protein quality control mechanism is causal, due either to the presence of an abnormal protein that overloads the degradation machinery and accumulates in the cell or to primary failure of the surveillance machinery itself, as in the case of chaperone-based disease. The autophagy-proteinopathy link has been best studied in Huntington's disease, where expression of glutamine repeat-containing Huntingtin protein is a robust inducer of autophagy. In this setting, autophagy functions in an adaptive manner, facilitating clearance of both intracellular aggregates as well as pre-aggregate, amyloid-like structures [49].

5. HspB8: A Direct Molecular Link between Chaperone Function and Autophagy

HspB8 (also known as Hsp22) is a small heat shock protein expressed at high levels in human skeletal muscle, heart, and central nervous system, with spinal cord and sensory neurons especially enriched in the protein [30]. Lower levels of expression are found in prostate, lung, and kidney [50, 51]. Within the cell, HspB8 has been localized to the cytoplasm, the nucleus, and in association with the inner surface of the plasma membrane [52]. Like other sHsp, HspB8 expression increases in response to stress, though its induction appears to be dependent on the specific nature of the stress and the cell type in question; for example, heat shock induces HspB8 in MCF-7 cells but not in HeLa cells [53]. In piglet heart, hypoxia does not induce HspB8, whereas it is up-regulated after transient ischemia [54].

HspB8 has been implicated as an important and adaptive response to multiple forms of stress. In heart, ischemia/reperfusion dramatically up-regulates HspB8, which then serves to regulate cytoprotection, cell growth, and attenuation of apoptosis [54]. HspB8 levels are elevated in hibernating myocardium, mechanically dysfunctional tissue whose contractile performance returns dramatically following revascularization [55]. *In vitro* studies demonstrate a specific mechanism wherein HspB8 can act as a chaperone to other chaperones that may themselves have assumed a non-native structure. This is seen, for example, in the case of an R→G missense mutation in the sHsp CryAB ($CryAB^{R120G}$). As discussed later, $CryAB^{R120G}$ elicits significant proteinopathy with devastating cardiac malfunction. Over-expression of HspB8 *in vitro* is sufficient to rescue $CryAB^{R120G}$ oligomeric organization, allowing the mutant CryAB protein to assume its native conformation and carry out its chaperone function [56]. Expression of a cardiac-specific HspB8 transgene inhibits the progression of cardiomyopathy in a transgenic mouse model of $CryAB^{R120G}$-mediated heart disease [57].

Insights into HspB8 function have emerged in studies addressing its role in Huntington's disease (HD). Analogous to other neurodegenerative conditions, pathogenesis in HD involves expression of a mutant protein with non-native conformation that overloads the chaperone machinery leading to extensive accumulation of intracellular aggregates. In this setting, autophagy has emerged as a functionally relevant cellular response. For example, pharmacological activation of autophagy attenuates the accumulation of non-native Huntingtin protein. Conversely, pharmacologic or genetic inhibition of autophagy results in enhanced aggregate accumulation and cellular toxicity [49]. An interesting finding in these studies was that enhancement or inhibition of autophagic activity had no discernable effect on levels of native Huntingtin protein, suggesting the existence of substrate specificity in autophagic clearance. Furthermore, autophagy is capable of clearing both aggregated as well as soluble 'monomeric' Huntingtin species [58]. This observation is of particular interest as it suggests the existence of molecular machinery capable of specifically identifying non-native substrate and targeting it toward autophagic degradation. Collectively, these studies demonstrate that autophagy is induced by the expression of mutant Huntingtin protein, which functions as an adaptive mechanism of aggregate clearance. Furthermore, specificity for mutant Huntingtin protein, as opposed to native protein, points to mechanisms of substrate selection in autophagic clearance.

The striking link between autophagy and selective clearance of mutant Huntingtin protein was clarified in recent studies of HspB8. Expression of Htt43Q, mutant Huntingtin harboring a 43-glutamine track, results in extensive, SDS-insoluble, perinuclear inclusions. Surprisingly, co-expression of HspB8 (but not Hsp27 or CryAB) was sufficient to prevent the accumulation of SDS-insoluble Htt43Q [59]. In fact, mutant protein accumulated exclusively in cells that were concurrently treated with proteasome and autophagy inhibitors, indicating that over-expression of HspB8 resulted in enhanced Htt43Q clearance, so much so that aggregates were prevented from forming [59]. This work uncovered a *bona fide* chaperone function of HspB8 *in vivo* to maintain Htt43Q in a non-aggregated state competent for degradation. Further, these studies pointed to a direct link between sHsp-mediated protein quality control and the autophagic pathway. Indeed, this paradigm has multiple similarities with that involved in protein-refolding, including interaction of multiple binding partners with the non-native protein-sHsp complex, recognition by HspB8 of the denatured conformation, and subsequent binding and stabilization of an intermediate state. HspB8 subsequently binds the cofactor Bag3 in a 2:1 molar ratio, and it is this interaction that links HspB8 to the autophagic machinery [60, 61].

Bag3 is a member of the family of co-chaperone, Bag domain-containing proteins. Bag3 has 3 putative protein binding domains: BAG, WW, and proline rich domains [62]. The proline rich domain is essential for Bag3-dependent clearance of mutant Huntingtin. Notably, deletion of the BAG domain, which is required for binding of Hsp70 and Bcl-2, had no effect on autophagic activity. The proline-rich domain interacts with LC3, a major structural component of the nascent autophagic vacuole, thus bringing the *de novo* autophagic vacuole into close proximity with HspB8-bound non-native protein [61]. In addition to LC3 binding, Bag3 expression is sufficient to induce elevated LC3 protein levels [60]. Interestingly, knockdown of Bag3 abolishes clearance of mutant Huntingtin, whereas knockdown of HspB8 decreases, but does not abolish, clearance. Thus, Bag3 is necessary for linking Htt43Q to the autophagosome, but HspB8 is expendable [60], suggesting redundancy in molecular chaperones. Although Hsp27 and CryAB do not facilitate Htt43Q clearance, it is possible that other sHsps are involved.

Mutations in HspB8 are associated with human disease. Consistent with its high-level expression in spinal cord and peripheral motor neurons, mutations in this protein present with severe neurological sequelae. To date, 4 independent families have been found with 2 different missense mutations in HspB8, *viz.* $HspB8^{K141N}$ and $HspB8^{K141E}$ [30]. Common to each is the loss of the positively charged amino acid lysine in a region that is central to the crystallin domain. *In vitro* expression of these mutant HspB8s results in formation of extensive intracellular aggregates which are largely comprised of the mutant HspB8 and its binding partner HspB1. In cultured N2a neurons, these events lead to decreased cell viability [30]. Clinically, these hereditary conditions present as distal motor neuropathies characterized by pure motor dysfunction and associated atrophy and wasting of distal limb musculature [63].

Thus, HspB8 participates in cellular protein quality control at multiple points. Under conditions of stress, HspB8 attenuates the formation and accumulation of non-native protein substrates, preventing their disruption of cellular pathways. In the presence of the $CryAB^{R120G}$ mutant, HspB8 chaperone activity facilitates refolding of mutant protein, promoting re-emergence of CryAB chaperone function. Lastly, recent evidence demonstrates that HspB8 participates in the direct targeting of toxic, non-native substrates toward autophagic degradation. It is notable that this targeting proceeds in a highly specific manner, where only toxic forms of Huntingtin protein are removed while levels of the native species remain unchanged. Together these findings provide important clues regarding mechanistic links between autophagy and sHsp biology, a fertile ground for future research.

6. CryAB and Autophagy

CryAB/HspB5 is expressed at high levels in the ocular lens, brain, skeletal muscle, and heart. By contrast, expression of the closely related αA-crystallin (CryAA/HspB4) is largely restricted to the lens. Indeed, under baseline conditions CryAB is the most abundant sHsp in the heart, and levels of this protein can increase to a dramatic 3% of total cellular protein under conditions of stress [64, 65], suggesting an important role for this protein in disease.

CryAB binds to both the intermediate filament protein desmin and cytoplasmic actin helping to maintain cytoskeletal integrity. It is also required for assembly of desmin at the Z-line of the sarcomere and thus is critical for skeletal muscle integrity. Evidence suggests that improper folding of desmin leads to toxic accumulations, which, in turn, inhibit the proteasome, resulting in a failure to clear the misfolded proteins. The so called desmin-related cardiomyopathies (DRM) are caused by mutations in either the desmin or CryAB genes [66]. Over-expression of CryAB in cultured cardiomyocytes and in transgenic mouse hearts protects against ischemia/reperfusion-induced cell death.

Parsing the exact role of any sHsp is difficult given the functional overlap among these related proteins. Surprisingly, inactivation of the gene coding for CryAB has no effect on lens development nor is there decreased resistance to cataract formation, findings suggestive of functional redundancy with other sHsp [67]. Interestingly, while CryAB knockout mice are developmentally normal, a phenotype emerges at later ages. Knockout animals manifest a general loss of body weight, muscle degeneration with extensive accumulation of amorphous electron dense material between myofibrils, and abnormal development of the skeleton with consequent kyphosis.

6.1. CryAB in Disease

Early evidence suggesting that CryAB plays a critical role in the heart's adaptive response to diverse forms of stress was uncovered in microarray analysis of myocardial tissue from patients with either familial hypertrophic cardiomyopathy [68] and cardiomyopathy due to DRM [69]. In each of these distinct forms of heart disease, CryAB transcripts were significantly elevated, suggestive of a protective role. Conversely, CryAB is down-regulated in specimens taken from patients with terminal, class IV congestive heart failure, suggesting that failure in this chaperone system is a component of cardiac decompensation. Experimentally, a protective role for CryAB has also been demonstrated. For example, forced over-expression of CryAB protects cardiomyocytes from ischemia/reperfusion injury in both cell culture and transgenic animals [70, 71].

As mentioned earlier, DRMs are sporadic and familial myopathies caused by mutations in either desmin or CryAB. To date, nine unique mutations in CryAB have been identified as causative in human disease. In all reported cases, mutation results in an inherited myofibrillar myopathy with extensive intracellular protein aggregation. Clinical phenotypes vary; some mutations present with early-onset cataracts, whereas others manifest as desmin-related myopathy/cardiomyopathy with severe peripheral muscle and heart problems. Interestingly, none of the reported mutations has been associated with neurological disease despite the high level of CryAB expression in this compartment. This is likely due either to a limited role of CryAB or the co-expression of molecular chaperones with overlapping function.

Considerable investigation has focused on an arginine-to-glycine missense mutation at amino acid 120 (CryAB^{R120G}). This sequence variant was originally discovered in a familial case of DRM [31], an autosomal dominant disease presenting with weakness of the proximal and distal limb muscles, early onset cataracts, and severe cardiomyopathy. Mechanistically this condition is interesting in that disease is caused by a mutated protein that is itself a molecular chaperone; meanwhile, wild-type CryAB readily associates with protein aggregates which are comprised, in part, of denatured CryAB. Additional important insights have emerged from the study of CryAB^{R120G}-induced myopathy. For example, CryAB^{R120G} is capable of disrupting the UPS, with subsequent induction of an alternative degradative pathway, autophagy.

CryAB^{R120G} causes disease by multiple mechanisms. On one hand, mutant CryAB^{R120G} protein possesses no functional chaperone-like activity. Loss of chaperone function can be explained by aberrant folding, where the R120G mutation disrupts secondary, tertiary, and quaternary structure, resulting in a final conformation susceptible to heat-induced denaturation [72]. Beyond that, non-native, mutant CryAB^{R120G} can also disrupt native sHsps, functioning in a dominant-negative fashion. Indeed, *in vitro* experiments demonstrate that the addition of CryAB^{R120G} to α-lactalbumin enhances the kinetics and extent of protein aggregation. In this setting CryAB^{R120G} becomes entangled with the unfolded α-lactalbumin and comprises a major component of the insoluble protein aggregate [73].

CryAB^{R120G} co-aggregates with multiple intermediate filament proteins, highlighting intermediate filaments as an important physiologic substrate for CryAB [72]. Given CryAB's role as a chaperone of intermediate filament proteins, it is not surprising that CryAB^{R120G} causes extensive aggregation of desmin. These intracellular aggregates and pre-aggregated, soluble amyloid together cause significant disruption of the UPS. This has been shown both *in vivo* and *in vitro* by expression of a GPF-tagged degron, a substrate that is specifically degraded by the UPS. Co-expression of CryAB^{R120G}, but not WT CryAB, results in dramatic accumulation of the GFP substrate, despite an overall increase in the cell's proteasome capacity as measured by proteolysis assays. Dysfunction results from inadequate delivery of substrate to the 20S proteolytic core and concurrent depletion of key components of the 19S complex [70]. In other words, accumulation of poly-ubiquitinated, aggregate-prone protein is amplified by a functional "clogging up" of the UPS system, despite an increase in overall UPS capacity.

6.2. Autophagy in CryAB-Associated Myopathies

The role of autophagy in crystallin-induced desmin-related cardiomyopathy was directly addressed in a series of recent studies. Numerous reports confirm that forced expression of CryAB^{R120G} in neonatal rat cardiomyocytes is sufficient to induce robust accumulation of intracellular aggregates which are actively processed into peri-nuclear aggresomes [31, 74, 75]. Consistent with a role of autophagy in aggregate clearance, ultrastructural analysis revealed extensive accumulation of electron dense autophagosomes directly surrounding the protein aggregate [75]. Furthermore, live-cell monitoring of autophagy using a GFP-LC3 reporter system demonstrated a 3-fold increase in the number of autophagosomes in cells expressing mutant crystallin as compared with wild-type protein or mock-infected cells [75]. These findings, then, demonstrated that $CryAB^{R120G}$ expression is sufficient to induce the accumulation of intracellular aggregates, leading to robust induction of autophagy in the immediate peri-aggregate region.

While the aforementioned data revealed a spatial and temporal relationship between aggregates and autophagy, alone they do not demonstrate a mechanistic link. To test this, cells expressing either WT or mutant crystallin were treated with 3MA [3-methyladenine, an inhibitor of autophagosome formation), or rapamycin (an activator of autophagy). Strikingly, blunting autophagy resulted in a dramatic increase in the accumulation of insoluble CryAB [75], findings consistent with imaging studies demonstrating a dramatic increase in aggregates in cells with blunted autophagy.

Given that blunting autophagy with 3MA enhanced aggregate accumulation, we hypothesized that rapamycin, an inducer of autophagy, would facilitate clearance. Surprisingly, this was not the case. Rapamycin had a negligible effect on aggregate formation and induction of autophagy. However, this finding was elucidated by recognition that mTOR, the autophagy-inhibiting protein target of rapamycin, can be sequestered within intracellular aggregates [75]; thus, mTOR's inhibitory actions on autophagy are blunted already, and rapamycin would be expected to have little or no additional effect. Consistent with this model, confocal imaging demonstrated that mTOR is sequestered within the CryAB-desmin based aggregates [75].

To test this model *in vivo*, mice engineered with cardiomyocyte-restricted over-expression of mutant CryAB (αMHC-$CryAB^{R120G}$) [76] were studied. Consistent with the *in vitro* findings, $CryAB^{R120G}$ expression resulted in extensive intracellular protein aggregation and aggresome formation, again with robust induction of autophagy in the immediate peri-aggregate region [75]. Skeletal muscle from αMHC-$CryAB^{R120G}$ transgenic animals remained unaffected, with no aggregates or myofibrillar disarray and only housekeeping levels of autophagy, demonstrating specificity of the mechanism. With time, these animals developed a cardiomyopathy consistent with the phenotype observed in affected patients, with slowly progressive heart failure characterized by systolic dysfunction, increased ventricular stiffness, and sudden cardiac death by midlife [75, 77].

Prior studies have demonstrated that pressure overload-induced cardiomyocyte autophagy is both robust and maladaptive [78]. Nothing, however, was known about the autophagic response occurring in $CryAB^{R120G}$ mice. To evaluate the effect of autophagy on structural remodeling in DRM, hearts from *αMHC-CryAB*^{R120G} mice were crossed into a *beclin 1* haploinsufficient background. *Beclin 1*, the mammalian homologue of the proautophagic gene *ATG6* in yeast, is required for recruitment of Atg12–Atg5 conjugates to preautophagosomal membranes and is required for autophagosome formation [79]. While B*eclin 1*$^{-/-}$ mice die *in utero*, heterozygous *beclin 1*$^{+/-}$ animals have a ~50% reduction in autophagic capacity and no observed cardiac phenotype through at least 1 year of age [78, 80]. Maintenance of basal levels of autophagy is typically required for cell survival, accomplishing important housekeeping functions, as numerous models of complete inhibition of autophagy manifest multi-system failure [81, 82]. Thus, crossing these two strains provided a model where mutant crystallin is expressed in the setting of attenuated autophagic capacity, affording a platform for testing the role of aggregate-induced autophagy.

Cardiac performance was monitored serially over a 12 month period. As expected, *αMHC-CryAB*^{R120G} mice developed progressive declines in left ventricular systolic performance as measured by percent fractional shortening (%FS). By 12 months of age, these mice manifested overt heart failure (%FS <40%). Then, to test the role of autophagic activity in *αMHC-CryAB*^{R120G} mice, *αMHC-CryAB*^{R120G}*;beclin 1*$^{+/-}$ mice were studied. Intriguingly, declines in cardiac performance were significantly accelerated in these mice. Indeed, these animals developed significant decrements in cardiac function by 6 months of age, with severe heart failure by 9 months of age [75]. Mortality was significantly increased ($p<0.05$) in *αMHC-CryAB*^{R120G}*;beclin 1*$^{+/-}$ mice, as well. By 9 months of age, *αMHC-CryAB*^{R120G}*;beclin 1*$^{+/-}$ mice manifested 25% mortality (two of eight animals), whereas no mortality was observed in WT (zero of seven), *beclin 1*$^{+/-}$ (zero of nine), or *αMHC-CryAB*^{R120G} (zero of eight) animals. Echocardiographic determination of posterior wall thickness and necropsy evaluation of heart mass both revealed similar degrees of hypertrophic growth in *αMHC-CryAB*^{R120G} and *αMHC-CryAB*^{R120G}*;beclin 1*$^{+/-}$ mice. Interestingly, in both *αMHC-CryAB*^{R120G} and *αMHC-CryAB*^{R120G}*;beclin 1*$^{+/-}$ mice, heart failure was systolic in nature, with significant increases developing in end-systolic diameter without ventricular dilatation as measured by end-diastolic dimensions [75].

Collectively, these studies demonstrate that autophagy and defective sHsp function in crystallin-induced DRCM are interrelated. Beyond that, they demonstrate that the cell's capacity for autophagy has a direct and significant impact on cell survival and proper function. Indeed, it has been suggested that autophagy may have two distinct, beneficial effects in protein conformation disease [75, 83]. First, this pathway functions to clear the primary toxin contributing to disease pathogenesis. Second, enhanced autophagy attenuates apoptotic responses to various insults, rendering the cell resistant to programmed cell death. Importantly, recent studies demonstrating that pharmacological up-regulation of autophagy is protective in a wide variety of disease models associated with intracellular protein aggregation raises the exciting prospect of autophagic activation as a novel therapeutic strategy [25].

6.3. A Model

The sHsps participate in and govern protein quality control at many points (Figure 1). Surveillance commences with nascent polypeptide translation at the ribosome and continues through on-going monitoring of proteome integrity. sHsp also play central roles in deciding the fate of non-native proteins; when non-native species are identified, the chaperone machinery binds them, maintaining them in an intermediate state until their ultimate outcome is determined through interaction with a complex system of co-chaperones. Traditionally substrates at this triage point were thought to undergo one of two fates -- either refolding or shunting towards the proteasome for degradation. As reviewed here, multiple lines of evidence now demonstrate the existence of cellular machinery linking sHsps with the autophagic pathway, providing an alternative fate for substrates unfit for refolding but too large to fit within the proteasome core.

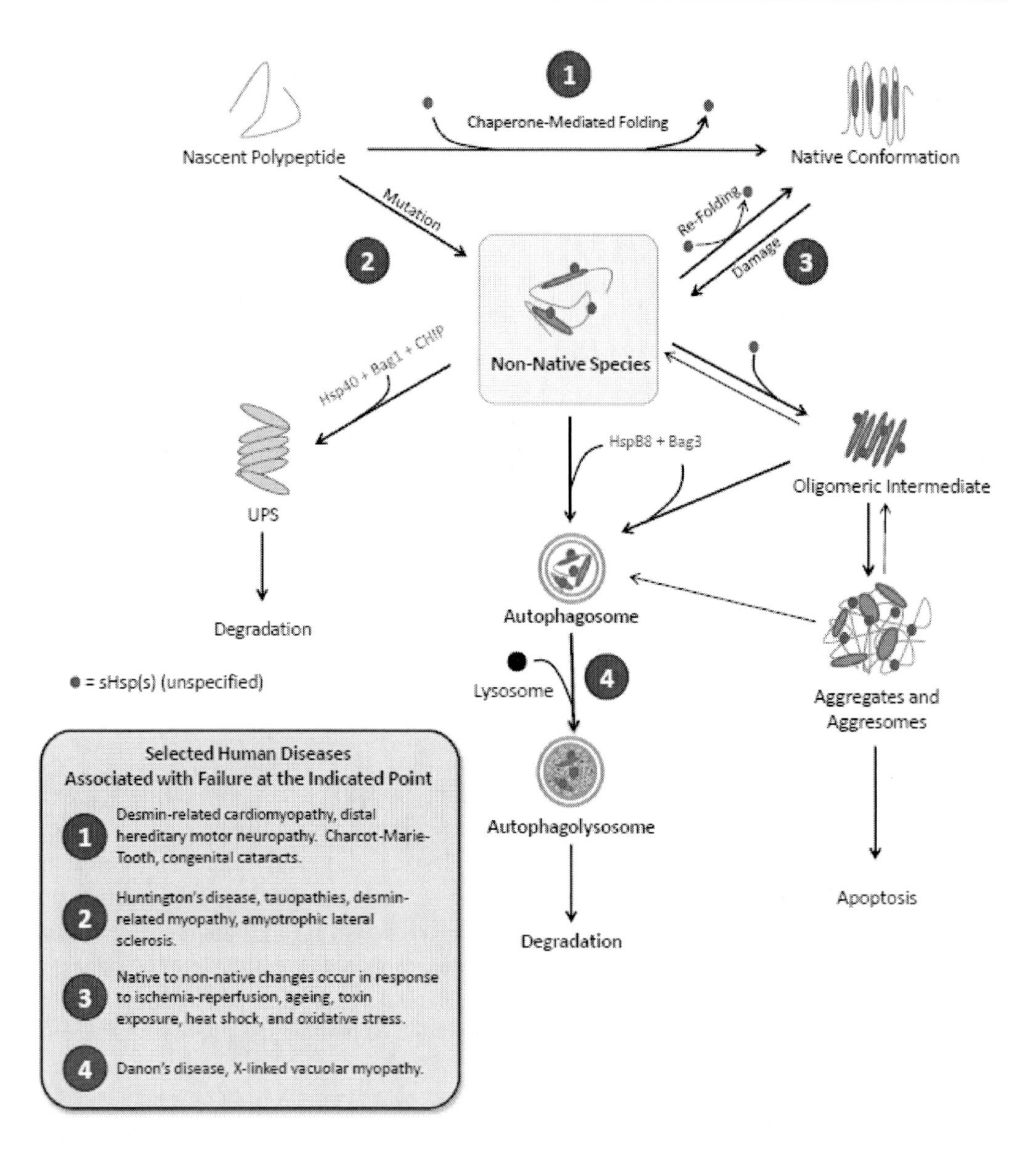

Figure 1. Working model of sHsps and protein quality control. sHsps participate in every step of a protein's life cycle. Initial binding of sHsps to the nascent polypeptide facilitates folding into the native conformation. Accumulation of non-native species results from mutations that alter the polypeptide sequence, or following damage and denaturation of an existing native protein. The ultimate fate of the denatured species is determined by a complex interaction with various sHsps and their binding partners. Ultimately the non-native species can undergo either re-folding back to the native state, targeting toward proteasome-mediated or autophagic degradation, or oligomerization into insoluble substrates that are then processed into aggregates and aggresomes. An error at any point in these pathways can result in clinical disease. Some of the more common diseases, and where the causative lesion exists, are highlighted.

In the case of HspB8, this sHsp is able to recognize a non-native protein, and then through interaction with specific co-factors, the non-native protein is fated for refolding, proteasomal degradation, or direct targeting to the autophagosome. These facts demonstrate how through interaction with various protein cofactors, sHsps provide some degree of substrate specificity to the autophagic process. Conversely, in the case of CryAB, devastating consequences result from failure in chaperone-mediated protein quality control, particularly in tissues that express high levels of CryAB such as cardiac and skeletal muscle. Here, autophagy is activated, functioning in an adaptive manner to facilitate aggregate clearance, thereby delaying the onset of heart failure.

CONCLUSION

Protein quality control is essential for proper functioning and longevity of multiple cell types, and failure of this system results in devastating disease. In the case of heart failure, enormous effort has been directed at identifying novel strategies with long-term therapeutic efficacy. The broad category of proteinopathies is categorized by severe and progressive disease, and therapeutic options are presently limited. Recently, however, autophagy is increasingly appreciated as an important proteolytic response to the presence of non-native protein substrates, and as demonstrated in DRM and various neurodegenerative diseases, the autophagic response can play a significant role in attenuating disease progression. This combination of demonstrated specificity for non-native protein substrates combined with a pathophysiologically important and adaptive role for autophagy is significant for a number of reasons: (*i*) it indicates that autophagy is a mechanism in proteinopathy suitable for consideration as a therapeutic intervention, (*ii*) it suggests that inter-individual variations in autophagic capacity or responsiveness may contribute to the heterogeneous presentation of clinical disease, and (*iii*) it suggests that these tenets may hold for proteinopathies of diverse molecular etiology. Finally, given that drugs that alter the process of autophagy are already in widespread clinical use, we envision an era where targeted activation of autophagy may be a therapeutically viable option for patients suffering from a variety of diseases. Major challenges remain to dissect mechanisms underlying these events, but patients with heart disease are likely to benefit from their elucidation.

ACKNOWLEDGMENTS

This work was supported by grants from the Donald W. Reynolds Cardiovascular Clinical Research Center (JAH), NIH (HL-72016, BAR; HL-075173, JAH; HL-080144, JAH; HL-006296, JAH; HL-090842, JAH), AHA (0655202Y, BAR; 0640084N, JAH), and by an award from the American Heart Association-Jon Holden DeHaan Foundation (JAH).

CONFLICTS OF INTEREST DISCLOSURES

None.

REFERENCES

[1] Hill JA, Olson EN. Cardiac plasticity. *New Engl J Med.* 2008;358(13):1370-80.

[2] Chien KR. Stress pathways and heart failure. *Cell*, 1999 1999///;98:555-8.

[3] Frey N, Katus HA, Olson EN, Hill JA. Hypertrophy of the heart: a new therapeutic target? *Circulation,* 2004 Apr 6;109(13):1580-9.

[4] Wang X, Robbins J. Heart failure and protein quality control. *Circ Res.* 2006 Dec 8;99(12):1315-28.

[5] Wang X, Su H, Ranek MJ. Protein quality control and degradation in cardiomyocytes. *J Mol Cell Cardiol.* 2008 Jul;45(1):11-27.

[6] Shemetov AA, Seit-Nebi AS, Gusev NB. Structure, properties, and functions of the human small heat-shock protein HSP22 (HspB8, H11, E2IG1): a critical review. J Neurosci Res2008 Feb 1;86(2):264-9.

[7] Haslbeck M. sHsps and their role in the chaperone network. Cell Mol Life Sci2002 Oct;59(10):1649-57.

[8] Li GC, Laszlo A. Amino acid analogs while inducing heat shock proteins sensitize CHO cells to thermal damage. J Cell Physiol1985 Jan;122(1):91-7.

[9] Benjamin IJ, McMillan DR. Stress (heat shock) proteins: molecular chaperones in cardiovascular biology and disease. Circ Res1998 Jul 27;83(2):117-32.

[10] Morimoto RI. Regulation of the heat shock transcriptional response: cross talk between a family of heat shock factors, molecular chaperones, and negative regulators. Genes Dev1998 Dec 15;12(24):3788-96.

[11] Pockley AG. Heat shock proteins, inflammation, and cardiovascular disease. *Circulation*, 2002 Feb 26;105(8):1012-7.

[12] Sun Y, MacRae TH. The small heat shock proteins and their role in human disease. *FEBS,* J2005 Jun;272(11):2613-27.

[13] Taylor RP, Benjamin IJ. Small heat shock proteins: a new classification scheme in mammals. *J Mol Cell Cardiol.* 2005 Mar;38(3):433-44.

[14] Westerheide SD, Morimoto RI. Heat shock response modulators as therapeutic tools for diseases of protein conformation. *J Biol Chem.* 2005 Sep 30;280(39):33097-100.

[15] Sreedhar AS, Csermely P. Heat shock proteins in the regulation of apoptosis: new strategies in tumor therapy: a comprehensive review. Pharmacol Ther2004 Mar;101(3):227-57.

[16] Haslbeck M, Franzmann T, Weinfurtner D, Buchner J. Some like it hot: the structure and function of small heat-shock proteins. Nat Struct Mol Biol2005 Oct;12(10):842-6.

[17] Chung E, Leinwand LA. Rescuing cardiac malfunction: the roles of the chaperone-like small heat shock proteins. *Circ Res.* 2008 Dec 5;103(12):1351-3.

[18] Toko H, Minamino T, Komuro I. Role of heat shock transcriptional factor 1 and heat shock proteins in cardiac hypertrophy. *Trends Cardiovasc Med.* 2008 Apr;18(3):88-93.

[19] Zou Y, Zhu W, Sakamoto M, Qin Y, Akazawa H, Toko H, et al. Heat shock transcription factor 1 protects cardiomyocytes from ischemia/reperfusion injury. *Circulation,* 2003 Dec 16;108(24):3024-30.

[20] Sakamoto M, Minamino T, Toko H, Kayama Y, Zou Y, Sano M, et al. Upregulation of heat shock transcription factor 1 plays a critical role in adaptive cardiac hypertrophy. *Circ Res.* 2006 Dec 8;99(12):1411-8.

[21] Arndt V, Daniel C, Nastainczyk W, Alberti S, Hohfeld J. BAG-2 acts as an inhibitor of the chaperone-associated ubiquitin ligase CHIP. *Mol Biol Cell,* 2005 Dec;16(12):5891-900.

[22] Dai Q, Qian SB, Li HH, McDonough H, Borchers C, Huang D, et al. Regulation of the cytoplasmic quality control protein degradation pathway by BAG2. *J Biol Chem.* 2005 Nov 18;280(46):38673-81.

[23] Lewandowska A, Gierszewska M, Marszalek J, Liberek K. Hsp78 chaperone functions in restoration of mitochondrial network following heat stress. Biochimica et Biophysica Acta (BBA) - *Molecular Cell Research,* 2006;1763(2):141-51.

[24] McDonough H, Patterson C. CHIP: a link between the chaperone and proteasome systems. Cell Stress Chaperones2003 Winter;8(4):303-8.

[25] Williams A, Jahreiss L, Sarkar S, Saiki S, Menzies FM, Ravikumar B, et al. Aggregate-prone proteins are cleared from the cytosol by autophagy: therapeutic implications. Curr Top Dev Biol2006;76:89-101.

[26] Verhoef LG, Lindsten K, Masucci MG, Dantuma NP. Aggregate formation inhibits proteasomal degradation of polyglutamine proteins. Hum Mol Genet2002 Oct 15;11(22):2689-700.

[27] Chen Q, Liu JB, Horak KM, Zheng H, Kumarapeli AR, Li J, et al. Intrasarcoplasmic amyloidosis impairs proteolytic function of proteasomes in cardiomyocytes by compromising substrate uptake. *Circ Res.* 2005 Nov 11;97(10):1018-26.

[28] Carra S, Brunsting JF, Lambert H, Landry J, Kampinga HH. HspB8 participates in protein quality control by a non-chaperone-like mechanism that requires eIF2{alpha} phosphorylation. *J Biol Chem.* 2009 Feb 27;284(9):5523-32.

[29] Evgrafov OV, Mersiyanova I, Irobi J, Van Den Bosch L, Dierick I, Leung CL, et al. Mutant small heat-shock protein 27 causes axonal Charcot-Marie-Tooth disease and distal hereditary motor neuropathy. *Nat Genet.* 2004 Jun;36(6):602-6.

[30] Irobi J, Van Impe K, Seeman P, Jordanova A, Dierick I, Verpoorten N, et al. Hot-spot residue in small heat-shock protein 22 causes distal motor neuropathy. Nat Genet2004 Jun;36(6):597-601.

[31] Vicart P, Caron A, Guicheney P, Li Z, Prevost MC, Faure A, et al. A missense mutation in the alphaB-crystallin chaperone gene causes a desmin-related myopathy. Nat Genet1998 Sep;20(1):92-5.

[32] Litt M, Kramer P, LaMorticella DM, Murphey W, Lovrien EW, Weleber RG. Autosomal dominant congenital cataract associated with a missense mutation in the human alpha crystallin gene CRYAA. *Hum Mol Genet.* 1998 Mar;7(3):471-4.

[33] McCampbell A, Taylor JP, Taye AA, Robitschek J, Li M, Walcott J, et al. CREB-binding protein sequestration by expanded polyglutamine. *Hum Mol Genet.* 2000 Sep 1;9(14):2197-202.

[34] Maloyan A, Gulick J, Glabe CG, Kayed R, Robbins J. Exercise reverses preamyloid oligomer and prolongs survival in alphaB-crystallin-based desmin-related cardiomyopathy. *Proc Natl Acad Sci U S A.* 2007 Apr 3;104(14):5995-6000.

[35] Fan GC, Zhou X, Wang X, Song G, Qian J, Nicolaou P, et al. Heat shock protein 20 interacting with phosphorylated Akt reduces doxorubicin-triggered oxidative stress and cardiotoxicity. *Circ Res.* 2008 Nov 21;103(11):1270-9.

[36] Heimrath O, Oxner A, Myers T, Legare JF. Heat shock treatment prior to myocardial infarction results in reduced ventricular remodeling. *J Invest Surg*. 2009 Jan-Feb;22(1):9-15.
[37] Johnston JA, Ward CL, Kopito RR. Aggresomes: a cellular response to misfolded proteins. *J Cell Biol*. 1998 Dec 28;143(7):1883-98.
[38] Massey AC, Zhang C, Cuervo AM. Chaperone-mediated autophagy in aging and disease. *Curr Top Dev Biol*. 2006;73:205-35.
[39] Massey A, Kiffin R, Cuervo AM. Pathophysiology of chaperone-mediated autophagy. Int *J Biochem Cell Biol*. 2004 Dec;36(12):2420-34.
[40] Sugie K, Yamamoto A, Murayama K, Oh SJ, Takahashi M, Mora M, et al. Clinicopathological features of genetically confirmed Danon disease. *Neurology,* 2002 Jun 25;58(12):1773-8.
[41] Ahlberg J, Glaumann H. Uptake--microautophagy--and degradation of exogenous proteins by isolated rat liver lysosomes. Effects of pH, ATP, and inhibitors of proteolysis. *Exp Mol Pathol*. 1985 Feb;42(1):78-88.
[42] Mortimore GE, Lardeux BR, Adams CE. Regulation of microautophagy and basal protein turnover in rat liver. Effects of short-term starvation. J Biol Chem1988 Feb 15;263(5):2506-12.
[43] Levine B, Klionsky DJ. Development by self-digestion: molecular mechanisms and biological functions of autophagy. Dev Cell2004 Apr;6(4):463-77.
[44] Pandey UB, Nie Z, Batlevi Y, McCray BA, Ritson GP, Nedelsky NB, et al. HDAC6 rescues neurodegeneration and provides an essential link between autophagy and the UPS. *Nature,* 2007 Jun 14;447(7146):859-63.
[45] Tannous P, Zhu H, Nemchenko A, Berry JM, Johnstone JL, Shelton JM, et al. Intracellular protein aggregation is a proximal trigger of cardiomyocyte autophagy. *Circulation,* 2008 Jun 17;117(24):3070-8.
[46] Sandri M, Sandri C, Gilbert A, Skurk C, Calabria E, Picard A, et al. Foxo transcription factors induce the atrophy-related ubiquitin ligase atrogin-1 and cause skeletal muscle atrophy. *Cell,* 2004 Apr 30;117(3):399-412.
[47] Mammucari C, Milan G, Romanello V, Masiero E, Rudolf R, Del Piccolo P, et al. FoxO3 controls autophagy in skeletal muscle in vivo. Cell Metab2007 Dec;6(6):458-71.
[48] Winslow AR, Rubinsztein DC. Autophagy in neurodegeneration and development. Biochim Biophys Acta2008 Dec;1782(12):723-9.
[49] Sarkar S, Rubinsztein DC. Small molecule enhancers of autophagy for neurodegenerative diseases. *Mol Biosyst*. 2008 Sep;4(9):895-901.
[50] Benndorf R, Sun X, Gilmont RR, Biederman KJ, Molloy MP, Goodmurphy CW, et al. HSP22, a new member of the small heat shock protein superfamily, interacts with mimic of phosphorylated HSP27 ((3D)HSP27). J Biol Chem2001 Jul 20;276(29):26753-61.
[51] Gober MD, Smith CC, Ueda K, Toretsky JA, Aurelian L. Forced expression of the H11 heat shock protein can be regulated by DNA methylation and trigger apoptosis in human cells. *J Biol Chem*. 2003 Sep 26;278(39):37600-9.
[52] Chowdary TK, Raman B, Ramakrishna T, Rao Ch M. Interaction of mammalian Hsp22 with lipid membranes. *Biochem*. J2007 Jan 15;401(2):437-45.

[53] Chowdary TK, Raman B, Ramakrishna T, Rao CM. Mammalian Hsp22 is a heat-inducible small heat-shock protein with chaperone-like activity. *Biochem J.* 2004 Jul 15;381(Pt 2):379-87.
[54] Depre C, Tomlinson JE, Kudej RK, Gaussin V, Thompson E, Kim SJ, et al. Gene program for cardiac cell survival induced by transient ischemia in conscious pigs. *Proc Natl Acad Sci U S A.* 2001 Jul 31;98(16):9336-41.
[55] Depre C, Kim SJ, John AS, Huang Y, Rimoldi OE, Pepper JR, et al. Program of cell survival underlying human and experimental hibernating myocardium. *Circ Res.* 2004 Aug 20;95(4):433-40.
[56] Chavez Zobel AT, Loranger A, Marceau N, Theriault JR, Lambert H, Landry J. Distinct chaperone mechanisms can delay the formation of aggresomes by the myopathy-causing R120G alphaB-crystallin mutant. *Hum Mol Genet.* 2003 Jul 1;12(13):1609-20.
[57] Sanbe A, Daicho T, Mizutani R, Endo T, Miyauchi N, Yamauchi J, et al. Protective effect of geranylgeranylacetone via enhancement of HSPB8 induction in desmin-related cardiomyopathy. PLoS ONE2009;4(4):e5351.
[58] Rubinsztein DC. The roles of intracellular protein-degradation pathways in neurodegeneration. *Nature*, 2006 Oct 19;443(7113):780-6.
[59] Carra S, Sivilotti M, Chavez Zobel AT, Lambert H, Landry J. HspB8, a small heat shock protein mutated in human neuromuscular disorders, has in vivo chaperone activity in cultured cells. *Hum Mol Genet.* 2005 Jun 15;14(12):1659-69.
[60] Carra S, Seguin SJ, Lambert H, Landry J. HspB8 chaperone activity toward poly(Q)-containing proteins depends on its association with Bag3, a stimulator of macroautophagy. *J Biol Chem.* 2008 Jan 18;283(3):1437-44.
[61] Carra S, Seguin SJ, Landry J. HspB8 and Bag3: a new chaperone complex targeting misfolded proteins to macroautophagy. Autophagy2008 Feb 16;4(2):237-9.
[62] Kabbage M, Dickman MB. The BAG proteins: a ubiquitous family of chaperone regulators. Cell Mol Life Sci2008 May;65(9):1390-402.
[63] Irobi J, De Jonghe P, Timmerman V. Molecular genetics of distal hereditary motor neuropathies. *Hum Mol Genet.* 2004 Oct 1;13 Spec No 2:R195-202.
[64] Iwaki T, Kume-Iwaki A, Goldman JE. Cellular distribution of alpha B-crystallin in non-lenticular tissues. J Histochem Cytochem1990 Jan;38(1):31-9.
[65] Longoni S, James P, Chiesi M. Cardiac alpha-crystallin. I. Isolation and identification. *Mol Cell Biochem.* 1990 Dec 3;99(1):113-20.
[66] Wang X, Osinska H, Klevitsky R, Gerdes AM, Nieman M, Lorenz J, et al. Expression of R120G-alphaB-crystallin causes aberrant desmin and alphaB-crystallin aggregation and cardiomyopathy in mice. Circ Res2001 Jul 6;89(1):84-91.
[67] Brady JP, Garland DL, Green DE, Tamm ER, Giblin FJ, Wawrousek EF. AlphaB-crystallin in lens development and muscle integrity: a gene knockout approach. *Invest Ophthalmol Vis Sci.* 2001 Nov;42(12):2924-34.
[68] Hwang JJ, Allen PD, Tseng GC, Lam CW, Fananapazir L, Dzau VJ, et al. Microarray gene expression profiles in dilated and hypertrophic cardiomyopathic end-stage heart failure. *Physiol Genomics,* 2002 Jul 12;10(1):31-44.
[69] Arbustini E, Pasotti M, Pilotto A, Pellegrini C, Grasso M, Previtali S, et al. Desmin accumulation restrictive cardiomyopathy and atrioventricular block associated with desmin gene defects. Eur J Heart Fail2006 Aug;8(5):477-83.

[70] Martin JL, Mestril R, Hilal-Dandan R, Brunton LL, Dillmann WH. Small heat shock proteins and protection against ischemic injury in cardiac myocytes. *Circulation,* 1997 Dec 16;96(12):4343-8.
[71] Ray PS, Martin JL, Swanson EA, Otani H, Dillmann WH, Das DK. Transgene overexpression of alphaB crystallin confers simultaneous protection against cardiomyocyte apoptosis and necrosis during myocardial ischemia and reperfusion. FASEB J2001 Feb;15(2):393-402.
[72] Perng MD, Muchowski PJ, van Den IP, Wu GJ, Hutcheson AM, Clark JI, et al. The cardiomyopathy and lens cataract mutation in alphaB-crystallin alters its protein structure, chaperone activity, and interaction with intermediate filaments in vitro. *J Biol Chem.* 1999 Nov 19;274(47):33235-43.
[73] Bova MP, Yaron O, Huang Q, Ding L, Haley DA, Stewart PL, et al. Mutation R120G in alphaB-crystallin, which is linked to a desmin-related myopathy, results in an irregular structure and defective chaperone-like function. *Proc Natl Acad Sci U S A.* 1999 May 25;96(11):6137-42.
[74] Sanbe A, Osinska H, Saffitz JE, Glabe CG, Kayed R, Maloyan A, et al. Desmin-related cardiomyopathy in transgenic mice: a cardiac amyloidosis. *Proc Natl Acad Sci U S A.* 2004 Jul 6;101(27):10132-6.
[75] Tannous P, Zhu H, Johnstone JL, Shelton JM, Rajasekaran NS, Benjamin IJ, et al. Autophagy is an adaptive response in desmin-related cardiomyopathy. *Proc Natl Acad Sci U S A.* 2008 Jul 15;105(28):9745-50.
[76] Rajasekaran NS, Connell P, Christians ES, Yan LJ, Taylor RP, Orosz A, et al. Human alpha B-crystallin mutation causes oxido-reductive stress and protein aggregation cardiomyopathy in mice. *Cell,* 2007 Aug 10;130(3):427-39.
[77] Goldfarb LG, Olive M, Vicart P, Goebel HH. Intermediate filament diseases: desminopathy. *Adv Exp Med Biol.* 2008;642:131-64.
[78] Zhu H, Tannous P, Johnstone JL, Kong Y, Shelton JM, Richardson JA, et al. Cardiac autophagy is a maladaptive response to hemodynamic stress. *J Clin Invest.* 2007 Jul;117(7):1782-93.
[79] Kihara A, Noda T, Ishihara N, Ohsumi Y. Two distinct Vps34 phosphatidylinositol 3-kinase complexes function in autophagy and carboxypeptidase Y sorting in Saccharomyces cerevisiae. *J Cell Biol.* 2001 Feb 5;152(3):519-30.
[80] Qu X, Yu J, Bhagat G, Furuya N, Hibshoosh H, Troxel A, et al. Promotion of tumorigenesis by heterozygous disruption of the beclin 1 autophagy gene. *J Clin Invest.* 2003 Dec;112(12):1809-20.
[81] Kuma A, Hatano M, Matsui M, Yamamoto A, Nakaya H, Yoshimori T, et al. The role of autophagy during the early neonatal starvation period. *Nature*, 2004 Dec 23;432(7020):1032-6.
[82] Komatsu M, Waguri S, Ueno T, Iwata J, Murata S, Tanida I, et al. Impairment of starvation-induced and constitutive autophagy in Atg7-deficient mice. *J Cell Biol.* 2005 May 9;169(3):425-34.
[83] Rothermel BA, Hill JA. Autophagy in load-induced heart disease. *Circ Res.* 2008 Dec 5;103(12):1363-9.

In: Small Stress Proteins and Human Diseases
Editors: Stéphanie Simon et al.
ISBN: 978-1-61470-636-6

Chapter 1.7

THE TWO-FACED NATURE OF SMALL HEAT-SHOCK PROTEINS: AMYLOID FIBRIL ASSEMBLY AND THE INHIBITION OF FIBRIL FORMATION. RELEVANCE TO DISEASE STATES

Heath Ecroyd[1, 2#], Sarah Meehan[3#] and John A. Carver[2*]

[1] School of Biological Sciences, University of Wollongong, Wollongong, NSW 2522, Australia

[2] School of Chemistry & Physics, The University of Adelaide, Adelaide, SA 5005, Australia

[3] Department of Chemistry, University of Cambridge, Cambridge, CB2 1EW, UK.

ABSTRACT

The ability of small heat-shock proteins (sHsps) such as αB-crystallin to inhibit the amorphous (disordered) aggregation of varied target proteins in a chaperone-like manner has been well described. The mechanistic details of this action are not understood. Amyloid fibril formation is an alternative off-folding pathway that leads to highly ordered β-sheet-containing aggregates. Amyloid fibril formation is associated with a broad range of protein conformational diseases such as Alzheimer's, Parkinson's and Huntington's and sHsp expression is elevated in the protein deposits that are characteristic of these disease states. The ability of sHsps to prevent fibril formation has been less well characterised. It has been shown, however, that sHsps are potent inhibitors of fibril formation of a range of target proteins. In this chapter, the disease-related significance of this observation is discussed. Interestingly, in addition to being effective

---

[#] These two authors contributed equally to this chapter.

[*] Corresponding Author: John A Carver

School of Chemistry & Physics
The University of Adelaide
Adelaide, SA 5005, Australia
john.carver@adelaide.edu.au

molecular chaperones, αA- and αB-crystallin themselves, along with some of their peptide fragments, readily form amyloid fibrils under slightly destabilising solution conditions. The implications of this observation in terms of protein conformational diseases, e.g. cataract, along with the potential nanotechnological applications of these fibrils, are discussed.

1. Protein Folding, Misfolding and Aggregation: The Role of Molecular Chaperones

Globular proteins do not exist as structured, folded entities for their entire life. For example, after synthesis, the polypeptide chain is unstructured and must fold to its correct three-dimensional, functional conformation. To cross a biological membrane or move from one cellular compartment to another, a protein must unfold from its native state. Furthermore, if a protein is subject to stress, e.g. elevated temperature, low pH, oxidative conditions etc., it may partially unfold to adopt intermediately folded states. Many of these intermediately folded states have characteristics of molten globule states, i.e. they have elements of secondary structure but little or no tertiary structure, and are dynamic species with an expanded structure relative to the native state and have their hydrophobic core exposed to solution. The greater exposure of hydrophobicity to solution encourages protein misfolding, mutual association (aggregation) and potentially, precipitation. The cell has various mechanisms available to counteract these processes of which, arguably the most important, is the utilization of molecular chaperones, a large group of structurally unrelated proteins which interact with destabilized proteins to prevent their aggregation and facilitate their correct fold [1]. Expression of molecular chaperones is increased under stress conditions, e.g. elevated temperature. Hence, they are often known as heat-shock proteins (Hsps). Thus, for a newly synthesized protein in the crowded environment of the cell, ATP-dependent molecular chaperones such as Hsp70 and Hsp60 ensure high fidelity of protein folding and assembly [1]. In this process, Hsp70 interacts initially with the unfolded, nascent polypeptide chain, specifically with extended regions of the chain. If the protein requires further encouragement to fold correctly, it is sequestered into the large cavity of the Hsp60 aggregate, where it can be isolated from the surrounding cellular milieu, and fold via repeated interactions with the wall of the Hsp60 cavity [1]. If a protein is irretrievably misfolded and therefore incapable of folding properly, e.g. as a result of mutation, it may be labeled with ubiquitin and thereby targeted for degradation into its constituent amino acids via the proteasome, a large multi-subunit ordered complex with a large central cavity into with the targeted protein is inserted and digested.

The folding pathway for a protein proceeds via a series of intermediately folded states and is relatively rapid (Figure 1). Most newly synthesized proteins fold in a few milliseconds from their unstructured state so that the intermediates are only transiently present. However, if they linger for too long, e.g. because of difficulty in folding, mutation, or due to cellular stress, these states may associate and enter the slow off-folding pathways which, if not recognized by the ubiquitin/proteasome system, eventually lead to precipitated amorphous or amyloid fibril aggregates (Figure 1). In many cases, aggregation and precipitation of proteins are highly deleterious to cell viability and are hallmarks of protein misfolding diseases (more below). In all these diseases, with the possible exception of cataract [2], the precipitated state of the proteins is highly ordered, existing in a cross β-sheet array arranged into long amyloid fibrils (more below) [3-5]. By contrast, as the name implies, the amorphous aggregation pathway is composed of cellular aggregates that are relatively disordered in arrangement. Both pathways arise from association of intermediately folded states of proteins via a nucleation-dependent mechanism (Figure 1). In the case of the amorphous aggregation pathway, this leads to the formation of aggregates that when a critical size (a nucleus) is reached, readily sequester other intermediates and form large aggregates that are intrinsically unstable and precipitate [6, 7].

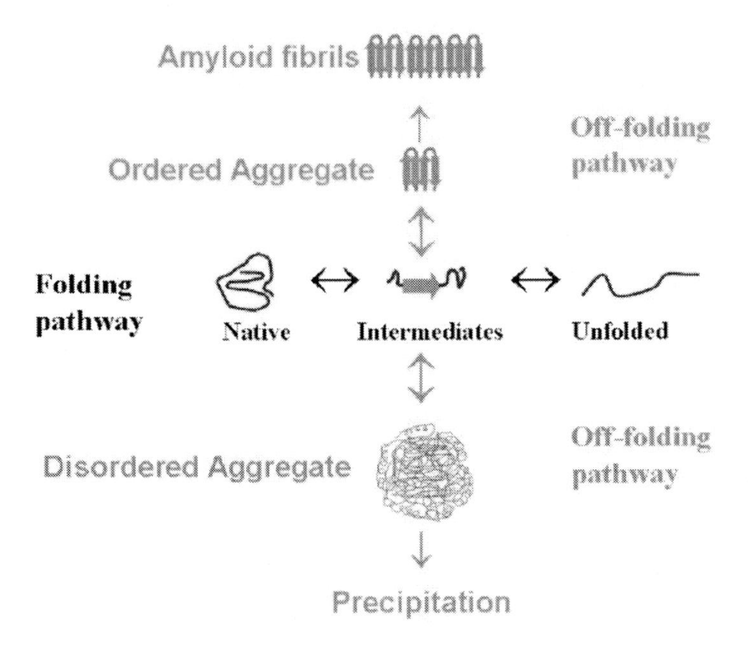

Figure 1. Amyloid fibrils arise from the association of a partially structured intermediate that deviates from the protein folding pathway. In this pathway, proteins move from their unfolded to folded/native states via an intermediate or intermediates. Amorphous aggregates arise from these states via relatively rapid association that leads to disordered precipitates whereas the formation of fibrils occurs over a longer timeframe via a nucleation and ordered growth mechanism involving a soluble precursor.

Intracellularly, a unique class of ubiquitous and abundant ATP-independent molecular chaperones, the small heat-shock proteins (sHsps), interact specifically with partially folded proteins that enter the off-folding protein aggregation pathways (Figure 2). The ability of sHsps to prevent amorphous protein aggregation was described over 15 years ago [8] and led to major research activity in relation to the principal lens protein α-crystallin, a sHsp, and its involvement in the prevention of protein aggregation in the lens and hence cataract formation [6]. A mechanism for this action has been proposed (see Figure 2 for a description) [6, 9], although specific details, at the atomic level, about the mechanism of sHsp chaperone action are not known.

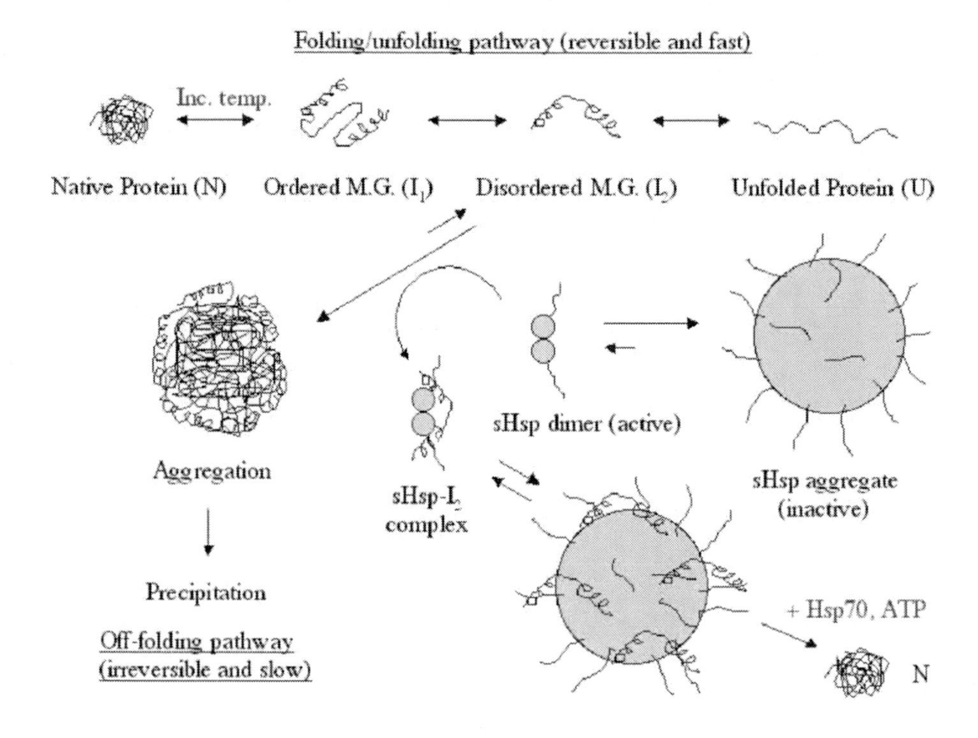

Figure 2. Schematic of sHsp chaperone action with amorphously aggregating proteins [6, 112]. Along the rapid folding pathway, the intermediately folded states of proteins are transiently present. The off-folding pathway is slow by comparison. As a result, the more disordered intermediate or molten globule (M.G.) state of the target protein (I_2), which is highly dynamic and displays significant hydrophobicity to solution, is present for a significant period of time and therefore has the propensity to aggregate along the off-folding pathway via a nucleation-dependent mechanism. sHsps exist as large, dynamic aggregates which are in constant exchange with dissociated, probably dimeric species. The dissociated sHsp may be more chaperone-active than the aggregated species, potentially via exposure of its chaperone binding site region during subunit dissociation. The aggregation of I_2 is also a dynamic process involving equilibria between various aggregated species. The dynamic nature of both protein species may be the trigger that facilitates sequestration of I_2 by the sHsp dimer. In support of this, sHsps specifically interact with target proteins that are aggregating via a nucleation-dependent mechanism [113]. Following binding to I_2, the complex is incorporated into a high-molecular mass aggregate. Refolding of I_2 to the native state (N) can occur via the action of another chaperone protein, e.g. Hsp70, which requires the hydrolysis of ATP. The squiggly lines on the surface of the sHsp oligomer and dimer represent the highly mobile and unstructured C-terminal extensions [6, 49, 50, 58].

2. Amyloid Fibrils: Mechanism of Formation, Structure and their Association with Disease

The reasons for a protein entering either the amorphous or amyloid fibril-forming, off-folding pathways are not entirely understood but may involve one or more of the following factors:

1. The amino acid sequence of the protein: some sequences are more prone to fibril formation since the physiochemical properties of the amino acids are highly influential on fibril-forming propensity of a polypeptide sequence [10-12]. For example, factors such as charge, hydrophobicity and propensity to adopt a β-sheet configuration are likely to play important roles. There are a variety of algorithms available to predict the fibril-forming propensity of amino acid sequences [12-15].
2. The kinetics of aggregation, with a slow aggregation rate favoring ordered, amyloid fibril formation.
3. The 'nature' of the intermediate state from which aggregation occurs, i.e. what elements of secondary structure (e.g. β-sheet in the case of amyloid fibril formation) are in place within the intermediate to facilitate protein association and aggregation along either pathway.

The misfolding, and subsequent formation of amyloid fibrils, of a normally soluble protein or peptide precursor, is associated with a variety of debilitating diseases, including Alzheimer's disease (AD), Parkinson's disease, Creutzfeldt-Jakob and Mad-Cow disease, Huntington's disease and type II-diabetes [5, 7, 16-18]. The incidence of many of these diseases will grow over the next 20-30 years as the population ages. In each case, the proteinaceous deposits or plaques that are hallmarks of the disease predominately consist of one protein or peptide, for example amyloid-β peptides (Aβ) in AD, α-synuclein in Parkinson's, prion protein in Creutzfeldt-Jakob, huntingtin in Huntington's and amylin in type II-diabetes. The severity of the disease, however, does not necessarily correlate with the extent of amyloid fibril deposition [16], and precursors to amyloid fibrils, or small oligomeric species in particular, are believed to be especially toxic to cells [19, 20], although, the mature fibril can also be toxic [21, 22], and the most cytotoxic species may vary depending on the fibril-forming protein.

Amyloid fibrils are long thread-like protein aggregates typically 2-10 nm in diameter and ranging from one to many microns in length, as visualized by transmission electron microscopy (TEM) [23] and atomic force microscopy (AFM) [24, 25] (Figure 3). Amyloid fibrils are characterized by a 'cross-β' X-ray fiber diffraction pattern arising from a predominantly β-sheet structure, comprising two major anisotropic reflections: a meridional reflection at 4.7 Å, corresponding to the distance between β-strands, and an equatorial reflection, ca. 9-11 Å, corresponding to the distance between β-sheets [26, 27] (Figure 3). The anisotropy arises since the two aforementioned distances are perpendicular to each other; the β-strands align to form β-sheets, stabilized by extensive hydrogen bonding that runs throughout the length of the fibrils, and these β-sheets stack to form protofilaments, which in turn assemble in a helical fashion to comprise a 'mature' amyloid fibril [28-30]. Fibrils exhibit a positive interaction with the dyes Congo red and thioflavin T (ThT) and hence these dyes are frequently used to monitor amyloid fibril assembly [31, 32]. It has been proposed that intercalation of the dye molecules into the extended β-sheet structure of the fibrils gives rise to either a characteristic spectral spectral shift for Congo red or a change in the fluorescence properties/emission intensities of ThT [33]. A range of biophysical techniques (i.e. TEM, AFM, X-ray fibre diffraction, and dye binding) are useful in characterizing amyloid fibril assembly.

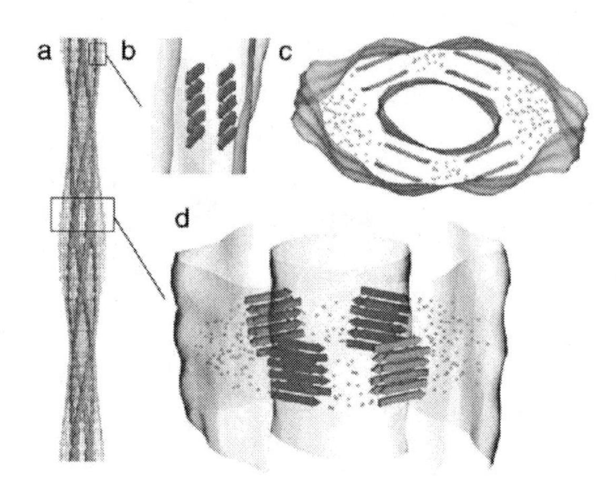

Figure 3. Structural model of an amyloid fibril, formed by the SH3 domain derived from cryo-EM and X-ray fiber diffraction data (reproduced from [29]). An amyloid fibril is a highly structured arrangement of the polypeptide chain that is characterized by an array of stacked cross β-sheets (b,d). The individual β-strands are oriented perpendicular to the long axis of the fibril (b,d) and stack to form protofilaments (b). The fibril may comprise more than one twisted protofilament, i.e. four in this case (a,c,d) which helically assemble into the overall fibril structure (a). Fibrils can be very long (hundreds of nm in length) (c), and may be highly stable and insoluble when sufficiently large (Figure 1). The amyloid fibril is proposed to be a generic structure that is accessible to all polypeptide chains irrespective of their sequence [3].

The ability to assemble into amyloid fibrils has been proposed to be a generic property of polypeptide chains [4], although the propensity of proteins to aggregate to form fibrils can vary widely depending on the factors listed above and solution conditions [34]. The amyloid pathway proceeds via a soluble protofibril precursor (nucleus) that acts as a template to sequester other intermediates, via a nucleation-dependent mechanism, which eventually leads to the insoluble amyloid fibril [5, 34]. This mechanism is consistent with the observed kinetics of fibril formation, as monitored using ThT or Congo red. Thus, both the length of the lag phase (i.e. the time taken to form a stable nucleus) and the rate of elongation are highly dependent on the protein concentration through their reliance on the concentration of partially folded intermediates present at any given time [35]. Whilst this nucleation-dependent mechanism holds for most *in vitro* amyloid fibril forming species studied to date, alternative mechanisms do exist [36].

3. Amyloid Fibril Formation by α-Crystallin: *In Vitro* Studies

Although renowned for their impressive chaperone ability, i.e. the capability to prevent target proteins from aggregating amorphously or to form amyloid fibrils [37], sHsps themselves can self-assemble into amyloid fibrils [38-41]. α-Crystallin is the principal lens protein and is comprised of two closely related subunits, αA- and αB-crystallin (HspB4 and HspB5). Amyloid fibril formation is observed for a variety of α-crystallin isoforms *in vitro* (Figure 4), as summarized by Table 1, which lists the protein source and the relevant references describing their fibril formation. For example, both wild-type and mutant human recombinant αA- and αB-crystallins form aggregates displaying the characteristic properties of amyloid fibrils, including long filamentous structures as observed by TEM (Figure 4B and C), a positive interaction with Congo red, and cross-β X-ray fiber patterns [39].

Table 1. Summary of *in vitro* studies of amyloid fibril formation by α-crystallins

sHsps	*Protein Source*	*Ref.*
αT-Crystallin, mixture of αA + αB-crystallin	Bovine lenses (purified)	[38]
αT-Crystallin, mixture of αA + αB-crystallin	Bovine lenses (crude extracts)	[41]
Wild-type αA-crystallin, 173 aa	Human recombinant	[39]
αA-Crystallin, range of peptide fragments	Human recombinant	[40]
Wild-type αB-crystallin, 175 aa	Human recombinant	[39]
R120G αB-crystallin, 175 aa	Human recombinant	[39]

Amyloid fibrils can be produced from α-crystallins obtained from a variety of sources, including bovine (Figure 4A) [38], deer, and sheep eye lenses [41]. Both purified α-crystallin fractions and crude protein extracts, containing a mixture of crystallin proteins together with other unrelated crystallins (β and γ) as minor components, form fibrils [41]. Although the major component from bovine lens crude protein extracts is α-crystallin, the exact protein composition of fibrils formed from this mixture remains to be explored, and could include incorporation of some β- and γ-crystallin subunits within the fibrils [41], both of which form fibrils on their own *in vitro* [38, 42, 43]. Recently, the highly-ordered nature of amyloid fibrils has generated significant research interest within the nanotechnology industry as self-assembling nanofibres, with potentially wide ranging applications as gelation agents, viscosifiers, nanowires and scaffolds for cellular support and for enzyme immobilization [44-46]. Crude bovine crystallin protein extracts represent an industrial waste product from the meat industry. It has been suggested that the ability to form fibrils from this crude mixture, without the need for time-consuming and expensive purification procedures, could be advantageous in the large-scale production of fibrils as bionanomaterials [41].

3.1. Solution Conditions that Induce α-Crystallin Amyloid Fibril Assembly *In Vitro*

Fibril formation by α-crystallin is favored when it is subjected to partially denaturing conditions and incubation at elevated temperature, i.e. sufficient destabilization of the native state is required to give the protein the opportunity to rearrange and self-assemble into amyloid fibrils [38], as has been observed in other amyloid-forming systems [47, 48]. For example, bovine α-crystallin (as isolated from the lens and containing αA- and αB-crystallin subunits) forms amyloid fibrils when incubated at elevated temperature (60°C) with 10% (v/v) trifluoroethanol at acidic pH or with 1 M guanidine hydrochloride at neutral pH (Figure 4A; lower concentrations of guanidine hydrochloride were found insufficient to induce amyloid formation within 24 hours) [38]. Real-time 1D ^{1}H NMR spectroscopy was employed to monitor the time course for fibril formation by αB-crystallin and confirmed that major unfolding of the native protein precedes self-assembly into fibrils [39]. The ^{1}H NMR spectrum of the native αB-crystallin oligomer is well-resolved and arises solely from a solvent-exposed, flexible and unstructured 12-residue C-terminal extension. Resonances from the remaining part of the protein are not observed by ^{1}H NMR spectroscopy because the large native aggregate (mass of ~650 kDa) tumbles slowly [49, 50]. During amyloid fibril assembly by αB-crystallin, however, ^{1}H NMR resonances in the aromatic region of the spectrum became visible for a brief period during the early stages of aggregation [39]. Given that all the aromatic residues for αB-crystallin are located in the N-terminal and central (α-crystallin) domains, the implication is that these regions partially unfold prior to fibril formation [39].

3.2. Fibril Formation by Peptides Derived From αA-Crystallin

Cross-linking studies have identified regions of αA-crystallin responsible for binding to proteins undergoing amorphous aggregation [51]. Synthesized peptides corresponding to these regions inhibit both amorphous target protein aggregation [51] and amyloid fibril formation [40]. For example, residues 71-88, (αAC(71-88)), corresponding to a putative chaperone target protein binding site, inhibit amyloid fibril formation by Aβ(1-40) [40]. Interestingly, αAC(71-88) itself assembles into amyloid fibrils *in vitro* at pH 7.5 and 37°C [40]. The ability of this αA-crystallin peptide to form amyloid fibrils is, however, suppressed when it is mixed with Aβ(1-40), suggesting that the chaperone activity and the propensity to form amyloid fibrils may be closely linked. The link between chaperone action and amyloid fibril formation is further supported by the observation that residue F71 in αA-crystallin is important for both properties; removal of F71 from αAC(71-88) stifled both of these activities [40].

The close links between these apparently opposing behaviors, i.e. the ability to inhibit aggregation and also to self-aggregate, may be rationalized when the mechanism of sHsp chaperone activity is taken into account. The aptitude of sHsps to be effective chaperones arises, at least in part, from their sequences containing hydrophobic regions, such as amino acids 71-88 in αA-crystallin, that facilitate their interaction with exposed hydrophobic patches on partially unfolded target proteins to prevent their aggregation [52]. Whereas on one hand this property is useful for chaperone activity, on the other hand it may have detrimental consequences, since hydrophobic sequences also have a high propensity to aggregate and are prone to forming amyloid fibrils [10], as was observed for αAC(71-88) [40]. Full-length sHsp sequences may have evolved to counteract this issue, for example by flanking hydrophobic sequences with residues likely to suppress aggregation. This could include charged residues or flexible hydrophilic regions, that are likely to hinder aggregation [11, 12, 53]. Indeed, αAC(70-88), which has an extra lysine residue at the N-terminus, did not form fibrils [40], indicating that the charged K70 residue may play an important role in suppressing amyloid fibril formation in full-length αA-crystallin. The hydrophilic and highly flexible 10-12 amino acid C-terminal extensions of αA- and αB-crystallin [49, 50] may also play a role in hindering amyloid fibril formation by these proteins since they are likely to discourage formation of inter-molecular contacts that could lead to fibril formation [53].

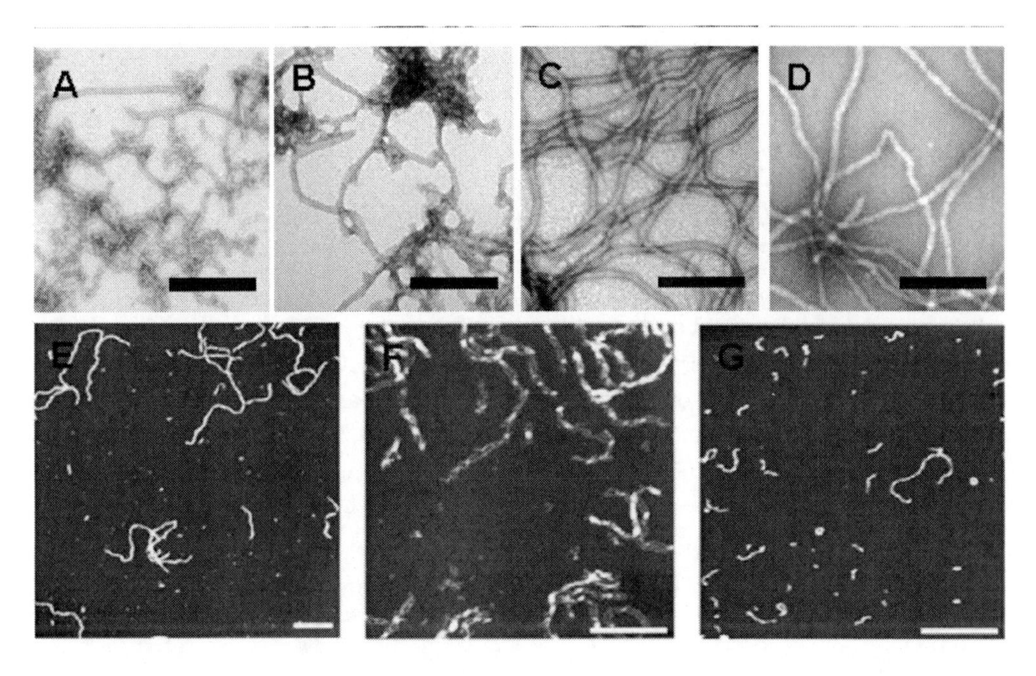

Figure 4. A-D, Transmission electron microscopy (TEM) images of amyloid fibrils formed from A, bovine α-crystallin (purified); B, human recombinant αA-crystallin; C, human recombinant αB-crystallin; and D, human recombinant R120G αB-crystallin. Amyloid fibrils were formed by dissolving the protein at 1 mg/ml in the presence of 1 M guanidine hydrochloride (GdnHCl) (pH 7.4) and incubating at 60 °C for 2-24 h. E-G, Atomic force microscopy (AFM) images of human recombinant αB-crystallin fibril solutions prepared at pH 7.4 with GdnHCl, then, following a ten-fold (F), or one 100-fold (G), dilution into a pH 2.0 solution. In A-D, the scale bars represent 200 nm and in E-G, the scale bars represent 500 nm.

3.3. Structural Properties of Fibrils Formed From α-Crystallins

α-Crystallin forms fibrils with a variety of morphologies, depending on the experimental conditions employed. For example, the level of purity of α-crystallin, prepared from bovine eye lenses, was highly influential on the resultant fibril morphology as observed by TEM; short and curly fibrils were formed from crude crystallin protein extracts, and long, straight fibrils arose from purified bovine α-crystallin [41]. Imaging studies of human recombinant αB-crystallin fibrils reveal unusual and apparently branched morphologies at physiological pH in the presence of guanidine hydrochloride (Figure 4E) [39]. These 'mature' fibrils, comprising multiple protofilament assemblies, were found to unravel when the pH of solution containing the fibrils was altered (Figure 4E-G) [39]. For example, changing the pH from 7 to 2 makes the overall charge of αB-crystallin more positive and leads to significant repulsion between molecules, shifting the equilibrium between the protofilament assemblies and resulting in dissociation and large morphological changes (as assessed by AFM, Figure 4E-G) [39]. The dissociated fibrils at acidic pH displayed 'protofibrillar' morphologies and possessed a relatively low elastic modulus, weak intermolecular interactions, and less regular structures in comparison to a range of 'mature' amyloid fibrils formed from a variety of proteins (Figure 4F) [24]. These flexible protofibrillar morphologies readily assembled into amyloid pores (Figure 4G), similar to those reported for other amyloid systems, e.g. apolipoprotein C-II [54, 55] and α-synuclein [56]. Interestingly, this morphology has been proposed to represent a toxic species [56, 57].

Overall, a combination of biophysical studies, including NMR spectroscopy and AFM imaging, has revealed important insights into the structural properties of α-crystallin amyloid fibrils formed *in vitro*. For example, ^{1}H NMR studies of αB-crystallin amyloid fibrils show that the flexible C-terminal extension is not incorporated into the structured cross-β sheet fibrillar core, but instead protrudes into solution [39], i.e. the extension is flexible in both the fibrillar [39] and native αB-crystallin aggregate [49]. The C-terminal extension plays an important role in maintaining solubility of the native protein and the oligomeric complexes it forms with target proteins in preventing their aggregation [49, 50, 58], and therefore may also influence the solubility of the αB-crystallin fibrils.

4. Amyloid Fibril Formation by the Small Heat-Shock Proteins: A Disease Role *In Vivo*?

Amyloid fibril formation by sHsps may be linked to disease formation, in particular lens cataract, desmin-related myopathy (DRM) and cardiomyopathy. The following sections discuss these associations in detail.

4.1. Amyloid Fibril Formation in the Eye Lens: Cataract and α-Crystallin

Cataract is defined as opacity of the eye lens. In the healthy eye lens, transparency is thought to be maintained by a liquid-like, short-range order that is present in highly concentrated solutions of the crystallins [59]. The crystallins are organized in a stable supramolecular β-sheet structure within the healthy eye lens and α-crystallin plays an important role in chaperoning the other lens crystallins (β and γ), by suppressing aberrant aggregation that would otherwise cause light scattering and impair vision [2]. It has been generally considered that the opacity present in cataractous lenses arises from amorphous aggregation of the crystallin proteins. There is increasing evidence to suggest, however, that amyloid fibril formation by crystallin proteins within the lens could be the cause of certain forms of cataract. For example, in a murine model for cataract, a direct causal relationship was reported between the disease state and amyloid fibril formation by a natural deletion mutant of γ-crystallin [43]. Additionally, a connection has been reported between AD and supranuclear cataract [60]. *Ex vivo* investigation of lenses from individuals with AD revealed the accumulation of Aβ peptides colocalized with αB-crystallin in electron dense deposits that exhibited birefringence with Congo red staining and a positive interaction with thioflavin S, both of which bind strongly to amyloid fibrils [60]. These electron dense deposits correlated with the same areas of the lens where cataract was identified by slit lamp examination. *In vitro*, Goldstein et al. [60] demonstrated that Aβ bound αB-crystallin with high affinity, promoting the formation of aggregates and amyloid protofibrils. Overall, it was suggested that the presence in the lens of Aβ may promote protein aggregation, amyloid fibril formation and supranuclear cataract [60].

There is no direct *in vivo* evidence, however, to show that fibril formation by α-crystallin occurs within the eye lens. Nevertheless, intuitively the lens environment may be favorable to amyloid fibril assembly. For instance, the concentration of α-crystallin is high (approximately 150 mg/ml) and there is very little protein turnover within the avascular lens; it contains proteins as old as the individual [2]. With time, there is a mounting strain of longevity exerted on the crystallins as they degrade and undergo an increasing level of post-transitional modifications in old age (e.g. truncation, phosphorylation, glycation and deamidation) that could lead to destabilization of their native state [61]. Furthermore, crystallin protein aggregation occurs slowly. As a result of these factors, an opportunity may arise for the crystallins to rearrange and self-assemble into amyloid fibrils. Furthermore, Frederikse reported that healthy mammalian eye lenses stain positively with the amyloidophilic dyes Congo red and Thioflavin [62], implying that crystallins in the lens reside in a 'amyloid-like' β-sheet supramolecular structure [62], and thus it is conceivable that the conformational change required to form amyloid fibrils could be relatively minor. Additionally, as summarized above, the sHsp α-crystallin readily forms amyloid fibrils *in vitro* when subjected to partially destabilizing conditions [38].

Further work is required, however, to determine whether any forms of cataract can be attributed to amyloid fibril aggregation, and additionally the role α-crystallin may play in this process, i.e. whether it suppresses amyloid assembly by the other lens crystallins (i.e. β- and γ-crystallin), and/or whether it forms amyloid fibrils itself in the eye lens.

4.2. Amyloid Fibril Formation by R120G αB-Crystallin Leads to Desmin-Related Cardiomyopathy

The naturally occurring missense mutant R120G αB-crystallin causes desmin-related cardiomyopathy, an autosomal dominant disorder presenting in late adulthood with symptoms of cardiomyopathy, cataract, and desmin aggregation in muscles [63, 64]. The exact nature and causes of the pathogenicity involved in R120G αB-crystallin-induced cardiomyopathy remain to be fully understood, although reductive stress is likely to play an important role [65]. Parallels have been drawn, however, between desmin-related cardiomyopathy and neurodegenerative diseases such as AD [66], and the pathogenic processes associated with desmin-related cardiomyopathy may additionally involve the formation of toxic amyloid oligomers by R120G αB-crystallin [67]. Electron dense deposits have been identified containing R120G αB-crystallin which stain positively with the dye Congo red and are also immunoreactive with conformationally dependent epitope antibodies that recognize amyloid oligomers [67]. In a cardiac-specific transgenic mouse model of this disease, R120G αB-crystallin amyloid oligomer formation produced mitochondrial dysfunction and apoptosis [68]. Terminating R120G αB-crystallin expression resulted in a decrease in amyloid oligomer formation and rescued the transgenic animals from premature death [69]. Prolonged voluntary exercise of the mice was also shown to reduce amyloid oligomer levels and slow the progression to heart failure [66].

In comparison to the wild-type protein, R120G αB-crystallin has a significantly destabilized structure that has the tendency to aggregate [70]. It readily forms amyloid fibrils *in vitro*, and possesses structural properties akin to fibrils formed from the wild-type protein, including an apparently identical morphology (as visualized by TEM) and a similar cross-β X-ray fiber diffraction pattern [39]. Amyloid formation by R120G αB-crystallin occurs with a slower growth rate than the wild-type protein *in vitro*. Given early amyloid oligomers are toxic to cells [20], the ability of R120G αB-crystallin to persist in this form was suggested to be consistent with its potentially pathogenic role *in vivo* [39].

Sanbe *et al.* [71] showed that recombinant R120G αB-crystallin forms toxic amyloid oligomers *in vitro*, approximately 240-480 kDa in mass. The addition of other sHsps, Hsp22 (HspB8) and Hsp25, blocked oligomer formation by R120G αB-crystallin and recovered cell viability in a cardiomyocyte-based based study, whereas wild-type αB-crystallin enhanced cell toxicity, suggesting that the mutant and wild-type αB-crystallin co-aggregate as toxic oligomers [71]. Thus, one therapeutic approach in the treatment of desmin-related cardiomyopathy may involve elevating cellular levels of Hsp22 and Hsp25.

5. sHsps are Associated with Amyloid Plaques and Deposits in Protein Conformational Diseases

Apart from their ability to form amyloid fibrils when placed under destabilizing conditions (see above), there is a growing body of evidence to show that the chaperone activity of sHsps prevents other proteins from forming amyloid fibrils and that sHsps are intimately associated with protein conformational diseases [37]. For example, sHsps have been found within amyloid fibril deposits and plaques [72-76] and the expression of some sHsps is dramatically up-regulated in response to pathological conditions associated with fibril formation, e.g. in Dementia with Lewy bodies [75, 77], Alexander's disease [78-80], Creutzfeldt–Jakob disease [72], AD [73, 81] and other neurological conditions [82, 83]. The up-regulation of sHsps expression is presumably a stress response, at the cellular level, to the aggregating species and is thought to reflect an attempt by the cell to mitigate fibril formation.

It remains to be established what effect sHsps have on the toxicity associated with fibril formation. In transgenic *Caenorhabditis elegans* worms, the expression of human Aβ(1-42) led to the induction of Hsp16 proteins (which are homologs of αB-crystallin in vertebrates) [84]. Moreover, the Hsp16 proteins co-localized and co-immunoprecipitated with Aβ(1-42) in this model [84] and increasing the expression of Hsp16 partially suppressed the Aβ-mediated toxicity to these worms [85]. Similarly, the toxicity associated with ataxin-3 aggregation in a *Drosophila* model of spinocerebellar ataxia type-3 (a poly-glutamine mediated protein conformational disease) was reduced when αB-crystallin was over-expressed in this system

[86]. Cell culture models of α-synuclein aggregation have shown that the mammalian sHsp, Hsp27, inhibits both the aggregation and toxicity associated with this process [77, 87]. It was concluded from both the Aβ and α-synuclein studies that sHsps inhibit fibril formation by acting early during the aggregation process to prevent the formation of toxic species [77, 85, 87]. This finding is supported by *in vitro* studies into the ability of sHsps to inhibit amyloid fibril formation (see below).

A possible mechanism by which sHsps may prevent cytotoxicity associated with fibril formation is suggested by studies showing that Hsp27 and αB-crystallin increase the resistance of cultured cells [88-90] and mice [65, 91] to oxidative stress. This may be significant with regard to amyloid fibril formation since it has been reported that the toxicity of prefibrillar amyloid species is due to the production of reactive oxygen species by the aggregating fibril-forming protein, which are generated as a consequence of fibril formation [92]. Therefore, sHsps may protect cells from the cytotoxic effects of fibril formation both through their ability to inhibit the process directly and to mediate the redox resistance of cells.

5.1. sHsps Inhibit Amyloid Fibril Formation *In Vitro*

Despite the up-regulation of sHsp expression and their co-localization with amyloid fibril deposits in various protein conformational diseases, the precise role of sHsps in interacting with the aggregating, fibril-forming target protein remains to be established [37]. The *in vitro* inhibition of fibril formation by sHsps of a variety of target proteins has now been demonstrated. For example, αB-crystallin, the most studied sHsp in this context, inhibits the fibrillation of α-synuclein [93-95], β2-microglobulin [96] (Esposito, Carver, et al., unpublished results), κ-casein [97, 98], insulin [99], the synthetic ccβ-Trp peptide [97], apolipoprotein C-II [100], and the prion protein (Ecroyd, unpublished results). Whilst there has been some conjecture over whether sHsps are able to inhibit fibril formation by the Aβ peptides [96, 101-103], our recent results suggest that αB-crystallin is a potent inhibitor of fibril formation by Aβ1-40 and Aβ1-42 and, in doing so also prevents the cytotoxicity associated with this process (Dehle, Ecroyd, Carver, unpublished results). Thus, along with the well-established activity of sHsps to inhibit the disordered (amorphous) forms of protein aggregation, it is evident that sHsps also have a generic ability to inhibit ordered amyloid fibril protein aggregation. However, the mechanism(s) by which sHsps inhibit fibril formation appear to vary, not only from the one(s) utilized to inhibit amorphous aggregation, but also between different fibril-forming target proteins and under various solution conditions [97].

5.2. Mechanisms by which sHsps Inhibit Amyloid Fibril Formation

αB-Crystallin acts at very low sub-stoichiometric ratios to prevent fibril formation by apolipoprotein C-II and does so by forming a transient complex with the partially folded form of the protein, which stabilizes it and enables it to refold back to its native state via the reversible on-folding pathway [100]. In contrast, when αB-crystallin prevents fibril formation by α-synuclein [93, 94] and κ-casein [97], it does so by forming a stable, soluble, chaperone-target protein complex with the target protein (i.e. a 'reservoir of intermediates') in a manner presumed to be analogous to the mechanism of chaperone action used by sHsps against amorphously aggregating proteins [9]. In addition, in the case of α-synuclein, αB-crystallin redirects the protein from the amyloid fibril-forming pathway to the amorphous aggregation pathway [93], a process that is similar to the mechanism by which the polyphenol (-)-epigallocatechin-3-gallate (EGCG) inhibits fibril formation [104]. Redirecting the aggregation of a target protein from an ordered (amyloid fibril) to disordered (amorphous) pathway would benefit the cell since amorphous aggregates are typically non-toxic and more easily degraded via the ubiquitin-proteasome system.

There are also mechanistic differences in the manner by which sHsps interact with and inhibit protein aggregation leading to amyloid fibril formation as compared to amorphous aggregation. This conclusion is based on studies which have compared the chaperone activity of mutant and/or chimeric forms of sHsps against amorphous and amyloid fibril-forming target proteins. For example, Raman *et al.* (2005) [96], using chimeric α-crystallin proteins in which the N-terminal domain of αA-crystallin was fused with the C-terminal domain of αB-crystallin (αANBC) and vice-versa (αBNAC), showed that whilst αANBC has no ability to prevent the reduction-induced amorphous aggregation of insulin B-chain [105], it was as effective as wild-type αA-crystallin in inhibiting amyloid fibril formation by Aβ(1-40) and β2-microglobulin [96]. Similarly, the chaperone activity of the R120G αB-crystallin mutant (which, as discussed above, is linked to DRM and early onset cataract [64, 106, 107]) towards amorphously aggregating target proteins is significantly reduced compared to the wild-type protein [70, 108], however, its ability to prevent fibril formation by Aβ(1-40) is only slightly decreased [96]. A double mutant (E164A/E165A αB-crystallin) in the highly flexible and unstructured C-terminal extension of αB-crystallin [6, 9, 50, 58], had a markedly reduced ability to inhibit the reduction-induced amorphous aggregation of insulin and heat-induced amorphous aggregation of β_L-crystallin, but significantly increased capacity to prevent fibril formation by κ-casein and ccβ-Trp [109]. Together, these results indicate that there are distinct differences in the mechanism by which sHsps inhibit these two modes of protein aggregation. Interestingly, this raises the possibility that mutant forms of sHsps may be designed as more effective inhibitors of fibril formation compared to the wild-type protein and therefore may be an avenue for therapeutic potential in the future [109].

Recent work has shown that sHsps interact directly with amyloid fibrils by binding along their length, inhibiting their further aggregation and ability to 'seed' the formation of more fibrils [96, 99] (unpublished data). αB-Crystallin does not, however, have the ability to disassemble preformed fibrils [94]. The binding of αB-crystallin to fibrils is intriguing, as such a process may help to explain why sHsps are associated with amyloid deposits and plaques in protein conformational diseases (see above). Moreover, our recent results suggest that the interaction of αB-crystallin with preformed fibrils may result in them tangling and therefore the interaction of Hsps with fibrils may expedite plaque formation *in vivo*. Further studies in this exciting new area of sHsp-fibril research will hopefully shed more light on this interaction and the consequence for plaque assembly in diseases associated with fibril formation.

Conclusion

Overall, a range of studies has demonstrated that the sHsp α-crystallin readily assembles into amyloid fibrils *in vitro* when subjected to partially destabilizing conditions. The potential role of amyloid fibril formation by α-crystallin in the bio-nanomaterial industry, and also in diseases such as cataract and DRM has been discussed. Whilst amyloid fibril assembly by other sHsps has not been reported to date, given that the central α-crystallin domain (of approximately 90 amino acids in length) is conserved between all sHsps and is likely to adopt a β-sandwich structure, these proteins may also have the potential to form amyloid fibrils. Additionally, there may be a close relationship between sHsp chaperone action and the propensity to self-aggregate and form fibrils, since both these activities are favored by exposed stretches of hydrophobicity. It is interesting that α-crystallin (along with other crystallins) adopts an amyloid-like ordered structure within the healthy eye lens. Overall, therefore, there may be a delicate balance between the structure and chaperone function of the native α-crystallin oligomer and the formation of amyloid fibrils. With regard to their ability to inhibit protein aggregation, including amyloid fibril formation, sHsps have significant therapeutic potential [83, 110, 111]. For example, sHsps could be used directly to suppress protein aggregation whereby enhancing their expression and/or chaperone ability chemically or by mutagenesis would have therapeutic application in the treatment of cataract and other protein misfolding diseases.

Acknowledgments

We thank Tuomas P.J. Knowles (University of Cambridge) for acquisition and analysis of the AFM images shown in Figure 3. Our work described in this chapter has been supported by grants from the National Health and Medical Research Council (NHMRC) of Australia and the Australian Research Council. HE was supported by a NHMRC Peter Doherty Fellowship and SM is supported by a Royal Society Dorothy Hodgkin Fellowship.

Abbreviations

Aβ; amyloid-beta
AD; Alzheimer's disease.
AFM; atomic force microscopy
sHsp; small heat-shock protein
TEM; transmission electron microscopy
ThT; thioflavin T

References

[1] Horwich, A. (2002) Protein Folding in the Cell. Academic Press, San Diego, USA.

[2] Harding, J. J. (1991) Cataract: Biochemistry, epidemiology and pharmacology. Chapman and Hall, London.

[3] Dobson, C. M. (1999) Protein misfolding, evolution and disease. *Trends Biochem Sci.* 24, 329-332.

[4] Dobson, C. M. (2001) The structural basis of protein folding and its links with human disease. *Philos Trans R Soc Lond B Biol Sci.* 356, 133-145

[5] Chiti, F. and Dobson, C. M. (2006) Protein misfolding, functional amyloid, and human disease. *Annu Rev Biochem.* 75, 333-366.

[6] Treweek, T. M., Morris, A. M. and Carver, J. A. (2003) Intracellular protein unfolding and aggregation: the role of small heat-shock chaperone proteins. *Aust J Chem.* 56, 357-367.

[7] Ecroyd, H. and Carver, J. A. (2008) Unraveling the mysteries of protein folding and misfolding. *IUBMB Life* 60, 769-774

[8] Horwitz, J. (1992) Alpha-crystallin can function as a molecular chaperone. *Proc Natl Acad Sci U S A..* 89, 10449-10453

[9] Carver, J. A., Rekas, A., Thorn, D. C. and Wilson, M. R. (2003) Small heat-shock proteins and clusterin: intra- and extracellular molecular chaperones with a common mechanism of action and function? *IUBMB Life,* 55, 661-668.

[10] Chiti, F., Stefani, M., Taddei, N., Ramponi, G. and Dobson, C. M. (2003) Rationalization of the effects of mutations on peptide and protein aggregation rates. *Nature,* 424, 805-808

[11] DuBay, K. F., Pawar, A. P., Chiti, F., Zurdo, J., Dobson, C. M. and Vendruscolo, M. (2004) Prediction of the absolute aggregation rates of amyloidogenic polypeptide chains. *J Mol Biol.* 341, 1317-1326

[12] Pawar, A. P., Dubay, K. F., Zurdo, J., Chiti, F., Vendruscolo, M. and Dobson, C. M. (2005) Prediction of "aggregation-prone" and "aggregation-susceptible" regions in proteins associated with neurodegenerative diseases. *J Mol Biol.* 350, 379-392.

[13] Fernandez-Escamilla, A. M., Rousseau, F., Schymkowitz, J. and Serrano, L. (2004) Prediction of sequence-dependent and mutational effects on the aggregation of peptides and proteins. *Nat Biotechnol.* 22, 1302-1306

[14] Lopez de la Paz, M. and Serrano, L. (2004) Sequence determinants of amyloid fibril formation. *Proc Natl Acad Sci U S A.* 101, 87-92

[15] Tartaglia, G. G., Pawar, A. P., Campioni, S., Dobson, C. M., Chiti, F. and Vendruscolo, M. (2008) Prediction of aggregation-prone regions in structured proteins. *J Mol Biol.* 380, 425-436

[16] Lansbury, P. T., Jr. (1999) Evolution of amyloid: what normal protein folding may tell us about fibrillogenesis and disease. *Proc Natl Acad Sci U S A.* 96, 3342-3344.

[17] Stefani, M. and Dobson, C. M. (2003) Protein aggregation and aggregate toxicity: new insights into protein folding, misfolding diseases and biological evolution. *J Mol Med.* 81, 678-699

[18] Prusiner, S. B. (1998) Prions. *Proc Natl Acad Sci U S A.* 95, 13363-13383

[19] Bucciantini, M., Giannoni, E., Chiti, F., Baroni, F., Formigli, L., Zurdo, J., Taddei, N., Ramponi, G., Dobson, C. M. and Stefani, M. (2002) Inherent toxicity of aggregates implies a common mechanism for protein misfolding diseases. *Nature*, 416, 507-511.

[20] Bucciantini, M., Calloni, G., Chiti, F., Formigli, L., Nosi, D., Dobson, C. M. and Stefani, M. (2004) Prefibrillar amyloid protein aggregates share common features of cytotoxicity. *J Biol Chem.* 279, 31374-31382

[21] Ward, R. V., Jennings, K. H., Jepras, R., Neville, W., Owen, D. E., Hawkins, J., Christie, G., Davis, J. B., George, A., Karran, E. H. and Howlett, D. R. (2000) Fractionation and characterization of oligomeric, protofibrillar and fibrillar forms of beta-amyloid peptide. *Biochem J.* 348, 137-144

[22] Novitskaya, V., Bocharova, O. V., Bronstein, I. and Baskakov, I. V. (2006) Amyloid fibrils of mammalian prion protein are highly toxic to cultured cells and primary neurons. *J Biol Chem.* 281, 13828-13836.

[23] Serpell, L. C., Sunde, M. and Blake, C. C. (1997) The molecular basis of amyloidosis. *Cell Mol Life Sci.* 53, 871-887.

[24] Knowles, T. P., Fitzpatrick, A. W., Meehan, S., Mott, H. R., Vendruscolo, M., Dobson, C. M. and Welland, M. E. (2007) Role of intermolecular forces in defining material properties of protein nanofibrils. *Science,* 318, 1900-1903

[25] Smith, J. F., Knowles, T. P., Dobson, C. M., MacPhee, C. E. and Welland, M. E. (2006) Characterization of the nanoscale properties of individual amyloid fibrils. *Proc Natl Acad Sci U S A..* 103, 15806-15811.

[26] Sunde, M. and Blake, C. (1997) The structure of amyloid fibrils by electron microscopy and X-ray diffraction. *Adv Protein Chem.* 50, 123-159.

[27] Sunde, M., Serpell, L. C., Bartlam, M., Fraser, P. E., Pepys, M. B. and Blake, C. C. (1997) Common core structure of amyloid fibrils by synchrotron X-ray diffraction. *J Mol Biol.* 273, 729-739.

[28] Jimenez, J. L., Guijarro, J. I., Orlova, E., Zurdo, J., Dobson, C. M., Sunde, M. and Saibil, H. R. (1999) Cryo-electron microscopy structure of an SH3 amyloid fibril and model of the molecular packing. *EMBO J.* 18, 815-821

[29] Jimenez, J. L., Nettleton, E. J., Bouchard, M., Robinson, C. V., Dobson, C. M. and Saibil, H. R. (2002) The protofilament structure of insulin amyloid fibrils. *Proc Natl Acad Sci U S A.* 99, 9196-9201

[30] Zhang, R., Hu, X., Khant, H., Ludtke, S. J., Chiu, W., Schmid, M. F., Frieden, C. and Lee, J. M. (2009) Interprotofilament interactions between Alzheimer's A{beta}1-42 peptides in amyloid fibrils revealed by cryoEM. *Proc Natl Acad Sci U S A.*

[31] Klunk, W. E., Jacob, R. F. and Mason, R. P. (1999) Quantifying amyloid by congo red spectral shift assay. *Methods Enzymol.* 309, 285-305

[32] LeVine, H., 3rd (1999) Quantification of beta-sheet amyloid fibril structures with thioflavin T. *Methods Enzymol.* 309, 274-284

[33] Krebs, M. R., Bromley, E. H. and Donald, A. M. (2005) The binding of thioflavin-T to amyloid fibrils: localisation and implications. *J Struct Biol.* 149, 30-37

[34] Dobson, C. M. (2003) Protein folding and misfolding. *Nature,* 426, 884-890.

[35] Harper, J. D. and Lansbury, P. T., Jr. (1997) Models of amyloid seeding in Alzheimer's disease and scrapie: mechanistic truths and physiological consequences of the time-dependent solubility of amyloid proteins. *Annu Rev Biochem.* 66, 385-407

[36] Ecroyd, H., Koudelka, T., Thorn, D. C., Williams, D. M., Devlin, G., Hoffmann, P. and Carver, J. A. (2008) Dissociation from the oligomeric state is the rate-limiting step in amyloid fibril formation by kappa-casein. *J Biol Chem.* 283, 9012-9022

[37] Ecroyd, H. and Carver, J. A. (2009) Crystallin proteins and amyloid fibrils. *Cell Mol Life Sci.* 66, 62-81

[38] Meehan, S., Berry, Y., Luisi, B., Dobson, C. M., Carver, J. A. and MacPhee, C. E. (2004) Amyloid fibril formation by lens crystallin proteins and its implications for cataract formation. *J Biol Chem.* 279, 3413-3419

[39] Meehan, S., Knowles, T. P., Baldwin, A. J., Smith, J. F., Squires, A. M., Clements, P., Treweek, T. M., Ecroyd, H., Tartaglia, G. G., Vendruscolo, M., MacPhee, C. E., Dobson, C. M. and Carver, J. A. (2007) Characterisation of amyloid fibril formation by small heat-shock chaperone proteins human alphaA-, alphaB- and R120G alphaB-crystallins. *J Mol Biol.* 372, 470-484

[40] Tanaka, N., Tanaka, R., Tokuhara, M., Kunugi, S., Lee, Y. F. and Hamada, D. (2008) Amyloid fibril formation and chaperone-like activity of peptides from alphaA-crystallin. *Biochemistry*, 47, 2961-2967

[41] Garvey, M., Gras, S. L., Meehan, S., Meade, S. J., Carver, J. A. and Gerrard, J. A. (2009) Protein nanofibres of defined morphology prepared from mixtures of crude crystallins. *Int J Nanotech.* 6, 258-273.

[42] Papanikolopoulou, K., Mills-Henry, I., Thol, S. L., Wang, Y., Gross, A. A., Kirschner, D. A., Decatur, S. M. and King, J. (2008) Formation of amyloid fibrils in vitro by human gammaD-crystallin and its isolated domains. *Mol Vis.* 14, 81-89

[43] Sandilands, A., Hutcheson, A. M., Long, H. A., Prescott, A. R., Vrensen, G., Loster, J., Klopp, N., Lutz, R. B., Graw, J., Masaki, S., Dobson, C. M., MacPhee, C. E. and Quinlan, R. A. (2002) Altered aggregation properties of mutant gamma-crystallins cause inherited cataract. *EMBO J.* 21, 6005-6014

[44] MacPhee, C. E. and Woolfson, D. N. (2004) Engineered and designed peptide-based fibrous biomaterials. *Curr Opin Solid State Mat Sci.* 8, 141-149

[45] Baldwin, A. J., Bader, R., Christodoulou, J., MacPhee, C. E., Dobson, C. M. and Barker, P. D. (2006) Cytochrome display on amyloid fibrils. *J Am Chem Soc.* 128, 2162-2163

[46] Gras, S. L. (2007) Amyloid fibrils: From disease to design. New biomaterial applications for self-assembling cross beta-fibrils. *Aust J Chem.* 60, 333-342

[47] Kelly, J. W. (1998) The alternative conformations of amyloidogenic proteins and their multi-step assembly pathways. *Curr Opin Struct Biol.* 8, 101-106

[48] Uversky, V. N. and Fink, A. L. (2004) Conformational constraints for amyloid fibrillation: the importance of being unfolded. *Biochim Biophys Acta*, 1698, 131-153

[49] Carver, J. A., Aquilina, J. A., Truscott, R. J. and Ralston, G. B. (1992) Identification by ^{1}H NMR spectroscopy of flexible C-terminal extensions in bovine lens alpha-crystallin. *FEBS Lett.* 311, 143-149

[50] Carver, J. A. and Lindner, R. A. (1998) NMR spectroscopy of alpha-crystallin. Insights into the structure, interactions and chaperone action of small heat-shock proteins. *Int J Biol Macromol.* 22, 197-209

[51] Sharma, K. K., Kumar, R. S., Kumar, G. S. and Quinn, P. T. (2000) Synthesis and characterization of a peptide identified as a functional element in alphaA-crystallin. J *Biol Chem.* 275, 3767-3771

[52] Reddy, G. B., Kumar, P. A. and Kumar, M. S. (2006) Chaperone-like activity and hydrophobicity of alpha-crystallin. *IUBMB Life,* 58, 632-641
[53] Hall, D., Hirota, N. and Dobson, C. M. (2005) A toy model for predicting the rate of amyloid formation from unfolded protein. *J Mol Biol.* 351, 195-205
[54] Hatters, D. M., MacPhee, C. E., Lawrence, L. J., Sawyer, W. H. and Howlett, G. J. (2000) Human apolipoprotein C-II forms twisted amyloid ribbons and closed loops. *Biochemistry,* 39, 8276-8283
[55] Hatters, D. M., MacRaild, C. A., Daniels, R., Gosal, W. S., Thomson, N. H., Jones, J. A., Davis, J. J., MacPhee, C. E., Dobson, C. M. and Howlett, G. J. (2003) The circularization of amyloid fibrils formed by apolipoprotein C-II. *Biophys J.* 85, 3979-3990.
[56] Lashuel, H. A., Hartley, D., Petre, B. M., Walz, T. and Lansbury, P. T., Jr. (2002) Neurodegenerative disease: amyloid pores from pathogenic mutations. *Nature,* 418, 291
[57] Lashuel, H. A. and Lansbury, P. T., Jr. (2006) Are amyloid diseases caused by protein aggregates that mimic bacterial pore-forming toxins? *Q Rev Biophys.* 39, 167-201
[58] Carver, J. A. (1999) Probing the structure and interactions of crystallin proteins by NMR spectroscopy. *Prog Retin Eye Res.* 18, 431-462
[59] Delaye, M. and Tardieu, A. (1983) Short-range order of crystallin proteins accounts for eye lens transparency. *Nature*, 302, 415-417
[60] Goldstein, L. E., Muffat, J. A., Cherny, R. A., Moir, R. D., Ericsson, M. H., Huang, X., Mavros, C., Coccia, J. A., Faget, K. Y., Fitch, K. A., Masters, C. L., Tanzi, R. E., Chylack, L. T., Jr. and Bush, A. I. (2003) Cytosolic beta-amyloid deposition and supranuclear cataracts in lenses from people with Alzheimer's disease. *Lancet*, 361, 1258-1265
[61] Derham, B. K. and Harding, J. J. (1999) Alpha-crystallin as a molecular chaperone. *Prog Retin Eye Res.* 18, 463-509
[62] Frederikse, P. H. (2000) Amyloid-like protein structure in mammalian ocular lenses. *Curr Eye Res.* 20, 462-468.
[63] Fardeau, M., Godet-Guillain, J., Tome, F. M., Collin, H., Gaudeau, S., Boffety, C. and Vernant, P. (1978) A new familial muscular disorder demonstrated by the intra-sarcoplasmic accumulation of a granulo-filamentous material which is dense on electron microscopy (author's translation). *Rev Neurol (Paris)* 134, 411-425
[64] Vicart, P., Caron, A., Guicheney, P., Li, Z., Prevost, M. C., Faure, A., Chateau, D., Chapon, F., Tome, F., Dupret, J. M., Paulin, D. and Fardeau, M. (1998) A missense mutation in the alphaB-crystallin chaperone gene causes a desmin-related myopathy. *Nat Genet.* 20, 92-95.
[65] Rajasekaran, N. S., Connell, P., Christians, E. S., Yan, L. J., Taylor, R. P., Orosz, A., Zhang, X. Q., Stevenson, T. J., Peshock, R. M., Leopold, J. A., Barry, W. H., Loscalzo, J., Odelberg, S. J. and Benjamin, I. J. (2007) Human alpha B-crystallin mutation causes oxido-reductive stress and protein aggregation cardiomyopathy in mice. *Cell,* 130, 427-439
[66] Maloyan, A., Gulick, J., Glabe, C. G., Kayed, R. and Robbins, J. (2007) Exercise reverses preamyloid oligomer and prolongs survival in alphaB-crystallin-based desmin-related cardiomyopathy. *Proc Natl Acad Sci U S A..* 104, 5995-6000.

[67] Sanbe, A., Osinska, H., Saffitz, J. E., Glabe, C. G., Kayed, R., Maloyan, A. and Robbins, J. (2004) Desmin-related cardiomyopathy in transgenic mice: a cardiac amyloidosis. *Proc Natl Acad Sci U S A.* 101, 10132-10136

[68] Maloyan, A., Sanbe, A., Osinska, H., Westfall, M., Robinson, D., Imahashi, K., Murphy, E. and Robbins, J. (2005) Mitochondrial dysfunction and apoptosis underlie the pathogenic process in alpha-B-crystallin desmin-related cardiomyopathy. *Circulation,* 112, 3451-3461

[69] Sanbe, A., Osinska, H., Villa, C., Gulick, J., Klevitsky, R., Glabe, C. G., Kayed, R. and Robbins, J. (2005) Reversal of amyloid-induced heart disease in desmin-related cardiomyopathy. *Proc Natl Acad Sci U S A.* 102, 13592-13597

[70] Treweek, T. M., Rekas, A., Lindner, R. A., Walker, M. J., Aquilina, J. A., Robinson, C. V., Horwitz, J., Perng, M. D., Quinlan, R. A. and Carver, J. A. (2005) R120G alphaB-crystallin promotes the unfolding of reduced alpha-lactalbumin and is inherently unstable. *FEBS J.* 272, 711-724

[71] Sanbe, A., Yamauchi, J., Miyamoto, Y., Fujiwara, Y., Murabe, M. and Tanoue, A. (2007) Interruption of CryAB-amyloid oligomer formation by HSP22. *J Biol Chem.* 282, 555-563

[72] Renkawek, K., de Jong, W. W., Merck, K. B., Frenken, C. W., van Workum, F. P. and Bosman, G. J. (1992) alpha B-crystallin is present in reactive glia in Creutzfeldt-Jakob disease. *Acta Neuropath.* 83, 324-327

[73] Shinohara, H., Inaguma, Y., Goto, S., Inagaki, T. and Kato, K. (1993) Alpha B crystallin and HSP28 are enhanced in the cerebral cortex of patients with Alzheimer's disease. *J Neurol Sci.* 119, 203-208

[74] McLean, P. J., Kawamata, H., Shariff, S., Hewett, J., Sharma, N., Ueda, K., Breakefield, X. O. and Hyman, B. T. (2002) TorsinA and heat shock proteins act as molecular chaperones: suppression of alpha-synuclein aggregation. *J Neurochem.* 83, 846-854

[75] Pountney, D. L., Treweek, T. M., Chataway, T., Huang, Y., Chegini, F., Blumbergs, P. C., Raftery, M. J. and Gai, W. P. (2005) Alpha B-crystallin is a major component of glial cytoplasmic inclusions in multiple system atrophy. *Neurotox Res.* 7, 77-85

[76] Wilhelmus, M. M., Otte-Holler, I., Wesseling, P., de Waal, R. M., Boelens, W. C. and Verbeek, M. M. (2006) Specific association of small heat shock proteins with the pathological hallmarks of Alzheimer's disease brains. *Neuropathol Appl Neurobiol.* 32, 119-130

[77] Outeiro, T. F., Klucken, J., Strathearn, K. E., Liu, F., Nguyen, P., Rochet, J. C., Hyman, B. T. and McLean, P. J. (2006) Small heat shock proteins protect against alpha-synuclein-induced toxicity and aggregation. *Biochem Biophys Res Commun.* 351, 631-638

[78] Iwaki, T., Kume-Iwaki, A., Liem, R. K. and Goldman, J. E. (1989) Alpha B-crystallin is expressed in non-lenticular tissues and accumulates in Alexander's disease brain. *Cell,* 57, 71-78

[79] Ochi, N., Kobayashi, K., Maehara, M., Nakayama, A., Negoro, T., Shinohara, H., Watanabe, K., Nagatsu, T. and Kato, K. (1991) Increment of alpha B-crystallin mRNA in the brain of patient with infantile type Alexander's disease. *Biochem Biophys Res Commun.* 179, 1030-1035

[80] Kato, K., Inaguma, Y., Ito, H., Iida, K., Iwamoto, I., Kamei, K., Ochi, N., Ohta, H. and Kishikawa, M. (2001) Ser-59 is the major phosphorylation site in alphaB-crystallin accumulated in the brains of patients with Alexander's disease. *J Neurochem.* 76, 730-736

[81] Renkawek, K., Voorter, C. E., Bosman, G. J., van Workum, F. P. and de Jong, W. W. (1994) Expression of alpha B-crystallin in Alzheimer's disease. *Acta Neuropath.* 87, 155-160.

[82] Iwaki, T., Wisniewski, T., Iwaki, A., Corbin, E., Tomokane, N., Tateishi, J. and Goldman, J. E. (1992) Accumulation of alpha B-crystallin in central nervous system glia and neurons in pathologic conditions. *Am J Pathol.* 140, 345-356

[83] Clark, J. I. and Muchowski, P. J. (2000) Small heat-shock proteins and their potential role in human disease. *Curr Opin Struct Biol.* 10, 52-59

[84] Fonte, V., Kapulkin, V., Taft, A., Fluet, A., Friedman, D. and Link, C. D. (2002) Interaction of intracellular beta amyloid peptide with chaperone proteins. *Proc Natl Acad Sci U S A..* 99, 9439-9444.

[85] Fonte, V., Kipp, D. R., Yerg, J., 3rd, Merin, D., Forrestal, M., Wagner, E., Roberts, C. M. and Link, C. D. (2008) Suppression of in vivo beta-amyloid peptide toxicity by overexpression of the HSP-16.2 small chaperone protein. *J Biol Chem.* 283, 784-791

[86] Bilen, J. and Bonini, N. M. (2007) Genome-wide screen for modifiers of ataxin-3 neurodegeneration in Drosophila. *PLoS Genet.* 3, 1950-1964

[87] Zourlidou, A., Payne Smith, M. D. and Latchman, D. S. (2004) HSP27 but not HSP70 has a potent protective effect against alpha-synuclein-induced cell death in mammalian neuronal cells. *J Neurochem.* 88, 1439-1448

[88] Mehlen, P., Schulze-Osthoff, K. and Arrigo, A. P. (1996) Small stress proteins as novel regulators of apoptosis. Heat shock protein 27 blocks Fas/APO-1- and staurosporine-induced cell death. *J Biol Chem.* 271, 16510-16514

[89] Baek, S. H., Min, J. N., Park, E. M., Han, M. Y., Lee, Y. S., Lee, Y. J. and Park, Y. M. (2000) Role of small heat shock protein HSP25 in radioresistance and glutathione-redox cycle. *J Cell Physiol.* 183, 100-107

[90] Arrigo, A. P. (2001) Hsp27: novel regulator of intracellular redox state. *IUBMB Life,* 52, 303-307

[91] Yan, L. J., Christians, E. S., Liu, L., Xiao, X., Sohal, R. S. and Benjamin, I. J. (2002) Mouse heat shock transcription factor 1 deficiency alters cardiac redox homeostasis and increases mitochondrial oxidative damage. *EMBO J.* 21, 5164-5172

[92] Tabner, B. J., El-Agnaf, O. M., German, M. J., Fullwood, N. J. and Allsop, D. (2005) Protein aggregation, metals and oxidative stress in neurodegenerative diseases. *Biochem Soc Trans.* 33, 1082-1086

[93] Rekas, A., Adda, C. G., Aquilina, J. A., Barnham, K. J., Sunde, M., Galatis, D., Williamson, N. A., Masters, C. L., Anders, R. F., Robinson, C. V., Cappai, R. and Carver, J. A. (2004) Interaction of the molecular chaperone alphaB-crystallin with alpha-synuclein: effects on amyloid fibril formation and chaperone activity. *J Mol Biol.* 340, 1167-1183

[94] Rekas, A., Jankova, L., Thorn, D. C., Cappai, R. and Carver, J. A. (2007) Monitoring the prevention of amyloid fibril formation by alpha-crystallin. Temperature dependence and the nature of the aggregating species. *FEBS J.* 274, 6290-6304

[95] Wang, J., Martin, E., Gonzales, V., Borchelt, D. R. and Lee, M. K. (2008) Differential regulation of small heat shock proteins in transgenic mouse models of neurodegenerative diseases. *Neurobiol Aging* 29, 586-597

[96] Raman, B., Ban, T., Sakai, M., Pasta, S. Y., Ramakrishna, T., Naiki, H., Goto, Y. and Rao Ch, M. (2005) AlphaB-crystallin, a small heat-shock protein, prevents the amyloid fibril growth of an amyloid beta-peptide and beta2-microglobulin. *Biochem J.* 392, 573-581.

[97] Ecroyd, H., Meehan, S., Horwitz, J., Aquilina, J. A., Benesch, J. L., Robinson, C. V., MacPhee, C. E. and Carver, J. A. (2007) Mimicking phosphorylation of alphaB-crystallin affects its chaperone activity. *Biochem J.* 401, 129-141

[98] Ecroyd, H. and Carver, J. A. (2008) The effect of small molecules in modulating the chaperone activity of alphaB-crystallin against ordered and disordered protein aggregation. *FEBS J.* 275, 935-947

[99] Knowles, T. P., Shu, W., Devlin, G. L., Meehan, S., Auer, S., Dobson, C. M. and Welland, M. E. (2007) Kinetics and thermodynamics of amyloid formation from direct measurements of fluctuations in fibril mass. *Proc Natl Acad Sci U S A.* 104, 10016-10021

[100] Hatters, D. M., Lindner, R. A., Carver, J. A. and Howlett, G. J. (2001) The molecular chaperone, alpha-crystallin, inhibits amyloid formation by apolipoprotein C-II. *J Biol Chem.* 276, 33755-33761

[101] Stege, G. J., Renkawek, K., Overkamp, P. S., Verschuure, P., van Rijk, A. F., Reijnen-Aalbers, A., Boelens, W. C., Bosman, G. J. and de Jong, W. W. (1999) The molecular chaperone alphaB-crystallin enhances amyloid beta neurotoxicity. *Biochem Biophys Res Commun.* 262, 152-156

[102] Liang, J. J. (2000) Interaction between beta-amyloid and lens alphaB-crystallin. *FEBS Lett.* 484, 98-101

[103] Wilhelmus, M. M., Boelens, W. C., Otte-Holler, I., Kamps, B., de Waal, R. M. and Verbeek, M. M. (2006) Small heat shock proteins inhibit amyloid-beta protein aggregation and cerebrovascular amyloid-beta protein toxicity. *Brain Res.* 1089, 67-78.

[104] Ehrnhoefer, D. E., Bieschke, J., Boeddrich, A., Herbst, M., Masino, L., Lurz, R., Engemann, S., Pastore, A. and Wanker, E. E. (2008) EGCG redirects amyloidogenic polypeptides into unstructured, off-pathway oligomers. *Nat Struct Mol Biol.* 15, 558-566

[105] Kumar, L. V. and Rao, C. M. (2000) Domain swapping in human alpha A and alpha B crystallins affects oligomerization and enhances chaperone-like activity. *J Biol Chem.* 275, 22009-22013

[106] Simon, S., Fontaine, J. M., Martin, J. L., Sun, X., Hoppe, A. D., Welsh, M. J., Benndorf, R. and Vicart, P. (2007) Myopathy-associated alphaB-crystallin mutants: abnormal phosphorylation, intracellular location, and interactions with other small heat shock proteins. *J Biol Chem.* 282, 34276-34287

[107] Litt, M., Kramer, P., LaMorticella, D. M., Murphey, W., Lovrien, E. W. and Weleber, R. G. (1998) Autosomal dominant congenital cataract associated with a missense mutation in the human alpha crystallin gene CRYAA. *Hum Mol Genet.* 7, 471-474

[108] Kumar, L. V., Ramakrishna, T. and Rao, C. M. (1999) Structural and functional consequences of the mutation of a conserved arginine residue in alphaA and alphaB crystallins. *J Biol Chem.* 274, 24137-24141

[109] Treweek, T. M., Ecroyd, H., Williams, D. M., Meehan, S., Carver, J. A. and Walker, M. J. (2007) Site-directed mutations in the C-terminal extension of human alphaB-crystallin affect chaperone function and block amyloid fibril formation. *PLoS ONE,* 2, e1046

[110] Arrigo, A. P., Simon, S., Gibert, B., Kretz-Remy, C., Nivon, M., Czekalla, A., Guillet, D., Moulin, M., Diaz-Latoud, C. and Vicart, P. (2007) Hsp27 (HspB1) and alphaB-crystallin (HspB5) as therapeutic targets. *FEBS Lett.* 581, 3665-3674

[111] Muchowski, P. J. (2002) Protein misfolding, amyloid formation, and neurodegeneration: a critical role for molecular chaperones? *Neuron.* 35, 9-12

[112] Carver, J. A., Lindner, R. A., Lyon, C., Canet, D., Hernandez, H., Dobson, C. M. and Redfield, C. (2002) The interaction of the molecular chaperone alpha-crystallin with unfolding alpha-lactalbumin: a structural and kinetic spectroscopic study. *J Mol Biol.* 318, 815-827

[113] Devlin, G. L., Carver, J. A. and Bottomley, S. P. (2003) The selective inhibition of serpin aggregation by the molecular chaperone, alpha-crystallin, indicates a nucleation-dependent specificity. *J Biol Chem.* 278, 48644-48650.

PART II : MUTATIONS IN SMALL STRESS PROTEINS RESPONSIVE OF HUMAN DISEASES

In: Small Stress Proteins and Human Diseases
Editors: Stéphanie Simon et al.
ISBN: 978-1-61470-636-6

Chapter 2.1

From Sequence to Pathology: The Required Integrity of the Structure-Dynamics-Function Relationships at the Molecular Level

Annette Tardieu[1], Magalie Michiel[2], Fériel Skouri-Panet[1], Elodie Duprat[1], Stéphanie Simon[3] and Stéphanie Finet[1*]

[1] Institut de Minéralogie et de Physique des Milieux Condensés, UMR 7590, CNRS-UPMC- IPGP-Université Paris Diderot, 140 rue de Lourmel 75015 Paris, France

[2] Université d'Evry Val d'Essonne (UEVE), CEA, DSV, Institut de Génomique (IG), CNRS, UMR8030, 2 rue Gaston Crémieux, 91057 Evry cedex 07, France

[3]Université Lyon 1 - CGMC, UMR5534, Equipe 'Stress, Chaperons et Morts Cellulaires', 16 rue Raphael Dubois, 69622 Villeurbanne Cedex, France

Abstract

The understanding of the small Heat Shock Proteins (sHSP) properties is becoming of outmost importance for society. Indeed, it is now well recognized that sHSPs participate in a variety of degenerative or inflammation related pathologies. The sHSPs were also found implicated in bad cancer prognosis and resistance to anti-cancer drugs. sHSP also participated at different levels of cell regulation (differentiation, apoptosis, autophagy …). Since the sHSPs are believed to function through a variety of associations and interactions, between them and with substrates, pathologies may originate from dysfunction at different levels of structural organization and regulation. Moreover, an increasing number of point mutations have been identified, leading to cardiomyopathies, neuropathies or cataracts. All those aspects are treated in this book. This chapter deals with the mutation-induced structural modifications underlying the pathological states.

* Corresponding Author: Stéphanie Finet
Institut de Minéralogie et de Physique des Milieux Condensés,
UMR 7590, CNRS-UPMC- IPGP-Université Paris Diderot,
140 rue de Lourmel 75015 Paris, France
stephanie.finet@upmc.fr

INTRODUCTION

The small Heat Shock Proteins (sHSPs) are ubiquitous proteins, specialized in cell protection against stress [1-3]. The family is characterized by a highly conserved domain of around 80 amino acids, the α-crystallin domain or ACD (see Figure 1) [4]. The sHSPs usually form large, yet highly variable, assemblies (2-50 subunits of 12-50kDa), either monodisperse or polydisperse, i.e. with a fixed or with a variable number of subunits (SU). The oligomers are able to rapidly exchange SU. The anti-stress function relies on an exceptional ability, known as the "chaperone-like activity", to associate a variety of partially unfolded substrates and prevent their stress-induced aggregation and precipitation. The chaperone-like activity is tightly associated with quaternary structure and SU exchange (see introduction of this book).

The archetype R120G mutation of the HSPB5 induces a myofibrillar myopathy which can be associated with cardiomyopathy and cataract [5]; these facts told us for a long time that point mutations may have drastic effects on the overall complex structure and stability of the sHSP assemblies. Indeed, large aggregates of mutant sHSPs have been observed *in vivo* and *in cellulo* [6-7]. *In vitro*, the assembly size, polydispersity and stability were found deeply modified [8-20].

After describing the concepts and ideas underlying our present knowledge of the sHSP structure, dynamics and function, the challenge of this chapter is to understand how little modifications like point mutations may perturb these normal relationships at different levels of integration and organization and lead to pathologies. A model of their relationships is presented at the end.

1. FROM SEQUENCE TO ASSEMBLIES: THE INPUT OF 3D STRUCTURES

1.1. A Conserved ACD

Nowadays, the sequence information is abundant, with more than 350 sequences, covering almost all life forms [21-22]. Several sHSPs are possibly present in one organism: in some species (typically in plants) more than 15 sHSPs have been observed. The sHSP evolutionary aspects were first analyzed by W. de Jong and his group. It was shown that, at the monomer level, the highly conserved ACD [4] was surrounded by more variable N-terminal domains (N-td) and C-terminal extensions (C-te). Then, extensive sequence analyses led to the identification of 10 human sHSPs, called HSPB1-10 [21, 23]. The highly expressed HSP27, αA- and αB-crystallin, HSP20, HSP22, became respectively HSPB1, B4, B5, B6 and B8. The sequences of these human sHSPs and of the sHSPs for which structural information is available are schematically represented in figure 1. A relative, human HSP16.2, was recently found [24], as discussed in the introduction of this book.

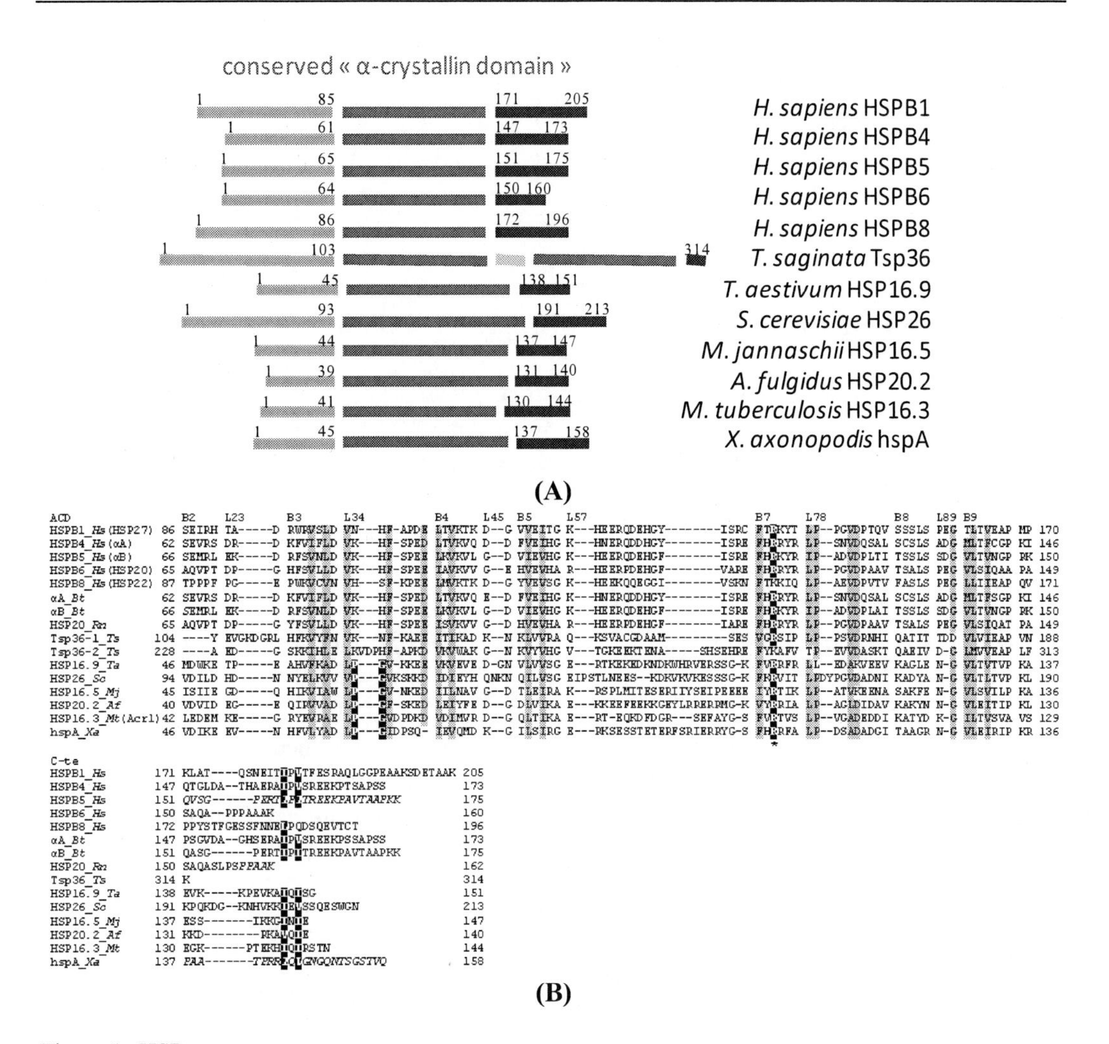

Figure 1.sHSP sequences

A-Sequences of 12 sHSPs from animals (*Homo sapiens* HSPB1, HSPB4, HSPB5, HSPB6 and HSPB8; *Taenia saginata* Tsp36), plant (*Triticum aestivum* HSP16.9), fungi (*Saccharomyces cerevisiae* HSP26), archae (*Methanococcus jannaschii* HSP16.5 and *Archaeglobus fulgidus* HSP20.2), and bacteria (*Mycobacterium tuberculosis* HSP16.3 and *Xanthomonas axonopodis* hspA). The N-terminal domain (N-td), ACD, and C-terminal extension (C-te) are represented in green, red and blue, respectively. The sequence numbering of amino acids at start and end of each N-td and C-te is indicated.

B-Multiple sequence alignment of the ACD and C-te of 15 sHSPs : human HSPB1 (or HSP27) (Swiss-Prot accession number: P04792), human HSPB4 (αA-crystallin) (P02489), human HSPB5 (αB-crystallin) (P02511; PDB code: 2WJ7), human HSPB6 (HSP20) (O14558), human HSPB8 (HSP22) (Q9UJY1), rat HSP20 (P97541; 2WJ5), bovine αA-crystallin (P02470), bovine αB-crystallin (P02510), *Taenia saginata* Tsp36 (Q7YZT0, 2BOL), *Triticum aestivum* HSP16.9 (Q41560, 1GME), *Saccharomyces cerevisiae* HSP26 (P15992; 2H50 and 2H53), *Methanococcus jannaschii* HSP16.5 (Q57733; 1SHS), *Archaeoglobus fulgidus* HSP20.2 (O28308), *Mycobacterium tuberculosis* HSP16.3 (Acr1) (P0A5B7; 2BYU), *Xanthomonas axonopodis* hspA (Q8PNC2; 3GLA). *Taenia saginata* Tsp36 comprises one N-ter region, two ACDs separated by a connecting peptide (sequence not shown), followed by one C-ter extension; Tsp36-1 and Tsp36-2 refer to the first and second ACD sequences, respectively. In italic are the amino acids absent from the 3D structure. The ACD secondary structures of *Triticum aestivum* HSP16.9 (1GME, chain A) are indicated by B (beta-strands) or L (loops). The most conserved amino acid sites (exact identity or similar physicochemical properties) are highlighted in gray. The PG pattern (highly conserved in the non-animal sHSPs [Fu and Chang, 2006]) is indicated by white letters and highlighted in black; almost all of these sHSPs also share the IXI pattern [Pasta et al, 2004], located in the C-te. The position 120 (in the human HSPB5 sequence), and the equivalent ones for the other sequences, are indicated by a black star; at this site, the R residues are indicated by white letters and highlighted in black.

1.2. High or Medium Resolution 3D Structures Available

Structural information at high or even medium resolution is limited. Among the monodisperse sHSP assemblies, only a few were crystallized, probably because of their flexibility or dynamical properties, yet cryo-electron microscopy (cryo-EM) reconstruction has been developed. The polydisperse sHSPs are deeply lacking since polydispersity precludes crystallization and the assembly size generally limits NMR studies. The recent expression and purification of specific excised ACDs from human and rat sHSPs, after years of unsuccessful attempts, now open new ways for polydisperse sHSP structural analyses [25-26]. Other strategies are also emerging like solid state NMR [26]. The structures of wild-type monodisperse sHSPs and of the ACD homodimers of truncated sHSPs that have been published so far, at high or medium resolution, are listed in Table 1.

Only four wild-type assembly 3D structures have been determined so far on an atomic scale by X-rays: the hyperthermophile archae *Methanococcus jannaschii* HSP16.5 [27], the wheat *Triticum aestivum* HSP16.9 [28], the tapeworm *Taenia saginata* Tsp36 [29] and the fourth one from *Xanthomonas axonopodis pv. citri* has been recently released [30]. Among them, the N-td was either not resolved at all or was resolved in helical structure. It can be pointed out that Tsp 36 protein has an original feature since it is a dimer where each SU contains two ACDs connected by a linker region.

Two ACD homodimers have been analyzed: one from human, HSPB5, was studied by NMR and X-rays, the other, HSP20 from rat, by X-rays [25-26].

The other structures of monodisperse sHSPs determined at lower resolution by cryoEM include: yeast HSP26, under two forms, compact and expanded [31-32], *Mycobacterium tuberculosis* HSP16.3 [33], and *Methanococcus jannaschii* HSP16.5-P1, a HSP16.5 chimeric variant with an additional Ntd 14-residue 'P1 sequence' specific to HSPB1 N-td sequence [34-35]. Finally, HSP20.2 from *Archeoglobus fulgidus* was compared to HSP16.5 from *Methanococcus jannaschii* [36].

1.3. Comparison of the Monomer 3D Structures

The 3D structures of monodisperse assemblies have shown that the ACD adopts a two β-sheet (β sandwich) fold, reminiscent of an immunoglobulin-fold, yet with different connections. In all these structures the β sandwich fold, essentially made of 6-9 β-strands, is highly conserved, whereas the L57 loop is more variable, as may be seen on figure 2. The C-te, however, may adopt different orientations.

The ACDs obtained from truncated mammalian sHSPs present some differences in the 3D structure. The β sandwich fold is essentially preserved, yet the two β strands B6 and B7 are merged in a unique and extended β strand (B6+7), whereas the L56 loop is minimized. It is not possible at present to tell whether this new ACD fold is representative of all the ACDs from polydisperse native sHSPs.

Table 1. High or medium resolution sHSP 3D structures available.
(1) The corresponding sequence length: amino acid number (aa) are in bracket for truncated sequences. (2) Additional 14 residues of the variant *M. jannaschii* (see details in the text).

Kingdom	Species	Usual name	aa (1)	SU nb	technique	Pdb
Animals	*H. sapiens*	HSPB5 homodimer	[67-157]	2	X-rays	2WJ7
			[64-152]	2	NMR	
	R. norvegicus	HSP20 homodimer	[65-162]	2	X-rays	2WJ5
	T. saginata	Tsp36	1-314	2= 4ACD	X-rays	2BOL
Plants	*T. aestivum*	HSP16.9	1-151	12	X-rays	1GME
Fungi	*S. cerevisiae*	HSP26 compact HSP26 expanded	1-213	24 24	cryoEM-SAXS cryoEM	2H50 2H53
Archae	*M. jannaschii*	HSP16.5 HSP16.5-P1 (variant)	1-147 +14 (2)	24 48	X-rays cryoEM	1SHS
	A. fulgidus	HSP20.2 compact HSP20.2 expanded	1-140	24	cryoEM	
Bacteria	*M. tuberculosis*	HSP16.3-Acr1	1-144	12	cryoEM	2BYU
	X. Axonopodis pv.citri	HSPA	[40-139]	2	X-rays	3GLA

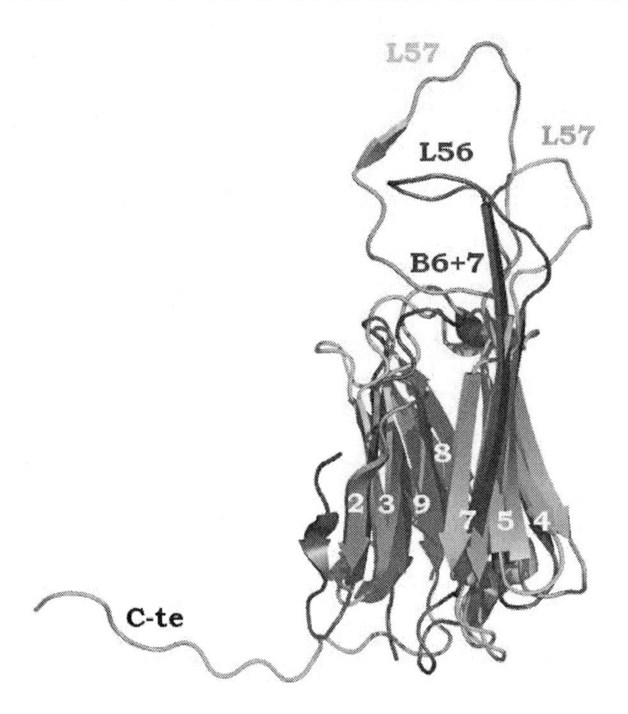

Figure 2. Structural superposition of *Triticum. aestivum* HSP16.9 (PDB code: 1GME, orange), *Taenia saginata* Tsp36 (2BOL, green), and *Homo sapiens* HSPB5 (2WJ7, blue) monomer. The beta-strands of ACD are indicated by numbers; the long loop (between the beta-strands B5 and B7 for HSP16.9 and Tsp36) is indicated by L57. For HSPB5, the extended β strand is noted B6+7 and the loop is then noted L56.

1.4. The Dimer as a Building Block of Monodisperse Assemblies but Distinct Inter SU Contacts

The high and medium resolution structural studies were, of course, limited to monodisperse sHSPs. A strong resemblance exists at the dimer level for HSP16.5, HSP16.9 and *X. Axonopodis* HspA. The dimeric substructures involve ACD/ACD intermolecular interactions, as shown on figure 3 where the HSP16.9 dimer (1GME) is represented with the N-td in green, the ACD in red, and the C-te in blue. These interactions are mediated by the long loop, L57, (comprising the small β-strand B6), and a small strand, B2. These data have been used by the community to construct the EM models of the other sHSPs [32, 34]. It was found that, in all cases, the same dimeric ACD motif could be fit within the electron density.

Dimeric building blocks have been proposed to favor monodispersity [28]. In fact, from the 3D structures available up to now, the occurrence of dimers as building blocks may be considered essentially valid for polydisperse assemblies. The mammalian truncated homodimers recently published present, however, different ACD/ACD interactions [25]. Indeed, the structural studies demonstrated that the interfaces in the new dimeric substructures were now mediated by the extended B6+7 strand. It remains unknown so far whether the new dimeric motif is responsible for polydispersity, or associated with the occurrence of a shorter L57 loop, or with something else in the sequence.

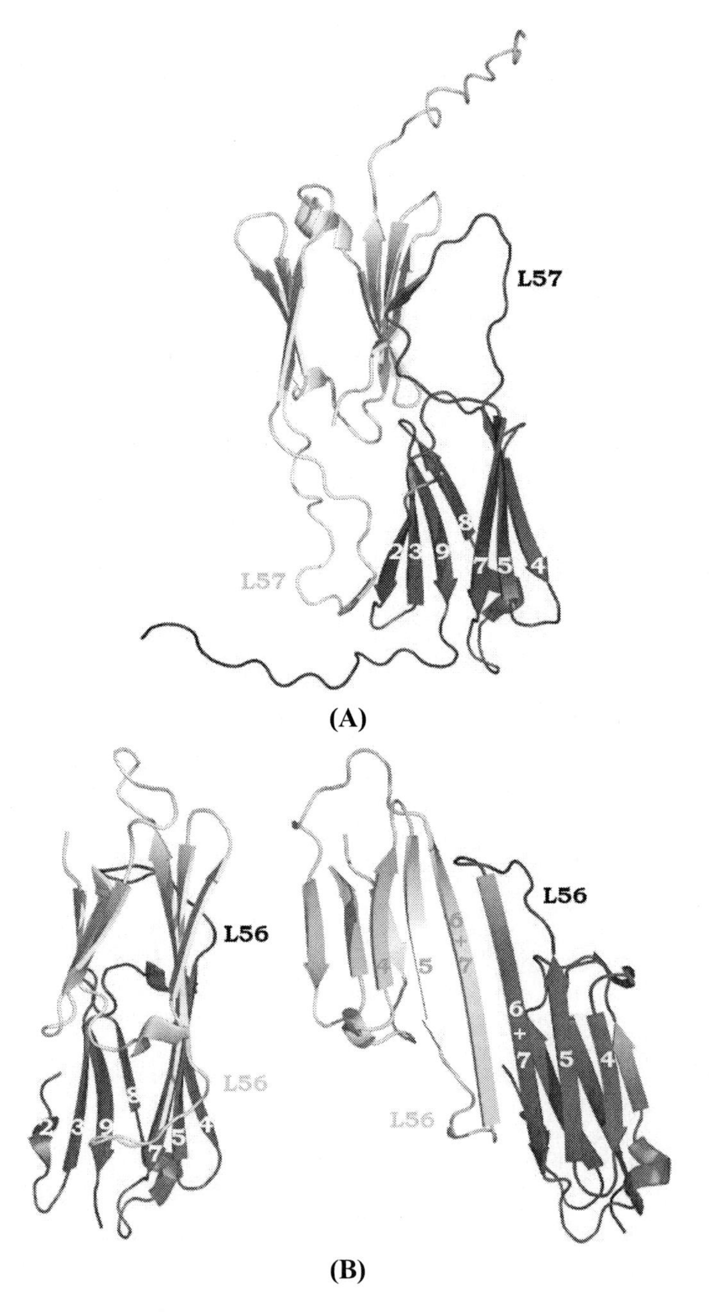

Figure 3. sHSP dimeric interface. The N-td, ACD and C-te of one monomer are represented in green, red and blue, respectively; the second monomer is represented in gray. **A-***Triticum aestivum* HSP16.9 (PDB code: 1GME); similar interface (not shown) is observed for *Methanococcus jannaschii* HSP16.5 (1SHS), *Taenia saginata* Tsp36 (2BOL) and *Xanthomonas axonopodis* hspA (3GLA). The N-td (resolved for one monomer of the dimer) is not shown. **B-***Homo sapiens* HSPB5 (2WJ7): lateral and front views. In the lateral view, the extended beta-strand B6+7 of the first monomer is partially hidden by the second monomer (in gray). Similar interface (not shown) is observed for *Rattus norvegicus* HSP20 (2WJ5).

1.5. The Building-up of Assemblies: What we Know From Monodisperse sHSPs

The 3D structures available told us that the monodiserse sHSPs display a variety of quaternary structures, either in solution or in crystals. Assemblies with either 2, 12, 24, or 48 SU have already been observed (see Table I). In addition, compact and expanded forms were found in several cases (see Table I also).

To stabilize these assemblies, each sHSP (or groups of sHSPs) seems to have developed its own series of inter SU contacts. The analysis of the known 3D structures indicates, however, which regions are particularly involved in the formation and stabilization of the various assemblies. Some of these structures are shown in figure 4.

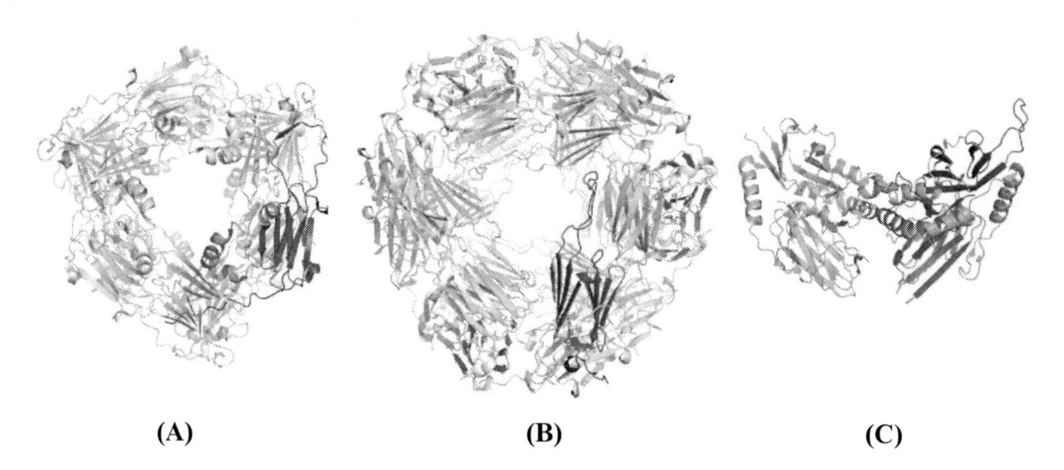

Figure 4. sHSP 3D structures determined by X-rays. The one from *Xanthomonas axonopodis* (PDB code: 3GLA) is not shown. Among them, the N-td was either not resolved at all or was resolved in helical structure. The N-td, ACD and C-te are in green, red and blue, respectively. **A-***Triticum aestivum* HSP16.9 (1GME): 12 SU organized in 2 rings of 3 dimers. Note that only one of the two N-td was resolved. **B-***Methanococcus jannaschii* HSP16.5 (PDB code: 1SHS): 24 SU organized in 2 rings of 6 dimers. **C-***Taenia saginata* Tsp36 (2BOL): a dimer where each SU contains 2 ACDs (in red) connected by a linker region (in yellow).

Three regions seem of particular interest. The L57 loop that stabilizes the dimer interface has already been mentioned. The dimer represents the first level of organization. Then, both the C-te and the N-td are obviously playing an important role in the stabilization of higher order assemblies. Indeed, the C-te and its highly conserved IXI/V motif [Pasta et al. 2004] are encountered in two different structures to interact with a groove of an adjacent dimer and stabilize the higher order assembly. In addition, since its orientation may vary from one sHSP to the other, as can be seen on figures 2 and 3, the C-te is compatible with the building up of structures of different sizes and symmetries. Finally the much more variable N-td, able to fold in two ways within the same structure [28] and organized in different ways in the different assemblies, could well be a major determinant of the final assembly and symmetry.

2. From Structure to Function: The Relevant Concepts

2.1. Mammalian sHSPs: Some Landmarks

The main objective was, however, to understand at the molecular level the sHSP structure - dynamics - function relationships in mammalian sHSPs, i.e. in polydisperse assemblies. The initial questions addressed, mainly focused on lens crystallins and transparency, first prompted the community to develop biophysical *in vitro* approaches. Since α-crystallin control eye lens transparency, many studies were devoted, either to native bovine (bαN) crystallin, made of a mixture of αA and αB crystallin in a 3:1 ratio, or recombinant homo-oligomers human HSPB4 and HSPB5 (see reviews in [37-39]). The α-crystallin repulsive interactions were first shown to account for the physical phenomenon of eye lens transparency [40]. Then, α-crystallin were shown to also possess, like the other sHSPs, a chaperone-like activity toward their physiological substrates, β- and γ-crystallin, enabling them to protect eye lens against cataract [41]. Later on, with the understanding of the importance of sHSPs for human health, and despite the difficulty to express and purify large, complex and polydisperse objects, a large range of tools was developed to analyze sHSPs (see below). Essentially, three original sHSP characteristics have emerged from these extensive studies: i) the structural sHSP flexibility, demonstrated by the sHSP ability to undergo conformational changes and structural transitions as a function of the environment, ii) the dynamic aspect of sHSP quaternary structure, illustrated by SU exchange, iii) the sHSP chaperone-like activity toward a large spectrum of client proteins. It now remains to fill the gaps between all the pieces of information gathered so far, that originated either from monodisperse or polydisperse assembly analyses.

2.2. A Variety of Approaches

Size exclusion chromatography - multi-angle light scattering (SEC-MALS) was found particularly adequate to provide us with precise molecular weights of the polydisperse populations [38], whereas dynamic light scattering (DLS), small angle X-ray scattering (SAXS) and electron microscopy (EM) were appropriate for the study of the assembly sizes and quaternary structure transitions (e.g. [42-43]). Mass spectrometry was able to reveal the relative populations of oligomers in bovine refolded αB-crystallin [44]. A variety of spectroscopic techniques, including EPR [45] and NMR [26, 46] were extensively used to follow secondary and tertiary structures whereas fluorescence, mass spectrometry, isoelectrofocusing (IEF) gels or anion exchange chromatography were demonstrating SU exchange [15, 47-49]. A variety of spectrophotometric assays were designed to measure the chaperone-like activity [41, 50-55]. Recently, *in silico* approaches were called for contribution to overcome the shortage of 3D structures [15]. The 3D structures of monodisperse sHSPs, however, did not provide information about the location, at the atomic level, of the functional interactive surface patches of sHSPs (moreover, no structure of any sHSP-substrate complex has been available thus far). A number of residues and peptides, specifically involved in the associations and interactions essential for assembly and function, have been identified thanks to the construction of a variety of mutants [56-57] and the development of pin array analyses, coupled to the structural information available [58].

2.3. sHSP Structural Transitions as a Function of the Environment

The native bovine α-crystallin (bαN) was the first to be extensively studied, since it was available in large quantities from calf lenses, and served as a reference, rapidly followed by studies on recombinant human HSPB4 and HSPB5. In the early eighties the bαN molecular weight (MW) value was a matter of debate for a while, since different values could be found in the literature. Finally, it was clarified that the MW was varying with the preparation conditions and, once purified, with the environment, e.g. temperature, pH and ionic strength [59-61]. It is now recognized that, in physiological conditions, the average αN MW is of the order of 800 ± 150 kDa [39], whereas the average HSPB4 and HSPB5 MW were lower, of the order of 600 kDa. Lower MW, however, continue to be reported in non physiological environment, e.g. with preparation at pH 9 [62]. Variability of SU number and polydispersity were better understood when mass spectrometry was able to reveal the relative populations of individual oligomers in recombinant (rec) bαA-crystallin and human HSPB5 compared to the refolded αA and αB-crystallin from native bovine αN [63]. In solution, rec-αA, respectively rec-HSPB5, could possess either an even or an odd number of SU (from 22 to 30, respectively 24 to 33, SU) with a most populated assembly containing 26, respectively 28, SU. It was shown that reassembly of unfolded (e.g. in urea) crystallins results in lower MW particles with a more homogenous distribution of even and odd number of SU.

Later on, the temperature effect was more precisely analyzed and it was observed that the assembly was essentially unchanged until 60°C, but that the assembly size roughly doubled between 60°C and 66°C [42, 54]. The transition was not reversible. The size evolution of as a function of temperature was followed by different techniques, DLS, SAXS [43] and, thanks to the irreversibility, by SEC-MALS after incubation at the desired temperature. The later experiments are shown in figure 5. The transition is also schematically represented in the figure. At higher temperatures the formation of large aggregates, sedimenting with time, was progressively observed. The doubling in size was also observed under pressure, yet in that case the transition was reversible.

The main results obtained from the initial bαN analysis were confirmed in recent years to be of general validity for sHSPs. It has now been observed in a variety of systems, either monodisperse or polydisperse, that the assemblies are highly flexible, that the assembly size may vary as a function of temperature, pH, ionic strength..., and that structural transitions, often associated to normal functioning, may occur.

The nature of the observed changes varies, however, with the sHSP under study. The comparison of the HSP26 and bαN behaviors is particularly illuminating. The cryo-EM study of White et al. [32] observed that the HSP26 24-mer could exist in two forms at room temperature, one compact and one expanded. In addition, close to the temperature conditions where HSP26 plays in role *in vivo* [64], a structural transition occurs. The HSP26 24-mer dissociates into dimers [43]. With both HSP26 and bαN a transition is observed when increasing temperature, yet with HSP26 it is dissociation whereas with bαN it is a doubling in size.

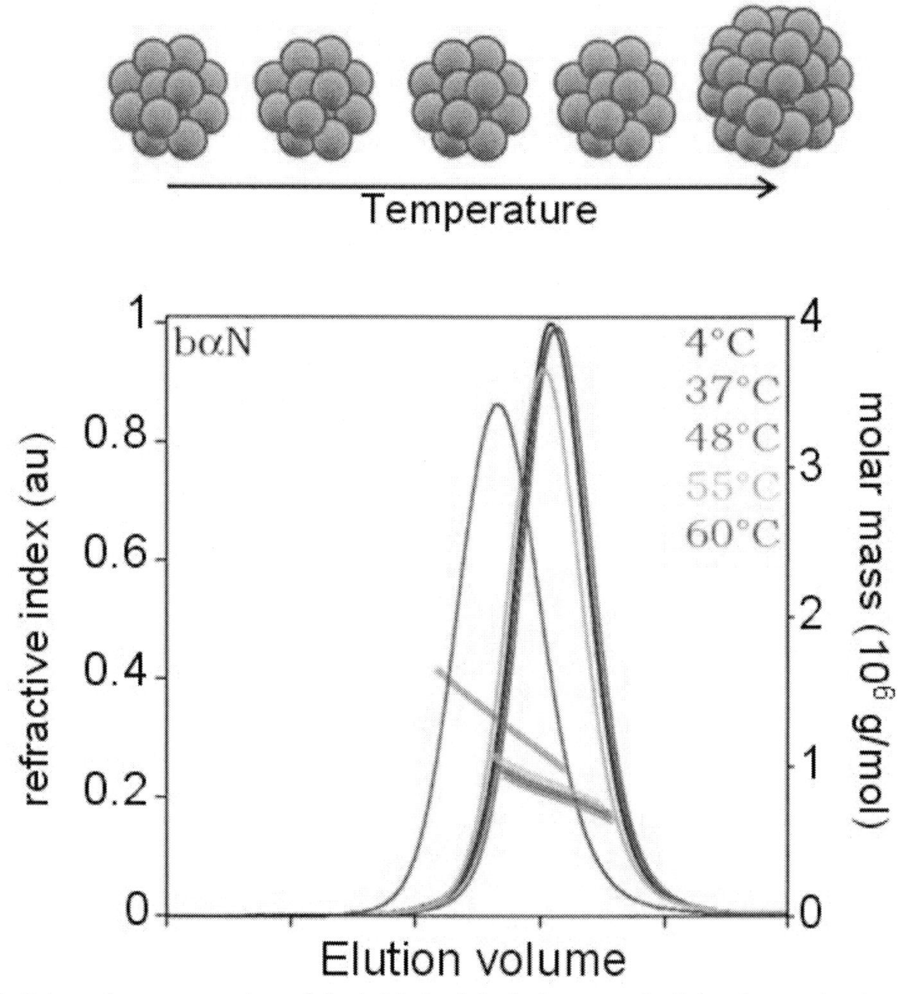

Figure 5. Schematic representation of the bαN physiological state and of the changes in size with increasing temperature and SEC-MALS measurements. The chromatography was done after incubation of the samples at the desired temperature for one hour.

HSP16.5 and HSP20.2 were also shown to adopt at least two conformations in solution [36]. It was hypothesized that these two states were associated to their anti-stress activity.

The most spectacular example, however, of the *in vivo* flexibility of sHSP assemblies is probably provided by HSP27, which is known for a long time [65] to change its oligomerization state after phosphorylation or dephosphorylation.

2.4. The Special Role of SU Exchange in sHSPS

Oligomeric enzymes in solution are, usually, in equilibrium with their dissociation products: either monomers, or dimers, or higher oligomers. The observation that sHSP oligomers could gain or lose SU had been made a long time ago. Such exchanges were necessary to explain the α-crystallin changes in oligomeric size as a function of pH, temperature or salt concentration, analyzed in the eighties [59-61]. Yet, with sHSPs, no isolated SU could ever be found in solution.

The SU exchange concept as proposed by Bova et al. [48] was different. The novelty was that SU exchange was shown to occur all the time, so that the mixture of e.g. initially differently labeled populations was rapidly leading to the formation of a unique final population where the labels were uniformly distributed. The initial study was performed on HSPB4. It was rapidly demonstrated that SU exchange was a general property of sHSPs, whatever their species, and could also take place between different, yet related, assemblies: between HSPB4 and HSPB5, between different archae, etc. But the question of the basic unit, dimeric and/or monomeric, associated to the SU exchange process is still under debate.

When SU exchange was analyzed as a function of temperature, it was observed that the exchange rate increased with increasing temperature. No exchange was observed at 4°C. The exchange rate started being significant around 37°C and then increased rapidly with temperature. The exchange was also shown to be reduced by the binding of various substrates, in a substrate dependent manner [48, 54]. The SU exchange is underlying the sHSP structural transitions and functional properties that are described below.

2.5. Dynamic Structure and Chaperone-Like Activity

α-crystallin was demonstrated by Horwitz [41] to exhibit chaperone properties *in vitro*. It was able to bind β- and γ-crystallins at the onset of their thermal unfolding, thus preventing further denaturation, aggregation and precipitation. Following this pioneering work, the sHSP chaperone-like activity has been the subject of numerous studies. Both *in vivo* and *in vitro* sHSPs function as molecular chaperones [41, 50-53, 55] whose main function is to associate, not refold, a variety of stress-unfolded proteins and prevent further aggregation and precipitation, in an ATP-independent process. SU exchange might control the chaperone-like activity, often associated with changes in oligomeric state [31, 42, 54]. Functional models have been suggested [34, 66-68]. Essentially, hydrophobic patches on the surface, either present in the native state or revealed after structural modifications or during SU exchange, would associate with a variety of partially denatured substrates (yet not all unfolded substrates) in a substrate dependent manner. Reciprocally, substrate fixation was observed to slow down the SU exchange rate [54].

Assays using unfolded proteins, e.g. oxidized insulin, temperature unfolded citrate synthase, alcohol dehydrogenase or α-lactalbumin, were designed to test the sHSP activity. The chaperone efficiency is usually measured by the ability to prevent the onset of light scattering generated by substrate aggregation. Interestingly, the α-crystallin chaperone-like activity toward the physiological substrates β/γ-cristallins was shown to require the temperature-induced transition previously described.

With further refinements of such studies, it has been shown that highly destabilized sHSPs were usually poor chaperones, but could retain some SU exchange capacity [63]. On the other hand, when the native oligomeric assembly was lost, chaperone-like activity could be partially retained by smaller entities, or even by peptides [51, 56].

Table 2. sHSP point mutations or deletions leading to pathologies.

DHMN stands for distal hereditary motor neuropathy, CMT for Charcot-Marie-Tooth, Myo for myopathy and Cardiomyo for cardiomyopathy. N-td domain mutations and ACD domain mutations are indicated in italic and in bold, respectively.

sHSP	Mutation	Associated pathology	Reference
HSPB1	*P39L*	DHMN	[69]
	G84R	DHMN	[7, 69]
	L99M	CMT	[69]
	R127W	DHMN / CMT 2F	[70]
	S135F	CMT 2F / DHMN	[69-70]
	R136W	CMT 2F	[70]
	R140G	DHMN	[70]
	K141Q	DHMN	[71]
	T151I	DHMN	[70]
	P182L	DHMN	[70]
	P182S	DHMN	[72]
HSPB4	*W9X*	Cataract	[73]
	R12C	Cataract	[74]
	R21L, R21W	Cataract	[73]
	R49C	Cataract	[75]
	R54C	Cataract	[76]
	F71L	Cataract	[9]
	G98R	Cataract	[77-78]
	R116C, R116H	Cataract	[11, 79]
HSPB5	*P20S*	Cataract	[13]
	R56W	Cataract	[80]
	R120G	Myo, Cardiomyo, Cataract	[5]
	D140N	Cataract	[14]
	K150X (450delA*)	Cataract	[81]
	Q151X (464delCT*)	Myo	[17]
	R157H	Cardiomyo	[82]
HSPB6	*P20L*	Cardiomyo	[83]
HSPB8	**K141E, K141N**	DHMN-CMT	[84]

*indicates nucleotidic frameshift mutation.

3. Mutations Leading to Pathologies

3.1. Prevalence of Point Mutations in Humans

In recent years, a number of point mutations or deletions were identified, that lead to various diseases, such as cataracts, cardiomyopathies, neuropathies. They are summarized in Table 2.

It can be seen that mutations are localized in the three sHSP identified regions: N-td, ACD and C-te (Figure 1). Among those located in the ACD, the first to be discovered and the most well-known is the HSPB5 R120G mutation [5]. The equivalent mutation also leads to pathologies in HSPB1, HSPB4 and HSPB8. Of all the pathological mutants presently known, only a few have been studied so far at the molecular level, and our present knowledge of the relationships between sHSP structure and pathology relies mainly on the studies performed on HSPB4 R116C and HSPB5 R120G pathological mutants. Moreover, in both cases other mutants were constructed at the same position for the sake of comparison: R116K and R116D for HSPB4 [8], R120K, R120D, and R120C for HSPB5 [15, 19].

3.2. Abnormal Distribution in Mammalian Cell Lines

The sHSP pathological mutants studied so far in mammalian cell lines, *e.g.* HSPB1 G84R [7], HSPB5 R120G [5-6], were observed to accumulate as aggregates at the exception of HSPB5 R151H. Wild-type (wt) HSPB5 and its R120 mutants were therefore used as a model system to compare *in cellulo* the cytoplasmic distribution of normal and pathological sHSPs [18-19]. Whereas wt HSPB5 always showed a diffuse cytoplasmic pattern, the R120G mutant was found initially as well defined multi-foci aggregates in the cytoplasm (Figure 6), that became bigger and perinuclear as a function of time. The mutant R120K showed the same pattern of distribution than the wt HSPB5. Instead, the mutants R120C and R120D had a tendency to form aggregates with sizes and location similar to those of R120G. Typical results are shown in figure 6. These results tend to indicate that charge preservation might be enough to preserve a normal behavior *in cellulo*.

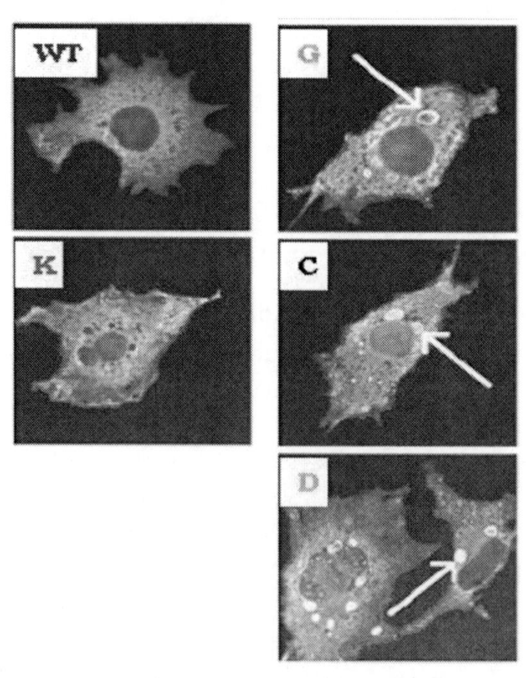

Figure 6. Cellular location of wt HSPB5 and its R120K, R120G, R120C, and R120D mutants in Cos-7 cells, that do not normally express HSPB5, observed with immunofluorescence 48 h post-transfection.

3.3. Impaired Solubility in Bacterial Hosts

In agreement with the findings described above, pathological mutant purification from bacterial host is usually difficult. The HSPB4 G98R mutant has been reported to form inclusion bodies in *E. coli* [77], which could be rescued with co-expression with the wild-type HSPB4 [78]. In the same way, whereas the HSPB5 and its R120K mutant were mainly found in the bacterial soluble fraction, a significant fraction of the R120G and R120D mutants remained in the insoluble one [19]. Moreover, the R120C mutant was found exclusively in the insoluble fraction, hindering any further study *in vitro;* indeed, it is known that refolding HSPB5 *in vitro* does not provide us with native oligomeric structures ([63] and unpublished observations).

3.4. Abnormal Interactions with Cellular Partners

The myofibrillar myopathy R120G is characterized by intracellular HSPB5 and desmin co-aggregates and it has been shown that this mutation is responsible for an increased affinity to desmin [85]. The mutated proteins interfere with normal structure of the cytoskeleton. In addition, expression of R120G HSPB5 in mice cardiomyocytes is linked to amyloid oligomers detection [86]. The correlation between mutated HSPB5 expression and amyloid oligomer high levels (cessation of R120G HSPB5 expression in transgenic mice is associated with significant decrease of amyloid oligomers) reveals an abnormal and negative correlation leading to severe pathology [87] (see chapter 1-7).

Recently, several abnormal interactions between R120G HSPB5 and diverse proteins have been reported as with members of apoptotic cascade (*i.e.* procaspase-3), with protein of the ubiquitin pathway (Fbx-4) or protein of the splicing components (SMN) [88-91]. Those data explained in part why the R120G HSPB5 has a dominant effect within cells.

Moreover, the mutant HSPB5 proteins (R120G, Q151X and 464delCT) also show abnormal interactions with most of the others sHSPs presents in muscle [18]. Interestingly, a similar phenomenon has been reported between the mutant HSPB8 proteins (K141E and K141N) and the others sHSPs presents in neurons [92] (see chapters 2-5 and 4-3). The observations of those modifications indicate that each mutant protein induced a specific break in the natural equilibrium between sHSPs within the cells.

4. From Impaired Structure to Dysfunction: Mutant Studies at the Molecular Level *In Vitro*

4.1. Abnormal Assemblies

HSPB4 and its R116 mutants [8], HSPB5 and its R120 mutants [15, 19] were first investigated with biophysical approaches for their size and polydispersity *in vitro*. It has to be emphasized once more that sHSP structure is, in general, highly sensitive to environmental conditions: temperature, pressure, pH, ionic strength and preparation conditions. The results mentioned below originate from experiments done at physiological pH and ionic strength and either at ambient or physiological temperature. The pathological HSPB4 R116C and HSPB5 R120G mutants were both found to display an increased SU number and polydispersity associated with a decreased stability. The stability of the HSPB5 R120G mutant was particularly impaired. Moreover, the other mutants at this special position, also presented abnormalities and increased MW, yet to a lesser extent. Only the HSPB4 R116K mutant seemed to be rescued. Increased molecular mass was also observed with the HSPB5 D140N mutant [14] whereas the HSPB8 K137E [93] and 141E [94] mutations seemed to increase the structure disorder without affecting the quaternary structure.

4.2. Abnormal Transitions of HSPB5 R120 Mutants

HSPB5 and its R120K, R120D and R120G mutants were investigated as a function of increasing temperature with DLS and SAXS [15]. Essentially, the mutants assembled into oligomers larger and more polydisperse than those formed by HSPB5. Size and polydispersity were minimal at physiological temperature and then increased as a function of temperature until precipitation occurs, at 60°C for HSPB5 and at lower temperatures for the mutants (in a mutant dependent manner). For HSPB5, the transition roughly corresponded to a doubling in size. Yet the change in size was less important for the mutants than for HSPB5. The average SU numbers measured as a function of temperature by SEC-MALS for HSPB5 and its R120K mutant are reported in the Table 3.

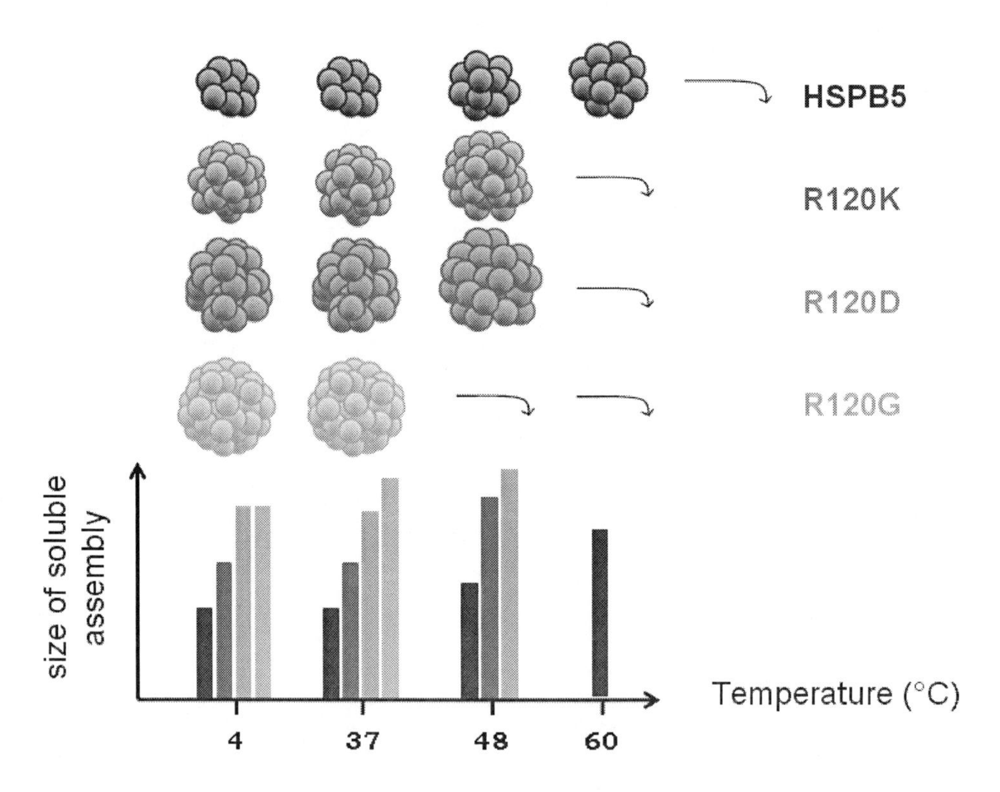

Figure 7. HSPB5 and its R120 mutants: schematic representation of the assembly state as a function of temperature.

Table 3. Average SU number as a function of incubation temperature as measured by SEC-MALS and, in parentheses, SU numbers at the limits of the peak

Temperature	4°C	37°C	48°C	55°C
HSPB5	32 [26-46]	31[27-41]	36 [31-47]	44 [35-60]
HSPB5 R120K	65 [63-82]	57 [50-75]	63 [52-93]	84 [58-161]

The pressure-induced structural modifications observed between ambient pressure and 300 MPa mimic those induced by temperature between room temperature and 60°C. The pressure-induced transition was, however, fully reversible with some hysteresis for HSPB5 and at least partially reversible for the mutants [15]. The whole of the results obtained with HSPB5 and its R120 mutants are schematically represented in figure 7.

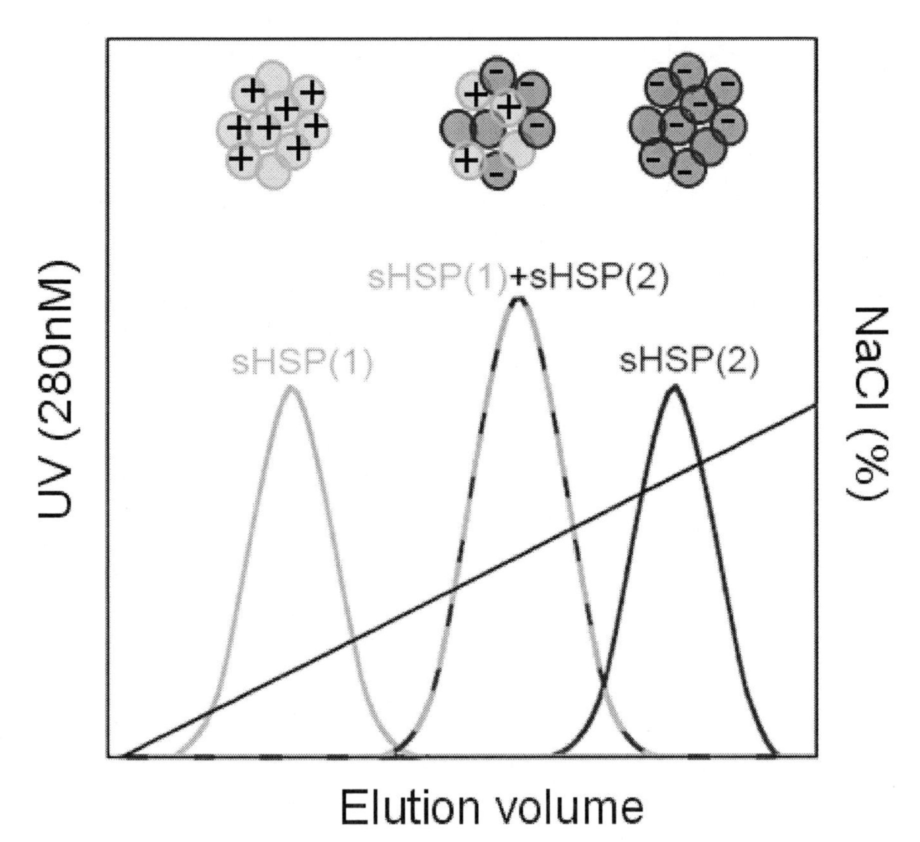

Figure 8. Principle of the SU exchange experiment. Mixtures of wt HSPB5 or its R120 mutants/bαN in a 1:1 ratio, are incubated at 37°C for different times. Then, the products present in solution are analyzed on an affinity chromatography column. The bαN was chosen as a reference for its high stability and charge difference with HSPB5 and the mutants. Indeed HSPB5 and bαN elute respectively at the beginning and at the end of the affinity gradient. After incubation at 37°C where SU exchange occurs, a new mixed population is formed, that becomes the only one visible after a few hours. SU exchange was found to occur faster between bαN and the mutants than with HSPB5: with R120K and R120G, only one peak was visible after 3 h incubation at 37°C, 1 h was sufficient for R120D. Instead 5 h were necessary for the wt HSPB5.

The stability of the different proteins was also followed as a function of time. The HSPB5, the R120K and R120D mutants were found to remain stable at 4°C for several days [15]. On the contrary, the R120G mutant was forming whitish precipitates at the bottom of the tube in a few days; it is known for a long time to be inherently unstable [20].

4.3. Modified SU Exchange of HSPB5 R120 Mutants

Since the building of an assembly implies a number of SU-SU interactions or associations, since in addition the environment-induced transitions require correct flexibility and dynamics, it may be anticipated that abnormal assemblies will also be impaired in chaperone-like activity and/or SU exchange. Only a few studies were devoted to analyze SU exchange between pathological mutants or between normal sHSPs and mutants (e.g. [95]). The result was that pathological mutations were modifying SU exchange, without stopping it. The HSPB5 R120 mutants were again chosen as a model system to further investigate the point [15]. The strategy adopted to compare HSPB5 and its R120 mutants is described in figure 8. SU exchange was achieved with a reference sample, the bαN, and the species formed were analyzed as a function of time by affinity chromatography. Once more, each R120 mutant was found to behave in its own and different way, yet all the HSPB5 R120 mutants were found to exchange SU more rapidly than HSPB5, which probably indicates a destabilization of the mutants as compared with HSPB5.

The most important observation, however, was that whatever the mutation, only one mixed population was observed after a few hours, which might indicate a possibility of rescuing pathological sHSPs *in vivo*, through SU exchange with the other cellular sHSPs. The point will be discussed below.

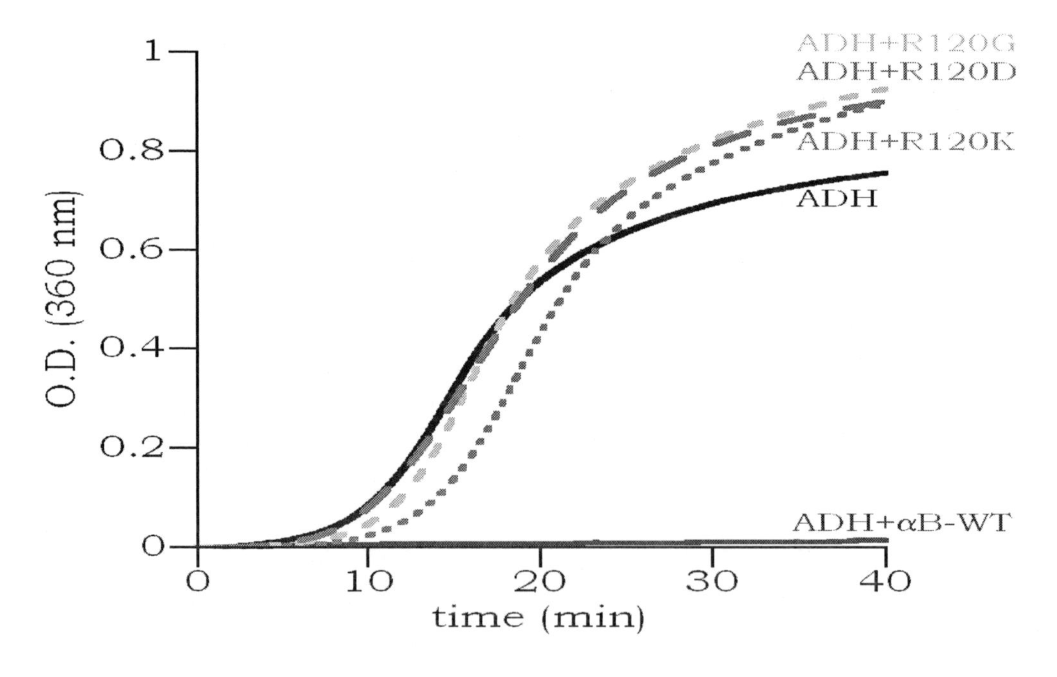

(A)

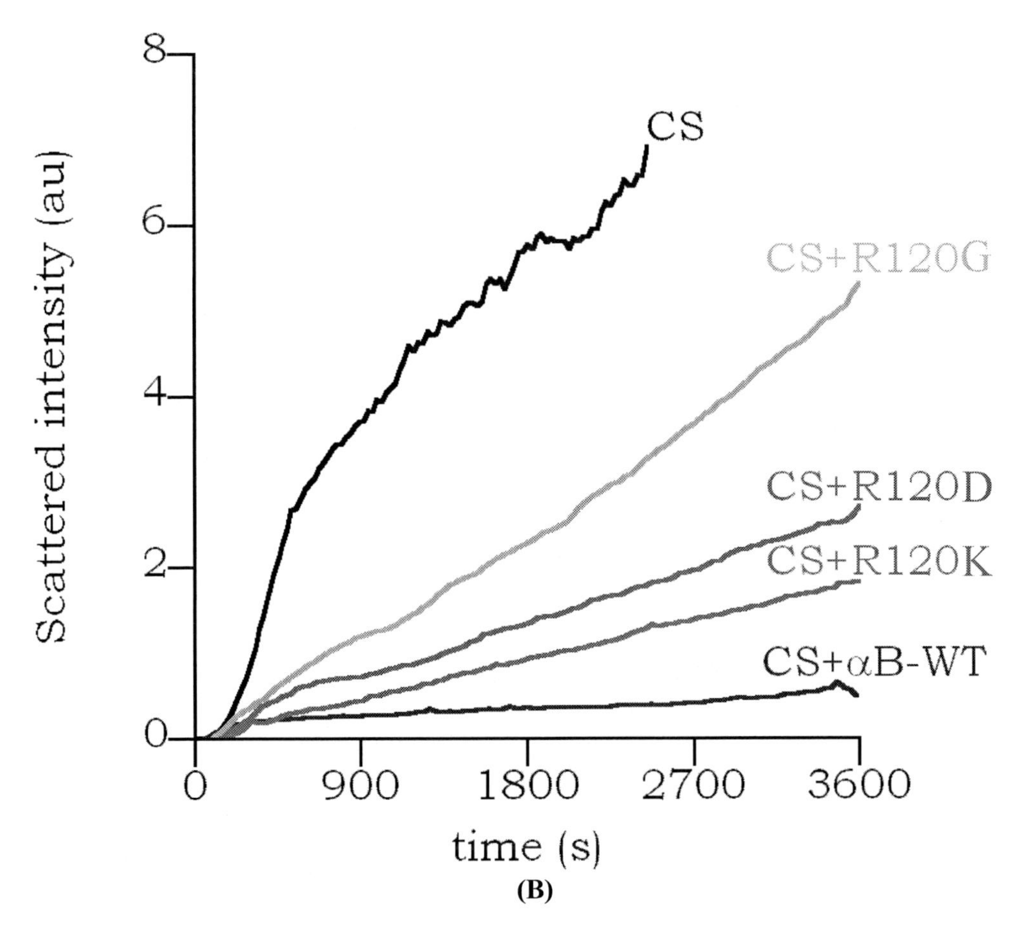

(B)

Figure 9. Chaperone-like activity of HSPB5 and its R120 mutants. **A**-toward ADH at 42°C. Aggregation was facilitated by addition of 1-10 phenanthroline and **B**-toward CS at 48°C.

4.4. Impaired Chaperone-Like Activity

Pathological sHSP behaviors were usually thought to result from, or at least to be closely associated with defective chaperone-like activity. Indeed, an impressive amount of data was collected, showing that pathological mutations are associated with impaired chaperone-like activity, *e.g.* HSPB4 F71L [9], G98R [77, 96], R116H [11], R116C [8], HSPB5 R120G [15, 19], D140N [13-14], HSPB8 K141E [94]. For the sake of illustration, Figure 9 shows the results obtained with HSPB5 and its mutants, toward two target proteins, alcohol dehydrogenase (ADH) and citrate synthase (CS) [19].

As can be seen on figure 9, the mutants had no effect on the ADH aggregation, indicating the *in vitro* chaperone-like activity was lost. All the mutants, however, retained a significant chaperone-like activity toward CS. Furthermore, this reduced activity could be different depending upon the mutant. Such results illustrate a target specificity of the sHSP pathological mutants and a direct connection between chaperone-like activity and mutant quaternary structure and stability. They suggest that a fine regulation of the sHSP dynamic properties is required for normal function.

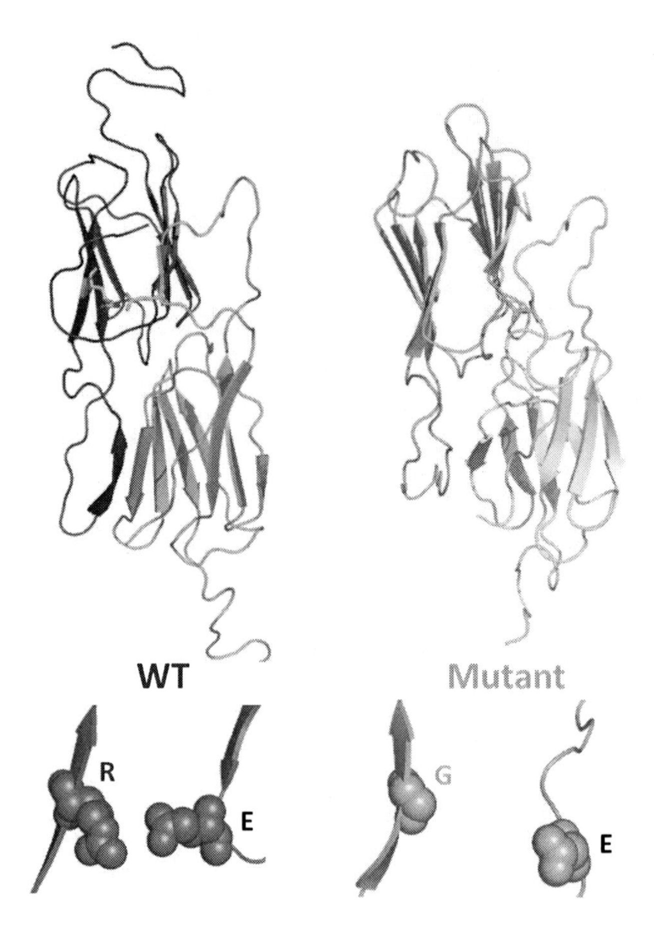

Figure 10. Structural consequences induced by the R120 homologous mutations: 3D dimeric substructures of WT and R120G homologous mutant of the *M. jannaschii* HSP16.5 (R107G) during *in silico* simulations. From left to right: the dimer structures are shown in the upper line, with the 107 (R of G) and E98 (within the L57 loop) residues represented as sticks. These structures correspond to the 1SHS pdb dimer (chains A and C in dark and light blue, respectively) and G mutant (orange), both after MD simulation (simulated annealing). Each 3D structure is partial, the main part of the N-ter region being absent (until residue 33). The lower line focuses on the local environment around the residue 107, at the dimeric interface. The interaction between the residues at position 107 and E98, which contribute to the stability of the dimer, is lost in the mutant; as a consequence, the L57 loop and the dimer are destabilized.

4.5. Abnormal sHSP Behavior Could Originate From Destabilization of the Dimeric Substructure

Among the new approaches able to precise the links between structure, dynamics and function, *in silico* approaches look promising. Whenever 3D structures are known, mutations can be generated *in silico* by amino acid replacement in the 3D coordinate files at critical positions. Molecular dynamics (MD) simulations can then be carried out for the wild-type and for the mutants. The approach was applied to dimeric substructures of HSP16.5 and HSP16.9, the coordinates of which were available. Mutations were generated at the R120 equivalent position, which was involved in interactions with the L57 loop of the adjacent monomer and therefore in the ACD/ACD interface, illustrated in figure 10. One unambiguous result of the

MD simulations was to show that any mutation at the R120 equivalent position modified the interaction between the residue at that position and the L57 loop, leading to a destabilization of the dimeric substructure [15].

The determination of the truncated HSPB5 dimeric 3D structure [25], confirmed that the R120 residue is involved in the dimeric interface, suggesting that mutations at this position is likely to affect the dynamics of the assembly. A locally more flexible (or less constrained) structure would be compatible with increased size and polydispersity and increased SU exchange capacity, as observed. The decreased chaperone-like activity would result from perturbations of the substrate interactive surface because of increased flexibility or decreased accessibility of the mutant structures.

4.6. Possible Rescue of Pathological Mutants via Hetero-Oligomer Formation with Normal sHSPs

In vivo, sHSPs are often found associated to form hetero-oligomers. In the eye lens, the functional entity is a complex made of HSPB4 and HSPB5. When analyzed *in vitro*, the complex was found much more stable than the homo-oligomers made of either HSPB4 or HSPB5. In other organs of the human body, HSPB1 and HSPB5 are often found in association, although it remains unclear whether they also form complexes [97]. HSPB2 and HSPB3 were recently found to form well defined hetero-oligomers in a 3:1 ratio [98] and HSPB8 was observed in complexes with HSPB5 and HSPB6 [99].

Since pathological mutants retain their ability to exchange SU, the probability is high for them to associate with other cellular sHSPs, either constitutively expressed or stress-induced. It has been shown that co-expression of wt HSPB4 and its G98R mutant [78] and co-expression of HSPB5 R120G and wild-type HSPB1 [100] suppress the formation of inclusion bodies induced by mutations, therefore partial rescue might be obtained *in vivo* through hetero-oligomer formation. Such a possibility to rescue abnormal sHSPs should perhaps be considered in some therapeutic strategies [97] (see chapter 4-3).

CONCLUSION

In this chapter, our present knowledge, at the molecular level, of the sHSP structure, dynamics and function relationships has been reviewed. We have seen that sHSP quaternary structures were either monodisperse or polydisperse, variable in size and number of SU, yet that the ACD core was highly conserved and the dimeric substructure likely to be of particular relevance for quaternary structure stability. The sizes and MW of the assemblies are controlled by the environmental conditions, through SU exchange. With changes of the environment, induced by a variety of stress, the sHSP structure is sufficiently flexible and dynamic to undergo structural transitions, reorganizations, changes in size, for which SU exchange is, again, essential. Finally, the sHSP chaperone-like capacity, that requires quaternary structure and dynamics integrity, is assumed to rescue deleterious stress-induced protein aggregation. This present understanding of sHSP structure-dynamics-function relationships is summarized in figure 11.

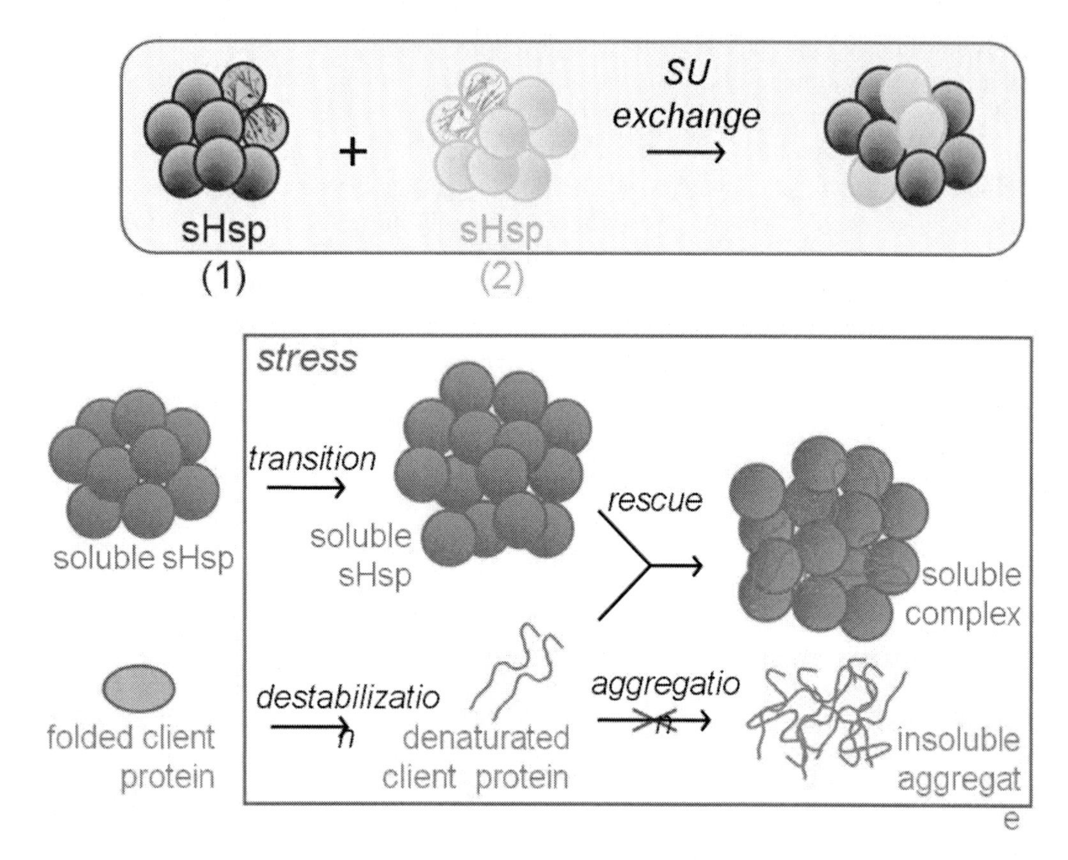

Figure 11. Illustration of the relationships between **A-**sHSP quaternary structure and SU exchange **B-** sHSP quaternary structure transitions and chaperone-like activity of the sHSPs.

Despite the advances already obtained, many original features of the sHSPs, illustrated on the figure, remain poorly understood and characterized. A major challenge for future will be to identify the structural determinants and characteristics leading either to polydisperse assemblies or monodisperse oligomers. Nothing is known of the SU exchange process *per se*. No individual SU has been isolated yet (whatever the technique), suggesting that the exchange mechanism can only take place at the moment of contact. The point remains to be demonstrated. Last but not least, the sites involved in client protein association during chaperone-like activity also require to be better defined, even if some surface areas have already been postulated as likely candidates. A deeper knowledge of these points would be particularly valuable in view of understanding pathological issues and designing therapeutic strategies.

The questions addressed in this chapter were focused on the point mutations leading to pathologies. It was shown that the pathological mutants known so far presented abnormalities at the molecular level. The most obvious abnormalities were bearing on the assembly size, on the stress-induced transitions and on the chaperone-like activity, i.e. on the flexibility and dynamics of the quaternary structure. All the sHSP pathological mutants were impaired in similar ways, although more or less severely, suggesting that abnormal assembly and transitions could be a pathology marker.

A good correlation was observed between *in cellulo* and *in vitro* studies since the most structurally impaired mutants were observed as large aggregates in the cell cytoplasm. It was therefore strongly suggested that a normal quaternary structure was a prerequisite for a right functioning.

In vitro studies on some sHSP and their pathological mutants were found particularly adequate to define which characteristics are essential for sHSPs to act as anti-stress proteins. In order to function as molecular chaperones, sHSPs require the right combination of quaternary structure and dynamics. Only in these conditions can sHSPs expose the interactive surfaces able to bind partially denatured substrates. Any failure or impairment at one stage or another may result in an incorrect balance of hydrophilicity and hydrophobicity and loss of chaperone-like activity. Of course, given the huge variety of sHSP assemblies, it may be anticipated that quaternary structure and dynamics will be combined in different ways in the different species.

The take-home message is probably that with sHSPs, pathology and large structural abnormalities are intimately associated. sHSPs therefore represent a good example of molecular structural biology. A right functioning of the different pieces is required at the different levels of organization for the whole system to play its role.

Acknowledgments

The authors gratefully acknowledge C. Férard for her constant help in the sample preparation and purification. AT, FSP, SF and ED were supported by the Centre National de la Recherche Scientifique (CNRS) and the Université Pierre et Marie Curie (UPMC-Paris 6) respectively. MM were supported by the Université Pierre et Marie Curie (UPMC-Paris 6) and the Université d'Evry Val d'Essonne (UEVE), and SS was supported by the French association against myopathies (AFM).

References

[1] Haslbeck M, Franzmann T, Weinfurtner D, Buchner J: Some like it hot: the structure and function of small heat-shock proteins. *Nat Struct Mol Biol* 2005, 12(10):842-846.

[2] Ingolia TD, Craig EA: Four small Drosophila heat shock proteins are related to each other and to mammalian alpha-crystallin. *Proc Natl Acad Sci USA* 1982, 79(7):2360-2364.

[3] Van Montfort R, Slingsby C, Vierling E: Structure and function of the small heat shock protein/alpha-crystallin family of molecular chaperones. *Adv Protein Chem* 2001, 59:105-156.

[4] de Jong WW, Caspers GJ, Leunissen JA: Genealogy of the alpha-crystallin small heat shock protein superfamily. *Int J Biol Macromol* 1998, 22(3-4):151-162.

[5] Vicart P, Caron A, Guicheney P, Li Z, Prevost MC, Faure A, Chateau D, Chapon F, Tome F, Dupret JM *et al*: A missense mutation in the alphaB-crystallin chaperone gene causes a desmin-related myopathy. *Nat Genet* 1998, 20(1):92-95.

[6] Chavez Zobel AT, Loranger A, Marceau N, Theriault JR, Lambert H, Landry J: Distinct chaperone mechanisms can delay the formation of aggresomes by the

myopathy-causing R120G alphaB-crystallin mutant. *Hum Mol Genet* 2003, 12(13):1609-1620.

[7] James PA, Rankin J, Talbot K: Asymmetrical late onset motor neuropathy associated with a novel mutation in the small heat shock protein HSPB1 (HSP27). *J Neurol Neurosurg Psychiatry* 2008, 79(4):461-463.

[8] Bera S, Thampi P, Cho WJ, Abraham EC: A positive charge preservation at position 116 of alpha A-crystallin is critical for its structural and functional integrity. *Biochemistry* 2002, 41(41):12421-12426.

[9] Bhagyalaxmi SG, Srinivas P, Barton KA, Kumar KR, Vidyavathi M, Petrash JM, Bhanuprakash Reddy G, Padma T: A novel mutation (F71L) in alphaA-Crystallin with defective chaperone-like function associated with age-related cataract. *Biochim Biophys Acta* 2009, 1792(10):974-981.

[10] Bova MP, Yaron O, Huang Q, Ding L, Haley DA, Stewart PL, Horwitz J: Mutation R120G in alphaB-crystallin, which is linked to a desmin-related myopathy, results in an irregular structure and defective chaperone-like function. *Proc Natl Acad Sci USA* 1999, 96(11):6137-6142.

[11] Gu F, Luo W, Li X, Wang Z, Lu S, Zhang M, Zhao B, Zhu S, Feng S, Yan YB *et al*: A novel mutation in AlphaA-crystallin (CRYAA) caused autosomal dominant congenital cataract in a large Chinese family. *Hum Mutat* 2008, 29(5):769.

[12] Huang Q, Ding L, Phan KB, Cheng C, Xia CH, Gong X, Horwitz J: Mechanism of cataract formation in alphaA-crystallin Y118D mutation. *Invest Ophthalmol Vis Sci* 2009, 50(6):2919-2926.

[13] Liu M, Ke T, Wang Z, Yang Q, Chang W, Jiang F, Tang Z, Li H, Ren X, Wang X *et al*: Identification of a CRYAB mutation associated with autosomal dominant posterior polar cataract in a Chinese family. *Invest Ophthalmol Vis Sci* 2006, 47(8):3461-3466.

[14] Liu Y, Zhang X, Luo L, Wu M, Zeng R, Cheng G, Hu B, Liu B, Liang JJ, Shang F: A novel alphaB-crystallin mutation associated with autosomal dominant congenital lamellar cataract. *Invest Ophthalmol Vis Sci* 2006, 47(3):1069-1075.

[15] Michiel M, Skouri-Panet F, Duprat E, Simon S, Ferard C, Tardieu A, Finet S: Abnormal assemblies and subunit exchange of alphaB-crystallin R120 mutants could be associated with destabilization of the dimeric substructure. *Biochemistry* 2009, 48(2):442-453.

[16] Perng MD, Muchowski PJ, van Den IP, Wu GJ, Hutcheson AM, Clark JI, Quinlan RA: The cardiomyopathy and lens cataract mutation in alphaB-crystallin alters its protein structure, chaperone activity, and interaction with intermediate filaments in vitro. *J Biol Chem* 1999, 274(47):33235-33243.

[17] Selcen D, Engel AG: Myofibrillar myopathy caused by novel dominant negative alpha B-crystallin mutations. *Ann Neurol* 2003, 54(6):804-810.

[18] Simon S, Fontaine JM, Martin JL, Sun X, Hoppe AD, Welsh MJ, Benndorf R, Vicart P: Myopathy-associated alphaB-crystallin mutants: abnormal phosphorylation, intracellular location, and interactions with other small heat shock proteins. *J Biol Chem* 2007, 282(47):34276-34287.

[19] Simon S, Michiel M, Skouri-Panet F, Lechaire JP, Vicart P, Tardieu A: Residue R120 is essential for the quaternary structure and functional integrity of human alphaB-crystallin. *Biochemistry* 2007, 46(33):9605-9614.

[20] Treweek TM, Rekas A, Lindner RA, Walker MJ, Aquilina JA, Robinson CV, Horwitz J, Perng MD, Quinlan RA, Carver JA: R120G alphaB-crystallin promotes the unfolding of reduced alpha-lactalbumin and is inherently unstable. *Febs J* 2005, 272(3):711-724.

[21] Franck E, Madsen O, van Rheede T, Ricard G, Huynen MA, de Jong WW: Evolutionary diversity of vertebrate small heat shock proteins. *J Mol Evol* 2004, 59(6):792-805.

[22] Fu X, Chang Z: Identification of a highly conserved pro-gly doublet in non-animal small heat shock proteins and characterization of its structural and functional roles in Mycobacterium tuberculosis Hsp16.3. *Biochemistry (Mosc)* 2006, 71 Suppl 1:S83-90.

[23] Kappe G, Franck E, Verschuure P, Boelens WC, Leunissen JA, de Jong WW: The human genome encodes 10 alpha-crystallin-related small heat shock proteins: HspB1-10. *Cell Stress Chaperones* 2003, 8(1):53-61.

[24] Bellyei S, Szigeti A, Pozsgai E, Boronkai A, Gomori E, Hocsak E, Farkas R, Sumegi B, Gallyas F, Jr.: Preventing apoptotic cell death by a novel small heat shock protein. *Eur J Cell Biol* 2007, 86(3):161-171.

[25] Bagneris C, Bateman OA, Naylor CE, Cronin N, Boelens WC, Keep NH, Slingsby C: Crystal structures of alpha-crystallin domain dimers of alphaB-crystallin and Hsp20. *J Mol Biol* 2009, 392(5):1242-1252.

[26] Jehle S, van Rossum B, Stout JR, Noguchi SM, Falber K, Rehbein K, Oschkinat H, Klevit RE, Rajagopal P: alphaB-crystallin: a hybrid solid-state/solution-state NMR investigation reveals structural aspects of the heterogeneous oligomer. *J Mol Biol* 2009, 385(5):1481-1497.

[27] Kim KK, Kim R, Kim SH: Crystal structure of a small heat-shock protein. *Nature* 1998, 394(6693):595-599.

[28] van Montfort RL, Basha E, Friedrich KL, Slingsby C, Vierling E: Crystal structure and assembly of a eukaryotic small heat shock protein. *Nat Struct Biol* 2001, 8(12):1025-1030.

[29] Stamler R, Kappe G, Boelens W, Slingsby C: Wrapping the alpha-crystallin domain fold in a chaperone assembly. *J Mol Biol* 2005, 353(1):68-79.

[30] Hilario E, Teixeira EC, Pedroso GA, Bertolini MC, Medrano FJ: Crystallization and preliminary X-ray diffraction analysis of XAC1151, a small heat-shock protein from Xanthomonas axonopodis pv. citri belonging to the alpha-crystallin family. *Acta Crystallogr Sect F Struct Biol Cryst Commun* 2006, 62(Pt 5):446-448.

[31] Haslbeck M, Walke S, Stromer T, Ehrnsperger M, White HE, Chen S, Saibil HR, Buchner J: Hsp26: a temperature-regulated chaperone. *Embo J* 1999, 18(23):6744-6751.

[32] White HE, Orlova EV, Chen S, Wang L, Ignatiou A, Gowen B, Stromer T, Franzmann TM, Haslbeck M, Buchner J *et al*: Multiple distinct assemblies reveal conformational flexibility in the small heat shock protein Hsp26. *Structure* 2006, 14(7):1197-1204.

[33] Kennaway CK, Benesch JL, Gohlke U, Wang L, Robinson CV, Orlova EV, Saibil HR, Keep NH: Dodecameric structure of the small heat shock protein Acr1 from Mycobacterium tuberculosis. *J Biol Chem* 2005, 280(39):33419-33425.

[34] McHaourab HS, Godar JA, Stewart PL: Structure and mechanism of protein stability sensors: chaperone activity of small heat shock proteins. *Biochemistry* 2009, 48(18):3828-3837.

[35] Shi J, Koteiche HA, McHaourab HS, Stewart PL: Cryoelectron microscopy and EPR analysis of engineered symmetric and polydisperse Hsp16.5 assemblies reveals determinants of polydispersity and substrate binding. *J Biol Chem* 2006, 281(52):40420-40428.

[36] Haslbeck M, Kastenmuller A, Buchner J, Weinkauf S, Braun N: Structural dynamics of archaeal small heat shock proteins. *J Mol Biol* 2008, 378(2):362-374.

[37] Bloemendal H, de Jong W, Jaenicke R, Lubsen NH, Slingsby C, Tardieu A: Ageing and vision: structure, stability and function of lens crystallins. *Prog Biophys Mol Biol* 2004, 86(3):407-485.

[38] Horwitz J: Alpha-crystallin. *Exp Eye Res* 2003, 76(2):145-153.

[39] Horwitz J: Alpha crystallin: the quest for a homogeneous quaternary structure. *Exp Eye Res* 2009, 88(2):190-194.

[40] Delaye M, Tardieu A: Short-range order of crystallin proteins accounts for eye lens transparency. *Nature* 1983, 302(5907):415-417.

[41] Horwitz J: Alpha-crystallin can function as a molecular chaperone. *Proc Natl Acad Sci USA* 1992, 89(21):10449-10453.

[42] Burgio MR, Kim CJ, Dow CC, Koretz JF: Correlation between the chaperone-like activity and aggregate size of alpha-crystallin with increasing temperature. *Biochem Biophys Res Commun* 2000, 268(2):426-432.

[43] Skouri-Panet F, Quevillon-Cheruel S, Michiel M, Tardieu A, Finet S: sHSPs under temperature and pressure: the opposite behaviour of lens alpha-crystallins and yeast HSP26. *Biochim Biophys Acta* 2006, 1764(3):372-383.

[44] Aquilina JA, Benesch JL, Bateman OA, Slingsby C, Robinson CV: Polydispersity of a mammalian chaperone: mass spectrometry reveals the population of oligomers in alphaB-crystallin. *Proc Natl Acad Sci U S A* 2003, 100(19):10611-10616.

[45] Koteiche HA, McHaourab HS: Folding pattern of the alpha-crystallin domain in alphaA-crystallin determined by site-directed spin labeling. *J Mol Biol* 1999, 294(2):561-577.

[46] Carver JA, Aquilina JA, Truscott RJ, Ralston GB: Identification by 1H NMR spectroscopy of flexible C-terminal extensions in bovine lens alpha-crystallin. *FEBS Lett* 1992, 311(2):143-149.

[47] Aquilina JA, Benesch JL, Ding LL, Yaron O, Horwitz J, Robinson CV: Subunit exchange of polydisperse proteins: mass spectrometry reveals consequences of alphaA-crystallin truncation. *J Biol Chem* 2005, 280(15):14485-14491.

[48] Bova MP, Ding LL, Horwitz J, Fung BK: Subunit exchange of alphaA-crystallin. *J Biol Chem* 1997, 272(47):29511-29517.

[49] Sun TX, Liang JJ: Intermolecular exchange and stabilization of recombinant human alphaA- and alphaB-crystallin. *J Biol Chem* 1998, 273(1):286-290.

[50] Basha E, Lee GJ, Demeler B, Vierling E: Chaperone activity of cytosolic small heat shock proteins from wheat. *Eur J Biochem* 2004, 271(8):1426-1436.

[51] Bhattacharyya J, Padmanabha Udupa EG, Wang J, Sharma KK: Mini-alphaB-crystallin: a functional element of alphaB-crystallin with chaperone-like activity. *Biochemistry* 2006, 45(9):3069-3076.

[52] Ghosh JG, Estrada MR, Clark JI: Interactive domains for chaperone activity in the small heat shock protein, human alphaB crystallin. *Biochemistry* 2005, 44(45):14854-14869.

[53] Ghosh JG, Shenoy AK, Jr., Clark JI: N- and C-Terminal motifs in human alphaB crystallin play an important role in the recognition, selection, and solubilization of substrates. *Biochemistry* 2006, 45(46):13847-13854.

[54] Putilina T, Skouri-Panet F, Prat K, Lubsen NH, Tardieu A: Subunit exchange demonstrates a differential chaperone activity of calf alpha-crystallin toward beta LOW- and individual gamma-crystallins. *J Biol Chem* 2003, 278(16):13747-13756.

[55] Shashidharamurthy R, Koteiche HA, Dong J, McHaourab HS: Mechanism of chaperone function in small heat shock proteins: dissociation of the HSP27 oligomer is required for recognition and binding of destabilized T4 lysozyme. *J Biol Chem* 2005, 280(7):5281-5289.

[56] Santhoshkumar P, Murugesan R, Sharma KK: Deletion of (54)FLRAPSWF(61) residues decreases the oligomeric size and enhances the chaperone function of alphaB-crystallin. *Biochemistry* 2009, 48(23):5066-5073.

[57] Santhoshkumar P, Sharma KK: Conserved F84 and P86 residues in alphaB-crystallin are essential to effectively prevent the aggregation of substrate proteins. *Protein Sci* 2006, 15(11):2488-2498.

[58] Ghosh JG, Clark JI: Insights into the domains required for dimerization and assembly of human alphaB crystallin. *Protein Sci* 2005, 14(3):684-695.

[59] Siezen RJ, Bindels JG, Hoenders HJ: The quaternary structure of bovine alpha-crystallin. Effects of variation in alkaline pH, ionic strength, temperature and calcium ion concentration. *Eur J Biochem* 1980, 111(2):435-444.

[60] Tardieu A, Laporte D, Licinio P, Krop B, Delaye M: Calf lens alpha-crystallin quaternary structure. A three-layer tetrahedral model. *J Mol Biol* 1986, 192(4):711-724.

[61] Van den Oetelaar PJ, Clauwaert J, Van Laethem M, Hoenders HJ: The influence of isolation conditions on the molecular weight of bovine alpha-crystallin. *J Biol Chem* 1985, 260(26):14030-14034.

[62] Peschek J, Braun N, Franzmann TM, Georgalis Y, Haslbeck M, Weinkauf S, Buchner J: The eye lens chaperone alpha-crystallin forms defined globular assemblies. *Proc Natl Acad Sci U S A* 2009, 106(32):13272-13277.

[63] Benesch JL, Ayoub M, Robinson CV, Aquilina JA: Small heat shock protein activity is regulated by variable oligomeric substructure. *J Biol Chem* 2008, 283(42):28513-28517.

[64] Franzmann TM, Menhorn P, Walter S, Buchner J: Activation of the chaperone Hsp26 is controlled by the rearrangement of its thermosensor domain. *Mol Cell* 2008, 29(2):207-216.

[65] Mehlen P, Arrigo AP: The serum-induced phosphorylation of mammalian hsp27 correlates with changes in its intracellular localization and levels of oligomerization. *Eur J Biochem* 1994, 221(1):327-334.

[66] Ahmad MF, Raman B, Ramakrishna T, Rao Ch M: Effect of phosphorylation on alpha B-crystallin: differences in stability, subunit exchange and chaperone activity of homo and mixed oligomers of alpha B-crystallin and its phosphorylation-mimicking mutant. *J Mol Biol* 2008, 375(4):1040-1051.

[67] Aquilina JA, Benesch JL, Ding LL, Yaron O, Horwitz J, Robinson CV: Phosphorylation of alphaB-crystallin alters chaperone function through loss of dimeric substructure. *J Biol Chem* 2004, 279(27):28675-28680.

[68] Pasta SY, Raman B, Ramakrishna T, Rao Ch M: Role of the conserved SRLFDQFFG region of alpha-crystallin, a small heat shock protein. Effect on oligomeric size, subunit exchange, and chaperone-like activity. *J Biol Chem* 2003, 278(51):51159-51166.

[69] Houlden H, Laura M, Wavrant-De Vrieze F, Blake J, Wood N, Reilly MM: Mutations in the HSP27 (HSPB1) gene cause dominant, recessive, and sporadic distal HMN/CMT type 2. *Neurology* 2008, 71(21):1660-1668.

[70] Evgrafov OV, Mersiyanova I, Irobi J, Van Den Bosch L, Dierick I, Leung CL, Schagina O, Verpoorten N, Van Impe K, Fedotov V *et al*: Mutant small heat-shock protein 27 causes axonal Charcot-Marie-Tooth disease and distal hereditary motor neuropathy. *Nat Genet* 2004, 36(6):602-606.

[71] Ikeda Y, Abe A, Ishida C, Takahashi K, Hayasaka K, Yamada M: A clinical phenotype of distal hereditary motor neuronopathy type II with a novel HSPB1 mutation. *J Neurol Sci* 2009, 277(1-2):9-12.

[72] Kijima K, Numakura C, Goto T, Takahashi T, Otagiri T, Umetsu K, Hayasaka K: Small heat shock protein 27 mutation in a Japanese patient with distal hereditary motor neuropathy. *J Hum Genet* 2005, 50(9):473-476.

[73] Hejtmancik JF: Congenital cataracts and their molecular genetics. *Semin Cell Dev Biol* 2008, 19(2):134-149.

[74] Zhang LY, Yam GH, Tam PO, Lai RY, Lam DS, Pang CP, Fan DS: An alphaA-crystallin gene mutation, Arg12Cys, causing inherited cataract-microcornea exhibits an altered heat-shock response. *Mol Vis* 2009, 15:1127-1138.

[75] Andley UP: AlphaA-crystallin R49Cneo mutation influences the architecture of lens fiber cell membranes and causes posterior and nuclear cataracts in mice. *BMC Ophthalmol* 2009, 9:4.

[76] Khan AO, Aldahmesh MA, Meyer B: Recessive congenital total cataract with microcornea and heterozygote carrier signs caused by a novel missense CRYAA mutation (R54C). *Am J Ophthalmol* 2007, 144(6):949-952.

[77] Singh D, Raman B, Ramakrishna T, Rao Ch M: The cataract-causing mutation G98R in human alphaA-crystallin leads to folding defects and loss of chaperone activity. *Mol Vis* 2006, 12:1372-1379.

[78] Singh D, Raman B, Ramakrishna T, Rao Ch M: Mixed oligomer formation between human alphaA-crystallin and its cataract-causing G98R mutant: structural, stability and functional differences. *J Mol Biol* 2007, 373(5):1293-1304.

[79] Litt M, Kramer P, LaMorticella DM, Murphey W, Lovrien EW, Weleber RG: Autosomal dominant congenital cataract associated with a missense mutation in the human alpha crystallin gene CRYAA. *Hum Mol Genet* 1998, 7(3):471-474.

[80] Safieh LA, Khan AO, Alkuraya FS: Identification of a novel CRYAB mutation associated with autosomal recessive juvenile cataract in a Saudi family. *Mol Vis* 2009, 15:980-984.

[81] Berry V, Francis P, Reddy MA, Collyer D, Vithana E, MacKay I, Dawson G, Carey AH, Moore A, Bhattacharya SS *et al*: Alpha-B crystallin gene (CRYAB) mutation causes dominant congenital posterior polar cataract in humans. *Am J Hum Genet* 2001, 69(5):1141-1145.

[82] Inagaki N, Hayashi T, Arimura T, Koga Y, Takahashi M, Shibata H, Teraoka K, Chikamori T, Yamashina A, Kimura A: Alpha B-crystallin mutation in dilated cardiomyopathy. *Biochem Biophys Res Commun* 2006, 342(2):379-386.

[83] Nicolaou P, Knoll R, Haghighi K, Fan GC, Dorn GW, 2nd, Hasenfub G, Kranias EG: Human mutation in the anti-apoptotic heat shock protein 20 abrogates its cardioprotective effects. *J Biol Chem* 2008, 283(48):33465-33471.

[84] Irobi J, Van Impe K, Seeman P, Jordanova A, Dierick I, Verpoorten N, Michalik A, De Vriendt E, Jacobs A, Van Gerwen V *et al*: Hot-spot residue in small heat-shock protein 22 causes distal motor neuropathy. *Nat Genet* 2004, 36(6):597-601.

[85] Perng MD, Wen SF, van den IP, Prescott AR, Quinlan RA: Desmin aggregate formation by R120G alphaB-crystallin is caused by altered filament interactions and is dependent upon network status in cells. *Mol Biol Cell* 2004, 15(5):2335-2346.

[86] Sanbe A, Osinska H, Villa C, Gulick J, Klevitsky R, Glabe CG, Kayed R, Robbins J: Reversal of amyloid-induced heart disease in desmin-related cardiomyopathy. *Proc Natl Acad Sci U S A* 2005, 102(38):13592-13597.

[87] Ecroyd H, Carver JA: Crystallin proteins and amyloid fibrils. *Cell Mol Life Sci* 2009, 66(1):62-81.

[88] den Engelsman J, Bennink EJ, Doerwald L, Onnekink C, Wunderink L, Andley UP, Kato K, de Jong WW, Boelens WC: Mimicking phosphorylation of the small heat-shock protein alphaB-crystallin recruits the F-box protein FBX4 to nuclear SC35 speckles. *Eur J Biochem* 2004, 271(21):4195-4203.

[89] den Engelsman J, Gerrits D, de Jong WW, Robbins J, Kato K, Boelens WC: Nuclear import of {alpha}B-crystallin is phosphorylation-dependent and hampered by hyperphosphorylation of the myopathy-related mutant R120G. *J Biol Chem* 2005, 280(44):37139-37148.

[90] den Engelsman J, Keijsers V, de Jong WW, Boelens WC: The small heat-shock protein alpha B-crystallin promotes FBX4-dependent ubiquitination. *J Biol Chem* 2003, 278(7):4699-4704.

[91] Ikeda R, Yoshida K, Ushiyama M, Yamaguchi T, Iwashita K, Futagawa T, Shibayama Y, Oiso S, Takeda Y, Kariyazono H *et al*: The small heat shock protein alphaB-crystallin inhibits differentiation-induced caspase 3 activation and myogenic differentiation. *Biol Pharm Bull* 2006, 29(9):1815-1819.

[92] Fontaine JM, Sun X, Hoppe AD, Simon S, Vicart P, Welsh MJ, Benndorf R: Abnormal small heat shock protein interactions involving neuropathy-associated HSP22 (HSPB8) mutants. *FASEB J* 2006, 20(12):2168-2170.

[93] Kasakov AS, Bukach OV, Seit-Nebi AS, Marston SB, Gusev NB: Effect of mutations in the beta5-beta7 loop on the structure and properties of human small heat shock protein HSP22 (HspB8, H11). *Febs J* 2007, 274(21):5628-5642.

[94] Kim MV, Kasakov AS, Seit-Nebi AS, Marston SB, Gusev NB: Structure and properties of K141E mutant of small heat shock protein HSP22 (HspB8, H11) that is expressed in human neuromuscular disorders. *Arch Biochem Biophys* 2006, 454(1):32-41.

[95] Liang JJ, Liu BF: Fluorescence resonance energy transfer study of subunit exchange in human lens crystallins and congenital cataract crystallin mutants. *Protein Sci* 2006, 15(7):1619-1627.

[96] Murugesan R, Santhoshkumar P, Sharma KK: Cataract-causing alphaAG98R mutant shows substrate-dependent chaperone activity. *Mol Vis* 2007, 13:2301-2309.

[97] Arrigo AP, Simon S, Gibert B, Kretz-Remy C, Nivon M, Czekalla A, Guillet D, Moulin M, Diaz-Latoud C, Vicart P: Hsp27 (HspB1) and alphaB-crystallin (HspB5) as therapeutic targets. *FEBS Lett* 2007, 581(19):3665-3674.

[98] den Engelsman J, Boros S, Dankers PY, Kamps B, Vree Egberts WT, Bode CS, Lane LA, Aquilina JA, Benesch JL, Robinson CV *et al*: The Small Heat-Shock Proteins HSPB2 and HSPB3 Form Well-defined Heterooligomers in a Unique 3 to 1 Subunit Ratio. *J Mol Biol* 2009.

[99] Fontaine JM, Sun X, Benndorf R, Welsh MJ: Interactions of HSP22 (HSPB8) with HSP20, alphaB-crystallin, and HSPB3. *Biochem Biophys Res Commun* 2005, 337(3):1006-1011.

[100] Ito H, Kamei K, Iwamoto I, Inaguma Y, Tsuzuki M, Kishikawa M, Shimada A, Hosokawa M, Kato K: Hsp27 suppresses the formation of inclusion bodies induced by expression of R120G alpha B-crystallin, a cause of desmin-related myopathy. *Cell Mol Life Sci* 2003, 60(6):1217-1223.

In: Small Stress Proteins and Human Diseases
Editors: Stéphanie Simon et al.
ISBN: 978-1-61470-636-6

Chapter 2.2

ALPHA-CRYSTALLINS (HSPB4 AND HSPB5) AND CATARACT DISEASES

Usha P. Andley*
Department of Ophthalmology and Visual Sciences, Washington University School of Medicine, St. Louis, Missouri, U.S.

ABSTRACT

This chapter reviews the biological roles of two members of the small heat shock protein family, HspB4 (αA-crystallin) and HspB5 (αB-crystallin), in lens development, function, and disease. HspB4 and HspB5 are cytoplasmic, structural proteins that are essential for the optical properties of the lens. While the lens continues to grow throughout life, the bulk of the lens is composed of fiber cells that lack the cellular machinery for protein turnover. Thus, lens proteins must remain stable during a full lifespan in order to keep the lens transparent and competent for sight. The α-crystallins have been found to help maintain lens transparency by creating the proper refractive index in the lens that is necessary to refract light onto the retina. The crystallins are divided into two distinct classes, the α-crystallins and the βγ-crystallins, which comprise at least 16 different mammalian crystallin polypeptides that make up 90% of the water-soluble proteins in the lens. Most crystallins appear to be essential for lens transparency, raising the question of why so many different crystallins are required. This question may be partially answered by recent advances in the understanding of HspB4 and HspB5, α-crystallins. HspB4 and HspB5 have the chaperone-like property of binding unfolded proteins, allowing them to act as a sink to trap denatured proteins. This prevents unfolded protein aggregation which can reduce lens transparency and cause age-related lens damage. In addition, the study of mice harboring targeted deletions or mutations of the α-crystallin genes has demonstrated that the α-crystallins are required for lens development

* Corresponding Author: Usha Andley
Department of Ophthalmology and Visual Sciences,
Washington University School of Medicine,
660 S. Euclid Avenue,
Campus Box 8096,
St. Louis, Missouri 63110
andley@vision.wustl.edu

and function. In the past two decades, our understanding of α-crystallins structure and function has increased exponentially. Here we review the recent knowledge of the roles of HspB4 and HspB5 in the development of cataracts, the most common cause of blindness. This chapter will outline the current research and the critical goals of future research on α-crystallin functions in cataract disease.

INTRODUCTION

The eye lens, together with the cornea, focuses light on the retina to allow vision to occur. Lens transparency is required for vision and is maintained by a high concentration of proteins called lens crystallins. Nearly 90% of the water-soluble proteins in vertebrate lenses are crystallins [1, 2] which are expressed at high concentrations (up to 500 mg/ml) in lens fiber cells. When the concentration of most proteins reaches 10 mg/ml, they aggregate and scatter light. In contrast, crystallins remain soluble up to 500 mg/ml, creating instead a transparent refractive index gradient via short-range interactions [3, 4]. However, this gradient can be disrupted by crystallin structure modifications due to post-translational modification, degradation, or gene mutation. Structural changes can alter the packing and short-range interactions of crystallins, thereby altering the refractive index gradient, leading to increased light scattering by lens opacities that are commonly known as cataracts. Some crystallins, including HspB4 (αA-crystallin) and HspB5 (αB-crystallin), are also expressed at abundant levels in the lens epithelium where they play important roles in cell growth, differentiation and development of the ocular lens [5-7].

The lens crystallins are composed of structurally distinct α and βγ families. While the β- and γ-crystallins are related solely by structure, the α-crystallins are related by structure and function. The α-crystallin family consists of HspB4 (αA-crystallin) and HspB5 (αB-crystallin) which share a highly conserved α-crystallin domain sequence of ~90 amino acids. The α-crystallins are also members of the small heat shock protein family, HspB1 through HspB10. Lens α-crystallin is composed of two approximately 20 kDa subunits, HspB4 and HspB5, which aggregate into species of varying sizes and undefined quaternary structure. Point-mutations in crystallin genes cause structural and functional crystallin changes that have been shown to cause human cataracts at birth or in early childhood. In addition, age-related crystallin protein modifications and structural changes have been shown to contribute to age-related cataract development (Table 1). Thus, understanding the development of the lens in the presence of mutant crystallin genes is greatly relevant to understanding the mechanisms of age-related cataracts. Recent reviews of the genetics of crystallins, regulation of α-crystallin gene expression, and evolution of α-crystallins are available [8-10].

The main lens pathology that is caused by crystallin defects is cataract. Cataract is loosely defined as lens opacification which reduces the amount of light reaching the retina and thereby obstructs vision. Cataracts develop over time and are especially prevalent in the elderly. The α-crystallins mutations that increase the risk of early onset cataracts have been identified over the past 12 years. At present, the mechanism whereby α-crystallin mutation causes cataract formation is poorly understood for two reasons. First, few human cataracts that are known to be caused by α-crystallin mutations have been documented or photographed which means there is little detailed information about the geographic distribution, etiology, and severity of congenital cataracts. Second, models to study the effect of mutations *in vivo* have only recently been developed, thus limited data is currently available from these systems. In contrast, numerous studies have examined the structure-function relationships of HspB4 and HspB5 *in vitro*. Studies of protein solutions have addressed the effects of mutations on protein structure and substrate binding properties, but have not dealt with the localization of opacities or the sequelae of changes in the lens that lead to its opacification.

1. The Unique Architecture of the Ocular Lens

In order to understand the biochemical, developmental, and structural processes that lead to cataract formation, it is important to consider lens structual organization. The lens is the transparent asymmetric oblate spheroid structure that lies beneath the cornea and the aqueous humor [11]. It is an avascular tissue which is derived from surface ectoderm and is exposed to multiple inductive factors during the gastrulation stage of the embryo. Thickening of the presumptive lens ectoderm leads to the formation of the lens placode, which then invaginates to form the lens pit and the lens vesicle. Cells in the anterior of the lens vesicle proliferate to form the lens epithelium, while those in the posterior of the lens vesicle elongate to form primary lens fiber cells. Secondary lens fiber cells are formed by the proliferation and differentiation of a small number of epithelial cells near the lens equator. The anterior of the fully formed lens is lined by a monolayer of cuboidal epithelial cells. These cells differentiate into secondary lens fiber cells in the equatorial region to form the bulk of the lens. The fiber cells are morphologically elongated with a width of only a few micrometers wide and a length of hundreds of micrometers. Primary lens fiber cells form the core of the lens. The entire lens is encapsulated in a basement membrane called the capsule. Secondary fiber cell differentiation occurs at a diminishing rate throughout the lifetime of an organism. Maturing primary and secondary fiber cells elongate progressively in the developing lens, adding shells of new fiber cells to overlay older fiber cells as the lens grows. Details of the growth and differentiation process have been recently reviewed [7, 12-15]. Human cataract development involves specific topographic regions of the lens, which are described below.

The Lens Epithelium

All cells in the lens are derived by proliferation and differentiation of lens epithelial cells. The three distinct regions of the lens epithelium are the central, germinative, and transitional zones. Cell division in the lens epithelium initially occurs throughout the epithelium, but becomes restricted to a narrow band of epithelial cells in the germinative zone during development. Daughter cells that shift below the lens equator enter the post-mitotic region of the epithelium called the transitional zone. These cells are in transition to become differentiated lens fiber cells. Historically, lens epithelial cells were believed not to express crystallins, since their crystallin levels are well below the concentrations present in the lens fiber cells. However, recent work has demonstrated that lens epithelial cells express both HspB4 and HspB5, and several members of the βγ-crystallin family [6, 7, 16, 17]. Knockout mouse models have provided convincing evidence that HspB4 and HspB5 are essential for lens epithelial cell function. Targeted gene deletion of HspB4 leads to decreased lens growth and increased epithelial cell death. Gene deletion studies of HspB5 suggest that it is necessary for maintaining the stability of the epithelial cell genome, at least in cell culture [18, 19].

The Lens Cortex

The outermost, more recently synthesized fiber cells are in the lens cortex [11]. In these distinctly elongated cells, α-, β-, and γ-crystallins function as structural proteins. The cortical fiber cells accumulate high concentrations of crystallins in order to generate the refractive index that is necessary for lens function. Cortical fiber cells also lack the common cellular organelles which would block the path of light to the retina.

The Lens Nucleus

The innermost cells of the lens are formed at the earliest stages of lens development and comprise the lens nucleus or core. Cataracts commonly form in the nucleus, where they inhibit the passage of light to the retina. Nuclear cataract formation has been associated with inherited mutations in the HspB4 and HspB5 genes.

Lens Zonule or Lamella

The mammalian lens is suspended by zonular fibers that attach to the ciliary muscles. The ciliary muscles contract and relax, modifying the tension on the zonular fibers and altering the shape of the lens to accommodate the depth of vision [20]. Zonular and lamellar cataracts refer to cataracts that form in this same region.

Anterior and Posterior Poles

The front edge of the lens faces the cornea and is called the anterior pole. The back end faces the lens capsule and is called the posterior pole. Cataracts of the anterior pole have not been associated with α-crystallin mutations, while cataracts of the posterior pole have [21].

Anterior and posterior sutures. Lens fibers are formed during differentiation of epithelial cells. They elongate and migrate bidirectionally until their tips reach a point where they encounter fiber cells from the opposite hemisphere of the lens. Fiber cells eventually detach from the lens capsule and overlap with the tips of opposing fiber cells, forming a seam known as the suture line [22]. Improper or irregular migration of fiber cell ends results in an irregular suture, causing reduced optical quality and lens opacity [23].

2. Type of Human Catacts

Human age-related cataracts are divided into three distinct categories defined by the location of the opacity within the lens.

(i) *Nuclear cataracts* are among the most commonly observed cataracts and occur in the fully-differentiated innermost fiber cells of the central region of the lens. This region is completely devoid of cellular organelles, including nuclei and contains primary fiber cells, which are the oldest lens fiber cells that originated during embryonic development. As nuclear cataracts advance, they enlarge to sequentially include the fetal lens nucleus, the adult lens nucleus, and the deep cortex. In humans, nuclear cataracts are characterized by a general hardening, often including coloration, and containing aggregates of proteins of mass greater than five million Daltons. These protein aggregates are responsible for the increased light scattering observed in nuclear cataracts. Hereditary human cataracts that are linked with the HspB4 (R116C) and HspB4 (R49C) mutations have been described as nuclear cataracts.

(ii) *Cortical cataracts* are defined by spoke-like opacities in the fully differentiated lens fiber cells peripheral to the lens nucleus. These opacities in humans are typically inferonasally localized during the early stages of cataract formation. The opacities are found both in the anterior and posterior cortex and often cause asymmetric sutures with variable branching. The cortical opacities and changes in suture branches appear to be correlated, which suggests that abnormal lens growth with suture asymmetry predisposes the lens to cortical cataracts.

(iii) Defects in the terminal differentiation of cells in the transitional zone of the lens epithelium result in *posterior subcapsular cataracts*. Transitional zone epithelial cells are columnar cells, which are aligned in orderly meridional rows at the lens equator. As the transitional zone epithelial cells begin to differentiate, they rotate 180° around the polar axis and elongate bidirectionally towards the anterior and posterior poles. Early posterior subcapsular cataracts have disordered meridional rows and irregularly shaped transitional zone cells. These cells fail to rotate completely, and do not begin the normal process of bidirectional elongation. The abnormal transitional zone cells enlarge into ovate-shaped cells, and may be surrounded by normal elongating fibers as they migrate towards the posterior pole. These cells are considered to be dysplastic fiber cells having degenerate cell nuclei and few other organelles. However, they also upregulate crystallin synthesis, like lens fiber cells, and contain an extensive cytoskeleton. The accumulation of these aberrant cells and nuclei at the posterior pole of the lens causes light scattering that is responsible for the posterior subcapsular cataracts.

3. Structure of HspB4 (αA-Crystallin) and HspB5 (αB-crystallin)

α-crystallin is a heteroaggregate of two related polypeptides, HspB4 and HspB5, which share about 57% sequence identity. These subunits arose from a gene duplication event. In humans the HspB4 gene (*CRYAA*) is on chromosome 21 and encodes a 173 amino acid polypeptide [24]. The human HspB5 gene (*CRYAB*) is on chromosome 11 and encodes a 175 amino acid polypeptide. The aggregate size of HspB4 and HspB5 is 300 kDa to 1 million kDa, with an average size of 800 kDa. When isolated from human, rat, and bovine lenses, the ratio of HspB4 to HspB5 in α-crystallin aggregates is 3:1. The exact structure of α-crystallin oligomers has not been determined, because attempts to crystallize them have failed. However, studies on x-ray crystal structure of the microbial and eukaryotic small heat shock proteins Hsp16.5 and Hsp16.9, respectively, have provided insights into the general organization and assembly of small heat shock protein oligomers [25, 26]. Comparison to the crystal structures of other small heat shock proteins indicates that the conserved C-terminal α-crystallin domain forms a solvent-exposed outer shell on the oligomer. The variable-sequence containing N-terminal domain of the small heat shock proteins forms the core of the oligomer. It is also known that α-crystallin contains about 50% β-sheet and 10-15% α-helix structures. Cryoelectron microscopy suggests that HspB5 aggregates contain approximately 32 subunits that form a globule with a hollow cavity. The subunits making up this cavity freely exchange in a temperature-dependent manner (see chapter 2-1).

4. Transcriptional Regulation

Mechanisms of lens-preferred expression of HspB4 (mouse αA-crystallin/cryaa) and HspB5 (mouse αB-crystallin/cryab) have been studied extensively [27]. The expression of the HspB4 gene is relatively low in the lens pit and lens vesicle [28], and is mediated by lens epithelial transcription factors including abundant Pax 6 and low abundant MafB, and possibly CREB. These factors interact with two distal enhancers (DCR1 and DCR3) and the promoter [29]. In differentiating lens fiber cells, Pax6 is reduced while expression of c-Maf is strongly increased [30, 31]. The promoter region of HspB4/Cryaa shows increased binding of c-Maf. In addition, binding of CREB is increased in the promoter region and its 3'-enhancer, DCR3 [29]. This results in a significant upregulation of HspB4/Cryaa gene expression. The lens-specific chromatin domain of the HspB4/Cryaa locus also contains histone acetyltransferases p300 and CBP, and ATP-dependent chromatin remodelers Brg1 and Snf2h [29, 32].

In contrast to the lens-restricted expression of HspB4, murine HspB5 is highly expressed in the lens, but also in the significant levels in the heart, skeletal muscle and lung, and appreciable levels in kidney,brain, eye ciliary body, cornea, and retina [33]. HspB5 expression is also elevated in primary lens fiber cells and in secondary lens fibers [27]. However, HspB5 expression persists longer than HspB4 expression in the epithelium. Mammalian HspB5/Cryab 5'-promoter region contains two lens-specific regions (LSR), LSR1 and LSR2 [34]. Both LSR1 and LSR2 can be activated via Pax6 and large Maf transcription factors (MafA, MafB, c-Maf and NRL) [32, 35]. In addition, a heat shock responsive element has been found in the LSR2 region of the HspB5 gene promoter [36].

5. HSPB4 (αA-CRYSTALLIN) AND CATARACTS

HspB4 expression is largely confined to the lens, but low levels of this protein are also detectable in the spleen, thymus, cornea, and retina. HspB4 expression in the lens is regulated at the transcriptional level, but unlike many heat shock proteins, is not upregulated by stress conditions [37]. HspB4 and HspB5 are thought to maintain transparency in the aging lens and delay lens opacification. Lens proteins do not turnover, so to maintain transparency over a lifespan, α-crystallin acts as a buffer or sink to bind partially unfolded β- and γ-crystallins that would otherwise aggregate, denature, and cause lens opacification. Once the supply of α-crystallin is exhausted in the lens core fiber cells, the concentration of irreversibly denatured crystallins rises and forms nuclear cataracts. By the same pathway, the loss of α-crystallin function due to hereditary mutations in lens epithelial cells is hypothesized to cause cortical cataracts [2, 38]. Several HspB4 mutations have been reported to date (Table 2). The α-crystallin isolated from the lens exists as a complex of HspB4 and HspB5 in a 3:1 stoichiometry. HspB5 is the first crystallin to be expressed in the developing mouse embryonic lens beginning around day 9.5. HspB4 transcript expression becomes detectable in the lens vesicle on embryonic day 10-10.5 [31]. Thus, it can be envisioned that mutations in HspB4 or HspB5 genes could lead to expression of mutant proteins in the embryo and cataract development early in life.

The first HspB4 mutation that was reported to be associated with human hereditary cataract development is the R116C mutation, which is due to a C to T transition in exon 3 of *CRYAA* [39]. This mutation was first identified in a cohort of ten patients, some of whom also developed posterior and cortical opacities in their thirties. A second family also had this mutation but developed both cataracts and a micro-cornea defect [40]. The *in vivo* effect of the HspB4 (R116C) mutation was recently described in a transgenic mouse model [47]. When mutant R116C protein subunits are expressed in a wild type endogenous mouse α-crystallin background, lens opacities and sutural defects occur even in cells expressing low levels of the mutant protein. In human carriers of the R116C mutation, some of the defects such as zonular central nuclear opacities occur in newborns, while cortical and posterior subcapsular defects are common in adult humans [39]. In contrast, the transgenic mice did not exhibit a high frequency of the central nuclear opacities at any age. The reason for these differences has yet to be determined.

The second HspB4 mutation to be identified is the HspB4(R49C) mutation which causes nuclear cataracts [41-42]. The R49C mutation is of special interest as it lies outside the conserved C-terminal α-crystallin domain of HspB4. This mutation introduces a new disulfide linkage between cysteine 49 and a second sulfhydryl group of βB1-crystallin, and therefore may alter the structure and function of HspB4 and/or higher order α-crystallin complexes [49]. Consistent with this hypothesis, studies have shown that replacement of arginine 49 with alanine dramatically reduces the *in vitro* chaperone activity of HspB4 [50]. The R49C mutation is in exon 1 of the *CRYAA* gene and produces a protein that causes cell death when exogenously expressed *in vitro* without extraneous stress conditions [42]. The protein was found primarily in the nuclei of lens epithelial cells, with low levels in the cytoplasm. This protein has recently been expressed *in vivo* in mice in which the mutation was knocked-in [48].

A third cataract-associated HspB4 mutation was identified as a C to A missense mutation causing substitution of the amino acid arginine with glycine at codon 98 (G98R) of HspB4 [43]. A fourth mutation of G to A which creates a HspB4 (W9X) mutant protein causes autosomal recessive cataracts [44]. No clinical descriptions of this cataract type have been published, although the premature chain termination of the W9X mutation may functionally mimic the changes observed in the *Cryaa* homozygous mouse knockout lens. A fifth cataract-associated mutation identified in a four generation family was an autosomal dominant (AD), G to A mutation in exon 3 of *CRYAA* which changed a conserved arginine to histidine [45]. Thirteen of the 28 family members with this mutation developed cataracts. This is one of only two reports where the cataract morphology was described in detail. The cataracts were described as diverse and variable, both between family members and inter-ocularly (between the two eyes of one individual). Embryonic nuclear opacities were evident, with some extending to the anterior of the lens or to the anterior pole, and some accompanied by cortical punctate opacities. A small cornea defect (micro-cornea) was evident in some of the patients. Interestingly, elevated cataract density did consistently correlate with age. Correlations between age and cataract density are rarely described in studies of genetic cataracts because of the small number of patients that do not undergo artificial lens replacement surgery. However, documenting the variations within patients and families is critical to understand the mechanism of cataract formation and the influence of age, other genes, and environmental factors on cataract development.

Of the human cataract-associated *CRYAA* mutations that have been reported, four are in codon 116 [38, 39, 41, 48]. Interestingly, none of the *CRYAA* mutations cause *isolated* cortical or posterior opacities. The amino acid substitution results in replacement of the charged arginine residue at amino acid 116 of *CRYAA* protein by a neutral amino acid in three unrelated families [39, 41, 45]. This mutation is required for proper structure and function of the HspB4 protein and is known to reduce the survival of lens epithelial cells in the presence of stress [46]. These reports underscore the importance of the hydrophobic and charged arginine at position 116 of HspB4.

6. HSPB5 (αB-CRYSTALLIN) AND CATARACTS

HspB5 and HspB4 are 57% identical at the amino acid level, and share many properties including anti-apoptotic functions, autokinase activity, interactions with cytoskeletal proteins, and the ability to act as a molecular chaperone. However, there are also many differences between the two proteins. Unlike HspB4, HspB5 is a stress-inducible protein, although recent work has identified HspB4 in diseases, such as in age-related macular degeneration and experimental uveitis [51, 52]. Another difference is that HspB5 is expressed in numerous tissues, whereas HspB4 expression is primarily lens-restricted. In particular, HspB5 is highly expressed in adult heart and muscle cells, although it is most highly expressed in the lens [37]. In mice, HspB5 is expressed in the heart at an earlier embryonic stage than in the lens [53]. Elevated HspB5 expression is also associated with multiple diseases such as scrapie-infected brain tissue. Its upregulation in numerous diseases has been reviewed in another chapter of this book.

Unexpectedly, the targeted deletion of the HspB5 gene (*CRYAB*) in mice does not produce changes in lens development and only mildly affects lens transparency throughout life [54]. HspB5 null mice also lacked the related small heat shock protein HspB2, which is not expressed in the lens and is not expected to contribute to lens phenotypes, but the lack of HspB5 and HspB2 did cause degeneration of skeletal muscle tissue. Primary cultures of lens epithelial cells derived from HspB5-/- mice contain cells with vastly enhanced proliferative ability; however, these hyperproliferative cells were not tumorigenic. The HspB5-/- cultures contained many tetraploid cells, had abnormal microtubule structures in the anaphase midzone, and had compromised p53 checkpoint function. These findings support the conclusion that HspB5 may be required for proper mitosis and cytokinesis, and like some other chaperones, may be necessary for the proper assembly of microtubules [19, 55].

Three HspB5 mutations have been shown to cause human cataract development. The first is a C to T substitution in exon 3 of *CRYAB* which causes mutation of arginine 120 to glycine, and corresponds to arginine 116 of HspB4. This mutation is associated with AD cataract formation and desmin-related myopathy [56]. A second cataract mutation is the HspB5 (450delA) frameshift mutation which causes AD posterior cataracts in the absence of myopathy [21]. This suggests that the aberrant 184 amino acid protein formed by the mutation has a toxic effect on lens cells. The third cataract-associated HspB5 mutation is a G to A substitution in *CRYAB* that causes a D140N mutation and AD lamellar (zonular) cataracts [57].

7. Mouse Models with HspB4 and HspB5 Mutations

Three HspB4 mutations, two in codon 49 and one in codon 188, are associated with cataract development. The HspB4 (R54C) mutation causes recessive cataracts in mice and humans [58]. The lenses of HspB4 (R54C) homozygous mice displayed dense nuclear opacities and microphthalmia. Histological analysis of the R54C homozygous mice showed severely altered lens epithelial cells and degenerated lens fiber cells. HspB4 (R54H) (lop18) homozygous mice also have severe cataracts. Heterozygous mice in each cases have transparent lenses. Homozygous HspB4 (R54H) mice had less severe nuclear cataracts than homozygous HspB4 (R54C) mice. These phenotypic differences may be due to the different properties of the residues that replace the positively charged arginine of codon 54. Histidine is a weakly positive-charged amino acid, whereas cysteine is a neutral amino acid residue with the ability to form intramolecular and intermolecular disulfide bonds. In contrast to codon 49 mutations, the autosomal dominant HspB4 (Y118D) mutation does not affect epithelial or fiber cell morphology but results in elevated insoluble and degraded proteins which would be expected to contribute to cataract development. The dominant nuclear cataract in Y118D mice is associated with a significant decrease in the amount of HspB4, leading to a reduction in total chaperone capacity needed for maintaining lens transparency [59].

Mice in which one wild type copy of the HspB4 gene has been replaced by the HspB4 (R49C) mutant have been generated by using stem cell-based gene targeting technologies. In these knock-in mice, it is possible to compare the gene dosage effect of the HspB4 (R49C) mutant by comparing the phenotypes of heterozygous and homozygous mice. These studies showed that the presence of one copy of the mutant protein was sufficient to cause cataracts in mice by 2-3 months of age, which progress to a complete opacity as the mice grow older. The opacity begins with posterior changes, progresses to nuclear changes, and eventually involves the entire cortex and lens as heterozygous mice age. In contrast, homozygous mice carrying two copies of the mutant gene had a more drastic phenotype characterized by a small eyes (microphthalmia) and lenses, and nuclear cataracts at birth. These homozygous mutant mice had fully formed cataracts at the relatively young age of 2 months.

Further analysis showed that the mechanism of cataract formation was due to the insolubility of the HspB4 (R49C) mutant protein [60]. Lens protein solubility was reduced by 80% in HspB4 (R49C) homozygous knock-in lenses. When HspB4 (R49C) protein begins to aggregate, it becomes insoluble. In contrast, other human cataract-causing HspB4 mutant proteins including HspB4 (G98R), HspB4 (R116C), HspB4 (R54C), and HspB4 (Y118D) can form higher molecular weight water-soluble complexes. This difference may be explained by the ability of HspB4 (R49C) purified from lenses to stably bind to β and γ-crystallins which do not bind to wild type HspB4. This binding may be the result of the increased surface hydrophobicity of the mutant or disulfide-crosslinks which can form between βB1-crystallin and HspB4 (R49C) [49]. Whatever the mechanism, the tendency of the HspB4 (R49C) mutant to become unstable and form aggregates with unfolded substrate proteins in the presence of wild type HspB4 in heterozygous lenses would be expected to reduce the rate of HspB4 (R49C) mutant incorporation into high molecular weight insoluble species, thus reducing the rate of cataract formation. Subsequent disulfide cross-linking may also have additional deleterious functional consequences. Much remains to be explored in terms of the oxidizing potential and the functional consequence of the disulfide bonds on the structure and chaperone activity of HspB4. More research is necessary to identify the structure, *in vivo* substrates, and mechanism of insolubilization of the mutant HspB4 proteins.

During the generation of mice carrying the HspB4 (R49C) mutation, another cataract model was generated. Mice carrying the neomycin gene in the intron of the HspB4 gene also had lens defects. The expression of the neomycin gene suppresses the severity of the cataracts in both HspB4-R49C^{neo} heterozygous and homozygous mice [61]. This effect is due to a decrease in the R49C protein expression in the presence of the neomycin gene. When the neomycin gene is deleted, the mice get a much more severe lens phenotype with the cataracts developing at birth [48]. However, HspB4-R49C^{neo} homozygous mice have posterior cataracts with a posterior migration of lens epithelial cells around 2 months of age and develop nuclear cataracts. In addition, as early as three weeks, the lenses show fiber cell membrane defects, prominent vacuoles, and intracellular gaps.

8. Mechanisms of Cataract Disease Caused by HspB4 and HspB5 Mutations

The mechanisms that have been suggested for congenital or hereditary cataract formation will be considered in this section. The congenital cataract formation in mice carrying HspB4 and HspB5 mutations is a convenient and powerful model that allows an age-related process that takes decades in humans to be studied in a time-compressed manner. For example, oxidation and formation of disulfide-linked aggregates is one of the causes of age-related cataracts which also occurs in HspB4 (R49C) mutant protein [49]. Formation of high molecular weight water-insoluble protein aggregates found in human cataract also occurs in HspB4 (R49C) mutant mice [60]. The mechanisms of binding of substrate proteins with HspB4 has been extensively studied [62]. The increased affinity of the HspB4 (R49C) mutant for substrate proteins, causes complex formation and cataract formation soon after birth, without the need for age-related modifications. From the dominant pattern of cataract development in heterozygous HspB4 mutant mice we can surmise that mutant HspB4 actively damages the lens epithelial and fiber cells. Alternatively, it may inhibit the function of wild type HspB4 [63, 64]. The role of reduced chaperone activity in cataract formation is further supported by the finding that the cataract-associated HspB4 (G98R) mutant also affects chaperone activity *in vitro* [65]. The HspB5 (R120G) mutant is inherently unstable [66]. The R120 position is critical to conserve proper intra- and intersubunit interactions [67]. The arginine at position 120 in the HspB5 sequence is critical for quaternary structure and functional integrity of HspB5 [69, 70]. Additional mechanisms are suggested by HspB5 (R120G) mutant mice. Studies on the mechanisms of HspB5 (R120G) mutant-induced changes in heart muscle cells indicates that cardiomyopathy is enhanced due to increased cycling of reduced glutathione and reductive stress [71]. HspB5 also has a phosphorylation-dependent role in the ubiquitination of a component of SC35 speckles [72]. Nuclear speckle localization of HspB5 is inhibited by the R120G mutation [73]. In addition, the HspB5 (R120G) mutation also promotes vimentin aggregation [68]. If and how these mechanisms contribute to cataract formation is unknown.

Evidence suggests several mechanisms may contribute to the high molecular protein aggregation that is caused by mutations in HspB4 and HspB5. A decrease in the chaperone activity of the mutant α-crystallin might allow damaged protein substrates to accumulate and aggregate in the lens cytoplasm, causing an increase in light scattering and lens opacity [70, 74, 75]. Alternatively, evidence suggests that mutant α-crystallin itself may form high molecular weight aggregates, which results in an increase in co-precipitation of substrate and non-substrate proteins [60]. Other studies suggest that oligomers of α-crystallin may dissociate into smaller multimers, which enhances their capacity to bind substrate proteins and cause aggregation of proteins [76]. The dominant model for cataract formation is that mutation-induced enhancement of substrate binding causes substrate-saturated α-crystallin to form high molecular weight aggregates that precipitate and form cataracts. This is supported by the finding that HspB5 (R120G) coprecipitates with substrates *in vitro* and *in vivo* [70]. This model is also supported by the observation that the α-crystallin fraction from HspB4 (R49C) mutant knock-in mouse lenses contains stable β-crystallin and γ-crystallin protein aggregates [60].

CONCLUSION

HspB4 and HspB5 are proteins abundant in the lens and loss of their function leads to cataract formation or epithelial cell proliferation defects in the lens. These proteins are the major determinants of health and disease in this tissue. HspB4 and HspB5 form heteroaggregates with one another, prevent protein aggregation *in vitro* and *in vivo*, and maintain the necessary refractive index gradient essential to maintain lens transparency. HspB5 is expressed very early in lens development, when vision is not a requirement, suggesting that at this stage it participates in the development of the fully formed lens. Clinical and functional genetic studies have provided evidence that HspB4 and HspB5 are essential for the primary function of the lens, which is to focus light onto the retina. Mutations in both the N- and C-terminal regions of HspB4 and HspB5 result in human cataracts inherited most often by dominant autosomal inheritance. These abundant proteins are prime candidates for age-related cataracts. It is likely that future work will find additional mutations in HspB4 and HspB5 genes that cause age-related cataracts. Thus, the development of methods to circumvent the deleterious effects of these mutant proteins is a critical goal for the prevention of age-related cataract development.

Molecular analysis of the biochemical mechanisms that contribute to congenital human cataract formation and the study of mouse models of cataract formation hold great promise for advancing our understanding of lens development and cataract disease caused by HspB4 and HspB5 crystallin mutations. The observation that loss of HspB4 or HspB5 causes cataract formation establishes their critical roles in cellular physiology and growth. Identification of the HspB4 and HspB5 lens substrates that contribute to cataract disease is critical for the characterization of the variability of cataract severity and location within the lens. How factors, such as mutant protein concentrations in different fiber cell regions, modulate cataract severity and localization also will contribute to the characterization of human cataract disease.

Acknowledgment: Work in the author's laboratory is supported by the National Institutes of Health (USA) Grant EY05681.

Table 1. Lens HspB4 and HspB5 modifications

Protein	Modification	Effect on protein	Ref.
HspB4	Truncation	Higher molecular weight	[77-80]
	Deamidation	Higher molecular weight	[77]
	Oxidation, S-S bonding	Higher molecular weight	[77]
	Phosphorylation	Lower molecular weight	[77, 78]
	Glycation	Higher molecular weight	
HspB5	Truncation		[79]
	Deamidation		[77]
	Oxidation		[77]
	Phosphorylation	Aggregation, disruption of dimeric substructure	[78, 79, 81]

Table 2. Mutation in HspB4 and HspB5 genes in mouse and human cataracts

Protein	Species	Mutation	Effect on Lens
HspB4	**Mouse**		
	Cryaa lop18	G→A R54H	Recessive nuclear and cortical cataract
	Cryaa Aey7	T → A V124E	Dominant nuclear cataract
	Cryaa R116C	C → T R116C	Posterior cortical cataract
	Cryaa	C → T R54C	Recessive nuclear and cortical cataract
	Cryaa	T →G Y118D	Dominant nuclear cataract
	Cryaa	C →T R49C	Dominant nuclear, posterior and cortical cataract
	Cryaa	Cryaa-/-	Nuclear cataract
HspB4	**Human**		
	CRYAA	C → T R116C	Dominant cataract
	CRYAA	C → T R49C	Dominant cataract
	CRYAA	G →A W9X	Recessive cataract
	CRYAA	G →A G98R	Dominant cataract
	CRYAA	G →A R116H	Dominant cataract
HspB5	**Mouse**		
	Cryab, HspB2	Cryab-/-	No ocular phenotype, but cultured cells show genomic instability
HspB5	**Human**		
	CRYAB	A →G R120G	Dominant nuclear cataract, myopathy
		450delA (frameshift in codon 150)	Dominant posterior polar cataracts
		G →A D140N	Dominant lamellar cataract

References

[1] Wistow, GJ; Piatigorsky, J. Lens crystallins: the evolution and expression of proteins for a highly specialized tissue. *Annu Rev Biochem*, 1988. 57: p. 479-504.

[2] Bloemendal, H;de Jong, W;Jaenicke, R;Lubsen, NH;Slingsby, C; Tardieu, A. Ageing and vision: structure, stability and function of lens crystallins. *Prog Biophys Mol Biol*, 2004. 86: p. 407-85.

[3] Delaye, M; Tardieu, A. Short-range order of crystallin proteins accounts for eye lens transparency. *Nature*, 1983. 302: p. 415-7.

[4] Benedek, GB. Theory of transparency of the eye. *Appl. Optics*, 1971. 10: p. 459-473.

[5] Andley, UP. The lens epithelium: focus on the expression and function of the alpha-crystallin chaperones. *Int J Biochem Cell Biol*, 2008. 40: p. 317-23.

[6] Wang, X;Garcia, CM;Shui, YB; Beebe, DC. Expression and regulation of alpha-, beta-, and gamma-crystallins in mammalian lens epithelial cells. *Invest Ophthalmol Vis Sci*, 2004. 45: p. 3608-19.

[7] Bhat, S. The ocular lens epithelium. *Bioscience Reports*, 2002. 21: p. 537-563.

[8] Graw, J. Genetics of crystallins: cataract and beyond. *Exp Eye Res*, 2009. 88: p. 173-89.

[9] Kappe, G;Franck, E;Verschuure, P;Boelens, WC;Leunissen, JA; de Jong, WW. The human genome encodes 10 alpha-crystallin-related small heat shock proteins: HspB1-10. *Cell Stress Chaperones*, 2003. 8: p. 53-61.

[10] Cvekl, A; Duncan, MK. Genetic and epigenetic mechanisms of gene regulation during lens development. *Prog Retin Eye Res*, 2007. 26: p. 555-97.

[11] Kuszak, J, Costello, MJ. The structure of the vertebrate lens. 1 ed. Development of the Ocular Lens, ed. Lovicu, F, Robinson, ML. 2004, Cambridge, U.K.: Cambridge University Press. 71-118.

[12] Sue Menko, A. Lens epithelial cell differentiation. *Exp Eye Res*, 2002. 75: p. 485-90.

[13] Bassnett, S. Lens organelle degradation. *Exp Eye Res*, 2002. 74: p. 1-6.

[14] Robinson, ML. An essential role for FGF receptor signaling in lens development. *Semin Cell Dev Biol*, 2006. 17: p. 726-40.

[15] Piatigorsky, J. Lens differentiation in vertebrates. A review of cellular and molecular features. *Differentiation*, 1981. 19: p. 134-53.

[16] Bhat, SP. Transparency and non-refractive functions of crystallins--a proposal. *Exp Eye Res*, 2004. 79: p. 809-16.

[17] Andley, UP. Crystallins in the eye: Function and pathology. *Prog Retin Eye Res*, 2007. 26: p. 78-98.

[18] Andley, UP;Song, Z;Wawrousek, EF; Bassnett, S. The molecular chaperone alphaA-crystallin enhances lens epithelial cell growth and resistance to UVA stress. *J Biol Chem*, 1998. 273: p. 31252-61.

[19] Andley, UP;Song, Z;Wawrousek, EF;Brady, JP;Bassnett, S; Fleming, TP. Lens epithelial cells derived from alphaB-crystallin knockout mice demonstrate hyperproliferation and genomic instability. *Faseb J*, 2001. 15: p. 221-229.

[20] Robinson M.L. and Lovicu, FJ. The Lens: Historical and Comparative Perspectives. First ed. Development of the Ocular Lens, ed. Lovicu, FJaR, ML. 2004, New York: Cambridge University Press. 3-23.

[21] Berry, V;Francis, P;Reddy, MA;Collyer, D;Vithana, E;MacKay, I;Dawson, G;Carey, AH;Moore, A;Bhattacharya, SS; Quinlan, RA. Alpha-B crystallin gene (CRYAB) mutation causes dominant congenital posterior polar cataract in humans. *Am J Hum Genet*, 2001. 69: p. 1141-5.

[22] Kuszak, JR;Zoltoski, RK; Sivertson, C. Fibre cell organization in crystalline lenses. *Exp Eye Res*, 2004. 78: p. 673-87.

[23] Al-ghoul, KJ;Novak, LA; Kuszak, JR. The structure of posterior subcapsular cataracts in the Royal College of Surgeons (RCS) rats. *Exp Eye Res*, 1998. 67: p. 163-77.

[24] Piatigorsky, J. Crystallin genes: specialization by changes in gene regulation may precede gene duplication. *J Struct Funct Genomics*, 2003. 3: p. 131-7.

[25] Kim, KK;Kim, R; Kim, SH. Crystal structure of a small heat-shock protein. *Nature*, 1998. 394: p. 595-9.

[26] van Montfort, RL;Basha, E;Friedrich, KL;Slingsby, C; Vierling, E. Crystal structure and assembly of a eukaryotic small heat shock protein. *Nat Struct Biol*, 2001. 8: p. 1025-30.
[27] Cvekl, A;Yang, Y;Chauhan, BK; Cveklova, K. Regulation of gene expression by Pax6 in ocular cells: a case of tissue-preferred expression of crystallins in lens. *Int J Dev Biol*, 2004. 48: p. 829-44.
[28] Wolf, L;Yang, Y;Wawrousek, E; Cvekl, A. Transcriptional regulation of mouse alpha A-crystallin gene in a 148 kb Cryaa BAC and its derivatives. *BMC Dev Biol*. 8: p. 88.
[29] Yang, Y;Stopka, T;Golestaneh, N;Wang, Y;Wu, K;Li, A;Chauhan, BK;Gao, CY;Cveklova, K;Duncan, MK;Pestell, RG;Chepelinsky, AB;Skoultchi, AI; Cvekl, A. Regulation of alphaA-crystallin via Pax6, c-Maf, CREB and a broad domain of lens-specific chromatin. *Embo J*, 2006. 25: p. 2107-18.
[30] Ring, BZ;Cordes, SP;Overbeek PA;Barsh GS. Regulation of mouse lens fiber cell development and differentiation by the Maf gene. *Development*, 2000. 127: p. 307-17.
[31] Robinson, ML; Overbeek, PA. Differential expression of alpha A- and alpha B-crystallin during murine ocular development. *Invest Ophthalmol Vis Sci*, 1996. 37: p. 2276-84.
[32] Yang, Y;Chauhan, BK;Cveklova,K; Cvekl, A.Transcriptional regulation of mouse alpha B- and gamma F-crystallin genes in lens: opposite promoter-specific interactions between Pax6 and large Maf transcription factors. *J Mol Biol*, 2004. 344: p. 351-68.
[33] Dubin, RA;Wawrousek, EF; Piatigorsky, J. Expression of the murine alpha B-crystallin gene is not restricted to the lens. *Mol Cell Biol*, 1989. 9: p. 1083-91.
[34] Gopal-Srivastava, R;Cvekl, A; Piatigorsky, J. Pax-6 and alphaB-crystallin/small heat shock protein gene regulation in the murine lens. Interaction with the lens-specific regions, LSR1 and LSR2. *J Biol Chem*, 1996. 271: p. 23029-36.
[35] Chauhan, BK;Zhang, W;Cveklova, K;Kantorow, M; Cvekl, A. Identification of differentially expressed genes in mouse Pax6 heterozygous lenses. *Invest Ophthalmol Vis Sci*, 2002. 43: p. 1884-90.
[36] Somasundaram, T; Bhat, SP. Canonical heat shock element in the alpha B-crystallin gene shows tissue-specific and developmentally controlled interactions with heat shock factor. *J Biol Chem*, 2000. 275: p. 17154-9.
[37] Sax, CM; Piatigorsky, J. Expression of the alpha-crystallin/small heat-shock protein/molecular chaperone genes in the lens and other tissues. *Adv Enzymol Relat Areas Mol Biol*, 1994. 69: p. 155-201.
[38] Horwitz, J. Alpha-crystallin. *Exp Eye Res*, 2003. 76: p. 145-53.
[39] Litt, M;Kramer, P;LaMorticella, DM;Murphey, W;Lovrien, EW; Weleber, RG. Autosomal dominant congenital cataract associated with a missense mutation in the human alpha crystallin gene CRYAA. *Hum Mol Genet*, 1998. 7: p. 471-4.
[40] Vanita, V;Singh, JR;Hejtmancik, JF;Nuernberg, P;Hennies, HC;Singh, D; Sperling, K. A novel fan-shaped cataract-microcornea syndrome caused by a mutation of CRYAA in an Indian family. *Mol Vis*, 2006. 12: p. 518-22.
[41] Beby, F;Commeaux, C;Bozon, M;Denis, P;Edery, P; Morle, L. New phenotype associated with an Arg116Cys mutation in the CRYAA gene: nuclear cataract, iris coloboma, and microphthalmia. *Arch Ophthalmol*, 2007. 125: p. 213-6.
[42] Mackay, DS;Andley, UP; Shiels, A. Cell death triggered by a novel mutation in the alphaA-crystallin gene underlies autosomal dominant cataract linked to chromosome 21q. *Eur J Hum Genet*, 2003. 11: p. 784-93.

[43] Santhiya, ST;Soker, T;Klopp, N;Illig, T;Prakash, MV;Selvaraj, B;Gopinath, PM; Graw, J. Identification of a novel, putative cataract-causing allele in CRYAA (G98R) in an Indian family. *Mol Vis*, 2006. 12: p. 768-73.

[44] Pras, E;Frydman, M;Levy-Nissenbaum, E;Bakhan, T;Raz, J;Assia, EI;Goldman, B; Pras, E. A nonsense mutation (W9X) in CRYAA causes autosomal recessive cataract in an inbred Jewish Persian family. *Invest Ophthalmol Vis Sci*, 2000. 41: p. 3511-5.

[45] Richter, L;Flodman, P;Barria von-Bischhoffshausen, F;Burch, D;Brown, S;Nguyen, L;Turner, J;Spence, MA; Bateman, JB. Clinical variability of autosomal dominant cataract, microcornea and corneal opacity and novel mutation in the alpha A crystallin gene (CRYAA). *Am J Med Genet A*, 2008. 146: p. 833-42.

[46] Andley, UP;Patel, HC; Xi, JH. The R116C mutation in alpha A-crystallin diminishes its protective ability against stress-induced lens epithelial cell apoptosis. *J Biol Chem*, 2002. 277: p. 10178-86.

[47] Hsu, CD;Kymes, S; Petrash, JM. A transgenic mouse model for human autosomal dominant cataract. *Invest Ophthalmol Vis Sci*, 2006. 47: p. 2036-44.

[48] Xi, JH;Bai, F;Gross, J;Townsend, RR;Menko, AS; Andley, UP. Mechanism of small heat shock protein function in vivo: a knock-in mouse model demonstrates that the R49C mutation in alpha A-crystallin enhances protein insolubility and cell death. *J Biol Chem*, 2008. 283: p. 5801-14.

[49] Kumar, MS;Koteiche, HA;Claxton, DP; McHaourab, HS. Disulfide cross-links in the interaction of a cataract-linked alphaA-crystallin mutant with betaB1-crystallin. *FEBS Lett*, 2009. 583: p. 175-9.

[50] Biswas, A;Miller, A;Oya-Ito, T;Santhoshkumar, P;Bhat, M; Nagaraj, RH. Effect of site-directed mutagenesis of methylglyoxal-modifiable arginine residues on the structure and chaperone function of human alphaA-crystallin. *Biochemistry*, 2006. 45: p. 4569-77.

[51] Rao, NA;Saraswathy, S;Wu, GS;Katselis, GS;Wawrousek, EF; Bhat, S. Elevated retina-specific expression of the small heat shock protein, alphaA-crystallin, is associated with photoreceptor protection in experimental uveitis. *Invest Ophthalmol Vis Sci*, 2008. 49: p. 1161-71.

[52] Crabb, JW;Miyagi, M;Gu, X;Shadrach, K;West, KA;Sakaguchi, H;Kamei, M;Hasan, A;Yan, L;Rayborn, ME;Salomon, RG; Hollyfield, JG. Drusen proteome analysis: an approach to the etiology of age-related macular degeneration. *Proc Natl Acad Sci U S A*, 2002. 99: p. 14682-7.

[53] Benjamin, IJ;Shelton, J;Garry, DJ; Richardson, JA. Temporospatial expression of the small HSP/alpha B-crystallin in cardiac and skeletal muscle during mouse development. *Dev Dyn*, 1997. 208: p. 75-84.

[54] Brady, JP;Garland, DL;Green, DE;Tamm, ER;Giblin, FJ; Wawrousek, EF. AlphaB-crystallin in lens development and muscle integrity: a gene knockout approach. *Invest Ophthalmol Vis Sci*, 2001. 42: p. 2924-34.

[55] Bai, F;Xi, JH;Wawrousek, EF;Fleming, TP; Andley, UP. Hyperproliferation and p53 status of lens epithelial cells derived from alphaB-crystallin knockout mice. *J Biol Chem*, 2003. 278: p. 36876-86.

[56] Vicart, P;Caron, A;Guicheney, P;Li, Z;Prevost, MC;Faure, A;Chateau, D;Chapon, F;Tome, F;Dupret, JM;Paulin, D; Fardeau, M. A missense mutation in the alphaB-

crystallin chaperone gene causes a desmin-related myopathy. *Nat Genet*, 1998. 20: p. 92-5.

[57] Liu, Y;Zhang, X;Luo, L;Wu, M;Zeng, R;Cheng, G;Hu, B;Liu, B;Liang, JJ; Shang, F. A Novel {alpha}B-Crystallin Mutation Associated with Autosomal Dominant Congenital Lamellar Cataract. *Invest Ophthalmol Vis Sci*, 2006. 47: p. 1069-75.

[58] Xia, CH;Liu, H;Chang, B;Cheng, C;Cheung, D;Wang, M;Huang, Q;Horwitz, J; Gong, X. Arginine 54 and Tyrosine 118 residues of {alpha}A-crystallin are crucial for lens formation and transparency. *Invest Ophthalmol Vis Sci*, 2006. 47: p. 3004-10.

[59] Huang, Q;Ding, L;Phan, KB;Cheng, C;Xia, CH;Gong, X; Horwitz, J. Mechanism of cataract formation in alphaA-crystallin Y118D mutation. *Invest Ophthalmol Vis Sci*, 2009. 50: p. 2919-26.

[60] Andley, UP;Hamilton, PD; Ravi, N. Mechanism of insolubilization by a single-point mutation in alphaA-crystallin linked with hereditary human cataracts. *Biochemistry*, 2008. 47: p. 9697-706.

[61] Andley, U.P. alphaA-crystallin R49C neo mutation influences the architecture of lens fiber cell membranes and causes posterior and nuclear cataracts in mice. *BMC Ophthalmology*, 2009. 9: p.4.

[62] Claxton, DP;Zou, P; McHaourab, HS. Structure and orientation of T4 lysozyme bound to the small heat shock protein alpha-crystallin. *J Mol Biol*, 2008. 375: p. 1026-39.

[63] Bateman, BJ; Harley, RD. Genetics of eye disease. 4th / [edited by] Leonard B. Nelson. ed. Harley's pediatric ophthalmology, ed. Nelson, LB. 1998, Philadelphia: W.B. Saunders. 1-59.

[64] Shiels, A; Hejtmancik, JF. Genetic origins of cataract. *Arch Ophthalmol*, 2007. 125: p. 165-73.

[65] Murugesan, R;Santhoshkumar, P; Sharma, KK. Cataract-causing alphaAG98R mutant shows substrate-dependent chaperone activity. *Mol Vis*, 2007. 13: p. 2301-9.

[66] Treweek, TM;Rekas, A;Lindner, RA;Walker, MJ;Aquilina, JA;Robinson, CV;Horwitz, J;Perng, MD;Quinlan, RA; Carver, JA. R120G alphaB-crystallin promotes the unfolding of reduced alpha-lactalbumin and is inherently unstable. *Febs J*, 2005. 272: p. 711-24.

[67] Michiel, M;Skouri-Panet, F;Duprat, E;Simon, S;Ferard, C;Tardieu, A; Finet, S. Abnormal assemblies and subunit exchange of alphaB-crystallin R120 mutants could be associated with destabilization of the dimeric substructure. *Biochemistry*, 2009. 48: p. 442-53.

[68] Song, S;Hanson, MJ;Liu, BF;Chylack, LT; Liang, JJ. Protein-protein interactions between lens vimentin and alphaB-crystallin using FRET acceptor photobleaching. *Mol Vis*, 2008. 14: p. 1282-7.

[69] Simon, S;Michiel, M;Skouri-Panet, F;Lechaire, JP;Vicart, P; Tardieu, A. Residue R120 is essential for the quaternary structure and functional integrity of human alphaB-crystallin. *Biochemistry*, 2007. 46: p. 9605-14.

[70] Bova, MP;Yaron, O;Huang, Q;Ding, L;Haley, DA;Stewart, PL; Horwitz, J. Mutation R120G in alphaB-crystallin, which is linked to a desmin-related myopathy, results in an irregular structure and defective chaperone-like function. *Proc Natl Acad Sci U S A*, 1999. 96: p. 6137-42.

[71] Rajasekaran, NS;Connell, P;Christians, ES;Yan, LJ;Taylor, RP;Orosz, A;Zhang, XQ;Stevenson, TJ;Peshock, RM;Leopold, JA;Barry, WH;Loscalzo, J;Odelberg, SJ;

Benjamin, IJ. Human alpha B-crystallin mutation causes oxido-reductive stress and protein aggregation cardiomyopathy in mice. *Cell*, 2007. 130: p. 427-39.

[72] den Engelsman, J;Bennink, EJ;Doerwald, L;Onnekink, C;Wunderink, L;Andley, UP;Kato, K;de Jong, WW; Boelens, WC. Mimicking phosphorylation of the small heat-shock protein alphaB-crystallin recruits the F-box protein FBX4 to nuclear SC35 speckles. *Eur J Biochem*, 2004. 271: p. 4195-203.

[73] van den, IP;Wheelock, R;Prescott, A;Russell, P; Quinlan, RA. Nuclear speckle localisation of the small heat shock protein alpha B-crystallin and its inhibition by the R120G cardiomyopathy-linked mutation. *Exp Cell Res*, 2003. 287: p. 249-61.

[74] Shroff, NP;Cherian-Shaw, M;Bera, S; Abraham, EC. Mutation of R116C results in highly oligomerized alpha A-crystallin with modified structure and defective chaperone-like function. *Biochemistry*, 2000. 39: p. 1420-6.

[75] Derham, BK; Harding, JJ. Effects of modifications of alpha-crystallin on its chaperone and other properties. *Biochem J*, 2002. 364: p. 711-7.

[76] Koteiche, HA; McHaourab, HS. Mechanism of a hereditary cataract phenotype. Mutations in alphaA-crystallin activate substrate binding. *J Biol Chem*, 2006. 281: p. 14273-9.

[77] Hanson, SR;Hasan, A;Smith, DL; Smith, JB. The major in vivo modifications of the human water-insoluble lens crystallins are disulfide bonds, deamidation, methionine oxidation and backbone cleavage. *Exp Eye Res*, 2000. 71: p. 195-207.

[78] Han, J; Schey, KL. MALDI Tissue Imaging of Ocular Lens {alpha}-Crystallin. *Invest Ophthalmol Vis Sci*, 2006. 47: p. 2990-6.

[79] Ueda, Y;Duncan, MK; David, LL. Lens proteomics: the accumulation of crystallin modifications in the mouse lens with age. *Invest Ophthalmol Vis Sci*, 2002. 43: p. 205-15.

[80] Ueda, Y;Fukiage, C;Shih, M;Shearer, TR; David, LL. Mass measurements of C-terminally truncated alpha-crystallins from two-dimensional gels identify Lp82 as a major endopeptidase in rat lens. *Mol Cell Proteomics*, 2002. 1: p. 357-65.

[81] Aquilina, JA;Benesch, JLP;Ding, LL;Yaron, O;Horwitz, J; Robinson, CV. Phosphorylation of {alpha}B-Crystallin Alters Chaperone Function through Loss of Dimeric Substructure. *J. Biol. Chem.*, 2004. 279: p. 28675-28680.

In: Small Stress Proteins and Human Diseases
Editors: Stéphanie Simon et al.
ISBN: 978-1-61470-636-6

Chapter 2.3

αB-CRYSTALLIN (HSPB5) AND MYOFIBRILLAR MYOPATHIES

Isidre Ferrer**, *Dolores Moreno and Anna Janué
Institut de Neuropatologia, Servei Anatomia Patològica, IDIBELL-Hospital Universitari de Bellvitge, Universitat de Barcelona, Hospitalet de Llobregat, CIBERNED, Spain

ABSTRACT

Myofibrillar myopathies (MFMs) are a group of muscular diseases characterized by the presence of foci of myofibril dissolution, accumulation of protein aggregates and ectopic expression of several proteins. Mutations in αB-crystallin gene (*CRYAB)* are causative of certain cases of MFM and cardiomyopathy. Mutant αB-crystallin (HSPB5) is prone to form abnormal protein aggregates in transgenic models and transfected cells and it is a principal component, together with desmin, of aberrant protein aggregates in the subgroup of MFMs named αB-crystallinopathies. Yet other cases of MFMs are linked to mutations in desmin (*DES*), myotilin (*MYOT, TTID*), ZASP (*LDB3)* and filamin C (*FLNC*). Expression levels of αB-crystallin are increased and the protein is accumulated in abnormal and in apparently less affected fibers in other subgroups of MFMs as well. αB-crystallin responses in MFMs occur in parallel with increased expression of other small heat shock proteins as HSP27, and high molecular weight HSPs as HSP70 and HSP72/73. These responses also take place in parallel with increased expression of endoplasmic reticulum stress markers. Importantly, αB-crystallin in MFMs is phosphorylated, at least, at Ser59 thus probably resulting in modified αB-crystallin binding capacities with other small HSP and with different chaperone targets, mainly desmin and other components of the Z-disk. Finally, αB-crystallin aggregates co-localize partially with desmin and myotilin in desminopathies and myotilinopathies. Therefore,

* Corresponding Author: Isidre Ferrer
Institut de Neuropatologia,
Servei Anatomia Patològica,
IDIBELL-Hospital Universitari de Bellvitge,
Universitat de Barcelona, Hospitalet de Llobregat,
CIBERNED, Spain
8082ifa@gmail.com

αB-crystallin is largely phosphorylated and deficiently degraded in muscular fibers thus accumulating in MFMs. Based on experimental evidence in other paradigms lost of αB-crystallin chaperone function can be expected in MFMs thus compromising muscular cytoskeletal stabilization.

INTRODUCTION

αB-crystallin is a member of the small heat shock protein (sHSP) family that has an N-hydrophobic domain, a conserved 80- to 100-amino acid sequence, called the crystallin core domain and found in all members of sHSP, at their C-terminal domain, and a C-terminal extension [22, 28, 57, 80, 100, 116, 132]. Ten sHPS are known [40, 66, 80, 81]: HSPB1 or HSP27, HSP2 or MKBP, HSPB3, HSPB4 or αA-crystallin, HSPB5 or αB-crystalin, HSPB6 or HSP20, HSPB7 or cvHSP, HSPB8 or HSP22, HSPB9 and HSPB10 or ODFP. HSPB2, HSPB3 and HSPB7 are mainly expressed in skeletal muscle and heart; HSPB1, HSPB5, HSPB6 and HSPB8 are encountered in skeletal muscle, heart and other organs, whereas HSPB9 and HSPB10 are restricted to germ cells in the testes and sperm cells, respectively [32, 40, 66, 72, 104]. αB-crystallin is encoded by the gene *CRYAB*; it encompasses three exons and codes for 175 amino acids [14, 116]. αB-crystallin has a molecular weight of 22 kDa, but may form large homodimers of about 200-800 kDa with variable quaternary structure, forming spheres measuring between 8 and 18 nm and showing a central hole [59]. Small HSPs act as chaperones by interacting with misfolded proteins in response to several stressful stimuli, thereby preventing protein aggregation [60, 82, 86, 103, 137]. αB-crystallin and other small HSPs protect actin, desmin and other filament assemblies from stressful stimuli [45, 96, 115]. Together with high-molecular-mass HSPs like HSP60, HSP70 and HSP72 [88, 91], sHSPs constitute the principal chaperone response in the face of physiological and pathological stress conditions in different tissues, including heart and skeletal muscle.

αB-crystallin is crucial for the development of the skeletal muscle. Skeletal muscles degenerate in knockout αB-crystallin mice [13]. Ultrastructural examination of moderately degenerated muscles in knockout αB-crystallin mice discloses accumulation of amorphous, flocculent, electron-opaque material intermingled with partially damaged myofibrils showing variable streaming and disorganization of the Z disc [13]. Interestingly, accumulation of flocculent material in KO mutants resembles the material accumulated in human desminopathies and crystallinopathies. Furthermore, double knockout of αB-crystallin and MKBP increases necrosis and apoptosis induced by ischemia/reperfusion [76].

Small HSPs play non-redundant roles under physiological and pathological conditions [39, 113, 144]. Yet small HSPs may form heterodimers and hetero-oligomers. αB-crystallin interacts with HSP20 [78, 114], HSP27 [44, 87, 146] and HSP22 [41]. HSP22 is also found in fractions containing high molecular complexes composed of αB-crystallin and HSP20 [41]. αB-crystallin oligomer formation is inhibited by HSP22, indicating that blockade of oligomer formation by small HSPs may have therapeutical implications [123]. Groups of dimers form different complexes. One of them is formed by HSP27, HSP20 and αB-crystallin while the other is composed of MKBP, cvHSP and HSPB3; HSP22 may interact with both groups [41, 133]. MKBP and HSPB3 complex participates in myogenic differentiation [133]. αB-crystallin is also induced during skeletal myogenesis and plays a role in preventing caspase-3-

mediated apoptosis during differentiation [67, 133]. αB-crystallin immunoreactivity is also observed in small cells consistent with myoblasts in regenerating muscles, lending support to the idea that αB-crystallin also participates in skeletal muscle responses reminiscent of those encountered during normal development.

Small HSPs interact with different proteins. αB-crystallin interacts with FGF-2, NGF-β, VEGF, insulin and β-catenin, suggesting that αB-crystallin prevents abnormal folding of these regulatory proteins under stress conditions [48]. Better known is the interaction of small HSPs with filaments, enabling the stabilization of cytoskeletal filament networks. In skeletal and cardiac muscle, αB-crystallin binds to desmin, actin, tubulin and titin [8, 11, 16, 45, 46, 47, 54, 55, 61, 62, 85, 96, 103, 110, 111, 117, 121]. αB-crystallin also maintains myosin enzymatic activity and prevents its aggregation under stress conditions [93, 131].

Interactions of small HSPs are principally regulated by phosphoylation which induces a change in their tertiary structure and modifies their capacity to form homo- and hetero-oligomers, and to interact with other proteins [2, 70, 84, 92]. HSP27 is phosphorylated at different sites by MAPK-activated protein kinase-2 (MK2), MK3, p38 and Δ isoform of protein kinase [145]. First studies indicated that αB-crystallin is phosphorylated at serines 19, 45 and 59 in response to various stressful stimuli [69, 71]. Phosphorylation at Ser45 is carried out by extracellular signal-regulated kinase-1/2 (MAPK/ERK 1/2) and phosphorylation at Ser59 by the p38 MAPK substrate MK2 [79]. More recently, the mechanisms and the functional implications of αB-crystallin phosphorylation have been examined in more detail. Oxidative stress and calpain inhibition induce αB-crystallin phosphorylation, as revealed with antibodies that recognize phosphorylation sites Ser 19, 45 and 59, via activation of MAPK/ERK 1/2 and p38 [1]. αB-crystallin is phosphorylated by p38 MAKP, but not by MAPK/ERK 1/2, during myocardial infarction [129] and in ischemically preconditioned rabbit cardiomyocytes [3], as a general cardiac stress response [63]. Phosphorylation at Ser 59 appears to have a protective role associated with a loss of chaperone activity [2, 70, 144]. αB-crystalin and also HSP27 are phosphorylated by active p38 in estrogen-activated cardioprotection following trauma-haemorrhage [65]. Moreover, overexpression of αB-crystallin with Ser to Glu substitution on Ser 59, which mimics phosphorylation at Ser59, protects cardiomyocytes from apoptosis following ischaemia-reperfusion [94]. The mechanisms of cardioprotection are not clearly understood but it seems that phosphorylated αB-crystallin is translocated after ischaemia [54, 55, 56, 144] and after cardioplegia and cardiopulmonary bypass [24]. In addition to the myofilament redistribution, it has been suggested that αB-crystallin phosphorylation at Ser59 translocates to mitochondria, in so doing providing myocardial mitochondria protection during ischaemia-reperfusion [75].

1. Myofibrillar Myopathies

Myofibrillar myopathies (MFMs) are a group of muscular diseases characterized by the presence of foci of myofibril dissolution, accumulation of protein aggregates and ectopic expression of several proteins [27, 98, 99]. MFMs are also known as desmin-related myopathies because desmin is one of the main components of abnormal protein aggregates [49, 50, 51, 107, 124, 128].

Genetic studies have demonstrated the causative role of distinct mutations in genes encoding components of the Z-disc and proteins engaged in the maintenance of sarcomere assembly. These include desmin (*DES*) [25, 26, 51, 52, 53, 97, 105], αB-crystallin (*CRYAB*) [126, 136], myotilin (*MYOT, TTID*) [106, 126], ZASP (*LDB3)* [127] and filamin C (*FLNC*) [139]. In addition, mutations in selenoprotein N (*SEPN1*) are the cause of desmin-related myopathy with Mallory body-like inclusions [38]. Finally, inclusion body myopathy associated with Paget disease of the bone and frontotemporal dementia (IBMPFD), which is caused by mutations in the valosin-containing protein gene (*VCP*), is also considered as a form of MFM on the basis of morphological, molecular and ultrastructural characteristics of damaged fibers [64, 142, 143]. MFMs are closely related to sporadic inclusion body myositis (sIBM), a sporadic muscular disease with abnormal protein aggregates, autophagic vacuoles and occasional inflammatory infiltrates [4, 7, 18, 77].

MFMs caused by desmin mutations are manifested in young adults as distal muscle weakness with later involvement of the proximal muscles or as limb girdle myopathy. Cardiomyopathy is common and, frequently, accompanied by atrioventricular conduction abnormalities (see chapter 2-4). Myotilinopathies may be manifested as limb girdle muscular dystrophy or as distal myopathy in older individuals. Peripheral neuropathy has been reported to be common but cardiomyopathy occurs in a minority of cases. Patients with MFM caused by mutations in ZASP suffer from distal leg weakness slowly progressing to involve intrinsic hand muscles, and mild proximal weakness. Cardiopathy and peripheral neuropathy are additional features in some patients. In addition, mutations in exon 4 and exon 6 of ZASP gene have been identified in patients suffering from dilated cardiomyopathy. Muscular involvement in filaminopathies follows a pattern of limb girdle muscular dystrophy. About one-third of patients suffer from cardiac abnormalities comprising conduction blocks, tachycardia, diastolic dysfunction and left ventricular hypertrophy. Respiratory muscle weakness and respiratory insufficiency are common complications in the clinical progression of MFMs [51, 53, 105, 106, 107, 109, 126].

Electron microscopic studies have shown that abnormalities begin at or close to the Z disc and are characterized by disintegration of myofibrils and streaming of the Z line, followed by focal accumulations of degraded myofibrillar components, increased numbers of mitochondria at the periphery of aberrant protein aggregates and the presence of autophagic vacuoles, occasionally trapping abnormal mitochondria. More refined immunohistochemical and ultrastructural studies have revealed particular features in desminopathies and myotilinopathies. Flocculent, amorphous, and granulofilamentous electrondense material is common in desminopathies whereas tubulofilamentous accumulations predominate in myotilinopathies; these accumulations are accompanied by the typical morphology of rubbed out fibres in desminopathies [20, 21, 37].

2. Abnormal Accumulation of Multiple Proteins in MFM

In addition to desmin in desminopathies and myotilin in myotilinopathies, a wide variety of proteins have been found to be in excess or aberrantly expressed in muscle cells in MFMs. These include cytoskeletal and myofibrillar proteins, intermediate filaments, nuclear proteins, proteins connected with the ubiquitin-proteasome system, kinases, proteins associated with oxidative and nitrosative stress, Alzheimer disease-linked proteins, neuronal proteins, and others (see [37] for review). The mechanisms gearing to the expression of ectopic proteins, such as certain neuronal proteins, are probably interrelated with the abnormal regulation of certain transcription factors such as neuron restrictive silencer factor (REST), which is a neuronal gene repressor in non-neural cells. Down-regulation of REST in MFMs is accompanied by aberrant expression of the neuronal proteins ubiquitin C-terminal hydrolase 1 (UCHL1), SNAP25, synaptophysin and α-internexin [10]. Other transcription factors are abnormally expressed in MFMs; one of these is TDP-43 [108], which is involved in the control of gene transcription, transcription of selected splicing processes, and maintenance of mRNA stability [17]. TDP-43 is associated with the Microprocessor complex Drosha/DGR8 [58], a major complex in the processing of microRNAs. TDP-43 in MFM is phosphorylated and translocated from the nucleus to the cytoplasm, thus impeding its normal function [108].

The reasons for aberrant protein accumulation in MFM are not clearly understood, although two factors are probably involved. First, accumulated proteins are abnormal. This includes a. primarily abnormal proteins such as mutant proteins resulting from causative mutations in the different MFMs (e.g., mutant desmin in animal models and ransfected cell lines); and b. secondarily altered proteins modified by oxidative stress [37]. One of these proteins is desmin, which is oxidized in desminopathies and myotilinopathies [73], but many other proteins may also be targets of oxidative damage considering the extension of oxidative stress damage and oxidative stress responses in MFMs [74]). The amount of mutant protein also represents a threshold of abnormal protein accumulation [23]. Furthermore, several components of the aggresome and the ubiquitin-proteasome system (UPS) are altered in MFM [35, 36, 37, 109]. Impaired UPS has also been demonstrated in sIBM [5, 6, 42]. Interestingly, mutant ubiquitin, resulting from RNA misreading and leading to abnormal aberrant UPS function, occurs in both sIBM and MFM [43, 109].

3. MUTANT αB-CRYSTALLIN AS A CAUSE OF MYOFIBRILLAR MYOPATHY

The first reported family with αB-crystallinopathy presented with proximal and distal myopathy, involvement of the velo-pharyngeal muscles, respiratory failure, cardiopathy, and cataracts. The age of onset was in the thirties. The disease was transmitted with an autosomal dominant inheritance [33, 34, 136]. This first discovered mutation was the missense mutation R120G in *CRYAB*, (^{R120G}αBC) also named CryabR120G [136]. Mutant αB-crystallin R120G has altered structure [12, 83, 112, 135] and shows impaired chaperone functions [12]. Two further cases of *CRYAB* mutations associated with MFM were later identified. One of them presented with cervical and limb girdle weakness and respiratory failure. The second patient had proximal and distal limb weakness. Both patients had peripheral neuropathy. Cardiac involvement and lens opacities were absent [125]. The harboured new mutations were the nonsense mutation Q151X (^{Q151X}αBC) and the frameshift mutation 464delCT (^{464}αBC) [125]. Another patient bearing the R151H mutation in *CRYAB* suffered from dilated cardiomyopathy without apparent skeletal muscle involvement [68].

Optical and electron microscopical findings in muscle biopsies in αB-crystallinopathy are indistinguishable from desminopathy, including the presence of desmin and αB-crystallin-immunoreactive inclusion, and granulofilamentous material [125, 136].

Early studies showed that muscle cells transfected with mutant αB-crystallin exhibited abnormal aggregates containing αB-crystallin and desmin [136]. Two independent transgenic mouse models were later produced [118, 141]. Cardiac-specific transgenic CryabR120G expression in the mouse is characterized by abnormal αB-crystallin and desmin aggregates in cardiomyocytes that cause heart failure by the age of 5-7 months [141]. Interestingly, protein aggregates in aggresomes contain large amounts of toxic oligomers [122, 140]. To further understand the mechanisms of cell damage, convergent approaches have been performed *in vitro* and *in vivo* [89]. Transfection of adult cardiomyocytes with CryabR120G-expressing adenovirus results in altered contractile mechanics, whereas CryabR120G expression *in vivo* produces alterations in mitochondria, reduction of oxygen consumption, impaired complex I function, alterations in the permeability transition pore, and altered inner mitochondrial membrane potential [89]. This is followed by impaired UPS function, and activation of apoptotic pathways and cardiomyocyte death [19, 89]. Therefore, at least two different mechanisms appear to be at work in cells bearing the CryabR120G mutation. On the one hand, impaired binding of mutant αB-crystallin to proteins of the cytoskeleton and of the Z disc induces protein aggregation. On the other, toxic oligomers impair mitochondrial function and trigger an apoptotic cascade leading to cardiomyocyte death [89]. Whether these two mechanisms are connected is a matter of speculation. In this line, it may be proposed that altered mitochondrial distribution resulting from a disorganized desmin network may impact on mitochondrial vulnerability and function. Yet this is probably not specific to αB-crystallinopathy or even desminopathies, as protein aggregates in aggresomes are usually accompanied by increased numbers of mitochondria and occasional mitochondrial abnormalities at the periphery of these aggregates. Mitochondrial dysfunction is not uncommon in myofibrillar myopathy [120].

In fact, the three mutations R120G, Q151X and 464delCT in *CRYAB* expressed in transfected cells have the capacity to induce protein aggregation, which is accompanied by an increase in the levels of αB-crystalin phosphorylation at serine residues 19, 45 and 59 [130]. Furthermore, interactions of mutant αB-crystallins with small heat shock proteins HSP20, HSP22, and possibly HSP27, are altered [130].

4. αB-Crystallin is Also Abnormally Expressed in MFM Not Associated with Mutations in CRYAB, and in Other Myopathies with Protein Aggregates

Increased muscular activity is accompanied by increased expression levels of αB-crystallin [30, 101, 102]. αB-crystallin expression levels increase with age as well [31]. Aging skeletal muscle shows enhanced αB-crystallin immunoreactivity co-localizing with the surface marker laminin [31]. Moreover, aging is associated with increased small HSP phosphorylation, which is accompanied by increased activity of corresponding kinases and increased ubiquitination of proteins that accumulate in the insoluble fractions of aged muscles [145]. The reasons for increased αB-crystallin expression in muscular activity and in aging may be linked with muscular protection. In this line, αB-crystallin overexpression protects cardiomyocytes against ischemic damage [90, 95, 119]. Therefore, it can be speculated that aging and muscular exercise are conditions in which increased αB-crystallin may play a protective role. However, it is important to stress that increased expression levels in the two situations are not accompanied by abnormal protein aggregates.

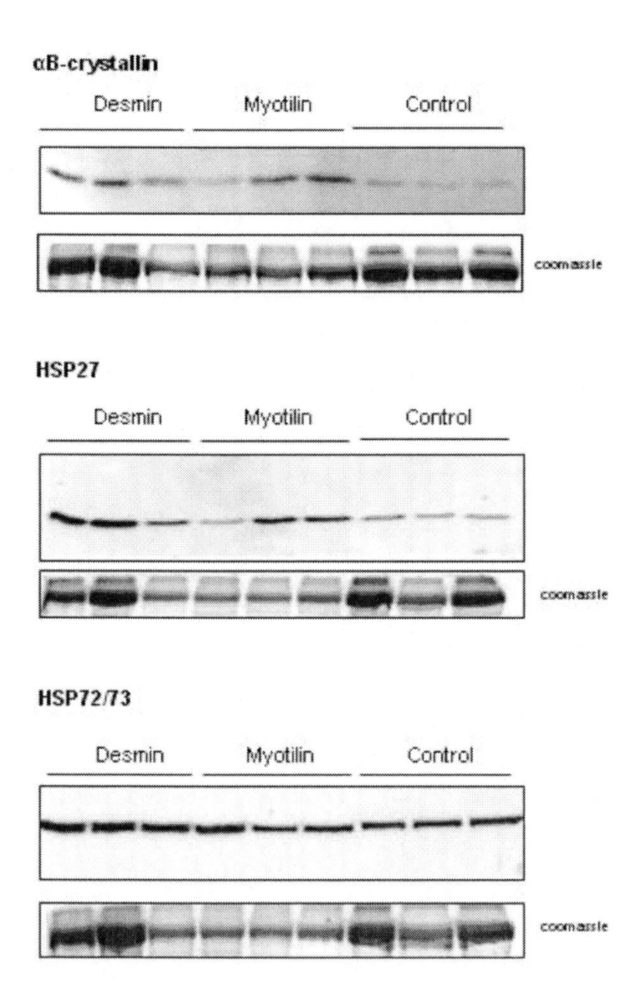

Figure 1. Gel electrophoresis and western blotting to αB-crystallin, HSP27 and HSP72/73 in muscle homogenates of desminopathies, myotilinopathies and controls. Significant increased expression levels are seen in diseased cases when compared with controls ($p < 0.05$). The densitometric quantification of western blot bands was carried out with Total Lab v2.01 software, and the data obtained were analyzed using Statgraphics Plus v5.1 software. Differences between control and diseased samples were analyzed with the Student T-test.

In striated muscle, increased and abnormal αB-crystallin immunoreactivity was first reported in IBM [9], and, later, in central and minicore lesions in congenital myopathies, as well as in target fibers of denervated muscles [39]. Immuno-electronmicroscopic studies have shown abnormal αB-crystallin associated with desmin in core myopathies and neurogenic atrophy [39]. αB-crystallin in IBM is mainly associated with abnormal protein aggregates although increased αB-crystallin also occurs in apparently normal fibers [9].

αB-crystallin accumulates in MFM (see ref [37] for review). αB-crystallin is equally present in the group of desminopathies/αB-crystallinopathies, and in the group of myotilinopathies/ZASPopathies. Further details have been obtained from present new studies of biopsies of myotilinopathies, desminopathies and sIBM, the general characteristics of which have been reported elsewhere [105, 106, 107]. Tissue samples were processed for gel electrophoresis and western blotting, and for immunohistochemistry, double-labelling immunofluorescence and confocal microscopy. Western blots show significantly increased αB-crystallin levels in total homogenates of striated muscle in desminopathies and myotilinopaties, indicating elevated expression levels of the protein (Figure 1). In addition, immunohistochemical studies have demonstrated abnormal αB-crystallin distribution and localization in affected muscles (Figure 2). αB-crystallin accumulates in protein inclusions, as well as in the sarcolemma and in rimmed vacuoles. Many atrophic fibers also are strongly stained wth anti-αB-crystallin antibodies.

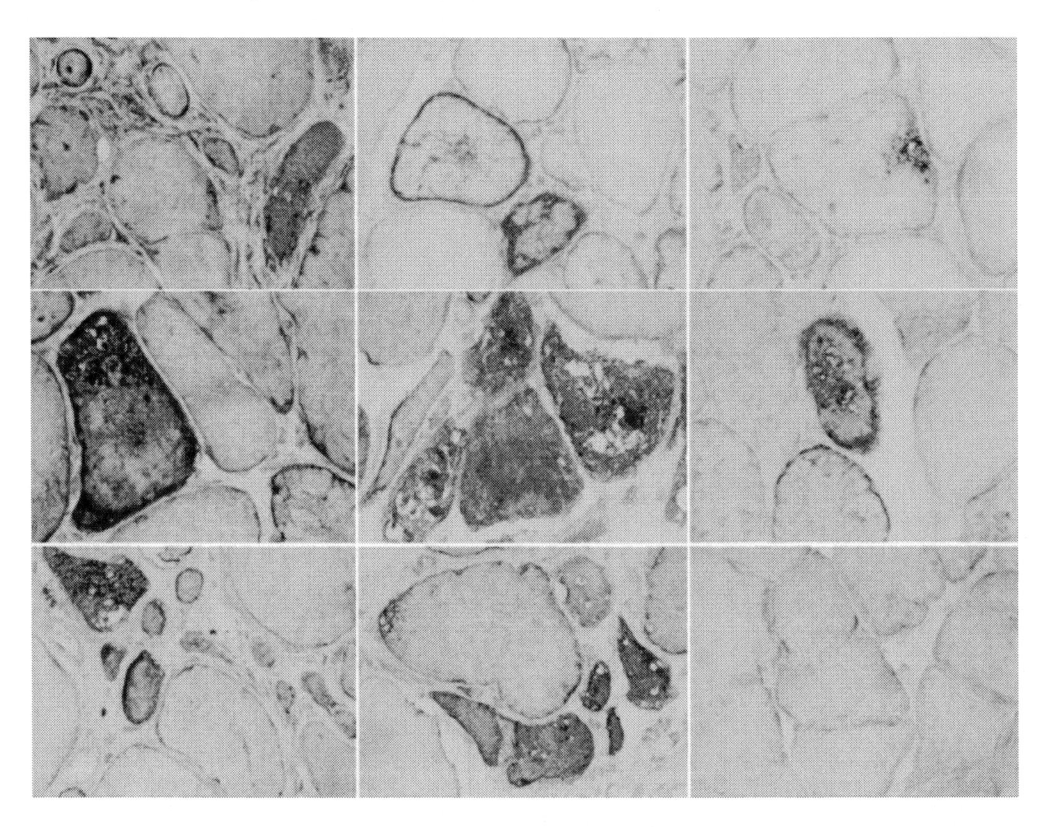

Figure 2. αB-crystallin immunohistochemistry in desminopathy (A-C), myotilinopathy (D-F) and sIBM (G-H), shown here for comparative purposes. Accumulation of αB-crystrallin occurs in fibers with abnormal protein aggregates and in fibres with various types of inclusions including those related with rimmed vacuoles (C, E, F). Normal αB-crystallin immunoreactivity is seen in a control fiber (I).

Double-labelling immunofluorescence and confocal microscopy have disclosed partial co-localization of αB-crystallin, desmin and myotilin in MFM (Figure 3 and 4). This is not a rare situation, as for example ubiquitin does not co-localize completely with desmin and myotilin in aberrant protein aggregates, and TDP-43 is not always ubiquitinated in protein inclusions [37, 108].

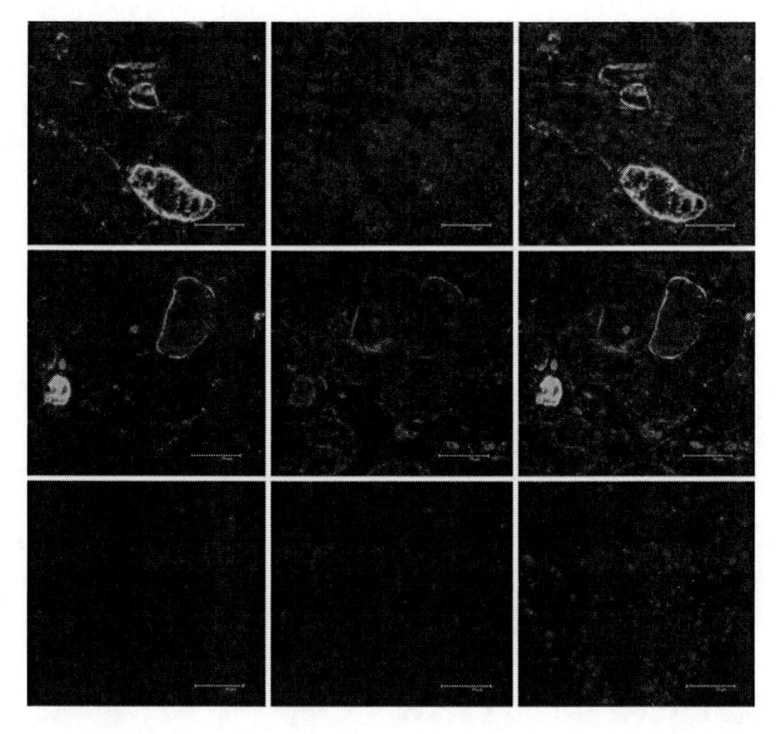

Figure 3. αB-crystallin (A, D; green) and desmin (B, red) or myotilin (E, red) in desminopathy showing weak co-localization of αB-crystallin with desmin and myotilin (C, F, merge). G-I: control sections incubated without the primary antibodies to test the specificity of the immunoreaction.

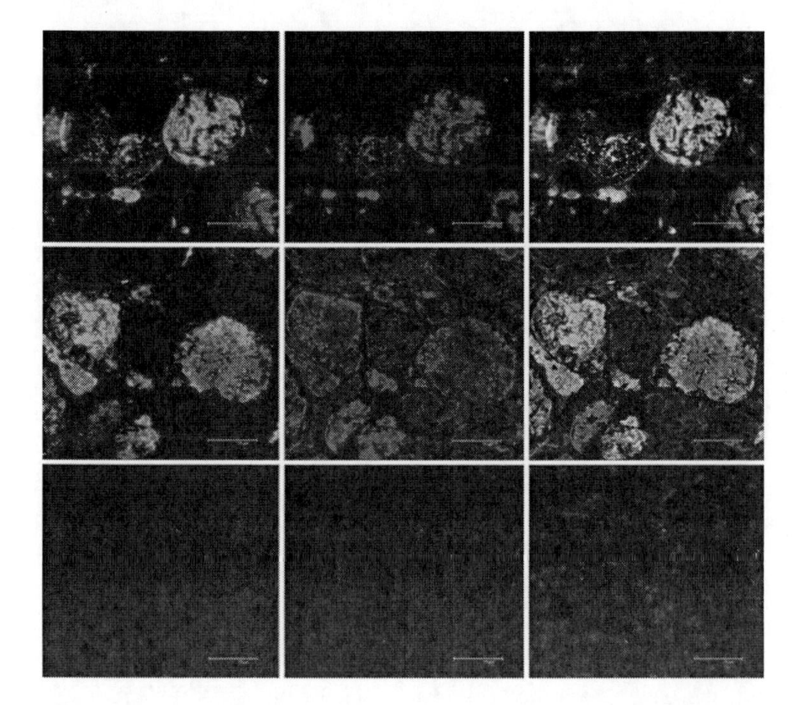

Figure 4. αB-crystallin (A, D; green) and desmin (B, red) or myotilin (E, red) in myotilinopathy showing co-localization of αB-crystallin with desmin and myotilin (C, F, merge). G-I, sections incubated without the primary antibodies to test the specificity of the immunoreaction.

Accumulation of mutant αB-crystallin is associated with αB-crystallin phosphorylation in the three known αB-crystallin mutations [130]. αB-crystallin is also phosphorylated in desminopathies and myotilinopathies (Figure 5). Since phosphorylation of crystallin is associated with abnormal dimer formation with other small HSPs in several paradigms, it is feasible that impaired small HSP binding with αB-crystallin also takes place in MFMs linked to *DES* and *MYOT* mutations.

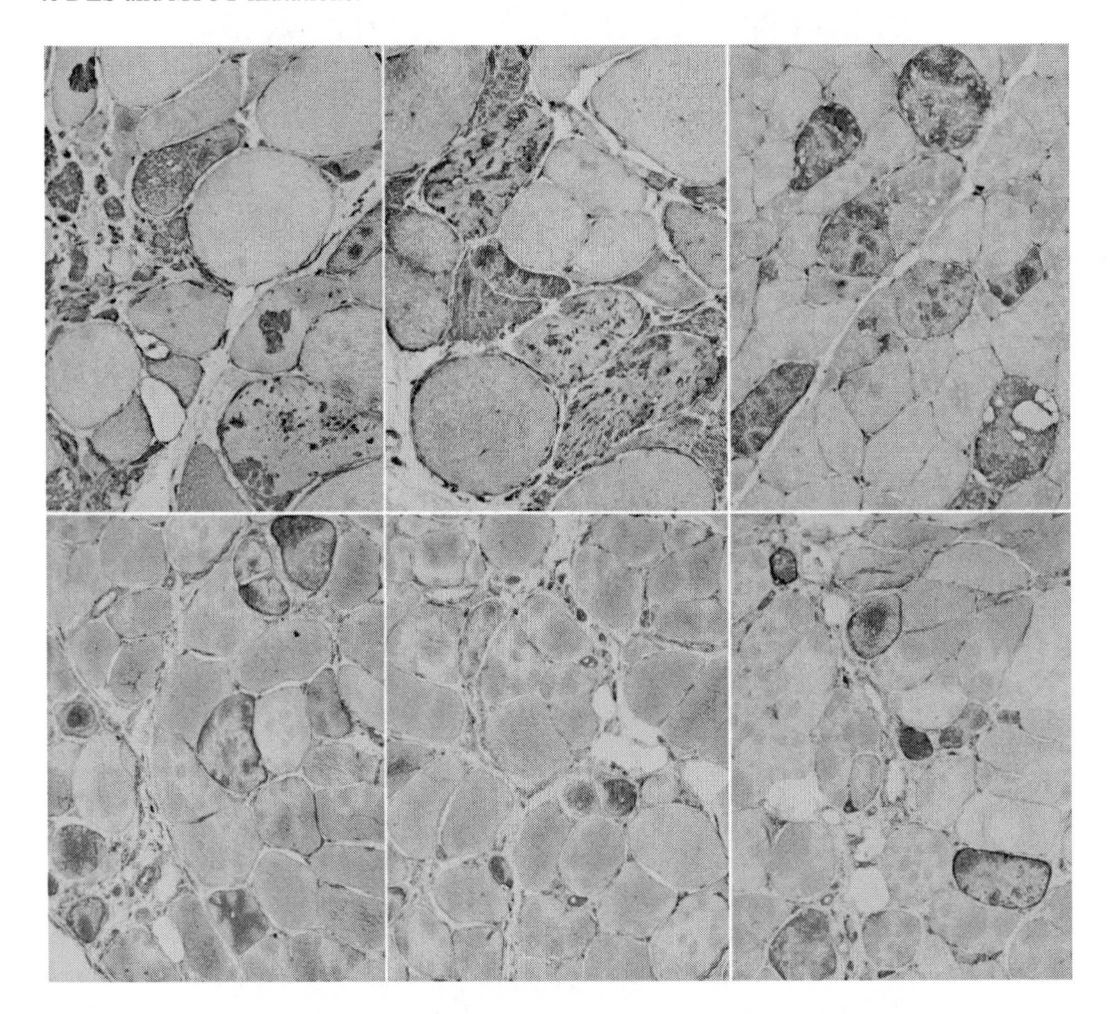

Figure 5. Phosphorylated αB-crystallin Ser59 in myotilinopathies (A-C) and desminopathies (D-F). Note the abundance of rubbed out fibers in desminopathy.

Other small HSPs also accumulate in altered fibers together with HSPs of high molecular weight such as HSP70 and HSP72 in desminopathies and myotilinopathies, as revealed by gel electrophoresis and western blotting, (Figure 1) and immunohistochemistry. Serial sections indicate that small HSPs and large HSPs do not always co-localize in the same lesions. Moreover, double-labelling immunofluorescence with confocal microscopy discloses co-localization of crystalline and HSP-27 only in a percentage of damaged muscle fibers which is variable from one case to another (Figure 6). These observations suggest that a complex and extensive HSP response occurs in myopathies with abnormal protein aggregates. Presumably, this response is geared to stop aberrant protein aggregation in the face of distinct stimuli including abnormal protein assembly resulting from post-translational modifications induced by oxidative stress. In this line, previous studies have shown that increased HSP70 expression is accompanied by enhanced protection of muscle fibres, including oxidative stress-induced cell damage [15, 88, 91]. As with αB-crystallin, other HSPs are also sequestered in aberrant protein aggregates, indicating abnormal HSP protein turnover.

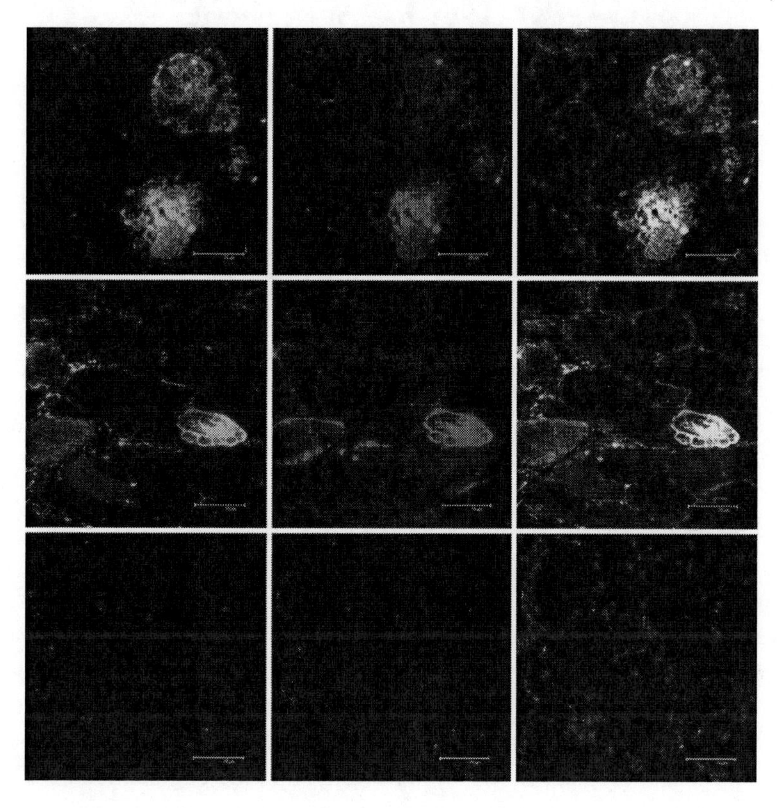

Figure 6. Double-labelling immunofluorescence and confocal microscopy showing partial co-localization of B-crystalin (A, D, green) and HSP27 (B, E, red) in myotilinopathy (A-C) and desminopathy (D-F). C, F: merge. D-F, sections stained only with the secondary antibodies to test the specificity of the immunoreaction.

5. Increased HSP Expression Parallels Endoplasmic Reticulum Stress Responses

Accumulation of misfolded proteins can trigger a reticulum stress response which is manifested by the activation of certain markers such as pancreatic endoplasmic reticulum kinase (PERK) and eukaryotic initiation factor 2 α (eIF2α). Although the total levels of eIF2α are not modified in MFMs, phosphorylated eIF2α is found in many damaged fibers with abnormal protein aggregates in desminopathies and myotilinopahies. Increased phosphorylated eIF2α immunoreactivity co-localizes with desmin and myotilin deposits, indicating that endoplasmic reticulum stress occurs in fibers with abnormal protein aggregates (Figure 7).

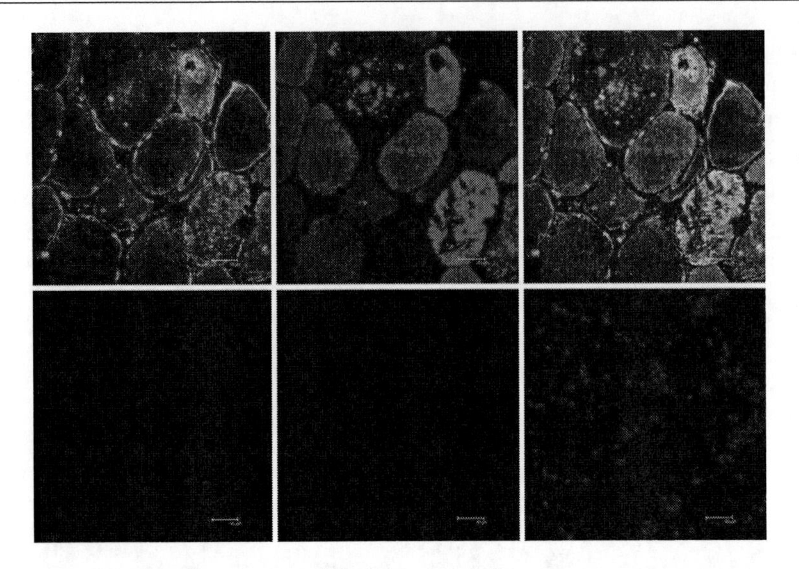

Figure 7. Double-labelling immunofluorescence and confocal microscopy showing partial co-localization of phosphorylated eiF2α (A, green) and myotilin (B, red) in myotilinopathy (C, merge). D-F, sections stained only with the secondary antibodies to test the specificity of the immunoreaction.

6. αB-Crystallin Degradation in MFM

Proteasome inhibition induces small HSP with chaperone function, including αB-crystallin [29, 138]. Since abnormal UPS function has been observed in MFM [36, 37] one of the mechanisms dealing with increased αB-crystallin expression in MFMs can be related with impaired UPS activity. However, αB-crystallin is not only increased but is also incorporated into abnormal protein aggregates in MFM, indicating impaired degradation of αB-crystallin.

Recent studies in CryabR120G transgenic mice have shown indirect evidence of increased autophagic activity in cardiomyocytes and direct evidence of abnormal autophagy in CryabR120G-expressing cells [134]. Moreover, inactivation of beclin1, an activator of autophagia, dramatically increases intracellular aggregates in cardiomyocytes of CryabR120G transgenic mice [134]. Unfortunately, no determination of αB-crystallin expression was carried out in these transgenic models. Therefore, the role of autophagy in αB-crystallin degradation in physiological and pathological conditions is not known.

Conclusion

Mutations in αB-crystallin gene are causative of a small percentage of cases affected by MFM. Mutant αB-crystallin accumulates together with other proteins in aberrant protein aggregates in damaged fibres. Yet accumulation of αB-crystallin and phosphorylated αB-crystallin Ser59 also occurs in MFM not associated with mutations in αB-crystallin gene. Such accumulation partially co-localizes with other proteins in abnormal protein aggregates, and also with high molecular weight HSP27 and HSPs. Therefore, two main mechanisms are altered in relation to αB-crystallin in MFMs. First, there is an accumulation of αB-crystallin which is largely phosphorylated (and mutated in inherited bearing mutations in *CRYAB*). Second, abnormal αB-crystallin is sequestered in protein aggregates and is not properly degraded by the UPS. HSP responses, including increased expression of αB-crystallin, are associated with endoplasmic reticulum stress in damaged fibres.

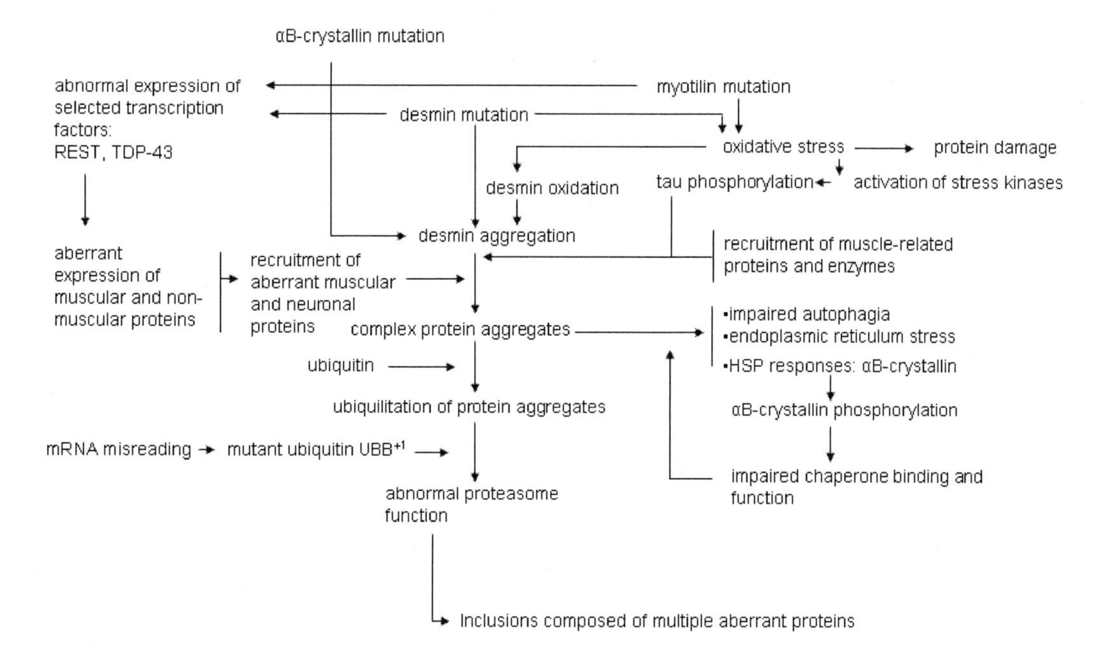

Figure 8. Summary of principal features linked to α B-crystallin in MFMs.

Acknowledgments

We wish to thank Dr. M. Olive for her collaboration in the selection of cases used in the present review, and T. Yohannan for editorial help.

References

[1] Aggeli, I-O., Beis, I. & Gaitanaki, C. (2008) Oxidative stress and calpain inhibition induce alpha B-crystallin phosphorylation via p38-MAPK and calcium signalling pathways in H9c2 cells. *Cell Sign, 20,* 1292-1302.

[2] Aquilina, J.A., Benesch, L.L., Ding, O., Yaron, J., Horwitz, C.V. & Robinson, J. (2004) Phosphorylation of alphaB-crystallin alters chaperone function through loss of dimeric substructure. *J Biol Chem, 279,* 28675-28680.

[3] Armstrong, S.C., Shivell, C. & Ganote, C.E. (2000) Differential translocation or phosphorylation of alpha B crystallin cannot be detected in ischaemically preconditioned rabbit cardiomyocytes. *J Mol Cell Cardiol, 32,* 1301-1314.

[4] Askanas, V. & Engel, W.K. (2001) Inclusion-body myositis: newest concepts of pathogenesis and relation to aging and Alzheimer's disease. *J Neuropathol Exp Neurol, 60,* 1-14.

[5] Askanas, V. & Engel, W.K. (2005) Sporadic inclusion-body myositis: a proposed key pathogenetic role of the abnormalities of the ubiquitin-proteasome system, and protein misfolding and aggregation. *Acta Myol, 24,* 17-24.

[6] Askanas, V. & Engel, W.K. (2006) Inclusion-body myositis: a myodegenerative conformational disorder associated with Abeta, protein misfolding, and proteasome inhibition. *Neurology, 24,* S39-48.

[7] Askanas, V., Engel, W.K. & Mirabella, M. (1994) Idiopathic inflammatory myopathies: inclusion-body myositis, polymyositis and dermatomyositis. *Curr Opin Neurol, 7,* 448-456.

[8] Atomi, Y., Yamada, S., Strohman, R. & Nonomura, Y. (1991) AlphaB-crystallin in skeletal muscle: purification and localization. *J Biochem (Tokyo), 110,* 812-822.

[9] Banwell, B.L. & Engel, A.G. (2000) αB-crystallin immunolocalization yields new insights into inclusion body myositis. *Neurology, 54,* 1033-1041.

[10] Barrachina, M., Moreno, J., Juvés, S., Moreno, D., Olivé, M. & Ferrer, I. (2007) Target genes of neuron-restrictive silencer factor are abnormally up-regulated in human myotilinopathy. *Am J Pathol, 171,* 1312-1323.

[11] Bennardini, F., Wrzosek, A. & Chiesi, M. (1992) αB-crystallin in cardiac tissue. Association with actin and desmin filaments. *Circ Res, 71,* 288-294.

[12] Bova, M.P., Yaron, O., Huang, Q., Ding, L., Haley, D.A., Sttewart, P.L., & Horwitz, J (1999) Mutation R120G in αB-crystallin, which is linked to a desmin-related myopathy, results in an irregular structure and defective chaperone-like function. *Proc Natl Acad Sci USA, 96,* 6137-6142.

[13] Brady, J.P., Garland, D.L., Green, D.E., Tamm, E.R., Giblin, F.J. & Wawrousek, E.F. (2001) αB-crystallin in lens development and muscle integrity: a gene knockout approach. *Invest Ophtalmol Vis Sci, 42,* 2924-2934.

[14] Brakenhoff, R.H., Guerts van Kessel, A.H., Oldenburg, M., Wijnen, J.T., Bloemendal, H., Meera Khan, P. & Schoenmakers, J.G. (1990) Human αB-crystallin (CRYA2) gene mapped to chromosome 11q12-q23. *Hum Genet, 85,* 237-240.

[15] Broome, C.S., Kayani, A.C., Palomero, J., Dillmann, W.H., Mestril, R., Jackson, M.J. & McArdle, A. (2006) Effect of lifelong overexpression of HSP70 in skeletal muscle on age-related oxidative stress and adaptation after nondamaging contractile activity. *FASEB J, 20,* 1549-1551.

[16] Bullard, B., Ferguson,C., Minajeva, A., Leake, M.C., Gautel, M., Labeit, D., Ding, L., Labeit, S., Horwitz, J., Leonard, K.R. & Linke, W.A. (2004) Association of the chaperone αB-crystallin with titin in heart muscle. *J Biol Chem, 279,* 7917-7924.

[17] Buratti, E. & Baralle, F.E. (2001) Characterization and functional implications of the RNA binding properties of nuclear factor TDP-43, a novel splicing regulator of CFTR exon 9. *J Biol Chem, 276,* 36337-36343.

[18] Carpenter, S. (1996) Inclusion body myositios, a review. *J Neuropathol Exp Neurol, 55,* 1105-1114.

[19] Chen, Q., Liu, J.B., Horak, K.M., Zheng, H., Kumarapeli, A.R., Li, J., Li, F., Gerdes, A.M., Wawrousek, E.F. & Wang, X. (2005) Intrasarcoplasmic amyloidosis impairs

proteolytic function of proteasomes in cardiomyocytes by compromising substrate uptake. *Circ Res, 97,* 1018-1026.

[20] Claeys, K.G., van der Ven, P.F., Behin, A., Stojkovic, T., Eymard, B., Dubourg, O., Laforêt, P., Faulkner, G., Richard, P., Vicart, P., Romero, N.B., Stoltenburg, G., Udd, B., Fardeau, M., Voit, T. & Fürst, D.O. (2009) Differential involvement of sarcomeric proteins in myofibrillar myopathies: a morphological and immunohistochemical study. *Acta Neuropathol, 117,* 293-307.

[21] Claeys, K.G., Fardeau, M., Schröder, R., Suominen, T., Tolksdorf, K., Behin, A., Dubourg, O., Eymard, B., Maisonobe, T., Stojkovic, T., Faulkner, G., Richard, P., Vicart, P., Udd, B., Voit, T. & Stoltenburg G. (2008) Electron microscopy in myofibrillar myopathies reveals clues to the mutated gene. *Neuromuscul Disord, 18,* 656-666.

[22] Clark, J.I. & Muchowski, P.J. (2000) Small heat-shock proteins and their potential role in human disease. *Curr Opin Struct Biol, 10,* 52-59.

[23] Clemen, C.S., Fischer, D., Reimann, J., Eichinger, L., Müller, C.R., Müller, H.D., Goebel, H.H. & Schröder, R. (2009) How much mutant protein is needed to cause a protein aggregate myopathy in vivo? Lessons from an exceptional desminopathy. *Hum Mutat, 30,* E490-E499.

[24] Clements, R.T., Sodha, N.R., Feng, J., Mieno, S., Boodhani M., Ramlawi, B., Bianchi, C. & Sellke, F.W. (2007) Phosphorylation and translocation of heat shock protein 27 and αB-crystallin in human myocardium after cardioplegia and cardiopulmonary bypass. *J Thor Cardiovasc Surg, 134,* 1461-1470.

[25] Dagvadorj, A., Olivé, M., Urtizberea, J.A., Halle, M., Shatunov, A., Bönnemann, C., Park, K.Y., Goebel, H.H., Ferrer, I., Vicart, P., Dalakas, M.C. & Goldfarb, L.G. (2004) A series of West European patients with severe cardiac and skeletal myopathy associated with a de novo R406W mutation in desmin. *J Neurol 251,*143-149.

[26] Dalakas, M.C., Park, K.Y., Semino-Mora, C., Lee, H.S., Sivakumar, K., & Goldfarb, L.G. (2000) Desmin myopathy, a skeletal myopathy with cardiomyopathy caused by mutations in the desmin gene. *N Engl J Med 342,* 770-780.

[27] De Bleecker, J.L., Engel, A.G. & Ertl, B. (1996) Myofibrillar myopathy with foci of desmin positivity. II. Immunocytochemical analysis reveals accumulation of multiple other proteins. *J Neuropathol Exp Neurol, 55,* 563-577.

[28] De Jong, W.W., Leunissen, J.A., & Voorter, C.E. (1993) Evolution of the α-crystallin/small heat-shock protein family. *Mol Biol Evol, 10,* 103-126.

[29] Doll, D., Sarikas, A., Krajcik, R. & Zolk, O. (2007) Proteomic expression analysis of cardiomyocytes subjected to proteasome inhibition. *Biochem Biophys Res Commun, 353,* 436-442.

[30] Donoghue, P., Doran, P., Dowling, P. & Ohlendieck, K. (2005) Differential expression of the fast skeletal muscle proteome following chronic low-frequency stimulation. *Biochem. Biophys Acta, 1752,* 166-176.

[31] Doran, P., Gannon, J., O'Connell, K. & Ohlendieck, K. (2007) Aging in skeletal muscle shows a drastic increase in the small heat shock proteins αB-crystallin/HspB5 and cvHsp/HspB7. *Eur J Cell Biol, 86,* 629-640.

[32] Dubin, R.A., Wawrousek, E.F. & Piatigorsky, J. (1989) Expression of the αB-crystallin gene is not restricted to the lens. *Mol Cell Biol, 9,* 10883-10891.

[33] Fardeau, M., Godet-Guillain, J., Tome, F.M., Collin, H., Gaudeau, S., Boffety, C. & Vernant, P. (1978) Une nouvelle affection musculaire familiale, definie par l'acumulation intra-sarco-plasmique d'un materiel granulo-filamentaire dense en microscopie electronique. *Rev Neurol 134*, 411-425.

[34] Fardeau, M., Vicart, P., Caron, A., Chateau, D., Chevallay, M., Collin, H., Chapon, F., Duboc, D., Eymard, B., Tomé, F.M., Dupret, J.M., Paulin, D. & Guicheney P. (2000). Myopathie familiale avec surcharge en desmine, sous forme de matériel granulo-filamentaire dense en microscopie électronique, avec mutation dans le gène de l'αB-cristalline. *Rev Neurol, 156,* 497-504.

[35] Ferrer, I., Carmona, M., Blanco, R., Moreno, D., Torrejón-Escribano, B. & Olivé, M. (2005) Involvement of clusterin and the aggresome in abnormal protein deposits in myofibrillar myopathies and inclusion body myositis. *Brain Pathol, 15,* 101-108.

[36] Ferrer, I, Martín, B., Castaño, J.G., Lucas, J.J., Moreno, D. & Olivé, M. (2004) Proteasomal expression and activity, and induction of the immunoproteasome in myofibrillar myopathies and inclusion body myositis. *J Neuropathol Exp Neurol, 63,* 484-498.

[37] Ferrer, I. & Olivé, M. (2008) Molecular pathology of myofibrillar myopathies. *Expert Rev Mol Med 10, e25,* 1-20.

[38] Ferreiro, A., Ceuterick-de Groote, C., Marks, J.J., Goemans, N., Schreiber, G., Hanefeld, F., Fardeau, M., Martin, J.J., Goebel, H.H., Richard, P., Guicheney, P. & Bönnemann C.G. (2004) Desmin-related myopathy with Mallory body-like inclusions is caused by mutations of the selenoprotein N gene. *Ann Neurol, 55,* 676-686.

[39] Fischer, D., Matten, J., Reimann, J., Bönneman, C. & Schröder, R. (2002) Expression, localization and functional divergence of αB-crystallin and heat shock protein 27 in core myopathies and neurogenic atrophy. *Acta Neuropathol, 104,* 297-304.

[40] Fontaine, J.M., Rest, J.S., Welsh, M.J. & Benndorf, R. (2003) The sperm outer dense fiber protein is the 10th member of the superfamily of mammalian small stress proteins. *Cell Stress Chaperones, 8,* 4610-4621.

[41] Fontaine, J-M., Sun, X., Benndorf, R. & Welsh, M.J. (2005) Interactions of HSP22 (HSPB8) with HSP20, αB-crystallin, and HSP23. *Biochem Biophys Res Commun, 337,* 1006-1011.

[42] Fratta, P., Engel, W.K., McFerrin, J., Davies, K.J., Lin, S.W. & Askanas, V. (2005) Proteasome inhibition and aggresome formation in sporadic inclusion-body myositis and in amyloid-beta precursor protein-overexpressing cultured human muscle fibers. *Am J Pathol, 167,* 517-526.

[43] Fratta, P., Engel, W.K., Van Leeuwen, F.W., Hol, E.M., Vattemi, G. & Askanas, V. (2004) Mutant ubiquitin UBB+1 is accumulated in sporadic inclusion-body myositis muscle fibers. *Neurology, 63,* 1114-1117.

[44] Fu, L. & Liang, J.J. (2003) Enhanced stability of alpha B-crystallin in the presence of small heat shock protein Hsp27. *Biochem Biophys Res Commun, 302,* 710-714.

[45] Fujita, Y., Ohto, E., Katayama, E. & Atomi, Y. (2004) αB-crystallin-coated MAP microtubule resists nocodazole and calcium-induced disassembly. *J Cell Sci, 117,* 1719-1726.

[46] Ghosh, J.G., Houck, S.A. & Clark, J.I. (2007) Interactive sequences in the stress protein and molecular chaperone human alphaB crystallin recognize and modulate the assembly of filaments. *Int J Biochem Cell Biol, 39,* 1804-1815.

[47] Ghosh, J.G., Houck, S.A. & Clark, J.I. (2007) Interactive domains in the molecular chaperone human alphaB crystallin modulate microtubule assembly and disassembly. *PLoS ONE. 2,* e498.

[48] Ghosh, J.G., Shenoy, A.K. & Clark, J.I. (2007) Interactions between important regulatory proteins and human αB-crystallin. *Biochemistry, 46,* 6308-6317.

[49] Goebel, H.H. (1995) Desmin-related neuromuscular disorders. *Muscle Nerve 18,* 1306-1320.

[50] Goebel, H.H. & Goldfarb, L. (2002) Desmin-related myopathies. In Structural and molecular basis of skeletal muscle diseases (Karpati, K., ed), pp. 70-73, Basel, ISN Neuropath Press.

[51] Goldfarb, L.G., Vicart, P., Goebel, H.H. & Dalakas, M.C. (2004) Desmin myopathy. *Brain, 127,* 723-734.

[52] Goldfarb, L.G., Park, K.Y., Cervenáková, L., Gorokhova, S., Lee, H.S., Vasconcelos, O., Nagle, J.W., Semino-Mora, C., Sivakumar, K. & Dalakas, MC. (1998) Missense mutations in desmin associated with familial cardiac and skeletal myopathy. *Nat Genet, 19,* 402-403.

[53] Goldfarb, L.G. Olivé, M., Vicart, P. & Goebel, H.H. (2008) Intermediate filament diseases: desminopathy. *Adv Exp Med Biol, 642,* 131-164.

[54] Golenhofen, N., Htun, P., Ness, W., Koob, R., Schapeer, W. & Drenckhahn, D. (1999) Binding of the stress protein αB-crystallin to cardiac myofibrils correlates with the degree of myocardial damage during ischemia/reperfusion in vivo. *J Mol Cell Cardiol, 31,* 569-580.

[55] Golenhofen, N., Ness., Koob, R. & Drenckhahn, D. (1998) Ischaemia-induced phosphorylation and translocation of stress protein αB-crystalin to Z lines of myocardium. *Am J Physiol, 274,* 457-464.

[56] Golenhofen, N., Perng, M.D., Quinlan, R.A. & Drenckhahn, D. (2004) Comparison of the small heat shock proteins αB-crystallin, MKBP, HSP25, HSP20, and vCHSP in heart and skeletal muscle. *Histochem Cell Biol, 122,* 415-425.

[57] Graw, J. (1997) The crystallins: genes, proteins and diseases. *Biol Chem, 378,* 1331-1348.

[58] Gregory, R.I., Yan, K.P., Amuthan, G., Chendrimada, T., Doratotaj, B., Cooch, N. & Shiekhattar, R. (2004) The Microprocessor complex mediates the genesis of microRNAs. *Nature, 432,* 235-240.

[59] Haley, D.A., Horwitz, J. & Stewart, P.L. (1998) The small heat shock protein, αB-crystallin, has a variable quaternary structure. *J Mol Biol, 277,* 27-35.

[60] Hasbeck, M., Fanzmann, T., Weintfurtner, D. & Buchner, J. (2005) Some like it hot: the structure and function of small heat-shock proteins. *Nat Struct Mol Biol, 12,* 842-846.

[61] Head, M.W., Corbin, E. & Goldman, J.E. (1994) Coordinate and independent regulation of αB-crystallin and hsp27 expression in response to physiological stress. *J Cell Physiol, 159,* 41-50.

[62] Head, M.W., Hurwitz, L., Kegel, K. & Goldman, J.E. (2000) αB-crystallin regulates intermediate filament organization in situ. *NeuroReport, 11,* 361-365.

[63] Hoover, H.E., Thuereauf, D.J., Martindale, J.J. & Glembotski, C.C. (2000) αB-crystallin gene induction and phosphorylation by MKK6-activated p38. A potential role

for αB-crystallin as a target of of the p38 branch of the cardiac stress response. *J Biol Chem, 275,* 23825-23833.

[64] Hübbers, C.U., Clemen, C.S., Kesper, K., Böddrich, A., Hofmann, A., Kämäräinen, O., Tolksdorf, K., Stumpf, M., Reichelt, J., Roth, U., Krause, S., Watts, G., Kimonis, V., Wattjes, M.P., Reimann, J., Thal, D.R., Biermann, K., Evert, B.O., Lochmüller, H., Wanker, E.E., Schoser, B.G., Noegel, A.A. & Schröder, R. (2007) Pathological consequences of VCP mutations on human striated muscle. *Brain, 130,* 381-393.

[65] Hsu, J-T., Hsieh, Y-C., Kan W.H., Chen, J.G., Choudry, M.A., Schwacha, M.G., Bland, K.I. & Chaudry, I.H. (2007) Role of p38 mitogen-activated protein kinase pathway in estrogen-mediated cardioprotection following trauma-hemorrhage. *Am J Physiol Heart Circ Physiol, 292,* 2982-2987.

[66] Hu, Z., Yang, B., Lu, W., Zhou, W., Zeng, L., Li, T. & Wang, X. (2008) HSPB2/MKBP, a novel and unique member of the small heat-shock protein family. *J Neurosci Res, 86,* 2125-2133.

[67] Ikeda, R., Yoshida, K., Ushiyama, M., Yamaguchi., T, Iwashita, K., Futagawa, T., Shibayama, Y., Oiso, S., Takeda, Y., Kariyazono, H., Furukawa, T., Nakamura, K., Akiyama, S., Inoue, I. & Yamada, K. (2006) The small heat shock protein alphaB-crystallin inhibits differentiation-induced caspase 3 activation and myogenic differentiation. *Biol Pharm Bull, 29,* 1815-1819.

[68] Inagaki, N., Hayashi, T. & Arimura, T. (2006) αB-crystalin mutation in dilated cardiomyopathy. *Biochem Biophys Res Commun, 342,* 379-386.

[69] Ito, H., Kamei, K., Iwamto, I., Inaguma, Y., Garcia-Mata, R., Sztul, E. & Kato, K. (2002) Inhibition of proteasomes induces accumulation, phosphorylation, and recruitment of HSP27 and αB-crystalin in aggresomes. *J Biochem (Tokyo), 131,* 593-603.

[70] Ito, H., Kamei, K., Iwamoto, I., Inaguma, Y., Nohara, D. & Kato, K. (2001) Phosphorylation-induced change of the oligomerization state of αB-crystallin. *J Biol Chem, 276,* 5346-5352.

[71] Ito, H., Okamoto, K., Nakayama, H., Isobe, T. & Kato, K. (1997) Phosphorylation of αB-crystallin in response to various types of stress. *J Biol Chem, 272,* 29934-29941.

[72] Iwaki, T., Kume-Iwaki, A. & Goldman, J.E. (1990) Cellular distribution of αB-crystallin in non-lenticular tissues. *J Histochem Cytochem, 38,* 31-39.

[73] Janué, A., Odena, M.A., Oliveira, E., Olivé, M. & Ferrer, I. (2007) Desmin is oxidized and nitrated in affected muscles in myotilinopathies and desminopathies. *J Neuropathol Exp Neurol, 66,* 711-723.

[74] Janué, A., Olivé, M. & Ferrer, I. (2007) Oxidative stress in desminopathies and myotilinopathies: a link between oxidative damage and abnormal protein aggregation. *Brain Pathol, 17,* 377-388.

[75] Jin, J.K., Whittaker, R., Glassy, M.S., Barlow, S.B., Gottlieb, R.A. & Glembotski, C.C. (2008) Localization of phosphorylated αB-crystallin to heart mitochondria during ischaemia-reperfusion. *Am J Physiol Heart Circ Physiol, 294,* 337-344.

[76] Kadono, T., Zhang, X.Q., Srinivasan, S., Ishida, H., Barry, W.H. & Benjamin, I.J. (2006) CRYAB and HSPB2 deficiency increases myocyte mitochondrial permeability transition and mitochondrial calcium uptake. *J Mol Cell Cardiol, 40,* 783-789.

[77] Karpati, G. & Hohlfeld, R. (2000) Biologically stressed muscle fibres in sporadic IBM: a clue for the enigmatic etiology? *Neurology, 54,* 1020-1021.

[78] Kato, K., Goto, S., Inaguma, Y., Hasegawa, K., Morishita, R. & Asano, T. (1994) Purification and characterization of a 20-kDa protein that is highly homologous to alpha B-crystallin. *J Biol Chem, 269,* 15302-15309.

[79] Kato, K., Ito, H., Kamei, K., Inaguna, Y., Iwamoto, I. & Saga, S. (1998) Phosphorylation of αB-crystallin in mitotic cells and identification of enzymatic activities responsible for phosphorylation. *J Biol Chem, 273*, 28346-28354.

[80] Kappe, G., Franck, E., Verschuure, P., Boelens, W.C., Leunissen, J.A. & de Jong, W.W. (2003) The human genome encodes 10 alpha-crystallin-related small heat shock proteins: HspB1-10. *Cell Stress Chaperones, 8*, 53-61.

[81] Kappe, G., Verschuure, P., Philipsen, R.L., Staalduinen, A.A., van de Boogaart, P., Boelens, W.C. & de Jong, W.W. (2001) Characterization of two novel small heat shock proteins: protein kinas-related HSPB8 and testis-specific HSPB9. *Biochem Biophys Acta, 1520,* 1-6.

[82] Koh, T.J. & Escobedo, J. (2004) Cytoskeletal disruption and small heat shock protein translocation immediately after lengthening contractions. *Am J Physiol Cell Physiol, 286,* 713-722.

[83] Kumar, L.V., Ramakrishna, T. & Rao, C.M. (1999) Structural and functional consequences of the mutation of a conserved arginine residue in alphaA and alphaB crystallins. *J Biol Chem, 274*, 24137-24141.

[84] Lambert, H., Charettet, S.J., Bernier, A.F., Guimond, A. & Landry, J. (1999) HSP27 multimerization mediated by phosphorylation-sensitive intermolecular interactions at the amino terminus. *J Biol Chem, 274,* 9378-9385.

[85] Liang, P. & MacRae, T.H. (1997) Molecular chaperones and the cytoskeleton. *J Cell Sci, 110,* 1431-1440.

[86] Lindquist, S. & Craig, E.A. (1998) The heat-shock proteins. *Annu Rev Genet, 22,* 631-677.

[87] Liu, C & Welsh, M.J. (1999) Identification of a site of Hsp27 binding with Hsp27 and alpha B-crystallin as indicated by the yeast two-hybrid system. *Biochem Biophys Res Commun, 255,* 256-261.

[88] Liu, Y., Gampert, L., Nething, K., & Steinacker, J.M. (2006) Response and function of skeletal muscle heat shock protein 70. *Front Biosci, 11,* 2802-2827.

[89] Maloyan, A., Sanbe, A., Osinska, H., Westfall, M., Robinson, D., Imahashi, K., Murphy, E., & Robbins, J. (2005) Mitochondrial dysfunction and apoptosis underlie the pathogenic process in alpha B-crystallin desmin-related cardiomyopathy. *Circulation 112,* 3451-3461.

[90] Martin, J.L., Mestril, R., Hilla-Dandan, R., Brunton, L.L. & Dillmann, W.H. (1997) Small heat shock proteins and protection against ischemic injury in cardiac myocytes. *Circulation, 96,* 4343-4348.

[91] McArdle, A., Dillmann, W.H., Mestril, R., Faulkner, J.A., & Jackson, M.J. (2004) Overexpression of HSP70 in mouse skeletal muscle protects against muscle damage and age-related muscle dysfunction. *FASEB J, 18,* 355, 357.

[92] Mehlen, P. & Arrigo, A.P. (1994) The serum-induced phosphorylation of mammalian hsp27 correlates with changes in its intracellular localization and levels of oligomerization. *Eur J Biochem, 221,* 327-334.

[93] Melkani, G.C., Cammarato, A., Bernstein, S.I. (2006) αB-crystallin maintains skeletal muscle myosin enzymatic activity and prevents its aggregation under heat-shock stress. *J Mol Biol, 358,* 635-645.

[94] Morrison, L.E., Hoover, H.E., Thuereauf, D.J. & Glembotski, C.C. (2003) Mimicking phosphorylation of αB-crystallin on serine-59 is necessary and sufficient to provide maximal protection of cardiac myocytes from apoptosis. *Circ Res, 92,* 203-211.

[95] Morrison, L.E., Whittaker, R.J., Klepper, E.F., Wawrousek, E.F. & Glembotski, C.C. (2004) Roles for alphaB-crystallin and HSPB2 in protecting the myocardium from ischemia-reperfusion-induced damage in a KO mouse model. *Am J Physiol Heart Circ Physiol, 286,* 847-855.

[96] Mounier, N. & Arrigo, A.P. (2002) Actin cytoskeleton and small heat shock proteins: How do they interact? *Cell Stress Chaperones, 7,* 167-176.

[97] Muñoz-Mármol, A.M., Strasser, G., Isamat, M., Coulombe, P.A., Yang, Y., Roca, X., Vela, E., Mate, J.L., Coll, J., Fernández-Figueras, M.T., Navas-Palacios, J.J., Ariza, A. & Fuchs E. (1998) A dysfunctional desmin mutation in a patient with severe generalised myopathy. *Proc Natl Acad Sci USA, 95,* 11312-11317.

[98] Nakano, S., Engel, A.G., Waclawik, A.J., Emslie-Smith, A.M. & Busis, N.A. (1996) Myofibrillar myopathy with abnormal foci of desmin positivity. I. Light and electron miscroscopy analysis of 10 cases. *J Neuropathol Exp Neurol, 55,* 549-562.

[99] Nakano, S., Engel, A.G., Akiguchi, I. & Kimura, J. (1997) Myofibrillar myopathy. III. Abnormal expression of cyclin-dependent kinases and nuclear proteins. *J Neuropathol Exp Neurol 56,* 850-856.

[100] Narberhaus, F. (2002) Alpha-crystallin-type heat shock proteins: socializing minichaperones in the context of a multichaperone network. *Microbiol Mol Biol Rev, 66,* 64-93.

[101] Neufer, P.D. & Benjamin, I.J. (1996) Differential expression of α-crystallin and Hsp27 in skeletal muscle during continuous contractile activity. Relationship to myogenic regulatory actors. *J Biol Chem, 271,* 24089-24095.

[102] Neufer, D., Ordway, G.A. & Williams, R.S. (1998) Transient regulation of c-fos, αB-crystallin and hsp70 in muscle during recovery from contractile activity. *Am J Physiol, 274,* 341-346.

[103] Nichol, I.D. & Quinlan, R.A. (1994) Chaperone activity of crystallins modulates intermediate filament assembly. *EMBO J,* 13, 945-953.

[104] Oertel, M.F., May, C.A., Bloemendal, H. & Lütjen-Drecoll, E. (2000) αB-crystallin expression in tissues derived from different age groups. *Ophthalmologica, 214,* 13-23.

[105] Olivé, M., Armstrong, J., Miralles, F., Pou, A., Fardeau, M., Gonzalez, L., Martínez, F., Fischer, D., Martínez-Matos, J.A., Shatunov, A., Goldfarb, L. & Ferrer, I. (2007) Phenotypic patterns of desminopathy associated with three novel mutations in the desmin gene. *Neuromusc Disord, 17,* 443-450.

[106] Olivé, M., Goldfarb, L.G., Shatunov, A., Fischer, D. & Ferrer, I. (2005) Myotilinopathy: refining the clinical and myopathological phenotype. *Brain, 128,* 2315-2326.

[107] Olivé, M., Goldfarb, L., Moreno, D., Laforet, E., Dagvadorj, A., Sambuughin, N., Martínez-Matos, J.A., Martínez, F., Alió, J., Farrero, E., Vicart, P. & Ferrer, I. (2004) Desmin-related myopathy: clinical, electrophysiological, radiological, neuropathological and genetic studies. *J Neurol Sci, 219,* 125-137.

[108] Olivé, M., Janué, A., Moreno, D., Gámez, J., Torrejón-Escribano, B. & Ferrer, I. (2009) TAR DNA-binding protein 43 accumulation in protein aggregate myopathies. *J Neuropathol Exp Neurol, 68,* 262-273.

[109] Olivé, M., van Leeuwen, F.W., Janué, A., Moreno, D., Torrejón-Escribano, B. & Ferrer, I. (2008) Expression of mutant ubiquitin (UBB+1) and p62 in myotilinopathies and desminopathies. *Neuropathol Appl Neurobiol, 34,* 76-87.

[110] Parcellier, A., Schmitt, E., Brunet, M., Hammann, A., Solary, E. & Garrido, C. (2005) Small heat shock proteins HSP27 and αB-crystallin: cytoprotective and oncogenic functions. *Antioxid Redox Signal, 7,* 404-413.

[111] Perng, M.D., Cairns, L., van Den Ijssel, P., Prescott, A., Hutchenson, A.M. & Quinlan, R.A. (1999) Intermediate filament filament interactions can be altered by HSP27 and αB-crystallin. *J Cell Sci,* 112, 2099-2112.

[112] Perng, M.D., Muchowski, P.J., van Den Ijssel, P., Wu, G.H., Hutchenson, A.M., Clark, J.I., & Quinlan, R.A. (1999) The cardiomyopathy and lens cataract mutation in alphaB-crystallin alters its protein structure, chaperone activity, and interactions with intermediate filaments in vitro. *J Biol Chem, 274,* 33235-33243.

[113] Pinz, I., Robbins, J., Rajasekaran, N.S., Benjamin, I.J. & Ingwall, J.S. (2008) Unmasking different mechanical and energetic roles for the small heat shock proteins CryAB and HSPB2 using genetically modified mouse hearts. *FASEB J, 22,* 84-92.

[114] Pipkin, W., Johnson, J.A., Creazo, T.L., Burch, J., Komalavilas, P. & Brophy, C. (2003) Localization, macromolecular associations, and function of the small heat shock protein HSP20 in rat heart. *Circulation, 107,* 469-476.

[115] Pivovarova, A.V., Mikkhailova, D.I., Chernik, I.S., Chebotareva, N.A., Levitsky, D.I. & Gusev, N.B. (2005) Effects of small heat shock proteins on the thermal denaturation and aggregation of F-actin. *Biochem Biophys Res Commun, 331,* 1548-1553.

[116] Quax-Jeuken, Y., Quax, W., Van Rens, G., Khan, P.M. & Bloemendal, H. (1985) Complete structure of the alpha B-crystallin gene: conservation of the exonintron distribution in the two nonlinked alpha-crystallin genes. *Proc Natl Acad Sci USA, 82,* 5919-5923.

[117] Quinlan, R. (2002) Cytoskeletal competence requires protein chaperones. *Prog Mol Subcell Biol, 28,* 219-233.

[118] Rajasekaran, N.S., Connell, P., Christians, E.S., Yan, L.J., Taylor, R.P., Orosz, A., Zhang, X.Q., Stevenson, T.J., Peshock, R.M., Leopold, J.A., Barry, W.H., Loscalzo, J., Odelberg, S.J. & Benjamin, I.J. (2007) Human αB-crystallin mutation causes oxido-reductive stress and protein aggregation cardiomyopathy in mice. *Cell, 130,* 427-439.

[119] Ray, P.S., Martin, J.L., Swanson, E.A., Otani, H., Dillman, W.H. & Das, D.K. (2001) Transgene overexpression of αB-crystallin confers simltaneous protection against cardiomyocyte apoptosis and necrosis during myocardial ischemia and reperfusion. *FASEB J, 15,* 393-402.

[120] Reimann, J., Kuntz, W.S., Vielhaber S., Capees-Horn, K. & Schroeder, R. (2003) Mitochondrial dysfunction in myofibrillar myopathy. *Neuropathol Appl Neurobiol, 29,* 45-51.

[121] Sakurai, T., Fujita, Y., Ohto, E., Oguro, A. & Atomi, Y. (2005) The decrease of the cytoskeleton tubulin follows the decrease of the associating molecular chaperone αB-crystallin in unloaded soleus muscle atrophy without stretch. *FASEB J, 19,* 1199-1201.

[122] Sanbe, A., Osinska, H., Safitz, J.E., Glabe, C.G., Kayed, R., Maloyan, A. & Robbins, J. (2004) Desmin-related cardiomyopathy in transgenic mice: a cardiac amyloidosis. *Proc Natl Acad Sci USA, 101,* 10132-10136.

[123] Sanbe, A., Yamauchi, J., Miyamoto Y., Fujiwara, Y., Murabe, M. & Tanoue, A. (2007) Interruption of CryAB-amyloid oligomer formation by HSP22. *J Biol Chem, 282,* 555-563.

[124] Selcen, D. (2008) Myofibrillar myopathies. *Curr Opin Neurol, 21,* 585-589.

[125] Selcen, D. & Engel, A.G. (2003) Myofibrillar myopathy caused by novel dominant negative alpha B-crystallin mutations. *Ann Neurol, 54*, 804-810.

[126] Selcen, D. & Engel, A.G. (2004) Mutations in myotilin cause myofibrillar myopathy. *Neurology, 62,* 1363-1371.

[127] Selcen, D. & Engel, A.G. (2005) Mutations in ZASP define a novel form of muscular dystrophy in humans. *Ann Neurol, 57,* 269-276.

[128] Selcen, D., Ohno, K. & Engel, A.G. (2004) Myofibrillar myopathy: clinical, morphological and genetic studies in 63 patients. *Brain, 127,* 439-451.

[129] Shu, E., Matsuno, H., Akamatsu, S., Kanno, Y., Suga, H., Nakajima, K., Ishisaki, A., Takai, S., Kato, K., Kitajima, Y. & Kozawa, O. (2005) αB-crystallin is phosphorylated during myocardial infarction: involvement of platelet-derived growth factor-BB. *Arch Biochem Biophys, 438,* 111-118.

[130] Simon, S., Fontaine, J.M., Martin, J.L., Sun, X., Hoppe, A.D., Welsh, M.J., Benndorf, R. & Vicart, P. (2007) Myopathy-associated alphaB-crystallin mutants: abnormal phosphorylation, intracellular location, and interactions with other small heat shock proteins. *J Biol Chem, 282,* 34276-34287.

[131] Srikakulam, R. & Winkelmann, D. (2004) Chaperone-mediated folding and assembly of myosin in striated muscle. *J Cell Sci, 117,* 641-671.

[132] Stamler, R., Kappe, G., Boelens, W. & Slingsby, C. (2005) Wrapping the α-crystallin domain fold in chaperone assembly. *J Mol Biol, 353,* 68-79.

[133] Sugiyama, Y., Suzuki, A., Kishikawa, M., Akutsu, R., Hirose, T., Waye, M.M., Tsui, S.K., Yoshida, S. & Ohno, S. (2000) Muscle develops a specific form of small heat shock protein complex composed of MKBP/HSPB2 and HSP23 during myogenic differentiation. *J Biol Chem, 275*, 1095-1104.

[134] Tannous, P., Zhu, H., Johnstone, J.L., Shelton, J.M., Rajasekaran, N.S., Benjamin, I.J., Nguyen, L., Gerard, R.D., Levine, B., Rothermel, B.A. & Hill, J.A. (2008) Autophagy is an adaptative response in desmin-related cardiomyopathy. *PNAS, 105,* 9745-9750.

[135] Treweek, T.M., Rekas, A., Lindner, R.A., Walker, M.J., Aquilina, J.A., Robinson, C.V., Horwitz, J., Perng, M.D., Quinlan, R.A. & Carver, J.A. (2005) R120G alphaB-crystallin promotes the unfolding of reduced alpha-lactalbumin and is inherently unstable. *FEBS J, 272,* 711-724.

[136] Vicart, P., Caron, A., Guicheney, P., Li, Z., Prevost, M.C., Faure, A., Chateau, D., Chapon, F., Tome F., Dupret, J.M., Paulin, D. & Fardeau, M. (1998) A missense mutation in the αB-crystallin chaperone gene causes a desmin-related myopathy. *Nat Genet, 20*, 92-95.

[137] Voellmy, R.,& Boellmann, F (2007) Chaperone regulation of the heat shock protein response. *Adv Exp Med Biol, 594,* 89-99.

[138] Vojcik, S., Engel, W.K., McFerrin, J., Paciello, O. & Askanas, V. (2006) A PP overexpresssion and proteasome inhibition increase B-Crystallin in cultured human muscle: relevance to inclusion-body myositis. *Neuromuscul Disord, 16,* 839-844.

[139] Vorgerd, M. van der Ven, P.F., Bruchertseifer, V., Löwe, T., Kley, R.A., Schröder, R., Lochmüller, H., Himmel, M., Koehler, K., Fürst, D.O. & Huebner, A. (2005) A mutation in the dimerization domain of filamin causes a novel type of autosomal dominant myofibrillar myopathy. *Am J Hum Genet, 77,* 297-304.

[140] Wang, X., Osinska, H., Klevitsky, R., Gerdes, A.M., Nieman, M., Lorenz, J., Hewett, T. & Robbins, J. (2001) Expression of R120G-αB-crystallin causes aberrant desmin and αB-crystallin aggregation and cardiomyopathy in mice. *Circ Res, 89,* 84-91.

[141] Wang, X., Osinska, H., Dorn, G.W., Nieman, M., Lorenz, J.N., Gerdes, A.M., Witt, S., Kimball, T., Gulik, J. & Robbins, J. (2001) Mouse model of desmin-related cardiomyopathy. *Circulation 103*, 2402-2407.

[142] Watts, G.D., Thomasova, D., Ramdeen, S.K., Fulchiero, E.C., Mehta, S.G., Drachman, D.A., Weihl, C.C., Jamrozik, Z., Kwiecinski, H., Kaminska, A. & Kimonis, V.E. (2007) Novel VCP mutations in inclusion body myopathy associated with Paget disease of bone and frontotemporal dementia. *Clin Genet, 72,* 420-426.

[143] Watts, G.D., Wymer, J., Kovach, M.J., Mehta, S.G., Mumm, S., Darvish, D., Pestronk, A., Whyte, M.P. & Kimonis, V.E. (2004) Inclusion body myopathy associated with Paget disease of bone and frontotemporal dementia is caused by mutant valosin containing protein. *Nat Genet, 36,* 377-381.

[144] White, M.Y., Hambly, B.D., Jeremy, R.W. & Cordwell, S.J. (2006) Ischemia-specific phosphorylation and myofilament translocation of heat shock protein 27 precedes alpha B-crystallin and occurs independently of reactive oxygen species in rabbit myocardium. *J Mol Cell Cardiol, 40,* 761-774.

[145] Yamaguchi, T., Arai, H., Katayama, N., Ishikawa, T., Kikumoto, K. & Atomi, Y. (2007) Age-related increase of insoluble, phosphorylated small heat shock proteins in human skeletal muscle. *J Gerontol, 62,* 481-489.

[146] Zantema, A., Verlaan-De Vries, M., Maasdam, D., Bol, S. & van der Eb, A. (1992) Heat shock protein 27 and lpha B-crystallin can form a complex, which dissociates by heat shock. *J Biol Chem 267,* 12936-12941.

In: Small Stress Proteins and Human Diseases
Editors: Stéphanie Simon et al.
ISBN: 978-1-61470-636-6

Chapter 2.4

HSPB5 (*CRYAB*; αB-CRYSTALLIN) AND PROTEIN AGGREGATION CARDIOMYOPATHY

Huali Zhang and Ivor J. Benjamin*
Center for Cardiovascular Translational Biomedicine, Division of Cardiology, Department of Internal Medicine, University of Utah School of Medicine, Salt Lake City, UT 84132, U.S.

ABSTRACT

For decades, the biological functions of the evolutionarily conserved heat shock proteins have garnered widespread attention from cell biologists to physiologists alike. Among the most abundant proteins expressed in the heart (1-3% soluble homogenates), gain-of-function model systems have been established the pivotal roles for HSPB5 (CryAB, αB-crystallin) to speed up physiological recovery after ischemia/reperfusion injury. Remarkably, loss of function studies in mice indicated CryAB was dispensable in the heart under basal conditions, attributed to the plasticity among the members of the small MW HSP family. In contrast, the missense mutation of CryAB (R120G) linked genetically to the multisystem diseases including cardiomyopathy has underlying protein misfolding and aggregate formation in humans. Similar to numerous inheritable diseases, intense efforts to establish causal mechanisms are more demanding and elusive. Notwithstanding, recent progress using genetically engineered models has uncovered mechanisms involving alterations in mitochondrial organization and architecture, activation of autophagy and new surprises related to oxido-reductive stress.

* Ivor J. Benjamin
Center for Cardiovascular Translational Biomedicine,
Division of Cardiology,
Department of Internal Medicine,
University of Utah School of Medicine,
Salt Lake City, UT 84132, USA
Ivor.benjamin@hsc.utah.edu

INTRODUCTION

Various (patho)physiological stresses evoke profound shifts in the protein folding landscape that jeopardize cell survival if prompt counterbalancing measures do not restore cellular homeostasis. Both constitutive and inducible expression of ubiquitous genes encoding heat shock proteins such as HSP70 and HSPB5 (CryAB, αB-crystallin) have been established to speed up physiological recovery and to afford effective cardioprotection after noxious stimuli including ischemia/reperfusion injury [1-3] (see chapter 1-5). We have summarized this literature in several reviews [4, 5] but the present chapter focuses on the expression and consequences of mutant stress (HSP) proteins that are pathogenically linked to inheritable conditions characterized by protein aggregation, reduced functional capacity, and multisystem diseases in humans.

Small heat shock proteins (sHSP) such as HSPB5 are abundantly expressed in the lens yet selectively found in striated muscles, approaching 1-3% of the soluble homogenates in the heart [6]. Most sHSPs like CryAB/HSPB5 share the highly conserved α-crystallin domain and form heter-oligomeric cytosolic complexes with other sHSPs whose interactions and differing intracellular distribution are hypothesized to confer distinct intracellular functions. At the transcriptional level, the promoters for CryAB and HSPB2 are bidirectionally oriented but share regulatory elements that control their coordinate expression [7, 8], suggesting potential common functional roles.

During physiological perturbations induced by heat shock, ischemia and exercise, CryAB/HSPB5 translocates from the cytosol to the cytoskeleton, where it binds to titin, actin and desmin to prevent stress-induced protein aggregates and, perhaps, myocardial dysfunction [9-12].

In recent studies, the requirements for CryAB/HSPB5 expression have been addressed using both gain-of-function and loss-of-function model systems in mice. Overexpressing CryAB in mouse hearts protects against ischemic myocardial damage and improves post-ischemic contractile performance [13]. Using mice lacking both CryAB/HSPB5 and HSPB2 sHSPs (DKO) generated by Brady and coworkers [14], Morrison and colleagues first reported severe contractile dysfunction and increased myocardial injury in isolated hearts after ischemia/reperfusion *ex vivo* [15]. Golenhofen and colleagues then reported that isolated papillary muscles of DKO after ischemia/reperfusion exhibited an earlier rise in resting tension and increased post-ischemic contracture compared with control animals [12]. Our group has similarly demonstrated that an increased susceptibility after simulated ischemic conditions of DKO compared to wild-type controls [16].

To provide further insights about specific roles, our laboratory recently used available genetic tools to unmask distinct roles for CryAB and HSPB2 in cardiac mechanics and energetics respectively. DKO mice were intercrossed with a CryAB overexpressor line, enabling testing partial or complete rescue of the CryAB phenotype, in genetically engineered mice solely lacking HSPB2. We found that restoring CryAB in the DKO background protected these hearts against ischemia-induced diastolic dysfunction whereas the persistence of HSPB2 deficiency affected both normal systolic performance and interesting cardiac energetics [17]. While these studies implicate new roles of these sHSPs in ischemic cardioprotection, an unambiguous role for either HSPB2 or CryAB could not be assigned owing the inherent limitations related to the DKO experimental model. Tissue specific knockout animals, currently in development by independent laboratories, are needed to definitely resolve these questions for the field. Without doubt, the importance of sHSPs, in general, and CryAB, in particular, was provided in the late 90s when Vicart and coworkers first reported on the CryAB mutation (Arg120Gly) that was linked genetically to the multisystem protein aggregation diseases including cardiomyopathy in humans [18].

1. Genetic Causes of Cardiomyopathy Include CryAB

Cardiomyopathy encompasses multiple cardiac conditions affecting the heart's muscle with manifestations of congestive heart failure, a form of cardiac arrhythmia termed sudden death, and increased mortality. Such events are dramatically found especially in young male elite athletes who succumb to such conditions in their teens during bouts of intensive exercise [19, 20]. At the anatomical level, cardiomyopathy has been conveniently classified on the basis on chamber size and left ventricular wall thickening: namely, hypertrophic cardiomyopathy (HCM), dilated cardiomyopathy (DCM), and restrictive cardiomyopathy (RCM). Besides left ventricular wall thickening, extensive pathologic studies have documented other characteristics of HCM include myocyte disarray and cardiac fibrosis. In 1990, the seminal studies from the Seidmans' laboratories established the causal role of mutations in genes encoding sarcomeric proteins for hypertrophic cardiomyopathy (21)]. This discovery ushered similar findings for numerous autosomal dominant and recessive mutations for cardiac diseases now numbering more than 450 different mutations. An extensive network of candidates with diverse biological functions including sarcolemmal, membrane, nuclear envelope, intermediate filament, and the molecular chaperone CryAB/HSPB5 has been identified (Table 1) [22, 23].

Table 1. Genetic heterogeneity of hypertrophic cardiomyopathy

Encoded protein	Gene symbol	Chromosome locus	Sarcomere component
Myosin heavy chain 7	MYH7	14q12	Thick filament
Myosin-binding protein c	*MYBPC3*	11p11.2	Thick filament
Troponin t2, cardiac	*TNNT2*	1q32	Thin filament
Troponin I, cardiac	*TNNI3*	19q13.4	Thin filament
Tropomyosin 1	*TPM1*	15q22.1	Thin filament
Actin, alpha, cardiac muscle	*ACTC*	15q14	Thin filament
Myosin regulatory light chain	*MYL2*	12q23-q24.3	Thin filament
Myosin essential light chain, cardiac	*MYL3*	3p	Thin filament
Titin	*TTN*	2q31	Thick filament/Z-Disc
Telethonin	*TCAP*	17q12	Z-Disc
Myozenin 2	*MYOZ2*	4q26	Z-Disc
Vinculin	VCL	10q22.2	Intercalated disc
Muscle LIM protein	CSRP3	11p15.1	Z-Disc
Lysosomal associated membrane protein 2	*LAMP2*	Xq24	no
5-AMP-activated protein kinase, gamma-2 subunit	*PRKAG2*	7q35-q36	no
Alpha-galactosidase	*GLA*	Xq22	no

Dilated cardiomyopathy (DCM), the most common form, is characterized by cardiac dilation, reduced systolic function, heart failure, thromboembolic events, all of which account for the substantial morbidity and premature mortality. Heterogeneous etiologies of DCM also include myocarditis, coronary artery disease, systemic diseases, myocardial toxins and genetic factors [24]. Pertaining to genetic influences, idiopathic dilated cardiomyopathy accounts for approximately one-half of all cases with a prevalence of about 36.5 per 100,000 and is one of the leading indications for cardiac transplantation in the US. A predominant familial occurrence of autosomal dominant DCM accounts for 20 to 25%, is characterized by age-related penetrance and can be transmitted as an autosomal recessive, X-linked, or matrilinear (mitochondrial) trait. Mutations in many genes, which overlap with HCM, have been found to cause different forms of dilated cardiomyopathy (Table 2).

Table 2. Genetic heterogeneity of dilated cardiomyopathy

Encoded protein	**Gene symbol**	**Chromosome locus**
Troponin t2, cardiac	*TNNT2*	1q32
Myosin heavy chain 7	MYH7	14q12
Actin, alpha, cardiac muscle	*ACTC*	15q14
Lamin A/C	LMNA	1q21
Desmin	DES	2q35
ATP-BINDING CASSETTE, SUBFAMILY C, MEMBER 9	ABCC9	12p12.1
LIM DOMAIN-BINDING 3	LDB3, ZASP	10q22.2-q23.3
SODIUM CHANNEL, VOLTAGE-GATED, TYPE V, ALPHA SUBUNIT	SCN5A	3p21
Titin	*TTN*	2q31
Muscle LIM protein	CSRP3	11p15.1
Telethonin	TCAP	17q12
Tropomyosin 1	*TPM1*	15q22.1
Delta-sarcoglycan	SGCD	5q33
Phospholamban	PLN	6q22.1
Thymopoietin	TMPO	12q22
presenilin-1	PSEN1	14q24.3
presenilin-2	PSEN2	1q31-q42
Metavinculin	VCL	10q22-q23
Fukutin	FCMD	9q31
Tropomyosin	TPM1	15q22.1
Troponin C, slow	TNNC1	3p21.3-p14.3
Actinin, Alpha-2	ACTN2	1q42-q43
Desmoglein	DSG2	18q12.1-q12.2

2. PROTEIN AGGREGATION CARDIOMYOPATHY

Among mutant proteins linked to familial dilated cardiomyopathy is the intermediate filament protein, desmin, which mediates attachments between the terminal Z disc and the junctional membrane [25]. In addition to autosomal dominant dilated cardiomyopathy [26], the mutations in *DES* (Desmin) gene have been found to cause familial myofibrillar myopathy (MFM), which is often termed "'Desmin-related myopathy" or "Desminopathy" [27-30]. Desminopathy is a distinct form of protein aggregate myopathy characterized by skeletal muscle weakness which could be associated with cardiac conduction blocks, arrhythmias, and restrictive heart failure. Disorganization of the desmin filament network and sarcomeric disorganization of striated muscles might accompany the accumulation of insoluble desmin-containing aggregates [28, 31] (see chapter 2-3).

Table 3. Human protein aggregation myopathy

Disease		Disease gene	Phenotype
Primary Desminopathy		Desmin	Distal myopathy, cardiac arrhythmias, cardiomyopathy onset between the second and fourth decades of life
Secondary Desminopathy	αB crystallinopathy	αB-crystallin	Weakness of skeletal muscles, cardiomyopathy, and lens opacities
	Myotilinopathy	Myotilin	distal muscle weakness , peripheral neuropathy with hyporeflexia
	ZASP-related MFM	ZASP	muscle weakness
	filaminopathy	g-filamin	slowly progressive skeletal muscle weakness
	Rigid spine syndrome	selenoprotein N1	Childhood onset of marked limitation in flexion of the whole dorsolumbar and cervical spine
	BAG3-related MFM	BAG3	late-childhood onset of rapidly progressive myopathy affecting both skeletal and cardiac muscle with severe respiratory insufficiency

On the basis of morphological hallmarks of desmin-positive protein aggregates with myofibillar abnormalities, desminopathies have expanded beyond the desmin to include genes encoding desmin-associated proteins such as αB-crystallin, myotilin, ZASP, γ-filamin, selenoprotein N1 or Bag 3 (Table 3) [18, 32-35]. Additional secondary desminopathies include crystallinopathy, myotilinopathy, ZASP-related MFM, filaminopathy and Rigid spine syndrome. Although protein aggregation is a well-recognized pathological feature in desminopathy, there is increasing evidence that such immuno-morphological classification developed in the pre-genomic era might both inadequate and inaccurate for the field. Accordingly, we have opted for the more general term protein aggregation disease and, in current context for the R20G isoform in the heart, protein aggregation cardiomyopathy. A burgeoning series of diverse proteins shown to be associated with cellular aggregates is listed in Table 4.

Table 4. Multiprotein complexes associate with Desmin-related diseases

Cytoskeletal proteins	**Sarcomeric proteins**	**Nuclear proteins**
Desmin Vimentin Nestin Dystrophin β-Spectrin	Nebulin Titin Actin a-Actinin Myosin fast Myosin slow	Emerin Lamin B Nuclear matrix-associated protein
Cyclin-dependent kinases	Chaperone proteins	Other proteins
CDK1 CDK2 CDK3 CDK4 CDK5 CDK7 CDK2 kinase p21(CDK inhibitor)	Ubiquitin α-B Crystallin HSPB1	Cathepsin B Calpain Gelsolin Utrophin α-1 Antichymotrypsin N-CAM

3. CRYAB/HSPB5 MUTATIONS CAUSE MULTISYSTEM DISEASES

As a *bona fide* molecular chaperone, the small MW heat shock protein αB-crystallin (CryAB, HSPB5) abrogates aggregation formation of client proteins such as desmin thus maintaining muscle integrity and stress tolerance. Following the identification by Vicart *et al* of an Arg120-to-Gly (R120G) substitution in the CryAB gene, of desmin-positive cytoplasmic or perinuclear aggregates, functional expression studies in muscle cells, there was considerable excitement for understanding the causal mechanisms for a condition originally described by Fardeau *et al* as autosomal dominant desmin-related myopathy [18, 36]. Two additional mutations identified with desmin-related myofibrillar myopathy were a 2-bp deletion (464delCT) in the C terminus of the CRYAB gene, resulting in a truncated protein of 162 amino acids instead of the normal 175 and a 451C-T transition in the CRYAB gene, resulting in a Gln151-to-Ter (Q151X) mutation [37]. Two missense mutations of CRYAB, Arg157His and Gly54Ser, which affect evolutionary conserved residues, were associated with late-onset in DCM patients without cataracts. Functional analysis revealed that Arg157His mutation decreased the binding to titin/connectin heart-specific N2B domain but without aggregates in cardiomyocytes [38, 39]. Berry and coworkers described familial congenital cataracts with an exon 3 deletion mutation, 450delA, of CryAB arising from a frameshift mutation in codon 150 that produced an aberrant protein consisting of 184 residues [40]. All these mutations were predicted to impair CryAB's chaperone-like properties to inhibit heat-induced protein aggregation of unfolded and denatured proteins, resulting in aberrant accumulation of proteins in muscle fibers [41, 42]. However, such *in vitro* analyses do not always faithfully recapitulate the *in vivo* conditions linked to disease pathogenesis for which more robust experimental model systems are increasingly becoming essential (see chapter 2-1).

4. Small Animal Models of Human Diseases

Genetically engineered mice created by conventional pronuclear injections have been invaluable experimental tools for investigating cardiac diseases when candidate genes are placed under the control of the cardiac-specific αMHC promoter [43]. A milestone forward towards understanding the molecular mechanisms of mutant CryAB expression was the generation, by two independent laboratories, of mice harboring the mouse and human R120G mutations, respectively [44, 45]. Such transgenic mouse models of cardiomyocyte-specific R12OG CryAB developed hypertrophy that progressed into cardiomyopathy and decompensated heart failure. Parallel studies revealed a very benign course for mice with WT CryAB overexpression, whereas modest levels of R120G-CryAB protein caused aberrant desmin and CryAB aggregation, disruption of the desmin network, perturbation of myofibril alignment, and compromised muscle function [44]. A gene 'knockin' strategy is an alternative but time-consuming approach that achieves mutant protein expression under the control of its endogenous promoter. Notwithstanding, current models have been a tremendous boast for the field as new insights into mechanisms have emerged from the detailed characterization of protein misfolding diseases associated with aberrant oligomerization and aggregation.

5. Mechanisms of Cell Death and Heart Failure in R120G CryAB Mice

The etiology of heart failure involves multiple biochemical, physiological and molecular pathways but the progressive loss of cardiac myocyte is among the most important pathogenic hallmarks. The collapse of the desmin network in R120G CryAB transgenic mice undoubtedly contributes to the developing disease, yet similar structural changes of the cytoskeleton in another animal model expressing mutant desmin mutation resulted in significantly less morbidity and mortality [46]. Therefore, it is becoming more widely appreciated that mutant CryAB expression has other effects beyond alterations of the desmin network. In particular, recent attention has focused on the integrity of the cytoskeleton on mitochondrial organization in both skeletal and cardiac myocytes [47]. In R120G CryAB transgenic mice, alterations in both mitochondrial organization and architecture, reduction in the maximal rate of oxygen consumption with substrates that utilize complex I activity, increased opening of the permeability transition pore, and compromised inner membrane potential have been reported [48]. Whereas these etiologic factors have been associated with apoptotic pathways, they do not *per se* reveal the precise mechanisms by which R120G contributes to cardiomyocyte death, dilation, and heart failure. Recent studies from our laboratory are among the first to provide such mechanistic insights, uncovering an entirely new paradigm for disease pathogenesis.

6. Reductive Stress as a Causal Mechanism for R120G Cardiomyopathy

Hearts of CryAB deficient mice were remarkably devoid of pathogenic protein aggregates [49] but based on initial biochemical analyses of R120G and similar mutant isoforms some have proposed for loss of function mechanism(s) leading to R120G pathology [41, 42]. Our laboratory, however, has challenged this paradigm by using a mouse heart model to express a mutant human CryAB that causes protein misfolding cardiomyopathy[45]. Our results show that these heart muscles are under 'reductive stress' from an over-activated antioxidative system, indicating R120G CryAB expression induces a novel 'gain-of- toxic' function mechanism linked to increased activity of glucose 6-phosphate dehydrogenase (G6PD) mimicking 'reductive stress.' Reductive stress refers to pathogenic excesses of reducing equivalents (e.g., glutathione, NADPH), which has been demonstrated in lower eukaryotes but uncommonly in mammals and/or disease states [50, 51]. Our studies revealed an increased reduced glutathione (GSH) to oxidized glutathione (GSSG) ratio arising from increased specific enzymatic activities of glucose-6-phosphate dehydrogenase (G6PD). G6PD generates NADPH, an essential co-factor for numerous biochemical pathways, for glutathione reductive whose hyperactivity contributes to the recycling of oxidized GSSG to reduced GSH. Since R120G has profound effects on HSPB1/Hsp25 expression, these findings are strikingly reminiscent of earlier findings reported from the laboratory of Arrigo who associated HSPB1/Hsp25 expression with upholding the glutathione redox state [52]. Proof of principle for this causal pathway was obtained by genetic studies in which hR120GCryAB Tg cardiomyopathic animals intercrossed with G6PD-deficient (20% normal G6PD activity) mice completely rescues the pathology of protein aggregation cardiac hypertrophy, and heart failure [45].

Because the prevention from dysregulation of G6PD activity dramatically improves the pathology, our studies of R120GCryAB cardiomyopathy raise several important questions for redox regulation, causal mechanisms of protein aggregation, and disease susceptibility pathways in humans. The significance of work for future studies is to unambiguously clarify whether or not similar pathogenic mechanisms such as hyperactivity of antioxidative pathways, which can protect the intact heart against ischemic injury, can play similar pathogenic roles for other protein aggregation diseases involving the swath of inherited or acquired mutant proteins.

7. Oxidative Stress in the Pathogenesis of R120G Cardiomyopathy and Heart Failure

Oxidative stress is widely implicated in pathogenic states including aging and heart failure. A large body of evidence suggests that excessive production of reactive oxygen species (ROS) is an important hallmark of heart failure leading to clinical trials of antioxidants treatment to prevent the progression of heart failure in terms of cardiac dysfunction. However, targeted inhibition of xanthine oxidase (XO) by allopurinol or oxypurinol can rescue cardiac structure and function and substantially improve the heart failure phenotype in a canine model of pacing-induced cardiomyopathy [53]. Maloyan and coworkers recently administered this strategy, the XO inhibitor oxypurinol, to R120G CryAB mice for a period of 1 or 3 months and found that mitochondrial function was dramatically improved in treated animals. Specifically, XO therapy improved complex I activity, conservation of mitochondrial membrane potential and restored normal mitochondrial morphology but that contractility was not improved because of mechanical deficits in passive cytoskeletal stiffness [54]. These data are consistently with the hypothesis that R120G triggers "early-onset" oxidative stress and that compensatory antioxidative pathways become dysregulated after a transition phase leading to "late-onset" pathogenic reductive stress (Figure 1). In this model, the introduction of antioxidants would be beneficial during the compensatory phase but would be deleterious during cardiac decompensation when we hypothesize an 'anti-reductant' might be administered.

8. Role of Autophagy in R120G Protein Aggregation Cardiomyopathy

Autophagy has evolved as a conserved physiological process for lysosomal degradation and recycling of cytoplasmic components, such as long-lived proteins and damaged organelles (see chapter 1-6). In this process, the autophagosome, a double-membrane structure, sequesters cytosolic constituents such as ubiquitinated protein aggregates or organelles including mitochondria, peroxisomes and endoplasmic reticulum. Autophagy is controlled by autophagy-related genes (Atgs). Two conjugation pathways, Atg12-Atg5 and Atg8 (microtubule-associated protein light chain 3)-phosphatidylethanolamine are involved in autophagy process [55, 56]. In general, autophagy occurs constitutively to maintain cellular homeostasis. However, autophagy can be stimulated by extrinsic stressors and intrinsic factors such as ATP depletion, reactive oxygen species and mitochondrial permeability transition pore opening. The activation of autophagy plays a critical role in cell survival mechanism. Blocking of autophagy by pharmacological inhibitors or inactivation of Atg genes can increase cell death during *in vivo* experiments [57]. In hR120G Tg mice, Tannous and coworkers test the hypothesis that autophagic activation represents an adaptive response to mutant protein aggregates in vivo [58]. Mutant CryAB (CryAB(R120G)) correlated with activation of cardiomyocyte autophagy and aggregate formation varied inversely with the requirements for beclin 1, a gene required for autophagy, and the development of heart failure. These findings led us to conclude that autophagy can remove damaged organelles and misfolded proteins as a cytoprotective mechanism under stress condition, but excessive autophagic activity induced by severe stimuli might destroy cytosol constituents and organelles and lead to cell death.

CONCLUSION

Enormous strides have made been in the field of stress protein biology from its inception with the seminal discovery that all livings cells are endowed with evolutionarily conserved capacity, termed 'heat shock response (HSR)', for mounting robust stress tolerance. Evidence for the multiple families of genes encoding heat shock protein (HSPs) were quickly matched with the discovery of families of heat shock transcription factors, which governed the transcription regulation in diverse biological processes. Legions of investigators using multiple model organisms (e.g., cells, yeast, flies, fish, and mice) have provided the rich tapestry of cellular and molecular knowledge now being focus on the pathogenic mechanisms. Without doubt, the future dividends on this investment would be targeted therapies of the critical pathways for protein aggregation diseases in the emerging era of genomic and personalized medicine.

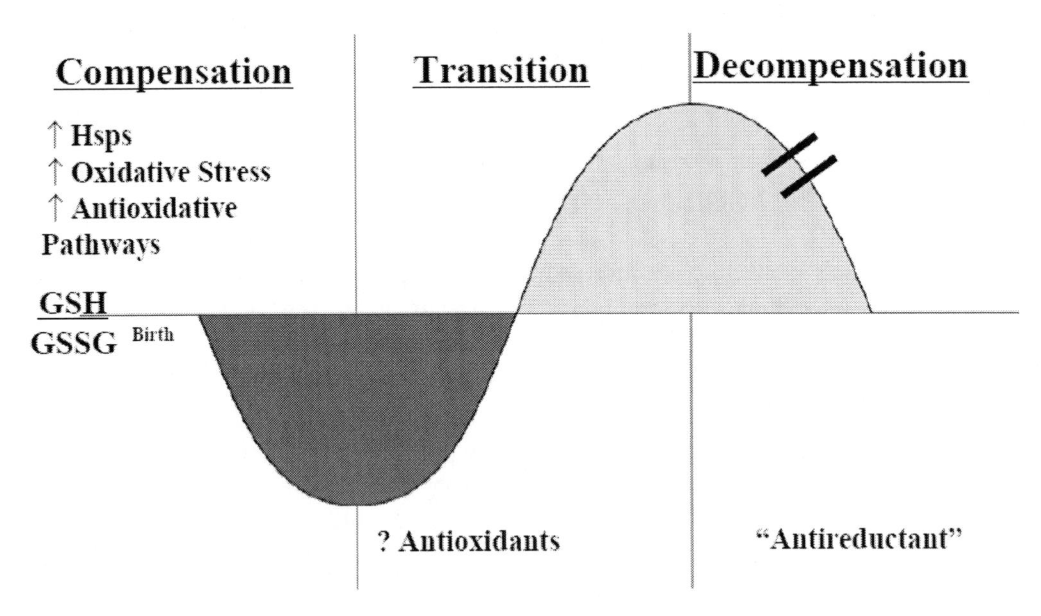

Figure 1. R120G triggers oxidative stress and that compensatory antioxidative pathways become dysregulated after a transition phase leading to pathogenic reductive stress.

REFERENCES

[1] Plumier, J.C., Ross, B.M., Currie, R.W., Angelidis, C.E., Kazlaris, H., Kollias, G. and Pagoulatos, G.N. (1995) Transgenic mice expressing the human heat shock protein 70 have improved post-ischemic myocardial recovery. *J Clin Invest,* 95, 1854-60.

[2] Radford, N.B., Fina, M., Benjamin, I.J., Moreadith, R.W., Graves, K.H., Zhao, P., Gavva, S., Wiethoff, A., Sherry, A.D., Malloy, C.R. *et al.* (1996) Cardioprotective effects of 70-kDa heat shock protein transgenic mice. *Proc Natl Acad Sci USA,* 93, 2339-2342.

[3] Marber, M.S., Mestril, R., Chi, S.H., Sayen, M.R., Yellon, D.M. and Dillmann, W.H. (1995) Overexpression of the rat inducible 70-kD heat stress protein in a transgenic mouse increases the resistance of the heart to ischemic injury. *J Clin Invest,* 95, 1446-56.

[4] Benjamin, I.J. and McMillan, D.R. (1998) Stress (Heat Shock) Proteins: Molecular Chaperones in Cardiovascular Biology and Disease. *Circ Res,* 83, *117-132.*

[5] Christians, E.S., Yan, L.J. and Benjamin, I.J. (2002) Heat shock factor 1 and heat shock proteins: Critical partners in protection against acute cell injury. *Crit Care Med,* 30, S43-50.

[6] Kato, K., Shinohara, H., Kurobe, N., Inaguma, Y., Shimizu, K. and Ohshima, K. (1991) Tissue distribution and developmental profiles of immunoreactive alpha B crystallin in the rat determined with a sensitive immunoassay system. *Biochim Biophys Acta,* 1074, 201-8.

[7] Iwaki, A., Nagano, T., Nakagawa, M., Iwaki, T. and Fukumaki, Y. (1997) Identification and characterization of the gene encoding a new member of the alpha-crystallin/small hsp family, closely linked to the alphaB- crystallin gene in a head-to-head manner. *Genomics,* 45, 386-94.

[8] Suzuki, A., Sugiyama, Y., Hayashi, Y., Nyu-i, N., Yoshida, M., Nonaka, I., Ishiura, S., Arahata, K. and Ohno, S. (1998) MKBP, a novel member of the small heat shock protein family, binds and activates the myotonic dystrophy protein kinase. *J Cell Biol,* 140, 1113-1124.

[9] Barbato, R., Menabo, R., Dainese, P., Carafoli, E., Schiaffino, S. and Di Lisa, F. (1996) Binding of cytosolic proteins to myofibrils in ischemic rat hearts. *Circ Res,* 78, 821-8.

[10] Golenhofen, N., Ness, W., Koob, R., Htun, P., Schaper, W. and Drenckhahn, D. (1998) Ischemia-induced phosphorylation and translocation of stress protein alpha B-crystallin to Z lines of myocardium. *Am J Physiol,* 274, H1457-64.

[11] Golenhofen, N., Arbeiter, A., Koob, R. and Drenckhahn, D. (2002) Ischemia-induced Association of the Stress Protein alpha B-crystallin with I-band Portion of Cardiac Titin. *J Mol Cell Cardiol,* 34, 309-19.

[12] Golenhofen, N., Redel, A., Wawrousek, E.F. and Drenckhahn, D. (2006) Ischemia-induced increase of stiffness of alphaB-crystallin/HSPB2-deficient myocardium. *Pflugers Arch,* 451, 518-25.

[13] Ray, P.S., Martin, J.L., Swanson, E.A., Otani, H., Dillmann, W.H. and Das, D.K. (2001) Transgene overexpression of alphaB crystallin confers simultaneous protection against cardiomyocyte apoptosis and necrosis during myocardial ischemia and reperfusion. *Faseb J,* 15, 393-402.

[14] Brady, J.P., Garland, D.L., Green, D.E., Tamm, E.R., Giblin, F.J. and Wawrousek, E.F. (2001) AlphaB-crystallin in lens development and muscle integrity: a gene knockout approach. *Invest Ophthalmol Vis Sci,* 42, 2924-34.

[15] Morrison, L.E., Whittaker, R.J., Klepper, R.E., Wawrousek, E.F. and Glembotski, C.C. (2004) Roles for alphaB-crystallin and HSPB2 in protecting the myocardium from ischemia-reperfusion-induced damage in a KO mouse model. *Am J Physiol Heart Circ Physiol,* 286, H847-55.

[16] Kadono, T., Zhang, X.Q., Srinivasan, S., Ishida, H., Barry, W.H. and Benjamin, I.J. (2006) CRYAB and HSPB2 deficiency increases myocyte mitochondrial permeability transition and mitochondrial calcium uptake. *J Mol Cell Cardiol,* 40, 783-9.

[17] Pinz, I., Robbins, J., Rajasekaran, N.S., Benjamin, I.J. and Ingwall, J.S. (2007) Unmasking different mechanical and energetic roles for the small heat shock proteins CryAB and HSPB2 using genetically modified mouse hearts. *FASEB J,* in press.

[18] Vicart, P., Caron, A., Guicheney, P., Li, Z., Prevost, M.C., Faure, A., Chateau, D., Chapon, F., Tome, F., Dupret, J.M. *et al.* (1998) A missense mutation in the alphaB-crystallin chaperone gene causes a desmin-related myopathy. *Nat Genet,* 20, 92-5.

[19] Pelliccia, A., Maron, B.J., Culasso, F., Spataro, A. and Caselli, G. (1996) Athlete's heart in women. Echocardiographic characterization of highly trained elite female athletes. *Jama,* 276, 211-5.

[20] Maron, B.J., Shirani, J., Poliac, L.C., Mathenge, R., Roberts, W.C. and Mueller, F.O. (1996) Sudden death in young competitive athletes. Clinical, demographic, and pathological profiles [see comments]. *Jama,* 276, 199-204.

[21] Tanigawa, G., Jarcho, J.A., Kass, S., Solomon, S.D., Vosberg, H.P., Seidman, J.G. and Seidman, C.E. (1990) A molecular basis for familial hypertrophic cardiomyopathy: an alpha/beta cardiac myosin heavy chain hybrid gene. *Cell,* 62, 991-8.

[22] Alcalai, R., Seidman, J.G. and Seidman, C.E. (2008) Genetic basis of hypertrophic cardiomyopathy: from bench to the clinics. *J Cardiovasc Electrophysiol,* 19, 104-10.

[23] Lind, J.M., Chiu, C. and Semsarian, C. (2006) Genetic basis of hypertrophic cardiomyopathy. *Expert Rev Cardiovasc Ther,* 4, 927-34.

[24] Olson, T.M. and Keating, M.T. (1996) Mapping a cardiomyopathy locus to chromosome 3p22-p25. *J Clin Invest,* 97, 528-32.

[25] Tidball, J.G. (1992) Desmin at myotendinous junctions. *Exp Cell Res,* 199, 206-12.

[26] Li, D., Tapscoft, T., Gonzalez, O., Burch, P.E., Quinones, M.A., Zoghbi, W.A., Hill, R., Bachinski, L.L., Mann, D.L. and Roberts, R. (1999) Desmin mutation responsible for idiopathic dilated cardiomyopathy. *Circulation,* 100, 461-4.

[27] Dalakas, M.C., Park, K.Y., Semino-Mora, C., Lee, H.S., Sivakumar, K. and Goldfarb, L.G. (2000) Desmin myopathy, a skeletal myopathy with cardiomyopathy caused by mutations in the desmin gene. *N Engl J Med,* 342, 770-80.

[28] Goldfarb, L.G., Park, K.Y., Cervenakova, L., Gorokhova, S., Lee, H.S., Vasconcelos, O., Nagle, J.W., Semino-Mora, C., Sivakumar, K. and Dalakas, M.C. (1998) Missense mutations in desmin associated with familial cardiac and skeletal myopathy. *Nat Genet,* 19, 402-3.

[29] Munoz-Marmol, A.M., Strasser, G., Isamat, M., Coulombe, P.A., Yang, Y., Roca, X., Vela, E., Mate, J.L., Coll, J., Fernandez-Figueras, M.T. *et al.* (1998) A dysfunctional desmin mutation in a patient with severe generalized myopathy. *Proc Natl Acad Sci U S A,* 95, 11312-7.

[30] Sjoberg, G., Saavedra-Matiz, C.A., Rosen, D.R., Wijsman, E.M., Borg, K., Horowitz, S.H. and Sejersen, T. (1999) A missense mutation in the desmin rod domain is associated with autosomal dominant distal myopathy, and exerts a dominant negative effect on filament formation. *Hum Mol Genet,* 8, 2191-8.

[31] Schroder, R., Vrabie, A. and Goebel, H.H. (2007) Primary desminopathies. *J Cell Mol Med,* 11, 416-26.

[32] Selcen, D. and Engel, A.G. (2004) Mutations in myotilin cause myofibrillar myopathy. *Neurology,* 62, 1363-71.

[33] Selcen, D. and Engel, A.G. (2005) Mutations in ZASP define a novel form of muscular dystrophy in humans. *Annals of neurology,* 57, 269-76.

[34] Vorgerd, M., van der Ven, P.F., Bruchertseifer, V., Lowe, T., Kley, R.A., Schroder, R., Lochmuller, H., Himmel, M., Koehler, K., Furst, D.O. *et al.* (2005) A mutation in the dimerization domain of filamin c causes a novel type of autosomal dominant myofibrillar myopathy. *Am J Hum Genet,* 77, 297-304.

[35] Selcen, D., Muntoni, F., Burton, B.K., Pegoraro, E., Sewry, C., Bite, A.V. and Engel, A.G. (2009) Mutation in BAG3 causes severe dominant childhood muscular dystrophy. *Annals of neurology,* 65, 83-9.

[36] Fardeau, M., Godet-Guillain, J., Tome, F.M., Collin, H., Gaudeau, S., Boffety, C. and Vernant, P. (1978) [A new familial muscular disorder demonstrated by the intra-sarcoplasmic accumulation of a granulo-filamentous material which is dense on electron microscopy (author's transl)]. *Rev Neurol (Paris),* 134, 411-25.

[37] Selcen, D. and Engel, A.G. (2003) Myofibrillar myopathy caused by novel dominant negative alpha B-crystallin mutations. *Annals of neurology,* 54, 804-10.

[38] Inagaki, N., Hayashi, T., Arimura, T., Koga, Y., Takahashi, M., Shibata, H., Teraoka, K., Chikamori, T., Yamashina, A. and Kimura, A. (2006) Alpha B-crystallin mutation in dilated cardiomyopathy. *Biochem Biophys Res Commun,* 342, 379-86.

[39] Pilotto, A., Marziliano, N., Pasotti, M., Grasso, M., Costante, A.M. and Arbustini, E. (2006) alphaB-crystallin mutation in dilated cardiomyopathies: low prevalence in a consecutive series of 200 unrelated probands. *Biochem Biophys Res Commun,* 346, 1115-7.

[40] Berry, V., Francis, P., Reddy, M.A., Collyer, D., Vithana, E., MacKay, I., Dawson, G., Carey, A.H., Moore, A., Bhattacharya, S.S. *et al.* (2001) Alpha-B crystallin gene (CRYAB) mutation causes dominant congenital posterior polar cataract in humans. *Am J Hum Genet,* 69, 1141-5.

[41] Perng, M.D., Muchowski, P.J., van Den, I.P., Wu, G.J., Hutcheson, A.M., Clark, J.I. and Quinlan, R.A. (1999) The cardiomyopathy and lens cataract mutation in alphaB-crystallin alters its protein structure, chaperone activity, and interaction with intermediate filaments in vitro. *J Biol Chem,* 274, 33235-43.

[42] Bova, M.P., Yaron, O., Huang, Q., Ding, L., Haley, D.A., Stewart, P.L. and Horwitz, J. (1999) Mutation R120G in alphaB-crystallin, which is linked to a desmin- related myopathy, results in an irregular structure and defective chaperone-like function. *Proc Natl Acad Sci U S A,* 96, 6137-42.

[43] Gulick, J., Subramaniam, A., Neumann, J. and Robbins, J. (1991) Isolation and characterization of the mouse cardiac myosin heavy chain genes. *J Biol Chem,* 266, 9180-5.

[44] Wang, X., Osinska, H., Klevitsky, R., Gerdes, A.M., Nieman, M., Lorenz, J., Hewett, T. and Robbins, J. (2001) Expression of R120G-alphaB-crystallin causes aberrant desmin and alphaB-crystallin aggregation and cardiomyopathy in mice. *Circ Res,* 89, 84-91.

[45] Rajasekaran, N.S., Connell, P., Christians, E.S., Yan, L.J., Taylor, R.P., Orosz, A., Zhang, X.Q., Stevenson, T.J., Peshock, R.M., Leopold, J.A. *et al.* (2007) Human alphaB-crystallin mutation causes oxido-reductive stress and protein aggregation cardiomyopathy in mice. *Cell,* 130, 427-39.

[46] Wang, X., Osinska, H., Dorn, G.W., 2nd, Nieman, M., Lorenz, J.N., Gerdes, A.M., Witt, S., Kimball, T., Gulick, J. and Robbins, J. (2001) Mouse model of desmin-related cardiomyopathy. *Circulation,* 103, 2402-7.

[47] Reimann, J., Kunz, W.S., Vielhaber, S., Kappes-Horn, K. and Schroder, R. (2003) Mitochondrial dysfunction in myofibrillar myopathy. *Neuropathol Appl Neurobiol,* 29, 45-51.

[48] Maloyan, A., Sanbe, A., Osinska, H., Westfall, M., Robinson, D., Imahashi, K., Murphy, E. and Robbins, J. (2005) Mitochondrial dysfunction and apoptosis underlie the pathogenic process in alpha-B-crystallin desmin-related cardiomyopathy. *Circulation,* 112, 3451-61.

[49] Benjamin, I.J., Guo, Y., Srinivasan, S., Boudina, S., Taylor, R.P., Rajasekaran, N.S., Gottlieb, R., Wawrousek, E.F., Abel, E.D. and Bolli, R. (2007) CRYAB and HSPB2 deficiency alters cardiac metabolism and paradoxically confers protection against myocardial ischemia in aging mice. *Am J Physiol Heart Circ Physiol,* 293, H3201-9.

[50] Gores, G.J., Flarsheim, C.E., Dawson, T.L., Nieminen, A.L., Herman, B. and Lemasters, J.J. (1989) Swelling, reductive stress, and cell death during chemical hypoxia in hepatocytes. *The American journal of physiology,* 257, C347-54.

[51] Trotter, E.W. and Grant, C.M. (2002) Thioredoxins are required for protection against a reductive stress in the yeast Saccharomyces cerevisiae. *Molecular microbiology,* 46, 869-78.

[52] Mehlen, P., Hickey, E., Weber, L.A. and Arrigo, A.P. (1997) Large unphosphorylated aggregates as the active form of hsp27 which controls intracellular reactive oxygen species and glutathione levels and generates a protection against TNFalpha in NIH-3T3-ras cells. *Biochem Biophys Res Commun,* 241, 187-92.

[53] Amado, L.C., Saliaris, A.P., Raju, S.V., Lehrke, S., St John, M., Xie, J., Stewart, G., Fitton, T., Minhas, K.M., Brawn, J. *et al.* (2005) Xanthine oxidase inhibition ameliorates cardiovascular dysfunction in dogs with pacing-induced heart failure. *J Mol Cell Cardiol,* 39, 531-6.

[54] Maloyan, A., Osinska, H., Lammerding, J., Lee, R.T., Cingolani, O.H., Kass, D.A., Lorenz, J.N. and Robbins, J. (2009) Biochemical and mechanical dysfunction in a mouse model of desmin-related myopathy. *Circ Res,* 104, 1021-8.

[55] Levine, B. and Klionsky, D.J. (2004) Development by self-digestion: molecular mechanisms and biological functions of autophagy. *Dev Cell,* 6, 463-77.

[56] Ohsumi, Y. (2001) Molecular dissection of autophagy: two ubiquitin-like systems. *Nat Rev Mol Cell Biol,* 2, 211-6.

[57] Boya, P., Gonzalez-Polo, R.A., Casares, N., Perfettini, J.L., Dessen, P., Larochette, N., Metivier, D., Meley, D., Souquere, S., Yoshimori, T. *et al.* (2005) Inhibition of macroautophagy triggers apoptosis. *Mol Cell Biol,* 25, 1025-40.

[58] Tannous, P., Zhu, H., Johnstone, J.L., Shelton, J.M., Rajasekaran, N.S., Benjamin, I.J., Nguyen, L., Gerard, R.D., Levine, B., Rothermel, B.A. *et al.* (2008) Autophagy is an adaptive response in desmin-related cardiomyopathy. *Proc Natl Acad Sci U S A,* 105, 9745-50.

In: Small Stress Proteins and Human Diseases
Editors: Stéphanie Simon et al.
ISBN: 978-1-61470-636-6

Chapter 2.5

HspB1 and HspB8 Mutations in Neuropathies

Rainer Benndorf*

Center for Clinical and Translational Research, The Research Institute at Nationwide Children's Hospital, and Department of Pediatrics, Ohio State University, Columbus, Ohio, U.S.

Abstract

Several mutations in the two related small Heat Shock Protein (sHSP) genes, *HSPB1* and *HSPB8,* are associated with and cause the two inherited peripheral motor neuron diseases, distal hereditary motor neuropathy and Charcot-Marie-Tooth disease. Most mutations have an autosomal dominant or semi-dominant inheritance, are of the missense type, and are positioned in the conserved α-crystallin domain of these proteins. The role of sHSPs in neurons is only poorly understood. Likewise, the slow death specifically of motor neurons resulting from these mutations is insufficiently understood.

The studied mutant sHSPs exhibit a disparate pattern of abnormal properties and functions, including structural instabilities and impaired chaperon-like and anti-apoptotic activities. Cell-biological consequences that are common to most if not all mutant sHSPs are their propensity to form intracellular inclusion bodies (aggresomes) and to disrupt cytoskeletal networks. Both phenomena are likely to affect intracellular transport and eventually the viability of neurons. Given the involvement of sHSPs in many cellular processes, arguments are summarized for possible adverse effects of mutant sHSPs on the proteasome system, the RNA processing machinery, and the redox homeostasis. Additionally, arguments will be provided for the possibility that mutant sHSP mRNA may exert adverse effects on its own without protein involvement.

---

* Rainer Benndorf
Center for Clinical and Translational Research
The Research Institute at Nationwide Children's Hospital
Room WA2109, Research Building II
700 Children's Drive
Columbus, OH 43205, USA
Phone: 614-722-2672
Fax: 614-722-5892
Rainer.Benndorf@Nationwidechildrens.org

It is proposed that most (semi-)dominant mutations affect neurons by one or more gain-of-function mechanisms including "classic" dominant-negative effects (through abnormal sHSP interactions), cytotoxic gene products (through formation of aggresomes or through "toxic" mRNAs), altered structural proteins (through a disruption of the cytoskeleton), and/or by increased protein activity (through a downstream redox imbalance), all known mechanisms of genetic dominance which are consistent with experimental findings. On the other hand, haploinsufficiency due to a simple loss of sHSP functions (e.g. of their chaperone-like or anti-apoptotic activities) seems less likely to be the primary defect in the case of the (semi-)dominant mutant sHSPs.

Given the (semi-)dominant nature of most sHSP mutations, specific inhibition of expression of the mutant sHSP genes in the affected neurons should be the preferred approach of future therapeutic strategies. Alternative strategies include the stimulation of the proteasome system and autophagy, the stabilization of the redox homeostasis, and application of substances that inhibit protein aggregation.

INTRODUCTION

In recent years a number of mutations in the two related small heat shock protein (sHSP)[1] genes, *HSPB1* (protein: HspB1, Hsp27, Hsp25) and *HSPB8* (protein: HspB1, Hsp22, H11, E2IG1) were identified as a cause of the two inherited peripheral motor neuron diseases (MND)[2,3], distal hereditary motor neuropathy (dHMN) and Charcot-Marie-Tooth disease (CMT) [25, 37, 39, 42, 44, 48, 57, 86]. These discoveries came not entirely unexpected as earlier work had revealed the importance of HspB1, HspB8, and also of the related HspB5 (αB-crystallin) for the biology of neurons, both in health and disease. Initially, elevated levels of HspB1 and HspB5 were found in patients with a number of neurodegenerative diseases as diverse as amyotrophic lateral sclerosis (ALS), multiple sclerosis, and Alexander, Parkinson, and Alzheimer diseases. A hallmark of these diseases is the deposition of unfolded or denatured proteins (amyloids) in the affected neuronal tissues that typically contain high amounts of HspB1 and HspB5 [34].

Later sHSPs were identified as important players in normal neuron function and development [53, 55, 58, 70, 94, 96]. HspB1, for example, plays a critical role in neurite extension and branching. HspB1 is induced in neurons by ischemia, heat, and apoptosis-inducing withdrawal of growth factors. Acquired stress tolerance of neurons largely correlates with the induction of heat shock proteins including sHSPs. More direct evidence of the neuroprotective effect of sHSPs was obtained from overexpression experiments in various experimental systems including transgenic mice. HspB1 and HspB5 exhibited a potent general cytoprotective function, typically by inhibition of apoptosis, and this function was verified for neurons [98] (see also reviews [22, 55] and chapters 1-3, 1-4 and 3-4). In spite of these insights, the precise role of the single wild-type sHSPs (^{wt}sHSP) or of the entire cohort in neurons and in particular in motor neurons is poorly understood, and much less is the role of mutant sHSPs (^{mu}sHSP).

In this chapter, the knowledge on MND-associated mutant forms of HspB1 (^{mu}HspB1) and HspB8 (^{mu}HspB8) will be summarized and candidate pathological mechanisms will be considered. Since much of our knowledge on the cellular functions of sHSPs originates from studies on both wild-type and mutant HspB5 (^{wt}HspB5, ^{mu}HspB5), findings related to this sHSP will be also included where appropriate, in spite of the fact that all identified mutations in the *HSPB5* gene are associated with the muscular disorders myofibrillar (or desmin-related) myopathy and dilated cardiomyopathy and/or with cataracts in the lens of the eye (see chapters 2-3 and 2-4).

1. Survey on the Molecular Etiology of the Peripheral Neuropathies dHMN and CMT

Peripheral neuropathy denotes a wide array of symptoms that can develop when nerves are injured. More than 100 types of peripheral neuropathies have been identified, each with its own characteristic symptoms, patterns of development, and prognosis[4]. Peripheral neuropathy may be either inherited or acquired. dHMN is a heterogeneous group of inherited peripheral neuropathies that affect primarily the lower motor nerves. Motor neurons degenerate beginning in the distal parts of the limbs. Weakness and wasting of the extensor muscles of the feet frequently result in foot deformity. As the disease progresses, distal upper limb muscles are also involved, and other symptoms may occur like paralysis of the vocal cords and diaphragm. CMT has similar symptoms, with additional involvement of the peripheral sensory nerves. The clinical symptoms in dHMN and CMT vary considerably and overlap partially. Symptoms in dHMN and CMT may also overlap with other MNDs such as certain forms of ALS [18], although this disease affects both upper and lower motor neurons.

In recent years, mutations in many genes were identified that are associated with and cause dHMN and CMT. Mutations affect primarily the myelin sheath around the axons (demyelinating forms), the axons themselves (axonal forms), or both (intermediate forms). Excellent reviews on the molecular genetics in dHMN and CMT were published earlier [43, 63, 78, 100]. A stunning observation is that mutations in very different groups of genes result in similar or identical disease phenotypes, although some of the affected genes can be grouped based on their known or putative functions in cells, tentatively suggesting involvement in the same pathomechanism(s). This genetic heterogeneity has hindered our understanding of the pathology of MND and also the development of therapeutic concepts. In axonal or intermediate forms of dHMN and CMT, mutated proteins are largely involved in cell architecture (cytoskeleton) and intracellular transport (e.g. neurofilament light chain, gigaxonin, dynactin, kinesin1B, lamininA, ras-related protein Rab-7), in mitochondrial fusion and fragmentation (mitofusin2, ganglioside-induced differentiation-associated protein-1), and in metabolism and processing of RNA (glycyl- and tyrosyl-tRNA synthetases, putative RNA helicase senataxin). HspB1 and HspB8 form a further group of proteins that are affected by mutations [8, 22]. With a typical late onset age of 15 years or older, ^{mu}sHSP-associated MNDs share this feature with many other neurodegenerative diseases.

2. Neuropathy- and Myopathy-Associated Mutations in HspB1, HspB8, and HspB5

The positions and nature of the known mutations in HspB1 and HspB8 related to MND are shown in Figure 1. For comparison, mutations in HspB5 associated with myopathy are also included. Overall, these mutations have no obvious common denominator, although

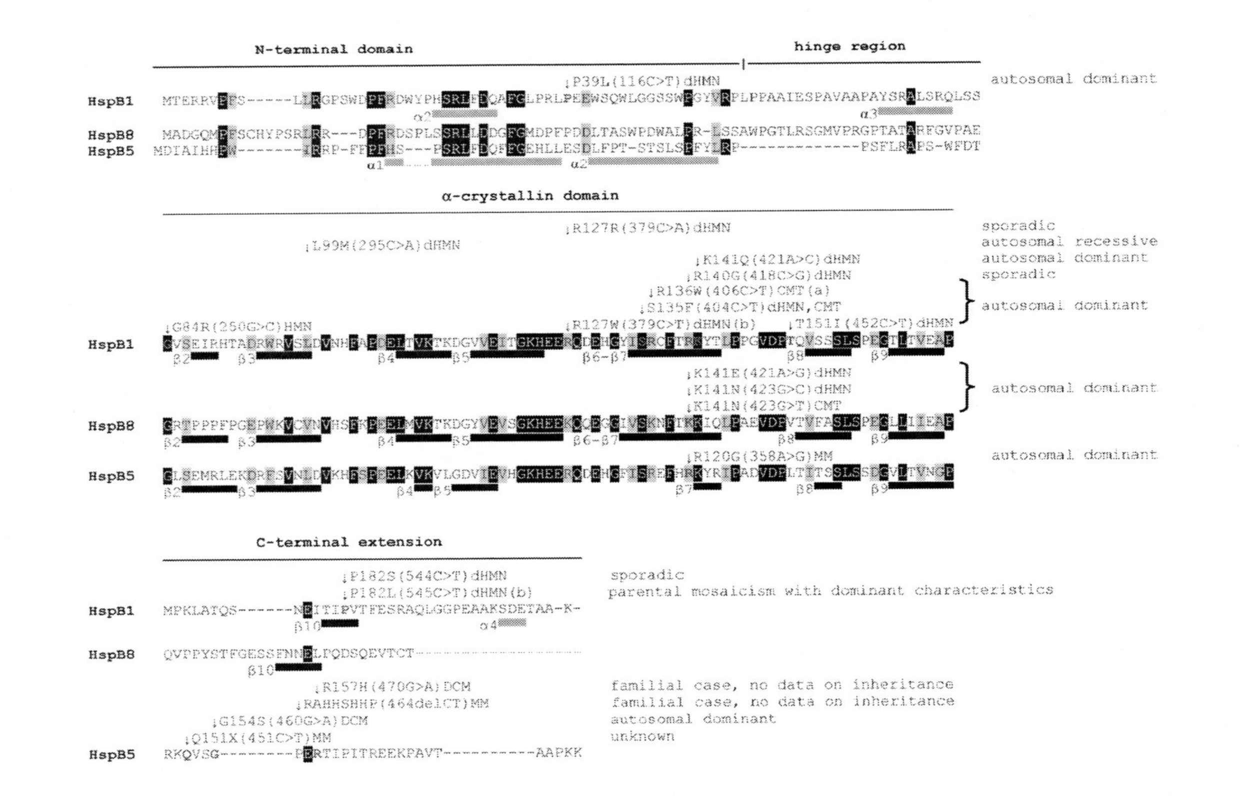

Figure 1. Positions and types of neuropathy- and myopathy-related mutations in the aligned sequences of HspB1, HspB8, and HspB5. Alignment and domain designation was taken from the alignment of all 10 human sHSPs [27]. Identical and similar (E/D, A/G, H/F/W/Y, S/T, I/L/V, H/R/K) amino acid residues in all 3 sequences are highlighted in black and gray, respectively. Mutations were taken from these reports: HspB1: R127W, S135F, R136W, T151I, P182L [25]; P182S [48]; R127W [57]; P39L, G84R, L99M, R127W, R127R, S135F, R140G [37; 44]; K141Q [39]. HspB8: K141E, K141N [42]; K141N [86]. HspB5: R120G [91]; Q151X, 464delCT [75]; G154S [69]; R157H [40]. The predicted positions of the α-helices α1-α4 (gray bars) and β-strands β2-β10 (black bars) are indicated below each sequence as published previously for HspB1 [88] and HspB5 [47], or as predicted by the JPRED program for HspB8. Abbreviations: dHMN, distal hereditary motor neuropathy; CMT, Charcot-Marie-Tooth disease; MM, myofibrillar or desmin-related myopathy; DCM, dilated cardiomyopathy. (a) Personal communication on inheritance by Dr. J. Irobi, Antwerp, Belgium; (b) Inheritance described in [20].

several observations are noteworthy: Identical mutations can cause both dHMN and CMT. Among the mutations with known inheritance, most are dominant (or semi-dominant, see below), while only one mutation is recessive. Most are missense mutations, and only one mutation each is a nonsense, frameshift, or silent mutation. Mutations are clustered in the conserved α-crystallin domains - in particular in the regions of the predicted β-strand(s) β6-β7 - and also in the highly variable C-terminal extensions. Only 1 mutation is located in the moderately conserved N-terminal domain. Most mutations affect well conserved residues, while some positions are more variable. Positions P39, T151, and P182 in HspB1, for example, are only moderately conserved among the human sHSPs [27], albeit these residues are highly conserved among the mammalian orthologs (not shown). One homologous hot spot region is mutated in all 3 sequences: ^{R140G}HspB1, ^{K141Q}HspB1, ^{K141E}HspB8, ^{K141N}HspB8, and ^{R120G}HspB5 have in common the elimination of the positively charged amino acid residues R or K. Whether the silent mutation ^{R127R}HspB1(379C>A) is associated with disease or represents a polymorphism remains to be determined (see below).

3. Normal and Abnormal Properties of Wild-Type and Mutant HspB1 and HspB8 with Relevance for MND

Phylogenetic analysis revealed that human wild-type HspB1 (^{wt}HspB1) and HspB8 (^{wt}HspB8) are most similar among all proteins [85]. Thus, the fact that mutations in either gene cause identical or similar disease phenotypes is not a surprise. Both genes are expressed in mammals in most tissues and organs including neuronal tissues [41, 70, 90]. Their proximal promoter regions contain heat-responsive elements and consequently the expression can be induced by heat stress in addition to other stress factors [17, 36]. Neuronal tissues typically contain, in addition to HspB1 and HspB8, other sHSPs such as HspB5 and HspB6, and this fact obscures the function of a particular sHSP as these proteins interact with one another with the degree of functional redundancy or specificity being unknown. Another complicating factor for understanding the role of sHSPs in neurons is their involvement in many different cellular processes.

In this section, normal and abnormal properties and functions of wild-type and mutant forms of HspB1 and HspB8 are summarized that are of potential relevance for the pathology of the associated MND. Those aspects of sHSP functions that are covered by other chapters of this book will be mentioned only briefly. See also recent reviews on sHSP functions [3, 76, 92].

3.1. sHSP Structure and Interactions

Because none of the mammalian sHSPs has been crystallized to date, the structure of mammalian HspB1 and HspB5 was modeled by using crystallographic data of *Methanococcus* and *Triticum* sHSPs and other approaches [47, 88; and references therein]. According to these models, the prevailing secondary structural element of HspB1 and HspB5 are β-strands, and 7 or 8 β-strands in the α-crystallin domain are organized in two β-sheets. Prediction of the secondary structure of HspB8 using the JPRED program[5] indicates a β-strand distribution which is very similar to that of HspB1 (Figure 1). These β-strands play a critical role in the structure of the sHSP monomers and in interactions between the monomers. Thus, it is not surprising that most of the MND-associated mutations in HspB1 and HspB8 are positioned in or adjacent to the β-strands in the α-crystallin domain (notably in or adjacent to strands β6/β7), and indeed abnormalities and instabilities of the secondary, tertiary and quaternary structures were demonstrated *in vitro* and/or *in vivo* for several ^{mu}sHSP species including ^{S135F}HspB1, ^{P182L}HspB1, ^{R120G}HspB5, ^{K141E}HspB8, and hamster ^{R148G}HspB1 (homologous to human ^{R140G}HspB1) [14, 47, 49, 52, 62, 67].

Both HspB1 and HspB8 have the capacity to interact with themselves and with other sHSPs resulting in the formation of homo- and hetero-dimers [85]. HspB1 and possibly also HspB8 dimers can assemble in oligomeric complexes with >600 kDa size. The existence of HspB1-HspB8 hetero-dimers provides another explanation why mutations in either protein result in similar disease phenotypes as the same cellular processes may be affected. At the level of dimers, abnormal interaction properties were reported for all studied ^{mu}sHSPs including ^{S135F}HspB1, ^{R120G}HspB5, ^{Q151X}HspB5, ^{464delCT}HspB5, ^{K141E}HspB8, and ^{K141N}HspB8 [26, 79]. These interaction abnormalities are likely to result in abnormally structured and malfunctioning sHSP complexes. Such abnormal complexes may well be at the base of the associated pathologies as they potentially can affect all described sHSPs properties and functions including chaperoning, regulation of apoptotic activities and of redox homeostasis, organization of the cytoskeleton, and formation of aggresomes with subsequent adverse consequences e.g. for the axonal transport and eventually for neuron viability.

3.2. Aggresome Formation

Proteins affected by mutations frequently form cytotoxic amyloid fibrils which precipitate into aggresomes. In experimental systems, aggresomes develop in a time-dependent process starting with formation of multiple foci that subsequently merge into large perinuclear aggregates. As prefibrillary precursor forms exhibit a much higher cytotoxicity than the aggresomes themselves [82], it is not clear whether aggresome formation is part of the pathological processes or of a detoxifying mechanism. Most if not all studied mutant forms of HspB1 and HspB8 tended to form aggresomes, including ^{G84R}HspB1, ^{R127W}HspB1, ^{S135F}HspB1, ^{R136W}HspB1, ^{T151I}HspB1, ^{P182L}HspB1, ^{K141E}HspB8, ^{K141N}HspB8, and hamster ^{R148G}HspB1 (homologous to human ^{R140G}HspB1) [1, 14, 25, 26, 42, 44, 52]. See also the chapter 1-7 in this book. In some cases these aggresomes were shown to recruit intermediate filament proteins.

On the other hand, ^{wt}sHSPs have the capacity to attenuate the formation of aggresomes caused by precipitation of ^{mu}sHSPs or of other proteins. ^{wt}HspB1 and ^{wt}HspB5, for example, prevented aggresome formation by co-oligomerizing with ^{R120G}HspB5 [15]. Similarly, ^{wt}HspB1 or ^{wt}HspB8 interrupted the formation of ^{R120G}HspB5 aggresomes in virus-infected cardiomyocytes [73]. In transfected Cos-7 cells, aggresome formation of ^{K141E}HspB8- and ^{K141N}HspB8-citrine fusion proteins was attenuated, although not prevented, by co-expression of various ^{wt}sHSPs [26].

This formation of aggresomes, or more likely of their highly toxic precursor forms, can be assumed to be critical for the pathology of the associated MND. Theoretically, ^{mu}sHSPs-caused aggresome formation can result either from a loss of the chaperone-like activity leading to initial precipitation of client proteins with possible subsequent recruitment of the ^{mu}sHSPs into these aggresomes (loss of function), or from acquisition of a new and toxic property of the ^{mu}sHSPs themselves leading to protein precipitation (gain of function), or from both (see below).

3.3. Chaperone-like Properties

HspB1 has a chaperone-like property *in vitro* and *in vivo*, albeit it does not actively fold proteins on its own. Large oligomeric forms of HspB1 bind non-native and partially denatured client proteins and trap them in a folding-competent state. Eventually these client proteins are released by ATP-dependent chaperones (such as Hsp70) if refolding occurs, or by the proteasome system if degradation occurs [92]. Thus, HspB1 acts at the branch point that destines the fate of partially denatured proteins either towards recycling or degradation. It is widely believed that the involvement of HspB1 in chaperoning is its primary function, and that this is the basis for its well-described cytoprotective and anti-apoptotic activity *in vivo*. Similar to HspB1, HspB8 exhibits chaperone-like activity *in vitro* [17, 50], and this role was verified for the *in vivo* situation when its expression prevented huntingtin-caused inclusion body formation in association with Bag3 [13].

Some of the tested ^{mu}sHSP species (^{P182L}HspB1; ^{R120G}HspB5, ^{K141E}HspB8) lost their chaperone-like properties at least partially, while other studied ^{mu}sHSP species (^{R127W}HspB1, ^{S135F}HspB1, ^{R136W}HspB1, ^{T151I}HspB1) were unchanged in this property [49, 52, 67]. Thus, loss of the chaperone-like property apparently is not a common denominator among the various ^{mu}sHSP species.

3.4. Anti- and Pro-Apoptotic Activities of HspB1 and HspB8

The cytoprotective and anti-apoptotic activity of HspB1 and of other sHSPs was demonstrated in numerous studies [reviewed in 3], and this function of HspB1 was also verified for neurons [7]. See also the chapters 1-1, 1-3, 1-4 and 1-5 in this book. HspB8 also modulates cell death, although apparently in a pro-apoptotic way. In cardiomyocytes, high level expression of HspB8 caused apoptosis involving inhibition of casein kinase-2 [33]. Similarly, a pro-apoptotic role of this sHSP was reported in human melanoma cells [30]. As both HspB1 and HspB8 are similar proteins, this discrepancy is somewhat surprising. The pro-apoptotic activity of ^{wt}HspB8 is of potential interest as MND-associated mutations may enhance this intrinsic property (see below).

While all tested disease-associated ^{mu}HspB1 species (^{R127W}HspB1, ^{S135F}HspB1, ^{R136W}HspB1, ^{T151I}HspB1, ^{P182L}HspB1) reduced cell viability if expressed in neuronal cells, only a subset (^{R136W}HspB1, ^{T151I}HspB1, ^{P182L}HspB1) apparently induced apoptosis [25, 52]. The ^{mu}HspB8 species (^{K141E}HspB8, ^{K141N}HspB8) also reduced cell viability, however their apoptotic activities were not studied [42].

Theoretically, reduced cell viability can be achieved either by loss of cytoprotective (or anti-apoptotic) properties or by acquisition of cytotoxic (or pro-apoptotic) properties resulting from mutations (see below). Whatever the mechanism, reduced cell viability - but not necessarily pro-apoptotic activity - appears to be a common denominator of ^{mu}sHSP species. In case of the ^{mu}HspB8 species, this cytotoxic property may be subjoined to the basal pro-apoptotic property of ^{wt}HspB8.

3.5. Maintenance of the Redox Homeostasis

Another line of experimentation has implicated HspB1 in the regulation of homeostasis of the intracellular redox state and in the resistance of cells to oxidative injuries. Large oligomers of HspB1 both decreased the levels of reactive oxygen species and increased the level of reduced glutathione through their ability to increase the glucose-6-phosphate dehydrogenase activity [4]. HspB1 also maintained the mitochondrial membrane potential thus supporting ATP production in stress situations. The redox homeostasis function of HspB1 is thought to contribute to its cytoprotective and anti-apoptotic activities.

As myopathy-associated ^{R120G}HspB5 caused oxido-reductive stress in mice by abnormally increasing enzymatic activities of glucose-6-phosphate dehydrogenase and of other enzymes of the redox metabolism [71], impaired redox homeostasis function of ^{mu}HspB1 and ^{mu}HspB8 species is a realistic scenario for the damage of motor neurons in the associated diseases, although this remains to be substantiated experimentally.

3.6. Organization and Stabilization of the Cytoskeleton

In many studies, HspB1 and HspB5 were found to be associated with components of the cytoskeleton, notably microfilaments and intermediate filaments (see chapter 1-2). Both sHSPs were implicated in the organization and stabilization of these cytoskeletal systems, a role which is thought to contribute to their cytoprotective function.

As some of the ^{mu}sHSPs are known to affect the organization of the cytoskeleton, this aspect of their malfunction can be expected to contribute to the pathology of the associated diseases. Expression of ^{S135F}HspB1 in adrenal carcinoma cells disrupted the neurofilaments and caused the formation of aggresomes which did not happen in the presence ^{wt}HspB1 [25]. Similar effects of ^{S135F}HspB1 and ^{P182L}HspB1 were observed in cultured motor neurons and in primary neuronal cells, respectively [1, 99]. Overexpressing ^{R127W}HspB1, ^{S135F}HspB1, ^{R136W}HspB1, ^{T151I}HspB1, or ^{P182L}HspB1 in a neuronal cell line also resulted in formation of aggresomes [52]. Since aggresomes usually recruit also cytoskeletal proteins, disruption of the neurofilament network and possibly of other cytoskeletal systems can be assumed to be a consequence of most if not all ^{mu}HspB1 species.

Given the association of HspB1 with the neurofilament network, it is not surprising that several mutations in the neurofilament light chain (NF-L) are also associated with axonal forms of CMT [99, and references therein]. Mutations in either HspB1 (^{S135F}HspB1) or NF-L disrupted the neurofilament assembly and caused aggregation of NF-L protein. ^{S135F}HspB1 had dominant effects on the disruption of neurofilament assembly and the aggregation of NF-L protein. Co-expression of ^{wt}HspB1 diminished and reversed aggregation of mutant NF-L and increased cell viability. Together these data support the notion that disruption of neurofilament network is a common and critical event in the pathology of both ^{mu}HspB1- and mutant NF-L-associated MND.

Interestingly, there is a similar relationship between HspB5 and the intermediate filament protein desmin. HspB5 is thought to play a role in desmin filament network assembly, and mutations in HspB5 or desmin cause desmin-related (or myofibrillar) myopathy characterized by a disruption of the desmin filament network in the affected muscles [31, 91].

3.7. Involvement of HspB8 in Autophagy and Translational Arrest

While a chaperone-like activity for HspB8 was reported *in vivo*, it actually may act in a non-canonical manner which is un-related to the classical chaperone model [12]. In complex with Bag3, HspB8 induced phosphorylation of eIF2α resulting in stimulation of autophagy and translational arrest. The significance of autophagic activity as a protective mechanism in aggresome-prone diseases has been demonstrated for ^{R120G}HspB5-associated myopathy [87]. The fact that a mutation in the *BAG3* gene causes severe dominant muscular dystrophy with aggresome formation further supports the significance of the HspB8/Bag3 complex for clearing of cells from aggresomes [74]. Why mutations in Bag3 and HspB8 cause myopathy and MND, respectively, is not known. The fact that HspB8/Bag3 induced phosphorylation of eIF-2α is highly interesting as also HspB1 was implicated in regulation of phosphorylation of eIF-2α in the context of stress granule function, although with different consequences (see below).

Together these data suggest a role of HspB8/Bag3 in autophagy which is critical for clearing of cells from aggresomes and their precursors. Accordingly, ^{mu}HspB8 species may affect this mechanism thus contributing to the disease, although the evidence remains to be provided.

3.8. Involvement of sHSPs in Proteasome-Mediated Protein Degradation

Growing evidence suggests a critical role of ^{wt}HspB1, ^{wt}HspB8, and also of ^{wt}HspB5 in the proteasome-mediated protein degradation. As proteasomes are key elements in the cell's ability to cope with cytotoxic amyloid proteins, the potential consequences of mutations in HspB1 and HspB8 for the proteasome function are of interest in the context of the associated MNDs.

The binding of ^{wt}HspB1 was required for both activation of 26S proteasomes and promotion of degradation of ubiquitinated proteins such as phosphorylated I-κBα [66]. This function of ^{wt}HspB1 could contribute, through increased NF-κB activity, to its anti-apoptotic properties. Similarly, ^{wt}HspB1 stimulated p27Kip1 ubiquitination and degradation in stress situations [65].

In a mouse model of cardiac hypertrophy induced by overexpression of ^{wt}HspB8, increased abundance and activity of proteasomes was reported [35]. Proteasome inhibition reversed the cardiac hypertrophy suggesting an abnormally increased proteasome activity as a critical event downstream of ^{wt}HspB8. In the presence of ^{wt}HspB8, aggresome formation by myopathy-associated ^{R120G}HspB5 was prevented, probably involving proteasomal activity [73]. This role of ^{wt}HspB8 in activating proteasome activity may involve the E3-ubiquitin-protein ligase seven-in-absentia-homolog 1 (SIAH1) and the structurally related ring finger protein 10 (RNF10). Both proteins were identified as the main interacting partners of ^{wt}HspB8 (X. Sun, R. Benndorf, unpublished).

^{wt}HspB5 is an essential component of the Skp1-Cul1-F-box complex which functions as E3-ubiquitin-protein ligase [56]. In this system, rapid ubiquitination and degradation of the client proteins required the presence of ^{wt}HspB5.

The fact that several mutations in the ring finger protein and putative E3 protein-ubiquitin ligase SIMPLE causes CMT supports the importance of protein degradation for motor neuron function [72]. Thus, ^{mu}sHSPs could affect protein degradation in a similar way through interaction with components of the proteasome system.

Taken together, experimental data suggest a critical role of several sHSPs in the regulation of the proteasome activity. Impaired proteasome function resulting from mutations in HspB1 and HspB8 seems to be a realistic scenario for the lesion of neurons in MND.

3.9. The RNA Connection

This section summarizes data on RNA-related processes with potential consequences for the pathology of ^{mu}sHSP-associated and other MNDs.

The existence of the sporadic silent mutation ^{R127R}HspB1 (379C>A) indicates that the HspB1 mRNA may be the damaging agent [37]. However, as the inheritance of this mutation is not known, the association with the disease is not clear and thus this sequence variation may represent a polymorphism without relation to disease. On the other hand, the mutation ^{R127W}HspB1 (379C>T) which is clearly associated with disease affects the same position in the mRNA, albeit this mutation affects also the protein [20]. Thus, 379C may be a hot spot position in the HspB1 mRNA, and both mutations may result in a mRNA with abnormal and toxic properties causing injury of motor neurons, possibly without involving the HspB1 protein. Such a situation is not uncommon as single stranded mRNAs are known to be highly susceptible in their folding structures to single nucleotide substitutions which affect mRNA processing [45]. Applying this rationale to other sHSP mutations leads to the provocative hypothesis according to which the ^{mu}sHSP mRNAs would be the damaging agents with abnormal ^{mu}sHSP proteins being irrelevant. More likely, however, is a scenario in which both mutant mRNA and protein contribute to the death of neurons. Before final conclusions can be drawn, more work on the cellular effects of ^{mu}sHSP mRNAs is needed.

Apart from that, growing evidence supports a role of several sHSPs in various aspects of processing and regulation of RNA. Reports from ~2 decades ago identified sHSPs as components of so-called heat shock or stress granules [64]. Named neuronal granules in neurons, these cytoplasmic domains have emerged as important players in the posttranscriptional regulation of gene expression. In cellular stress situations, stalled translation initiation complexes are recruited and sorted to sites of storage, reinitiation, or decay [97]. The assembly of stress granules involves the recruitment of HspB1 [28] which may play a role in the restoration of translational activity after stress treatment by promoting dephosphorylation of eIF-2α [23]. In this respect, HspB1 has quite the opposite effect on phosphorylation of eIF-2α as compared to HspB8 (see above).

Later, both HspB1 and HspB5 were found to localize in nuclear storage domains for RNA-processing factors called nuclear speckles, and translocation to speckles depended on phosphorylation of these sHSPs [11, 19, 89]. Co-immunoprecipitation experiments suggested that the nuclear import of HspB5 is regulated by its interaction with the survival-of-motor-neurons (SMN) protein, a core component of the SMN complexes which are involved in small nuclear ribonucleoprotein assembly and RNA processing. Myopathy-associated ^{R120G}HspB5 was affected in its interaction with the SMN protein and in nuclear import [19]. In protein interaction experiments, ^{wt}HspB8 was found to interact with the DEAD box protein Ddx20 (gemin3, DP103), another core component of the SMN complexes, and the two disease-associated ^{mu}HspB8 forms showed abnormally increased binding to Ddx20 [84]. Thus, two independent lines of experimentation have connected the sHSPs with the SMN complexes. These findings are of potential interest as mutations in the *SMN1* gene cause spinal muscular atrophy (SMA), another MND and one of the most prevalent genetic causes of infant mortality [32]. These protein interaction studies have linked the etiologic factor sHSPs (HspB1, HspB8, HspB5) with the seemingly unrelated etiologic factor SMN protein to a common signaling pathway or protein complex, and mutations in any of their genes cause the various forms of MNDs or myopathies.

Other studies also reported a role of sHSPs in RNA processing and regulation. HspB1 bound to cell-death-inhibiting RNA CDIR and c-YES mRNA resulted in down-regulation of mRNAs [77, 81], and a mimic of phosphorylated HspB1 stabilized the cyclooxygenase-2 mRNA [54]. HspB1 enhanced recovery of splicing activity after heat shock by specifically regulating the splicing factor SRp38 [61]. HspB8 interacted with Sam68, an RNA-binding protein that has been implicated in cell proliferation and tumorigenesis [5].

Further support for RNA involvement in MND comes from the fact that a group of genes affected by mutations in dHMN and CMT are involved in RNA processing or metabolism. These genes include *GARS* (glycyl-tRNA synthetase), *YARS* (tyrosyl-tRNA synthetase), *IGHMBP2* (associated with the RNA processing machinery), and *SETX* (senataxin, a putative RNA helicase) [43, 46, 78].

In summary, available data tentatively suggest involvement ^{mu}sHSP mRNA in the pathology of the associated MNDs, as ^{mu}sHSP mRNA may have cytotoxicity on its own. Additionally, experimental data suggest involvement of several sHSPs, including HspB1, HspB5, and HspB8, in the RNA processing machinery which is of potential interest in the context of ^{mu}sHSP-associated MND.

3.10. Is HspB8 a Protein Kinase?

The reported protein kinase activity of ^{wt}HspB8 could be of potential interest for the pathology of mutant-associated MND [80]. However, as no convincing biochemical evidence for this activity has been provided to date, false-positive data cannot be excluded. Phylogenetic analysis did not reveal a significant relation of HspB8 to any of the protein kinases in the human kinome [60]. Instead, HspB8 clearly belongs to the superfamily of sHSPs [85]. While the conflicts on the role of HspB8 have not been resolved to date [cf. 29, 51], it seems unlikely that this claimed protein kinase activity of HspB8 contributes to the associated MNDs.

4. Genetic (Semi-)Dominance in Mutant HspB1- and Mutant HspB8-Associated MND

Most MND-associated mutations in HspB1 and HspB8 with known inheritance are autosomal-dominant (Figure 1) implying that the affected individuals have both mutant (MU) and wild-type (WT) alleles. Whether these MU/WT genotypes have true dominant or semi-dominant inheritance is unknown, because MU/MU genotypes have not been reported and comparisons cannot be made. This prevailing of genetic dominance is noteworthy as studies of mutagenesis in many organisms showed that ~90% of the wild-type alleles are dominant over mutant alleles, thus the recessive phenotype being the "default" state [95]. Therefore, a higher incidence of recessive ^{mu}HspB1 and ^{mu}HspB8 alleles in the population in symptom-free heterozygote individuals can be assumed, as compared to the (semi-)dominant alleles. To date, one dHMN-associated recessive missense mutation in HspB1 was reported (Figure 1).

A number of molecular mechanisms of genetic dominance were identified that seem to accommodate most situations [95]. Among them, haploinsufficiency is the only loss-of-function mechanism as opposed to the gain-of-function mechanisms which include i) dominant-negative effects, ii) toxic gene products, iii) altered structural proteins, and iv) increased protein activities.

For the (semi-)dominant mutations in sHSPs, haploinsufficiency is not supported by the available data: Mice with disrupted HspB1 or HspB5/HspB2 genes are viable and do not show symptoms that are related to the ^{mu}sHSP-associated neuro- or myopathies [10, 38]. Additionally, some of the studied ^{mu}HspB1 species - although not all - did not lose key sHSP functions such as chaperone-like and anti-apoptotic activities [52]. This disparate pattern of properties indicates that loss of these functions may not be critical for these diseases. Together these facts tentatively indicate that a simple loss of function may not be the primary lesion of the (semi-)dominant ^{mu}sHSP species, although it is premature to definitely exclude it for each single mutation. Recently a sporadic mutation in the promoter of HspB1 in an ALS patient was identified [21]. This mutation decreased the basal promoter activity to ~50%, and it also affected the cellular stress response. Provided that association of this mutation with the disease can be demonstrated, this report indicates that HspB1-related loss-of-function mechanisms may play a role in MND, albeit in this case the disease phenotype, ALS, is different.

For the (semi-)dominant ^{mu}sHSP-associated MNDs, one or more of the known gain-of-function mechanisms seem to be good candidates causing the diseases. 'Classic' dominant-negative effects could readily be explained by the formation of homo- and hetero-dimers and mixed oligomers. Incorporation of just one ^{mu}sHSP molecule could compromise the entire sHSP complex with adverse consequences for the cytoprotective or other cell functions that involve sHSPs. Such secondary loss of function(s) resulting from dominant-negative effects should be distinguished from a simple loss of function resulting from haploinsufficiency.

Toxic properties of the ^{mu}sHSPs are another plausible explanation for genetic dominance. In fact, the increased tendency of probably all MND-associated ^{mu}sHSPs to form aggresomes, including their highly toxic precursors, represents such a toxic property. ^{mu}sHSPs caused increased aggresome formation also in the presence of ^{wt}sHSPs, although ^{wt}sHSPs could attenuate aggresome formation in some of the experiments [26, 42]. In other studies, ^{wt}sHSPs prevented aggresome formation by ^{mu}sHSPs [15, 73]. Apparently, the actual ratio of

^{mu}sHSP/^{wt}sHSP seems to determine the degree of aggresome formation, and this may determine the severity of the disease. This situation was found in transgenic mouse lines expressing different levels of ^{R120G}HspB5 with the severity of the associated cardiomyopathy being gene dosage-dependent [93]. These considerations would support semi-dominant rather than dominant properties of the aggresome forming ^{mu}sHSP species. To be complete, a toxic gain-of-function mechanism could be caused also by 'toxic' ^{mu}sHSP mRNAs (see above).

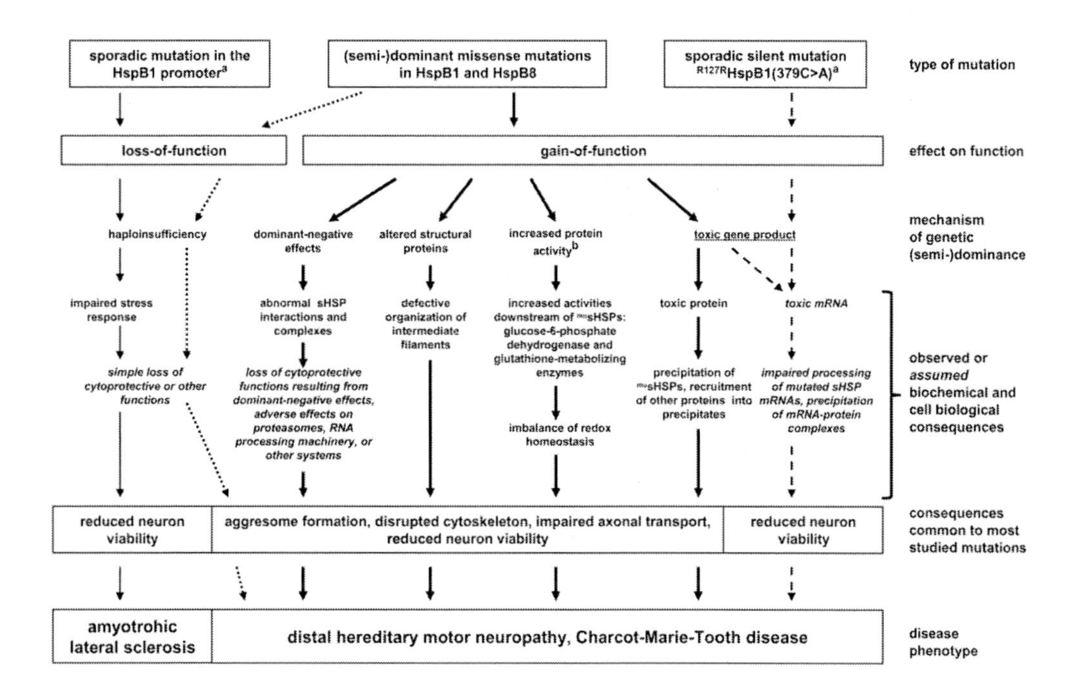

Figure 2. Proposed mechanisms of neuron damage in ^{mu}HspB1- and ^{mu}HspB8-related MND. Most mutations in dHMN and CMT are of the missense type and have dominant or semi-dominant inheritance. Available experimental data are consistent with the major gain-of-function mechanisms of genetic dominance including dominant-negative effects, altered structural proteins, increased protein activity, and toxic gene products (bold arrows). One identified sporadic mutation in the HspB1 promoter of an ALS patient suggests haploinsufficiency as a possible mechanism of motor neuron damage (thin arrows) [21]. One identified sporadic silent mutation in HspB1 in a dHMN patient tentatively suggests involvement of mRNA in motor neuron damage (dashed arrows) [37]. Although not supported by available data, simple loss-of-function mechanisms (haploinsufficiency) cannot be excluded for each single dHMN- and CMT-associated (semi-)dominant mutation (dotted arrows). Such simple loss of function should be distinguished from secondary loss of function resulting from dominant-negative effects (e.g. through malfunction of the entire sHSP complex). [a] Inheritance not known; [b] Experimental data are available only for myopathy-associated ^{R120G}HspB5 [71].

Yet another potential reason for the genetic (semi-)dominance of ^{mu}sHSPs is their adverse effect on the overall cell architecture. The negative impact of ^{S135F}HspB1, ^{P182L}HspB1, and ^{R120G}HspB5 on the intermediate filaments (neurofilaments, desmin filaments) is well documented [1, 25, 91, 99]. Genetic (semi-)dominance has been frequently seen with other mutations affecting various structural and associated proteins. Examples are CMT-associated mutations in NF-L and the kinesin-like protein KIF1B [78, 99], dHMN-associated mutations in dynactin [43], and myopathy-associated mutations in desmin [31].

Finally, increased enzymatic activities can also contribute to genetic dominance and this has been demonstrated for cardiomyopathy-causing ^{R120G}HspB5 [71]. Although ^{R120G}HspB5 itself has no known increased activity, its transgenic expression in mice led downstream to increased expression and activities of glucose-6-phosphate dehydrogenase and of glutathione-metabolizing enzymes resulting in a toxic reductive stress which was causatively involved in the disease phenotype.

In summary, available data favor one or several molecular gain-of-function mechanisms for the (semi-)dominant ^{mu}sHSP species (summarized in Figure 2). Accordingly, these ^{mu}sHSPs have acquired a new and cytotoxic property which results in adverse cell biological consequences including formation of aggresomes, collapse of the cytoskeleton, and possibly in other lesions such as impairment of the proteasome system, autophagy, RNA processing, or imbalance of the redox homeostasis. The various ^{mu}sHSP species may adopt more than one of these mechanisms, even with varying proportions thus contributing to the clinical variability seen among the patients. Based on these considerations, a simple loss of the chaperone-like or cytoprotective functions of ^{mu}sHSPs is unlikely to be the primary lesion, although a secondary loss of function may result from dominant-negative effects. What appears like a distinction without difference may have decided consequences for the design of therapeutic strategies.

5. Do Wild-Type sHSPs Exhibit also Cytotoxic and Pro-Apoptotic Properties?

Given the well documented cytoprotective and anti-apoptotic activity of HspB1 and HspB5, it was unexpected to learn about the pro-apoptotic activity of HspB8 in cardiomyocytes and melanoma cells [30, 33]. As HspB1 and HspB8 are most similar among all proteins [85], this discrepancy cannot be explained easily. While the chaperone-like activity of HspB1 provides a rationale for its anti-apoptotic activity, it is not clear how the chaperone-like activity of HspB8 can be reconciled with its pro-apoptotic activity. This pro-apoptotic activity of ^{wt}HspB8 may well contribute to the neurotoxic effect of ^{mu}HspB8 as seen in MND, e.g. the mutations may enhance this intrinsic property of ^{wt}HspB8. If correct, could this argumentation also apply to HspB1? Could ^{wt}HspB1 also carry a cytotoxic property, which might be normally suppressed? Contrary to the overwhelming number of studies reporting cytoprotection, a few studies indeed report cytotoxic effects of ^{wt}HspB1 and also of ^{wt}HspB5. Immortalized human KMST-6 fibroblasts stably expressing high levels of ^{wt}HspB1 were more sensitive to growth inhibition by H_2O_2 than transformants expressing low levels [2]. Also L929 cells overexpressing HspB1 exhibited increased sensitivity to TNFα, etoposide, and H_2O_2 [59]. In Alzheimer disease, ^{wt}HspB1 was suspected to contribute to the pathology by aggravating microtubule disruption [9]. ^{wt}HspB5 was reported to increase cytotoxicity as its expression increased the neurotoxicity of amyloid beta, the main constituent of amyloid plaques in the brains of Alzheimer disease patients, by keeping it in a highly toxic, non-fibrillar form [83]. Mice that systemically overexpress human HspB1 showed increased renal injury following ischemia/reperfusion by exacerbating renal and systemic inflammation [16].

This sHSP-mediated cytotoxicity may involve phosphorylation. Three forms of ^{mu}HspB5, $^{\mathrm{R120G}}$HspB5, $^{\mathrm{Q151X}}$HspB5, and $^{\mathrm{464delCT}}$HspB5 were hyperphosphorylated in all three phosphorylation sites [79], and it was this hyperphosphorylation of $^{\mathrm{R120G}}$HspB5 - and not the mutation *per se* - which caused abnormal cellular localization and abnormal interaction with SMN protein [19]. In this context it is highly interesting that activated p38 MAPK has been implicated in the damage of motor neurons in ALS [6]. As activated p38 MAPK phosphorylates HspB1 through its downstream protein kinases MK-2/3, increased phosphorylation of HspB1 can be expected to be associated with the death of motor neurons, at least in ALS. Whether this also applies to dHMN or CMT is not known. This finding also raises the question of the role of environmental factors as etiologic agents in sporadic forms of MND, as a number of environmental toxicants including toxic metals and chloro-organic compounds activate p38 MAPK signaling. Certain forms of MND without obvious genetic predisposition (e.g. the Guam-type of ALS) were suspected to be caused by environmental toxicants, although the scientific proof remains to be provided.

It is concluded that cytotoxic and pro-apoptotic properties of the ^{wt}sHSPs should not be excluded *a priori*. The view according to which overexpression of ^{wt}sHSPs would necessarily be beneficial in terms of increased resistance to any type of stress in any type of cells obviously is too simple. ^{wt}sHSPs may exhibit both cytoprotective and, under certain circumstances, also cytotoxic properties and thus play a more ambivalent cellular role than generally acknowledged. Which of the two sides prevails may depend on a variety of factors including harbored mutations, expression of a specific profile of sHSPs, the degree of their phosphorylation, the cell type, and the developmental and physiologic status of the cells and tissues.

CONCLUSION

MND-associated ^{mu}HspB1 and ^{mu}HspB8 species act probably through the same pathomechanism(s), as i) both sHSPs are interacting proteins, ii) both sequences are most similar among all proteins, and iii) mutations in either gene cause identical or similar disease phenotypes. As far as studied, some abnormal properties and cell-biological consequences resulting from mutation are shared by most if not all (semi-)dominant ^{mu}sHSPs. These include abnormal sHSP interaction patterns, increased tendency to form aggresomes, and disruption of the intermediate filament network. Eventually these abnormalities are likely to affect the axonal transport and cell viability.

Known properties and consequences of the (semi-)dominant ^{mu}sHSPs are concordant with major gain-of-function mechanisms of genetic dominance. A simple loss of function (e.g. chaperoning or anti-apoptotic activities) of the (semi-)dominant ^{mu}sHSPs resulting from haploinsufficiency is probably not the primary cause of the disease. This does not exclude, however, loss of function(s) resulting from dominant-negative effects. In line with this argumentation, formation of aggresomes or the collapse of the cytoskeleton would result primarily from new and toxic properties of the ^{mu}sHSPs and not from a simple loss of the chaperone-like activity. Such a distinction may have consequences for the development of future therapeutic concepts. Experiments tentatively favor semi-dominant over dominant properties of ^{mu}sHSPs.

There is growing evidence for a critical role of ^{wt}sHSPs in various cell processes including proteasome function, RNA processing, and redox homeostasis. ^{mu}sHSPs may affect these processes which is of potential interest in the context of the associated MNDs.

^{wt}sHSPs may play a more ambivalent role than generally acknowledged, as they can exhibit also cytotoxic properties in addition to their cytoprotective properties. The view according to which overexpression of ^{wt}sHSPs would necessarily be beneficial in terms of increased resistance to any type of stress in any type of cells obviously is too simple. This situation should be considered when designing future therapeutic strategies.

Given the (semi-)dominant nature of most of the mutations, a therapeutic strategy aiming to up-regulate the expression of any of the ^{wt}sHSPs in order to increase the cell's chaperone capacity is unlikely to succeed. Even if a semi-dominant mechanism is assumed, it appears impractical to permanently and effectively "dilute" the ^{mu}sHSPs in the motor neurons by increased expression of ^{wt}sHSPs. More logically appears to decrease the abundance of the ^{mu}sHSPs, e.g. by future RNAi-based technologies. This approach would be particularly useful as ^{mu}sHSP mRNA may also contribute to the pathology, and it would also counteract any potential toxic effects of the ^{mu}sHSP species on the cytoskeleton, the RNA processing machinery, or on the proteasome system. Since increased aggresome formation seems to be common to all ^{mu}sHSP species, measures to stimulate clearing mechanisms could be another promising strategy. Possibilities include the stimulation of the proteasome system and autophagy, or application of substances that reduce misfolding and aggregation of proteins. A promising candidate drug for this is the secondary plant metabolite (–)-epigallocatechin-3-gallate [24]. Finally, stabilization of the redox homeostasis is also a strategy which should be pursued in future.

ACKNOWLEDGMENT

Critical reading of the chapter by Dr. S. Simon, Lyon, France, is gratefully acknowledged.

ENDNOTES

[1]Abbreviations: ALS, amyotrophic lateral sclerosis; CMT, Charcot-Marie-Tooth-disease; dHMN, distal hereditary motor neuropathy; HspB1 (^{wt}HspB1, ^{mu}HspB1), HspB8 (^{wt}HspB8, ^{mu}HspB8), HspB5 (^{wt}HspB5, ^{mu}HspB5), small heat shock proteins 27, 22, and αB-crystallin with their wild-type and mutant forms; MND, motor neuron disease; NF-L, neurofilament light chain; sHSP (^{wt}sHSP, ^{mu}sHSP), small heat shock protein as general term with its wild-type and mutant forms; SMA, spinal muscular atrophy; SMN complex, SMN protein, survival-of-motor-neurons complex and protein, respectively.

[2]Herein, dHMN, CMT, ALS, and SMA are collectively referred to as motor neuron diseases (MND).

[3]Note added in proof: Recently a missense mutation in the N-terminal part of HspB3, another sHSP, was identified that was also associated with axonal peripheral motor neuropathy (Kolb, S.J., Snyder, P.J., Poi, E.J., Renard, E.A., Bartlett, A., Gu, S., Sutton, S.,

Arnold, W.D., Freimer, M.L., Lawson, V.H., Kissel, J.T. & Prior, T.W. (2010). Mutant small heat shock protein B3 causes motor neuropathy: utility of a candidate gene approach. *Neurology, 74*, 502-506.

[4]National Institute of Neurological Disorders and Stroke, peripheral neuropathy information page at http://www.ninds.nih.gov.

[5]Secondary structure prediction server at http://www.compbio.dundee.ac.uk/www-jpred/.

REFERENCES

[1] Ackerley, S., James, P.A., Kalli, A., French, S., Davies, K.E. & Talbot, K. (2006). A mutation in the small heat-shock protein HSPB1 leading to distal hereditary motor neuronopathy disrupts neurofilament assembly and the axonal transport of specific cellular cargoes. *Hum. Mol. Genet., 15*, 347-354.

[2] Arata, S., Hamaguchi, S. & Nose, K. (1995). Effects of the overexpression of the small heat shock protein, HSP27, on the sensitivity of human fibroblast cells exposed to oxidative stress. *Cell Physiol., 163*, 458-465.

[3] Arrigo, A.-P. (2007). The cellular "networking" of mammalian Hsp27 and its functions in the control of protein folding, redox state and apoptosis. *Adv. Exp. Med. Biol., 594*, 14-26.

[4] Arrigo, A.-P. (2001). Hsp27: novel regulator of intracellular redox state. *IUBMB Life, 52,* 303-307.

[5] Badri, K.R., Modem, S., Gerard, H.C., Khan, I., Bagchi, M., Hudson, A.P. & Reddy, T.R. (2006). Regulation of Sam68 activity by small heat shock protein 22. *J. Cell Biochem., 99*, 1353-1362.

[6] Bendotti, C., Bao Cutrona, M., Cheroni, C., Grignaschi, G., Lo Coco, D., Peviani, M., Tortarolo, M., Veglianese, P. & Zennaro, E. (2005). Inter- and intracellular signaling in amyotrophic lateral sclerosis: role of p38 mitogen-activated protein kinase. *Neurodegener. Dis., 2*, 128-134.

[7] Benn, S.C., Perrelet, D., Kato, A.C., Scholz, J., Decosterd, I., Mannion, R.J., Bakowska, J.C. & Woolf, C.J. (2002). Hsp27 upregulation and phosphorylation is required for injured sensory and motor neuron survival. *Neuron, 36*, 45-56.

[8] Benndorf, R. & Welsh, M.J. (2004). Shocking Degeneration. *Nat. Genet., 36,* 547-548.

[9] Björkdahl, C., Sjögren, M.J., Zhou, X., Concha, H., Avila, J., Winblad, B. & Pei, J.J. (2008). Small heat shock proteins Hsp27 or αB-crystallin and the protein components of neurofibrillary tangles: tau and neurofilaments. *J. Neurosci. Res., 86*, 1343-1352.

[10] Brady, J.P., Garland, D.L., Green, D.E., Tamm, E.R., Giblin, F.J. & Wawrousek, E.F. (2001). αB-crystallin in lens development and muscle integrity: a gene knockout approach. *Invest. Ophthalmol. Vis. Sci., 42*, 2924-2934.

[11] Bryantsev, A.L., Chechenova, M.B. & Shelden, E.A. (2007). Recruitment of phosphorylated small heat shock protein Hsp27 to nuclear speckles without stress. *Exp. Cell Res., 313*, 195-209.

[12] Carra, S., Brunsting, J.F., Lambert, H., Landry, J. & Kampinga, H.H. (2009) HSPB8 participates in protein quality control by a non chaperone-like mechanism that requires eIF2α phosphorylation. *J. Biol. Chem., 284,* 5523-5532,

[13] Carra, S., Seguin, S.J., Lambert, H. & Landry, J. (2008). HspB8 chaperone activity toward poly(Q)-containing proteins depends on its association with Bag3, a stimulator of macroautophagy. *J. Biol. Chem., 283,* 1437-1444.

[14] Chávez-Zobel, A.T., Lambert, H., Thériault, J.R. & Landry, J. (2005). Structural instability caused by a mutation at a conserved arginine in the α-crystallin domain of Chinese hamster heat shock protein 27. *Cell Stress Chaperones, 10*, 157-166.

[15] Chávez-Zobel, A.T., Loranger, A., Marceau, N., Thériault, J. R., Lambert H. & Landry, J. (2003). Distinct chaperone mechanisms can delay the formation of aggresomes by the myopathy-causing R120G αB-crystallin mutant. *Hum. Mol. Genet., 12*, 1609-1620.

[16] Chen, S.W., Kim, M., Kim, M., Song, J.H., Park, S.W., Wells, D., Brown, K., Belleroche, J., D'Agati, V.D. & Lee, H.T. (2009). Mice that overexpress human heat shock protein 27 have increased renal injury following ischemia reperfusion. *Kidney Int., 75,* 499-510.

[17] Chowdary, T.K., Raman, B., Ramakrishna, T. & Rao, C.M. (2004). Mammalian Hsp22 is a heat-inducible small heat-shock protein with chaperone-like activity. *Biochem. J.,* 381, 379-87.

[18] De Jonghe, P., Auer-Grumbach, M., Irobi, J., Wagner, K., Plecko, B., Kennerson, M., Zhu, D., De Vriendt, E., Van Gerwen, V., Nicholson, G., Hartung, H.P. & Timmerman, V. (2002). Autosomal dominant juvenile amyotrophic lateral sclerosis and distal hereditary motor neuronopathy with pyramidal tract signs: synonyms for the same disorder? *Brain, 125*, 1320-1325.

[19] den Engelsman, J., Gerrits, D., de Jong, W.W., Robbins, J., Kato, K. & Boelens, W.C. (2005). Nuclear import of αB-crystallin is phosphorylation-dependent and hampered by hyperphosphorylation of the myopathy-related mutant R120G. *J. Biol. Chem., 280*, 37139-37148.

[20] Dierick, I., Baets, J., Irobi, J., Jacobs, A., De Vriendt, E., Deconinck, T., Merlini, L., Van den Bergh, P., Rasic, V.M., Robberecht, W., Fischer, D., Morales, R.J., Mitrovic, Z., Seeman, P., Mazanec, R., Kochanski, A., Jordanova, A., Auer-Grumbach, M., Helderman-van den Enden, A.T., Wokke, J.H., Nelis, E., De Jonghe, P. & Timmerman, V. (2008). Relative contribution of mutations in genes for autosomal dominant distal hereditary motor neuropathies: a genotype-phenotype correlation study. *Brain, 131,* 1217-1227.

[21] Dierick, I., Irobi, J., Janssens, S., Theuns, J., Lemmens, R., Jacobs, A., Corsmit, E., Hersmus, N., Van Den Bosch, L., Robberecht, W., De Jonghe, P., Van Broeckhoven, C. & Timmerman, V. (2007). Genetic variant in the HSPB1 promoter region impairs the HSP27 stress response. *Hum. Mutat., 28*, 830.

[22] Dierick, I., Irobi, J., De Jonghe, P. & Timmerman, V. (2005). Small heat shock proteins in inherited peripheral neuropathies. *Ann. Med., 37*, 413-422.

[23] Doerwald, L., van Genesen, S.T., Onnekink, C., Marín-Vinader, L., de Lange, F., de Jong, W.W., & Lubsen, N.H. (2006). The effect of αB-crystallin and Hsp27 on the availability of translation initiation factors in heat-shocked cells. *Cell. Mol. Life Sci., 63*, 735-743.

[24] Ehrnhoefer, D.E., Duennwald, M., Markovic, P., Wacker, J.L., Engemann, S., Roark, M., Legleiter, J., Marsh, J.L., Thompson, L.M., Lindquist, S., Muchowski, P.J. & Wanker, E.E. (2006). Green tea (-)-epigallocatechin-gallate modulates early events in

huntingtin misfolding and reduces toxicity in Huntington's disease models. *Hum. Mol. Genet., 15,* 2743-2751.

[25] Evgrafov, O.V., Mersiyanova, I., Irobi, J., Van Den Bosch, L., Dierick, I., Leung, C.L., Schagina, O., Verpoorten, N., Van Impe, K., Fedotov, V., Dadali, E., Auer-Grumbach, M., Windpassinger, C., Wagner, K., Mitrovic, Z., Hilton-Jones, D., Talbot, K., Martin, J.J., Vasserman, N., Tverskaya, S., Polyakov, A., Liem, R.K., Gettemans, J., Robberecht, W., De Jonghe, P. & Timmerman, V. (2004). Mutant small heat-shock protein 27 causes axonal Charcot-Marie-Tooth disease and distal hereditary motor neuropathy. *Nat. Genet., 36*, 602-606.

[26] Fontaine, J.-M., Sun X., Hoppe, A.D., Simon, S., Vicart, P., Welsh, M.J. & Benndorf, R. (2006). Abnormal small heat shock protein interactions involving neuropathy-associated HSP22 (HSPB8) mutants. *FASEB J., 20*, 2168-2170.

[27] Fontaine, J.M., Rest, J.S., Welsh, M.J. & Benndorf, R. (2003). The sperm outer dense fiber protein is the 10th member of the superfamily of mammalian small stress proteins. *Cell Stress Chaperones, 8*, 62-69.

[28] Gilks, N., Kedersha, N., Ayodele, M., Shen, L., Stoecklin, G., Dember, L.M. & Anderson P. (2004). Stress granule assembly is mediated by prion-like aggregation of TIA-1. *Mol. Biol. Cell, 15*, 5383-5398.

[29] Gober, M.D., Depre, C. & Aurelian L. (2004) Correspondence regarding M.V. Kim et al. "Some properties of human small heat shock protein Hsp22 (H11 or HspB8)". *Biochem. Biophys. Res. Commun., 321*, 267-268.

[30] Gober, M.D., Smith, C.C., Ueda, K., Toretsky, J.A. & Aurelian L. (2003). Forced expression of the H11 heat shock protein can be regulated by DNA methylation and trigger apoptosis in human cells. *J. Biol. Chem., 278,* 37600-37609.

[31] Goldfarb, L.G., Vicart, P., Goebel, H.H. & Dalakas, M.C. (2004). Desmin myopathy. *Brain, 127,* 723-734.

[32] Gubitz, A.K., Feng, W. & Dreyfuss, G. (2004). The SMN complex. *Exp. Cell Res., 296,* 51-56.

[33] Hase, M., Depre, C., Vatner, S.F. & Sadoshima, J. (2005). H11 has dose-dependent and dual hypertrophic and proapoptotic functions in cardiac myocytes. *Biochem J., 388*, 475-483.

[34] Head, M. W. & Goldman, J.E. (2000). Small heat shock proteins, the cytoskeleton, and inclusion body formation. *Neuropathol. Appl. Neurobiol., 26*, 304-312.

[35] Hedhli, N., Wang, L., Wang, Q., Rashed, E., Tian, Y., Sui, X., Madura, K. & Depre, C. (2008). Proteasome activation during cardiac hypertrophy by the chaperone H11 Kinase/Hsp22. *Cardiovasc. Res., 77*, 497-505.

[36] Hickey, E., Brandon, S.E., Potter, R., Stein, G., Stein, J. & Weber, L.A. (1986). Sequence and organization of genes encoding the human 27 kDa heat shock protein. *Nucleic Acids Res. 14,* 4127-4145. Erratum in: *Nucleic Acids Res., 14,* 8230 (1986).

[37] Houlden, H., Laura, M., Wavrant-De Vrièze, F., Blake, J., Wood, N. & Reilly, M.M. (2008). Mutations in the HSP27 (HSPB1) gene cause dominant, recessive, and sporadic distal HMN/CMT type 2. *Neurology, 71*, 1660-1668.

[38] Huang, L., Min, J.N., Masters, S., Mivechi, N.F. & Moskophidis, D. (2007). Insights into function and regulation of small heat shock protein 25 (HSPB1) in a mouse model with targeted gene disruption. *Genesis, 45*, 487-501.

[39] Ikeda, Y., Abe, A., Ishida, C., Takahashi, K., Hayasaka, K. & Yamada, M. (2009). A clinical phenotype of distal hereditary motor neuronopathy type II with a novel HSPB1 mutation. *J. Neurol. Sci., 277,* 9-12.

[40] Inagaki, N., Hayashi, T., Arimura, T., Koga, Y., Takahashi, M., Shibata, H., Teraoka, K., Chikamori, T., Yamashina, A. & Kimura, A. (2006). αB-crystallin mutation in dilated cardiomyopathy. *Biochem. Biophys. Res. Commun., 342*, 379-386.

[41] Inaguma, Y., Hasegawa, K., Goto, S., Ito, H. & Kato, K. (1995). Induction of the synthesis of hsp27 and αB-crystallin in tissues of heat-stressed rats and its suppression by ethanol or an alpha 1-adrenergic antagonist. *J. Biochem* (Tokyo), *117,* 1238-1243.

[42] Irobi, J., Van Impe, K., Seeman, P., Jordanova, A., Dierick, I., Verpoorten, N., Michalik, A., De Vriendt, E., Jacobs, A., Van Gerwen, V., Vennekens, K., Mazanec, R., Tournev, I., Hilton-Jones, D., Talbot, K., Kremensky, I., Van Den Bosch, L., Robberecht, W., Van Vandekerckhove, J., Van Broeckhoven, C., Gettemans, J., De Jonghe, P. & Timmerman, V. (2004). Hot-spot residue in small heat-shock protein 22 causes distal motor neuropathy. *Nat. Genet., 36*, 597-601.

[43] Irobi, J., De Jonghe, P. & Timmerman, V. (2004). Molecular genetics of distal hereditary motor neuropathies. *Hum. Mol. Genet., 13 (Suppl. 2)*, R195-R202.

[44] James, P.A., Rankin, J. & Talbot, K. (2008). Asymmetrical late onset motor neuropathy associated with a novel mutation in the small heat shock protein HSPB1 (HSP27). *J. Neurol. Neurosurg. Psychiatry, 79,* 461-463.

[45] Johnson, A.D., Wang, D. & Sadee, W. (2005). Polymorphisms affecting gene regulation and mRNA processing: broad implications for pharmacogenetics. *Pharmacol. Ther., 106,* 19-38.

[46] Jordanova, A., Irobi, J., Thomas, F.P., Van Dijck, P., Meerschaert, K., Dewil M,, Dierick, I., Jacobs, A., De Vriendt, E., Guergueltcheva, V., Rao, C.V., Tournev, I., Gondim, F.A., D'Hooghe, M., Van Gerwen, V., Callaerts, P., Van Den Bosch, L., Timmermans, J.P., Robberecht, W., Gettemans, J., Thevelein, J.M., De Jonghe, P., Kremensky, I. & Timmerman V. (2006). Disrupted function and axonal distribution of mutant tyrosyl-tRNA synthetase in dominant intermediate Charcot-Marie-Tooth neuropathy. *Nat. Genet., 38,* 197-202.

[47] Kasakov, A.S., Bukach, O.V., Seit-Nebi, A.S., Marston, S.B. & Gusev, N.B. (2007). Effect of mutations in the beta5-beta7 loop on the structure and properties of human small heat shock protein HSP22 (HspB8, H11). *FEBS J. 274,* 5628-5642.

[48] Kijima, K., Numakura, C., Goto, T., Takahashi, T., Otagiri, T., Umetsu, K. & Hayasaka, K. (2005). Small heat shock protein 27 mutation in a Japanese patient with distal hereditary motor neuropathy. *J. Hum. Genet., 50*, 473-476.

[49] Kim, M.V., Kasakov, A.S., Seit-Nebi, A.S., Marston, S.B. & Gusev, N.B. (2006). Structure and properties of K141E mutant of small heat shock protein HSP22 (HspB8, H11) that is expressed in human neuromuscular disorders. *Arch. Biochem. Biophys. 454,* 32-41.

[50] Kim, M.V., Seit-Nebi, A.S., Marston, S.B. & Gusev, N.B. (2004). Some properties of human small heat shock protein Hsp22 (H11 or HspB8). *Biochem. Biophys. Res. Commun., 315*, 796-801.

[51] Kim, M,V,, Seit-Nebi, A.S. & Gusev, N.B. (2004). The problem of protein kinase activity of small heat shock protein Hsp22 (H11 or HspB8). *Biochem. Biophys. Res. Commun, 325,* 649-652.

[52] Krishnan, J., D'Ydewalle, C., Dierick, I., Irobi, J., van den Berghe, P., Janssen, P., Timmerman, V., Robberecht, W. & van den Bosch, L. (2008). Pathogenic mechanisms of HSPB1 mutations associated with distal hereditary motor neuropathies. *38th Annual Meeting of the Society for Neuroscience,* Washington, DC, Nov. 15-19, 2008.

[53] Krueger-Naug, A.M., Plumier, J.C., Hopkins, D.A. & Currie, R.W. (2002). Hsp27 in the nervous system: expression in pathophysiology and in the aging brain. *Prog. Mol. Subcell. Biol., 28,* 235-251.

[54] Lasa, M., Mahtani, K.R., Finch, A., Brewer, G., Saklatvala, J. and Clark, A.R. (2000). Regulation of cyclooxygenase 2 mRNA stability by the mitogen-activated protein kinase p38 signaling cascade. *Mol. Cell. Biol., 20*, 4265-4274.

[55] Latchman, D.S. Protection of neuronal and cardiac cells by HSP27. in: Arrigo, A.-P. & Müller W.E.G. (eds), Small Stress Proteins, Springer-Verlag Berlin. 2002, pp. 253-265.

[56] Lin, D.I., Barbash, O., Kumar, K.G., Weber, J.D., Harper, J.W., Klein-Szanto, A.J., Rustgi, A., Fuchs, S.Y. & Diehl, J.A. (2006) Phosphorylation-dependent ubiquitination of cyclin D1 by the SCF(FBX4-αB-crystallin) complex. *Mol. Cell, 24*, 355-366.

[57] Liu, X.M., Tang, B.S., Zhao, G.H., Xia, K., Zhang, F.F., Pan, Q., Cai, F., Hu, Z.M., Zhang, C., Chen, B., Shen, L., Zhang, R.X. & Jiang, H. (2005). Mutation analysis of small heat shock protein 27 gene in Chinese patients with Charcot-Marie-Tooth disease. *Zhonghua Yi Xue Yi Chuan Xue Za Zhi, 22*, 510-513.

[58] Loones, M.T., Chang, Y., Morange, M. (2000). The distribution of heat shock proteins in the nervous system of the unstressed mouse embryo suggests a role in neuronal and non-neuronal differentiation. *Cell Stress Chaperones, 5*, 291-305.

[59] Mairesse, N., Bernaert, D., Del Bino, G., Horman, S., Mosselmans, R., Robaye, B. & Galand P. (1998). Expression of HSP27 results in increased sensitivity to tumor necrosis factor, etoposide, and H_2O_2 in an oxidative stress-resistant cell line. *J. Cell. Physiol., 177*, 606-617.

[60] Manning, G., Whyte, D.B., Martinez, R., Hunter, T. & Sudarsanam, S. (2002). The protein kinase complement of the human genome. *Science. 298*, 1912-1934.

[61] Marin-Vinader, L., Shin, C., Onnekink, C., Manley, J.L. & Lubsen, N.H. (2006). Hsp27 enhances recovery of splicing as well as rephosphorylation of SRp38 after heat shock. *Mol. Biol. Cell, 17*, 886-894.

[62] Michiel, M., Skouri-Panet, F., Duprat, E., Simon, S., Férard, C., Tardieu, A. & Finet, S. (2009). Abnormal assemblies and subunit exchange of αB-crystallin R120 mutants could be associated with destabilization of the dimeric substructure. *Biochemistry, 48,* 442-453.

[63] Niemann, A., Berger, P. & Suter, U. (2006). Pathomechanisms of mutant proteins in Charcot-Marie-Tooth disease. *Neuromolecular Med., 8*, 217-242.

[64] Nover, L., Scharf, K.D. & Neumann, D. (1989). Formation of cytoplasmic heat shock granules in tomato cell cultures and leaves. *Mol. Cell. Biol., 3*, 1648-1655.

[65] Parcellier, A., Brunet, M., Schmitt, E., Col, E., Didelot, C., Hammann, A., Nakayama, K., Nakayama, K.I., Khochbin, S., Solary, E. & Garrido, C. (2006). HSP27 favors ubiquitination and proteasomal degradation of p27Kip1 and helps S-phase re-entry in stressed cells. *FASEB J.,* 20, 1179-1181.

[66] Parcellier, A., Schmitt, E, Gurbuxani, S., Seigneurin-Berny, D., Pance, A., Chantôme, A., Plenchette, S., Khochbin, S., Solary, E. & Garrido, C. (2003). HSP27 is a ubiquitin-

binding protein involved in I-kappaBalpha proteasomal degradation. *Mol. Cell Biol., 23*, 5790-5802.

[67] Perng, M.D., Muchowski, P.J., van Den IJssel, P., Wu, G.J., Hutcheson, A.M., Clark, J.I. & Quinlan, R.A.(1999). The cardiomyopathy and lens cataract mutation in αB-crystallin alters its protein structure, chaperone activity, and interaction with intermediate filaments in vitro. *J. Biol. Chem., 274,* 33235-33243.

[68] Perng, M.D., Cairns, L., van den IJssel, P., Prescott, A., Hutcheson, A.M. & Quinlan, R.A. (1999). Intermediate filament interactions can be altered by HSP27 and αB-crystallin. *J. Cell Sci., 112 (Pt 13),* 2099-2112.

[69] Pilotto, A., Marziliano, N., Pasotti, M., Grasso, M., Costante, A.M. & Arbustini, E. (2006). αB-crystallin mutation in dilated cardiomyopathies: low prevalence in a consecutive series of 200 unrelated probands. *Biochem. Biophys. Res. Commun., 346*, 1115-1117.

[70] Quraishe, S., Asuni, A., Boelens, W.C., O'Connor, V. & Wyttenbach, A. (2008). Expression of the small heat shock protein family in the mouse CNS: Differential anatomical and biochemical compartmentalization. *Neuroscience, 153*, 483-491.

[71] Rajasekaran, N.S., Connell, P., Christians, E.S., Yan, L.J., Taylor, R.P., Orosz, A., Zhang, X.Q., Stevenson, T.J., Peshock, R.M., Leopold, J.A., Barry, W.H., Loscalzo, J., Odelberg, S.J. & Benjamin, I.J. (2007). Human αB-crystallin mutation causes oxido-reductive stress and protein aggregation cardiomyopathy in mice. *Cell, 130,* 427-439.

[72] Saifi, G.M., Szigeti, K., Wiszniewski, W., Shy, M.E., Krajewski, K., Hausmanowa-Petrusewicz, I., Kochanski, A., Reeser, S., Mancias, P., Butler, I. & Lupski, J.R. (2005). SIMPLE mutations in Charcot-Marie-Tooth disease and the potential role of its protein product in protein degradation. *Hum. Mutat., 25*, 372-383.

[73] Sanbe, A., Yamauchi, J., Miyamoto, Y., Fujiwara, Y., Murabe, M. & Tanoue, A. (2007). Interruption of CryAB-amyloid oligomer formation by HSP22. *J. Biol. Chem., 282*, 555-563.

[74] Selcen, D., Muntoni, F., Burton, B.K., Pegoraro, E., Sewry, C., Bite, A.V. & Engel, A.G. (2009). Mutation in BAG3 causes severe dominant childhood muscular dystrophy. *Ann. Neurol., 65,* 83-89.

[75] Selcen, D. & Engel, A.G. (2003). Myofibrillar myopathy caused by novel dominant negative αB-crystallin mutations. *Ann. Neurol., 54*, 804-810.

[76] Shemetov, A.A., Seit-Nebi, A.S. &, Gusev, N.B. (2008) Structure, properties, and functions of the human small heat-shock protein HSP22 (HspB8, H11, E2IG1): a critical review. *J. Neurosci. Res., 86,* 264-269

[77] Shchors, K., Yehiely, F. & Deiss, L.P. (2002). Cell death inhibiting RNA (CDIR) derived from a 3'-untranslated region binds AUF1 and heat shock protein 27. *J. Biol. Chem., 277*, 47061-47072.

[78] Shy, M.E. (2004). Charcot-Marie-Tooth disease: an update. *Curr. Opin. Neurol., 17*, 579-585.

[79] Simon, S., Fontaine, J.-M., Martin, J.L., Sun, X., Hoppe, A.D., Welsh, M.J., Benndorf, R. & Vicart, P. (2007). Myopathy-associated αB-crystallin mutants: abnormal phosphorylation, intracellular location, and interactions with other small heat shock proteins. *J. Biol. Chem., 282*, 34276-34287.

[80] Smith, C.C., Yu, Y.X., Kulka, M. & Aurelian, L. (2000). A novel gene homologous to PK coding domain of the large subunit of herpes simplex virus type 2 ribonucleotide

reductase (ICP10) codes for a serine-threonine PK and is expressed in melanoma cells. *J. Biol. Chem.*, 275, 25690-25699.

[81] Sommer, S., Cui, Y., Brewer, G. & Fuqua, S.A. (2005). The c-Yes 3'-UTR contains adenine/uridine-rich elements that bind AUF1 and HuR involved in mRNA decay in breast cancer cells. *J. Steroid. Biochem. Mol. Biol.*, 97, 219-229.

[82] Stefani, M. & Dobson, C.M. (2003). Protein aggregation and aggregate toxicity: new insights into protein folding, misfolding diseases and biological evolution. *J. Mol Med., 81*, 678-699.

[83] Stege, G.J., Renkawek, K., Overkamp, P.S., Verschuure, P., van Rijk, A.F., Reijnen-Aalbers, A., Boelens, W.C., Bosman, G.J. & de Jong, W.W. (1999). The molecular chaperone αB-crystallin enhances amyloid beta neurotoxicity. *Biochem. Biophys. Res. Commun., 262*, 152-156.

[84] Sun, X., Fontaine, J.-M., Hoppe, A.D., Carra, S., DeGuzman, C., Martin, J.L., Simon, S., Vicart, P., Welsh, M.J., Landry, J. & Benndorf, R. (2010). Abnormal interaction of motor neuropathy-associated mutant HspB8 (Hsp22) forms with the RNA helicase Ddx20 (gemin3). *Cell Stress Chaperones* (in press).

[85] Sun, X., Fontaine, J.-M., Rest, J.S., Shelden, E.A., Welsh, M.J. & Benndorf, R. (2004). Interaction of human HSP22 (HSPB8) with other small heat shock proteins. *J. Biol. Chem., 279*, 2394-2402.

[86] Tang, B.S., Zhao, G.H., Luo, W., Xia, K., Cai, F., Pan, Q., Zhang, R.X., Zhang, F.F., Liu, X.M., Chen, B., Zhang, C., Shen, L., Jiang, H., Long, Z.G. & Dai, H.P. (2005). Small heat-shock protein 22 mutated in autosomal dominant Charcot-Marie-Tooth disease type 2L. *Hum. Genet., 116*, 222-224.

[87] Tannous, P., Zhu, H., Johnstone, J.L., Shelton, J.M., Rajasekaran, N.S., Benjamin, I.J., Nguyen, L., Gerard, R.D., Levine, B., Rothermel, B.A. & Hill, J.A. (2008). Autophagy is an adaptive response in desmin-related cardiomyopathy. *Proc. Natl. Acad. Sci. USA., 105,* 9745-9750.

[88] Thériault, J.R., Lambert, H., Chávez-Zobel, A.T., Charest, G., Lavigne, P. & Landry, J. (2004). Essential role of the NH2-terminal WD/EPF motif in the phosphorylation-activated protective function of mammalian Hsp27. *J. Biol. Chem., 279,* 23463-23471.

[89] van den Ijssel, P., Wheelock, R., Prescott, A., Russell, P. & Quinlan, R.A. (2003). Nuclear speckle localisation of the small heat shock protein αB-crystallin and its inhibition by the R120G cardiomyopathy-linked mutation. *Exp. Cell Res., 287*, 249-261.

[90] Verschuure, P., Tatard, C., Boelens, W.C., Grongnet, J.F. & David, J.C. (2003). Expression of small heat shock proteins HspB2, HspB8, Hsp20 and cvHsp in different tissues of the perinatal developing pig. *Eur. J. Cell Biol., 82*, 523-530.

[91] Vicart, P., Caron, A., Guicheney, P., Li, Z., Prevost, M.C., Faure, A., Chateau, D., Chapon, F., Tome, F., Dupret, J.M., Paulin, D. & Fardeau, M. (1998). A missense mutation in the αB-crystallin chaperone gene causes a desmin-related myopathy. *Nat. Genet., 20*, 92-95.

[92] Vos, M.J., Hageman, J., Carra, S. & Kampinga, H.H. (2008). Structural and functional diversities between members of the human HSPB, HSPH, HSPA, and DNAJ chaperone families. *Biochemistry, 47*, 7001-7011.

[93] Wang, X., Osinska, H., Klevitsky, R., Gerdes, A.M., Nieman, M., Lorenz, J., Hewett, T. & Robbins, J. (2001). Expression of R120G-αB-crystallin causes aberrant desmin and αB-crystallin aggregation and cardiomyopathy in mice. *Circ. Res., 89,* 84-91.

[94] Wilhelmus, M.M., Boelens, W.C., Otte-Höller, I., Kamps, B., de Waal, R.M., Verbeek, M.M. (2006). Small heat shock proteins inhibit amyloid-beta protein aggregation and cerebrovascular amyloid-beta protein toxicity. *Brain Res., 17,* 1089, 67-78.

[95] Wilkie, A.O.M. (1994). The molecular basis of genetic dominance. *J. Med. Genet., 31*, 89-98.

[96] Williams, K.L., Rahimtula, M. & Mearow, K.M. (2006) Heat shock protein 27 is involved in neurite extension and branching of dorsal root ganglion neurons in vitro. *J. Neurosci. Res., 84,* 716-723.

[97] Yamasaki, S. & Anderson, P. (2008). Reprogramming mRNA translation during stress. *Curr. Opin. Cell Biol., 20*, 222-226.

[98] Ying, X., Zhang, J., Wang, Y., Wu, N., Wang, Y. & Yew, D.T. (2008). α-crystallin protected axons from optic nerve degeneration after crushing in rats. *J. Mol. Neurosci., 35*, 253-258.

[99] Zhai, J., Lin, H., Julien, J.P., & Schlaepfer, W.W. (2007). Disruption of neurofilament network with aggregation of light neurofilament protein: a common pathway leading to motor neuron degeneration due to Charcot-Marie-Tooth disease-linked mutations in NFL and HSPB1. *Hum. Mol. Genet., 16,* 3103-3116.

[100] Züchner, S. & Vance, J.M. (2006). Molecular genetics of autosomal-dominant axonal Charcot-Marie-Tooth disease. *Neuromolecular Med., 8*, 63-74.

Part III: Small Heat Shock Proteins in Cancer Diseases

In: Small Stress Proteins and Human Diseases
Editors: Stéphanie Simon et al.
ISBN: 978-1-61470-636-6

Chapter 3.1

SMALL STRESS PROTEINS, BIOMARKERS OF CANCER

Daniel R. Ciocca*, Mariel A. Fanelli, F. Darío Cuello-Carrión and Gisela N. Castro

Laboratory of Oncology, Institute of Experimental Medicine and Biology of Cuyo, Scientific and Technological Centre, National Research Council (CONICET) and Argentine Foundation for Cancer Research. Casilla de Correo 855, 5500 Mendoza, Argentina

ABSTRACT

Many oncogenic agents/events generate in cancer cells stress-responsive proteins, also known as heat shock proteins (HSPs). These proteins are implicated in the genesis as well as in the progression of cancer; the small HSPs share the chaperone activity of the other HSPs and thus the small HSPs should also be described as chaperones of tumorigenesis. Since the small HSPs are found at relatively high levels in tumor cells, the study of the expression of these proteins is of high interest as biomarkers of cancer. In this chapter we have analysed the evidence regarding its usefulness: 1) to detect in fluids and tissues the early changes in cell transformation, 2) to help in the diagnosis of the disease, 3) in predicting the recurrence, aggressiveness and the development of metastasis, and 4) to predict the response to therapies and to monitor the efficacy/safety of therapeutic agents. In breast cancer, earlier studies and more recent proteomic studies reveal that the small HSPs are mainly implicated in the response to therapies, HSPB1 (HSP27) appears as a biomarker of resistance to different anticancer therapies. In other tumors, like in prostate cancer, they appear as biomarkers of disease prognosis, while in uterine cervical cancer HSPB1 appears as a marker of cell differentiation. The molecular pathways implicated in the small HSPs activation, and the association of these proteins

* Daniel Ciocca
Laboratory of Oncology,
Institute of Experimental Medicine and Biology of Cuyo
Scientific and Technological Centre, National Research Council and
Argentine Foundation for Cancer Research.
Casilla de Correo 855
5500 Mendoza, Argentina
dciocca@lab.cricyt.edu.ar

with other proteins, results in a unique molecular context in each cancer cell type (and ultimately in each patient) appearing as biomarkers of different situations. Overall, it is evident that more studies on the topic of biomarkers of anticancer therapies are needed to change the way we treat cancer patients, but so far it seems that the small HSPs, in conjunction with other biomarkers, have a great future to achieve tailored-treatment strategies.

INTRODUCTION

Stress-responsive proteins, also known as heat shock proteins (HSPs), include the group of small HSPs (due to their relatively lower molecular weight) that at present consists of 11 members characterized by a signature conserved crystallin domain flanked by variable N- and C-termini [1]. These authors have unified the nomenclature of the HSPs and it will be used in this chapter. In cancer, the best studied member of the small HSPs is HSPB1 (previously known as HSP27).

HSPs are implicated in the genesis and progression of cancer, many oncogenic agents/events generate in the transformed cells a HSP response facilitating viral protein assembly and virus transportation, causing stabilization of mutated/damaged proteins implicated in signaling for proliferation, overriding cell cycle checkpoints, improving DNA repair and cell survival, and even provoking immune-modulation [2, 3]. The increased transcription of HSPs in tumor cells is due in part to loss of p53 function and to higher expression of the proto-oncogene HER-2 and c-Myc; moreover, cancer cells are exposed to many stressful situations like hypoxia, changes in the pH, exposure to radiation and to chemotherapeutic agents all of which could trigger a HSP response [reviewed in 4]. Therefore, the expression of the HSPs is in general higher in cancer tissues than in their normal counterparts [5]. Coming back to the small HSPs, they share the chaperone activity of the other HSPs and thus the small HSPs should also be described as chaperones of tumorigenesis. HSPB1 is a marker of metaplasia (one of the first steps in cell transformation) in the uterine cervix, it is overexpressed in all high-grade gliomas, it has been associated with a more aggressive phenotype in several cancer types, and with resistance to certain chemotherapies [5]; while HSPB5 (crystallin alpha B) has been described as an oncoprotein due to its expression in basal-like breast carcinomas (poor prognosis), and to its capacity to transform and produce neoplastic-like changes in immortalized human mammary epithelial cells [6].

Since the small HSPs are found at relatively high levels in tumor cells, the study of the expression of these proteins is of high interest as biomarkers of cancer. The need for newer biomarkers of cancer has been stressed in recent publications [7, 8], ideally they can be useful: 1) to detect in fluids (i.e., blood, urine) and tissues (biopsies) the early changes in cell transformation, 2) in the diagnosis of the disease, 3) in predicting the recurrence, aggressiveness and the development of metastasis, and 4) to predict the response to therapies and to monitor the efficacy/safety of therapeutic agents.

Gene expression profiles have contributed to identify tumor subclasses, disease prognosis and have been correlated to drug sensitivity. However, most therapeutic drugs are designed to target proteins, and the gene expression analysis does not correlate well with protein expression, as a consequence oncoproteome profiles offer advantages, allowing the analysis of post-translational modifications, subcellular distribution, and protein-protein interactions in a functional context [7]. In recent articles exploring the oncoproteomics to discover biomarkers of cancer, HSPB1 has emerged as one of the identified proteins. This phosphoprotein appeared in the signal transduction pathway of the Epstein-Barr virus encoded latent membrane protein 1 (LMP1) [8]. Moreover, HSPB1 was also identified as one of the biomarker proteins expressed at high levels in invasive ductal carcinomas of the breast (using nonequilibrium pH gradient electrophoresis and mass spectrometry) [9], and more specifically in subcohorts of patients with luminal A breast cancer subtype with high expression of estrogen receptor (using protein chip arrays) [10]. These findings confirm previous immunohistochemical studies reporting the high expression of HSPB1 in cancer tissues including breast carcinomas [5]. Here the authors reviewed the clinical significances of HSPs in cancer, regarding the diagnostic, prognostic, predictive, and treatment implications, containing HSPB1 and HSPB5. In this chapter we will update the data focusing on the small HSPs.

1. Genital and Urinary Tract

Since HSPB1 is present in several tumor types, it is not useful in diagnostic immunopathology to determine the cell of origin of a given tumor. In summary [5], in early studies HSPB1 was found in breast cancer cells and tissues, the expression of this protein correlated with that of estrogen receptors alpha, in fact HSPB1 is under estrogen regulation, however, HSPB1 should not be considered by itself as a biomarker of hormonodependency since it is also found in some estrogen receptor-negative tumors. The presence of HSPB1 in the serum of cancer patients is not useful as a biomarker of the disease; CA15-3 is more reliable for this purpose.

Regarding the prognostic significance of HSPB1 in breast cancer, this protein tends to be at higher levels in patients with poor prognosis but the association is not very strong, it seems more useful in patients with lymph node metastases than in patients with lymph node-negative tumors [5]. In more recent proteomic studies we mentioned that HSPB1 appeared as a biomarker of luminal A subtype [10], however, it is also a molecular feature of basal-like breast cancer together with HSPB5, cytokeratins 5 and 17, c-KIT, caveolins, P-cadherin, nuclear factor–kB (NF-kB) and many others [11]. Therefore, there are contradictory results (good *versus* poor prognosis) and we believe that this is due to the molecular pathways implicated in HSPB1 activation and the association of HSPB1 with other proteins, making a unique molecular context (Figure 1, with ref. 12-18).

In a recent study, we have reported that in breast cancer cells and tissues HSPB1 interacts with proteins involved with the cadherin-catenin cell adhesion system that is required for both the establishment and maintenance of cell-cell adhesion, in signaling pathways and in tumor cell invasion [19]. β-catenin interacted with HSPB1 and HSF1 but not with HSPA1A (Hsp70-1), and other HSP family members, and this interaction depended on the cellular localization of β-catenin. When this protein was expressed in the cytoplasm of the tumor cells it appeared, together with P-cadherin, as a biomarker of poor prognosis. Murine breast cancer cells transfected with HSPB1 showed a redistribution of β-catenin from the cell membrane to the cytoplasm. This study points to the importance of HSPB1 in influencing β-catenin, a protein important not only in the organization of cell-cell contacts but also in the downstream Wnt/wingless signalling pathways. In this context it is possible to suggest that HSPB1 overexpression might stabilize β-catenin in the cytoplasm of the tumor cells, impeding its normal function on cell-cell adhesion facilitating both metastasis formation and the β-catenin-mediated activation of genes in concert with TCF/LEF proteins. The role of HSPB1 in cell migration is further supported by a recent study showing that when HSPB1 was decreased by gene silencing the migratory capability of the highly metastatic murine 4T1 breast cancer cells was eliminated [20].

On the other hand, HSPB1 appears as a biomarker of resistance to different anticancer therapies: 1) tamoxifen (and perhaps other forms of hormone-therapies), 2) chemotherapies, 3) immunotherapy with trastuzumab (Herceptin), and 4) radiotherapy (Figure 1). We already mentioned that in Her-2/neu positive breast cancer cells this protein increases the oncoprotein stability reducing the susceptibility to trastuzumab [15]. Moreover, the presence of the phosphorylated form at Ser^{78} of HSPB1 has been reported correlated with Her-2/neu positive tumors and with lymph node positivity [21]. In a previous study our laboratory group reported that breast cancer patients with Her-2/neu + tumors showed more resistance to tamoxifen and to chemotherapy, and the possible molecular mechanisms implicated [17, 18, 22, 23]. We have also showed that in breast cancer patients with locally advanced disease, neoadjuvant chemotherapy was more ineffective in patients with high HSPB1 expression levels [24]. In more recent studies, in breast cancer cells HSPB1 has been found again associated with chemo-resistance [25, 26], while the chemotherapeutic drug paclitaxel inhibited HSPB1 expression causing topoisomerase (target for doxuribicin) up-regulation, thus Shi et al [26] proposed that the sequence paclitaxel-doxorubicin is more effective in cells with HSPB1 overexpression. In contrast, HSPB1 and other molecules were found down-regulated in MCF-7 human breast cancer cells treated with cisplatin, and in this proteomic analysis the down-regulation of these proteins has been associated with cisplatin resistance [27]. The three above mentioned chemotherapeutic drugs have different mechanisms of action, cisplatin mainly acts producing DNA adducts, and HSPB1 has been described as participating in the DNA repair mechanisms mediated by mismatch repair proteins [28] and base excision repair [29]; therefore, if these mechanisms are deficient due to less HSBP1 (and other components of the repairing complex) the cancer cells accumulate DNA damages but are not killed by the drug. Finally, HSPB1 has been associated with radioprotection in submandibular glands of rats [30]. The implications of HSPB1 in apoptosis and senescence can be found in other chapters of this book. Overall, it is evident that more studies on the topic of biomarkers of anticancer therapies are needed to change the way we treat breast cancer patients, but so far it seems that HSPB1 in conjunction with other biomarkers have a great future to achieve tailored-treatment strategies.

There are fewer studies on the role of HSPB1 in other cancer types. In a recent proteomic analysis performed on squamous carcinomas of the uterine cervix, HSPB1 emerged at higher expression levels than in the normal mucosa, appearing as a marker of various stages of cervical intraepithelial neoplasia [31]. This confirms in part previous immunohistochemical studies where HSPB1 was described at higher levels in cancer tissues, as a marker of cell differentiation in endometrial and endocervical adenocarcinomas and in squamous carcinomas of the uterine cervix, but was not useful to identify the different degrees of cervical intraepithelial neoplasia [32, 33].

In prostate cancer HSF1 is generally found at higher levels than in the normal gland, and it has been found at increased levels in cancer cells with metastatic ability [34]. In this study, HSPB1 was consistently noted at higher levels in both nonmetastatic and metastatic cell lines. The recurrence free-survival in patients with strong HSPB1 expression was shorter than in those with weak expression [together with HSPD1 (HSP60)] in biopsies analysed by tissue microarray [35]. This is consistent with a recent study of prostate cancer patients where HSPB1 expression was found correlated with several conventional clinicopathological prognostic factors of recurrence [36], and with previous studies reviewed earlier [5].

Finally, in earlier studies on urinary bladder cancer the relationships of HSPB1 with disease status has not been conclusive [5], however, in a more recent study HSPB1 expression (together with that of HSPA1A) has been associated with poor prognosis in schistosomiasis-associated bladder carcinoma patients [37]. Moreover, antisense oligonucleotides targeting HSPB1 administered intravesically showed antitumor activity in an orthotopic model of high-grade bladder cancer suggesting its effectiveness in non-muscle-invasive tumors [38].

2. Head and Neck Cancers

In head and neck cancers the altered expression of HSPB1 has been documented by proteomic studies, either increasing or decreasing, however its prognostic significance remains unclear [39-41]. In oral squamous cell carcinomas reduced HSPB1 expression has been associated with more aggressive and poorly differentiated tumors [42]. This is consistent with a retrospective analysis of 57 tumours performed by the same research group where reduced expression of HSPB1 in squamous carcinomas was found as an independent biomarker of the poorest overall survival rate [43]. Regarding the response to therapy, HSPB1 is one of the candidate proteins for chemoresistance. It has been demonstrated in *in vitro* studies carried out in Hep-2 laryngeal cancer cells that the overexpression of human HSPB1 induced resistance to several cytotoxic compounds (cisplatin, staurospin, H_2O_2) but no resistance against irradiation or serum starvation [44]. This chemoresistance was associated with retarded cell proliferation, HSPB1 seems to have a citoprotective activity. Although, the stabilization of F-actin filaments may explain, at least in part, the delay in cell progression, further studies will be necessary to elucidate this point. In esophageal squamous cell carcinomas, the expression of HSPB1, according to a multivariate analysis, was a good predictor of the response to chemo-radiotherapy and radiotherapy, those patients with tumours with negative HSPB1 (as well as with negative HSPA1A and p21) were good responders to therapy [45]. However, a recent study carried out in esophageal adenocarcinoma patients, showed a significant association between low HSPB1 expression and non-response to neoadjuvant chemotherapy [46]. The critical mechanisms by which HSPB1 positive tumor cells exhibit chemoresistance have not been clearly elucidated, perhaps some of them may be via the prevention of necrosis and apoptosis (see chapters in this book). Some authors have related chemoresistance via cell-cell adhesion in colorectal cancer cells and in larynx carcinoma cells, where the resistance to chemotherapeutic drugs in confluent cultures correlated with an increase in the expression of the HSPB1 and a decrease in topoisomerase 2 content [47]. Head and neck squamous cell carcinomas represent a major indication for radiotherapy, but intrinsic radioresistance of the tumor often leads to local recurrence, so by this manner HSPB1 could be a useful target in the treatment of patients with cancers resistant to radiotherapy. In a recent study, it was shown in a squamous cell carcinoma cell line that HSPB1 gene silencing produced radiosensitization effects indicating a protective role of HSPB1 against gamma radiation-induced apoptosis [48]. Clearly, the roles of HSPB1 in head and neck cancers warrants further investigations but this protein appears here, like in other cancer types, as protecting the tumor cells against anticancer therapies.

3. Digestive Tract

Hepatocellular carcinoma (HCC) is a common and aggressive malignant tumor worldwide, after resection there is a high rate of metastases or recurrence. Consequently, metastases remain one of the major obstacles for the treatment of HCC. Its molecular basis is still poorly known and there are few effective diagnostic biomarkers and therapeutic targets. Some reports have evaluated HSPB1 expression in HCC and have found that elevated levels could be used not only as a diagnostic marker but also as a valuable prognostic factor associated with poor disease outcome [49-51]. In vitro studies carried out in HCC-derived HuH7 cells have described that phosphorylated HSPB1 stopped cell growth via inhibition of extracellular signals regulated by kinases. So, the control of phosphorylated HSPB1 levels may be a new therapeutic strategy [52]. In intrahepatic cholangiocarcinomas, patients with HSPB1 expression showed a high mitotic index, tumor greatest dimension, capsular and vascular invasion and the worst mean survival [53].

Unfortunately, patients with pancreatic adenocarcinomas have a poor prognosis due to late clinical evidences, frequency of metastasis, and the tumor´s aggressive nature. Only 4% of patients survive for more than 5 years after diagnosis. Surgical resection is the sole curative treatment that is currently available, but only 10-15% of patients are free from metastasis at diagnosis. The existing biomarkers, such as CA 19-9 and CEA, are not suitable as early detection markers of pancreatic cancer. Therefore, there is a great need of new biomarkers to reveal this pathology. In a recent study, performing protein profiling of microdissected cryostat section of 9 pancreatic adenocarcinomas and 10 healthy pancreatic tissue samples using 2-DE followed by SELDI MS, the authors identified two proteins, DJ-1 and HSPB1, which were up-regulated in pancreatic cancer and discriminated well between central tumor and tumor margin vs healthy tissue and between central tumor and healthy tissue [54]. They also identified HSPB1 levels in serum by ELISA, indicating a sensitivity of 100% and specificity of 84% for the recognition of pancreatic cancer. Of course, the clinical relevance of these discovers needs further studies. Gemcitabine appears to be the only clinically effective drug for pancreatic cancer, however, the median survival time is only 6.3 months. Thus, it is urgent to understand the cellular and molecular mechanism of gemcitabine resistance. Obviously, increased HSPB1 expression in tumor specimens was related to higher resistibility to gemcitabine in pancreatic cancer patients, therefore HSPB1 could be a possible biomarker for predicting the response of pancreatic cancer patients to gemcitabine treatment [55].

Gastric cancer takes the first place among the malignant tumors of the digestive tract. Due to the relatively asymptomatic nature in the early stages of the disease and the lack of adequate screening tests, the majority of patients are diagnosed at an advanced stage of the disease. Hence, to search and identify specific disease-associated proteins as potential biomarkers for the early diagnosis and new therapeutic targets is a necessary topic to offer effective treatment. In this manner, in a study involving the serum proteomic spectra in patients with gastric cancer before surgery, HSPB1 appeared (among others such as glucose-regulated protein, prohibitin, protein disulfide isomerase A3) overexpressed and was obviously down-regulated in patients after surgical resection, allowing to distinguish gastric cancer patients from healthy people with 95.7% sensitivity and 92.5% specificity, respectively [56]. Then, they could use the proteomic spectra to diagnose gastric cancer quickly and exactly. According to the different regions of the stomach there exist different patterns of recurrence and outcome. It has been demonstrated co-up-regulation of HSPB1, HSPA1A and HSPD1 (and Prx-2) in gastric tumor tissues from the cardia region [57]. Interestingly, this stomach region tumor (gastric cardia adenocarcinoma) is far more prevalent with a higher incidence of lymph node metastasis and a poorer prognosis than distal gastric adenocarcinoma [58]. With respect to multidrug resistance, it has been shown that the suppression of HSPB1 expression by antisense oligonucleotides could increase vincristine chemosensitivity, in a vincristine resistant human gastric cancer cell line, where HSPB1 was overexpressed [59].

The pattern of HSPB1 expression and its role in colorectal cancer has been documented. As was described in a proteome analysis of human colon carcinoma cell lines with different potential mestastasis (SW480, SW620), the overexpression of HSPB1 was associated with metastasis and progression of colorectal carcinoma [60]. It is known that HSPB1 has several roles within the cells, one of these include its association with actin. In this manner, an *in vitro* wound closure assay showed that HSPB1 was involved in cell motility interacting with actin in human colorectal carcinoma [61]. Recently, it has been described that HSPB1 overexpression is closely connected with 5-fluorouracil (5-FU) resistance in human colorectal carcinoma cell lines seeing that HSPB down-regulation suppressed 5-FU resistance. In addition, HSPB1 may be a clinical target in patients who do not respond to this drug [62, 63].

4. LUNG

Lung cancer is the leading cause of cancer death in men, and in women it has surpassed even breast cancer [64], obviously metastasis is the major problem. There is little information about the roles played by HSPs in lung cancer. Previously, evaluating the immunoreactivity of HSPA1A and HSPB1 in tumor tissues from 60 patients with non-small cell lung cancer (NSCLC), the authors suggested that HSPA1A expression was a powerful and significant prognostic indicator; it was related to histopathological differentiation, lymph node metastasis, clinical stages, and smoking history, whereas HSPB1 expression was not [65]. However, a more recent study has demonstrated that there was an obvious correlation between HSPB1 overexpression and survival of patients with NSCLC: 70% of patients with HSPB1-negative tumors died within one year after the surgery. Therefore, HSPB1 (and HSPA1A) positivity may represent a favorable prognostic factor in NSCLC [66]. Few epidemiologic studies have demonstrated the associations between lymphocyte HSP levels and lung cancer risk. Recently has been suggested that lower lymphocyte HSPB1 levels might be associated with an increase risk of lung cancer (evaluated in 263 cases), this finding will need to be validated in a large prospective study [67]. With respect to chemotherapy response, there is a first study of HSPB1 expression in clinical NSCLC samples. The results pointed out that HSPB1 expression may be useful as a predictor of response to single-agent vinorelbine chemotherapy, there was a trend for patients with HSPB1-positive tumours to possess disease progression [68]. *In vitro* studies carried out in lung carcinoma cells NCI-HI299 overexpressing HSPB1, showed that the treatment with small interference RNA specifically targeting HSPB1 resulted in inhibition of their resistance to radiation or cisplatin. Perhaps some of these effects on resistance can be explained through direct interaction with PKC delta. Interaction between HSPB1 and PKC delta inhibited PKC delta kinase activity which resulted in suppression of PKC delta mediated cell death. Treatment of cancer cells with a novel heptapeptide that contains the binding sequence for PKC delta interaction with HSPB1 induced chemosensitization or radiosensitization because this heptapeptide restored the PKC delta kinase activity that had been inhibited by HSPB1. These results offer a new strategy for selective neutralization of HSPB1[69].

5. Leukemias

The term leukemia comprises a group of malignant diseases of the bone-marrow and blood, characterized by the production of abnormal white blood cells. Leukemia is classified in four categories: myelogenous or lymphocytic, and each of them can be acute or chronic. Thus, the four major types of leukemia are: Acute Lymphocytic Leukemia (ALL), Chronic Lymphocytic Leukemia (CLL), Acute Myelogenous Leukemia (AML) and Chronic Myelogenous Leukemia (CML).

With the purpose to identify prognostic factors or protein markers of drug-resistance and apoptosis, Thomas et al. [70] studied the expression of several HSPs in 98 newly diagnosed AML. These authors found that HSPB1 was expressed in 39% of the cases and that the positive expression appeared as a poor prognostic factor in patients with unfavorable karyotypes. Previously, it- has been reported that HSPB1 predicts a poor response to chemotherapy in leukemia patients [Reviewed in 5]. Diverse studies have pointed out the central role of HSPB1 in the inhibition of apoptosis and resistance to chemotherapeutic drugs in leukemic cells. U-937 promonocytes, HL-60 and NB4 human myeloid leukemia cell lines treated with arsenic trioxide at low doses showed apoptosis, and this effect was potentiated when PIK3 inhibitors, LY294002 and wortmannin, and Akt inhibitor Akt(i)5 were co-administrated with the arsenic compound [71]. This strenghtent of As_2O_3-provoked apoptosis involved the attenuation of HSPB1 expression, among other effects. In another work, Schepers et al. [72] have been used RNA interference for HSPB1 in leukemic TF-1 cells treated with VP-16 and CD95/Fas to induce apoptosis. HSPB1 RNA interference resulted in a twofold increase in VP-16-evoked apoptosis. DAXX co-immunoprecipitated with HSPB1 suggesting an inhibitory role of this small HSP in VP-16-mediated activation of ASK1/p38/JNK pathway. Moreover, CD95/Fas-mediated apoptosis, was unaffected by siRNA, probably, due to an up-regulation of HSPB1. Although, phosphorylated HSPB1 was over-expressed in primitive monocytic AML blasts (M4-M5, 91%, n=11) and was undetectable in myeloid blasts (M1-M2, n=5), VP-16-induced apoptosis moderately correlated with expression of HSPB1. Perhaps, owing to the co-expression of p21/Waf1/Cip1, which cytoplasmic over-expression inhibited the enhance p38 phosphorylation after HSPB1 RNAi treatment, suggesting a prevailing anti-apoptotic role of p21 over HSPB1. In conclusion: HSPB1 inhibits VP-16-mediated phosphorylation of p38 and c-Jun, cytochrome c release, and apoptosis; HSPB1 is expressed and activated in monocytic AML blasts; and p21 cytoplasmic expression equilibrates the deficiency of HSPB1. Navas et al. [73] demonstrated that continue treatment of multiple myeloma cells (MMC) with bortezomid, a proteasome inhibitor, leading to SCIO-469 (p38αinhibitor)-enhanced down-regulation of HSPB1 and to elevated MMC apoptosis. Previously, these authors have shown that p38 inhibition with SCIO-469 enforces MMC cytotoxicity of bortezomid by inhibiting the transient expression and phosphorylation of HSPB1, a downstream target of p38. Furthermore, p38 inhibition enhances the bortezomid-induced MMC apoptosis by upregulation of p53 and downregulation of BCL-X_L and Mcl-1. In a recent work, Mandal et al. [74] have been using Withaferin A (WA) to induce apoptosis in a dose-dependent manner on several human leukemic cell lines, and on primary cells from patients with lymphoblastic and myeloid leukemia. They found that in addition to the typical events of apoptosis (phosphatidylserine externalization, a time-dependent increase in Bax/Bcl-2 ratio, loss of mitochondrial transmembrane potential, cytochrome c release, caspases 9 and 3 activation, etc.), the apoptosis-WA induced was mediated by an increase of phosphorylated p38MAPK, which further activated downstream signaling by phosphorylating ATF-2 and HSPB1 in leukemic cells.

In an intent to elucidate the mechanisms of multidrug resistance (MDR) in leukemic cells, Zhang et al. [75] using proteomic tools, have screened MDR-related proteins in vincristine (VCR)-resistant leukemia cell line K562 and analyzed the probable mechanism of MDR in leukemia cells. This study identified a differential expression of HSPB1 between K562/VCR and K562 cells. HSPB1 was highly expressed in K562/VCR cells, and the blockade of its expression by antisense oligonucleotide could enhance chemosensitivity of K562/VCR cells to VCR. In a proteomic analysis of NOD/SCID mouse xenograft model of ALL, it has been found that alterations in the actin and tubulin cytoskeleton are involved in *in vivo* VCR resistance [76], and that altered expression of HSPB1, between others proteins, was associated with *in vivo* VCR resistance. This work provides the first evidence for a role of the actin cytoskeleton in intrinsic and acquired *in vivo* anti-microtubule drug resistance in childhood leukemia. These studies stand out the important role of HSPB1 in apoptosis and MDR mechanisms in leukemic cells.

In a recent and interesting investigation Tajeddine et al. [77] performed a gene profiling study by analyzing Preferentially Expressed Antigen of Melanoma (PRAME) shRNA-silenced leukemic cells on high density microarrays in 28 AML pediatric patients. They demonstrated that PRAME over-expression induces the repression of three genes: HSPB1, S100A4 and p21, associated with an unfavorable prognosis in leukemia. The results of this study suggest that PRAME is a favorable prognostic factor in leukemic patients, and this effect could be mediated, at least in part, by the modified expression of genes such as HSPB1, S100A4, p21, IL-8 and IGFBP. Previously, these authors [78] have suggested that PRAME can induce caspase-independent apoptosis via down regulation of HSPB1 and S100A4.

6. Lymphomas

Lymphoma is a general term for a group of cancers that originates in the lymphatic system. This malignancy appears when a lymphatic cell undergoes an aberrant transformation and begins to multiply, eventually crowding out healthy cells and creating tumors which enlarge the lymph nodes or other sites in the body (53% of blood cancers diagnosed are lymphomas). Non-Hodgkin lymphoma (NHL) represents a diverse group of cancers, with the distinctions between types based on the characteristics of the cancerous cells. Generally, the groups are classified as indolent or aggressive, low, intermediate and high grade. Each group is diagnosed and treated differently, and each has prognostic factors that categorize it as more or less favorable. On the other hand, Hodgkin lymphoma is a specialized form of lymphoma and represents about 11.1% of all lymphomas diagnosed in 2008. This tumor has characteristics that distinguish it from the other lymphatic cancers: including the presence of an abnormal cellular type called the Reed-Sternberg cell (a large, malignant cell found in Hodgkin lymphoma tissues), incidence rates higher in adolescents and young adults, and long-term survival rates of more than 86%.

In several B-malignancies, defective apoptosis signaling is a typical feature and the transcription factor nuclear factor-kappaB (NF-κB) is a critical mediator of resistance to apoptosis and oncogenic growth. Thomas et al. [79] have studied the effect of different inhibitors of NF-κB on classic Hodgkin´s lymphoma, multiple myeloma and activated B-cell-like diffuse large B-cell lymphoma cell lines with different defects in apoptosis signaling, both quantitatively and qualitatively. The cyclopentenone prostaglandin, 15-deoxi-Δ12,14-prostaglandin J (2) induced down-regulation of X-linked inhibitor of apoptosis protein and HSPB1, and led to breakdown of the mitochondrial membrane potential an cleavage of caspase-3 irrespective of IkappaBalpha (IκBα) mutational status. The results of this treatment suggest an important role of HSPB1 in the regulation of apoptosis in B-cell malignancies and its likely therapeutic role joined to therapeutic approach targeting IκBkinases. Chauhan et al. [80], treated lymphoma cells with PS-341, a selective proteasome inhibitor, and evaluated the levels of HSPB1 through microarray analysis. These authors found that a differential level- of RNA in DHL4 versus DHL6 cells correlated with HSPB1 expression. This was the first evidence that HSPB1 confers PS-431 resistance in lymphoma cells.

7. Sarcomas

Sarcomas are relatively uncommon cancers arising in connective tissues (fat, muscle, blood vessels, deep skin tissues, nerves, bones and cartilage) resulting in mesoderm proliferation. They are subdivided into two main categories: soft-tissue sarcomas (such as leiomyosarcoma) and non soft-tissue sarcomas (such as the bone cancer, osteosarcoma). Ewing's bone sarcoma contains both soft-tissue and non soft-tissue elements. The term soft tissue sarcoma is used to describe tumors of soft-tissue, which includes elements that are *in* connective tissue, but not derived from it (such as muscles and blood vessels). There are over 50 subtypes of sarcoma. About 1% of all adult cancers are sarcomas and between 15-20% of all children's cancers are sarcomas. Pediatric bone and soft-tissue malignant tumors are common tumors in children. It is also called musculoskeletal sarcoma, which means a cancer of mesenchymal tissues, such as the bone, soft tissues, and connective tissue. This kind of cancer is highly malignant and harmful to children.

HSPB1 over-expression showed a strong negative prognostic value in osteosarcoma and a decreased expression of this protein was found in chondrosarcoma [reviewed in 5]. Debes et al. [81] have shown that the bioflavonoid quercetin may be useful for optimizing the efficacy of hyperthermia in combination with chemotherapy, quercetin alone or combined with thermochemotherapy inhibited the expression of HSPB1 and HSPA in two Ewing´s tumor cells lines, SK-ES-1 and RD-ES. Suehara et al. [82] used 2-D difference gel electrophoresis to perform a global protein expression in soft-tissue sarcomas, in order to detect novel diagnostic biomarkers and to achieve molecular classification. These authors identified five proteins, including HSPB1, which could differentiate between malignant fibrous histiocytomas and leiomyosarcomas in grade III into low- and high-risk groups, which have different survival rates. Ha et al. [83] have studied gene expression profile in Hs701.T synovial sarcoma (SS) cell line under IL-1β stimulation, this cytokine has been found in the tumor microenvironment playing an important role in the pathogenesis of SS. The microarray data revealed that the most up-regulated genes were related to tumor progression, with HSPB1 and DAXX being involved in mediation of apoptosis. In a recent paper Zanini et al. [84] analyzed neuroblastoma (NB) and Ewing´s sarcoma (ES) cell lines by 2D gel electrophoresis searching for a new diagnostic/prognostic markers, and they found that HSPB1 could be used as a marker of neural differentiation in NB but not in ES.

8. Skin Cancer and Melanoma

Previous studies have demonstrated that HSPB1 expression is present in melanocytes in normal skin, as well as in melanomas, nevi and non-malignant melanocytes in culture [92]. Over-expression of HSPB1, among others HSPs, has been observed in metastatic malignant melanoma in comparison with the primary lesion [93]. In a recent proteomic analysis of human melanoma cells [94], HSPB1 has been identified as a downstream mediator of Secreted Protein Acidic and Rich Cysteine (SPARC) activity. The expression of this secreted protein has been related with increased aggressiveness in several human cancers. Antisense-mediated down-regulation of SPARC resulted in decreased levels of N-cadherin (N-CAD) and clusterin (CLU) and increased amounts of HSPB1. Transient re-expression of SPARC produced similar protein expression of N-CAD, CLU and HSPB1 to those in control cells. This work is the first evidence that SPARC, N-CAD, CLU and HSPB1 share a same molecular network in melanoma cells. In addition, HSPB1 plays a main role in protection of epidermal melanocytes under UV exposure. To arise this protective capacity is crucial the presence of mono- or bi-phosphorylated isoform of the protein. The UVB-induced HSPB1 phosphorylation was inhibited in melanocytes treated with an inhibitor to p38MAPKinase prior to irradiation. This fact suggests that UVB-induced HSPB1 phosphorylation through reactive oxygen species/p38MAPKinase pathway and this protein may confer a protective capacity against UVB-induced cell damage in the skin [95]. In the murine melanoma cell line K1735-C123, a cytoplasmatic over-expression of HSPB1 was associated with a larger proportion of DX-5+ NK cells-mediated lysis. This HSPB1 over-expression could be used as a marker of enhanced susceptibility to NK cytotoxicity in rats [96]. In humans, HSPB1 overexpression of transfected-A375 melanoma cell line yielded a decreased in cell invasiveness and metastatic potential *in vitro* [97]. Berhane et al. [98] have been studied the progression of actinic keratosis (AK) to squamous cell carcinoma (SCC). AK can be the first step leading to SCC and is therefore known as a "precancerous." Although the vast majority of AK remains benign, some studies report that up to ten percent may advance to SCC. Skin tumors from 50 patients with asymptomatic AK, inflamed AK or SSC were examined for studies of differentiation using HSPB1. The authors found a stepwise loss of differentiation as the tumors progressed from asymptomatic AK, through inflamed AK to SSCs. In other work, the expression of HSPB1 was associated with the onset of skin keratinocyte differentiation, but not with the progression of SCC. Using an immunostaining method of cutaneous tumors samples induced by UVB-irradiation, the authors showed that the level of HSPB1 decreases significantly as epithelial carcinoma growth progress upon UVB-exposure [99]. Previously, Trautinger et al. [100], have been studied the HSPB1 expression in 62 biopsy samples from normal human skin and from inflammatory and neoplastic skin diseases. The authors found a differentiation-related pattern of HSPB1 and they suggested that the absence of this HSP in the upper epidermal layers can be used a marker of epidermal malignancy. HSPB1 could play an inhibitory-regulatory function in tumor progression in melanoma such as suggested in a study carry out in A375 melanoma cell line HSPB1-transfected [101]. HSPB1 has previously been found as a differentiation marker in keratinocytes.

In canine skin samples, HSPB1 expression seems to correlate directly with cellular differentiation. In contrast, the participation of HSPA1 and HSPA8 in proliferation and differentiation of tumor cells still stay unclear. In normal epidermis, HSPB1 exhibited cytoplasmatic localization, whereas in neoplastic tissues it was detected in areas showing squamous differentiation. However, the pattern and intensity of immunostaining of HSPB1 and others HSPs did not show significant differences between intracutaneuos cornifying epithelioma and SCC, therefore they do not seem to be useful differential factors in the diagnosis of these canine tumors [102]. These authors, in a recent work have studied the expression of HSPB1, HSPA1 and HSPA8 in canine infundibular keratinizing achantomas and SCCs. The absence of correlation between the TUNEL Index and caspase-3 activated expression, an insufficient active-caspase-3-positive cells and HSPA1 over-expression were an indicative of apoptosis inhibition, suggesting that cell death repress- plays a role in oncogenesis and in progression of canine skin neoplasms [103].

9. HSPB5

HSPB5 is a major protein of the eye together with alpha acidic-crystalline (αA-crystalline), both proteins play an important role in maintaining crystalline transparency. In other tissues of the human body these proteins are expressed constitutively but at much lower levels [104]. HSPB5 can be phosphorylated and therefore it is under the control of several signaling pathways. The elevated expression of HSPB5 has been correlated with resistance to apoptosis induced by DNA-damaging agents, tumor necrosis factor alpha and fas ligation, and HSPB5 also plays a major role in tumor cell survival to stressful situations induced by external stimuli. Such protection may result through modulation of the activity of various apoptotic proteins (Bax, Bcl-xs, p53) [105]. Recently, oncogenic properties have been proposed for HSPB5, and this protein has been described as a novel regulator of tumor angiogenesis [104, 106].

HSPB5 is expressed in many tumors: gliomas, renal carcinomas, head and neck carcinomas and breast cancers [105]. Few studies exist regarding the expression of this protein in renal tumors and gliomas, most of them are from the decade of the 90's. Elevated HSPB5 levels in glial tumors (astrocytomas, glioblastomas multiforme and oligodendrogliomas) are associated with more aggressive disease stages. This relationship to the degree of glial differentiation suggests that this polypeptide is an important biochemical marker for studying human brain tumors [107]. In renal carcinomas, the HSPB5 immunostaining pattern is not absolutely specific but may be included in an array of markers to aid in the diagnosis of these tumors [108]. In head and neck carcinomas (62 patients with a follow-up of 5 years), the expression of the HSPB5 correlated with poor clinical outcomes [109].

There are several studies about the expression the HSPB5 in breast cancer. We already mentioned that Moyano and colleagues in a recent work [6] demonstrated oncogenic properties of HSPB5 in breast cancer specifically, overexpression of WT HSPB5 in several cell lines induced abnormalities changes such as EGF- and anchorage-independent growth, increased proliferation, loss of polarity, disorganized acinar structure, diminished luminal apoptosis and increased migration and invasion, all then are defining the character of a malignant neoplasm. This transformation in human mammary epithelial cell lines induced the expression and phosphorylation of ERK1/2 AKT and P38. The authors showed that the neoplastic changes were dependent on the ERK/MAPK pathway, and it also appears to be dependent on the phosphorylation status of HSPB5. Clearly, the mechanism by which HSPB5 activates ERK has yet to be elucidated. It is possible to speculate that HSPB5 regulates the activation of key proteins of the ERK/MAPK pathway such as Ras GTPases or Rafs [6]. Five molecular subtypes of breast cancer have been identified; the basal-like subtype expresses basal cytokeratins and is ER-negative, PR-negative, and Her2-negative (triple negative). This subtype currently has not targeted therapies, in contrast with other subtypes [110]. The small heat shock protein HSPB5 is expressed in 80% of the basal-like tumors and in 86% of the metaplastic carcinomas; it was not expressed in the no basal-like group. This finding indicates that HSPB5 may be a sensitive and specific marker for basal-like breast carcinomas. Moreover its overexpression contributes to aggressive phenotype and predicts poor survival [110]. Another study revealed a strong association between high expression levels of HSPB5 in primary breast cancer and lymph node involvement [111]. The antiapoptotic function of HSPB5 can explain the resistance to neoadjuvant chemotherapy of breast cancers expressing HSPB5 [112]. In this study, 112 patients who received neoadjuvant chemotherapy and had post-chemotherapy tumor specimens available for study were included.

In a recent work [113], a possible regulator of HSPB5 expression in basal-like breast cancers has been found by a bioinformatics method: the protein Est1. This is an oncogenic transcription factor with a specific ETS-binding union site (EBS) in the promotional region of the *HSPB5* (CRYAB) gene. These authors also demonstrated that the overexpression of Est1 in breast cancer cells increases HSPB5 levels, whereas its silencing reduces HSPB5 levels. In addition, they demonstrated that Est1 is expressed in basal-like breast cancer associated with poor survival. These tumours are associated with poor prognosis because they are highly proliferative and invasive. This may be due to the HSPB5 regulation by Ets1, possibly promoting metastasis by initiating invasion and apoptosis suppression. The co-expression of these two proteins needs to be investigated. Consistent with all these characteristics, HSPB5 was identified recently as a novel regulator of tumour angiogenesis in endothelial cells using a tubular morphogenesis assay in a 3D collagen matrix [106]. The new investigations should provide important data to support the potential role of HSPB5 as a target for therapy.

10. HspB8

Little is known about HSPB8, it is involved in the regulation of cell proliferation, cardiac hypertrophy, apoptosis and carcinogenesis. Depending on the cell type and its expression, HSPB8 might have either pro- or antiapoptotic effects. The chaperone activity of this protein seems to be relevant for the antiapoptotic capacity of HSPB8, while the proapoptotic effects are related with the activation of different protein kinases [114, 115].

In breast cancer, the overexpression of HSPB8 has been associated with the expression of cyclin D1 and ERs [116]. It is known that elevated cyclin D1 levels are related with poor prognosis and its expression distinguishes malignant breast carcinomas from benign and premalignant lesions. However, cyclin D1 expression is also associated with enhance radiation sensitivity. Then, what is the relationship of cyclin D1 with HSPB8? HSPB8 has been identified as a target candidate of CDK-independent cyclin D1; HSPB8 seems to mediate the effects of cyclin D1. Then, the expression of HSPB8 enhances the radiation sensitivity of tumour cells, and its loss in cyclin D1-overexpressing cells seems to block some of cyclin D1-dependent events.

Estrogens can induce HSPB8 expression in ER-positive MCF-7 breast cancer cells. In this cell line HSPB8 interacted with HSPB1, a related small HSP with important effects in breast cancers. Cadmium was also a potent inducer of HSPB8 in MCF-7 cells; this metalloestrogen has been implicated in breast cancer [117]. The induction of HSPB8 by estrogens and Cd in ER+ breast cancer cells represent a novel aspect of the estrogenic effects and of the aberrant response to an environmental factor with estrogenic properties. Further studies will be necessary to define the specific role(s) of HSPB8 in breast cancer.

On the other hand, HSPB8 is silenced in melanoma cell lines but not in benign melanocytic lesions or in normal skin melanocytes, the mechanism by which this lack of expression is produced is due to the aberrant methylation of the *HSPB8* gene that occurs in 60-75% of melanomas and atypical nevi [118]. Recently it has been shown that HSPB8 overload induces apoptosis in 55% of melanoma cells in culture [119]. These authors reported that HSPB8 overexpression induces both apoptosis through TAK1 activation and melanoma growth arrest, suggesting that HSPB8 activation can be used as a promising molecular target for therapy.

Conclusion

Several stressful situations and molecular events induce the small HSP response in several tumor types; therefore, these proteins are not useful as biomarkers to determine the cell of origin of a given tumor. However, the small HSPs appear implicated in several steps of tumorigenesis from the beginning to the "end" and therefore they are important as biomarkers

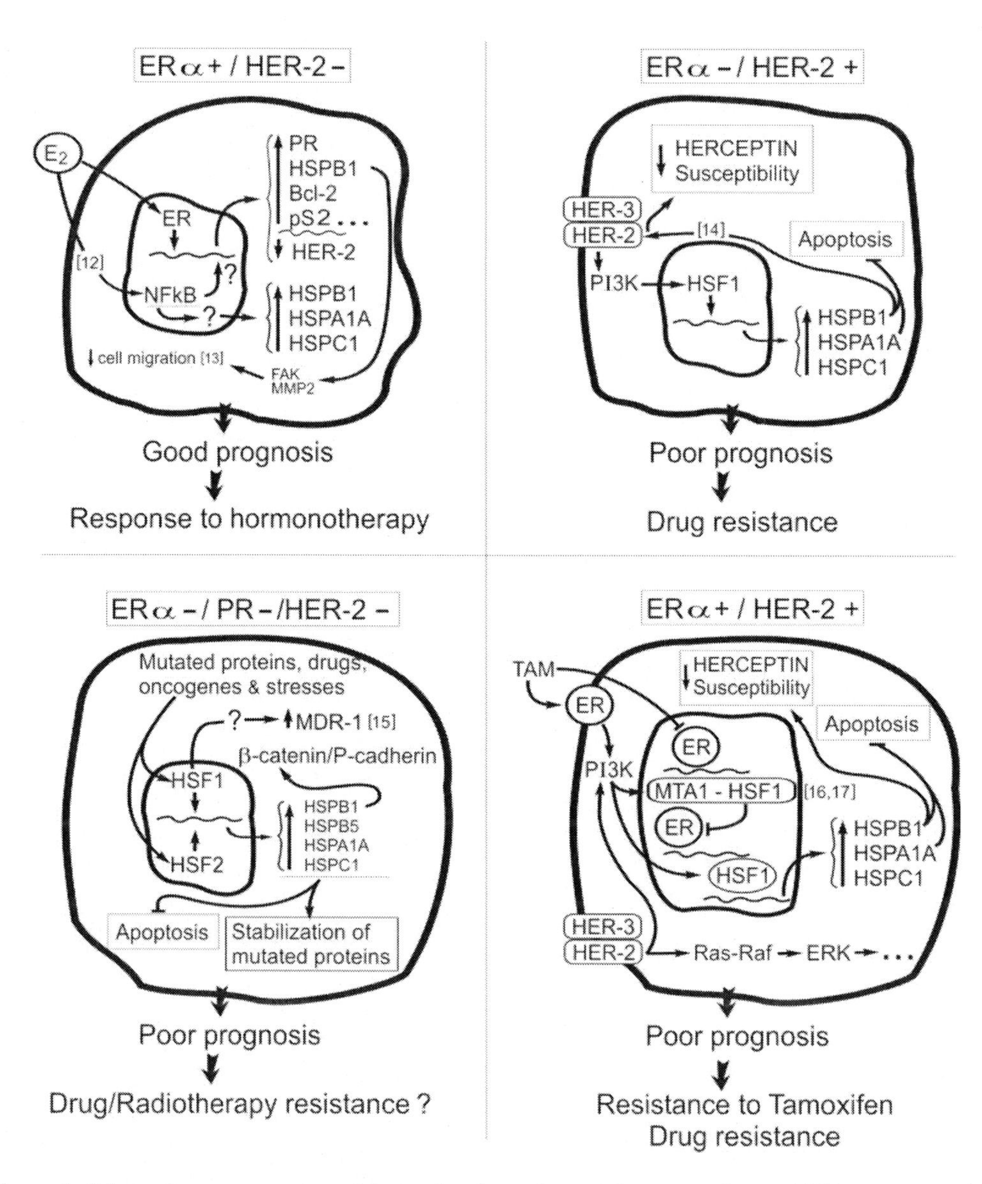

Figure 1. Schematic representation of the molecular pathways that can activate a HSP response, taken into account the presence or absence of hormone receptors (mainly estrogen receptor alpha: ERα) and Her-2/neu (c-erbB-2, human epidermal growth factor receptor-2). The four cases represent the different subtypes of breast cancer: 1) luminal A, 2) Her-2/neu +, 3) triple negative, and 4) luminal B. In the hypothetical cases here presented, HSPB1 is induced by the cancer cells, however, it is also possible to have the phenotypes presented but with absence or very low levels of HSPB1, and therefore the consequences will be different (not shown). In the first case (ERα+ and Her-2/neu negative tumours) HSPB1 is under estrogen regulation, the cancer cells still have similitude to the luminal normal mammary cells, and HSPB1 performs many of the normal constitutive housekeeping functions. Here HSPB1 is related with the luminal A subtype [10] being a biomarker of relatively good prognosis. In contrast, in the triple negative tumors, HSPB1 could be a biomarker of poor prognosis when characterizing a basal-like breast cancer [11]. In the case of ERα- and Her-2/neu+ tumors, it is of interest to mention the pathways of HSP activation related with Her-2/neu, HSPB1 can stabilize Her-2/neu decreasing the susceptibility of the cancer cells to the monoclonal antibody Herceptin [15].

Finally, in ERα+/Her-2/neu+ tumors the complex molecular mechanisms leading to tamoxifen (TAM) resistance have been described in a recent review [17].

of cell transformation, they could be useful to identify certain cancer cells (like Reed-Sternberg cells) and cancer cell types/subtypes. The main challenge is that the small HSP response is very versatile, changing in different cancer cell tissues and in each individual cancer cell, this in conjunction to their intrinsic chaperone activities can modify the molecular context of a given cancer cell. As a consequence sometimes there is confusion about the role(s) that they are playing, they can appear as biomarkers of cell differentiation (good and poor), or associated with disease prognosis (good and poor), or as biomarker of resistance to different anticancer therapies. The biomarker utility of these proteins should therefore be established when studied in the context of the proteome considering the molecular interrelations created in a cancer cell.

ACKNOWLEDGMENTS

This work was supported by the National Research Council (CONICET) PIP2428 of Argentina, the National Agency for Research (Préstamo BID, PICT 1047) and the Argentine Foundation for Cancer Research.

REFERENCES

[1] Kampinga, HH; Hageman, J; Vos, MJ; Kubota, H; Tanguay, RM; Bruford, EA; Chetham, ME; Chen, B; Hightower, LE. (2009). Guidelines for the nomenclature of the human heat shock proteins. *Cell Stress and Chaperones, 14*:105-111.

[2] Ciocca, DR; Fanelli, MA; Cuello-Carrión, FD; Calderwood, SK. (2007). Implications of heat shock proteins in carcinogenesis and cancer progression. In: S.K. Calderwood, M.Y. Sherman; D.R. Ciocca (Eds.), *Heat Shock Proteins in Cancer*. (Heat Shock Proteins, Vol 2. Series Ed.: A.A.A. Asea and S.K. Calderwood, pp. 31-51). The Netherlands:Springer.

[3] Calderwood, SK; Khaleque MA; Sawyer, DB; Ciocca DR. (2007). Heat shock proteins in the progression of cancer. In: S.K. Calderwood (Ed.). *Protein Reviews* (Cell Stress Proteins, Vol 7. Series Ed.: M. Z. Atassi, pp. 422-450). New York:Springer Science.

[4] Calderwood, SK; Khaleque, A; Sawyer, DB; Ciocca, DR. (2006). Heat shock proteins in cancer: chaperones of tumorigenesis. *Trends Biochem Sci, 31*:164-172.

[5] Ciocca, DR; Calderwood, SK. (2005). Heat shock proteins in cancer: diagnostic, prognostic, predictive, and treatment implications. *Cell Stress and Chaperones, 10*:86-103.

[6] Moyano, JV; Evans, JR; Chen, F; Lu, M; Werner, ME; Yehiely, F; Diaz, LK; Turbin, D; Karaca, G; Wiley, E; Nielsen, TO; Perou, CM; Cryns, V. (2006). αB-Crystallin is a novel oncoprotein that predicts poor clinical outcome in breast cancer. *J Clin Invest, 116*:261-270.

[7] Verrils, NM. (2006). Clinical proteomics: present and future prospects. *Clin Biochem Rev, 27*:99-116.

[8] Yan, G; Li, L; Tao, Y; Liu, S; Liu, Y; Luo, W; Wu, Y; Tang, M; Dong, Z; Cao, Y. (2006). Identification of novel phosphoproteins in signaling pathways triggered by latent membrane protein 1 using functional proteomics technology. *Proteomics, 6*:1810-1821.

[9] Cho, WCS. (2007). Contribution of oncoproteomics to cancer biomarker discovery. *Molecular Cancer 6*:25-37.

[10] Kabbage, M; Chahed, K; Hamrita, B; Guillier, CL; Trimeche, M; Remadi, S; Hoebeke, J; Chouchane, L. (2008). Protein alterations in infiltrating ductal carcinomas of the breast as detected by nonequilibrium pH gradient electrophoresis and mass spectrometry. *J Biomed Biotechnol, 2008*: ID#564127 (10 pages).

[11] Brozkova, K; Budinska, E; Bouchal, P; Hernychova, L; Knoflickova, D; Valik, D; Vyzula, R; Vojtesek, B; Nenutil, R. (2008). Surface-enhanced laser desorption/ionisation time-of-flight proteomic profiling of breast carcinomas identifies clinicopathologically relevant groups of patients similar to previously defined clusters from cDNA expression. *Breast Cancer Res 10*:R48.

[12] Rakha, EA; Reis-Filho, JS, Ellis, IO. (2008). Impact of basal-like breast carcinoma determination for a more specific therapy. *Pathobiology, 75*:95-103.

[13] Stice, JP; Knowlton, AA. (2008). Estrogen, NFkB, and the heat shock response. *Mol Med, 14*:517-527.

[14] Lee, JW; Kwak, HJ; Lee, JJ; Kim, YN; Lee, JW; Park, MJ; Jung, SE; Hong, SI; Lee, JH; Lee, JS. (2008). HSP27 regulates cell adhesion and invasion via modulation of focal adhesion kinase and MMP-2 expression. *Eur J Cell Biol 87*:377-387.

[15] Kang, SH; Kang, KW; Kim, K-H; Kwon, B; Kim, S-K; Lee, H-Y; Kong, S-Y; Lee, ES; Jang, S-G; Yoo, BC. (2008). Upregulated HSP27 in human breast cancer cells reduces Herceptin susceptibility by increasing Her2 protein stability. *BMC Cancer, 8*:286.

[16] Tchénio, T; Havard, M; Martinez, LA; Dautry, F. (2006). Heat shock-independent induction of multidrug resistance by heat shock factor 1. *Mol Cell Biol, 26*:580-591.

[17] Ciocca, DR; Gago, FE; Fanelli, MA; Calderwood, SK. (2006). Co-expression of steroid receptors (estrogen receptor alpha and/or progesterone receptors) and Her-2/neu: clinical implications. *J Steroid Biochem Mol Biol, 102*:32-40.

[18] Khaleque, A; Bharti, A; Gong, J; Ciocca, D; Stati, A; Fanelli, M; Calderwood, SK. (2008) Heat shock factor 1 represses estrogen dependent transcription through association with MTA1. *Oncogene 27*:1886-1893.

[19] Fanelli, MA; Montt-Guevara, M; Diblasi, AM; Gago, FE; Tello, O; Cuello-Carrión, FD; Callegary, E; Bausero, MA; Ciocca, DR. (2008). P-cadherin and β-catenin are useful prognostic markers in breast cancer patients; β-catenin interacts with heat shock protein Hsp27. *Cell Stress and Chaperones 13*:207-220.

[20] Bausero, MA; Bharti, A; Page, DT; Perez, KD; Eng, JWL; Jantschitsch, C; Kindas-Muegge, I; Ciocca, D; Asea, A. (2006). Silencing the hsp25 gene eliminates migration capability of the highly metastatic murine 4T1 breast adenocarcinoma cell. *Tumor Biol, 27*:17-26.

[21] Zhang, D; Wong, LL; Koay, ESC. (2007). Phosphorylation of Ser^{78} of Hsp27 correlated with HER-2/*neu* status and lymph node positivity in breast cancer. *Mol Cancer, 6*:52.

[22] Gago, FE; Fanelli, MA; Ciocca DR (2006). Co-expression of steroid hormone receptors (estrogen receptor α and/or progesterone receptors) and Her2/neu (c-erbB-2) in breast

cancer: clinical outcome following tamoxifen-based adjuvant therapy. *J Steroid Biochem Mol Biol, 98*:36-40.

[23] Vargas-Roig, LM; Gago, FE; Tello, O; Martin de Civetta, MT; Ciocca, DR. (1999). c-erbB-2 (HER-2/neu) protein and drug resistance in breast cancer patients treated with induction chemotherapy. *Int J Cancer (Pred Oncol), 84*:129-134.

[24] Vargas-Roig, LM; Gago, FE; Tello, O; Aznar, JC; Ciocca, DR (1998). Heat shock protein expression and drug resistance in breast cancer patients treated with induction chemotherapy. *Int J Cancer (Pred Oncol), 79*:468-475.

[25] Huthapisith, S; Layfield, R; Kerr, ID; Hughes, C; Eremin, O. (2007). Proteomic profiling of MCF-7 breast cancer cells with chemoresistance to different types of anticancer drugs. *Int J Oncol, 30*:1545-1551.

[26] Shi, P; Wang, MM; Jiang, LY; Liu, HT; Sun, JZ. (2008). Paclitaxel-doxorubicin sequence is more effective in breast cancer cells with heat shock protein 27 overexpression. *Chin Med J (Engl), 121*:1975-1979.

[27] Smith, L; Welham, KJ; Watson, MB; Drew, PJ; Lind, MJ; Cawkwell, L. (2007). The proteomic analysis of cisplatin resistance in breast cancer cells. *Oncol Res, 16*:497-506.

[28] Nadin, SB; Vargas-Roig, LM; Drago, G; Ibarra, J; Ciocca, DR. (2007). Hsp27, Hsp70 and mismatch repair proteins hMLH1 and hMSH2 expression in peripheral blood lymphocytes from healthy subjects and cancer patients. *Cancer Lett, 252*:131-146.

[29] F.Mendez, F; Sandigursky, M; Franklin, WA; Kenny, MK; Kureekattil, R; Bases, R. (2000). Heat-shock proteins associated with base excision repair enzymes in HeLa cells. *Radiat Res, 2*:186–195.

[30] Lee, H-J; Lee, Y-J; Kwon, H-C; Bae, S; Kim, S-H; Min, J-J; Cho, C-K; Lee, Y-S. (2006). Radioprotective effect of heat shock protein 25 on submandibular glands of rats. *Am J Pathol, 169*:1601-1611.

[31] Ono, A; Kumai, T; Koizumi, H; Nishikawa, H; Kabayashi, S; Tadokoro, M. (2009). Overexpression of heat shock protein 27 in squamous cell carcinoma of the uterine cervix: a proteomic analysis using archival formalin-fixed, paraffin-embedded tissues. *Human Pathol, 40*:41-49.

[32] Ciocca, DR; Puy, LA; Fasoli, LC. (1989). Study of estrogen receptor, progesterone receptor, and the estrogen-regulated Mr 24,000 protein in patients with carcinomas of endometrium and cervix. *Cancer Res, 49*:4298-4304.

[33] Puy, LA; Lo Castro, G; Olcese, JE; Lotfi, HO; Brandi, HR; Ciocca, DR. (1989). Analysis of a 24-kilodalton (KD) protein in the human uterine cervix during abnormal growth. *Cancer, 64*:1067-1073.

[34] Hoang, AT; Huang, J; Rudra-Ganguly, N; Zheng, J; Powell, WC; Rabindran, SK; Wu, C; Roy-Burman, P: (2000). A novel association between the human heat shock transcription factor 1 (HSF1) and prostate adenocarcinoma. *Am J Pathol, 156*:857-864.

[35] Glaessgen, A; Jonmarker, S; Lindberg, A; Nilsson, B; Lewensohn, R; Ekman, P; Valdman, A; Egevad, L. (2008). Heath shock proteins 27, 60 and 70 as prognostic markers of prostate cancer. *APMIS, 116*:888-895.

[36] Miyake, H; Muramaki, M; Kurahashi, T; Takenaka, A; Fujisawa, M. (2008). Expression of potential molecular markers in prostate cancer: correlation with clinicopathological outcomes in patients undergoing radical prostatectomy. *Urol Oncol* (Epub ahead of print).

[37] El-Meghawry El-Kenawy, A; El-Kott, AF; Hasan, MS. (2008). Heat shock protein expresión independently predicts survival outcome in schistosomiasis-associated urinary bladder cancer. *Int J Biol Markers, 23*:214-218.

[38] Hadaschik, BA; Jackson, J; Fazli, L; Zoubeidi, A, Burt, HM; Gleave, ME; So, AI. (2008). Intravesically administered antisense oligonucleotides targeting heat-shock protein-27 inhibited the growth of non-muscle-invasive bladder cancer. *BJU Int, 102*:610-616.

[39] He, QY; Chen, J; Kung, HF; Yuen, AP; Chiu, JF. (2004). Identification of tumor-associated proteins in oral tongue squamous cell carcinoma by proteomics. *Proteomics, 4*:271-278.

[40] Miyarek, AM; Balys, RL; Su, J; Hier, MP; Black, MJ; Alaoui-Jamali, MA. (2007). A cell proteomic approach for the diction of secretable biomarkers of invasiveness in oral squamous cell carcinoma. *Arch Otolarynol Head Neck Surg, 133*:910-8.

[41] Lo, W-Y; Tsai, M-H; Tsai, Y; Hua, Ch-H; Tsai, F-J; Huang, S-Y; Tsai, Ch-H; Lai, Ch-Ch. (2007). Identification of over-expressed proteins in oral squamous cell carcinoma (OSCC) patients by clinical proteomic analysis. *Clinica Chimica Acta, 376*:101-107.

[42] Lo Muzio, L; Leonardi, R; Mariggiò, MA; Mignogna, MD; Rubini, C; Vinella, A; Pannone, G; Giannetti, L; Serpico, R; Testa, NF; De Rosa, G; Staibano, S. (2004). HSP27 as possible prognostic factor in patients with oral squamous cell carcinoma. *Histol Histopathol, 19*:119-128.

[43] Lo Muzio, L; Campisi, G; Farina, A; Rubini, C; Ferrari, F; Falaschini, S; Leonardi, R; Carinci, F; Stalbano, S; De Rosa, G. (2006). Prognostic value of hsp27 in head and neck squamous cell carcinoma: a retrospective analysis of 57 tumours. *Anticancer Res, 26*:1343-1349.

[44] Lee, JH; Sun, D; Cho, KJ; Kim, MS; Hong, MH; Kim, IK; Lee, JS; Lee, JH. (2007). Overexpression of human 27 kDa heat shock protein in laryngeal cancer cells confers chemoresistance associated with cell growth delay. *J Cancer Res Clin Oncol, 133*:37-46.

[45] Miyazaki, T; Kato, H; Faried, A; Sohda, M; Nakajima, M; Fukai, Y; Masuda, N; Manda, R; Fukuchi, M; Ojima, H; Tsukada, K; Kuwano, H. (2005). Predictors of response to chemo-radiotherapy and radiotherapy for esophageal squamous cell carcinoma. *Anticancer Res, 25*:2749-2755.

[46] Langer, R; Ott, K; Specht, K; Becker, K; Lordick, F; Burian, M; Herrmann, K; Schrattenholz, A; Cahill, MA; Schwaiger, M; Hofler, H; Wester, HJ. (2008). Protein expression profiling in esophageal adenocarcinoma patients indicates association of heat-shock protein 27 expression and chemotherapy response. *Clin Cancer Res, 14*:8279-8287.

[47] Mazurov, VV; Solovieva, ME; Leshchenko, VV; Kruglov, AG; Edelweiss, EF; Yakubovskaya, RI; Akatov, VS. (2003). Small Heat shock protein hsp27 as a possible mediator of intercellular adhesion-induced drug resistance in human larynx carcinoma HRp-2 cells. *Bioscience Reports, 23*:187-197.

[48] Aloy, M-T; Hadchity, E; Bionda, C; Diaz-Latud, Ch; Claude, L; Rousson, R; Arrigo, A-P; Rodríguez-Lafrasse, C. (2008). Protective role of Hsp27 protein against gamma radiation-induced apoptosis and radiosensitization effects of Hsp27 gene silencing in different human tumor cells. *Int J Radiation Oncology Biol Phys, 70*:543-553.

[49] Li, AF; Chau, GY; Chi, CW; Wu, CW; Huang, CL; LUI, WY. (2000). Prognostic significances of heat shock protein 27expression in hepatocellular carcinoma and its relation to histologic grading and survival. *Cancer, 88*:2464-2470.

[50] Luck, JM; Lam, CT; Siu, AF; Lam, BY; Ng, IO; HU MY; Che, CM; Fan, ST. (2006) Proteomic profiling of hepatocellular carcinoma in chinese cohort reveals heat shock proteins (Hsp27, Hsp70, GRP78) up-regulation and their associated prognostic values. *Proteomics, 6*:1049-1057.

[51] 50.- Song, HY; Liu, YK; Feng, JT; Cui, JF; Dai, Z; Zhang, LJ; Feng, JX; Shen, HL; Tang, ZY. (2006). Proteomic analysis on metastasis-associated proteins of human hepatocellular carcinoma tissues. *J Cancer Res Clin Oncol, 132*:92-98.

[52] Matsushima-Nishiwaki, R; Takai S; Adachi, S; Minamitani, C; Yasuda, E; Noda, T; Kato, K; Toyoda, H; Kaneoka, Y; Yamaguchi, A; Kumada, T; Kozawa, O. (2008). Phosphorylated heat shock protein 27 represses growth hepatocellular carcinoma via inhibition of extracellular signal-regulated kinase. *J Biol Chem, 283*:18852-18860.

[53] Romani, AA; Crafa, P; Desenzani, S; Graiani, G; Lagrasta, C; Sianesi, M; Soliani, P; Borghetti, AF. (2007). The expression of HSP27 is associated with poor clinical outcome in intrahepatic cholangiocarcinoma. *BMC Cancer, 7*:232.

[54] Melle, Ch; Ernst, G; Escher, K; Hartmann, D; Schimmel, B; Bleul, A; Thieme, H; et al. (2007). Protein profiling of microdissected pancreas carcinoma and identification of HSP27 as a potential serum marker. *Clinical Chemistry, 53*:629-635.

[55] Mori-Iwamot, S; Kuramitsu, Y; Ryozawa, S; Mikuria, K; Fijumoto, M; Maehara, Y; Okita K; Nakamura, K; Saokaida, I. (2007). Proteomics finding of heat shock protein 27 as a biomarker for resistance of pancreatic cancer cells to gemcitabine. *International J Oncol, 31*: 1345-1350.

[56] Ren, H; Du, N; Liu, G; Hu, H-T; Tian, WDeng, Z-P; Shi J-S. (2006). Analysis of variabilities of serum proteomic spectra in patients with gastric cancer before and after *operation World J Gastroenterol, 12*:2789-2792.

[57] Cheng, Y; Zhang, J; Li, Y; Wang, Y; Gong, J. (2007). Proteome analysis of human gastric cardia adenocarcinoma by laser capture microdissection. *BMC Cancer, 7*:191-200.

[58] Saito, H; Fukumoto, Y; Osaki, T; Fukuda, K; Tatebe, A; Tsujitani, S; Ikeguchi, M. (2006). Distinct recurrence pattern and outcome of adenocarcinoma of the gastric cardia in comparison with carcinoma of other regions of the stomach. *World J Surg, 30*:1864-1869.

[59] Yang, YX; Xiao, ZQ; Chen, ZC; Zhang, GY; Yi, H; Zhang, PF; Li, JL; Zhu, G. (2006). Proteome analysis of multidrug resistance in vincristine-resistant human gastric cancer cell line SGC7901/VCR. *Proteomics, 6*:2009-2021.

[60] Zhao, L; Liu, LI; Wang, Shuang; Zhang, Y-F; Yu, L; Ding Y-Q. (2007). Differential proteomic analysis of human colorectal carcinoma cell lines metastasis-associated proteins. *J Cancer Res Clin Oncol, 133*: 771-782.

[61] Doshi, BM; Hightower, LE; Lee, J. (2009). The role of Hsp27 and actin in the regulation of movement in human cancer cells responding to heat shock. *Cell Stress Chaperones*, (epub ahead of print).

[62] Wong, CS; Wong, VW; Chan, CM; Ma, BB; Hui, EP; Wong, MC; Lam, MY; Au, TC; Chan, WH; Cheuk, W; Chan, AT. (2008). Identification of 5-fluorouracil response

proteins in colorectal carcinoma cell line SW480 by two-dimensional electrophoresis and MALDI-TOF mass spectrometry. *Oncol Rep, 20*:89-98.

[63] Tsuruta, M; Nishibori, H; Hasegawa, H; Ishii, Y; Kubota, T; Kitajima, M; Kitagawa, Y. (2008). Heat shock protein 27, a novel regulator of 5-fluorouracil resistance in colon cancer. *Oncol Rep, 20*:1165-1172.

[64] Hoffman, PC; Mauer, AM; Vokes, EE. (2000). Lung Cancer. *Lancet, 355*:479-485.

[65] Huang, Q; Zu, Y; Fu, X; Wu, T. (2005). Expression of heat shock protein 70 and 27 in non-small cell lung cancer and its clinical significance. *J Huazhong Univ Sci Technolog Med Sci, 25*:693-695.

[66] Malusecka, E; Krzyzowska-Gruca, S; Gawrychowski, J; Fiszer-Kierzkowska, A; Kolosza, Z; Krawczyk, Z. (2008). Stress proteins Hsp27 and Hsp70i predict survival in non-small cell lung carcinoma. *Anticancer Res, 28*:501-506.

[67] Wang, F; Feng, M; Xu, P; Xiao, H; Niu, P; Yang, X; Bai, Y; Peng, Y; Yao, P; Tan, H; Tanguay, RM; Wu, T. (2008). The level of Hsp27 in lymphocytes is negatively associated with a higher risk of lung cancer. *Cell Stress & Chaperones* (Epub ahead of print).

[68] Berrieman, HK; Cawkwell, L; O¨Kane, SL; Smith, L; Lind, MJ. (2006). Hsp27 may allow prediction of the response to single agent vinorelbine chemotherapy in non-small cell lung cancer. *Oncology Reports, 15*: 283-286.

[69] Kim, EH; Lee, HJ; Lee, DH; Bae, S; Soh, JW; Jeoung, D; Kim, J; Cho, CK; Lee, YJ; Lee, YS. (2007). Inhibition of heat shock protein 27 mediated resistance to DNA damaging agents by a novel PKCdelta-V5 heptapeptide. *Cancer Res, 67*:6333-6341.

[70] Thomas, X; Campos, L; Mounier, C; Cornillon, J; Flandrin, P; Le, QH; Piselli, S; Guyotat, D. (2005). Expression of heat-shock proteins is associated with major adverse prognostic factors in acute myeloid leukemia. *Leuk Res, 29*:1049-1058.

[71] Ramos, AM; Fernández, C; Amrán, D; Sancho, P; de Blas, E; Aller, P. (2005). Pharmacologic inhibitors of PI3K/Akt potentiate the apoptotic action of the antileukemic drug arsenic trioxide via glutathione depletion and increased peroxide accumulation in myeloid leukemia cells. *Blood, 105*:4013-4020.

[72] Schepers, H; Geugien, M; van der Toorn, M; Bryantsev, AL; Kampinga, HH; Eggen, BJ; Vellenga, E. (2005). HSP27 protects AML cells against VP-16-induced apoptosis through modulation of p38 and c-Jun. *Exp Hematol. 33*:660-670.

[73] Navas, TA; Nguyen, AN; Hideshima, T; Reddy, M; Ma, JY; Haghnazari, E; Henson, M; Stebbins, EG; Kerr, I; O'Young, G; Kapoun, AM; Chakravarty, S; Mavunkel, B; Perumattam, J; Luedtke, G; Dugar, S; Medicherla, S; Protter, AA; Schreiner, GF; Anderson, KC; Higgins, LS. (2006). Inhibition of p38alpha MAPK enhances proteasome inhibitor-induced apoptosis of myeloma cells by modulating Hsp27, Bcl-X(L), Mcl-1 and p53 levels in vitro and inhibits tumor growth in vivo. *Leukemia, 20*:1017-27.

[74] Mandal, C; Dutta, A; Mallick, A; Chandra, S; Misra, L; Sangwan, RS; Mandal, C. (2008). Withaferin A induces apoptosis by activating p38 mitogen-activated protein kinase signaling cascade in leukemic cells of lymphoid and myeloid origin through mitochondrial death cascade. *Apoptosis, 13*:1450-1464.

[75] Zhang, ZX; Wen, FQ; Liu, ZP; Cheng, YD. (2008). Correlation of high expression of HSP27 to multidrug resistance of leukemia cell line K562/VCR. *Ai Zheng, 27*:348-353.

[76] Verrills, NM; Liem, NL; Liaw, TY; Hood, BD; Lock, RB; Kavallaris, M. (2006). Proteomic analysis reveals a novel role for the actin cytoskeleton in vincristine resistant childhood leukemia--an in vivo study. *Proteomics, 6*:1681-1694.

[77] Tajeddine, N; Louis, M; Vermylen, C; Gala, JL; Tombal, B; Gailly, P. (2008). Tumor associated antigen PRAME is a marker of favorable prognosis in childhood acute myeloid leukemia patients and modifies the expression of S100A4, Hsp27, p21, IL-8 and IGFBP-2 in vitro and in vivo. *Leuk Lymphoma, 49*:1123-1131.

[78] Tajeddine, N; Gala, JL; Louis, M; Van Schoor, M; Tombal, B; Gailly, P. (2005). Tumor associated antigen preferentially expressed antigen of melanoma (PRAME) induces caspase-independent cell death in vitro and reduces tumorigenicity in vivo. *Cancer Res, 65*:7348-7355.

[79] Thomas, RK; Sos, ML; Zander, T; Mani, O; Popov, A; Berenbrinker, D; Smola-Hess, S; Schultze, JL; Wolf, J. (2005). Inhibition of nuclear translocation of nuclear factor-kappaB despite lack of functional IkappaBalpha protein overcomes multiple defects in apoptosis signaling in human B-cell malignancies. *Clin Cancer Res, 11*:8186-8194.

[80] Chauhan, D; Li, G; Shringarpure, R; Podar, K; Ohtake, Y; Hideshima, T; Anderson, KC. (2003). Blockade of Hsp27 overcomes Bortezomib/proteasome inhibitor PS-341 resistance in lymphoma cells. *Cancer Res, 63*:6174-6177.

[81] Debes, A; Oerding, M; Willers, R; Göbel, U; Wessalowski, R. (2003). Sensitization of human Ewing's tumor cells to chemotherapy and heat treatment by the bioflavonoid quercetin. *Anticancer Res, 23*:3359-3366.

[82] Suehara, Y; Kondo, T; Fujii, K; Hasegawa, T; Kawai, A; Seki, K; Beppu, Y; Nishimura, T; Kurosawa, H; Hirohashi, S. (2006). Proteomic signatures corresponding to histological classification and grading of soft-tissue sarcomas. *Proteomics, 6*:4402-4409.

[83] Ha, WY; Li, XJ; Yue, PY; Wong, DY; Yue, KK; Chung, WS; Zhao, L; Leung, PY; Liu, L; Wong, RN. (2006). Gene expression profiling of human synovial sarcoma cell line (Hs701.T) in response to IL-1beta stimulation. *Inflamm Res, 55*:293-299.

[84] Zanini, C; Pulerà, F; Carta, F; Giribaldi, G; Mandili, G; Maule, MM; Forni, M; Turrini, F. (2008). Proteomic identification of heat shock protein 27 as a differentiation and prognostic marker in neuroblastoma but not in Ewing's sarcoma. *Virchows Arch, 452*:157-167.

[85] Assimakopoulou, M; Sotiropoulou-Bonikou, G; Maraziotis, T; Varakis I. (1997). Prognostic significance of HSP-27 in astrocytic brain tumors: An immunohistochemical study. *Anticancer Res, 17*:2677-2682.

[86] Golembieski, WA; Thomas, SL; Schultz, CR; Yunker; CK; McClung; HM; Lemke, N; Cazacu, S; Barker, T; Sage, EH; Brodie, C; Rempel, SA.(2008). HSP27 mediates SPARC-induced changes in glioma morphology, migration, and invasion._*Glia, 56*:1061-1075.

[87] Aloy, MT; Hadchity, E; Bionda, C; Diaz-Latoud, C; Claude, L; Rousson, R; Arrigo, AP; Rodriguez-Lafrasse, C. (2008). Protective role of Hsp27 protein against gamma radiation-induced apoptosis and radiosensitization effects of Hsp27 gene silencing in different human tumor cells. *Int J Radiat Oncol Biol Phys, 70:*543-553.

[88] Hauser, P; Hanzély, Z; Jakab, Z; Oláh, L; Szabó, E; Jeney, A; Schuler, D; Fekete, G; Bognár, L; Garami, M.(2006). Expression and prognostic examination of heat shock

proteins (HSP 27, HSP 70, and HSP 90) in medulloblastoma. *J Pediatr Hematol Oncol, 28*:461-466.

[89] Kase S, Parikh JG, Rao N. (2008). Expression of heat shock protein 27 and alpha-crystallins in human retinoblastoma after chemoreduction. *Br J Ophthalmol*, [Epub ahead of print].

[90] Mitsiades, N; Mitsiades, CS; Poulaki, V; Chauhan, D; Fanourakis, G; Gu, X; Bailey, C; Joseph, M; Libermann, TA; Treon, SP; Munshi, NC; Richardson, PG; Hideshima, T; Anderson, KC. (2002). Molecular sequelae of proteosome inhibition in human multiple myeloma cells. *Proc Natl Acad Sci USA,* 99:14374–14379.

[91] Combaret, V; Boyault, S; Iacono, I; Brejon, S; Rousseau, R; Puisieux, A. (2008). Effect of bortezomib on human neuroblastoma: analysis of molecular mechanisms involved in cytotoxicity. *Mol Cancer, 7*:50.

[92] Kang, SH; Fung, MA; Gandour-Edwards, R; Reilly, D; Dizon, T; Grahn, J; Isseroff, RR. (2004). Heat shock protein 27 is expressed in normal and malignant human melanocytes in vivo. *J Cutan Pathol, 31*:665-671.

[93] Carta, F; Demuro, PP; Zanini, C; Santona, A; Castiglia, D; D'Atri, S; Ascierto, PA; Napolitano, M; Cossu, A; Tadolini, B; Turrini, F; Manca, A; Sini, MC; Palmieri, G; Rozzo, AC. (2005). Analysis of candidate genes through a proteomics-based approach in primary cell lines from malignant melanomas and their metastases. *Melanoma Res, 15*:235-244.

[94] Sosa, MS; Girotti, MR; Salvatierra, E; Prada, F; de Olmo, JA; Gallango, SJ; Albar, JP; Podhajcer, OL; Llera, AS. (2007). Proteomic analysis identified N-cadherin, clusterin, and HSP27 as mediators of SPARC (secreted protein, acidic and rich in cysteines) activity in melanoma cells. *Proteomics, 7*:4123-4134.

[95] Shi, B; Grahn, JC; Reilly, DA; Dizon, TC; Isseroff, RR. (2008). Responses of the 27 kDa heat shock protein to UVB irradiation in human epidermal melanocytes. *Exp Dermatol, 17*:108-114.

[96] Jantschitsch, C; Trautinger, F; Klosner, G; Gsur, A; Herbacek, I; Micksche, M; Kindås-Mügge, I. (2002). Overexpression of Hsp25 in K1735 murine melanoma cells enhances susceptibility to natural killer cytotoxicity. *Cell Stress & Chaperones. 7:*107-117.

[97] Aldrian, S; Trautinger, F; Fröhlich, I; Berger, W; Micksche, M; Kindas-Mügge, I. (2002). Overexpression of Hsp27 affects the metastatic phenotype of human melanoma cells in vitro. *Cell Stress & Chaperones 7*:177-185.

[98] Berhane, T; Halliday, GM; Cooke, B; Barnetson, RS. (2002). Inflammation is associated with progression of actinic keratoses to squamous cell carcinomas in humans. *Br J Dermatol, 46*:810-815.

[99] Kiriyama, MT; Oka, M; Takehana, M; Kobayashi, S. (2001). Expression of a small heat shock protein 27 (HSP27) in mouse skin tumors induced by UVB-irradiation. *Biol Pharm Bull, 24*:197-200.

[100] Trautinger, F; Kindas-Mügge, I; Dekrout, B; Knobler, RM; Metze, D. (1995). Expression of the 27-kDa heat shock protein in human epidermis and in epidermal neoplasms: an immunohistological study. *Br J Dermatol, 133*:194-202.

[101] Aldrian, S; Kindas-Mügge, I; Trautinger, F; Fröhlich, I; Gsur, A; Herbacek, I; Berger, W; Micksche, M. (2003). Overexpression of Hsp27 in a human melanoma cell line: regulation of E-cadherin, MUC18/MCAM, and plasminogen activator (PA) system. *Cell Stress & Chaperones 8*:249-257.

[102] Romanucci, M; Bongiovanni, L; Marruchella, G; Marà, M; di Guardo, G; Preziosi, R; della Salda, L. (2005). Heat shock proteins expression in canine intracutaneous cornifying epithelioma and squamous cell carcinoma. *Vet Dermatol, 16*:108-116.

[103] Bongiovanni, L; Romanucci, M; Fant, P; Lagadic, M; Della Salda, L. (2008). Apoptosis and anti-apoptotic heat shock proteins in canine cutaneous infundibular keratinizing acanthomas and squamous cell carcinomas. *Vet Dermatol, 19*:271-279.

[104] Gruvberger-Saal, SK; Parsons, R. (2006). Is the small heat shock protein alphaB-crystallin an oncogene?javascript:AL_get(this, 'jour', 'J Clin Invest.'); *J Clin Invest, 116*:30-32.

[105] Arrigo, AP; Simon, A; Gibert, B; Kretz-Remy, C; Nivon, M; Czekalla, A; Guillet, D; Moulin, M; Diaz-Latoud, C; Vicart, P. (2007). Hsp27 (HspB1) and αB-crystallin (HspB5) as therapeutic targets. *FEBS Lett, 581*:3665-3674.

[106] Dimberg, A, Rylova, S; Dieterich, LC; Olsson, A; Schiller, P; Wikner, C; Bohman, S; Botling, J; Lukinius, A; Wawrousek, EF; Claesson-Welsh, L (2008). αB-crystallin promotes tumor angiogenesis by increasing vascular survival during tube morphogenesis. *Blood, 111*:2015–2023.

[107] Aoyama, A, Steiger, R; Fröhli, E; Schäfer, R; von Deimling, A; Wiestler, O; Klemenz, R; (1993). Expression of alpha B-crystallin in human brain tumors. *Int J Cancer, 55*:760-764.

[108] Pinder, SE; Balsitis, M; Ellis, IO; Landon, M; Mayer; RJ; Lowe, J. (1994). The expression of alpha B-crystallin in epithelial tumours: a useful tumour marker? *J Pathol, 174*:209-215.

[109] Chin, D; Boyle, GM; Williams, RM; Ferguson, K; Pandeya, N; Pedley, J; Campbell, CM; Theile, DR; Parson, PG; Coman, WB. (2005). αB-crystallin, a new independent marker for poor prognosis in head and neck cancer. *Laryngoscope, 115*:1239–1242.

[110] Sitterding, SM; Wiseman, WR; Schiller, CL; Luan, C; Chen, F; Moyano, JV; Watkin, WG; Wiley, EL; Cryns, VL; Diaz, LK. (2008). αB-crystallin: A novel marker of invasive basal-like and metaplastic breast carcinomas. *Ann Diagn Pathol, 1*:33-40.

[111] Chelouche-Lev, D; Kluger, HM; Berger, AJ; Rimm, DL; Price, JE. (2004). αB-crystallin as a marker of lymph node involvement in breast carcinoma. *Cancer, 100*:2543–2548.

[112] Ivanov, O; Chen, F; Wiley, EL; Keswani, A; Diaz, LK. (2008). αB-crystallin is a novel predictor of resistance to neoadjuvant chemotherapy in breast cancer. *Breast Cancer Res Treat, 111*:411–417.

[113] Bosman, JD; Yehiely, F; Evans, JR; Cryns, VL. (2009). Regulation of *a*B-crystallin gene expression by the transcription factor Ets1 in breast cancer. *Breast Cancer Res Treat.* [Epub ahead of print].

[114] Gober, MD; Smith, CC; Ueda, K; Toretsky, JA; Aurelian, L. (2003). Forced expression of the H11 heat shock protein can be regulated by DNA methylation and trigger apoptosis in human cells. *J Biol Chem, 278*:37600-37609.

[115] Shemetov, AA; Seit-Nebi, AS; Gusev, NB.(2008) Structure, properties, and functions of the human small heat-shock protein HSP22 (HspB8, H11, E2IG1): a critical review. *J Neurosci Res, 86:*264-269.

[116] Trent; S; Yang, C; Li; C; Lynch, M; Schmidt, EV. (2007). Heat shock protein B8, a cyclin-dependent kinase-independent cyclin D1 target gene, contributes to its effects on radiation sensitivity. *Cancer Res, 67*:10774-10781.

[117] Sun, X; Fontaine, JM; Bartl, I; Behnam, B; Welsh, MJ; Benndorf, R. (2007). Induction of Hsp22 (HspB8) by estrogen and the metalloestrogen cadmium in estrogen receptor-positive breast cancer cells. *Cell Stress & Chaperones, 12*:307-319.

[118] Sharma, BK; Smith, CC; Laing, JM; Rucker, DA; Burnett, JW; Aurelian, L. (2006). Aberrant DNA methylation silences the novel heat shock protein H11 in melanoma but not benign melanocytic lesions. *Dermatology, 213*:192-199.

[119] Li, B; Smith, CC; Laing, JM; Gober, MD; Liu, L; Aurelian, L. (2007). Overload of the heat-shock protein H11/HspB8 triggers melanoma cell apoptosis through activation of transforming growth factor-beta-activated kinase 1. *Oncogene, 26*:3521-3531.

In: Small Stress Proteins and Human Diseases
Editors: Stéphanie Simon et al.
ISBN: 978-1-61470-636-6

Chapter 3.2

SMALL HEAT SHOCK PROTEINS AS POTENTIAL MESSENGERS IN THE PATHWAYS TO CANCER

Anastassiia Vertii[1] and Matthias Gaestel[2*]
[1]Program in Molecular Medicine, University of Massachusetts Medical School, Worcester, MA 01605, U.S.
[2]Institute of Biochemistry, Hannover Medical School, 30625 Hannover, Germany

ABSTRACT

The multifunctional role of small heat shock proteins (sHsps) in cellular response to a wide spectrum of stress stimuli, including treatment with anti-cancer drugs, provides a platform to identify sHsps as one of the messengers in the cancer pathways. While anti-apoptotic function of sHsps had been extensively studied for many years, their effect on mechanisms and signaling cascades essential for tumorgenesis is relatively poorly understood so far. This chapter attempts to summarize knowledge and recent developments in understanding the role of sHsps in signal pathways known to be of importance in cell proliferation, cell cycle control, cancerogenesis and cancer progression.

INTRODUCTION

Oncogenic transformation and tumor development are a complex multistep process which displays different features dependent on its cellular origin. Six essential alterations which cooperatively lead to malignant growth (reviewed in [1]) are shared by the majority of human tumors: self-sufficiency in growth signals, insensitivity to growth-inhibitory signals, evasion of programmed cell death (apoptosis), limitless replicative potential, sustained angiogenesis,

* Mathias Gaestel
Institute of Biochemistry
Hannover Medical School
30625 Hannover, Germany
Gaestel.Matthias@mh-hannover.de

and tissue invasion and metastasis [1]. Signaling networks which contribute to these characteristics are potential targets for anti-cancer therapy, but due to their complexity and importance for the function of normal cells, specificity of targeting is often a problem. In this regard, heat shock proteins (Hsps) could be of practical interest, since this group of proteins is getting actively involved in cellular regulation under stress conditions, and is known to be upregulated in many cancers. While some families of Hsps, such as Hsp70 and Hsp90, are important for providing folding of newly synthesized polypeptides under physiological conditions; others, like several members of the group of small Hsps (sHsps), are present in the cells and seem to participate redundantly in essential cellular processes during non-stressed conditions. Therefore their depletion from normal tissues may have minimal side effects in comparison to other vitally important targets. This situation is demonstrated by the fact that genetic deletion of the prominent sHsp in mice, HspB1 or Hsp25, does not lead to a detectable phenotype under non-stressful conditions [2].

Exposure to inflammation, cancer, variety of stresses including anti-cancer therapy, results in rapid covalent modifications, such as phosphorylation, of some sHsps and often also to an increase in their expression, which leads to protection of stressed cells and increased survival. Since sHsps, such as HspB1, can contribute to rescuing cancer cells from drug-induced apoptosis, understanding of their precise role in cancer pathways may help to increase the drug-sensitivity of tumor cells in anti-cancer therapy. The function of tumor-related overexpression and modification of sHsps and their participation in signaling networks, which mediate the six steps to malignancy mentioned above, remains one of the major challenges in this field today.

1. Small Hsps and Phosphorylation

1.1. sHSPs

Cellular response to stress includes rapid activation of a specific protein kinases as well as chaperoning function of heat shock proteins (Hsps). Hsps are present in nearly all living organisms and are widely expressed in mammalians tissues. Although their synthesis is induced by stress, a basal level of Hsps exists in non-stressed cell as well. Heat shock factor 1 (HSF1) is responsible for the stress-mediated induction of transcription of the genes coding for heat shock proteins [3, 4]. Classification of Hsps is based on their molecular mass and on their ATP-dependence. Hsps can cooperate in order to chaperone their substrates, and some of them, like Hsp70 require co-chaperones. ATP-dependent Hsps, such as Hsp60, Hsp70 and Hsp90, are necessary for providing a proper folding for newly synthesized polypeptide chains in non-stressed cells [5, 6]. Small Hsps are a well-known group of ATP-independent chaperones which are able to rescue cells from a variety of stresses. However, among the 10 members of sHsps in mammalian organism - HspB1 (mouse: Hsp25, human: Hsp27), HspB2 (MKBP-myotonic dystrophy protein kinase-binding protein), HspB3, HspB4 (αA-crystallin), HspB5 (αB-crystallin), HspB6 (Hsp20), HspB7 (cv-cardiovascular Hsp), HspB8 (Hsp22), HspB9, HspB10 [7] - only HspB1 is extensively studied in terms its anti-apoptotic function and pro-survival properties. Expression of many sHsps family members is tissue specific (for example, HspB4 in eye lens, HspB7 in heart, HspB9 and HspB10- in male germ cells), while others, such as HspB1 and HspB5 (αB-crystallin), as well as HspB6 and HspB8 are ubiquitously expressed [7]. In the proposed new classification, ubiquitously expressed sHsps belong to Class I sHsp family, while tissue-restricted Hsps are related to Class II [7].

1.2. Signal-Dependent Phosphorylation of sHsps

One of the first reactions of the cell on stress which clearly precedes induction of Hsp expression is sHsp phosphorylation. This process is mediated by different protein kinases within seconds and minutes and is of importance for the regulation of the oligomeric structure and properties of sHsps [8-14]. HspB1 (Hsp25/27) phosphorylation is the result of activation of the stress-triggered signalling cascade containing p38 MAPK and its downstream kinases MAPK-activated protein kinase (MAPKAPK or MK) 2 and 3[14, 15]. The MK family of protein kinases contains three members, MK2, MK3 and MK5 [16]. MK2 is the main protein kinase responsible for HspB1 (Hsp25/27) phosphorylation and MK3, which is also able to phosphorylate HspB1 [17], shows lower expression and activity compared to MK2 [18]. Cells lacking MK5 do not display reduced Hsp25 phosphorylation [19]. Murine Hsp25 is phosphorylated at two different sites, S15 and S86 [20], while human Hsp27 has three phosphorylation sites, S15, S78 and S82 [21]. In addition, T143 of HspB1 (Hsp27) can be phosphorylated by cGMP-dependent protein kinase [22], and further kinases were described to phosphorylate HspB1 (Hsp27) at least *in vitro*, such as MK5 [23], PKC [24] and PKD at S82 [25]. It has been shown that S15 is located close to the WDGF motif in the N-terminal part of HspB1 (Hsp27) and mutation of this part of HspB1 (Hsp27) is critical for oligomerisation of HspB1 (Hsp27) [8], while the C-terminus was shown to contribute to full chaperone activity of Hsp25 [26]. Stress-dependent phosphorylation and disruption of large oligomer complexes is mediated by MK2 [15, 19, 27-30]. The role of these changes in HspB1-mediated cellular stress resistance has been studied extensively. While reports from some groups clearly showed that phosphorylation of HspB1 (Hsp27) is required for thermoprotection [31, 32], others postulate that large oligomeric forms of HspB1 (Hsp27) are more efficient in its chaperoning functions and protection against oxidative stress [12, 30, 33]. Phosphorylation of HspB8 (Hsp22) was demonstrated at S14 and T63 by PKC and at S27 and T87 by p44 MAPK, and is MK2-independent [34]. HspB6 (Hsp20) is phosphorylated in cyclic nucleotide-dependent manner at S16 by PKA, which results in disruption of non-phosphorylated oligomers of about 230kDa and accumulation of smaller complexes with molecular mass between 160 and 67kDa [35]. In contrast, phosphorylation of HspB5 (αB-crystallin) is cyclic AMP-independent, but occurs in response to a variety of stresses, like heat, arsenite, phorbol 12-myristate 13-acetate (PMA), okadaic acid, H_2O_2, anisomycin, and high concentrations of NaCl or sorbitol [10, 11]. HspB5 is phosphorylated at three serine residues, S19, S45, and S59, by p44/42MAP kinase and MK2, respectively [11, 36, 37]. In addition, the phosphorylation at S21, S43, S53, S76 had been reported [38]. Another member of α-crystallins, HspB4 (αA-crystallin) which is expressed predominantly in eye lens, is phosphorylated at T13, S45, S122, T140 during cataract development [38].

1.3. sHsps – Phosphorylation-Regulated Chaperones and More?

Initially mammalian sHsps, as well as other Hsps, were discovered as a group of proteins which is getting upregulated in response to heat shock [39], where an increasing pool of partially unfolded proteins requires chaperones. As a logic consequence, analysis of sHsps in cells was primarily focused on their chaperone properties. The chaperone function of HspB1 has been demonstrated by *in vitro* experiments and the correlation between cellular stress resistance and presence of sHsps has been studied intensively [12, 13, 26, 32, 33]. *In vitro* the chaperone activity of HspB6 (Hsp20) was reported to be dependent on its oligomeric state [40-42]. Similar to HspB1 and HspB6, chaperone properties of HspB5 (αB-crystallin) were phosphorylation-dependent displaying weaker chaperone properties of the phosphorylated form [10]. HspB4 (αA-crystallin) forms large oligomers of about 800kDa *in vitro* and exhibits UVA-sensitive chaperone activity which increases with temperature [43-47, 48]. Because sHsps are ATP-independent and, therefore, cannot actively refold substrate proteins, it has been proposed that, upon stress, sHsps stabilize a pool of partially unfolded proteins, thus preventing them from irreversible aggregation before ATP-dependent Hsps will assist these proteins to be refolded [49, 50]. However, some questions regarding sHsp functions *in vivo* are left open so far. One of the main questions regards the role of phosphorylation in chaperoning function of sHsps: if large oligomeric complexes act efficiently to prevent protein aggregation, as demonstrated *in vitro* [12, 13], why is phosphorylation-induced dis-aggregation and dimer formation of sHsps necessary *in vivo*? Is this to clear the chaperoning complexes from irreversibly misfolded proteins? Or does this activate other functions of sHsps? Besides being chaperones, sHsps are described to act as anti-apoptotic agents in order to protect cells against multiple stress factors: HspB1 prevents cytochrome C release from mitochondria and caspase activation [51-53]; (also reviewed by Garrido [54]) and interacts with DAXX thus affecting DAXX-mediated apoptosis [55, 56]. The cysteine residue Cys 137 within the crystalline domain of HspB1, which is important for dimer formation, was shown to be also critical for the interaction with cytochrome C [57]. For its interaction with DAXX, dimers seem to be more effective than the non-phosphorylated larger oligomers of HspB1 [55, 56].

2. sHsps in Cancer

Involvement of sHsps in cancer was noticed more than a decade ago [58-67]. Changes in expression of human HspB1 (Hsp27) were repeatedly associated with different varieties of cancers, such as breast cancer [58, 63, 66, 68-70], ovarian cancer [71], non-small cell lung cancer [72], acute leukemia [73], gastric cancer [74, 75], pancreas carcinoma [76, 77], bladder [78, 79], and prostate cancer [80-82] (see chapter 3-1). Until now, less is known about other members of the sHsp family in solid tumors or cell lines: for example, some evidence for altered expression of HspB5 [83] and HspB8 in breast cancer and melanoma was obtained [84-87] and, very recently, a number of studies alert the importance of HspB5 (αB-crystallin) in numerous cancers and in vascularization of solid tumors [88-90]. While both, HspB1 and HspB5, are becoming attractive targets for anti-cancer therapy [91, 92] (see chapters 4-2 and 4-3); data from analysis of HspB1 and HspB5 expression and regulation in thyroid carcinoma suggest that expression, and probably, function of these proteins, might differ from each other [93]. In addition, a proteomic study of expression of sHsps in cancer cells lines reported that HspB1 (Hsp27) is the only sHsp detected in SK-N-SH, HCT, A549, HL-60, MCF-7 and Hela cell lines [94]. Obviously, stimuli-dependent phosphorylation and changes in oligomeric form of different members of sHsp family does not necessarily reflect similarity in their function. Instead, other probably efficient ways of sHsps to influence the survival of cancer cells include their non-chaperone properties and implies participation in different signaling pathways in their phospho-forms. The present chapter aims to summarize knowledge of how sHsps are participating in cancer signaling cascades. Here, we focus mainly on HspB1 as the most well-studied member of the sHsp family and a potential target for anti-cancer treatment.

2.1. Induction of sHsps and Hormone Receptors

Estrogen receptor (ER)-dependent gene expression proceeds via ER-binding to estrogen-responsive promoter elements [95]. Regulation of expression of sHsps in breast and endometrial tumor tissue and cancer cell lines by ER was demonstrated early, when experiments from independent groups reported the increase of HspB1 expression on both, mRNA and protein level [59, 64]. Five years later, analysis of primary tumors as well as cancer cells lines revealed that estrogen-regulated increase in HspB1 expression may have strong correlation with tumor aggressiveness and resistance towards chemotherapy in case of ovarian cancer [62]. However, it seems that expression of HspB1 in other cancer types, as squamous carcinomas of the uterine cervix and in the endometrial adenocarcinomas with squamous cells, is ER-independent, since about 80% of studied tumors had increased HspB1 expression while most of them were lacking ER [96]. A recent comprehensive study which employed mice models, human cancer cell lines and transformation assays, demonstrated the wide role of classical transcription factor for heat shock proteins, HSF1, in cancerogenensis [97]. In this regard, regulation of sHsps expression in cancer may involve activation at least two different transcription factors, HSF1 and ER, as a result of multiple pathways.

Upregulation of HspB1 in cancer cells may change tumor development and resistance, and, hence, could be critical for chemotherapy: in testis cancer cells, where no or little HspsB1 upregulation is detected, a high percentage of cured cases is observed and, furthermore, overexpression of HspB1 leads to decrease in cellular response towards chemotherapy [98]. Although HSF1 is exclusively responsible for heat shock–related expression of HspB1 [3], it seems that regulatory mechanisms may change accordingly to stress type: in case of estrogen receptor-dependent upregulation of HspB1, there is a possible feedback loop, since it was consistently reported that HspB1 could be detected in the complex with both, estradiol- and estrogen-response element-binding protein and shuttles between cytoplasm and nuclei in response to estrogen treatment [99, 100]. Recently, indication for the existence of yet another protein complex, which is potentially relevant in tumorgenesis and which might involve regulatory feedback loop, had been reported: it was shown by immunoprecipitation, that HspB1 (Hsp27) may co-exist in the same complex with HSF1 and β-catenin [101]. Whether this complex is tumor-specific remains to be investigated, although the reported co-expression of these proteins in human breast cancer samples as well as in a murine breast cancer model [101] suggests a role of this interaction in breast cancer. On the other hand, all three proteins seem to have much wider potential to influence tumor development than the limited role of this complex to one type of cancer and, therefore, careful examination of the affinity and dynamics of these protein-protein interactions will be necessary.

2.2. Cancer-Relevant Signal Transduction Pathways Which Act Upstream of sHsp

Existence of sHsps in large oligomeric form and different pathways, which are responsible for phosphorylation-dependent oligomer disruption and changes in their chaperone function, have been studied extensively as a result of stress-mediated activation of signaling cascades, and we discussed this briefly in the introduction. In regard to subcellular localization of sHsps, it is known that stress exposure leads to, at least partial, cytoplasmic-nuclear translocation of HspB1. In the nucleus, HspB1 may participate in formation of stress granules [30, 102] - a process which depends on phosphorylation of HspB1 [30, 103, 104]. The molecular mechanism of this translocation is still unclear; however, one possibility is that the large oligomeric complexes prevent HspB1 from entering the nucleus due to size limitations. Interestingly, severe stress leads to nearly complete translocation of HspB1 to the nucleus, or, as more often investigated, to the insoluble fraction (which may include some cytoskeleton and cytoplasmic granules), while mild stresses, resulting in similar HspB1 phosphorylation, lead only to partial translocation and formation of distinct nuclear granules [30]. The presence of sHsps is crucial for cell survival [92], indicating a role of sHsps as mediator of stress-related signal transduction pathways. Some of these functions are performed most probably in the nucleus, and therefore characterization of nuclear interactions and functions of HspB1 are required.

Stress-mediated phosphorylation of sHsps and activation of upstream kinases, such as the p38-MK2 signaling module (reviewed by Gaestel [16]), was repeatedly reported to play significant role in cancer [105-113]. p38 is activated by the MAP3Ks, MEKK 1 to 4, MLK2 and -3, DLK, apoptosis-signal regulating kinase 1 (ASK1), and TGFβ-activated protein kinase (TAK1), the MAP2Ks MEK3 and MEK6 (also termed MKK3 and MKK6, respectively), which display high specificity for p38 MAPK. MKK4, which has specificity towards JNK, is also able to activate p38 under certain conditions [114, 115]. There are four known isoforms of p38 MAPK (α, β, γ, and δ), of these p38α and p38β are ubiquitously expressed. While MKK6 activates all p38 isoforms, MKK3 preferentially phosphorylates the p38α and p38β isoforms (reviewed in [115]). Activation of the p38 isoforms results from the MKK3/6-catalyzed phosphorylation of a conserved Thr-Gly-Tyr (TGY) motif in their activation loop, where ERK1/2 posses the TEY motif [116]. The structures of inactive and active (phosphorylated) p38 have been solved by X-ray crystallography and the size of the activation loop was found to differ from that of ERK2 and JNK. This probably contributes to the substrate specificity of p38 [115, 117]. Another factor that contributes to substrate specificity is the common docking (CD) motif of MAPK, which is located C-terminal outside of catalytic domain, involved in D motif interactions. The CD motif contains acidic and hydrophobic residues which provide hydrophobic and electrostatic combination and are important for upstream activators and downstream substrate interactions (reviewed in [118]).

Activity of a protein kinase can be blocked by specific inhibitors that often compete with ATP-binding to the catalytic domain. There exist several pyridinyl imidazole compounds, which are known as suppressors of cytokine synthesis and inhibitors of p38 [119]. One of these, SB203580, is a widely used inhibitor for the activity of p38 α and β and their downstream targets, such as MK2 and MK3. Stress-induced phosphorylation of p38 α,β leads to activation of a number of downstream targets including distinct families of kinases, such as MKs, MAPK kinase-interacting kinases (MNK) 1 and 2 [120, 121], mitogen- and stress-activated kinase-1 (MSK1) [122], as well as transcription factors such as cAMP response element-binding proteins (CREB) and ATF1 [123]. Therefore, p38 has an ability to influence multiple aspects of cellular functions, such as apoptosis, inflammation, cell cycle, and senescence (reviewed by [124]). Activation of p38 may lead to phosphorylation of p53 [125, 126] and stabilization of p21 [127], yet another key protein which regulates cell cycle and triggers cell cycle arrest [127-131]. Localization of p38 and phosphorylated p53 to the centrosome as a response to depletion of 14 centrosomal proteins was shown to be critical for G1-S cell cycle arrest [130]. Recently, a link between p38 and p53-dependent apoptosis via a novel regulatory protein was demonstrated [132].

The first substrate of p38 identified was MAPKAPK2 or MK2 [14, 27] and two years later another member of MKs, MK3, has also been shown to be specifically activated by p38 α,β [17]. MK2 not only interacts with p38 but has also a stabilizing effect on p38, since it was reported that in MK2-deficient mice the level of p38α was dramatically decreased [29]. Whereas p38α-deficient mice were generated and appeared to be embryonic lethal due to placental defects [133], MK2-deficient mice were viable, fertile and grew to a normal size [134]. Involvement of p38 in post-transcriptional regulation of inflammatory response had been narrowed to its substrate, MK2 [134], which in turn phosphorylates tristetraprolin (TTP), a protein that interacts with AU-rich elements of cytokine mRNA [135]. Interestingly, HspB1 has recently been described to modulate TAK1-mediated activation of the p38/MK2/3 module by IL-1 thus playing a feedback role in cytokine signaling as well [136].

As demonstrated above, HspB1 is not the only target of p38-MK2/3 cascade. Hence, the different functions of the p38-MK2/3 cascade, such as stabilization of mRNA of inflammatory cytokines by p38-MK2 [135], pro-apoptotic effects mediated via pro-inflammatory cytokines, as well as DNA damage check point control by CDC25 [108, 137-139], interfere with the function of HspB1-phosphorylation in apoptosis and tumorigenicity. Thus, on one hand, there is an upregulation of HspB1 in tumors, an on the other hand, there proceeds modulation of the HspB1 phosphorylation state by the p38-MK2/3 signaling cascade, which in general acts as a pro-apoptotic, and pro-inflammatory module. One more aspect of p38 makes the situation even more complex: p38 was shown to be important in cancer cell migration, oncogene-induced senescence, tumor suppression and invasion [106, 109, 140, 141], thus promoting cancerogenesis and masking the contribution to HspB1-mediated transduction of cancer pathways. It seems, however, that phosphorylated form of HspB1, which is lacking the capacity to build large oligomer complexes, was much less effective in prevention of apoptosis in cancer cells [57], therefore, implying that the p38-MK2-dependent phosphorylation of HspB1 aims to inactivate its chaperone and anti-apoptotic functions. Whether phosphorylated form is more efficient in cytoskeleton remodeling and tumor invasion at more advanced steps of cancerogenesis remains to be addressed experimentally. It is known, however, that expression and phosphorylation of HspB1 affects endothelial cells migration in response to growth factors [142, 143]. Both, the specific inhibitor of p38, SB203580, and transfection of non-phosphorylatable mutants of HspB1 reduced migration rate of endothelial and smooth muscle cells.

While activated p38 has a significant number of downstream targets which contribute to apoptosis and cancerogenesis, the specific role of p38-MK2 signaling module in HspB1 (Hsp27) phosphorylation in favor of survival of tumor cell should be approached by using HspB1 (Hsp27) phospho-site specific mutants. Questions about contribution of particular sites into cellular defense against apoptosis remain to be answered: and it is very much likely that pro- or anti-apoptotic role of sHsp phosphorylation may change to opposite depending on death stimuli type and dose.

Data about phospho-dependent interaction of HspB1 (Hsp27) with 14-3-3 proteins [30] and Hic5 [144, 145], which are multifunctional small proteins involved in different aspects of cellular responses and diseases, including cancer (reviewed in [146]) indicate that phospho-dimers of HspB1 (Hsp27) may be of importance for tumor development and invasion as well. At the moment, a comprehensive analysis of small Hsps oligomerization and subcellular localization at different stages of tumor development is lacking. Such data will probably help to narrow down the multifunctional role of small Hsps in cancer signal transduction in future.

2.3. Nuclear Localization of Small Hsps: Implications and Speculations

Nuclear translocation of HspB1 (Hsp27) is often described and the functional consequences are of special interest [102-104]. To approach this, the nuclear region where HspB1 is exactly located after translocation could be of importance. The mammalian nucleus is a complex organelle organized into chromatin territories and further discrete nuclear compartments and compact structures designated as bodies [147]. These compact structures include supramolecular complexes with mRNPs, matrix-associated deacetylase (MAD) bodies, Cajal bodies and promyelocytic leukemia protein (PML) bodies. PML bodies are also named POD (PML oncogenic domain), ND10 (nuclear domain 10) or Kremer bodies (Kr). There exist approximately 5-30 PML bodies per nucleus, ranging in size from 0.2 to 1μm. PML bodies function in cell differentiation, cell growth and apoptosis (reviewed in [148]). Among several protein components of PMLs, such as MDM2 and Sp100, a protein that is involved in Fas-mediated apoptosis, DAXX, had been identified [149]. So far, there is no evidence for co-localization between HspB1 (Hsp27) and PML bodies, however HspB1 (Hsp27) is known to interact with DAXX and affect its subcellular localization or with another member of PML granules and interacting partner for DAXX, p53 [56, 150-152]. Exposure to heat shock leads to release of DAXX and Sp100 from PML bodies and exposure to subtoxic concentrations of cadmium results in dissociation of DAXX and PML [153].

The idea that translocation of HspB1 (Hsp27) to the nucleus has functional consequences displays great potential. Recently shown association between nuclear HspB1 (Hsp27) and nuclear speckles, splicing factor SC35 and proteasomal degradation sites, where it was shown to be associated with overexpressed heat shock-sensitive luciferase [154] is interesting and requires further analysis of potential endogenous client proteins of HspB1 (Hsp27) in the nucleus. Co-localization between HspB1 (Hsp27), nuclear granules and splicing factor SC-35 is in consistence with the described function of phosphorylated HspB1 (Hsp27) in splicing during recovery from heat shock [155]. While rescue from heat-induced nuclear protein aggregation by HspB1 (Hsp27) [156] may result from chaperoning of both, specific as well as non-specific protein targets, described interactions of HspB1 (Hsp27) and DAXX [55, 56] or p53 [150-152] and regulation of their functions could represent a specific nuclear role of HspB1 (Hsp27). First evidence that HspB1 (Hsp27) may perform such client protein-specific function in the nucleus is shown for androgen receptor (AR), where HspB1 (Hsp27) replaces Hsp90 and regulates AR in phospho-dependent manner [157].

2.4. Role of sHsps in Cell Cycle Control and Checkpoint-Related Pathways

Activation of p53 in response to DNA damage, activated oncogenes and uncontrolled mitogenic signals, is known to mediate complex network in order to prevent survival of abnormal cells, which may lead to cancer, recently reviewed by Rodier [158]. The result of p53 activation is either cell cycle arrest, which is depending on DNA repair mechanisms either temporary or precedes cell death. Cell cycle arrest is mediated through p21, a downstream target of p53, while apoptosis is known to be mediated by BH3-only proteins which are transcriptional targets of p53 [158]. Another p53-regulated pathway, which may affect both, cell proliferation and apoptosis, is miR43a,b,c-a novel transcription target of p53 [159]. p53 is a central mediator of signals which are critical for at least two of the six features of cancer: evasion of apoptosis and limitless replicative potential [1]. In division 2.2 we already discussed the involvement of the p38-MK2 signaling module in the p53 pathway. In addition to that, proteomic analysis and immunoprecipitations indicate that HspB1 (Hsp27) may interact with p53 in cancer cell lines [160]. Furthermore, HspB1(Hsp27) regulates p53 and p21 levels in the breast cancer cells line MCF10A, in human mammary epithelial cells and in HCT116 human colon carcinoma cells [151]. Independent data suggest an interaction between p53 and HspB1 (Hsp27) in fibroblasts and cardiac cells, an effect on p21 and a possible contribution to doxorubicin-induced cell cycle arrest [152]. The earlier finding that overexpression of murine HspB1 (Hsp25) in fibroblasts regulates p21 level at different stages, by increasing p21 mRNA transcription and by stabilizing the protein implies other p53-independent mechanisms by which small Hsps may regulate p21 as well [161]. Thus HspB1 (Hsp27) participates in p53-mediated pathways. To what extend overexpression and phosphorylation of HspB1 affects DNA-damage-related checkpoint signaling cascades and prevents cell cycle arrest in tumors is still an open question. Also the regulation of p53 stabilisation by HspB1 remains to be investigated further. HspB1-mediated regulation of p53 should be different from p38-mediated one, since p38 is able to phosphorylate p53 [130], while HspB1 (Hsp27) forms a complex with p53. Phospho-dependent partial translocation of HspB1 (Hsp27) into the nucleus suggests that this might be one of the phosphorylation-regulated nuclear functions of HspB1. HspB1 (Hsp27) is also involved in stress-related regulation of cell cycle- by degradation of $p27^{Kip}$ [162]. Gel filtration and immunoprecipitation experiments showed that HspB1 interacts with $p27^{Kip}$ in form of small oligomers and facilitates degradation of $p27^{Kip}$ via ubiquitination, thus allowing progression through G_1/S cell cycle arrest. Maximal protein level of $p27^{Kip}$ during G_1 provides inhibition of cyclin E-Cdk2 complex, and the decrease in $p27^{Kip}$ leads to activation of cyclin E-CDK2 and cyclin A-CDK2 and entry to S phase [163]. While transcription of $p27^{Kip}$ is regulated by FOXO family of transcription factors, its degradation and subcellular localization is regulated by multiple pathways, including Akt/PKB (see also below) and miRNA [163]. Deregulation of $p27^{Kip}$ protein has poor prognosis in human cancers of epithelial origin [164-168].

Cytoplasmic mislocalization of p27 ^{Kip} as a result of phosphorylation-dependent interaction with 14-3-3 group of proteins contributes to cell motility [164, 169, 170]. Thus, by interacting with and destabilisation of p27^{Kip} protein, HspB1 (Hsp27) may significantly interfere with multiple signaling cascades, and influence cell decision about cell cycle progression during tumorgenesis.

2.5. Cytoplasmic Function of sHsps: HspB1 and Akt

The oncogenic protein kinase Akt (also designated PKB) is activated through the PI3-pathway and phosphorylates a variety of substrates in the cytoplasm as well as in the nucleus of the cell. Akt possesses a plekstrin homology (PH) domain on its N-terminus that is responsible for binding to PIP3 on the membrane, a process that is required for Akt activation by phosphorylation at two sites, T308 and S473. T308 lies within kinase domain, while S473 is located at a hydrophobic phosphorylation motif in the C-terminus [171]. HspB1 was shown to specifically interact with Akt [104, 172-174]. In these reports, data on stress-dependent Akt-Hsp27 binding are controversial: while in some studies heat shock and oxidative stress lead to increase in Akt-Hsp27 interaction [172] others studies detected a stress-dependent dissociation of Hsp27 from Akt-p38 complexes [104, 174].

Phosphorylation of numerous targets by Akt is aimed to contribute to cell survival, proliferation and migration. Akt phosphorylates BAD thus generating 14-3-3 binding sites on BAD, sequestering BAD from mitochondria membrane to the cytoplasm and acting anti-apoptotic (reviewed by Stern, 2004 [175]). Another important Akt substrate in the cytoplasm is GSK3-beta, a kinase that plays a role in Wnt signaling and insulin response. Nuclear targets of Akt include regulators of cells cycle p21 and p27, mTOR proteins and Forkhead transcription factors FOXO [176]. Akt-mediated p27 phosphorylation results in p27 binding to 14-3-3 protein and export from the nucleus to the cytoplasm where p27 may affect cytoskeleton rearrangement therefore allowing cell to proliferate [175]. Increase in Akt phosphorylation and its nuclear translocation had been associated with poor prognosis during cancerogenesis [175, 177, 178]. The Akt/HspB1 interaction could modulate some of the Akt functions described above.

2.6. Compensatory effects of HspB1 in case of Hsp90 inhibition

Hsp90 is an ATP-dependent chaperone [179] and, because of the long list of its client proteins (around 100 protein kinases [180]) includes oncogenes, such as Akt [181], is one of the important targets for anti-cancer therapy. Hsp90 may form a chaperone complex with Hsp70 [182, 183]. Regulation of activity of Hsp90-dependent protein kinases and sensitizing cancer cells is dependent on Hsp90 ATPase activity [184-187] and presence of co-chaperones [188, 189]. Therefore, inhibitors of Hsp90 are potential anti-cancer drugs, which undergo clinical trials [190, 191]. However, inhibition of Hsp90 by the geldanamycin-derivate 17-AAG, the first potent Hsp90 inhibitor which was analysed in clinical trials, often leads to formation of 17-AAG resistant cancer cells. The resistance of these cells was, at least partially acquired by overexpression of HspB1 (Hsp27) [192]. The interference with function of common interacting partners of Hsp90 and HspB1 (Hsp27), like Akt/PKB and p53 might be the reason for HspB1 (Hsp27)-mediated resistance of cancer cells with non-active Hsp90. Overlay between Hsp90 and HspB1 (Hsp27) function was demonstrated in processes like splicing [155] and hormone receptor signaling [157]. However, there is no evidence of a rescue of HspB1 (Hsp27)-mediated effects by Hsp90 overexpression so far.

2.7. sHsps and Invasion: MMPs, Beta-Catenin, Actin

Cell migration and invasion is one of the determinant of tumor progression and malignancy [1]. Migration and actin cytoskeleton rearrangements are regulated by a complex network, where MK2, an upstream kinase of HspB1 (Hsp27) plays a significant role and it was described that catalytic activity of MK2 is required for restoration of migration phenotype in cells lacking MK2, implying that phosphorylated form of HspB1 (Hsp27) is efficient in altering actin cytoskeleton [29]. These data are in consistence with experiments where p38 inhibitor, SB203580 and HspB1 (Hsp27) phospho-mutants drastically decreased cell migration [142]. *In vitro* non-phosphorylated HspB1 (Hsp27) is able to cap the actin fibers and to prevent their further polymerization [193]. Furthermore, phosphorylation of HspB1 (Hsp27) seems to be important for cytoskeleton dynamics in cultured cells (reviewed in [194]). Besides the possibility of direct stress-dependent interaction between HspB1 (Hsp27) and actin fibers, regulation of migration via focal adhesion kinase (FAK) and matrix metalloproteases MMP-2 and MMP-9 was proposed as another mechanisms [195-197]. In addition, co-expression and specific interaction between HspB1 (Hsp27) and β-catenin, a protein important in cell adhesion and in the Wnt signaling pathway leading to overexpression of oncogenes, was demonstrated in tissues from biopsies from breast cancer patients [101]. Therefore, downregulation of HspB1 (Hsp27) in cancer cells might be a promising tool to decrease metastatic potential of tumor [63, 198, 199].

2.8. HspB1 and Posttranscriptional Regulation of Gene Expression

The function of HspB1 (Hsp27) in recovery of translation machinery after stress exposure [200, 201] might be one of key points in regulating cellular survival. Stress-induced inhibition of cap-dependent translation is mediated by recruitment of translation initiation factors and mRNA to stress granules (SGs). Cytoplasmic mammalian SGs can be defined as cytoplasmic regions into which mRNAs are sorted dynamically in response to phosphorylation of eukaryotic initiation factor (eIF) 2α, a key regulator of translational initiation (reviewed in [202]). Studies with phospho-mimicking mutants of eIF2α showed that phosphorylation of eIF2α leads to SGs assembly. eIF2α is a component of eIF2-GTP-tRNAMet, the ternary complex that loads initiator methionine on to the small ribosomal subunit to initiate translation. The TIA proteins (TIA-1, T-cell internal antigen 1, and TIAR, TIA-1-related protein are required for SGs formation, which is blocked by expression of dominant-negative truncation mutant of TIA-1 [203]. Arsenite causes both TIA proteins to move from the nucleus to the cytoplasm and colocalise to form SGs - a process which takes about 15 min. The SGs continue to increase in size but decrease in number owing to fusion of smaller SGs for additional one to two hours, depending on the severity of the stress. Cytoplasmic SGs contain 40S small subunit of ribosome, translation factors eIF2, eIF3, eIF4E, PABP, and TIA-1, Staufen SMN and G3BP, proteins that regulate RNA stability, such as TTP, HuR/D and exonuclease XRN1, and the argonaute protein involved in siRNA regulation [204]. SGs also could contain proteins from processing bodies (P-bodies), structures distinct from SGs, which were described as highly motile cytoplasmic RNA-containing granules that contain components of mRNA decay machinery. Interestingly, P-bodies and SGs turn to share lot of components after stress treatment [204]. Thus, SGs are the start point of recovery of translation. Small Hsps, like HspB1 (Hsp27) and αB-crystallin, both are important for the availability of translation factors [205] and an association between HspB1 (Hsp27) and SGs upon heat shock treatment was reported [206]. So far it is not well understood at which stage of recovery small Hsps interact with translation initiation factors. Immunoprecipitation experiment demonstrated that HspB1 (Hsp27) interacts with eIF4G and mediates inhibition of cap-dependent translation upon heat shock [200]. This could mean that small Hsps will be important at the very beginning of stress-related inhibition of translation and not only during recovery period.

Members of SGs, like TTP (tristetraprolin) regulate mRNA stability via AU-rich elements of cytokine mRNA in a p38/MK2-dependent manner [134, 135]. In addition to Hsp70 as a well known component of mRNA decay mechanism [207], HspB1 was recently identified as a part of this decay machinery, acting to destabilize cytokine mRNA [208]. Knockdown of HspB1 (Hsp27) stabilized TNFα mRNA up to 10 fold, while treatment with 12-O-tetradecanoylphorbol-13-acetate (TPA) lead to decrease in HspB1 (Hsp27) efficiency [208]. Interestingly, TPA treatment results in PKC-mediated activation of p38, an upstream kinase of MK2, and consequent phosphorylation of HspB1 (Hsp27) [209], implying that MK2-mediated phosphorylation of HspB1 (Hsp27) may increase stability of mRNA, similar to what was reported earlier for MK2–dependent phosphorylation of TTP, which on turn, results in decrease in its destabilizing effect on mRNA[135]. These findings are in consistence with the stimulating function of HspB1 (Hsp27) in cytokine production [136].

Conclusion: Potential Role of sHsps in Mediating Cancer Pathways: Summary and Perspectives

Among the six essential alterations which are important hallmarks of cancer [1], small Hsps are implied in mediating pathways which lead to manifestation of at least four major features, namely escaping from apoptosis, reaching limitless replication, sustained angiogenesis, and tissue invasion and metastasis. The facts that sHsps may be critical for survival and development of tumor cells, and that their inhibition will not dramatically affect cellular physiology under non-stress conditions, qualify sHsps as potential targets for anticancer therapy. However, so far there are many unanswered questions which should be taken into consideration: sHsps are able to interact with a number of essential targets, mediating cancer pathways, such as Akt, p53/p21, $p27^{Kip}$, ER, HSF1, β-catenin, (Figure 1), but the tumor-specific regulation of these interactions and their relevance for tumorigenesis are not understood. Furthermore, molecular mechanisms and specificity of the described interactions are unclear. Taken together, the pleiothrophic functions of sHsps harbor both, promises as target in tumor therapy, but also inherently problems of lacking a specific mechanism of action. Without this mechanism, inhibitors of sHsps will probably not be tolerated in the clinics in future. The recently generated HspB1-deficient mice [2] could help to identify the relevance of some of the interactions described *in vivo* and should be suited to be challenged in established models of mouse tumorigenesis to monitor the contribution of HspB1.

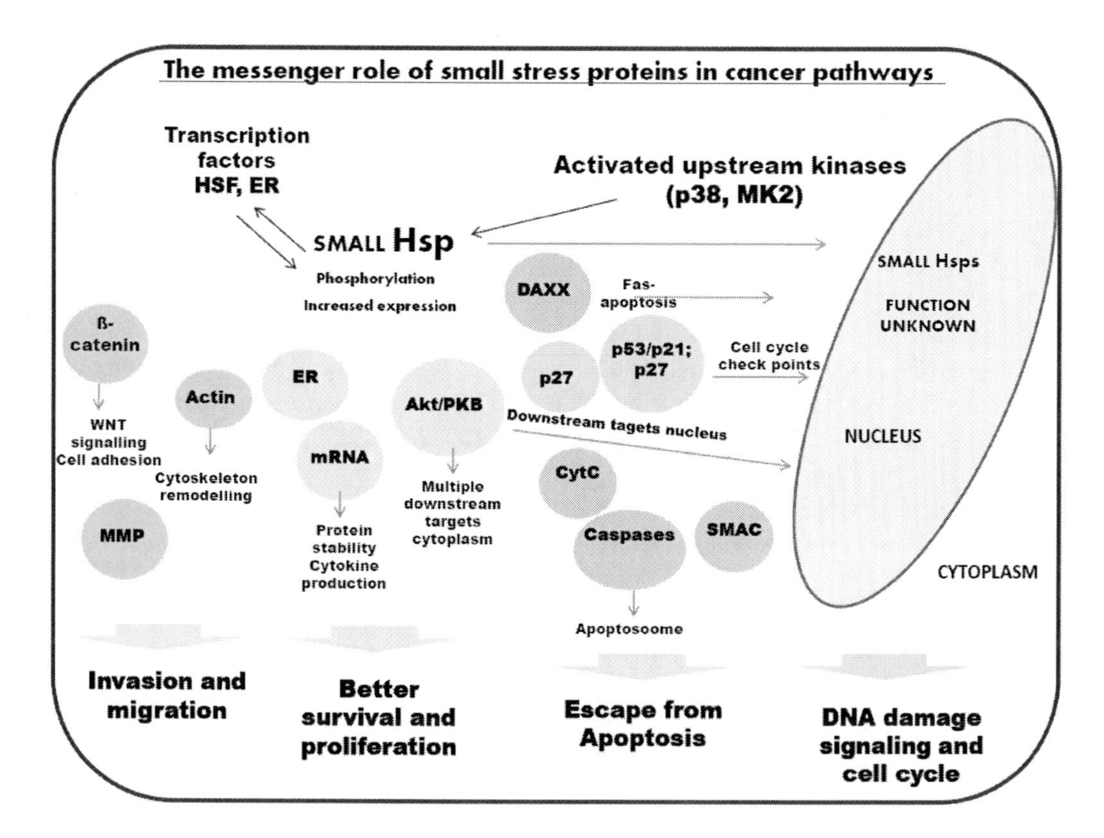

Figure 1.

References

[1] Hanahan, D, Weinberg, R. The hallmarks of cancer. *Cell,* 2000; 100:57.

[2] Huang, L, Min, J, Masters, S, et al. Insights into function and regulation of small heat shock protein 25 (HSPB1) in a mouse model with targeted gene disruption. *Genesis,* 2007; 45:487.

[3] McMillan, DR, Xiao, X, Shao, L, et al. Targeted Disruption of Heat Shock Transcription Factor 1 Abolishes Thermotolerance and Protection against Heat-inducible Apoptosis. *J. Biol. Chem.* 1998; 273:7523.

[4] Morimoto, RI. Regulation of the heat shock transcriptional response: cross talk between a family of heat shock factors, molecular chaperones, and negative regulators. *Genes Dev.* 1998; 12:3788.

[5] Hartl, FU, Hayer-Hartl, M. Converging concepts of protein folding in vitro and in vivo. *Nat Struct Mol Biol.* 2009; 16:574.

[6] Young, JC, Moarefi, I, Hartl, FU. Hsp90: a specialized but essential protein-folding tool. *J. Cell Biol.* 2001; 154:267.

[7] Taylor, R, Benjamin, I. Small heat shock proteins: a new classification scheme in mammals. J Mol Cell Cardiol 2005; 38:433.

[8] Lambert, H, Charette, SJ, Bernier, AF, et al. HSP27 multimerization mediated by phosphorylation-sensitive intermolecular interactions at the amino terminus. *J Biol Chem.* 1999; 274:9378.

[9] Dudich, I, Zav'yalov, V, Pfeil, W, et al. Dimer structure as a minimum cooperative subunit of small heat-shock proteins. Biochim Biophys Acta 1995; 1253:163.

[10] Ito, H, Kamei, K, Iwamoto, I, et al. Phosphorylation-induced Change of the Oligomerization State of alpha B-crystallin. J. Biol. Chem. 2001; 276:5346.

[11] Ito, H, Okamoto, K, Nakayama, H, et al. Phosphorylation of alpha B-Crystallin in Response to Various Types of Stress. J. Biol. Chem. 1997; 272:29934.

[12] Knauf, U, Jakob, U, Engel, K, et al. Stress- and mitogen-induced phosphorylation of the small heat shock protein Hsp25 by MAPKAP kinase 2 is not essential for chaperone properties and cellular thermoresistance. *EMBO J.* 1994; 13:54.

[13] Rogalla, T, Ehrnsperger, M, Preville, X, et al. Regulation of Hsp27 Oligomerization, Chaperone Function, and Protective Activity against Oxidative Stress/Tumor Necrosis Factor alpha by Phosphorylation 10.1074/jbc.274.27.18947. *J. Biol. Chem.* 1999; 274:18947.

[14] Rouse, J, Cohen, P, Trigon, S, et al. A novel kinase cascade triggered by stress and heat shock that stimulates MAPKAP kinase-2 and phosphorylation of the small heat shock proteins. Cell 1994; 78:1027.

[15] Stokoe, D, Engel, K, Campbell, DG, et al. Identification of MAPKAP kinase 2 as a major enzyme responsible for the phosphorylation of the small mammalian heat shock proteins. FEBS Lett 1992; 313:307.

[16] Gaestel, M. MAPKAP kinases - MKs - two's company, three's a crowd. *Nat Rev Mol Cell Biol.* 2006; 7:120.

[17] McLaughlin, MM, Kumar, S, McDonnell, PC, et al. Identification of mitogen-activated protein (MAP) kinase-activated protein kinase-3, a novel substrate of CSBP p38 MAP kinase. *J Biol Chem.* 1996; 271:8488.

[18] Ronkina, N, Kotlyarov, A, Dittrich-Breiholz, O, et al. The mitogen-activated protein kinase (MAPK)-activated protein kinases MK2 and MK3 cooperate in stimulation of tumor necrosis factor biosynthesis and stabilization of p38 MAPK. Mol Cell Biol 2007; 27:170.

[19] Shi, Y, Kotlyarov, A, Laass, K, et al. Elimination of Protein Kinase MK5/PRAK Activity by Targeted Homologous Recombination. Mol. Cell. Biol. 2003; 23:7732.

[20] Gaestel, M, Schroder, W, Benndorf, R, et al. Identification of the phosphorylation sites of the murine small heat shock protein hsp25. J Biol Chem 1991; 266:14721.

[21] Landry, J, Lambert, H, Zhou, M, et al. Human HSP27 is phosphorylated at serines 78 and 82 by heat shock and mitogen-activated kinases that recognize the same amino acid motif as S6 kinase II. J. Biol. Chem. 1992; 267:794.

[22] Butt, E, Immler, D, Meyer, HE, et al. Heat shock protein 27 is a substrate of cGMP-dependent protein kinase in intact human platelets: phosphorylation-induced actin polymerization caused by HSP27 mutants. J Biol Chem 2001; 276:7108.

[23] New, L, Jiang, Y, Zhao, M, et al. PRAK, a novel protein kinase regulated by the p38 MAP kinase. EMBO J 1998; 17:3372.

[24] Maizels, E, Peters, C, Kline, M, et al. Heat-shock protein-25/27 phosphorylation by the delta isoform of protein kinase C. Biochem J 1998; 332 (Pt 3):703.

[25] Doppler, H, Storz, P, Li, J, et al. A Phosphorylation State-specific Antibody Recognizes Hsp27, a Novel Substrate of Protein Kinase D. J. Biol. Chem. 2005; 280:15013.

[26] Lindner, RA, Carver, JA, Ehrnsperger, M, et al. Mouse Hsp25, a small heat shock protein: The role of its C-terminal extension in oligomerization and chaperone action. *Eur. J. Biochem.* 2000; 267:1923.

[27] Freshney, NW, Rawlinson, L, Guesdon, F, et al. Interleukin-1 activates a novel protein kinase cascade that results in the phosphorylation of Hsp27. Cell 1994; 78:1039.

[28] Kato, K, Hasegawa, K, Goto, S, Inaguma, Y. Dissociation as a result of phosphorylation of an aggregated form of the small stress protein, hsp27. J. Biol. Chem. 1994; 269:11274.

[29] Kotlyarov, A, Yannoni, Y, Fritz, S, et al. Distinct cellular functions of MK2. *Mol Cell Biol.* 2002; 22:4827.

[30] Vertii, A, Hakim, C, Kotlyarov, A, Gaestel, M. Analysis of Properties of Small Heat Shock Protein Hsp25 in MAPK-activated Protein Kinase 2 (MK2)-deficient Cells: MK2-dependent insolubilization of Hsp25 oligomers correlates with susceptibility to stress. J. Biol. Chem. 2006; 281:26966.

[31] Crete, P, Landry, J. Induction of HSP27 phosphorylation and thermoresistance in Chinese hamster cells by arsenite, cycloheximide, A23187, and EGTA. Radiat Res 1990; 121:320.

[32] Lavoie, JN, Lambert, H, Hickey, E, et al. Modulation of cellular thermoresistance and actin filament stability accompanies phosphorylation-induced changes in the oligomeric structure of heat shock protein 27. Mol Cell Biol 1995; 15:505.

[33] Mehlen, P, Hickey, E, Weber, LA, Arrigo, AP. Large unphosphorylated aggregates as the active form of hsp27 which controls intracellular reactive oxygen species and glutathione levels and generates a protection against TNFalpha in NIH-3T3-ras cells. *Biochem Biophys Res Commun.* 1997; 241:187.

[34] Benndorf, R, Sun, X, Gilmont, RR, et al. HSP22, a new member of the small heat shock protein superfamily, interacts with mimic of phosphorylated HSP27 (3DHSP27). *J. Biol. Chem.* 2001; 276:26753.

[35] Beall, A, Bagwell, D, Woodrum, D, et al. The small heat shock-related protein, HSP20, is phosphorylated on serine 16 during cyclic nucleotide-dependent relaxation. *J. Biol. Chem.* 1999; 274:11344.

[36] Kato, K, Inaguma, Y, Ito, H, et al. Ser-59 is the major phosphorylation site in alphaB-crystallin accumulated in the brains of patients with Alexander's disease. *J Neurochem.* 2001; 76:730.

[37] Kato, K, Ito, H, Kamei, K, et al. Phosphorylation of alpha B-crystallin in mitotic cells and identification of enzymatic activities responsible for phosphorylation. *J. Biol. Chem.* 1998; 273:28346.

[38] MacCoss, MJ, McDonald, WH, Saraf, A, et al. Shotgun identification of protein modifications from protein complexes and lens tissue. PNAS 2002; 99:7900.

[39] Arrigo, A, Welch, W. Characterization and purification of the small 28,000-dalton mammalian heat shock protein. J. Biol. Chem. 1987; 262:15359.

[40] Bukach, O, Seit-Nebi, A, Marston, S, Gusev, N. Some properties of human small heat shock protein Hsp20 (HspB6). Eur J Biochem 2004; 271:291.

[41] Lee, S, Carson, K, Rice-Ficht, A, Good, T. Hsp20, a novel {alpha}-crystallin, prevents A{beta} fibril formation and toxicity. Protein Sci. 2005; 14:593.

[42] van de Klundert, F, Smulders, R, Gijsen, M, et al. The mammalian small heat-shock protein Hsp20 forms dimers and is a poor chaperone. Eur J Biochem 1998; 258:1014.

[43] Andley, UP, Song, Z, Wawrousek, EF, Bassnett, S. The molecular chaperone alpha A-crystallin enhances lens epithelial cell growth and resistance to UVA stress. *J. Biol. Chem.* 1998; 273:31252.

[44] Bova, MP, Mchaourab, HS, Han, Y, Fung, BK-K. Subunit exchange of small heat shock proteins. Analysis of oligomer formation of alpha A-crystallin and Hsp27 by fluorescence resonance energy transfer and site-directed truncations. J. Biol. Chem. 2000; 275:1035.

[45] Das, KP, Choo-Smith, L-Pi, Petrash, JM, Surewicz, WK. Insight into the secondary structure of non-native proteins bound to a molecular chaperone alpha-crystallin. An isotope-edited infrared spectroscopic study. *J. Biol. Chem.* 1999; 274:33209.

[46] Merck, K, De Haard-Hoekman, W, Oude Essink, B, et al. Expression and aggregation of recombinant alpha A-crystallin and its two domains. Biochim Biophys Acta 1992; 1130:267.

[47] Merck, K, Horwitz, J, Kersten, M, et al. Comparison of the homologous carboxy-terminal domain and tail of alpha-crystallin and small heat shock protein. Mol Biol Rep 1993; 18:209.

[48] Reddy, GB, Das, KP, Petrash, JM, Surewicz, WK. Temperature-dependent chaperone activity and structural properties of human alpha A- and alpha B-crystallins. *J. Biol. Chem.* 2000; 275:4565.

[49] Ehrnsperger, M, Graber, S, Gaestel, M, Buchner, J. Binding of non-native protein to Hsp25 during heat shock creates a reservoir of folding intermediates for reactivation. *EMBO J.* 1997; 16:221.

[50] Ehrnsperger, M, Lilie, H, Gaestel, M, Buchner, J. The dynamics of Hsp25 quaternary structure. Structure and function of different oligomeric species. *J. Biol. Chem.* 1999; 274:14867.

[51] Fan, G-C, Chu, G, Mitton, B, et al. Small heat-shock protein Hsp20 phosphorylation inhibits {beta}-agonist-induced cardiac apoptosis. *Circ. Res.* 2004; 94:1474.

[52] Pandey, P, Farber, R, Nakazawa, A, et al. Hsp27 functions as a negative regulator of cytochrome c-dependent activation of procaspase-3. *Oncogene.* 2000; 19:1975.

[53] Paul, C, Manero, F, Gonin, S, et al. Hsp27 as a negative regulator of cytochrome c release. *Mol. Cell. Biol.* 2002; 22:816.

[54] Garrido, C, Schmitt, E, Cande, C, et al. HSP27 and HSP70: potentially oncogenic apoptosis inhibitors. *Cell Cycle,* 2003; 2:579.

[55] Charette, SJ, Landry, J. The interaction of HSP27 with daxx identifies a potential regulatory role of HSP27 in Fas-induced apoptosis. Ann. N.Y. Acad. Sci. 2000; 926:126.

[56] Charette, SJ, Lavoie, JN, Lambert, H, Landry, J. Inhibition of daxx-mediated apoptosis by heat shock protein 27. *Mol. Cell. Biol.* 2000; 20:7602.

[57] Bruey, J, Ducasse, C, Bonniaud, P, et al. Hsp27 negatively regulates cell death by interacting with cytochrome c. Nat Cell Biol 2000; 2:645.

[58] Ciocca, D, Luque, E. Immunological evidence for the identity between the hsp27 estrogen-regulated heat shock protein and the p29 estrogen receptor-associated protein in breast and endometrial cancer. Breast Cancer Res Treat 1991; 20:33.

[59] Ciocca, D, Stati, A, Amprino de Castro, M. Colocalization of estrogen and progesterone receptors with an estrogen-regulated heat shock protein in paraffin sections of human breast and endometrial cancer tissue. Breast Cancer Res Treat 1990; 16:243.

[60] Garrido, C, Fromentin, A, Bonnotte, B, et al. Heat shock protein 27 enhances the tumorigenicity of immunogenic rat colon carcinoma cell clones. Cancer Res. 1998; 58:5495.

[61] Garrido, C, Mehlen, P, Fromentin, A, et al. Inconstant association between 27-kDa heat-shock protein (Hsp27) content and doxorubicin resistance in human colon cancer cells. The doxorubicin-protecting effect of Hsp27. Eur J Biochem 1996; 237:653.

[62] Langdon, S, Rabiasz, G, Hirst, G, et al. Expression of the heat shock protein HSP27 in human ovarian cancer. Clin. Cancer Res. 1995; 1:1603.

[63] Lemieux, P, Oesterreich, S, Lawrence, J, et al. The small heat shock protein hsp27 increases invasiveness but decreases motility of breast cancer cells. Invasion *Metastasis*, 1997; 17:113.

[64] Manning, DL, Archibald, LH, Ow, KT. Cloning of estrogen-responsive messenger RNAs in the T-47D human breast cancer cell line. Cancer Res. 1990; 50:4098.

[65] Oesterreich, S, Hickey, E, Weber, L, Fuqua, S. Basal regulatory promoter elements of the hsp27 gene in human breast cancer cells. Biochem Biophys Res Commun 1996; 222:155.

[66] Oesterreich, S, Weng, C-N, Qiu, M, et al. The small heat shock protein hsp27 is correlated with growth and drug resistance in human breast cancer cell lines. *Cancer Res*. 1993; 53:4443.

[67] Trautinger, F, Kindas-Mugge, I, Dekrout, B, et al. Expression of the 27-kDa heat shock protein in human epidermis and in epidermal neoplasms: an immunohistological study. Br J Dermatol 1995; 133:194.

[68] Ioachim, E, Tsanou, E, Briasoulis, E, et al. Clinicopathological study of the expression of hsp27, pS2, cathepsin D and metallothionein in primary invasive breast cancer. *Breast*, 2003; 12:111.

[69] Kang, S, Kang, K, Kim, K, et al. Upregulated HSP27 in human breast cancer cells reduces Herceptin susceptibility by increasing Her2 protein stability. BMC Cancer 2008; 8:286.

[70] Zhang, D, Wong, L, Koay, E. Phosphorylation of Ser78 of Hsp27 correlated with HER-2/neu status and lymph node positivity in breast cancer. *Mol Cancer,* 2007; 6:52.

[71] Schneider, J, Jimenez, E, Marenbach, K, et al. Co-expression of the MDR1 gene and HSP27 in human ovarian cancer. Anticancer Res 1998; 18:2967.

[72] Berrieman, H, Cawkwell, L, O'Kane, S, et al. Hsp27 may allow prediction of the response to single-agent vinorelbine chemotherapy in non-small cell lung cancer. Oncol Rep 2006; 15:283.

[73] Tajeddine, N, Louis, M, Vermylen, C, et al. Tumor associated antigen PRAME is a marker of favorable prognosis in childhood acute myeloid leukemia patients and modifies the expression of S100A4, Hsp 27, p21, IL-8 and IGFBP-2 in vitro and in vivo. *Leuk Lymphoma*, 2008; 49:1123.

[74] Kapranos, N, Kominea, A, Konstantinopoulos, P, et al. Expression of the 27-kDa heat shock protein (HSP27) in gastric carcinomas and adjacent normal, metaplastic, and dysplastic gastric mucosa, and its prognostic significance. J Cancer Res Clin Oncol 2002; 128:426.

[75] Yang, Y, Sun, X, Cheng, A, et al. Increased expression of HSP27 linked to vincristine resistance in human gastric cancer cell line. J Cancer Res Clin Oncol 2008.

[76] Melle, C, Ernst, G, Escher, N, et al. Protein profiling of microdissected pancreas carcinoma and identification of HSP27 as a potential serum marker. Clin. Chem. 2007; 53:629.

[77] Yuan, J, Rozengurt, E. PKD, PKD2, and p38 MAPK mediate Hsp27 serine-82 phosphorylation induced by neurotensin in pancreatic cancer PANC-1 cells. *J Cell Biochem.* 2008; 103:648.

[78] Hadaschik, BA, Jackson, J, Fazli, L, et al. Intravesically administered antisense oligonucleotides targeting heat-shock protein-27 inhibit the growth of non-muscle-invasive bladder cancer. BJU Int 2008; 102:610.

[79] Kamada, M, So, A, Muramaki, M, et al. Hsp27 knockdown using nucleotide-based therapies inhibit tumor growth and enhance chemotherapy in human bladder cancer cells. *Mol. Cancer Ther.* 2007; 6:299.

[80] Cornford, PA, Dodson, AR, Parsons, KF, et al. Heat shock protein expression independently predicts clinical outcome in prostate cancer. Cancer Res 2000; 60:7099.

[81] Glaessgen, A, Jonmarker, S, Lindberg, A, et al. Heat shock proteins 27, 60 and 70 as prognostic markers of prostate cancer. APMIS 2008; 116:888.

[82] Xu, L, Bergan, RC. Genistein inhibits matrix metalloproteinase type 2 activation and prostate cancer cell invasion by blocking the transforming growth factor beta-mediated activation of mitogen-activated protein kinase-activated protein kinase 2-27-kDa heat shock protein pathway. Mol. Pharmacol. 2006; 70:869.

[83] Perou, C, Sorlie, T, Eisen, M, et al. Molecular portraits of human breast tumours. *Nature,* 2000; 406:747.

[84] Sharma, B, Smith, C, Laing, J, et al. Aberrant DNA methylation silences the novel heat shock protein H11 in melanoma but not benign melanocytic lesions. Dermatology 2006; 213:192.

[85] Sun, X, Fontaine, J, Bartl, I, et al. Induction of Hsp22 (HspB8) by estrogen and the metalloestrogen cadmium in estrogen receptor-positive breast cancer cells. *Cell Stress Chaperones*, 2007; 12:307.

[86] Trent, S, Yang, C, Li, C, et al. Heat shock protein B8, a cyclin-dependent kinase independent cyclin D1 target gene, contributes to its effects on radiation sensitivity. *Cancer Res*. 2007; 67:10774.

[87] Yang, C, Trent, S, Ionescu-Tiba, V, et al. Identification of cyclin D1- and estrogen-regulated genes contributing to breast carcinogenesis and progression. *Cancer Res*. 2006; 66:11649.

[88] Dimberg, A, Rylova, S, Dieterich, LC, et al. {alpha}B-crystallin promotes tumor angiogenesis by increasing vascular survival during tube morphogenesis. Blood 2008; 111:2015.

[89] Ivanov, O, Chen, F, Wiley, E, et al. alphaB-crystallin is a novel predictor of resistance to neoadjuvant chemotherapy in breast cancer. *Breast Cancer Res Treat*, 2008; 111:411.

[90] Stegh, AH, Kesari, S, Mahoney, JE, et al. Bcl2L12-mediated inhibition of effector caspase-3 and caspase-7 via distinct mechanisms in glioblastoma. PNAS 2008; 105:10703.

[91] Arrigo, A, Simon, S, Gibert, B, et al. Hsp27 (HspB1) and alphaB-crystallin (HspB5) as therapeutic targets. *FEBS Lett*. 2007; 581:3665.

[92] Parcellier, A, Schmitt, E, Brunet, M, et al. Small heat shock proteins HSP27 and alphaB-crystallin: cytoprotective and oncogenic functions. Antioxid Redox Signal 2005; 7:404.

[93] Mineva, I, Gartner, W, Hauser, P, et al. Differential expression of alphaB-crystallin and Hsp27-1 in anaplastic thyroid carcinomas because of tumor-specific alphaB-crystallin gene (CRYAB) silencing. *Cell Stress Chaperones,* 2005; 10:171.

[94] Myung, J, Afjehi-Sadat, L, Felizardo-Cabatic, M, et al. Expressional patterns of chaperones in ten human tumor cell lines. *Proteome Sci.* 2004; 2:8.

[95] Klein-Hitpass, L, Tsai, SY, Greene, GL, et al. Specific binding of estrogen receptor to the estrogen response element. *Mol. Cell. Biol*. 1989; 9:43.

[96] Ciocca, DR, Puy, LA, Fasoli, LC. Study of estrogen receptor, progesterone receptor, and the estrogen-regulated Mr 24,000 protein in patients with carcinomas of the endometrium and cervix. *Cancer Res*. 1989; 49:4298.

[97] Dai, C, Whitesell, L, Rogers, A, Lindquist, S. Heat shock factor 1 is a powerful multifaceted modifier of carcinogenesis. Cell 2007; 130:1005.

[98] Richards, EH, Hickey, E, Weber, L, Masters, JRW. Effect of overexpression of the small heat shock protein HSP27 on the heat and drug sensitivities of human testis tumor cells. *Cancer Res*. 1996; 56:2446.

[99] Chen, H, Hewison, M, Adams, JS. Control of estradiol-directed gene transactivation by an intracellular estrogen-binding protein and an estrogen response element-binding protein. *Mol. Endocrinol*. 2008; 22:559.

[100] Chen, H, Hewison, M, Hu, B, et al. An Hsp27-related, dominant-negative-acting intracellular estradiol-binding protein. J. Biol. Chem. 2004; 279:29944.

[101] Fanelli, M, Montt-Guevara, M, Diblasi, A, et al. P-Cadherin and beta-catenin are useful prognostic markers in breast cancer patients; beta-catenin interacts with heat shock protein Hsp27. Cell Stress Chaperones 2008; 13:207.

[102] Arrigo, AS, JP and Welch WJ. Dynamic changes in the structure and intracellular locale of the mammalian low-molecular-weight heat shock protein. *Mol. Cell. Biol.* 1988; 8:5059.

[103] Bryantsev, A, Chechenova, M, Shelden, E. Recruitment of phosphorylated small heat shock protein Hsp27 to nuclear speckles without stress. *Exp Cell Res*. 2007; 313:195.

[104] Mearow, K, Dodge, M, Rahimtula, M, Yegappan, C. Stress-mediated signaling in PC12 cells - the role of the small heat shock protein, Hsp27, and Akt in protecting cells from heat stress and nerve growth factor withdrawal. J Neurochem 2002; 83:452.

[105] Greenberg, AK, Basu, S, Hu, J, et al. Selective p38 Activation in Human Non-Small Cell Lung Cancer. Am. J. Respir. Cell Mol. Biol. 2002; 26:558.

[106] Han, Q, Leng, J, Bian, D, et al. Rac1-MKK3-p38-MAPKAPK2 pathway promotes urokinase plasminogen activator mRNA stability in invasive breast cancer cells. *J. Biol. Chem.* 2002; 277:48379.

[107] Hirose, Y, Katayama, M, Stokoe, D, et al. The p38 mitogen-activated protein kinase pathway links the DNA mismatch repair system to the G2 checkpoint and to resistance to chemotherapeutic DNA-methylating agents. Mol. Cell. Biol. 2003; 23:8306.

[108] Hsu, Y-L, Kuo, P-L, Lin, L-T, Lin, C-C. Asiatic acid, a triterpene, induces apoptosis and cell cycle arrest through activation of extracellular signal-regulated kinase and p38 mitogen-activated protein kinase pathways in human breast cancer cells. *J. Pharmacol. Exp. Ther.* 2005; 313:333.

[109] Huang, X, Chen, S, Xu, L, et al. Genistein inhibits p38 Map kinase activation, matrix metalloproteinase type 2, and cell invasion in human prostate epithelial cells. *Cancer Res.* 2005; 65:3470.

[110] Qi, X, Pramanik, R, Wang, J, et al. The p38 and JNK pathways cooperate to trans-activate vitamin D receptor via c-Jun/AP-1 and sensitize human breast cancer cells to vitamin D3-induced growth inhibition. J. Biol. Chem. 2002; 277:25884.

[111] Qi, X, Tang, J, Pramanik, R, et al. p38 MAPK activation selectively induces cell death in K-ras-mutated human colon cancer cells through regulation of vitamin D receptor. *J. Biol. Chem.* 2004; 279:22138.

[112] Quann, EJ, Khwaja, F, Djakiew, D. The p38 MAPK pathway mediates aryl propionic acid induced messenger RNA stability of p75NTR in prostate cancer cells. Cancer Res. 2007; 67:11402.

[113] Thoms, HC, Dunlop, MG, Stark, LA. p38-mediated inactivation of cyclin D1/cyclin-dependent kinase 4 stimulates nucleolar translocation of RelA and apoptosis in colorectal cancer cells. *Cancer Res.* 2007; 67:1660.

[114] Brancho, D, Tanaka, N, Jaeschke, A, et al. Mechanism of p38 MAP kinase activation in vivo. *Genes Dev.* 2003; 17:1969.

[115] Roux, PP, Blenis, J. ERK and p38 MAPK-activated protein kinases: a family of protein kinases with diverse biological functions. Microbiol Mol Biol Rev 2004; 68:320.

[116] Raingeaud, J, Gupta, S, Rogers, JS, et al. Pro-inflammatory cytokines and environmental stress cause p38 mitogen-activated protein kinase activation by dual phosphorylation on tyrosine and threonine. J Biol Chem 1995; 270:7420.

[117] Wang, XS, Diener, K, Manthey, CL, et al. Molecular cloning and characterization of a novel p38 mitogen-activated protein kinase. J Biol Chem 1997; 272:23668.

[118] Tanoue, T, Nishida, E. Molecular recognitions in the MAP kinase cascades. *Cell Signal*, 2003; 15:455.

[119] Lee, JC, Laydon, JT, McDonnell, PC, et al. A protein kinase involved in the regulation of inflammatory cytokine biosynthesis. Nature 1994; 372:739.

[120] Fukunaga, R, Hunter, T. MNK1, a new MAP kinase-activated protein kinase, isolated by a novel expression screening method for identifying protein kinase substrates. *EMBO J.* 1997; 16:1921.

[121] Waskiewicz, AJ, Flynn, A, Proud, CG, Cooper, JA. Mitogen-activated protein kinases activate the serine/threonine kinases Mnk1 and Mnk2. EMBO J 1997; 16:1909.

[122] Deak, M, Clifton, AD, Lucocq, LM, Alessi, DR. Mitogen- and stress-activated protein kinase-1 (MSK1) is directly activated by MAPK and SAPK2/p38, and may mediate activation of CREB. EMBO J 1998; 17:4426.

[123] Tan, Y, Rouse, J, Zhang, A, et al. FGF and stress regulate CREB and ATF-1 via a pathway involving p38 MAP kinase and MAPKAP kinase-2. EMBO J 1996; 15:4629.

[124] Zarubin, T, Han, J. Activation and signaling of the p38 MAP kinase pathway. *Cell Res.* 2005; 15:11.

[125] Bulavin, D, Saito, S, Hollander, M, et al. Phosphorylation of human p53 by p38 kinase coordinates N-terminal phosphorylation and apoptosis in response to UV radiation. *EMBO J.* 1999; 18:6845.

[126] Huang, C, Ma, W-Y, Maxiner, A, et al. p38 kinase mediates UV-induced phosphorylation of p53 protein at serine 389. J. Biol. Chem. 1999; 274:12229.

[127] Kim, G-Y, Mercer, SE, Ewton, DZ, et al. The stress-activated protein kinases p38alpha and JNK1 stabilize p21Cip1 by phosphorylation. J. Biol. Chem. 2002; 277:29792.

[128] Gaul, L, Mandl-Weber, S, Baumann, P, et al. Bendamustine induces G2 cell cycle arrest and apoptosis in myeloma cells: the role of ATM-Chk2-Cdc25A and ATM-p53-p21-pathways. *J Cancer Res Clin Oncol.* 2008; 134:245.

[129] Kim, H, Kokkotou, E, Na, X, et al. Clostridium difficile toxin A-induced colonocyte apoptosis involves p53-dependent p21(WAF1/CIP1) induction via p38 mitogen-activated protein kinase. *Gastroenterology,* 2005; 129:1875.

[130] Mikule, K, Delaval, B, Kaldis, P, et al. Loss of centrosome integrity induces p38-p53-p21-dependent G1-S arrest. *Nat Cell Biol.* 2007; 9:160.

[131] Stepniak, E, Ricci, R, Eferl, R, et al. c-Jun/AP-1 controls liver regeneration by repressing p53/p21 and p38 MAPK activity. *Genes & Dev.* 2006; 20:2306.

[132] Cuadrado, A, Lafarga, V, Cheung, PC, et al. A new p38 MAP kinase-regulated transcriptional coactivator that stimulates p53-dependent apoptosis. EMBO J 2007; 26:2115.

[133] Adams, RH, Porras, A, Alonso, G, et al. Essential role of p38alpha MAP kinase in placental but not embryonic cardiovascular development. Mol Cell 2000; 6:109.

[134] Kotlyarov, A, Neininger, A, Schubert, C, et al. MAPKAP kinase 2 is essential for LPS-induced TNF-alpha biosynthesis. *Nat Cell Biol.* 1999; 1:94.

[135] Hitti, E, Iakovleva, T, Brook, M, et al. Mitogen-activated protein kinase-activated protein kinase 2 regulates tumor necrosis factor mRNA stability and translation mainly by altering tristetraprolin expression, stability, and binding to adenine/uridine-rich element. Mol. Cell. Biol. 2006; 26:2399.

[136] Alford, KA, Glennie, S, Turrell, BR, et al. Heat shock protein 27 functions in inflammatory gene expression and transforming growth factor-beta-activated kinase-1 (TAK1)-mediated signaling. J Biol Chem 2007; 282:6232.

[137] Manke, I, Nguyen, A, Lim, D, et al. MAPKAP kinase-2 is a cell cycle checkpoint kinase that regulates the G2/M transition and S phase progression in response to UV irradiation. *Mol Cell.* 2005; 17:37.

[138] Mikhailov, A, Patel, D, McCance, D, Rieder, C. The G2 p38-mediated stress-activated checkpoint pathway becomes attenuated in transformed cells. Curr Biol 2007; 17:2162.

[139] Reinhardt, H, Aslanian, A, Lees, J, Yaffe, M. p53-deficient cells rely on ATM- and ATR-mediated checkpoint signaling through the p38MAPK/MK2 pathway for survival after DNA damage. *Cancer Cell*, 2007; 11:175.

[140] Chen, J, Baskerville, C, Han, Q, et al. alpha v integrin, p38 mitogen-activated protein kinase, and urokinase plasminogen activator are functionally linked in invasive breast cancer cells. *J. Biol. Chem.* 2001; 276:47901.
[141] Sun, P, Yoshizuka, N, New, L, et al. PRAK is essential for ras-induced senescence and tumor suppression. *Cell*, 2007; 128:295.
[142] Hedges, JC, Dechert, MA, Yamboliev, IA, et al. A role for p38(MAPK)/HSP27 pathway in smooth muscle cell migration. J Biol Chem 1999; 274:24211.
[143] Piotrowicz, RS, Hickey, E, Levin, EG. Heat shock protein 27 kDa expression and phosphorylation regulates endothelial cell migration. FASEB J 1998; 12:1481.
[144] Martin, JL, Bossuyt, J, Fontaine, J-M, et al. Inhibition of hsp27 phosphorylation increases interaction with Hic5 in vascular myocytes. FASEB J 2008; 22:1208.8.
[145] Jia, Y, Ransom, RF, Shibanuma, M, et al. Identification and characterization of hic-5/ARA55 as an hsp27 binding protein. J Biol Chem 2001; 276:39911.
[146] Wilker, E, Yaffe, M. 14-3-3 Proteins--a focus on cancer and human disease. *J Mol Cell Cardiol.* 2004; 37:633.
[147] Ching, RW, Dellaire, G, Eskiw, CH, Bazett-Jones, DP. PML bodies: a meeting place for genomic loci? J Cell Sci 2005; 118:847.
[148] Melnick, A, Licht, JD. Deconstructing a disease: RARalpha, its fusion partners, and their roles in the pathogenesis of acute promyelocytic leukemia. Blood 1999; 93:3167.
[149] Ishov, AM, Sotnikov, AG, Negorev, D, et al. PML is critical for ND10 formation and recruits the PML-interacting protein daxx to this nuclear structure when modified by SUMO-1. *J Cell Biol.* 1999; 147:221.
[150] Kim, E-J, Park, J-S, Um, S-J. Identification of Daxx interacting with p73, one of the p53 family, and its regulation of p53 activity by competitive interaction with PML. *Nucleic Acids Res.* 2003; 31:5356.
[151] O'Callaghan-Sunol, C, Gabai, VL, Sherman, MY. Hsp27 modulates p53 signaling and suppresses cellular senescence. *Cancer Res.* 2007; 67:11779.
[152] Venkatakrishnan, CD, Dunsmore, K, Wong, H, et al. HSP27 regulates p53 transcriptional activity in doxorubicin-treated fibroblasts and cardiac H9c2 cells: p21 upregulation and G2/M phase cell cycle arrest. *Am J Physiol Heart Circ Physiol.* 2008; 294:H1736.
[153] Nefkens, I, Negorev, DG, Ishov, AM, et al. Heat shock and Cd2+ exposure regulate PML and Daxx release from ND10 by independent mechanisms that modify the induction of heat-shock proteins 70 and 25 differently. J Cell Sci 2003; 116:513.
[154] Bryantsev, AL, Kurchashova, SY, Golyshev, SA, et al. Regulation of stress-induced intracellular sorting and chaperone function of Hsp27 (HspB1) in mammalian cells. *Biochem J.* 2007; 407:407.
[155] Marin-Vinader, L, Shin, C, Onnekink, C, et al. Hsp27 enhances recovery of splicing as well as rephosphorylation of SRp38 after heat shock. Mol Biol Cell 2006; 17:886.
[156] Kampinga, HH, Brunsting, JF, Stege, GJ, et al. Cells overexpressing Hsp27 show accelerated recovery from heat-induced nuclear protein aggregation. Biochem Biophys Res Commun 1994; 204:1170.
[157] Zoubeidi, A, Zardan, A, Beraldi, E, et al. Cooperative interactions between androgen receptor (AR) and heat-shock protein 27 facilitate AR transcriptional activity. Cancer Res 2007; 67:10455.

[158] Rodier, F, Campisi, J, Bhaumik, D. Two faces of p53: aging and tumor suppression. *Nucleic Acids Res.* 2007; 35:7475.

[159] He, L, He, X, Lim, LP, et al. A microRNA component of the p53 tumour suppressor network. *Nature*, 2007; 447:1130.

[160] Sun, Y, Yi, H, Zhang, P, et al. Identification of differential proteins in nasopharyngeal carcinoma cells with p53 silence by proteome analysis. FEBS Lett 2007; 581:131.

[161] Park, S-H, Lee, Y-S, Osawa, Y, et al. Hsp25 regulates the expression of p21(Waf1/Cip1/Sdi1) through multiple mechanisms. *J. Biochem.* 2002; 131:869.

[162] Parcellier, A, Brunet, M, Schmitt, E, et al. HSP27 favors ubiquitination and proteasomal degradation of p27Kip1 and helps S-phase re-entry in stressed cells. FASEB J 2006; 20:1179.

[163] Chu, IM, Hengst, L, Slingerland, JM. The Cdk inhibitor p27 in human cancer: prognostic potential and relevance to anticancer therapy. *Nat Rev Cancer*, 2008; 8:253.

[164] Alkarain, A, Slingerland, J. Deregulation of p27 by oncogenic signaling and its prognostic significance in breast cancer. *Breast Cancer Res.* 2004; 6:13.

[165] Belletti, B, Nicoloso, MS, Schiappacassi, M, et al. p27(kip1) functional regulation in human cancer: a potential target for therapeutic designs. Curr Med Chem 2005; 12:1589.

[166] Blain, SW, Scher, HI, Cordon-Cardo, C, Koff, A. p27 as a target for cancer therapeutics. *Cancer Cell*, 2003; 3:111.

[167] Brown, I, Shalli, K, McDonald, SL, et al. Reduced expression of p27 is a novel mechanism of docetaxel resistance in breast cancer cells. Breast Cancer Res 2004; 6:R601.

[168] Brunner, A, Verdorfer, I, Prelog, M, et al. Large-scale analysis of cell cycle regulators in urothelial bladder cancer identifies p16 and p27 as potentially useful prognostic markers. *Pathobiology,* 2008; 75:25.

[169] Fujita, N, Sato, S, Katayama, K, Tsuruo, T. Akt-dependent phosphorylation of p27Kip1 promotes binding to 14-3-3 and cytoplasmic localization. J. Biol. Chem. 2002; 277:28706.

[170] Wu, FY, Wang, SE, Sanders, ME, et al. Reduction of cytosolic p27(Kip1) inhibits cancer cell motility, survival, and tumorigenicity. Cancer Res 2006; 66:2162.

[171] Fayard, E, Tintignac, LA, Baudry, A, Hemmings, BA. Protein kinase B/Akt at a glance. *J. Cell Sci.* 2005; 118:5675.

[172] Konishi, H, Matsuzaki, H, Tanaka, M, et al. Activation of protein kinase B (Akt/RAC-protein kinase) by cellular stress and its association with heat shock protein Hsp27. FEBS Lett 1997; 410:493.

[173] Rane, MJ, Pan, Y, Singh, S, et al. Heat shock protein 27 controls apoptosis by regulating Akt activation. J. Biol. Chem. 2003; 278:27828.

[174] Zheng, C, Lin, Z, Zhao, ZJ, et al. MAPK-activated protein kinase-2 (MK2)-mediated formation and phosphorylation-regulated dissociation of the signal complex consisting of p38, MK2, Akt, and Hsp27. *J. Biol. Chem.* 2006; 281:37215.

[175] Stern, DF. More than a marker...Phosphorylated Akt in prostate carcinoma. Clin Cancer Res 2004; 10:6407.

[176] Burgering, BM, Medema, RH. Decisions on life and death: FOXO Forkhead transcription factors are in command when PKB/Akt is off duty. *J Leukoc Biol.* 2003; 73:689.

[177] Amaravadi, R, Thompson, CB. The survival kinases Akt and Pim as potential pharmacological targets. *J Clin Invest.* 2005; 115:2618.
[178] Shinohara, M, Chung, YJ, Saji, M, Ringel, MD. AKT in thyroid tumorigenesis and progression. *Endocrinology,* 2007; 148:942.
[179] Wandinger, SK, Richter, K, Buchner, J. The Hsp90 chaperone machinery. *J Biol Chem.* 2008; 283:18473.
[180] Citri, A, Harari, D, Shohat, G, et al. Hsp90 recognizes a common surface on client kinases. *J Biol Chem.* 2006; 281:14361.
[181] Sato, S, Fujita, N, Tsuruo, T. Modulation of Akt kinase activity by binding to Hsp90. PNAS 2000; 97:10832.
[182] Hernandez, MP, Sullivan, WP, Toft, DO. The assembly and intermolecular properties of the hsp70-Hop-hsp90 molecular chaperone complex. *J Biol Chem.* 2002; 277:38294.
[183] Schumacher, RJ, Hansen, WJ, Freeman, BC, et al. Cooperative action of Hsp70, Hsp90, and DnaJ proteins in protein renaturation. *Biochemistry,* 1996; 35:14889.
[184] Fumo, G, Akin, C, Metcalfe, DD, Neckers, L. 17-Allylamino-17-demethoxygeldanamycin (17-AAG) is effective in down-regulating mutated, constitutively activated KIT protein in human mast cells. Blood 2004; 103:1078.
[185] Grenert, JP, Johnson, BD, Toft, DO. The importance of ATP binding and hydrolysis by hsp90 in formation and function of protein heterocomplexes. J Biol Chem 1999; 274:17525.
[186] Scheibel, T, Weikl, T, Buchner, J. Two chaperone sites in Hsp90 differing in substrate specificity and ATP dependence. Proc Natl Acad Sci U S A 1998; 95:1495.
[187] Sullivan, W, Stensgard, B, Caucutt, G, et al. Nucleotides and two functional states of hsp90. J Biol Chem 1997; 272:8007.
[188] Holmes, JL, Sharp, SY, Hobbs, S, Workman, P. Silencing of HSP90 cochaperone AHA1 expression decreases client protein activation and increases cellular sensitivity to the HSP90 inhibitor 17-allylamino-17-demethoxygeldanamycin. *Cancer Res.* 2008; 68:1188.
[189] Smith, JR, Clarke, PA, de Billy, E, Workman, P. Silencing the cochaperone CDC37 destabilizes kinase clients and sensitizes cancer cells to HSP90 inhibitors. Oncogene 2009; 28:157.
[190] Brough, PA, Aherne, W, Barril, X, et al. 4,5-diarylisoxazole Hsp90 chaperone inhibitors: potential therapeutic agents for the treatment of cancer. J Med Chem 2008; 51:196.
[191] Sharp, S, Workman, P. Inhibitors of the HSP90 molecular chaperone: current status. *Adv Cancer Res.* 2006; 95:323.
[192] McCollum, AK, TenEyck, CJ, Sauer, BM, et al. Up-regulation of heat shock protein 27 induces resistance to 17-Allylamino-Demethoxygeldanamycin through a glutathione-mediated mechanism. *Cancer Res.* 2006; 66:10967.
[193] Benndorf, R, Hayess, K, Ryazantsev, S, et al. Phosphorylation and supramolecular organization of murine small heat shock protein HSP25 abolish its actin polymerization-inhibiting activity. *J Biol Chem.* 1994; 269:20780.
[194] Mounier, N, Arrigo, AP. Actin cytoskeleton and small heat shock proteins: how do they interact? *Cell Stress Chaperones,* 2002; 7:167.

[195] Hansen, R, Parra, I, Hilsenbeck, S, et al. Hsp27-induced MMP-9 expression is influenced by the Src tyrosine protein kinase yes. *Biochem Biophys Res Commun.* 2001; 282:186.

[196] Lee, J, Kwak, H, Lee, J, et al. HSP27 regulates cell adhesion and invasion via modulation of focal adhesion kinase and MMP-2 expression. *Eur J Cell Biol.* 2008; 87:377.

[197] Xu, L, Chen, S, Bergan, R. MAPKAPK2 and HSP27 are downstream effectors of p38 MAP kinase-mediated matrix metalloproteinase type 2 activation and cell invasion in human prostate cancer. Oncogene 2006; 25:2987.

[198] Bausero, M, Bharti, A, Page, D, et al. Silencing the hsp25 gene eliminates migration capability of the highly metastatic murine 4T1 breast adenocarcinoma cell. *Tumour Biol.* 2006; 27:17.

[199] Shin, KD, Lee, M-Y, Shin, D-S, et al. Blocking tumor cell migration and invasion with biphenyl isoxazole derivative KRIBB3, a synthetic molecule that inhibits Hsp27 phosphorylation. J. Biol. Chem. 2005; 280:41439.

[200] Cuesta, R, Laroia, G, Schneider, RJ. Chaperone Hsp27 inhibits translation during heat shock by binding eIF4G and facilitating dissociation of cap-initiation complexes. *Genes & Dev.* 2000; 14:1460.

[201] Doerwald, L, van Genesen, ST, Onnekink, C, et al. The effect of alphaB-crystallin and Hsp27 on the availability of translation initiation factors in heat-shocked cells. Cell Mol Life Sci 2006; 63:735.

[202] Anderson, P, Kedersha, N. Stressful initiations. *J Cell Sci.* 2002; 115:3227.

[203] Kedersha, N, Chen, S, Gilks, N, et al. Evidence that ternary complex (eIF2-GTP-tRNA(i)(Met))-deficient preinitiation complexes are core constituents of mammalian stress granules. *Mol Biol Cell,* 2002; 13:195.

[204] Anderson, P, Kedersha, N. RNA granules. *J Cell Biol.* 2006; 172:803.

[205] Doerwald, L, Onnekink, C, van Genesen, ST, et al. Translational thermotolerance provided by small heat shock proteins is limited to cap-dependent initiation and inhibited by 2-aminopurine. *J Biol Chem.* 2003; 278:49743.

[206] Kedersha, NL, Gupta, M, Li, W, et al. RNA-binding proteins TIA-1 and TIAR link the phosphorylation of eIF-2 alpha to the assembly of mammalian stress granules. *J Cell Biol.* 1999; 147:1431.

[207] Laroia, G, Cuesta, R, Brewer, G, Schneider, RJ. Control of mRNA decay by heat shock-ubiquitin-proteasome pathway. *Science,* 1999; 284:499.

[208] Sinsimer, KS, Gratacos, FM, Knapinska, AM, et al. Chaperone Hsp27, a novel subunit of AUF1 protein complexes, functions in AU-rich element-mediated mRNA *Decay Mol. Cell. Biol.* 2008; 28:5223.

[209] Takai, S, Matsushima-Nishiwaki, R, Tokuda, H, et al. Protein kinase C delta regulates the phosphorylation of heat shock protein 27 in human hepatocellular carcinoma. *Life Sci.* 2007; 81:585.

In: Small Stress Proteins and Human Diseases
Editors: Stéphanie Simon et al.
ISBN: 978-1-61470-636-6

Chapter 3.3

HSP16.2 IN CANCERS

***Szabolcs Bellyei*[1*]*, Eva Pozsgai*[2] *and Balazs Sumegi*[2]**
University of Pecs, Department of Oncotherapy[1], Department of Biochemistry and Medical Chemistry[2], Pecs, 7624, Hungary

ABSTRACT

Homology of HSP16.2 to αB-crystallin, the induction of its synthesis to heat stress and its ATP-independent chaperone activity suggested that HSP16.2 is a novel small Heat Shock Protein. Suppression of HSP16.2 sensitized cells to apoptotic stimuli, while the over-expressing of HSP16.2 protected cells against H_2O_2 and taxol-induced cell death. Under stress conditions, HSP16.2 inhibited the release of cytochrome c from the mitochondria, nuclear translocation of AIF and endonuclease G, and caspase-3 activation by protecting the integrity of the mitochondrial membrane system. It was demonstrated that HSP16.2 interacts with HSP90, and that the presence of functionally active HSP90 is a prerequisite of the cytoprotective action of HSP16.2. HSP16.2 also facilitated lipid rafts formation, which could serves as scaffolds for the activation of Akt. The inhibition of PI-3-kinase-Akt pathway by LY-294002, or wortmannin, significantly decreased the cytoprotective effect of HSP16.2. These data indicate that one of the main mechanisms by which HSP16.2 inhibits apoptosis is through its interaction with functionally active HSP90, promoting lipid raft formation and the activation of PI-3-kinase - Akt cytoprotective pathway. Expression of HSP16.2 varies in different types of brain tumors and a positive correlation can be found between the tumor grade and the quantity of HSP16.2 in the tumor cells' cytoplasm. Thus, HSP16.2 may become a valuable marker for brain cancer diagnosis, or a possible target of cancer therapy.

[*] Szabolcs Bellyei
University of Pecs
Department of Oncotherapy
Pecs, 7624
Edesanyak street 17.
Tel: +36-72-536-480
Fax: +36-72-536-481
szabolcs.bellyei@aok.pte.hu

Introduction

Small stress proteins are a group of Heat Shock Proteins that share an evolutionary conserved C-terminal region, called the α-crystallin domain, and whose molecular weight range between 15 and 43 kDa [1-5]. Small Heat Shock Proteins (sHSPs) are considered as ATP-independent chaperones that prevent protein aggregation and facilitate substrate refolding in conjunction with other molecular chaperones [6-9]. For this reason, it is considered that sHSPs can create a reservoir of denatured proteins, which could be refolded in the presence of ATP-dependent chaperones [10]. They are known to play a part in cell differentiation and in counteracting apoptosis [3]. Due to the anti-apoptotic activity of sHSPs, tumor cells expressing the protein highly may become increasingly resistant to chemo- or radiotherapy [2,11-13].

1. HSP16.2 is a sHSP

From its observed homology to αB-crystallin, we assumed that the protein encoded in C1ORF41 gene sequence belongs to the sHSPs family. Indeed, it was heat shock inducible in cell lines supporting this assumption. We named the protein small Heat shock protein 16.2, due to its weight of 16.2 kDa [14]. Our point of view has been confirmed as HSP16.2 has been included into the nomenclature of sHSPs as HspB11 [15]. HSP16.2 diminished CS aggregation. Furthermore, HSP16.2 suppressed the heat-induced inactivation of CS activity in a concentration-dependent manner unlike bovine serum albumin or IgG. According to these results, HSP16.2 recognized and bound CS that unfolded during thermal stress, protecting it from irreversible aggregation and the loss of enzyme activity [16], thus in this respect it behaved similarly to p26 and HSP90 [17]. These data together strongly suggest that HSP16.2 is indeed a sHSP.

2. HSP16.2 Has Anti-Apoptotic Properties

We over-expressed HSP16.2 in the NIH3T3 cell line, and suppressed its expression by dsiRNA technique in HeLa which naturally expresses it at a high extent. By using this model, we could establish that HSP16.2 has cytoprotective property since it protected the HSP16.2 over-expressing NIH3T3 cells against, and sensitized the HSP16.2 suppressed HeLa cells to either H_2O_2 or taxol induced oxidative stress [14].

It is very well established that destabilization of the mitochondrial membrane results in the release of pro-apoptotic signals from the mitochondrial inter-membrane space [18]. Indeed, we found the release of cytochrome c to the cytosol and the nuclear translocation of EndoG and AIF upon H_2O_2 stimulation that was significantly reduced by HSP16.2 over-expression. These differences in the release of pro-apoptotic proteins were reflected in caspase-3 activities, which is a good indicator of apoptosis [19]. These data suggested that HSP16.2 might have a direct stabilization effect on the mitochondrial membrane systems. We tested this hypothesis by adding recombinant HSP16.2 to isolated mitochondria undergoing calcium- and phosphate- induced collapse of the mitochondrial membrane potential. As revealed by the fluorescence quenching curves, HSP16.2 stabilized the mitochondrial membrane system in a concentration-dependent manner that strongly suggested a direct mitochondrial effect of the protein [14].

On the other hand, intracellular distribution of HSP16.2 in cells indicated that this protein did not exclusively bind to the mitochondria, but was more or less evenly distributed in the cytoplasm. For this reason, we looked for other intracellular targets. Immobilized recombinant HSP16.2 protein bound HSP90 specifically from the homogenate of HeLa and Panc-1 cells raising the possibility that the protective effect of HSP16.2 could partially be mediated by HSP90. To establish the physiological significance of the interaction of HSP16.2 and HSP90 in cytoprotection, HSP90 function was inhibited by geldanamycin in sham transfected wild type and HSP 16.2 over-expressing NIH3T3 cells, that were exposed or not to H_2O_2 or taxol. Geldanamycin completely abolished the cytoprotective effect of HSP16.2 over-expression indicating that the cytoprotection was mediated via its binding to HSP90 rather than via its direct mitochondrial effect [14]. It is known that HSP90 is involved in lipid raft stabilization [20-23], therefore, we investigated whether HSP16.2 over-expression can enhance the quantity of lipid rafts, which are cholesterol and sphingolipid enriched microdomains of the plasma membrane. HSP16.2 over-expression was found to increase lipid raft formation, the effect which was counteracted by geldanamycin. That is, HSP16.2 facilitated lipid raft microdomain formation by a HSP90 mediated mechanism.

Along with augmented lipid raft formation, we found increased long term Akt phosphorylation at serine 473 in HSP16.2 over-expressing NIH3T3 cells. Furthermore, inhibition of Akt activation by PI3-kinase inhibitors abolished the cytoprotective effect of HSP16.2 over-expression the same way as geldanamycin did in H_2O_2 or taxol treated HSP16.2 over-expressing NIH3T3 cells. Taken together, these data suggested that the cytoprotection by HSP16.2 over-expression was mediated by a HSP90 - lipid raft associated Akt 473 kinase - Akt pathway [14].

As we found, human cancer cell lines as well as the normal human liver line WRL-68 expressed HSP16.2 at a high extent, however, the mouse fibroblast line NIH3T3 expressed it at low concentrations. In different human cancer tissues particularly high levels of expression were seen in neuroectodermal cancers. Intracellular localization of HSP16.2 was found to be dominantly nuclear in tumors of low clinical grade, but cytoplasmic localization also appeared in tumors of higher clinical grades. Furthermore, its expression level increased with the clinical grading of the tumor. This finding turned our attention to examining sHSP16.2 expression in brain tumors differing in their grade and type [14,24].

3. HSP16.2 in Brain Cancer

A number of recent studies have been concerned with the possible role of sHSPs in the progression of brain tumors [11,25-27]. We began our examination of HSP16.2 expression using samples from a variety of brain tumor types. Since the fifty-one tumors did not significantly differ in their intra-nuclear labeling, but varied in their cytoplasmic labeling, it became apparent that the various tumors differed in the distribution and density of HSP16.2 in the cell cytoplasm. In accordance with earlier studies, astrocytomas (grade 1-2) as well as other benign brain tumors, such as meningothelial meningeoma, and oligodendroglioma (grade 1-2) exhibited low expression (+) of HSP16.2. The increase of anaplasia in the tumor cells resulted in moderate (++) expression (atypical meningeoma, malignant meningeoma, anaplastic astrocytoma, anaplastic oligodendroglioma) and high (+++) expression (glioblastoma, medulloblastoma, PNET) of the protein. Thus, it became evident that there is a direct correlation between the intensity of the staining and the histological grade of the brain tumor. Western blot analysis of twenty-one tumor samples gave similar results, indicating that HSP16.2 protein levels increase with the grade of the tumor [24].

CONCLUSION

In conclusion, homology of HSP16.2 to αB-crystallin, the induction of its synthesis to heat stress and its ATP-independent chaperone activity indicate that HSP16.2 is a novel small heat shock protein. Suppression of HSP16.2 sensitized cells to apoptotic stimuli, while over-expressing of HSP16.2 protected cells against H_2O_2 and taxol induced cell death. Under stress

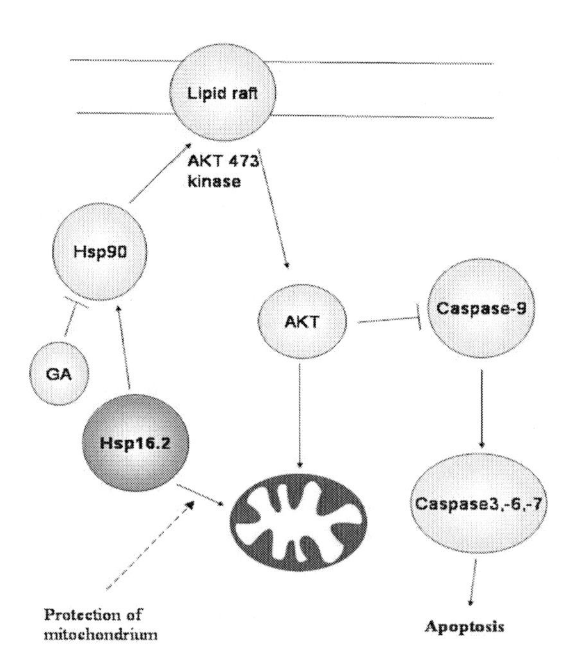

Figure 1. Possible molecular mechanism of the cytoprotective effect of HSP16.2 overexpression Lines with pointed ends denote activation, whereas the ones with flat ends indicate inhibition. HSP16.2 interferes with the cell death pathway at several points. It can protect the stability of the mitochondrial membrane system both in isolated mitochondria and in living cells. As a heat-shock protein, it protects cells from different types of stress. HSP16.2 binds to HSP90 and protects against Taxol and H_2O_2 induced stress by a HSP90 mediated way that can be disrupted by geldanamycin a well-established HSP90 inhibitor. HSP16.2 promotes the HSP90 mediated synthesis of lipid rafts, and likely through lipid raft-associated Akt 473 kinase activates Akt, which process seems mainly responsible for the anti-apoptotic and cytoprotective effect of HSP16.2 under our experimental conditions.

conditions, HSP16.2 inhibits the release of cytochrome c from the mitochondria, nuclear translocation of AIF and endonuclease G, and caspase 3 activation by protecting the integrity of mitochondrial membrane system. On the other hand, HSP16.2 binds to HSP90, and HSP16.2 mediated cytoprotection requires HSP90 activation. HSP16.2 over-expression facilitated lipid raft formation, and increased Akt phosphorylation (ser-473) supporting the idea that stabilization of lipid rafts is essential to Akt activation. The inhibition of PI-3-kinase-Akt pathway by LY-294002, or wortmannin, significantly decreased its protective effect. These data indicate that one of the main mechanisms by which HSP16.2 inhibits cell death is the activation of HSP90 followed by the activation of lipid raft formation and by the activation of PI-3-kinase - Akt cytoprotective pathway.

In addition, the expression of HSP16.2 in various brain tumors and the positive correlation shown with their increase of anaplasia and grade, indicates that high HSP16.2 content in tumor cells can be responsible for resistance to cytostatic treatment. Furthermore, HSP16.2 could become a valuable marker for primary brain tumor diagnosis and the anti-apoptotic activity of HSP16.2 could become the target of drug therapy.

Much effort needs yet to be made to reveal the exact role and function of HSP16.2 in normal and tumor cells and in the development of diseases. However, the significance of this novel small stress protein in medical, especially anti-cancer research seems certain.

REFERENCES

[1] Haslbeck M, Franzmann T, Weinfurtner D, Buchner J. Some like it hot: the structure and function of small heat-shock proteins. *Nat Struct Mol Biol.* 2005 Oct;12(10):842-6.

[2] Sun Y, MacRae TH. The small heat shock proteins and their role in human disease. *FEBS J.* 2005 Jun;272(11):2613-27.

[3] Arrigo AP. sHsp as novel regulators of programmed cell death and tumorigenicity. *Pathol Biol.* (Paris) 2000 Apr;48(3):280-8.

[4] Sun Y, MacRae TH. Small heat shock proteins: molecular structure and chaperone function. *Cell Mol Life Sci.* 2005 Nov;62(21):2460-76.

[5] Haslbeck M. sHsps and their role in the chaperone network. *Cell Mol Life Sci.* 2002 Oct;59(10):1649-57.

[6] Haslbeck M. sHsps and their role in the chaperone network. *Cell Mol Life Sci.* 2002 Oct;59(10):1649-57.

[7] Haslbeck M, Buchner J. Chaperone function of sHsps. *Prog Mol Subcell Biol.* 2002;28:37-59.

[8] Kamradt MC, Lu ML, Werner ME, Kwan T, Chen F, Strohecker A, et al. The small heat shock protein alpha B-crystallin is a novel inhibitor of TRAIL-induced apoptosis that suppresses the activation of caspase-3. *Journal of Biological Chemistry,* 2005 Mar 25;280(12):11059-66.

[9] Lee GJ, Roseman AM, Saibil HR, Vierling E. A small heat shock protein stably binds heat-denatured model substrates and can maintain a substrate in a folding-competent state. *EMBO J.* 1997 Feb 3;16(3):659-71.

[10] Sreedhar AS, Csermely P. Heat shock proteins in the regulation of apoptosis: new strategies in tumor therapy: a comprehensive review. *Pharmacol Ther.* 2004 Mar;101(3):227-57.
[11] Hitotsumatsu T, Iwaki T, Fukui M, Tateishi J. Distinctive immunohistochemical profiles of small heat shock proteins (heat shock protein 27 and alpha B-crystallin) in human brain tumors. *Cancer*, 1996 Jan 15;77(2):352-61.
[12] Hermisson M, Strik H, Rieger J, Dichgans J, Meyermann R, Weller M. Expression and functional activity of heat shock proteins in human glioblastoma multiforme. *Neurology,* 2000 Mar 28;54(6):1357-65.
[13] Strik HM, Weller M, Frank B, Hermisson M, Deininger MH, Dichgans J, et al. Heat shock protein expression in human gliomas. *Anticancer Research,* 2000 Nov;20(6B):4457-62.
[14] Bellyei S, Szigeti A, Boronkai A, Pozsgai E, Gomori E, Melegh B, et al. Inhibition of cell death by a novel 16.2 kD heat shock protein predominantly via Hsp90 mediated lipid rafts stabilization and Akt activation pathway. *Apoptosis,* 2007 Jan;12(1):97-112.
[15] Vos MJ, Hageman J, Carra S, Kampinga HH. Structural and functional diversities between members of the human HSPB, HSPH, HSPA, and DNAJ chaperone families. *Biochemistry,* 2008 Jul 8;47(27):7001-11.
[16] Bellyei S, Szigeti A, Pozsgai E, Boronkai A, Gomori E, Hocsak E, et al. Preventing apoptotic cell death by a novel small heat shock protein. *Eur J Cell Biol.* 2007 Mar;86(3):161-71.
[17] Jakob U, Lilie H, Meyer I, Buchner J. Transient Interaction of Hsp90 with Early Unfolding Intermediates of Citrate Synthase - Implications for Heat-Shock In-Vivo. *Journal of Biological Chemistry,* 1995 Mar 31;270(13):7288-94.
[18] Varbiro G, Toth A, Tapodi A, Bognar Z, Veres B, Sumegi B, et al. Protective effect of amiodarone but not N-desethylamiodarone on postischemic hearts through the inhibition of mitochondrial permeability transition. *Journal of Pharmacology and Experimental Therapeutics,* 2003 Nov;307(2):615-25.
[19] Mukai M, Kusama T, Hamanaka Y, Koga T, Endo H, Tatsuta M, et al. Cross talk between apoptosis and invasion signaling in cancer cells through caspase-3 activation. *Cancer Research,* 2005 Oct 15;65(20):9121-5.
[20] Chen S, Bawa D, Besshoh S, Gurd JW, Brown IR. Association of heat shock proteins and neuronal membrane components with lipid rafts from the rat brain. *J Neurosci Res.* 2005 Aug 15;81(4):522-9.
[21] Jain S, Li YN, Kumar A, Sehgal PB. Transcriptional signaling from membrane raft-associated glucocorticoid receptor. *Biochemical and Biophysical Research Communications,* 2005 Oct 14;336(1):3-8.
[22] hah M, Patel K, Fried VA, Sehgal PB. Interactions of STAT3 with caveolin-1 and heat shock protein 90 in plasma membrane raft and cytosolic complexes - Preservation of cytokine signaling during fever. *Journal of Biological Chemistry*, 2002 Nov 22;277(47):45662-9.
[23] Sreedhar AS, Mihaly K, Pato B, Schnaider T, Stetak A, Kis-Petik K, et al. Hsp90 inhibition accelerates cell lysis - Anti-Hsp90 ribozyme reveals a complex mechanism of Hsp90 inhibitors involving both superoxide- and Hsp90-dependent events. *Journal of Biological Chemistry*, 2003 Sep 12;278(37):35231-40.

[24] Pozsgai E, Gomori E, Szigeti A, Boronkai A, Gallyas F, Jr., Sumegi B, et al. Correlation between the progressive cytoplasmic expression of a novel small heat shock protein (Hsp16.2) and malignancy in brain tumors. *BMC Cancer*, 2007;7:233.

[25] Assimakopoulou M, Varakis J. AP-1 and heat shock protein 27 expression in human astrocytomas. *J Cancer Res Clin Oncol.* 2001 Dec;127(12):727-32.

[26] Khalid H, Tsutsumi K, Yamashita H, Kishikawa M, Yasunaga A, Shibata S. Expression of the small heat shock protein hsp27 in human astrocytomas correlates with histologic grades and tumor growth fractions. *Cell Mol Neurobiol.* 1995 Apr;15(2):257-68.

[27] Strik HM, Weller M, Frank B, Hermisson M, Deininger MH, Dichgans J, et al. Heat shock protein expression in human gliomas. *Anticancer Res.* 2000 Nov;20(6B):4457-62.

In: Small Stress Proteins and Human Diseases
Editors: Stéphanie Simon et al.
ISBN: 978-1-61470-636-6

Chapter 3.4

SMALL HEAT SHOCK PROTEIN HSP27 (HSPB1), αB-CRYSTALLIN (HSPB5) AND HSP22 (HSPB8): APOPTOTIC AND TUMORIGENIC FUNCTIONS

Aurelie de Thonel[1], Guillaume Wettstein[1], Anne-Laure Joly[1], Nathalie Decologne[1], Adonis Hazoumé[1], and Carmen Garrido[1,2,*]

[1]INSERM U-866, Faculty of Medicine and Pharmacy, 7 Boulevard Jeanne d'Arc, 21033 Dijon, France

[2]Clinical Hematology Unit, CHU le Bocage, BP1542, 21000 Dijon, France

ABSTRACT

HSP27, αB-crystallin and HSP22 are ubiquitous small heat shock proteins whose expression is induced in response to a wide variety of physiological and environmental insults. They allow the cells to survive to otherwise lethal conditions. The protective function of these HSP can largely be explained by their anti-apoptotic properties. Various mechanisms at the molecular level have been proposed to account for their function in apoptosis: 1) they are molecular chaperons that prevent aggregation of denatured proteins. 2) They regulate the activity of the apoptotic cysteine aspartate proteases called caspases. 3) They are involved in the regulation of the intracellular red-ox state. 4) They have a function in actin polymerization and cytoskeleton integrity and 5) they appear to play a role in the proteasome-mediated degradation of selected proteins. In cancer cells these small HSP are often overexpressed and associated with increased tumorigenicity, cancer cells metastatic potential and resistance to chemotherapy. Altogether, these

* Carmen Garrido
Faculty of Medicine
INSERM U866, IFR100
7, boulevard Jeanne d'Arc
21033 Dijon, France
Phone: 33 3 80 39 33 53
Fax: 33 3 80 39 34 34
cgarrido@u-bourgogne.fr

properties suggest that these small heat shock proteins are appropriate targets for modulating cell death pathways.

INTRODUCTION

Small heat shock proteins (sHSP) are a widespread and diverse class of proteins. These are low molecular range proteins (15-42 kDa) that exhibit a highly dynamic structure as a result of their ability to form oligomeric structures ranging from 9 to 50 subunits (see chapter 2-1). The mammalian sHSP family, now referred to as the HSPB family, includes 10 members: HSP27/HSPB1, MKBP (myotonic dystrophy protein kinase-binding protein)/HSPB2, HSPB3, αA-crystallin/HSPB4, αB-crystallin/HSPB5, HSP20/HSPB6, cvHSP (cardiovascular)/HSPB7, HSP22/HSPB8, HSPB9 and ODF1 (outer dense fiber protein)/HSPB10. They are chaperones that protect against protein aggregation [1]. These proteins share a common C-terminal motif, the so-called α-crystallin domain (figure 1), and the ability to associate to form globular structures of up to 800 KDa [2]. In humans, only three sHSP, namely HSP27, αB-crystallin and HSP22 are well characterized stress inducible heat shock proteins.

HSP27, αB-crystallin and HSP22 are ubiquitously expressed and have a cytosolic and nuclear localization. Under non-stressful conditions, their protein level in all tissues is low, with the noticeable exception of the eye lens in which αB-crystallin is highly expressed [3] and the proliferating keratinocytes where HSP22 is overexpressed [4]. However, after different stresses, their expression is strongly induced. Stresses that transiently induce these sHSP include heat shock, anticancer drugs, radiations and oxidative stress. For example, a shift of the redox status to an oxidative state due to loss of protein thiols or glutathione can trigger or enhance sHSP expression in the cell. On the other hand, overexpressed HSP27 modulates ROS (Reactive Oxygen Species) intracellular content. Stress-induced expression of these sHSP is often accompanied by a nuclear or perinuclear localization of the protein and by their transient phosphorylation. The effect of this transient translocation in sHSP function(s) remains unknown [5, 6].

The expression of HSP27, αB-crystallin and HSP22 also can vary during development, cell cycle and cell differentiation. These processes can be considered as physiological stress for the cells [7, 8]. For example, HSP27 has been observed to accumulate in leukemic cells undergoing differentiation [6]. HSP27 has been proposed as a pre-differentiation marker [9, 10]because its accumulation occurs early in the differentiation process, just when the cells stop proliferating [11].

The expression of HSP27, αB-crystallin and HSP22 is regulated at the transcriptional level. As observed with *HSP70* gene, binding of heat shock factor (HSF) to the regulatory heat shock element (HSE) found in the promoter region of their genes can induce their transcription [12, 13]. Another transcription factor involved in their transcriptional regulation is the cyclic AMP responsive element binding protein [14] whereas the presence of an estrogen responsive element (ERE) in *hsp27* gene promoter accounts for its strong induction by steroid hormones, e.g. by estrogens [15]. However, other mechanisms of regulation of HSP27, αB-crystallin and HSP22 expression must exist since these proteins can accumulate in the absence of transcriptional induction, e.g. in response to the cytotoxic drug, cisplatin [16].

The transient accumulation of HSP27, αB-crystallin and HSP22 induced after the stress is suppose to help the cell to cope with these adverse situations. In the last 15 years, many reports have demonstrated a role for these proteins in apoptosis or programmed cell death. They can play a role in apoptosis at different levels: at early stages, by interfering with apoptotic-inducing events and, later on, by their association with key apoptotic proteins. In this chapter, we will describe how these proteins, at the molecular level, can affect the apoptotic pathways and the repercussion in cancer cells.

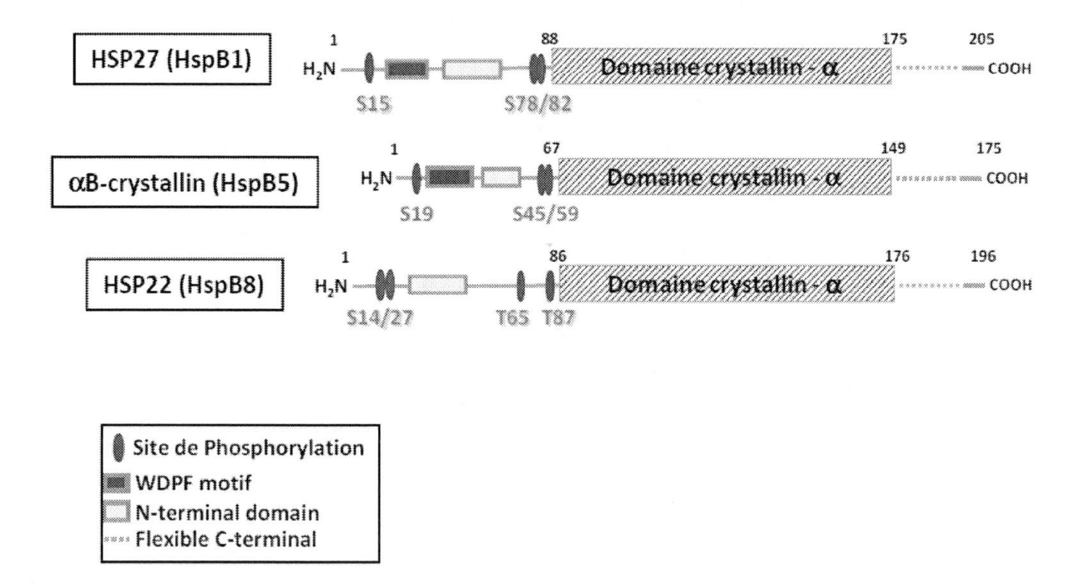

Figure 1. Structural domains of HSP27, HSP22 and αB-crystallin human proteins. Scheme indicates the N-terminal domain, flexible C-terminal domain, WDPF domain and the α-crystallin domain as well as the serine (S) and threonine (T) phosphorylation sites.

1. HSP27, αB-CRYSTALLIN AND HSP22 : REGULATION OF APOPTOTIC PATHWAYS

1.1. Apoptosis

Apoptosis or programmed cell death is responsible for the removal of unwanted or supernumerary cells during development, as well as for adult tissue homeostasis [17] (see also chapter 1-3). Apoptosis is also one of the cell death mechanisms triggered by cytotoxic drugs in tumor cells [18]. Two main pathways of apoptosis have been described both mediated by a family of cysteine proteases known as caspases: the intrinsic or mitochondrial pathway and the extrinsic or death receptors pathway. The two signal transducing cascades converge at the level of caspase-3, an effector caspase that leads to the typical morphologic and biochemical changes of the apoptotic cell (Figure 2).

The intrinsic pathway involves the production or activation of pro-apoptotic molecules upon intracellular stress signals. These molecules converge on the mitochondria to trigger the release of mitochondrial apoptogenic molecules under control of the Bcl-2 (B-cell lymphocytic-leukemia proto-oncogene) family of proteins. Bcl-2 proteins include anti-apoptotic members such as Bcl-2 and Bcl-xL, multi-domain pro-apoptotic members, mainly Bax and Bak [19, 20] as well as a series of BH3 domain-only pro-apoptotic proteins such as Bid that function upstream of Bax and Bak [21]. One of the released mitochondrial molecules is cytochrome *c*, which interacts with the cytosolic apoptotic protease activation factor-1 (Apaf-1) and pro-caspase-9 to form the apoptosome, the caspase-3 activation complex [22]. Two other mitochondrial proteins, Smac/Diablo and Htra2/Omi, activate apoptosis by neutralizing the inhibitory activity of the IAPs (Inhibitor of Apoptosis Proteins) that associate with and inhibit some of the activated caspases [23].

The extrinsic pathway is triggered through membrane proteins of the Tumor Necrosis Factor (TNF) receptor family known as death receptors and leads to the direct activation of the receptor-proximal caspase-8 or caspase-10 in the death-inducing signalling complex. Caspase-8 either directly activates the downstream cascade of caspases or cleaves Bid into an active truncated form, named tBid, that connects the extrinsic to the intrinsic apoptotic pathways through mitochondria permeabilization [24] (Figure 2).

1.1.1. HSP27: an Anti-apoptotic Protein

Overexpressed HSP27 protects against apoptotic cell death triggered by various stimuli, including hyperthermia, oxidative stress, staurosporine, ligation of the Fas/Apo-1/CD95 death receptor and cytotoxic drugs [6, 9, 25]. HSP27 has been shown to interact and inhibit components of both intrinsic and receptor induced apoptotic pathways (Figure 2). Experimental depletion of HSP27 suggests that HSP27 mainly functions as an inhibitor of caspase activation. Knock-down of HSP27 by siRNA (small interfering RNAs) induces apoptosis through caspase-3 activation [26, 27] (see chapter 4-2). This phenomenon can be explained by the reported ability of HSP27 to prevent the formation of the apoptosome and the subsequent activation of caspases [28] (Figure 2). Moreover, this may result from the capacity of HSP27 to sequester cytochrome *c* when released from the mitochondria into the cytosol, as demonstrated in leukemic U937, Jurkat T-lymhoma cells, and Pro-colon cancer cells treated with different apoptotic stimuli [29-31]. The heme group of cytochrome c is necessary but not sufficient for this interaction that involves amino-acids 51 and 141 of HSP27 and does not need the phosphorylation of HSP27 [30]. When expressed at high intracellular levels, HSPp27 can also interfere with caspase activation upstream of the mitochondria [31]. In L929 murine fibrosarcoma cells exposed to cytochalasin D or staurosporine, overexpressed HSP27 binds to F-actin preventing the cytoskeletal disruption and Bid intracellular redistribution that precede cytochrome c release [32, 31]. In multiple myeloma cells treated with dexamethasone, HSP27 has also been shown to inhibit the mitochondrial release of Smac [33] (Figure 2). HSP27 has been described to have important anti-oxidant properties. This is related to its ability to maintain glutathione in its reduced (non-oxidized) form, to decrease the abundance of ROS and nitric oxide. Consequently, in cells exposed to oxidative challenges, HSP27 expression reduces lipid peroxidation and neutralizes the toxic effects of oxidized proteins [5, 6, 34, 35]. In this way, HSP27 inhibits cisplatin-induced oxidation of thioredoxin by reversing the drug-induced inhibition of thioredoxin reductase activity [36]. The anti-oxidant properties of HSP27 seem particularly relevant for HSP27-mediated cytoprotection in neuronal cells and involve phosphorylated HSP27 [37].

HSP27 also inhibits apoptosis by regulating upstream signaling pathways. HSP27 has been described to interact with the protein kinase Akt, an association that is necessary for Akt activation in stressed cells (Figure 2). In turn, Akt can phosphorylate HSP27, thus leading to the disruption of HSP27-Akt complexes [38]. More recently, it has been shown in renal epithelial cells that HSP27, by promoting Akt activation through phosphatidylinositol 3-kinase, antagonizes Bax-mediated mitochondrial injury [39]. In mouse fibroblasts L929 cells, HSP27 has been shown to protect against cisplatin-induced apoptosis by enhancing Akt activation [36] and, in cardiomyocytes, HSP27 regulates ROS generation via Akt activation [40]. In acute myeloid leukemia HSP27 inhibits cytochrome release and subsequent apoptosis by blocking etoposide - induced phosphorylation of p38 and c-Jun [41].

The actin cytoskeleton is modulated by both the spatial arrangement as well as the polymerization dynamics of its different elements [42]. Overexpression of HSP27 increases the stability of F-actin microfilaments and prevents its aggregation during exposure to stress [32, 43, 44]. Although the exact mechanism and consequences of F-actin stabilization by HSP27 are not characterized, it is known that HSP27 directly binds to F-actin and this has been reported to affect mitochondria membrane structure and thereby the release of mitochondrial apoptogenic molecules such as cytochrome c [31].

HSP27 can form complexes with the estrogen and glucocorticoids receptors. Androgen binding to the androgen receptor induces rapid HSP27 phosphorylation that, in turn, facilitates genomic activity of the androgen receptor, thereby enhancing prostate cancer cell survival [45].

At the death receptor level, HSP27 also affects one of the downstream events elicited by the stimulation of CD95/Apo1/Fas (Figure 2). The phosphorylated form of HSP27 directly interacts with Daxx, which connects Fas signaling to the protein kinase Ask1 that mediates a caspase-independent cell death [46]. HSP27 has recently been reported to induce cytoprotection in prostate LNCaP tumor cells through the activator of the transcription-3 (STAT3), a cellular factor involved in oncogenic signaling pathways [47]. This transcription factor is constitutively active in most tumors and controls the expression of key genes involved in apoptosis inhibition such as those encoding Bcl-xL and survivine. HSP27 interacts with and inhibits STAT3, and the protective effect induced by HSP27 overexpression in those cells was attenuated by Stat3 knockdown [26]. Cytoprotection by HSP27 may also rely on its capacity to favour the proteasomal degradation of certain proteins under stress conditions. Two among the protein targets of HSP27 are the NF-κB (Nuclear factor-κB) inhibitor IκBα and the cyclin-dependent kinase inhibitor $p27^{kip1}$. Finally, HSP27 participates in the (de)phosphorylation reaction that controls the activity of the splicing regulator SRp38, thereby restoring the splicing activity inhibited after heat shock [48].

In conclusion, HSP27 is a powerful anti-apoptotic protein. Since several apoptotic stimuli induce in the cell the accumulation of misfolded aggregated proteins, HSP27 first protective role may therefore be related to its well known *in vitro* chaperone function: the inhibition of protein aggregation [49, 50]. Then, as summarised in Figure 2, HSP27 acts at different levels of the apoptotic pathways through the interaction with distinct proteins. The affinity of HPS27 for one partner seems to be defined by the phosphorylation and/or oligomerization status of HSP27 (and also probably its cell localization) that at its turn may depend on the particular stress stimuli that a given cell has received.

1.1.2. αB-crystallin: Strong Anti-apoptotic Properties

The overexpression of αB-crystallin confers protection against a large panel of apoptotic stimuli while its silencing sensitizes cells to apoptosis [51-53]. Several of these stimuli induce αB-crystallin overexpression, providing an example of how pro-apoptotic stimuli, delivered below a threshold level, can elicit protective responses. In glioma cells, microtubule disrupting drugs induce αB-crystallin expression via phosphorylation dependent reactions that are sensitive to staurosporine [54].

The anti-apoptotic effect of αB-crystallin overexpression has been demonstrated in epithelial lens cells and many other cell types. Caspase-3 processing inhibition by αB-crystallin appears to account for this anti-apoptotic effect in most cells [51, 53, 55, 56]. Addition of purified αB-crystallin to cell free extracts was shown to prevent pro-caspase-3 processing by either cytochrome c-dATP or caspase-8. This small stress protein interacts with the partially processed caspase-3, thus preventing further cleavage required for the protease to be activated. Pro-caspase-3 processing is inhibited by αB-crystallin after separation of the small subunit from the partially processed p24 (p20) fragment composed of the large subunit and the prodomain. αB-crystallin was shown to co-immunoprecipitate with this partially processed fragment of pro-caspase-3 [51, 53, 55]. This effect of αB-crystallin might be determinant to explain the anti-apoptotic function of some Bcl-2 family proteins as recently demonstrated in glioma multiforme, a highly aggressive brain tumor in which the nuclear oncoprotein Bcl2L12 (Bcl2-like 12 protein) is universally overexpressed. Stegh et al. demonstrated that αB-crystallin was an induced Bcl2L12 oncoprotein that, by selectively binding to pro-caspase 3 and its cleavage intermediates *in vitro* and *in vivo*, enabled Bcl2L12 to block the activation of this caspase [57].

In an ischemia/reperfusion mouse model, overexpresion of αB-crystallin not only protects myocardium cells from apoptosis but also prevents necrosis [56]. The phosphorylation of αB-crystallin on serine Ser59 appeared to be both necessary and sufficient to provide maximal protection of cardiac myocytes from apoptosis [56]. In cardiac myocytes, the hyperosmotic stressor sorbitol was shown to activate the p38/MAPK pathway through MKK6, thus leading to induction of αB-crystallin gene expression and phosphorylation of the protein on Ser59, two events that contribute to sorbitol-induced cell death inhibition [58]. Upstream of mitochondria, αB-crystallin has been shown to bind to pro-apoptotic Bax, Bcl-Xs [59] and p53 [60] to prevent their translocation to the mitochondria during staurosporine- and hydrogen peroxide-induced apoptosis.

αB-crystallin has been shown to protect cells also from death induced by proteasome inhibitors. Proteasome inhibition induces accumulation of unfolded protein which are toxic to the cells. In response to proteasome inhibitors and other stressful conditions, αB-crystallin relocalizes, like HSP27, to the cytoskeleton and binds to microtubules via the microtubule associated proteins (MAPs), which may protect the cells from damaged intracellular proteins by sequestering these proteins on the cytoskeleton [61].

Cell survival induced by αB-crystallin is associated to its role in maintenance of microfilaments integrity. One major target of the chaperone activity associated to αB-crystallin appears to be the type III intermediate filaments [62]. Bennardini et al. demonstrated that αB-crystallin could interact *in vitro* with desmin and actin filaments and their binding affinity increases after heat treatment [63]. *In vivo*, αB-crystallinopathies result from the misfolding and progressive aggregation of the mutated αB-crystallin to which subsequently associate with desmin filaments to form αB-crystallin/desmin aggregates [64] (see chapters 1-2 and 2-3). αB-crystallin regulates actin filaments dynamic *in vivo*. Reducing αB-crystallin using anti-sense strategy leads to disruption of the actin microfilament network [65]. The association of αB-crystallin with actin, that depends on its phosphorylation, in addition to protect the organization of the cytoskeletal network, helps the pinocytosis maintenance, a physiological function essential for cell survival [66].

We can conclude that anti-apoptotic functions of αB-crystallin depend mainly on its ability to block caspase-3 activation and to protect microfilaments integrity. These functions seem to be modulated by αB-crystallin phosphorylation and cellular redistribution.

1.1.3. HSP22: Anti- or Pro-apoptotic?

HSP22 has been suggested to trigger or block apoptosis in a cell type- and dose-dependent manner. In a melanoma and sarcoma cell lines, induction of a caspase- and p38MAPK-dependent apoptosis by the demethylating agent 5-aza-2-deoxine was associated with HSP22 overexpression [67]. However, HSP22 overexpression did not trigger apoptosis in HEK293 cells [68]. Moreover, HSP22, in keratinocytes [4] and hippocampal neurons, blocks apoptosis and this effect involves the activation of MEK/MAPK (Mitogen Extracellular Kinase/Mitogen Activated Protein kinase) survival pathway [69]. HSP22 is the eukaryotic homologue of the viral protein ICP10, an enzyme involved in the "immortalization" of cells infected by herpes simplex virus 2. Some reports have demonstrated that the protein kinase (PK) domain of ICP10 activates the Ras/MEK/MAPK and other survival pathways and thereby, can modulate apoptosis [70]. ICP10 PK inhibits apoptosis through the activation of the ERK (Extracellular Regulated signal Kinase) survival pathway which results in Bag1 upregulation and overrides the pro-apoptotic JNK/c-Jun signal induced by other viral proteins [70, 71]. Therefore, it has been suggested that, as for ICP10, HSP22 anti-apoptotic role requires a functional HSP22 PK activity [71] because protection from apoptosis was also associated with MEK/MAPK activation [4, 68, 72].

In a swine model and in human hibernating myocardium, HSP22 is upregulated and has been involved in cytoprotection by blocking apoptosis [73]. The protective effect of HSP22 can be at least partially explained by the direct binding and activation of phosphoglucomutase by HSP22 that results in stimulation of glycogen synthesis in the heart [74]. However, in cardiac myocytes, low level expression of HSP22 resulted in increased cell size and inhibition of apoptosis while a high level induced apoptosis [75]. This has been described as a dual-function kinase activity of HSP22 in cardiac cells. At low levels, HSP22 kinase independent activation of Akt results in cell survival [75, 76]. At higher levels HSP22, through a protein-kinase-dependent mechanism, particularly by binding and inhibiting casein kinase 2, this chaperone induces cell death. The HSP22 mutant K113G lacked pro-apoptotic activity because it does not interact with and thereby, does not inhibit casein kinase 2 [75]. Although this is an interesting explanation, the intrinsic protein kinase activity of HSP22 remains questionable, even at high levels. The kinase activity of HSP22 reported by some authors [71], has been argued that might be the result of a contamination [77].

Independently of this controversy concerning HSP22 intrinsic kinase activity, in most cancer cells HSP22 possesses anti-apoptotic and tumorigenic properties [78]. Expression of HSP22 induced anchorage independent and increased cell proliferation and protection to apoptosis of breast cancer cells [79]. In melanoma HEK 293 cells, overexpression of HSP22 is accompanied by cell resistance and anchorage-independent growth [80]. Further, in breast cancer MCF-7 cells, estradiol resulted in an accumulation of HSP22 [81] suggesting its association with cancer cells resistance to therapeutic agents.

Altogether we can conclude that HSP22, depending on the cellular model and experimental conditions can inhibit or induce apoptosis. A possible explanation, not involving the debatable HSP22 intrinsic kinase activity, is that HSP22 depending on its level of phosphorylation/oligomerization and maybe its hetero-oligomerization with other sHSP, like HSP27 or αB-crystallin, may bind to a distinct pattern of proteins (qualitative or quantitatively different) with a different result in the outcoming of the cell. Confirming this hypothesis it has recently been shown that HSP22 has a very flexible structure that enables its interaction with many cellular proteins [77] (see chapters 1-1 and 2-5).

1.2. HSP27, αB-crystallin and HSP22 and the Ubiquitin/Proteasome System

Another recently identified function of sHSP that could account for their protective effect towards cell death is its ability to facilitate the degradation of specific proteins by the ubiquitin/proteasome system, thus to play a role in the so-called "protein triage" that take place in cells recovering from stress or induced to differentiate [82] . The ubiquitination system labels proteins for degradation by the 26S proteasome, a multicatalytic protease composed by a catalytic 20S and two regulatory 19S subunits [83] and is essential for proper dissambly-assembly of protein complexes to prevent undesirable interactions and aggregation. In this respect, the lack of HSP27 or αB-crystallin during early differentiation induces aberrant cell differentiation and/or massive apoptosis [53, 84, 85].

We have demonstrated that HSP27 was able to enhance the catalytic activity of the 26S proteasome machinery and to increase the degradation of several ubiquitinated proteins in response to stressful stimuli. In contrast to HSP70 and HSP90, HSP27 directly interacts with ubiquitin [86]. This ability to directly interact with ubiquitin could account for the described co-localization of HSP27 with ubiquitinated proteins [87, 88] in cytoplasmic "aggresomes" that characterise neurodegenerative diseases [89]. A common feature of neurodegenerative disorders such as Alzheimer disease [90], Parkinson disease [91], Alexander disease [92] and Creutzfeld-Jacob syndrome [91, 93] is the deposition of improperly folded proteins in fibers, inclusion bodies and plaques in the nervous system [94]. This event has been associated with an increased expression of HSP27 and αB-crystallin that were typically found in association with insoluble protein-aggregates and ubiquitin [95].

αB-crystallin has been shown to interact with one of the 14 subunits of the 20S proteasome, namely the C8/α7 subunit [96]. However, this interaction does not seem to modulate the proteasome activity. When Ser19 and Ser45 of αB-crystallin are phosphorylated, the protein can interact with FBX4, which is a component of the ubiquitin-ligase SCF (SKP1/CUL1/F-box) [97].

αB-crystallin together with FBX4 governs ubiquitination substrate specificity. One of the substrates is the cell cycle protein, cyclin D1 [98, 99].

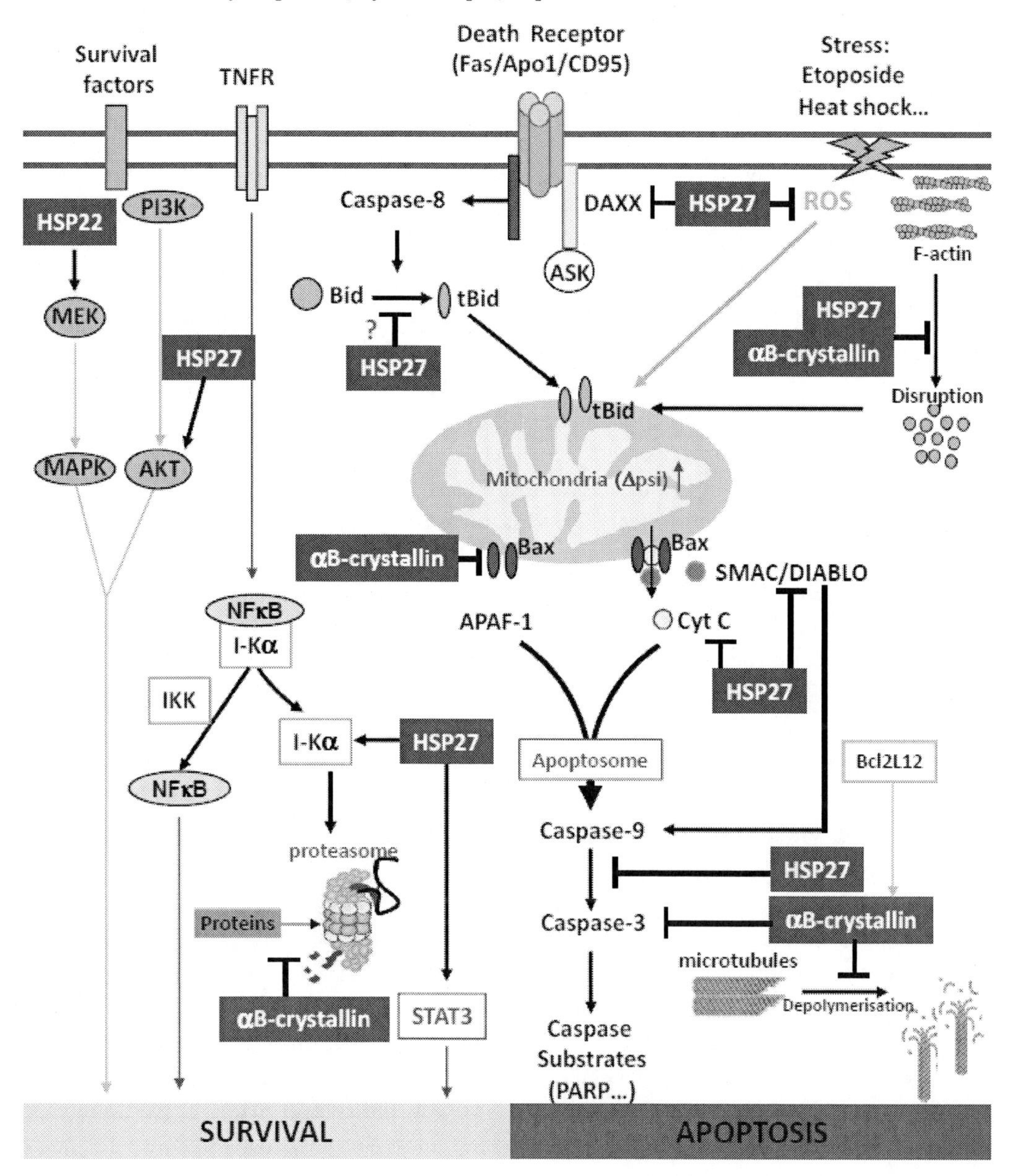

Figure 2. Regulation of apoptotic cell death by HSP27 and αB-crystallin.

- HSP27: Upstream of the mitochondria, HSP27 binds to Daxx preventing Ask1 activation and thereby, counteracting apoptosis induced by Fas/Apo1/CD95. HSP27 affects cytochrome c release from the mitochondria by inhibiting ROS production, F-actin disruption and thereby Bid redistribution. Downstream of the mitochondria, HSP27 prevents the apoptosome formation by direct interaction with cytochrome c in cytosol and inhibition of Smac/DIABLO release. HSP27 prevents also the caspase cascade by negatively modulating the caspase 3 activity. Furthermore, HSP27 can increase the resistance of microtubule against depolymerisation. Finally, HSP27 activates pro-survival pathways through Akt activation and the proteasome-mediated degradation of I-κBα, which result in increased NF-κB activity.

- αB-crystallin: αB-crystallin, at the mitochondria, inhibits pro-apoptotic Bcl2 members, like Bax, and subsequently the release of cytochrome c. At a post-mitochondria level, αB-crystallin interacts with pro-caspase 3 blocking its activation and can protect microtubules from depolymerization. Finally, αB-crystallin can enhance ubiquitination of yet unknown proteins that may contribute to its anti-apoptotic activity.

- HSP22: HSP22 can protect from apoptosis through the activation of the MEK/MAPK pathway.

The ability of overexpressed HSP27 to increase the degradation of specific proteins through the ubiquitin/proteasome pathway may account for its anti-apoptotic effect by modulating the expression of death regulatory proteins. In different cancer cells treated with etoposide or TNF-α, HSP27 favors the ubiquitination and subsequent degradation of IκBα (Figure 2). As a consequence, there is an increase in the activity of the survival factor NF-κB that contributes to the overall protective effect of HSP27 [100] (Figure 2). NF-κB is involved in the expression of several anti-apoptotic proteins such as Bcl-2, Bcl-xL ans c-IAPs. Under stressful conditions (serum depletion, staurosporine treatment), HSP27 also stimulates $p27^{kip1}$ ubiquitination/degradation. As a consequence, cells do not accumulate in the Go/G_1 phase of the cellular cycle but in the S-phase. Therefore, cells overexpressing HSP27 may be more ready to re-start proliferating once the stress conditions are over [11]. Another target of HSP27 is TRAF6 (TNF associated factor 6). HSP27 interact with TRAF6 and increases its ubiquitination thereby controlling IL-1β (Interleukin-1 beta) stimulation [101].

HSP22 also participates in protein quality control by a proteasome-unrelated pathway. HSP22, together with Bag3, induce the phosphorylation of the alpha subunit of the translation initiator factor, eIF2 which, in turn, causes a translational shut down and stimulates autophagy [102].

1.3. Phosphorylation and Oligomerization: Regulation of the Apoptotic Function of HSP27, αB-crystallin and HSP22

The chaperone function of high molecular weight HSP is regulated by ATP (Adenosine triphosphate). Concerning sHSP, it has been shown that αB-crystallin, like HSP27, binds *in vitro* several non-native proteins per oligomeric complex, thus representing the most efficient chaperones in terms of quantity of substrate binding [103, 104]. In some cases, the release of substrate proteins from the small heat shock protein complex is achieved in cooperation with HSP70 in an ATP-dependent reaction, suggesting that the role of sHSP in the network of chaperones is to create a reservoir of non-native refoldable protein. Although an influence of ATP on the chaperone function of αB-crystallin has been described [105], analysis of the chaperone function of HSP27, αB-crystallin and HSP22 *in vitro* was found to be completely independent of ATP binding or hydrolysis [106]. Therefore, other mechanisms to modulate these chaperones' function must exist (see chapters 2-1 and 4-3).

An intrinsic mechanism of regulation is the oligomerization of the protein [82]. The dynamic organization of HSP27, αB-crystallin and HSP22 oligomers appears to be a crucial factor which controls the activity of these proteins. These sHSP can form large oligomers of up to 800 kDa. The dimer seems to be the building block for these multimeric complexes. Only HSP22 has been reported to exist as a monomer [76, 107]. Oligomerization is a highly dynamic process that depends on the phosphorylation status of the protein. Phosphorylation induces modifications both in the oligomers size and the chaperone-like activity [106]. Human HSP27 is phosphorylated at serine Ser15, Ser78 and Ser82 [95, 108] and mouse at Ser15 and Ser86 [109] (Figure 1 and see chapter 3-2). This phosphorylation is a reversible process catalyzed by the kinase MAPKAP 2/3, downstream of p38 kinase and protein kinase C, more specifically its delta isoform [110]. Phosphorylation of HSP27 is observed in response to a variety of stimuli including differentiating agents, mitogens, inflammatory cytokines such as TNFα and IL-1β, hydrogen peroxide and other oxidants. Selectivity of phosphorylation sites in HSP27 for these enzymes is apparently absent [108, 110, 111]. Dephosphorylation of HSP27 *in vivo* is catalyzed by protein phosphatase 2A, although protein phosphatase 2B also demonstrated activity *in vitro* [112].

αB-crystallin is also phosphorylated at three serine residues, namely Ser19, Ser45 and Ser59 (Figure 1). This phosphorylation has been shown to occur in response to heat, arsenite, phorbol esters, okadaic acid, anisomycin, high concentration of sorbitol [113] and agents that result in disorganization of microfilaments or microtubules networks [114]. In the case of αB-crystallin, the *in vivo* phosphorylation of distinct serine residues appears to be regulated independently, i.e. p44/42 MAP kinase and MAPKAP kinase -2 are responsible for Ser45 and Ser59 phosphorylation, respectively [115]. The kinase responsible of the phosphorylation of Ser19 is still unknown.

HSP22 has also been shown to be phosphorylated . HSP22 is phosphorylated by protein kinase C (at residues Ser14 and Thr65), by p44 MAPK (at residues Ser27 and Thr87) and by casein kinase 2 but not by MAPKAP kinase -2 [67, 116, 117] (Figure 1). Moreover, HSP22, initially designed as H11 serine threonine protein kinase, was thought to have an autophosphorylation activity [118]. This is now a controversial issue, some authors considering that the kinase activity found was the result of a contamination [77].

The phosphorylating events are associated with changes in the oligomeric structure of the sHSP complexes. At least in the case of HSP27, phosphorylation induces a shift toward small oligomers [119, 120]. We have demonstrated that, while HSP27 phosphorylation is important for the regulation of its oligomerization in exponentially growing cells cultured *in vitro*, the formation of large oligomers occurs independently of the phosphorylation status of the protein in cells grown at confluence *in vitro* or grown *in vivo*. This indicates that cell-cell contact induces the formation of large oligomers independently of the phosphorylation status of the protein [121]. HSP27 biochemical functions may be modulated by its oligomeric (and phosphorylated) status since some biological activities of HSP27 are associated with small oligomers while others require the formation of large oligomers. The dissociated and phosphorylated form of HSP27 shows no chaperone activity [35] and its thermoprotective activity is strongly decreased [122]. We have demonstrated, *in vitro* and *in vivo,* that large non-phosphorylated oligomers of HSP27 are responsible for the caspase-dependent anti-apoptotic effect of this chaperone [121]. Other groups have demonstrated that large oligomers are needed for HSP27 anti-oxidant activity [35]. Large non-phosphorylated HSP27 oligomers bind to actin and the two proteins dissociate upon HSP27 phosphorylation [123]. In contrast, phosphorylated dimers of HSP27 directly associate with Daxx to negatively interfere with the extrinsic pathway of apoptosis [46]. Moreover, HSP27 in its small oligomeric form displays an affinity for ubiquitin chains and accelerates the degradation of certain proteins under stress conditions [11]. These results suggest that the state of oligomerization/phosphorylation of the protein alters HSP27 conformation and, hence, determines its capacity to interact with its different partners.

In contrast to HSP27, the protective effects of αB-crystallin appear to be mediated mostly by the phosphorylated form of the protein. In this way, the phosphorylated protein is involved in caspase-3 maturation [124] and interacts with actin, to help maintaining the functional integrity of the cells subjected to a stress [66].

We still do not have a deep knowledge about sHSP globular oligomeric structures, mainly because of their heterogeneity in size, dynamic properties and the ability of these sHSP to form hetero-complexes among them. These suggest that although there are only three proteins, many different structurally independent chaperone complexes may exist with distinct molecular targets which may be tissue specific. An example of the formation *in vivo* of this heteroligomeric structures connecting these three small chaperones is found in a familial form of desmin-related myopathy. This pathology is related to a missense mutation of αB-crystallin (R120G) that inhibits its chaperone function in desmin filament assembly and induces accumulation of both αB-crystallin mutant and desmin in inclusion bodies. It has been shown that wild-type αB-crystallin, HSP27 and HSP22 could prevent the formation of these abnormal structures by co-oligomerizing with the mutated αB-crystallin R120G, thus promoting an efficient folding of desmin filaments [125, 126].

Altogether we can hypothesize that sHSP modulate their interaction with various cellular partners, and thereby their cellular function, by shifting towards small or large homo- or hetero-oligomers. Phosphorylation, cell-cell contact and probably other factors regulate this equilibrium.

2. sHSP and Cancer

2.1. HSP27, αB-crystallin and HSP22: Tumorigenicity and Cancer Cells Resistance

2.1.1. Role of HSP27

HSP27 anti-apoptotic role is directly linked to its tumorigenic properties [30]. Rat colon cancer cells engineered to express human HSP27 were observed to form more aggressive tumors in syngeneic animals than control cells and the increase in tumorigenicity correlated with a reduced rate of tumor cell apoptosis [127]. Conversely, downregulation of HSP27 by anti-sense constructs enhanced apoptosis and delayed tumor progression [127, 128]. Overexpressed HSP27 did not increase the tumorigenicity of these rat colon cancer cells, nor did affect their survival rate *in vivo* when inoculated into immunodeficient animals. Hence, HSP27 is supposed to increase the ability of some cancer cells to resist to and evade from the apoptotic processes mediated by the immune system. However, HSP27 overexpression was reported to increase the metastatic potential of human breast cancer cells inoculated into athymic (nude) mice [43, 129]. Cell migration, a prerequisite for cancer invasion and metastasis, has also been shown to be affected by HSP27. Knockdown of HSP27 using small interfering RNA inhibited human breast cancer cells migration [130]. Han et al. have identified anti-migratory compounds by chemical library screening [130]. One of the inhibitors, KRIBB3, inhibited both tumor cell migration and invasion. They found that

KRIBB3 bound to HSP27 and inhibited cell migration by blocking PKC-dependent HSP27 phosphorylation [130]. That HSP27 binds and co-localizes with different components of the cytoskeleton may also contribute to its function in cell migration [43, 131]. In this way, HSP27 phosphorylation, through its function in the rearrangement of F-actin, plays a role in migration of vascular smooth muscle cells [132].

Clinically, in a number of cancers such as breast cancer, ovarian cancer, osteosarcoma, endometrial cancer and leukemias, an increased level of HSP27, relative to its level in non-transformed cells, has been detected [133]. In colorectal cancer patients, there is a clinical relationship between HSP27 expression levels and Irinotecan resistance, an inhibitor of topoisomerase I [134]. In ovarian tumors, HSP27 expression increases with the stage of the tumor [135]. Upregulated HSP27 in human breast cancer cells can reduce Herceptin susceptibility [136]. Expression of HSP27 is associated with poor clinical outcome in intra-hepatic cholangiocarcinoma [137] and with a more aggressive clinical behaviour in oligodendroglial tumors [138]. In addition, the pattern of HSP27 phosphorylation in tumor cells is different from that observed in primary non-transformed cells [139]. Consequently, the diversity of HSP27 phospho-isoforms may also represent a useful tumor predictor marker as demonstrated in human renal cell carcinomas and in myelodysplasic syndrome [140]. In prostate cancer cells, HSP27 participates in a phosphorylation-dependent manner, in survival through its role facilitating androgen receptor transactivation [45]. Confirming these results, anti-sense knockdown of HSP27 delays prostate cancer xenograft and androgen-independent progression [26, 128]. High levels of HSP27 are observed in metastatic tissues compared to non metastatic tissues suggesting the involvement of this chaperone in metastasis formation [141]. In this way, HSP27 expression is associated with metastatic disease and bad prognosis in breast and ovarian cancer [135, 142]. More detail can be found in chapter 3-1.

Apart from its protective role in the cytosol, HSP27 have been found in the extracellular space or on the plasma membrane and, by analogy with other HSP such as HSP70, HSP27 might play a role in the stimulation of the immune system. The immunogenic function of extracellular HSP is mainly described through their role chaperoning antigenic peptides. The word "chaperokine" has been evoked. In humans, the presence of HSP27 in the serum has been described associated with stress or pathological conditions including cancer [143]. Tumor cells were identified as a natural source for the extracellular location of HSP27. Previous studies identified both HSP27 and HSP70 expressed at the cancer cell surface during tumor growth and metastatic spread [144]. HSP27 has been found in the serum of breast cancer patients [82, 145]. Phosphorylated HSP27 has also been identified as being associated with cell membrane in migrating cells [146]. In different cancer cell lines (neuroblastoma, lung and colon adenocarcinoma) a high expression on the cell surface of different HSP, including HSP27, have been described [147]. However, although all these reports described the presence of HSP27 in the extracellular medium of cancer patients, none of them associate this presence with a good patient outcome. This suggests that if HSP27 has an immune stimulating function, its anti-apoptotic and tumorigenic properties somehow subvert the tumor-specific response.

2.1.2. Role of αB-crystallin

Overexpression of αB-crystallin has been observed in some human malignant tumors, for example in glial tumors such as astrocytoma, glioblastoma and oligodendroglioma [148], in oral squamous cell carcinomas [149] and in renal carcinoma tumors [150]. Abnormal high level of αB-crystallin has also been detected in pre-invasive ductal carcinoma that correlated with poor clinical outcome of the patients. In basal-like breast carcinomas, αB-crystallin is abundantly expressed. This abundant transcription of αB-crystallin seems due to the oncogenic transcription factor Ets-1, a member of the ETS family [151]. αB-crystallin has been described as a novel marker of invasive basal-like and metaplastic breast carcinomas [152], and as a predictor of resistance to neo-adjuvant chemotherapy in breast cancer [153]. Highly proliferating "normal" fibroblasts do not express αB-crystallin while proliferative neoplastic cells do express the protein, suggesting a transformation-associated deregulation of its expression [154]. Further, αB-crystallin is described as a novel oncoprotein that predicts poor clinical outcome in breast cancer [155, 156]. Phosphorylation of αB-crystallin inhibited neoplastic changes and invasive properties of breast cancer cells [157]. Indeed, Ser 59 phosphorylation reduces the oligomerization and anti-apoptotic properties of αB-crystallin [124, 155]. It can therefore be concluded that, as for HSP27, αB-crystallin may contribute to the aggressive behavior of cancer cells. However, the differential expression of HSP27 and αB-crystallin in anaplastic thyroid carcinomas and brain cancer suggests different involvement of theses sHSP at least in these two pathologies [158, 159]. An explanation would be the involvement of αB-crystallin in the ubiquitination and degradation of cyclin D1, a cell cycle protein overexpressed in human cancer [99]. In addition to the contribution of these sHSP in cancer, they are also responsible of cancer cell resistance to cytotoxic drugs. Chemotherapy, particularly cisplatin, vincristin and colchicines has been shown to enhance the expression of HSP27 and αB-crystallin in colon cancer [133], glioma [160] and neuroblastoma cells [161]. HSP27 has been shown to be induced by estrogens and glucocorticoids [162]. Collectively, HSP27 and/or αB-crystallin accumulation might impair the efficiency of the clinical treatment using chemotherapeutic agents.

2.1.3. Role of HSP22

In contrast to HSP27 and αB-crystallin, little is known about the role of HSP22 in cancer. HSP22 gene was detected among genes whose expression was increased in invasive lesions. Expression of HSP22 induced anchorage independent and increased cell proliferation and protection of breast cancer cells to apoptosis [79]. HSP22 has been involved in the biology of estrogen receptor-positive breast cancer cells. HSP22 is strongly induced by estrogen and the metallo-estrogen cadmium in estrogen receptor-positive breast cancer cells whereas it is not induced in estrogen receptor-negative cells [163]. HSP22 could play a role in tumorigenicity by its interaction with the protein Sam68, a partner of the oncogene c-Src [164].

Apart from its intracellular localization, HSP22 has also been shown to localize in the plasma membrane. HSP22 interacts with the membrane and this interaction leads to stable binding and conformational changes [165]. Whether, this membrane HSP22 has an immunological function, as reported for other externally expressed HSP, remains to be determined. Confirming this hypothesis, HSP22, together with HSP27 and αB-crystallin, have been reported to be TLR4 ligands (Toll-like receptor 4), that therefore, can activate dendritic cells [166].

2.2. The Inhibition of HSP27 and αB-crystallin in Cancer Therapy

HSP27 and αB-crystallin should be eliminated or their activity impaired in order to sensitize cancer cells to die. Indeed, until today, no report has described a positive role for these proteins in cancer cells, such as a better tumor-antigen recognition when expressed in the cell surface as already reported for HSP70 [167]. Hence, experiments have been performed using anti-sense or nucleotide based therapies in order to block HSP27 or αB-crystallin expression.

HSP27 anti-sense oligonucleotides, 2'-O-(2-methoxyethyl) (OGX 427, OncoGenex, Vancouver, Canada), particularly studied in prostate and bladder cancer, have been shown to enhance apoptosis and delay tumor progression in experimental rodent models. In prostate cancer xenografts, HSP27 anti-sense oligonucleotides delayed their growth [26, 45, 128]. This effect might be explained by the fact that HSP27 increases androgen receptor degradation and, thereby, increases prostate cancer cells apoptosis [45]. HSP27 is involved in resistance against anti-cancer agents. *In vitro*, in human bladder cancer UMUC-3 cells, OGX 427 enhanced sensitivity to paclitaxel and, in an orthotopic mouse model of high grade bladder cancer, intravesically administered OGX 427 showed an anti-tumor activity [168, 169]. Similarly, knockdown of HSP27 significantly increased paclitaxel-elicited caspase activation and apoptosis in different melanoma cell lines [170]. It has recently been demonstrated that HSP27 gene therapy offers a potential adjuvant to radiation based therapy. HSP27 depletion, by anti-sense or interfering RNA strategies, increases the sensitivity of different radio-resistant cancer cells to radiotherapy [171]. In dexamethasone-resistant cell lines, HSP27 is overexpressed. Its downregulation by siRNA restores the apoptotic response to dexamethasone of prostate cancer cells by triggering caspase activation [27]. Similarly, in colorectal cancer cells, siRNA-mediated downregulation of HSP27 increases Irinotecan sensitivity [172]. More detail about RNAi approache is given in the chapter 4.2. The degradation of still unknown tumorigenic or metastatic client proteins of HSP27 or αB-crystallin should also be considered. We have demonstrated that HSP27 participates in the process of ubiquitination/proteasomal degradation of certain proteins and this effect contributes to its protective functions by enhancing the activity of proteins, like NF-κB [86]. The proteasome inhibitor Velcade® (PS-341), currently tested in multiple myeloma clinical trials, has been shown *in vitro* to induce apoptosis in several cancer cell lines. HSP27 confers Velcade® resistance and an HSP27 antisense approach sensitizes cells to Velcade®-induced apoptosis [33, 173]. It is therefore tempting to conclude that a combinational therapy using Velcade® together with an inhibitor of HSP27 will increase the chemosensitization effect of both products. Antibodies-based inactivation of HSP27 may be an interesting approach. HSP27 antibodies enter in neuronal cells from human retina by an endocytic mechanism and, as a result, facilitate apoptotic cell death [174]. Finally, the peptide aptamers or drugs that bind a specific structural organization of HSP27 and αB-crystallin or block a particular function of these proteins (ie anti-oxidant function), may be very useful to block HSP27 or αB-crystallin ability to counteract the killing efficiency of many anti-cancer drugs and X-ray irradiation. Details on this strategy can be found in the chapter 4.3.

Conclusion

Recent evidences indicate several connections between the three small HSP, HSP27, αB-crystallin and HSP22 and the cell death machinery. These connections involve the chaperone functions of these proteins, their ability to interact with death regulatory proteins and probably their ability to connect the protein folding machinery to the protein degradation pathway. Ongoing studies will indicate whether these proteins are useful targets for therapeutic manipulation of the apoptotic pathways, for example to induce cancer cells death or to sensitize them to current therapeutic approaches. The structural complexity of these molecules makes difficult the search of therapeutical molecules that specifically neutralize them.

Acknowledgments

Our group is supported by the "Comité de la Nièvre de la Ligue contre le cancer" and the "Fondation de France contre la Leucémie."

Abbreviations

HSP: Heat Shock Proteins
sHSP: small Heat Shock Proteins
Apaf-1 : Apoptosis Protease Activating Factor-1
TNFα: Tumor Necrosis Factor alpha
IL1β: interleukine 1 beta
HSF1: Heat Shock Factor-1, Bcl-2 (B-cell lymphocytic-leukemia-2)
Bcl2L12 (Bcl2-like 12 protein)
HSE: Heat Shock Element
ERE: estrogen responsive element
TLR4: Toll-like Receptor-4
Caspase: Cyctein Aspartate Protease
PI-3K: Phosphoinositol-3 Kinase
NFkB: Nuclear Factor kappa B, MEK: Mitogen Extracellular Kinase
ERK: Extracellular signal-Regulated Kinase
MAPK: Mitogen Activated Protein Kinase
MKK: Mitogen-activated protein Kinase Kinase
STAT3: Signal Transducer and Activator of Transcription 3
IAP: Inhibitor of Apoptosis Proteins
TRAF6: TNF Receptor-Associated Factor 6
AMP: Adenosine Monophosphate
siRNA: small interfering Ribonucleic Acid
ATP: Adenosine Triphosphate
MAPs: Microtubule Associated Proteins.

References

[1] Markov DI, Pivovarova AV, Chernik IS, Gusev NB, Levitsky DI. Small heat shock protein Hsp27 protects myosin S1 from heat-induced aggregation, but not from thermal denaturation and ATPase inactivation. *FEBS letters,* 2008 Apr 30;582(10):1407-12.

[2] Fontaine JM, Rest JS, Welsh MJ, Benndorf R. The sperm outer dense fiber protein is the 10th member of the superfamily of mammalian small stress proteins. *Cell stress & chaperones,* 2003 Spring;8(1):62-9.

[3] Andley UP. The lens epithelium: focus on the expression and function of the alpha-crystallin chaperones. The international journal of biochemistry & cell biology. 2008;40(3):317-23.

[4] Aurelian L, Smith CC, Winchurch R, Kulka M, Gyotoku T, Zaccaro L, et al. A novel gene expressed in human keratinocytes with long-term in vitro growth potential is required for cell growth. The Journal of investigative dermatology. 2001 Feb;116(2):286-95.

[5] Arrigo AP, Virot S, Chaufour S, Firdaus W, Kretz-Remy C, Diaz-Latoud C. Hsp27 consolidates intracellular redox homeostasis by upholding glutathione in its reduced form and by decreasing iron intracellular levels. Antioxidants & redox signaling. 2005 Mar-Apr;7(3-4):414-22.

[6] Garrido C, Ottavi P, Fromentin A, Hammann A, Arrigo AP, Chauffert B, et al. HSP27 as a mediator of confluence-dependent resistance to cell death induced by anticancer drugs. *Cancer research.* 1997 Jul 1;57(13):2661-7.

[7] Arrigo AP. sHsp as novel regulators of programmed cell death and tumorigenicity. *Pathologie-biologie.* 2000 Apr;48(3):280-8.

[8] Wehmeyer N, Vierling E. The expression of small heat shock proteins in seeds responds to discrete developmental signals and suggests a general protective role in desiccation tolerance. Plant physiology. 2000 Apr;122(4):1099-108.

[9] Garrido C, Mehlen P, Fromentin A, Hammann A, Assem M, Arrigo AP, et al. Inconstant association between 27-kDa heat-shock protein (Hsp27) content and doxorubicin resistance in human colon cancer cells. The doxorubicin-protecting effect of Hsp27. *European journal of biochemistry / FEBS.* 1996 May 1;237(3):653-9.

[10] Gorman AM, Szegezdi E, Quigney DJ, Samali A. Hsp27 inhibits 6-hydroxydopamine-induced cytochrome c release and apoptosis in PC12 cells. *Biochemical and biophysical research communications.* 2005 Feb 18;327(3):801-10.

[11] Parcellier A, Brunet M, Schmitt E, Col E, Didelot C, Hammann A, et al. HSP27 favors ubiquitination and proteasomal degradation of p27Kip1 and helps S-phase re-entry in stressed cells. *Faseb J.* 2006 Jun;20(8):1179-81.

[12] Gaestel M, Gotthardt R, Muller T. Structure and organisation of a murine gene encoding small heat-shock protein Hsp25. *Gene.* 1993 Jun 30;128(2):279-83.

[13] Mosser DD, Kotzbauer PT, Sarge KD, Morimoto RI. In vitro activation of heat shock transcription factor DNA-binding by calcium and biochemical conditions that affect protein conformation. Proceedings of the National Academy of Sciences of the United States of America. 1990 May;87(10):3748-52.

[14] Choi HS, Li B, Lin Z, Huang E, Liu AY. cAMP and cAMP-dependent protein kinase regulate the human heat shock protein 70 gene promoter activity. *The Journal of biological chemistry*. 1991 Jun 25;266(18):11858-65.

[15] Fuqua SA, Blum-Salingaros M, McGuire WL. Induction of the estrogen-regulated "24K" protein by heat shock. *Cancer research*. 1989 Aug 1;49(15):4126-9.

[16] Edington BV, Hightower LE. Induction of a chicken small heat shock (stress) protein: evidence of multilevel posttranscriptional regulation. *Molecular and cellular biology*. 1990 Sep;10(9):4886-98.

[17] Jacobson MD, Weil M, Raff MC. Programmed cell death in animal development. *Cell*. 1997 Feb 7;88(3):347-54.

[18] Solary E, Droin N, Bettaieb A, Corcos L, Dimanche-Boitrel MT, Garrido C. Positive and negative regulation of apoptotic pathways by cytotoxic agents in hematological malignancies. *Leukemia*. 2000 Oct;14(10):1833-49.

[19] Wei MC, Zong WX, Cheng EH, Lindsten T, Panoutsakopoulou V, Ross AJ, et al. Proapoptotic BAX and BAK: a requisite gateway to mitochondrial dysfunction and death. *Science,* (New York, NY. 2001 Apr 27;292(5517):727-30.

[20] Zong WX, Lindsten T, Ross AJ, MacGregor GR, Thompson CB. BH3-only proteins that bind pro-survival Bcl-2 family members fail to induce apoptosis in the absence of Bax and Bak. *Genes & development*. 2001 Jun 15;15(12):1481-6.

[21] Cheng EH, Wei MC, Weiler S, Flavell RA, Mak TW, Lindsten T, et al. BCL-2, BCL-X(L) sequester BH3 domain-only molecules preventing BAX- and BAK-mediated mitochondrial apoptosis. *Molecular cell*. 2001 Sep;8(3):705-11.

[22] Li P, Nijhawan D, Budihardjo I, Srinivasula SM, Ahmad M, Alnemri ES, et al. Cytochrome c and dATP-dependent formation of Apaf-1/caspase-9 complex initiates an apoptotic protease cascade. *Cell*. 1997 Nov 14;91(4):479-89.

[23] Du C, Fang M, Li Y, Li L, Wang X. Smac, a mitochondrial protein that promotes cytochrome c-dependent caspase activation by eliminating IAP inhibition. Cell. 2000 Jul 7;102(1):33-42.

[24] Luo X, Budihardjo I, Zou H, Slaughter C, Wang X. Bid, a Bcl2 interacting protein, mediates cytochrome c release from mitochondria in response to activation of cell surface death receptors. *Cell*. 1998 Aug 21;94(4):481-90.

[25] Mehlen P, Schulze-Osthoff K, Arrigo AP. Small stress proteins as novel regulators of apoptosis. Heat shock protein 27 blocks Fas/APO-1- and staurosporine-induced cell death. *The Journal of biological chemistry*. 1996 Jul 12;271(28):16510-4.

[26] Rocchi P, Beraldi E, Ettinger S, Fazli L, Vessella RL, Nelson C, et al. Increased Hsp27 after androgen ablation facilitates androgen-independent progression in prostate cancer via signal transducers and activators of transcription 3-mediated suppression of apoptosis. *Cancer research*. 2005 Dec 1;65(23):11083-93.

[27] Rocchi P, Jugpal P, So A, Sinneman S, Ettinger S, Fazli L, et al. Small interference RNA targeting heat-shock protein 27 inhibits the growth of prostatic cell lines and induces apoptosis via caspase-3 activation in vitro. *BJU international*. 2006 Nov;98(5):1082-9.

[28] Garrido C, Bruey JM, Fromentin A, Hammann A, Arrigo AP, Solary E. HSP27 inhibits cytochrome c-dependent activation of procaspase-9. *Faseb J*. 1999 Nov;13(14):2061-70.

[29] Concannon CG, Orrenius S, Samali A. Hsp27 inhibits cytochrome c-mediated caspase activation by sequestering both pro-caspase-3 and cytochrome c. Gene expression. 2001;9(4-5):195-201.

[30] Bruey JM, Ducasse C, Bonniaud P, Ravagnan L, Susin SA, Diaz-Latoud C, et al. Hsp27 negatively regulates cell death by interacting with cytochrome c. *Nature cell biology*. 2000 Sep;2(9):645-52.

[31] Paul C, Manero F, Gonin S, Kretz-Remy C, Virot S, Arrigo AP. Hsp27 as a negative regulator of cytochrome C release. Molecular and cellular biology. 2002 Feb;22(3):816-34.

[32] Guay J, Lambert H, Gingras-Breton G, Lavoie JN, Huot J, Landry J. Regulation of actin filament dynamics by p38 map kinase-mediated phosphorylation of heat shock protein 27. *Journal of cell science*. 1997 Feb;110 (Pt 3):357-68.

[33] Chauhan D, Li G, Hideshima T, Podar K, Mitsiades C, Mitsiades N, et al. Hsp27 inhibits release of mitochondrial protein Smac in multiple myeloma cells and confers dexamethasone resistance. *Blood*. 2003 Nov 1;102(9):3379-86.

[34] Mehlen P, Kretz-Remy C, Preville X, Arrigo AP. Human hsp27, Drosophila hsp27 and human alphaB-crystallin expression-mediated increase in glutathione is essential for the protective activity of these proteins against TNFalpha-induced cell death. *The EMBO journal*. 1996 Jun 3;15(11):2695-706.

[35] Rogalla T, Ehrnsperger M, Preville X, Kotlyarov A, Lutsch G, Ducasse C, et al. Regulation of Hsp27 oligomerization, chaperone function, and protective activity against oxidative stress/tumor necrosis factor alpha by phosphorylation. *The Journal of biological chemistry*. 1999 Jul 2;274(27):18947-56.

[36] Zhang Y, Shen X. Heat shock protein 27 protects L929 cells from cisplatin-induced apoptosis by enhancing Akt activation and abating suppression of thioredoxin reductase activity. *Clin Cancer Res*. 2007 May 15;13(10):2855-64.

[37] Wyttenbach A, Sauvageot O, Carmichael J, Diaz-Latoud C, Arrigo AP, Rubinsztein DC. Heat shock protein 27 prevents cellular polyglutamine toxicity and suppresses the increase of reactive oxygen species caused by huntingtin. *Human molecular genetics*. 2002 May 1;11(9):1137-51.

[38] Rane MJ, Pan Y, Singh S, Powell DW, Wu R, Cummins T, et al. Heat shock protein 27 controls apoptosis by regulating Akt activation. *The Journal of biological chemistry*. 2003 Jul 25;278(30):27828-35.

[39] Havasi A, Li Z, Wang Z, Martin JL, Botla V, Ruchalski K, et al. Hsp27 inhibits Bax activation and apoptosis via a phosphatidylinositol 3-kinase-dependent mechanism. *The Journal of biological chemistry*. 2008 May 2;283(18):12305-13.

[40] Liu L, Zhang XJ, Jiang SR, Ding ZN, Ding GX, Huang J, et al. Heat shock protein 27 regulates oxidative stress-induced apoptosis in cardiomyocytes: mechanisms via reactive oxygen species generation and Akt activation. *Chinese medical journal*. 2007 Dec 20;120(24):2271-7.

[41] Schepers H, Geugien M, van der Toorn M, Bryantsev AL, Kampinga HH, Eggen BJ, et al. HSP27 protects AML cells against VP-16-induced apoptosis through modulation of p38 and c-Jun. *Experimental hematology*. 2005 Jun;33(6):660-70.

[42] Liang P, MacRae TH. Molecular chaperones and the cytoskeleton. *Journal of cell science*. 1997 Jul;110 (Pt 13):1431-40.

[43] Hino M, Kurogi K, Okubo MA, Murata-Hori M, Hosoya H. Small heat shock protein 27 (HSP27) associates with tubulin/microtubules in HeLa cells. *Biochemical and biophysical research communications*. 2000 Apr 29;271(1):164-9.

[44] Pivovarova AV, Chebotareva NA, Chernik IS, Gusev NB, Levitsky DI. Small heat shock protein Hsp27 prevents heat-induced aggregation of F-actin by forming soluble complexes with denatured actin. The FEBS journal. 2007 Nov;274(22):5937-48.

[45] Zoubeidi A, Zardan A, Beraldi E, Fazli L, Sowery R, Rennie P, et al. Cooperative interactions between androgen receptor (AR) and heat-shock protein 27 facilitate AR transcriptional activity. *Cancer research*. 2007 Nov 1;67(21):10455-65.

[46] Charette SJ, Lavoie JN, Lambert H, Landry J. Inhibition of Daxx-mediated apoptosis by heat shock protein 27. *Molecular and cellular biology*. 2000 Oct;20(20):7602-12.

[47] Song H, Ethier SP, Dziubinski ML, Lin J. Stat3 modulates heat shock 27kDa protein expression in breast epithelial cells. *Biochemical and biophysical research communications*. 2004 Jan 30;314(1):143-50.

[48] Marin-Vinader L, Shin C, Onnekink C, Manley JL, Lubsen NH. Hsp27 enhances recovery of splicing as well as rephosphorylation of SRp38 after heat shock. *Molecular biology of the cell*. 2006 Feb;17(2):886-94.

[49] Shashidharamurthy R, Koteiche HA, Dong J, McHaourab HS. Mechanism of chaperone function in small heat shock proteins: dissociation of the HSP27 oligomer is required for recognition and binding of destabilized T4 lysozyme. *The Journal of biological chemistry*. 2005 Feb 18;280(7):5281-9.

[50] Lelj-Garolla B, Mauk AG. Self-association of a small heat shock protein. *Journal of molecular biology*. 2005 Jan 21;345(3):631-42.

[51] Kamradt MC, Chen F, Cryns VL. The small heat shock protein alpha B-crystallin negatively regulates cytochrome c- and caspase-8-dependent activation of caspase-3 by inhibiting its autoproteolytic maturation. *The Journal of biological chemistry*. 2001 May 11;276(19):16059-63.

[52] Kamradt MC, Lu M, Werner ME, Kwan T, Chen F, Strohecker A, et al. The small heat shock protein alpha B-crystallin is a novel inhibitor of TRAIL-induced apoptosis that suppresses the activation of caspase-3. The Journal of biological chemistry. 2005 Mar 25;280(12):11059-66.

[53] Kamradt MC, Chen F, Sam S, Cryns VL. The small heat shock protein alpha B-crystallin negatively regulates apoptosis during myogenic differentiation by inhibiting caspase-3 activation. *The Journal of biological chemistry*. 2002 Oct 11;277(41):38731-6.

[54] Kato Y, Miyakawa T, Kurita J, Tanokura M. Structure of FBP11 WW1-PL ligand complex reveals the mechanism of proline-rich ligand recognition by group II/III WW domains. *The Journal of biological chemistry*. 2006 Dec 29;281(52):40321-9.

[55] Mao YW, Xiang H, Wang J, Korsmeyer S, Reddan J, Li DW. Human bcl-2 gene attenuates the ability of rabbit lens epithelial cells against H2O2-induced apoptosis through down-regulation of the alpha B-crystallin gene. *The Journal of biological chemistry*. 2001 Nov 16;276(46):43435-45.

[56] Morrison LE, Hoover HE, Thuerauf DJ, Glembotski CC. Mimicking phosphorylation of alphaB-crystallin on serine-59 is necessary and sufficient to provide maximal protection of cardiac myocytes from apoptosis. Circulation research. 2003 Feb 7;92(2):203-11.

[57] Stegh AH, Kesari S, Mahoney JE, Jenq HT, Forloney KL, Protopopov A, et al. Bcl2L12-mediated inhibition of effector caspase-3 and caspase-7 via distinct mechanisms in glioblastoma. Proceedings of the National Academy of Sciences of the United States of America. 2008 Aug 5;105(31):10703-8.

[58] Hoover HE, Thuerauf DJ, Martindale JJ, Glembotski CC. alpha B-crystallin gene induction and phosphorylation by MKK6-activated p38. A potential role for alpha B-crystallin as a target of the p38 branch of the cardiac stress response. *The Journal of biological chemistry*. 2000 Aug 4;275(31):23825-33.

[59] Mao YW, Liu JP, Xiang H, Li DW. Human alphaA- and alphaB-crystallins bind to Bax and Bcl-X(S) to sequester their translocation during staurosporine-induced apoptosis. *Cell death and differentiation*. 2004 May;11(5):512-26.

[60] Liu S, Li J, Tao Y, Xiao X. Small heat shock protein alphaB-crystallin binds to p53 to sequester its translocation to mitochondria during hydrogen peroxide-induced apoptosis. *Biochemical and biophysical research communications*. 2007 Mar 2;354(1):109-14.

[61] Verschuure P, Croes Y, van den IPR, Quinlan RA, de Jong WW, Boelens WC. Translocation of small heat shock proteins to the actin cytoskeleton upon proteasomal inhibition. *Journal of molecular and cellular cardiology*. 2002 Feb;34(2):117-28.

[62] Djabali K, de Nechaud B, Landon F, Portier MM. AlphaB-crystallin interacts with intermediate filaments in response to stress. Journal of cell science. 1997 Nov;110 (Pt 21):2759-69.

[63] Bennardini F, Wrzosek A, Chiesi M. Alpha B-crystallin in cardiac tissue. Association with actin and desmin filaments. Circulation research. 1992 Aug;71(2):288-94.

[64] Sanbe A, Osinska H, Villa C, Gulick J, Klevitsky R, Glabe CG, et al. Reversal of amyloid-induced heart disease in desmin-related cardiomyopathy. Proceedings of the National Academy of Sciences of the United States of America. 2005 Sep 20;102(38):13592-7.

[65] Iwaki T, Iwaki A, Tateishi J, Goldman JE. Sense and antisense modification of glial alpha B-crystallin production results in alterations of stress fiber formation and thermoresistance. *The Journal of cell biology*. 1994 Jun;125(6):1385-93.

[66] Singh D, Raman B, Ramakrishna T, Rao Ch M. Mixed oligomer formation between human alphaA-crystallin and its cataract-causing G98R mutant: structural, stability and functional differences. *Journal of molecular biology*. 2007 Nov 9;373(5):1293-304.

[67] Gober MD, Smith CC, Ueda K, Toretsky JA, Aurelian L. Forced expression of the H11 heat shock protein can be regulated by DNA methylation and trigger apoptosis in human cells. The Journal of biological chemistry. 2003 Sep 26;278(39):37600-9.

[68] Gober MD, Depre C, Aurelian L. Correspondence regarding M.V. Kim et al. "Some properties of human small heat shock protein Hsp22 (H11 or HspB8)". Biochemical and biophysical research communications. 2004 Aug 20;321(2):267-8.

[69] Perkins D, Pereira EF, Gober M, Yarowsky PJ, Aurelian L. The herpes simplex virus type 2 R1 protein kinase (ICP10 PK) blocks apoptosis in hippocampal neurons, involving activation of the MEK/MAPK survival pathway. *Journal of virology*. 2002 Feb;76(3):1435-49.

[70] Yu C, Wang S, Dent P, Grant S. Sequence-dependent potentiation of paclitaxel-mediated apoptosis in human leukemia cells by inhibitors of the mitogen-activated

protein kinase kinase/mitogen-activated protein kinase pathway. *Molecular pharmacology*. 2001 Jul;60(1):143-54.

[71] Perkins D, Pereira EF, Aurelian L. The herpes simplex virus type 2 R1 protein kinase (ICP10 PK) functions as a dominant regulator of apoptosis in hippocampal neurons involving activation of the ERK survival pathway and upregulation of the antiapoptotic protein Bag-1. *Journal of virology*. 2003 Jan;77(2):1292-305.

[72] Lebedeva IV, Su ZZ, Chang Y, Kitada S, Reed JC, Fisher PB. The cancer growth suppressing gene mda-7 induces apoptosis selectively in human melanoma cells. *Oncogene*. 2002 Jan 24;21(5):708-18.

[73] Depre C, Kim SJ, John AS, Huang Y, Rimoldi OE, Pepper JR, et al. Program of cell survival underlying human and experimental hibernating myocardium. *Circulation research*. 2004 Aug 20;95(4):433-40.

[74] Wang L, Zajac A, Hedhli N, Depre C. Increased expression of H11 kinase stimulates glycogen synthesis in the heart. Molecular and cellular biochemistry. 2004 Oct;265(1-2):71-8.

[75] Hase M, Depre C, Vatner SF, Sadoshima J. H11 has dose-dependent and dual hypertrophic and proapoptotic functions in cardiac myocytes. *The Biochemical journal*. 2005 Jun 1;388(Pt 2):475-83.

[76] Hu Z, Chen L, Zhang J, Li T, Tang J, Xu N, et al. Structure, function, property, and role in neurologic diseases and other diseases of the sHsp22. *Journal of neuroscience research*. 2007 Aug 1;85(10):2071-9.

[77] Shemetov AA, Seit-Nebi AS, Gusev NB. Structure, properties, and functions of the human small heat-shock protein HSP22 (HspB8, H11, E2IG1): a critical review. *Journal of neuroscience research*. 2008 Feb 1;86(2):264-9.

[78] Gober MD, Wales SQ, Aurelian L. Herpes simplex virus type 2 encodes a heat shock protein homologue with apoptosis regulatory functions. Front Biosci. 2005;10:2788-803.

[79] Yang C, Trent S, Ionescu-Tiba V, Lan L, Shioda T, Sgroi D, et al. Identification of cyclin D1- and estrogen-regulated genes contributing to breast carcinogenesis and progression. *Cancer research*. 2006 Dec 15;66(24):11649-58.

[80] Smith CC, Yu YX, Kulka M, Aurelian L. A novel human gene similar to the protein kinase (PK) coding domain of the large subunit of herpes simplex virus type 2 ribonucleotide reductase (ICP10) codes for a serine-threonine PK and is expressed in melanoma cells. *The Journal of biological chemistry*. 2000 Aug 18;275(33):25690-9.

[81] Charpentier AH, Bednarek AK, Daniel RL, Hawkins KA, Laflin KJ, Gaddis S, et al. Effects of estrogen on global gene expression: identification of novel targets of estrogen action. *Cancer research*. 2000 Nov 1;60(21):5977-83.

[82] Garrido C. Size matters: of the small HSP27 and its large oligomers. *Cell death and differentiation*. 2002 May;9(5):483-5.

[83] Ciechanover A. The ubiquitin-proteasome pathway: on protein death and cell life. *The EMBO journal*. 1998 Dec 15;17(24):7151-60.

[84] Duverger O, Paslaru L, Morange M. HSP25 is involved in two steps of the differentiation of PAM212 keratinocytes. The Journal of biological chemistry. 2004 Mar 12;279(11):10252-60.

[85] Mehlen P, Coronas V, Ljubic-Thibal V, Ducasse C, Granger L, Jourdan F, et al. Small stress protein Hsp27 accumulation during dopamine-mediated differentiation of rat

olfactory neurons counteracts apoptosis. Cell death and differentiation. 1999 Mar;6(3):227-33.

[86] Parcellier A, Schmitt E, Gurbuxani S, Seigneurin-Berny D, Pance A, Chantome A, et al. HSP27 is a ubiquitin-binding protein involved in I-kappaBalpha proteasomal degradation. *Molecular and cellular biology*. 2003 Aug;23(16):5790-802.

[87] Ito H, Iida K, Kamei K, Iwamoto I, Inaguma Y, Kato K. AlphaB-crystallin in the rat lens is phosphorylated at an early post-natal age. FEBS letters. 1999 Mar 12;446(2-3):269-72.

[88] Kato K, Ito H, Kamei K, Iwamoto I, Inaguma Y. Innervation-dependent phosphorylation and accumulation of alphaB-crystallin and Hsp27 as insoluble complexes in disused muscle. *Faseb J.* 2002 Sep;16(11):1432-4.

[89] Stumptner C, Fuchsbichler A, Heid H, Zatloukal K, Denk H. Mallory body--a disease-associated type of sequestosome. *Hepatology,* (Baltimore, Md. 2002 May;35(5):1053-62.

[90] Lowe J, Mayer RJ, Landon M. Ubiquitin in neurodegenerative diseases. *Brain pathology,* (Zurich, Switzerland). 1993 Jan;3(1):55-65.

[91] Iwaki T, Wisniewski T, Iwaki A, Corbin E, Tomokane N, Tateishi J, et al. Accumulation of alpha B-crystallin in central nervous system glia and neurons in pathologic conditions. *The American journal of pathology*. 1992 Feb;140(2):345-56.

[92] Iwaki T, Kume-Iwaki A, Liem RK, Goldman JE. Alpha B-crystallin is expressed in non-lenticular tissues and accumulates in Alexander's disease brain. Cell. 1989 Apr 7;57(1):71-8.

[93] Renkawek K, de Jong WW, Merck KB, Frenken CW, van Workum FP, Bosman GJ. alpha B-crystallin is present in reactive glia in Creutzfeldt-Jakob disease. *Acta neuropathologica*. 1992;83(3):324-7.

[94] Thomas LB, Gates DJ, Richfield EK, O'Brien TF, Schweitzer JB, Steindler DA. DNA end labeling (TUNEL) in Huntington's disease and other neuropathological conditions. Experimental neurology. 1995 Jun;133(2):265-72.

[95] Landry J, Lambert H, Zhou M, Lavoie JN, Hickey E, Weber LA, et al. Human HSP27 is phosphorylated at serines 78 and 82 by heat shock and mitogen-activated kinases that recognize the same amino acid motif as S6 kinase II. *The Journal of biological chemistry*. 1992 Jan 15;267(2):794-803.

[96] Boelens WC, Croes Y, de Jong WW. Interaction between alphaB-crystallin and the human 20S proteasomal subunit C8/alpha7. Biochimica et biophysica acta. 2001 Jan 12;1544(1-2):311-9.

[97] den Engelsman J, Keijsers V, de Jong WW, Boelens WC. The small heat-shock protein alpha B-crystallin promotes FBX4-dependent ubiquitination. *The Journal of biological chemistry*. 2003 Feb 14;278(7):4699-704.

[98] Lin DI, Barbash O, Kumar KG, Weber JD, Harper JW, Klein-Szanto AJ, et al. Phosphorylation-dependent ubiquitination of cyclin D1 by the SCF(FBX4-alphaB crystallin) complex. *Molecular cell*. 2006 Nov 3;24(3):355-66.

[99] Barbash O, Lin DI, Diehl JA. SCF Fbx4/alphaB-crystallin cyclin D1 ubiquitin ligase: a license to destroy. *Cell division*. 2007;2:2.

[100] Parcellier A, Gurbuxani S, Schmitt E, Solary E, Garrido C. Heat shock proteins, cellular chaperones that modulate mitochondrial cell death pathways. *Biochemical and biophysical research communications*. 2003 May 9;304(3):505-12.

[101] Wu PY, Lin YC, Chang CL, Lu HT, Chin CH, Hsu TT, et al. Functional decreases in P2X7 receptors are associated with retinoic acid-induced neuronal differentiation of Neuro-2a neuroblastoma cells. Cellular signalling. 2009 Jun;21(6):881-91.

[102] Carra S, Seguin SJ, Lambert H, Landry J. HspB8 chaperone activity toward poly(Q)-containing proteins depends on its association with Bag3, a stimulator of macroautophagy. The Journal of biological chemistry. 2008 Jan 18;283(3):1437-44.

[103] Ehrnsperger M, Graber S, Gaestel M, Buchner J. Binding of non-native protein to Hsp25 during heat shock creates a reservoir of folding intermediates for reactivation. *The EMBO journal.* 1997 Jan 15;16(2):221-9.

[104] Lee GJ, Roseman AM, Saibil HR, Vierling E. A small heat shock protein stably binds heat-denatured model substrates and can maintain a substrate in a folding-competent state. *The EMBO journal.* 1997 Feb 3;16(3):659-71.

[105] Muchowski PJ, Clark JI. ATP-enhanced molecular chaperone functions of the small heat shock protein human alphaB crystallin. Proceedings of the National Academy of Sciences of the United States of America. 1998 Feb 3;95(3):1004-9.

[106] Jakob U, Gaestel M, Engel K, Buchner J. Small heat shock proteins are molecular chaperones. *The Journal of biological chemistry,* 1993 Jan 25;268(3):1517-20.

[107] Chowdary TK, Raman B, Ramakrishna T, Rao CM. Mammalian Hsp22 is a heat-inducible small heat-shock protein with chaperone-like activity. *The Biochemical journal.* 2004 Jul 15;381(Pt 2):379-87.

[108] Stokoe D, Engel K, Campbell DG, Cohen P, Gaestel M. Identification of MAPKAP kinase 2 as a major enzyme responsible for the phosphorylation of the small mammalian heat shock proteins. FEBS letters. 1992 Nov 30;313(3):307-13.

[109] Gaestel M, Schroder W, Benndorf R, Lippmann C, Buchner K, Hucho F, et al. Identification of the phosphorylation sites of the murine small heat shock protein hsp25. The *Journal of biological chemistry,* 1991 Aug 5;266(22):14721-4.

[110] Maizels ET, Peters CA, Kline M, Cutler RE, Jr., Shanmugam M, Hunzicker-Dunn M. Heat-shock protein-25/27 phosphorylation by the delta isoform of protein kinase C. *The Biochemical journal.* 1998 Jun 15;332 (Pt 3):703-12.

[111] McLaughlin MM, Kumar S, McDonnell PC, Van Horn S, Lee JC, Livi GP, et al. Identification of mitogen-activated protein (MAP) kinase-activated protein kinase-3, a novel substrate of CSBP p38 MAP kinase. *The Journal of biological chemistry.* 1996 Apr 5;271(14):8488-92.

[112] Cairns J, Qin S, Philp R, Tan YH, Guy GR. Dephosphorylation of the small heat shock protein Hsp27 in vivo by protein phosphatase 2A. *The Journal of biological chemistry,* 1994 Mar 25;269(12):9176-83.

[113] Ito H, Okamoto K, Nakayama H, Isobe T, Kato K. Phosphorylation of alphaB-crystallin in response to various types of stress. *The Journal of biological chemistry.* 1997 Nov 21;272(47):29934-41.

[114] Launay N, Goudeau B, Kato K, Vicart P, Lilienbaum A. Cell signaling pathways to alphaB-crystallin following stresses of the cytoskeleton. *Experimental cell research.* 2006 Nov 1;312(18):3570-84.

[115] Kato K, Ito H, Kamei K, Inaguma Y, Iwamoto I, Saga S. Phosphorylation of alphaB-crystallin in mitotic cells and identification of enzymatic activities responsible for phosphorylation. *The Journal of biological chemistry.* 1998 Oct 23;273(43):28346-54.

[116] Benndorf R, Sun X, Gilmont RR, Biederman KJ, Molloy MP, Goodmurphy CW, et al. HSP22, a new member of the small heat shock protein superfamily, interacts with mimic of phosphorylated HSP27 ((3D)HSP27). *The Journal of biological chemistry*. 2001 Jul 20;276(29):26753-61.

[117] Kappe G, Verschuure P, Philipsen RL, Staalduinen AA, Van de Boogaart P, Boelens WC, et al. Characterization of two novel human small heat shock proteins: protein kinase-related HspB8 and testis-specific HspB9. Biochimica et biophysica acta. 2001 Jul 30;1520(1):1-6.

[118] Kim MV, Seit-Nebi AS, Gusev NB. The problem of protein kinase activity of small heat shock protein Hsp22 (H11 or HspB8). *Biochemical and biophysical research communications*. 2004 Dec 17;325(3):649-52.

[119] Arrigo AP, Simon S, Gibert B, Kretz-Remy C, Nivon M, Czekalla A, et al. Hsp27 (HspB1) and alphaB-crystallin (HspB5) as therapeutic targets. *FEBS letters*. 2007 Jul 31;581(19):3665-74.

[120] Kato K, Hasegawa K, Goto S, Inaguma Y. Dissociation as a result of phosphorylation of an aggregated form of the small stress protein, hsp27. *The Journal of biological chemistry*. 1994 Apr 15;269(15):11274-8.

[121] Bruey JM, Paul C, Fromentin A, Hilpert S, Arrigo AP, Solary E, et al. Differential regulation of HSP27 oligomerization in tumor cells grown in vitro and in vivo. Oncogene. 2000 Oct 5;19(42):4855-63.

[122] Martin JL, Hickey E, Weber LA, Dillmann WH, Mestril R. Influence of phosphorylation and oligomerization on the protective role of the small heat shock protein 27 in rat adult cardiomyocytes. *Gene expression*. 1999;7(4-6):349-55.

[123] During RL, Gibson BG, Li W, Bishai EA, Sidhu GS, Landry J, et al. Anthrax lethal toxin paralyzes actin-based motility by blocking Hsp27 phosphorylation. *The EMBO journal*. 2007 May 2;26(9):2240-50.

[124] Webster KA. Serine phosphorylation and suppression of apoptosis by the small heat shock protein alphaB-crystallin. *Circulation research*. 2003 Feb 7;92(2):130-2.

[125] Chavez Zobel AT, Loranger A, Marceau N, Theriault JR, Lambert H, Landry J. Distinct chaperone mechanisms can delay the formation of aggresomes by the myopathy-causing R120G alphaB-crystallin mutant. *Human molecular genetics*. 2003 Jul 1;12(13):1609-20.

[126] Ito H, Kamei K, Iwamoto I, Inaguma Y, Tsuzuki M, Kishikawa M, et al. Hsp27 suppresses the formation of inclusion bodies induced by expression of R120G alpha B-crystallin, a cause of desmin-related myopathy. *Cell Mol Life Sci*. 2003 Jun;60(6):1217-23.

[127] Garrido C, Fromentin A, Bonnotte B, Favre N, Moutet M, Arrigo AP, et al. Heat shock protein 27 enhances the tumorigenicity of immunogenic rat colon carcinoma cell clones. *Cancer research*. 1998 Dec 1;58(23):5495-9.

[128] Rocchi P, So A, Kojima S, Signaevsky M, Beraldi E, Fazli L, et al. Heat shock protein 27 increases after androgen ablation and plays a cytoprotective role in hormone-refractory prostate cancer. *Cancer research*. 2004 Sep 15;64(18):6595-602.

[129] Lemieux P, Oesterreich S, Lawrence JA, Steeg PS, Hilsenbeck SG, Harvey JM, et al. The small heat shock protein hsp27 increases invasiveness but decreases motility of breast cancer cells. *Invasion & metastasis*. 1997;17(3):113-23.

[130] Shin KD, Lee MY, Shin DS, Lee S, Son KH, Koh S, et al. Blocking tumor cell migration and invasion with biphenyl isoxazole derivative KRIBB3, a synthetic molecule that inhibits Hsp27 phosphorylation. The Journal of biological chemistry. 2005 Dec 16;280(50):41439-48.

[131] Doshi BM, Hightower LE, Lee J. The role of Hsp27 and actin in the regulation of movement in human cancer cells responding to heat shock. *Cell stress & chaperones*, 2009 Feb 18.

[132] Chen HF, Xie LD, Xu CS. Role of heat shock protein 27 phosphorylation in migration of vascular smooth muscle cells. *Molecular and cellular biochemistry*. 2009 Feb 4.

[133] Parcellier A, Schmitt E, Brunet M, Hammann A, Solary E, Garrido C. Small heat shock proteins HSP27 and alphaB-crystallin: cytoprotective and oncogenic functions. *Antioxidants & redox signaling*. 2005 Mar-Apr;7(3-4):404-13.

[134] Choi DH, Ha JS, Lee WH, Song JK, Kim GY, Park JH, et al. Heat shock protein 27 is associated with irinotecan resistance in human colorectal cancer cells. *FEBS letters*. 2007 Apr 17;581(8):1649-56.

[135] Langdon SP, Rabiasz GJ, Hirst GL, King RJ, Hawkins RA, Smyth JF, et al. Expression of the heat shock protein HSP27 in human ovarian cancer. *Clin Cancer Res*. 1995 Dec;1(12):1603-9.

[136] Kang SH, Kang KW, Kim KH, Kwon B, Kim SK, Lee HY, et al. Upregulated HSP27 in human breast cancer cells reduces Herceptin susceptibility by increasing Her2 protein stability. *BMC cancer*. 2008;8:286.

[137] Romani AA, Crafa P, Desenzani S, Graiani G, Lagrasta C, Sianesi M, et al. The expression of HSP27 is associated with poor clinical outcome in intrahepatic cholangiocarcinoma. *BMC cancer*. 2007;7:232.

[138] Yoshida T, Nakazato Y. Characterization of refractile eosinophilic granular cells in oligodendroglial tumors. Acta neuropathologica. 2001 Jul;102(1):11-9.

[139] Ciocca DR, Oesterreich S, Chamness GC, McGuire WL, Fuqua SA. Biological and clinical implications of heat shock protein 27,000 (Hsp27): a review. *Journal of the National Cancer Institute*, 1993 Oct 6;85(19):1558-70.

[140] Sarto C, Valsecchi C, Magni F, Tremolada L, Arizzi C, Cordani N, et al. Expression of heat shock protein 27 in human renal cell carcinoma. *Proteomics*, 2004 Aug;4(8):2252-60.

[141] Xu L, Chen S, Bergan RC. MAPKAPK2 and HSP27 are downstream effectors of p38 MAP kinase-mediated matrix metalloproteinase type 2 activation and cell invasion in human prostate cancer. Oncogene. 2006 May 18;25(21):2987-98.

[142] Eskenazi AE, Powers J, Pinkas J, Oesterreich S, Fuqua SA, Frantz CN. Induction of heat shock protein 27 by hydroxyurea and its relationship to experimental metastasis. *Clinical & experimental metastasis*. 1998 Apr;16(3):283-90.

[143] Kardys I, Rifai N, Meilhac O, Michel JB, Martin-Ventura JL, Buring JE, et al. Plasma concentration of heat shock protein 27 and risk of cardiovascular disease: a prospective, nested case-control study. Clinical chemistry. 2008 Jan;54(1):139-46.

[144] Bausero MA, Page DT, Osinaga E, Asea A. Surface expression of Hsp25 and Hsp72 differentially regulates tumor growth and metastasis. *Tumour Biol*. 2004 Sep-Dec;25(5-6):243-51.

[145] Bausero MA, Gastpar R, Multhoff G, Asea A. Alternative mechanism by which IFN-gamma enhances tumor recognition: active release of heat shock protein 72. *J Immunol.* 2005 Sep 1;175(5):2900-12.

[146] Piotrowicz RS, Hickey E, Levin EG. Heat shock protein 27 kDa expression and phosphorylation regulates endothelial cell migration. *Faseb J.* 1998 Nov;12(14):1481-90.

[147] Shin BK, Wang H, Yim AM, Le Naour F, Brichory F, Jang JH, et al. Global profiling of the cell surface proteome of cancer cells uncovers an abundance of proteins with chaperone function. The Journal of biological chemistry. 2003 Feb 28;278(9):7607-16.

[148] Aoyama A, Steiger RH, Frohli E, Schafer R, von Deimling A, Wiestler OD, et al. Expression of alpha B-crystallin in human brain tumors. *International journal of cancer.* 1993 Nov 11;55(5):760-4.

[149] Lo WY, Tsai MH, Tsai Y, Hua CH, Tsai FJ, Huang SY, et al. Identification of over-expressed proteins in oral squamous cell carcinoma (OSCC) patients by clinical proteomic analysis. Clinica chimica acta; international journal of clinical chemistry. 2007 Feb;376(1-2):101-7.

[150] Pinder SE, Balsitis M, Ellis IO, Landon M, Mayer RJ, Lowe J. The expression of alpha B-crystallin in epithelial tumours: a useful tumour marker? *The Journal of pathology.* 1994 Nov;174(3):209-15.

[151] Bosman JD, Yehiely F, Evans JR, Cryns VL. Regulation of alphaB-crystallin gene expression by the transcription factor Ets1 in breast cancer. Breast cancer research and treatment. 2009 Feb 11.

[152] Sitterding SM, Wiseman WR, Schiller CL, Luan C, Chen F, Moyano JV, et al. AlphaB-crystallin: a novel marker of invasive basal-like and metaplastic breast carcinomas. *Annals of diagnostic pathology.* 2008 Feb;12(1):33-40.

[153] Ivanov O, Chen F, Wiley EL, Keswani A, Diaz LK, Memmel HC, et al. alphaB-crystallin is a novel predictor of resistance to neoadjuvant chemotherapy in breast cancer. *Breast cancer research and treatment.* 2008 Oct;111(3):411-7.

[154] Bennardini F, Mattana A, Nossai EP, Mignano M, Franconi F, Juliano C, et al. Kinetic changes of alpha B crystallin expression in neoplastic cells and syngeneic rat fibroblasts at various subculture stages. *Molecular and cellular biochemistry.* 1995 Nov 8;152(1):23-30.

[155] Moyano JV, Evans JR, Chen F, Lu M, Werner ME, Yehiely F, et al. AlphaB-crystallin is a novel oncoprotein that predicts poor clinical outcome in breast cancer. *The Journal of clinical investigation.* 2006 Jan;116(1):261-70.

[156] Gruvberger-Saal SK, Parsons R. Is the small heat shock protein alphaB-crystallin an oncogene? *The Journal of clinical investigation.* 2006 Jan;116(1):30-2.

[157] Chelouche-Lev D, Kluger HM, Berger AJ, Rimm DL, Price JE. alphaB-crystallin as a marker of lymph node involvement in breast carcinoma. Cancer. 2004 Jun 15;100(12):2543-8.

[158] Hitotsumatsu T, Iwaki T, Fukui M, Tateishi J. Distinctive immunohistochemical profiles of small heat shock proteins (heat shock protein 27 and alpha B-crystallin) in human brain tumors. *Cancer.* 1996 Jan 15;77(2):352-61.

[159] Mineva I, Gartner W, Hauser P, Kainz A, Loffler M, Wolf G, et al. Differential expression of alphaB-crystallin and Hsp27-1 in anaplastic thyroid carcinomas because of tumor-specific alphaB-crystallin gene (CRYAB) silencing. Cell stress & chaperones. 2005 Autumn;10(3):171-84.

[160] Kato K, Ito H, Inaguma Y, Okamoto K, Saga S. Synthesis and accumulation of alphaB crystallin in C6 glioma cells is induced by agents that promote the disassembly of microtubules. *The Journal of biological chemistry*. 1996 Oct 25;271(43):26989-94.

[161] Ishiguro Y, Kato K, Akatsuka H, Iwata H, Nagaya M. Chemotherapy-induced expression of alpha B-crystallin in neuroblastoma. Medical and pediatric oncology. 1997 Jul;29(1):11-5.

[162] Barr CS, Dokas LA. Glucocorticoids regulate the synthesis of HSP27 in rat brain slices. *Brain research*. 1999 Nov 13;847(1):9-17.

[163] Sun X, Fontaine JM, Bartl I, Behnam B, Welsh MJ, Benndorf R. Induction of Hsp22 (HspB8) by estrogen and the metalloestrogen cadmium in estrogen receptor-positive breast cancer cells. Cell stress & chaperones. 2007 Winter;12(4):307-19.

[164] Badri KR, Modem S, Gerard HC, Khan I, Bagchi M, Hudson AP, et al. Regulation of Sam68 activity by small heat shock protein 22. *Journal of cellular biochemistry*. 2006 Dec 1;99(5):1353-62.

[165] Chowdary TK, Raman B, Ramakrishna T, Rao Ch M. Interaction of mammalian Hsp22 with lipid membranes. *The Biochemical journal*. 2007 Jan 15;401(2):437-45.

[166] Roelofs MF, Boelens WC, Joosten LA, Abdollahi-Roodsaz S, Geurts J, Wunderink LU, et al. Identification of small heat shock protein B8 (HSP22) as a novel TLR4 ligand and potential involvement in the pathogenesis of rheumatoid arthritis. J Immunol. 2006 Jun 1;176(11):7021-7.

[167] Wells AD, Malkovsky M. Heat shock proteins, tumor immunogenicity and antigen presentation: an integrated view. *Immunol Today,* 2000 Mar;21(3):129-32.

[168] Kamada M, So A, Muramaki M, Rocchi P, Beraldi E, Gleave M. Hsp27 knockdown using nucleotide-based therapies inhibit tumor growth and enhance chemotherapy in human bladder cancer cells. Mol Cancer Ther. 2007 Jan;6(1):299-308.

[169] Hadaschik BA, Jackson J, Fazli L, Zoubeidi A, Burt HM, Gleave ME, et al. Intravesically administered antisense oligonucleotides targeting heat-shock protein-27 inhibit the growth of non-muscle-invasive bladder cancer. BJU international. 2008 Aug 5;102(5):610-6.

[170] Lee SC, Sim N, Clement MV, Yadav SK, Pervaiz S. Dominant negative Rac1 attenuates paclitaxel-induced apoptosis in human melanoma cells through upregulation of heat shock protein 27: a functional proteomic analysis. *Proteomics*, 2007 Nov;7(22):4112-22.

[171] Aloy MT, Hadchity E, Bionda C, Diaz-Latoud C, Claude L, Rousson R, et al. Protective role of Hsp27 protein against gamma radiation-induced apoptosis and radiosensitization effects of Hsp27 gene silencing in different human tumor cells. *Int J Radiat Oncol Biol Phys*. 2008 Feb 1;70(2):543-53.

[172] Chiosis G, Timaul MN, Lucas B, Munster PN, Zheng FF, Sepp-Lorenzino L, et al. A small molecule designed to bind to the adenine nucleotide pocket of Hsp90 causes Her2 degradation and the growth arrest and differentiation of breast cancer cells. *Chem Biol*. 2001 Mar;8(3):289-99.

[173] Chauhan D, Li G, Shringarpure R, Podar K, Ohtake Y, Hideshima T, et al. Blockade of Hsp27 overcomes Bortezomib/proteasome inhibitor PS-341 resistance in lymphoma cells. Cancer research. 2003 Oct 1;63(19):6174-7.

[174] Tezel G, Wax MB. The mechanisms of hsp27 antibody-mediated apoptosis in retinal neuronal cells. *J Neurosci*. 2000 May 15;20(10):3552-62.

Part IV. Therapeutic Approaches

In: Small Stress Proteins and Human Diseases
Editors: Stéphanie Simon et al.
ISBN: 978-1-61470-636-6

Chapter 4.1

SMALL STRESS PROTEINS AND THEIR THERAPEUTIC POTENTIAL

Scott A. Houck[1], Joy G. Ghosh[2] and John I. Clark[1,3*]
Departments of Biological Structure[1] and Ophthalmology[3], University of Washington, Seattle, WA, U.S.
[2]Departments of Psychiatry, Ophthalmology, Neurology, Pathology,Laboratory Medicine, and Biomedical Engineering, Boston University School of Medicine and Photonics Center, Boston, MA, U.S.

ABSTRACT

Small stress or heat shock proteins (sHSP) are involved in protective activity against some of the most important pathologies affecting human health including cancer and neurodegeneration. The fundamental molecular mechanism for the function(s) of small stress proteins (sHSPs or small heat shock proteins) are just beginning to be understood. The archetype for small stress proteins is human αB-crystallin where multiple interactive sequences were identified using protein pin arrays and confirmed using site directed mutagenesis. A dynamic equilibrium between subunits of αB-crystallin with large polydisperse complexes has been described. The interactive domains were mapped to the surface of a homology model for the small stress proteins. The surface exposure of the interactive domains is dynamic and varies in response to protein unfolding and self assemby. Peptides were synthesized on the basis of the interactive sequences and were found to be active in assays for protection against proteins associated with protein unfolding diseases. The equilibrium between sHSP subunits and assembled complexes appears to be a dynamic mechanism for regulation of stress protein function.

* John I. Clark
University of Washington
357420 Biological Structure
Seattle, WA 98195-7420, USA
Phone: (206) 685-0950
Fax: (206) 543-1524
clarkji@u.washington.edu

Characterization of the functional mechanism of small stress protein action is expected to lead to novel therapies for diseases of aging (see also chapter 4-3).

INTRODUCTION

Small heat shock proteins, sHSPs, are transforming our ideas about endogenous protective mechanisms against molecular aging in cells. sHSPs are a family of stress proteins of molecular weight less than 43kD that are upregulated in response to pathologies characterized by protein unfolding and aggregation, the molecular phenotypes of neurodegeneration, cardiovascular disease, age related macular degeneration (AMD) and cataract, well known threats to the quality of life for aging humans [1-12]. In protein unfolding diseases, modest alterations in the distribution of surface charge or hydrophobicity can influence molecular conformation and protein – protein interactions to favor aggregation [13-23]. While not recognized as protein unfolding diseases, tumorogenesis and cancer are known to increase in prevalence with aging and are associated with the upregulation of sHSPs [24-29], an association which may account for the interactions between stress proteins and regulatory proteins involved in maintaining cellular homeostasis [30]. In the absence of pathological stress, protein – protein interactions are regulated by post translational modifications that can influence normal cell proliferation, cell shape, adhesion, signaling, migration or cell death (apoptosis), which are important dynamic processes in cells and tissues [31-38]. The special and unique activity of stress proteins in normal cell functions and in protective cell mechanism(s) makes the sHSPs ideal candidates for novel therapeutics against aggregation diseases of cell and molecular aging [39-41].

Important clues to the protective mechanism(s) of sHSPs and stress proteins can be found through a thorough analysis of the primary structure of αB-crystallin, a prominent lens protein, discovered to be a sHSP upregulated to protect against protein unfolding and aggregation diseases in aging. Our approach is to identify the specific interactive sequences responsible for the protective activity of αB-crystallin. We seek to characterize the recognition and response functions against protein unfolding that are necessary for protection against aggregate formation, for stabilization of the assembly of filaments and formation of protective protein complexes. The proteins that interact with αB-crystallin come from diverse families of proteins involved in dynamic cellular activities that may require functional stability for long time periods in living cells [39, 40]. The time periods can be as long as the lifetime of the individual organism for neurons, cardiomyocytes and transparent lens cells, where long term stability is required to protect against the abnormal effects of aging on filamentous cytoskeletal structures. The protective function of human αB-crystallin depends on high sensitivity in the recognition of protein unfolding and a dynamic response both to native and unfolding proteins. This review summarizes recent studies of the identification of interactive sequences in human αB-crystallin, mapping of their interactive surface domains, and modeling of their functions in the dynamic molecular mechanisms that regulate the self-assembly of αB-crystallin subunits, cytoskeletal proteins, and unfolding protein in protein aggregation diseases. We expect that defining the structural basis for the protective interactions responsible for the stress response of αB-crystallin to dysfunctional unfolding proteins will provide new insights for the design of novel therapeutics against protein aggregation disorders.

1. Human Diseases and Stress Proteins

There is compelling experimental support for the importance of protein – protein interactions in the protective cellular response of stress proteins (sHSPs) to diseases of aging from neurodegeneration to cancer (Table 1). Common protein targets for the protective response of the sHSPs are self assembling proteins in the cytoskeleton, amyloid fibrils, unfolding proteins, and regulatory molecules, such as β-catenin and growth factors [30, 42-45]. sHSP – unfolding protein interactions can contribute to the formation of disease causing protein aggregates. For example, in the ocular lens, mutations and post-translational modifications of αA and αB-crystallin lead to the aggregation of lens crystallins and cytoskeleton causing cataract [46-50]. In the retina, upregulation of αB-crystallin has been linked to age-related macular degeneration (AMD) and αB-crystallin is a component of Drusen [5, 8, 51, 52]. HSP27 and αB-crystallin are upregulated in the brain where they were associated with amyloid aggregates in Alzheimer's disease, Parkinson's disease, Creutzfeldt-Jakob disease, and familial amyloidotic polyneuropathy [53-57]. In amyotrophic lateral sclerosis (ALS), sHSP27 and αB-crystallin are upregulated and have been found in aggregates [3, 56]. Mutations in HSP27 and HSP22 are linked to Charcot-Marie-Tooth disease and distal hereditary motor neuropathy, possibly due to a defective interaction with neurofilaments [58-60]. Mutations in GFAP cause HSP27 and αB-crystallin to co-aggregate in Alexander's disease and mutations in αB-crystallin produce similar co-aggregates with desmin leading to cardiomyopathy and desmin-related myopathy [50, 58-60]. Our interest is in identifying the primary sequences of the interactive surface domains on human αB-crystallin, the archetype sHSP, and mapping the exposed interactive domains to the surface of the sHSP [30, 61-68]. Because the interactions with unfolding target proteins are dynamic they are expected to involve weak, non-covalent interactions. For this reason the mechanism of protective action can be expected to be unique to sHSPs and very different from the families of large HSPs. Advances in understanding the interactions between sHSPs and their target proteins under stress will provide opportunities for development of innovative therapies against protein unfolding and aggregation during aging.

While the list of diseases in Table 1 could suggest that sHSPs are involved nonspecifically in a wide range of diverse pathologies, the common elements of nearly all the diseases listed, are that they are the result of protein unfolding, and dynamic molecular mechanisms mediated by weak, non-covalent protein – protein interactions. Our approach to the development of innovative therapies in protein unfolding and aggregation diseases is to define the nature and mechanisms of the interactions between sHSPs and their target proteins. The first step was the identification of the interactive sequences. The next step was to observe the 3D structure of the interactive domains on the surface of the αB-crystallin molecule. Based on the 3D molecular structure of the interactive domains on the surface of human αB-crystallin, it is expected that novel protective reagents will be designed.

2. Molecular Structure of Human αB-Crystallin – Archetype for sHSP and Stress Proteins

The archetype for eleven sHSPs found in humans, is human αB-crystallin [35, 36, 61, 69], a highly interactive stress response protein characterized by three structural domains: the N-terminal domain, the α-crystallin core domain, and the C-terminal domain (Fig 1) [6, 70-73]. The N-terminal and C-terminal domains are variable in sequence (Fig 1) and, in the crystal structures of *Methanococcus jannaschii* (Mj) HSP16.3 and wheat HSP16.9, the N-terminal domains can contain α helical content but are largely unstructured. In sHSPs, the C-terminal extension protrudes from the α-crystallin core domain as an unstructured, flexible sequence. In contrast to the unstructured N- and C-terminal domains, the α-crystallin core domain is an immunoglobulin-like sandwich of β-strands that are stabilized by two anti-parallel β-sheets formed from six to nine β-strands connected by loops of variable lengths. The formation of dimers in wheat HSP16.9 is due to interactions between the β2 and β3 strands of one monomer with the β6 strand contained in the loop connecting β5 and β7 of another monomer. Extensive hydrogen bonding accounts for the stability of the α-crystallin core domain in sHSPs. The C-terminal extension contains a conserved I-X-I/V motif, where I is isoleucine, V is valine, and X is any natural amino acid [74]. In wheat HSP16.9, the I-X-I/V motif of one monomer interacts with residues of the β4 and β8 strands of another monomer to form the dodecameric quaternary structure observed using X-ray diffraction. While human αB-crystallin contains the same three structural domains found in Mj HSP16.5 and wheat HSP16.9, αB-crystallin forms large and polydisperse complexes. The heterogeneity in the size of the complex formed reflects the variation in the multiple interactive domains that respond to unfolding proteins and accounts for the fact that full-length αB-crystallin has not been crystallized for X-ray diffraction while HSP16.5 and HSP16.9 have [75-78]. This heterogeneity of αB-crystallin in complex size and in interactions with unfolding proteins is consistent with the fact that the multiple interactive domains in human αB-crystallin are more favorable for interactions with multiple unfolding proteins than the large HSPs. Proteolysis, deletion mutagenesis, single residue site-directed mutagenesis, and domain-swapping chimeric mutagenesis of αB-crystallin and *Caenorhabditis elegans* HSP12.2 were used to determine the sequences and specific residues in all three structural domains of sHSPs that were important for complex assembly and protective activity against protein unfolding and aggregation [63, 64]. While the homology model in Fig 1 is a computer simulation, the results of mutation studies and the use of synthetic peptides were consistent with the expected molecular structure of αB-crystallin and the interactive domains that respond to unfolding proteins. As will be described later, the structure of αB-crystallin provides a model for the design of novel therapeutics.

The software Molecular Operating Environment (Chemical Computing Group, Montreal, Quebec, Canada) was used to construct the 3D homology model of human αB-crystallin. MOE employs a number of algorithms including multiple sequence alignment, structure superposition, contact analyzer, and fold identification to develop homology models based on available high-resolution crystal and/or NMR structures of the template protein molecule [79]. The program analyzed the stereochemical quality of the predicted models and took into account planarity, chirality, phi/psi preferences, chi angles, nonbonded contact distances, unsatisfied donors, and acceptors [80]. The predicted secondary structure of human αB-crystallin (Fig 1) was aligned and found to be consistent with the secondary structure observed for wheat HSP16.9. The alignment was used by MOE to create a series of 10 energy-minimized 3D models. Each model was evaluated using the ModelEval module of MOE. The best model was selected and superimposed on the monomeric subunits of Mj HSP16.5 and wheat HSP16.9 and the 3D coordinates of the three structures were fit with a final RMSD of 3.25 Å [61]. The three structures have the same basic topology. The hydrophobic N terminus has helical elements. The α-crystallin core is an immunoglobulin-like β-sandwich that forms two anti-parallel β-sheets connected by a flexible loop. The C-terminal extension is charged and unstructured. The overall result was in good agreement with existing spectral and NMR data on the structure of αB-crystallin [81-83]. The homology model for human αB-crystallin provided the molecular structure for mapping the interactive sequences identified using protein pin arrays onto the surface of human αB-crystallin where the interactions with unfolding proteins occur.

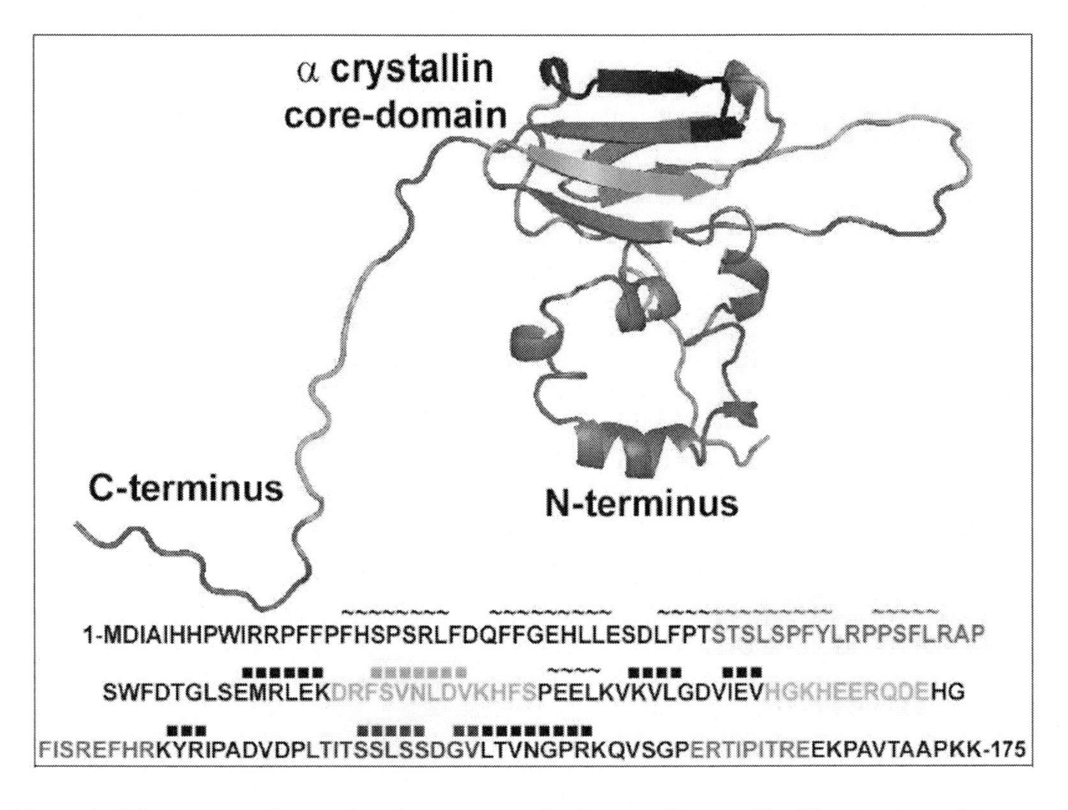

Figure 1. Primary, secondary and tertiary structure for human αB-crystallin. The structure of human αB-crystallin was modeled as previously described [61]. The amino acid sequence of human αB-crystallin is at the bottom of the figure. The secondary structure is indicated with squares (■) for β-strand and tildas (~) for α-helix. The color indicates locations of the interactive peptides, 41STSLSPFYLRPPSFLRAP58 (red), 73DRFSVNLDVKHFS85 (orange), 101HGKHEERQDE110 (light green), 113FISREFHR120 (dark green), 131LTITSSLSSDGV142 (blue), and 156ERTIPITRE164 (purple).

3. Identification of Interactive Sequences and the Function of sHSPs

Following the discovery of a 19 amino acid peptide which corresponded to the β3/β4 strands in human αA-crystallin and retained the protective activity against protein unfolding and aggregation [84, 85], we investigated the interactive sites in human αB-crystallin using protein pin array technology [61]. It was recognized quickly that αB-crystallin contained multiple interactive sequences that responded selectively, and rapidly to unfolding proteins with low affinity binding. The interactive sequences on αB-crystallin were sensitive to the subtle, dysfunctional modifications which, if left unprotected, could disrupt the cellular functions of proteins. Structurally αB-crystallin is a mosaic of interactive domains whose relative affinities must be coordinated to be effective, but not too effective, at binding unfolding proteins. For a multifunctional stress response protein, specific, high affinity binding may not be as effective as selective binding which suggests the importance of weak non-covalent interactions including hydrogen bonds, hydrophobic and electrostatic interactions. The goal was to identify the interactive sequences in human αB-crystallin and a protein pin array was designed to determine the interactive peptide sequences in αB-crystallin [61, 62, 66, 67]. In the assay, peptides corresponding to residues 1–175 of human αB-crystallin were synthesized employing a simultaneous peptide synthesis strategy developed by Geysen, called Multipin Peptide Synthesis [86]. The peptides were immobilized on derivatized polyethylene pins arranged in a microtiter plate format [61]. Each peptide was eight amino acids in length, and consecutive peptides were offset by two amino acids. All peptides were covalently bound to the surface plastic pins. The first peptide on the first pin of the array was $_{1}$MDIAIHHP$_{8}$, and the last peptide immobilized on the last pin of the array was $_{167}$PAVTAAPK$_{174}$ of human αB-crystallin. In total, eighty-four peptides corresponding to the 175-amino-acid primary sequence of human αB-crystallin were synthesized and fixed to sequential pins in the array. To screen for binding to the peptides, fixed concentrations of each target protein were added to each well in the microtiter plate and incubated for reaction with the peptides immobilized on the pins. Repeated washes removed unbound target protein before a labeling procedure measured the amount of target protein complexed with each αB-crystallin 8-mer sequence. The pin arrays can be regenerated up to 90 times for repeated use.

The target proteins used in the pin array assays can be categorized as (a) crystallins, (b) cytoskeletal proteins, (c) unfolding proteins and (d) regulators of growth and differentiation that participate in pathways of protection/stabilization of normal cell structure. These interactive proteins are important for fundamental cell functions during normal differentiation and aging and have been reported to interact with αB-crystallin. The pin array results determined that multiple interactive domains were important for the selectivity (not specificity) of the αB-crystallin in multifunctional mechanisms. For example, the β3 and β8 strands in αB-crystallin participate in self assembly of the αB-crystallin complex, filament interactions, protection against protein unfolding, growth factors and β catenin [61, 62, 66, 67]. In contrast, the β4 and β5 strands interact with signaling proteins and not in complex assembly or filament interactions. The β3-β8 strands form an interactive domain that is a key binding region where the effects of mutations on association/dissociations constants measured using SPR are expected to correlate closely with altered activity of αB-crystallin. Understanding the complex mechanism of the protective activity of sHSPs requires characterization of the specific functions of the individual sequences involved in interactions with unfolding proteins. The fact that sHSPs interact selectively with multiple proteins suggests that diversity in the multiple interactive domains rather than specificity is an important parameter in the protective mechanism(s) of sHSPs against protein unfolding and aggregation and accounts for the ability of αB-crystallin to recognize and bind a variety of unfolding proteins and other self assembling target proteins with different affinities. Once the multiple interactive sequences were identified, they were mapped to the 3D computer homology model of αB-crystallin based on the X-ray diffraction structures of Mj HSP16.5 and wheat HSP16.9 where it was possible to visualize, for the first time, the exposed surface domains that were most likely to be responsible for the interactive functions of sHSPs in cells. It was hypothesized that, in the presence of external stress, sHSPs detect protein unfolding with an extraordinary sensitivity that could be difficult to duplicate outside biological cells and tissues.

4. Exposed Surface Domains and Protective Interactions

To establish the relationship between αB-crystallin structure and biological activity, the multiple interactive sites on the surface of the αB-crystallin molecule needed to be defined and visualized in 3D structural models. Space-filling models of αB-crystallin were generated and the distributions of the exposed interactive domains were mapped to the surface, using the 3D molecular viewing software PyMol (Figure 2). In figure 2, all surface exposed side chains in the interactive regions are yellow and the surfaces not involved in protein – protein interactions are gray. While there are similarities in the surface exposed interactive domains, small variations are obvious. It is easily imagined that the surface variability is necessary to adapt to the differences in the surfaces of the diverse unfolding target proteins. Interactions occur on the exposed surface of the core domain of αB-crystallin, on the N-terminus and on the C-terminus which is consistent with reports of the involvement of the C-, N- termini and core domain in the variety of selective activities of sHSPs [87-92]. The results support the hypothesis that the molecular mechanisms for the multifunctional activity of sHSPs in disease involve diverse and multiple interactive domains. Visualization of the diverse domains permits the design of mutation experiments to confirm the precise residues involved in the recognition and binding of selected families of proteins, a major functions of human αB-crystallin. Point mutations in the surface exposed amino acids confirmed the models, and synthetic peptides based on the interactive domains were tested for functional activity in vitro [30, 61-64, 68]. The results determined that diversity in the interactive domains has functional importance and may account for the selectivity of αB-crystallin for several families of proteins and for the polydisperse dimensions of the complexes of αB-crystallin [75, 76, 78]. If there is functional significance to heterogeneity of the complex size, it is a significant mechanistic difference between the small and large HSPs [30, 66, 67, 93] that can be clarified through continued research on the dynamics of the functions of αB-crystallin.

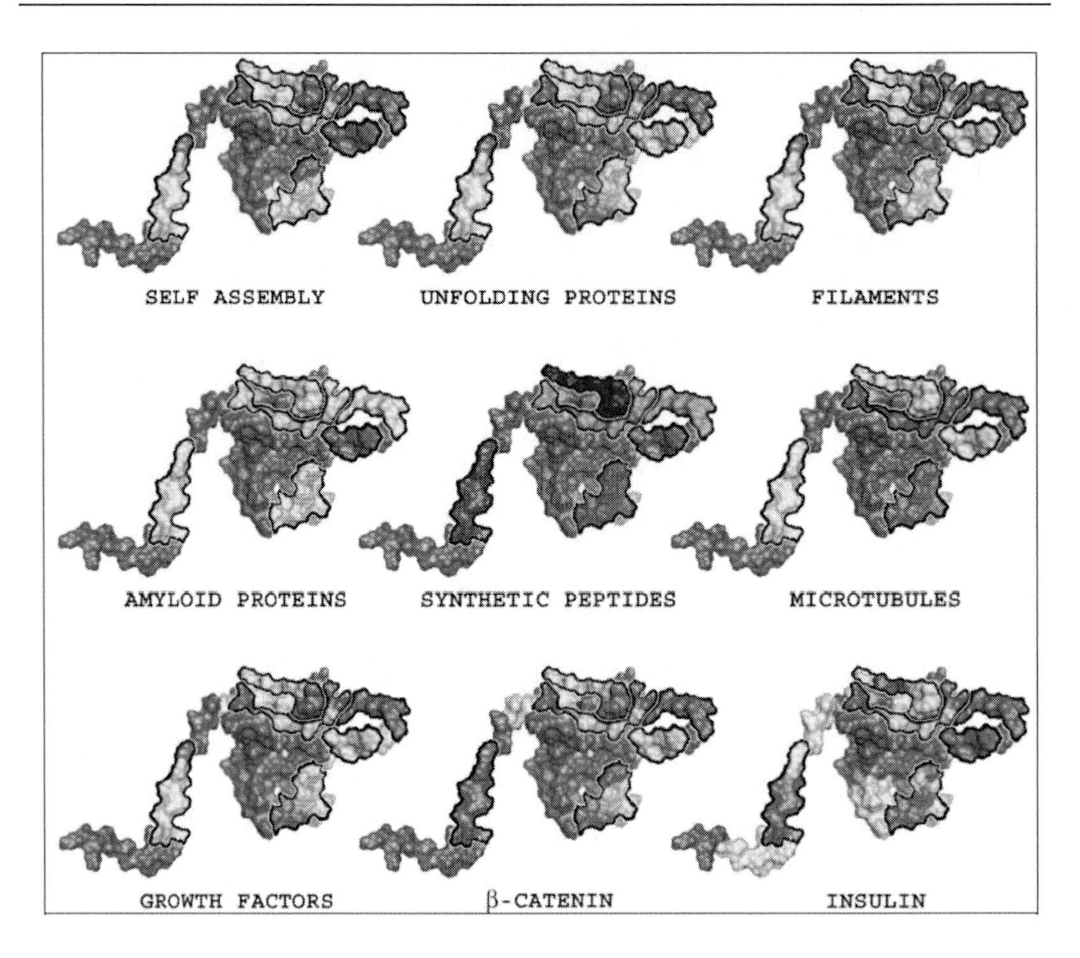

Figure 2. Diversity in the interactive surface domains of human αB crystallin. Synthetic peptides for potential use as therapeutics were based on the mapping of interactive domains to the surfaces of the 3D homology model of human αB crystallin and were colored (CENTER MODEL). The peptides are 41STSLSPFYLRPPSFLRAP58 (red), 73DRFSVNLDVKHFS85 (orange), 101HGKHEERQDE110 (light green), 113FISREFHR120 (dark green), 131LTITSSLSSDGV142(blue), and 156ERTIPITRE164(purple). The surrounding models represent the surface domains (yellow) for the interactive sequences required for self-assembly of αB crystallin, binding to unfolding proteins, binding to intermediate and micro-filaments, binding to microtubules, binding to insulin, binding toβ-catenin, binding to growth factors, and interactions with amyloid fibrils. All interactive sequences were identified using protein pin-arrays and functional assays with synthetic peptides. Surface domains for globular proteins were based on interactions with thermally unfolded alcohol dehydrogenase, citrate synthase, and β/γ crystallin. Surfaces domains for filaments were based on the interactions with actin, GFAP, vimentin, and desmin. Surfaces domains growth factors were defined by the interactions with NGF, VEGF, and FGF. Surface domains for amyloid proteins were based on interactions with β-amyloid 1-42, α-synuclein, β2-microglobulin, and transthyretin. The spatial distribution of the surface domains is diverse which may be necessary to adapt to the surface of the individual target proteins.

The results are consistent with the hypothesis that small variations in the interactive surfaces of a stress response protein can account for diverse selectivity in protein targets, an indication of the importance of surface exposed sequences in the stress response of αB-crystallin to unfolding proteins and self assembling systems. Detailed studies of the N-terminal interactive sequence LFPTSTSLSPFYLRPPSF determined that the exposed surface was 79% hydrophobic. The hydrophobic surface for the sequences FSVNLDVK, LTITSSLS, and GVLTVNGP that form the β3, β8, and β9 strands was calculated to be 65%, 68%, and 71% hydrophobic respectively, an average of 68% for the interface formed by all three β strands. The calculated hydrophobic surface for the C-terminal extension PERTIPTREEK was 58% hydrophobic [61]. The high proportion of hydrophobic surface in the interactive sequences is consistent with the potential importance of hydrophobic side chains in interactions between human αB-crystallin and its binding targets, α-crystallin subunits, cytoskeletal proteins, signaling proteins and regulators of normal growth [92]. Mapping of the interactive sequences to the 3D models of human αB-crystallin determined that the binding regions on the N-terminus, C-terminus, and α crystallin core domain were solvent-exposed on the surface of the monomeric subunit. The importance of understanding the variability in the exposure of the interactive sequences in αB-crystallin was the basis for the plan to synthesize interactive peptides for testing in amyloid formation assays *in vitro*.

5. Synthesis and Functional Evaluation of Bioactive Peptides

The αB-crystallin interactive peptides, DRFSVNLDVKHFS (DR), HGKHEERQDE (HG), STSLSPFYLRPPSFLRAP (ST), LTITSSLSSDGV (LT), ERTIPITRE (ER) and FISREFHR (FI), were synthesized or purchased for experimental tests using a thioflavin T fluorescence assay for amyloid fibril formation of β-amyloid (Aβ(1-42)), α synuclein, transthyretin (TTR), and β2-microglobulin (β2m) [68]. Thioflavin T is a cationic benzothiazole that has high affinity for amyloid fibrils and enhanced fluorescence when bound to aggregating amyloid fibrils. The effectiveness of individual interactive sequences against amyloid fibril formation was quantified using the Thioflavin T assay and was found to vary with the amyloid protein (Table 2). For example, the DR peptide inhibited the formation of α synuclein fibrils but increased the fibril formation of Aβ(1-42). In contrast, ST inhibited fibril formation of Aβ but increased formation of α synuclein fibrils. Truncated bioactive sequences as small as four amino acids retained activity. Similar to the large synthetic bioactive peptides, the effectiveness of the four amino acid peptides varied with the target amyloid. For example, the KHFS peptide was a very effective inhibitor of fibril formation of α synuclein and a poor inhibitor of fibril formation of β2 microglobulin (Table 2). The results were an impressive demonstration of the bioactivity of individual synthetic peptides on fibril formation of various amyloid proteins. The experimental results using synthetic bioactive peptides were important for several reasons and a number of conclusions can be made. First, the stress protein, human αB-crystallin, may have a more significant and direct function in the pathophysiology of amyloid fibril formation than previously thought. Second, the results were a clear demonstration that individual bioactive peptides in αB-crystallin have activity similar to that of the parent molecule. Third, individual bioactive peptides as small as four amino acids were effective inhibitors of fibril formation of amyloid proteins. Fourth, the multiple interactive domains on the surface of αB-crystallin had multiple effects on the formation of self assembling fibrils. The variation in response of fibril assembly to individual interactive peptides accounted for the adaptive capability of αB-crystallin in protection against multiple protein targets in aging diseases, stabilization of filaments and regulation of microtubule assembly during differentiation and development [66-68]. The collective effects of the multiple interactive surface domains on human αB-crystallin need to be clarified to understand the function of sHSPs in vivo. At this point in our understanding of the mechanism(s) of αB-crystallin activity, we can hypothesize that the activity of a single stress protein resembles a multiprotein complex in which a mosaic of interactive sites with variable affinities modulate selective and diverse interactions and the dynamics of protein self assembly. While it is unlikely that these results will translate immediately into new therapies for protein unfolding diseases including cataract or neurodegeneration, knowledge of the structure-function relationships will be useful for identification of specific activities of each interactive domain with selected target proteins and a better understanding of the molecular mechanisms for the protective activity of sHSPs.

6. The αB-crystallin Peptide – Amyloid Protein Complex

The protective effects of αB-crystallin against fibril formation in protein unfolding diseases that include Alzheimer's and Parkinson's are well documented [23, 33] (see chapter 1-7). The experiments using the bioactive peptides confirmed the activity of the individual interactive sequences in the function of αB-crystallin and provided a valuable model for determination of the importance of specific protein – protein interactions in sHSPs for recognition and binding to amyloid proteins. 3D molecular modeling was conducted to better characterize the structure activity relationships and the specific interactions responsible for the effects of the bioactive peptides on fibril formation of amyloid proteins. Three-dimensional co-ordinates for the interactive peptides were extracted from the computed homology model of human αB-crystallin (fig 1) [30, 61, 62]. The three-dimensional co-ordinates for human transthyretin (1DVQ) and human β2-microglobulin (1LDS) were obtained from their crystal structures in the protein databank [94, 95]. Co-ordinates for the peptides and the target proteins were used in the molecular docking program ClusPro [96, 97] which computed docking models for each protein–peptide combination (Fig 3) on the basis of free energy minimization, and filters for complex selection based on minimization of solvation and electrostatic energies to obtain the best structural fits between the bioactive peptides and the amyloid forming proteins. The 3D models characterized the interactions between transthyretin and β2-microglobulin and the αB-crystallin peptides DRFSVNLDVKHFS which forms the β3 strand and HGKHEERQDE which forms part of the loop connecting the β5 and β7 strands on the exposed surface of the αB-crystallin core domain (Fig 3A-C). Residues His-101, Lys-103, His-104, Glu-105, and Arg-107 from the 101HGKHEERQDE110 peptide interacted with residues Glu-51, His-52, Ser-53, and Leu-55 in β2-microglobulin (Fig 3A). Similarly, residues His-104, Glu-105, and Arg-107 from the 101HGKHEERQDE110 peptide interacted with residues His-79, His-81, Glu-83, and Val-85 in transthyretin (Figure3B). Residues Phe-75, Val-77, and Leu-79 from the 73DRFSVNLDVKHFS85 peptide interacted with residues Glu-51, Ser-53, Asp-54, Leu-55, and Phe-57 in β2-microglobulin (Figure 3C). Both αB crystallin peptides bound to the D-strand of β2-microglobulin and the F-strand of transthyretin. These β strands shared sequence homology and are known to be involved in critical intra-fibril interactions [68, 95, 98-100]. While homology models cannot be definitive, the interactions at the simulated sites of peptide - amyloid interactions are expected to be weak and non-covalent, consistent with the hypothetical mechanism proposed for protective activity of αB-crystallin against aggregation diseases. The models represent the direct interactions between residues of the αB-crystallin peptides and residues of amyloidogenic fibril forming β strands of transthyretin and β2-microglobulin that can be expected to prevent abnormal self-assembly of transthyretin and β2-microglobulin into β strand rich amyloid-like fibrils, as was observed in the thioflavin T assays.

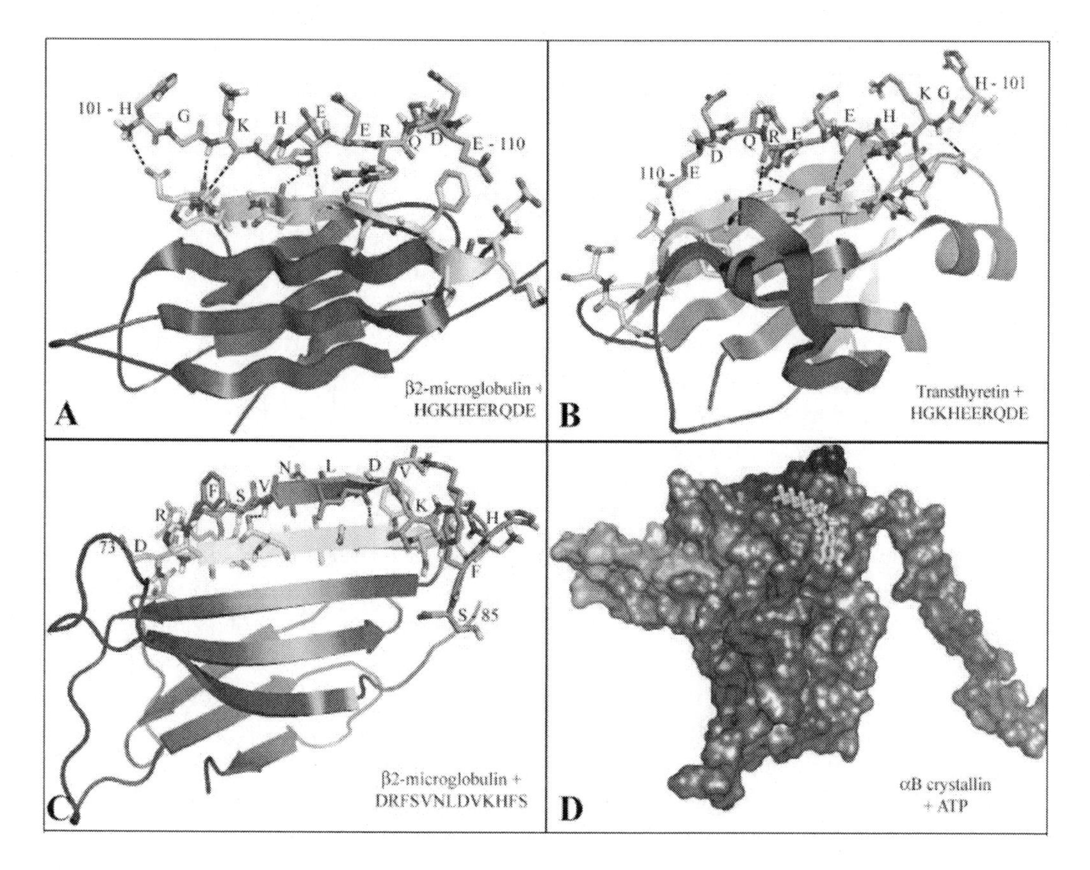

Figure 3. Models for molecular interactions between αB crystallin and amyloid proteins. The structures of αB crystallin peptides were docked with the amyloidogenic proteins, β2-microglobulin (PDB: 1LDS) and transthyretin (PDB:1DVQ) using the ClusPro molecular docking program. (A) The peptide $_{101}$HGKHEERQDE$_{110}$ (green) binding β2-microglobulin at the D strand (yellow). (B) The peptide $_{101}$HGKHEERQDE$_{110}$ (green) binding transthyretin at the critical amyloidogenic F strand(yellow). (C) The peptide $_{73}$DRFSVNLDVKHFS$_{85}$ (red) binding β2-microglobulin at the critical amyloidogenic D strand (yellow). The peptide has the potential to form a β-strand apposed to the β-sheet of β2-microglobulin, disrupting amyloid formation. (D) The surface domains of the interactive peptides in αB crystallin: $_{41}$STSLSPFYLRPPSFLRAP$_{58}$ (red), $_{73}$DRFSVNLDVKHFS$_{85}$ (orange), $_{101}$HGKH EERQDE$_{110}$ and $_{113}$FISREFHR$_{120}$ (green), $_{131}$LTITSSLSSDGV$_{142}$ (blue), and $_{156}$ERTIPITRE$_{164}$ (purple). Note the ATP (yellow) binding domain in a shallow groove at a Walker-B ATP binding motif between the β8-strand $_{131}$LTITSSLSSDGV$_{142}$ and the β4 strand [65].

For comparison, the interactive surface for ATP was mapped onto the 3D model for αB-crystallin (Fig 3D). Fluorescence resonance energy transfer (FRET) and mass spectrometry determined that the ATP interaction with αB-crystallin occurs in a shallow groove between the β4 and β8 strands of the exposed surface of the α-crystallin core domain. The ATP binding site is at the surface exposed interactive sequence in the β4 strand in αB-crystallin and is similar to a Walker B ATP binding motif [93]. This is not at the same surface site as the interactive domain for amyloidogenic proteins. The 3D molecular models generate new and detailed structure - activity relationships at the atomic level for the design of novel therapeutics based on the molecular structure and function of the small stress protein, human αB-crystallin.

7. Surface Accessible Interactive Domains and the Protective Mechanism of Human αB-Crystallin

One of the mysteries of stress proteins and sHSPs is the functional importance of the protein – protein interactions in the dynamic assembly of complexes in protection against protein unfolding (Fig 4). A well known characteristic of αB-crystallin and most sHSPs is the assembly of polydisperse complexes that are in dynamic equilibrium with the subunits. The dynamic equilibrium is sensitive to physical conditions in the environment that include temperature, pH, and ionic strength because subunit assembly is the result of weak non-covalent hydrogen bonds, electrostatic and hydrophobic interactions. The distribution of the multiple interactive domains on the surface of the space filled model of αB-crystallin is consistent with multiple interactions between subunits (see the colored domains in fig 4, top row left). The orange and blue areas are the exposed interfaces of the α crystallin core domain that contain the β3 (orange) and β8 (blue) strands for self-assembly of the αB crystallin complex [61]. In an assembled complex (top row right) the β3 (orange) and β8 strands (blue) are partially buried and inaccessible. In the absence of stress induced protein unfolding, αB crystallin subunits are in a dynamic equilibrium with the αB crystallin complex and interactions with native folded proteins are minimal (second row). In the presence of stress, unfolded proteins can shift the dynamic equilibrium to favor access to the β3 and β8 strands in the core domain of the dissociated αB crystallin subunits. Access to the interactive domains permits protective binding with unfolding proteins to inhibit the abnormal function and aggregation (middle row). The multiple interactive sites on the αB-crystallin subunits interact with unfolding proteins to form heterogeneous complexes that can grow to very large sizes [42, 78, 101, 102]. On the basis of a dynamic model, it can be hypothesized that the strength of the interactions between subunits and unfolding proteins influences the protective activity of the stress protein. For example, strong interactions between subunits could be expected to favor formation of homomeric complexes decreasing accessibility to the interactive domains on αB-crystallin subunits and reducing protective interactions with unfolding proteins. In contrast, weak interactions between subunits could be expected to favor dissociation of the complexes making the interactive domains on αB-crystallin subunits accessible for protective interactions with unfolding proteins. The model predicts that (a) the protective activity of the stress proteins is regulated dynamically by access for unfolding proteins to interactive domains on the surface of the stress response proteins and (b) the relative affinities between stress protein subunits and unfolding proteins have regulatory functions. In this dynamic model, the function of the complex is to sequester the protective interactive surface domains until stress shifts the dynamic subunit – complex equilibrium to favor dissociation of subunits

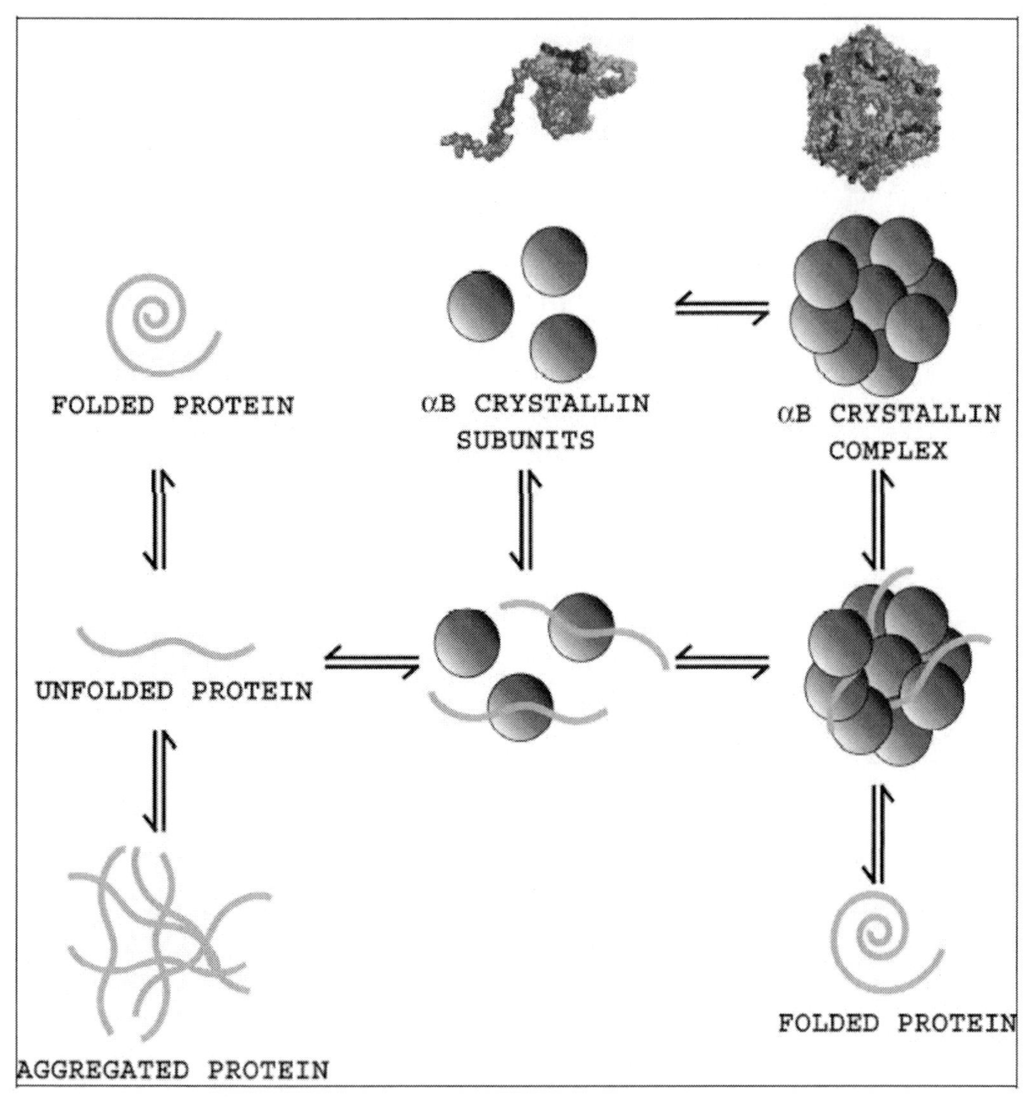

Figure 4. Model for the protection against protein unfolding and aggregation. In the first row, the colors label the exposed interactive sequences on the surface of an individual αB crystallin subunit (left image) which are partially buried in an assembled αB complex (right image). The surface domains for each interactive peptides are colored; 41STSLSPFYLRPPSFLRAP58 (red), 73DRFSVNLDVKHFS85 (orange), 101HGKHEERQDE110 (light green), 113FISREFHR120 (dark green), 131LTITSSLSSDGV142 (blue), and 156ERTIPITRE164 (purple). The second row models the dynamic equilibrium between the pool of αB subunits (grey ovals) and subunits assembled into complexes under normal conditions when interactions with native folded proteins (green) are minimal. Under conditions of cell stress, protein unfolding/misfolding increases the interactions between αB subunits and the unfolding/misfolding proteins (third row) and stronger subunit-unfolding protein interactions favor formation of αB crystallin-unfolded target protein complexes. The bottom row models the outcome of protein unfolding without (left) and with αB crystallin (right). The model suggests that the dynamics of complex assembly are integral to the function of the stress response protein, human αB crystallin.

for interaction with unfolding proteins. The emphasis of this model is on the dynamic nature of the subunit - complex equilibrium which can regulate access to the multiple interactive surfaces on αB-crystallin and other sHSPs. Detailed characterization of the relative strengths of the interactions between αB-crystallin subunits and unfolding proteins will be needed to test this hypothesis on the functional relationship(s) between the interactive surfaces on the stress protein, human αB-crystallin, and unfolding target proteins.

Table 1. Small heat-shock protein and stress protein associated diseases. Column 1 lists the name of the disease. Column 2 lists the organs affected. Column 3 lists the small heat-shock protein involved. Column 4 lists the names of the associated aggregated protein. Column 5 lists other details. Many of the diseases linked to sHSP are age related and most involve aggregate and filament formation. While cancer and tumors are not directly linked with protein unfolding and aggregation, modulation of cytoskeleton assembly or filament formation could account for the relationship between sHSP and these diseases which are increased in the aging populations.

Disease	Organ	sHSP	Aggregate proteins	Type	Notes
Alexander disease	Brain	HSP27, αB	GFAP	filament	HSP27 and αB crystallin coaggregate with GFAP filaments
Alzheimer's disease	Brain	HSP27,αB	β amyloid	amyloid	αB crystallin and HSP27 present in neurofibrillary tangles
Amyotrophic lateral sclerosis (ALS)	Brain	HSP27,αB	neurofilaments	filament	Presence of αB crystallin and HSP27 positive aggregates
Charcot-Marie-Tooth disease	Brain	HSP27, HSP22	neurofilaments	filament	Linked to mutations in HSP27 and HSP22
Creutzfeldt-Jakob disease	Brain	HSP27,αB	prion protein	amyloid	αB crystallin and HSP27 upregulated in disease
distal hereditary motor neuropathy	Brain	HSP27	neurofilaments	filament	Linked to mutations in HSP27
familial amyloidotic polyneuropathy	Brain	HSP27	transthyretin	amyloid	HSP27 upregulated in presence of aggregates
Parkinson's disease	Brain	HSP27,αB	α synuclein	amyloid	αB crystallin and HSP27 present in Lewy bodies
Age-related macular degeneration	Eye	αB	Drusen	unfolding proteins	αB crystallin present in Drusen aggregates
Cataract	Eye	αA, αB	β/γ crystallin, CP49, filensin, vimentin	unfolding proteins, filament	
Cardiomyopathy	Heart	αB	desmin	filament	αB crystallin coaggregates with desmin filaments
Desmin-related myopathy	skeletal muscle	αB	desmin	filament	αB crystallin coaggregates with desmin filaments
Cancer	prostate and mammary glands, brain, kidney, colon	HSP27, αB	-	-	αB crystallin and/or HSP27 are upregulated in disease. Some sHSP-positive tumors are drug resistant.

Table 2. Effects of αB crystallin peptides on amyloid fibril formation. The effects of the αB crystallin peptides on amyloid fibril formation of β-amyloid (A β (1-42), α synuclein, transthyretin (TTR), and β2-microglobulin (β2m) using Thioflavin-T fluorescence. The molar ratios of peptide:amyloid were 1:1 (1x) and 5:1 (5x). 100% is defined as the fibril formation of amyloid protein in the absence of peptide and 0% is defined as no fibril formation. The effectiveness of individual bioactive peptides derived from αB crystallin varies with the peptide sequence, the concentration and the protein target. The results suggest that different interactive sequences are specialized for different protein targets.

peptide	Residues	Aβ(1-42)		α synuclein		TTR		B2m	
		1x	5x	1x	5x	1x	5x	1x	5x
no peptide	-	100%	100%	100%	100%	100%	100%	100%	100%
STSLSPFYLRPPSFLRAP	41-58	0%	4%	446%	23%	92%	64%	62%	52%
DRFSVNLDVKHFS	73-85	49%	165%	0%	6%	55%	35%	38%	36%
DRFS	73-76	41%	26%	0%	9%	22%	38%	122%	217%
NLDV	78-81	0%	53%	0%	0%	25%	43%	50%	27%
KHFS	82-85	72%	71%	1%	0%	38%	45%	157%	196%
HGKHEERQDE	101-110	0%	0%	31%	34%	28%	29%	46%	42%
HGKH	101-104	100%	100%	2%	14%	13%	23%	13%	23%
HEER	104-107	151%	0%	43%	26%	8%	14%	8%	14%
RQDE	107-110	136%	114%	100%	161%	3%	22%	3%	22%
FISREFHR	113-120	96%	98%	100%	96%	93%	105%	229%	183%
LTITSSLSSDGV	131-142	168%	103%	30%	60%	86%	85%	37%	49%
ERTIPITRE	156-164	154%	140%	49%	59%	40%	93%	46%	59%

CONCLUSION

It is more important than ever before to understand the fundamental biological mechanisms responsible for the protective activity of stress response proteins and to maximize the development of therapeutics that can utilize these mechanisms. The multifunctional activity of human αB-crystallin can be accounted for by the diversity and variation in the interactive domains on the surface of the stress response protein. Similar to many sHSPs, αB-crystallin interacts selectively, not specifically, both with unfolding proteins and with proteins important for cell differentiation and development including proteins in regulatory pathways, apoptosis and filament assembly. The mechanisms for these activities distinguish the small from the large families of heat shock proteins. Similar to several sHSPs, αB-crystallin has special affinity for self-assembling protein systems in the formation of complexes, filaments, microtubules and amyloid fibrils. Under the stress of normal differentiation, growth and aging, αB-crystallin is a dynamic modulator of cell proliferation, migration and elongation which are sensitive to subtle changes in hydrophobicity, pH, ionic strength, charge, hydration, proteolysis and post-translational modifications which are important factors in protein – protein interactions. As stress response proteins, the protective function of αB-crystallin is beneficial against protein unfolding leading to the amyloidogenesis observed in a variety of aging diseases. The α crystallin core domain and the N- and C-termini contain interactive sequences that are critical for the normal activity of sHSPs and individually retain protective functions against the deleterious effects of unfolding proteins and unstable cytoskeletal assembly. Future experimental studies are expected to define the functional significance of the polydisperse complexes that can respond rapidly and dynamically to cellular stress. Our approach is to use the 3D molecular structures of the stress proteins and their interactions as the basis for identifying innovative therapies. Few proteins hold greater therapeutic promise than sHSPs for safe endogenous protection against protein unfolding and aggregation diseases (Table 1).

ACKNOWLEDGMENTS

Our work is supported by EY04542 and P30EY01730 from the National Eye Institute.

REFERENCES

[1] Head, M.W., Corbin, E., Goldman, J.E. (1993). Overexpression and abnormal modification of the stress proteins α B-crystallin and HSP27 in Alexander disease. *Am J Pathol 143*:1743-1753.

[2] Vicart, P., Caron, A., Guicheney, P., Li, Z., Prevost, M.C., et al. (1998). A missense mutation in the αB-crystallin chaperone gene causes a desmin-related myopathy. *Nat Genet 20*:92-95.

[3] Vleminckx, V., Van Damme, P., Goffin, K., Delye, H., Van Den Bosch, L., Robberecht, W. (2002). Upregulation of HSP27 in a transgenic model of ALS. *J Neuropathol Exp Neurol 61*:968-974.

[4] Arrigo, A.P., Muller, W.E.G. (2002). *Small Stress Proteins*. New York, NY: Springer Verlag.

[5] Crabb, J.W., Miyagi, M., Gu, X., Shadrach, K., West, K.A., et al. (2002). Drusen proteome analysis: an approach to the etiology of age-related macular degeneration. *Proc Natl Acad Sci U S A 99*:14682-14687.

[6] Bloemendal, H., de Jong, W., Jaenicke, R., Lubsen, N.H., Slingsby, C., Tardieu A. (2004). Ageing and vision: structure, stability and function of lens crystallins. *Prog Biophys Mol Biol.86*:407-485.

[7] Dabir, D.V., Trojanowski, J.Q., Richter-Landsberg, C., Lee, V.M., Forman, M.S. (2004). Expression of the small heat-shock protein αB-crystallin in tauopathies with glial pathology. *Am J Pathol 164*:155-166.

[8] Nakata, K., Crabb, J.W., Hollyfield, J.G. (2005). Crystallin distribution in Bruch's membrane-choroid complex from AMD and age-matched donor eyes. *Exp Eye Res 80*:821-826.

[9] Benjamin, I.J., Guo, Y., Srinivasan, S., Boudina, S., Taylor, R.P., et al. (2007). CRYAB and HSPB2 deficiency alters cardiac metabolism and paradoxically confers protection against myocardial ischemia in aging mice. *Am J Physiol Heart Circ Physiol. 293*:H3201-H3209.

[10] Kampinga, H.H., Henning, R.H., van Gelder, I.C., Brundel, B.J. (2007). Beat shock proteins and atrial fibrillation. *Cell Stress Chaperones12*:97-100.

[11] Quraishe, S., Asuni, A., Boelens, W.C., O'Connor, V., Wyttenbach, A. (2008). Expression of the small heat shock protein family in the mouse CNS: Differential anatomical and biochemical compartmentalization. *Neuroscience 153*:483-491.

[12] Vos M.J., Hageman, J., Carra, S., Kampinga, H.H. (2008). Structural and functional diversities between members of the human HSPB, HSPH, HSPA, and DNAJ chaperone families. *Biochemistry 47*:7001-7011.

[13] He, H.W., Zhang, J., Zhou, H.M., Yan, Y.B. (2005). Conformational change in the C-terminal domain is responsible for the initiation of creatine kinase thermal aggregation. *Biophys J 89*:2650-2658.

[14] Muchowski, P.J., Wacker, J.L. (2005). Modulation of neurodegeneration by molecular Chaperones. *Nat Rev Neurosci. 6*:11-22.

[15] Duennwald, M.L., Jagadish, S., Giorgini, F., Muchowski, P.J., Lindquist, S. (2006). A network of protein interactions determines polyglutamine toxicity. *Proc Natl Acad Sci U S A. 103*:11051-11056.

[16] Roberson, E.D., Mucke, L. (2006). 100 years and counting: prospects for defeating Alzheimer's disease. *Science 314*:781-784.

[17] Skovronsky, D.M., Lee, V.M., Trojanowski, J.Q. (2006). Neurodegenerative diseases: new concepts of pathogenesis and their Therapeutic implications. *Annu Rev Pathol.1*:151-170.

[18] Teplow, D.B., Lazo, N.D., Bitan, G., Bernstein, S., Wyttenbach,T., et al. (2006). Elucidating amyloid β-protein folding and assembly: A multidisciplinary approach. *Acc Chem Res 39*:635-645.

[19] Guo, Z., Eisenberg, D. (2007). The mechanism of the amyloidogenic conversion of T7 endonuclease I. *J Biol Chem. 282*:14968-14974.

[20] Ono, K., Condron, M.M., Ho, L., Wang, J., Zhao, W., et al.. (2008). Effects of grape seed-derived polyphenols on amyloid β-protein self-assembly and cytotoxicity. *J Biol Chem. 283*:32176-32187.

[21] Saibil, H.R. (2008). Chaperone machines in action. *Curr Opin Struct Biol. 18*:35-42.

[22] Cisse M., Mucke L. (2009). Alzheimer's disease: A prion protein connection. *Nature 457*:1090-1091.

[23] Ecroyd, H., Carver, J.A. (2009). Crystallin proteins and amyloid fibrils. *Cell Mol Life Sci. 66*:62-81.

[24] Aoyama, A., Steiger, R.H., Frohli, E., Schafer, R., von Deimling, A., et al. (1993). Expression of α B-crystallin in human brain tumors. *Int J Cancer 55*:760-764.

[25] Takashi, M., Katsuno, S., Sakata, T., Ohshima, S., Kato, K. (1998). Different concentrations of two small stress proteins, αB crystallin and HSP27 in human urological tumor tissues. *Urol Res 26*:395-399.

[26] Calderwood, S.K., Khaleque, M.A., Sawyer, D.B., Ciocca, D.R. (2006). Heat shock proteins in cancer: chaperones of tumorigenesis. *Trends Biochem Sci. 31*:164-172.

[27] Moyano, J.V., Evans, J.R., Chen, F., Lu, M., Werner, M.E., et al. (2006). AB-crystallin is a novel oncoprotein that predicts poor clinical outcome in breast cancer. *J Clin Invest 116*:261-270.

[28] Pozsgai, E., Gomori, E., Szigeti, A., Boronkai, A., Gallyas, F. Jr, et al. (2007). Correlation between the progressive cytoplasmic expression of a novel small heatshock protein (Hsp16.2) and malignancy in brain tumors. *BMC Cancer* 7:233.

[29] So, A., Hadaschik, B., Sowery, R., Gleave, M. (2007). The role of stress proteins in prostate cancer. *Curr Genomics 8*:252-261.

[30] Ghosh, J.G., Houck, S.A., Clark, J.I. (2007a). Interactive domains in the molecular chaperone human αB crystallin modulate microtubule assembly and disassembly. *PLoS ONE 2*:e498.

[31] Cherian, M. and Abraham, E.C. (1995). Decreased molecular chaperone property of α-crystallins due to posttranslational modifications. *Biochem BiophysRes Commun 208*: 675–679.

[32] Liu, J.P., Schlosser, R., Ma, W.Y., Dong, Z., Feng, H., et al. (2004). Human αA- and αB-crystallins prevent UVA-induced apoptosis through regulation of PKCα, RAF/MEK/ERK and AKT signaling pathways. *Exp Eye Res. 79*:393-403.

[33] Wyttenbach, A. (2004). Role of heat shock proteins during polyglutamine neurodegeneration: mechanisms and hypothesis. *J Mol Neurosci 23*:69-96.

[34] Gaestel, M. (2006). Molecular chaperones in signal transduction. *Handb Exp Pharmacol 172*:93-109.

[35] Bellyei, S., Szigeti, A., Boronkai, A., Pozsgai, E., Gomori, E., et al. (2007a). Inhibition of cell death by a novel 16.2 kD heat shock protein predominantly via Hsp90 mediated lipid rafts stabilization and Akt activation pathway. *Apoptosis 12*:97-112.

[36] Bellyei, S., Szigeti, A., Pozsgai, E., Boronkai, A., Gomori, E., et al. (2007b). Preventing apoptotic cell death by a novel small heat shock protein. *Eur J Cell Biol. 86*:161-171.

[37] Lanneau, D., Brunet, M., Frisan, E., Solary, E., Fontenay, M., Garrido, C. (2008). Heat shock proteins: essential proteins for apoptosis regulation. *J Cell Mol Med. 12*:743-761.

[38] Nicolaou, P., Knöll, R., Haghighi, K., Fan, G.C., Dorn, G.W. 2nd, et al. (2008). Human mutation in the anti-apoptotic heat shock protein 20 abrogates its cardioprotective effects. *J Biol Chem 283*:33465-33471.

[39] Haslbeck, M., Franzmann, T., Weinfurtner, D., Buchner, J. (2005). Some like it hot: the structure and function of small heat-shock proteins. *Nat Struct Mol Biol 12*:842-846.

[40] Nakamoto, H., Vigh, L. (2007). The small heat shock proteins and their clients. *Cell Mol Life Sci 64*:294-306.

[41] Arrigo, A.P., Simon, S., Gibert, B., Kretz-Remy, C., Nivon, M., et al. (2007). Hsp27 (HspB1) and αB-crystallin (HspB5) as therapeutic targets. *FEBS Lett 581*:3665-3674.

[42] Arai, H., Atomi, Y. (1997). Chaperone activity of α B-crystallin suppresses tubulin aggregation through complex formation. *Cell Struct Funct 22*:539-544.

[43] Liang, J.J. (2000). Interaction between β-amyloid and lens αB-crystallin. *FEBS Lett 484*:98-101.

[44] Perng, M.D., Cairns, L., van den, IJessel.P., Prescott, A., Hutcheson, A.M., Quinlan, R.A. (1999). Intermediate filament interactions can be altered by HSP27 and αB-crystallin. *J Cell Sci 112*:2099-2112.

[45] Xi, J.H., Bai, F., McGaha, R., Andley, U.P. (2006). A-crystallin expression affects microtubule assembly and prevents their aggregation. *Faseb J 20*:846-857.

[46] Devi, R.R., Yao, W., Vijayalakshmi, P., Sergeev, Y.V., Sundaresan, P., Hejtmancik , J.F. (2008). Crystallin gene mutations in Indian families with inherited pediatric cataract. *Mol Vis 14*:1157-1170.

[47] Litt, M., Kramer, P., LaMorticella, D.M., Murphey, W., Lovrien, E.W., Weleber, R.G. (1998). Autosomal dominant congenital cataract associated with a missense mutation in the human α crystallin gene CRYAA. *Hum Mol Genet 7*:471-474.

[48] Perng, M.D., Zhang, Q., Quinlan, R.A. (2007). Insights into the beaded filament of the eye lens. *Exp Cell Res 313*:2180-2188.

[49] Santhiya, S.T., Soker, T., Klopp, N., Illig, T., Prakash, M.V., et al. (2006). Identification of a novel, putative cataract-causing allele in CRYAA (G98R) in an Indian family. *Mol Vis 12*:768-773.

[50] Simon, S., Michiel, M., Skouri-Panet, F., Lechaire, J.P., Vicart, P., Tardieu, A. (2007). Residue R120 is essential for the quaternary structure and functional integrity of human αB-crystallin. *Biochemistry 46*:9605-9614.

[51] De, S., Rabin, D.M., Salero, E., Lederman, P.L., Temple, S., Stern, J.H. (2007). Human retinal pigment epithelium cell changes and expression of αB-crystallin: a biomarker for retinal pigment epithelium cell change in age-related macular degeneration. *Arch Ophthalmol 125*:641-645.

[52] Johnson, P.T., Brown, M.N., Pulliam, B.C., Anderson, D.H., Johnson, L.V. (2005). Synaptic pathology, altered gene expression, and degeneration in photoreceptors impacted by drusen. *Invest Ophthalmol Vis Sci 46*:4788-4795.

[53] Renkawek, K., Bosman, G.J., de Jong, W.W. (1994). Expression of small heat-shock protein hsp 27 in reactive gliosis in Alzheimer disease and other types of dementia. *Acta Neuropathol 87*:511-519.

[54] Renkawek, K., de Jong, W.W., Merck, K.B., Frenken, C.W., van Workum, F.P., Bosman G.J. (1992). α B-crystallin is present in reactive glia in Creutzfeldt-Jakob disease. *Acta Neuropathol 83*:324-327.

[55] Renkawek, K,, Stege, G.J., Bosman, G.J. (1999). Dementia, gliosis and expression of the small heat shock proteins hsp27 and α B-crystallin in Parkinson's disease. *Neuroreport 10*:2273-2276.

[56] Wang, J., Martin, E., Gonzales, V., Borchelt, D.R., Lee, M.K. (2008). Differential regulation of small heat shock proteins in transgenic mouse models of neurodegenerative diseases. *Neurobiol Aging 29*:586-597.

[57] Wilhelmus, M.M., Boelens, W.C., Otte-Holler, I., Kamps, B., Kusters, B., et al. (2006). Small heat shock protein HspB8: its distribution in Alzheimer's disease brains and its inhibition of amyloid-β protein aggregation and cerebrovascular amyloid-β toxicity. *Acta Neuropathol 111*:139-149.

[58] Evgrafov, O.V., Mersiyanova, I., Irobi, J., Van Den Bosch, L., Dierick, I., et al. (2004). Mutant small heat-shock protein 27 causes axonal Charcot-Marie-Tooth disease and distal hereditary motor neuropathy. *Nat Genet 36*:602-606.

[59] Irobi, J., Van Impe, K., Seeman, P., Jordanova, A., Dierick, I., et al. (2004). Hot-spot residue in small heat-shock protein 22 causes distal motor neuropathy. *Nat Genet 36*:597-601.

[60] Tang, B.S., Zhao, G.H., Luo, W., Xia, K., Cai, F., et al. (2005). Small heat-shock protein 22 mutated in autosomal dominant Charcot-Marie-Tooth disease type 2L. *Hum Genet 116*:222-224.

[61] Ghosh, J.G. & Clark, J.I. (2005). Insights into the domains required for dimerization and assembly of human αB-crystallin. *Protein Sci 14*:684-695.

[62] Ghosh, J.G., Estrada, M.R., Clark, J.I. (2005). Interactive domains for chaperone activity in the small heat shock protein, human αB crystallin. *Biochemistry 44*:14854-14869.

[63] Ghosh, J.G., Estrada, M.R., Clark, J.I. (2006a). Structure-based analysis of the β8 interactive sequence of human αB crystallin. *Biochemistry 45*:9878-9886.

[64] Ghosh, J.G., Estrada, M.R., Houck, S.A., Clark, J.I. (2006b). The function of the β3 interactive domain in the small heat shock protein and molecular chaperone, human αB crystallin. *Cell Stress Chaperones 11*:187-197.

[65] Ghosh, J.G., Houck, S.A., Doneanu, C.E., Clark, J.I. (2006c). The β4-β8 groove is an ATP-interactive site in the α crystallin core domain of the small heat shock protein, human αB crystallin. *J Mol Biol 364*:364-375.

[66] Ghosh, J.G., Houck, S.A., Clark, J.I. (2007b). Interactive sequences in the stress protein and molecular chaperone human αB crystallin recognize and modulate the assembly of filaments. *Int J Biochem Cell Biol 39*:1804-1815.

[67] Ghosh, J.G., Shenoy, A.K., Jr., Clark, J.I. (2007c). Interactions between important regulatory proteins and human αB crystallin. *Biochemistry 46*:6308-6317.

[68] Ghosh, J.G., Houck, S.A., Clark, J.I. (2008). Interactive sequences in the molecular chaperone, human αB crystallin modulate the fibrillation of amyloidogenic proteins. *Int J Biochem Cell Biol 40*:954-967.

[69] Kappé, G., Franck, E., Verschuure, P., Boelens, W.C., Leunissen, J.A., de Jong, W.W. (2003). The human genome encodes 10 α-crystallin-related small heat shock proteins: HspB1-10. *Cell Stress Chaperones 8*:53-61.

[70] Kim, K.K., Kim, R., Kim, S.H. (1998). Crystal structure of a small heat-shock protein. *Nature 394*:595-599.

[71] van Montfort,R.L., Basha, E., Friedrich, K.L., Slingsby, C., Vierling. E. (2001a). Crystal structure and assembly of a eukaryotic small heat shock protein. *Nat Struct Biol 8*:1025-1030.

[72] Van Montfort, R., Slingsby, C., Vierling, E. (2001b). Structure and function of the small heat shock protein/α-crystallin family of molecular chaperones. *Adv Protein Chem. 59*:105-156.

[73] Stamler, R., Kappé, G., Boelens, W., Slingsby, C. (2005). Wrapping the α-crystallin domain fold in a chaperone assembly. *J Mol Biol. 353*:68-79.

[74] Pasta, S.Y., Raman, B., Ramakrishna, T., Rao., Ch. M. (2004). The IXI/V motif in the C-terminal extension of α-crystallins: alternative interactions and oligomeric assemblies. *Mol Vis 10*:655-662.

[75] Haley, D.A., Horwitz, J., Stewart, P.L. (1998). The small heat-shock protein, αB-crystallin, has a variable quaternary structure. *J Mol Biol 277*:27-35.

[76] Haley, D.A., Bova, M.P., Huang, Q.L., McHaourab, H.S., Stewart, P.L. (2000). Small heat-shock protein structures reveal a continuum from symmetric to variable assemblies. *J Mol Biol 298*:261-272.

[77] Salerno, J.C., Eifert, C.L., Salerno, K.M., Koretz, J.F. (2003). Structural diversity in the small heat shock protein superfamily: control of aggregation by the N-terminal region. *Protein Eng. 16*:847-851.

[78] Horwitz, J. (2009). A crystallin: the quest for a homogeneous quaternary structure. *Exp Eye Res. 88*:190-194.

[79] Levitt, M. (1992). Accurate modeling of protein conformation by automatic segment matching. *J Mol Biol 226*:507-533.

[80] Fechteler, T., Dengler, U., Schomburg, D. (1995). Prediction of protein three-dimensional structures in insertion and deletion regions: a procedure for searching data

bases of representative protein fragments using geometric scoring criteria. *J Mol Biol 253*:114-131.

[81] Jehle, S., van Rossum, B., Stout, J.R., Noguchi, S.M., Falber, K., et al. (2009). AB-crystallin: a hybrid solid-state/solution-state NMR investigation reveals structural aspects of the heterogeneous oligomer. *J Mol Biol 385*:1481-1497.

[82] Koteiche, H.A., McHaourab, H.S. (1999). Folding pattern of the α-crystallin domain in αA-crystallin determined by site-directed spin labeling. *J Mol Biol 294*:561-577.

[83] Liang, J.J., Liu, B.F. (2006). Fluorescence resonance energy transfer study of subunit exchange in human lens crystallins and congenital cataract crystallin mutants. *Protein Sci 15*:1619-1627.

[84] Sharma, K.K., Kumar, R.S., Kumar. G.S., Quinn. P.T. (2000). Synthesis and characterization of a peptide identified as a functional element in αA-crystallin. *J Biol Chem 275*:3767-3771.

[85] Bhattacharyya, J., Sharma, K.K. (2001). Conformational specificity of mini-αA-crystallin as a molecular chaperone. *J Pept Res 57*:428-434.

[86] Geysen, H.M. (1990). Molecular technology: peptide epitope mapping and the pin technology. *Southeast Asian J Trop Med Public Health 21*:523-533.

[87] Derham, B.K., van Boekel, M.A., Muchowski, P.J., Clark, J.I., Horwitz, J., et al. (2001). Chaperone function of mutant versions of α A- and α B-crystallin prepared to pinpoint chaperone binding sites. *Eur J Biochem 268*:713-721.

[88] Muchowski, P.J., Wu, G.J., Liang, J.J., Adman, E.T., Clark, J.I. (1999). Site-directed mutations within the core "α-crystallin" domain of the small heat-shock protein, human αB-crystallin, decrease molecular chaperone functions. *J Mol Biol 289*:397-411.

[89] Pasta, S.Y., Raman, B., Ramakrishna, T., Rao, Ch. M. (2003). Role of the conserved SRLFDQFFG region of α-crystallin, a small heat shock protein. Effect on oligomeric size, subunit exchange, and chaperone-like activity. *J Biol Chem 278*:51159-51166.

[90] Saha, S., Das, K.P. (2004). Relationship between chaperone activity and oligomeric size of recombinant human αA- and αB-crystallin: a tryptic digestion study. *Proteins 57*:610-617.

[91] Sharma, K.K., Kaur, H., Kester, K. (1997). Functional elements in molecular chaperone α-crystallin: identification of binding sites in α B-crystallin. *Biochem Biophys Res Commun 239*:217-222.

[92] Shroff, N.P., Bera, S., Cherian-Shaw, M., Abraham, E.C. (2001). Substituted hydrophobic and hydrophilic residues at methionine-68 influence the chaperone-like function of αB-crystallin. *Mol Cell Biochem 220*:127-133.

[93] Ghosh, J.G., Shenoy, A.K., Jr., Clark, J.I. (2006d). N- and C-Terminal motifs in human αB crystallin play an important role in the recognition, selection, and solubilization of substrates. *Biochemistry 45*:13847-13854.

[94] Klabunde, T., Petrassi, H.M., Oza, V.B., Raman, P., Kelly, J.W., Sacchettini, J.C. (2000). Rational design of potent human transthyretin amyloid disease inhibitors. *Nat Struct Biol 7*:312-321.

[95] Trinh, C.H., Smith, D.P., Kalverda, A.P., Phillips, S.E., Radford, S.E. (2002). Crystal structure of monomeric human β-2-microglobulin reveals clues to its amyloidogenic properties. *Proc Natl Acad Sci U S A 99*:9771-9776.

[96] Comeau, S.R., Gatchell, D.W., Vajda, S., Camacho, C.J. (2004a). ClusPro: a fully automated algorithm for protein-protein docking. *Nucleic Acids Res 32*:W96-W99.

[97] Comeau, S.R., Gatchell, D.W., Vajda, S., Camacho, C.J. (2004b). ClusPro: an automated docking and discrimination method for the prediction of protein complexes. *Bioinformatics 20*:45-50.

[98] McParland, V.J., Kad, N.M., Kalverda, A.P., Brown, A., Kirwin-Jones, P., et al. (2000). Partially unfolded states of β(2)-microglobulin and amyloid formation in vitro. *Biochemistry 39*:8735-8746.

[99] Nelson, R., Eisenberg, D. (2006). Recent atomic models of amyloid fibril structure. *Curr Opin Struct Biol 16*:260-265.

[100] Schormann, N., Murrell, J.R., Benson, M.D. (1998). Tertiary structures of amyloidogenic and non-amyloidogenic transthyretin variants: new model for amyloid fibril formation. *Amyloid 5*:175-187.

[101] Friedrich, K.L., Giese, K.C., Buan, N.R., Vierling, E. (2004). Interactions between small heat shock protein subunits and substrate in small heat shock protein-substrate complexes. *J Biol Chem 279*:1080-1089.

[102] Lee, G.J., Roseman, A.M., Saibil, H.R., Vierling, E. (1997). A small heat shock protein stably binds heat-denatured model substrates and can maintain a substrate in a folding-competent state. *Embo J 16*:659-671.

In: Small Stress Proteins and Human Diseases
Editors: Stéphanie Simon et al.
ISBN: 978-1-61470-636-6

Chapter 4.2

ANTISENSE AND RNAI APPROACHES TO INTERFERE WITH HSP27 (HSPB1) ACTIVITY IN CANCER

Amina Zoubeidi and Martin E. Gleave***

The Vancouver Prostate Centre and Department of Urological Sciences, University of British Columbia, Vancouver, British Columbia, Canada

ABSTRACT

The 27 kDa heat shock protein (Hsp27/HspB1) is a stress-activated cytoprotective chaperone associated with poor prognosis in many malignancies including lung, breast, prostate, gastric, and liver. Hsp27 is up-regulated in an adaptive survival manner by various triggers to confer acquired treatment resistance after radiation, hormone or chemotherapy, and has therefore become an attractive strategy in cancer therapy. Hsp27 acts through an ATP-independent mechanism making this target less amenable to inhibition by small molecules, and so strategies to inhibit Hsp27 at the gene-expression level become appealing. Indeed, known nucleotide sequences of cancer-relevant genes offer the possibility to rapidly design antisense oligonucleotides (ASO) or short interfering RNA (siRNA) for loss-of-function and preclinical proof-of-principle studies. In this chapter we will review anti-apoptotic and cell survival pathways regulated by Hsp27 and the preclinical studies of using antisense and siRNA strategies in various cancers that support targeting Hsp27 in advanced and resistant cancers.

---

* Martin Gleave
University of British Columbia
The Vancouver Prostate Centre
2775 Laurel St, 6th Floor
Vancouver, BC V5Z 1M9
Phone: 604-875-5006
Fax: 604-875-5604
m.gleave@ubc.ca

Introduction

Acquired resistance to chemotherapeutic agents through inhibition of apoptosis has emerged as an important mechanism on tumor progression, metastasis and therapy outcome [1] (see chapter 3-4). Heat shock proteins (Hsps) were originally described for their role as chaperones induced by temperature shock [2, 3]. Hsps have subsequently been shown to be more associated more broadly with a stress response from many environmental (UV, radiation, heat shock, heavy metals and amino acids), pathological (bacterial, parasitic infections or fever, inflammation, malignancy or autoimmunity) or physiological (growth factors, cell differentiation, hormonal manipulation or tissue development) stressors that are all associated with remarkable increases in intracellular Hsp synthesis and/or activity. The essential chaperoning function of Hsps is subverted during oncogenesis to facilitate malignant transformation and rapid somatic evolution. They are biochemical buffers of numerous genetic lesions present within the tumors, allow mutated proteins to retain or gain their function while permitting cancer cells to tolerate imbalanced signaling that oncoprotein create [4]. Expression of small Hsps is essential to protect differentiating cells against apoptosis [5].

1. Small Heat Shock Protein 27

The 27-kDa Hsp (Hsp27) belongs to the family of small Hsp, a group of proteins that vary in size from 15 to 30 kDa and share sequence homologies and biochemical properties such as phosphorylation and oligomerization. Hsp27 protein is ubiquitously expressed and has been implicated in various biological functions. Hsp27 dimers function as building blocks for the multimeric complexes, and Hsp27 oligomers can form up to 1000 kDa.. Hsp27 oligomerization is a dynamic process that depends on the phosphorylation status of the protein and exposure to stress [6, 7]. Hsp27 can be phosphorylated at three serine residues, and its phosphorylation enhances oligomerization. This phosphorylation is a reversible process catalyzed by the MAPKAP kinases 2 and 3 in response to a variety of stresses, including differentiating agents, mitogens, inflammatory cytokines, such as tumor necrosis factor-α (TNF-α) and IL-1ß, hydrogen peroxide and other oxidants. Hsp27 is expressed in many cell types and tissues, at specific stages of development and differentiation [8]. A principle chaperone function of Hsp27 is protection against protein aggregation [9]. In contrast to large Hsp, Hsp27 acts through ATP-independent mechanisms, and in vivo, acts in concert with other chaperones by managing the reservoir of misfolded proteins, either assisting in their refolding or assisting their ubiquitination and proteasomal degradation [10].

2. Hsp27: A Powerful Survival and Lifeguard Protein

Hsp27 interacts with many key apoptosis-associated proteins to regulate a cell's apoptotic rheostat. Hsp27 overexpression protects against apoptotic cell death triggered by various stimuli, including hyperthermia, oxidative stress, staurosporine, ligation of the Fas/Apo-1/CD95 death receptor, and cytotoxic drugs [11, 12]. Hsp27 has been shown to interact and inhibit components of both stress- and receptor-induced apoptotic pathways. Hsp27 prevents activation of caspases by sequestering cytochrome C in the cytoplasm [13]. Cytochrome-c interacts with Apaf-1 and caspase-9 to form the "apoptosome" which activates caspase-3, leading to an activation cascade of downstream caspases, the so called "effectors" of cell death [14]. Hsp27 interacts with pro-caspase 3 interferes with caspase-3 activation, an effect related to the ability of Hsp27 to stabilize actin microfilaments [15]. Hsp27 binds to F-actin to prevent disruption of the cytoskeleton resulting from either heat shock, cytochalasin D, and other stresses [16]. In L929 murine fibrosarcoma cells exposed to cytochalasin D or staurosporine, overexpression of Hsp27 prevents cytoskeleton disruption and Bid intracellular redistribution that precede cytochrome *c* release [15]. Hsp27 also inhibits mitochondrial release of Smac and confers resistance to dexamethasone in multiple myeloma cells [17], and interferes with Bax activation via Akt [18]. Hsp27 can inhibit apoptosis induced by etoposide or TNF-α in different cancer cell lines by increasing the activity of the survival transcription factor nuclear factor-αB (NF-κB). Under stress conditions, Hsp27 increases IκBα ubiquitination/ degradation, which results in an increase in NF-κB activity and increased survival [19]. Hsp27 protects cells from TNF-alpha-mediated apoptosis by a mechanism that involves the downregulation of reactive oxygen species and neutralizes the toxic effects of oxidized proteins [5, 20, 21]. Hsp27 also inhibits apoptosis by regulating upstream signaling pathways. Survival factors, such as nerve growth factor or platelet-derived growth factor, inhibit apoptosis by activating the phosphatidylinositol 3-kinase pathway (PI3-K). Activated PI3-K phosphorylates inositol lipids in the plasma membrane that attract the serine/threonine kinase Akt/PKB. Akt targets multiple proteins of the apoptotic machinery, including Bad and caspase-9 [18, 22-24].

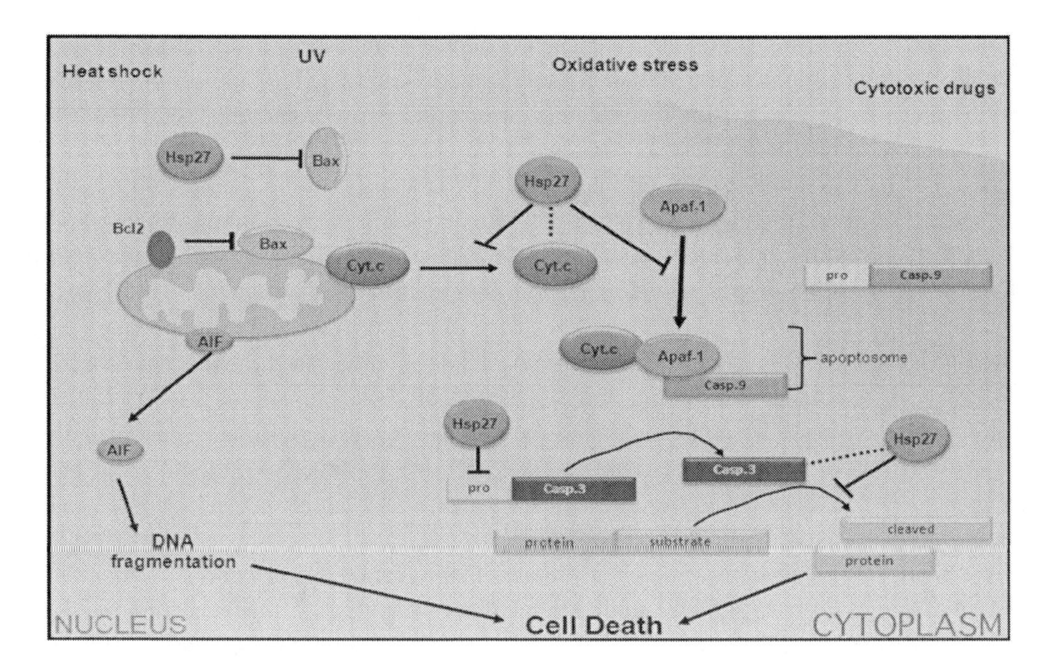

Figure 1. Effect of Hsp27 in suppressing stress induced apoptosis. Hsp27 prevent cytochrome c (cyt.c) release from mitochondria. Apoptosome formation is blocked by Hsp27 binding to cytochrome c.

Hsp27 prevents apoptosis downstream of caspase-3 activation by interacting with caspase-3. Adapted from Mosser and Morimoto [2].

Hsp27 interacts with the protein kinase Akt, an interaction that is necessary for Akt phosphorylation in stressed cells. In turn, Akt phosphorylates Hsp27, thus leading to the disruption of Hsp27-Akt complexes [25]. Hsp27 also affects Fas-mediated apoptotic pathways. The phosphorylated form of Hsp27 directly interacts with Daxx by connecting Fas signaling to the protein kinase Ask1 that mediates a caspase-independent cell death [26]. Phosphorylated Hsp27 directly interacts with Daxx and inhibits apoptosis [27] suggesting that Hsp27 oligomerization/phosphorylation status alters its conformation and hence determines its capacity to interact with different apoptotic proteins.

The preceding review illustrates that Hsp27 is a promiscuous protein that can interact with many varied and distinct partners involved in survival pathways (see chapters 1-3 and 3-4). Although the role of oligomerization/phosphorylation in the antiapoptotic functions of Hsp27 remains uncertain, it is believed that this ATP-independent chaperone, for which no co-chaperones have been described, seems to modulate its different protective properties by changing its oligomerization pattern (which is regulated by the phosphorylation status of the protein) (see also chapter 4-3). Depending on which stress or survival pathways are involved, phospho-Hsp27 will interact with various client proteins to inhibit intrinsic or extrinsic apoptotic activation, or facilitate survival signaling and transcriptional pathways (see chapter 3-2).

3. Hsp27 in Cancer and Treatments Resistance

Increase of Hsps transcription in cancer is coupled with basic oncogenic pathways. In normal cells, tumor suppressor genes such as p53 and p63 repress Hsps' transcription through their binding to the NF-Y response element present in Hsps promoters [28, 29]. During transformation, mutation of p53 occurs and reverses this mechanism and leads to the activation of Hsps promoters [29]. Of interest, Hsp27 is a target of the tumor suppressor p53 polypeptide, however, the nature of Hsp27-p53 interactions in the regulation of apoptosis are not yet fully defined [30]. Moreover, several of the genes that are upregulated by c-Myc in cells expressing wild-type p53 encode chaperones related to cell survival as Hsp27 [31]. *In vivo*, the anti-apoptotic property of Hsp27 has been demonstrated using virus bearing Hsp27 coding sequence [32, 33]. Hsp27 interacts with factors involved in the oncogenic signaling pathway such as Signal Transducer and Activator of Transcription (STAT3), which is constitutively active in different tumors and controls the expression of many genes involved in transformation or cell survival. Hsp27-STAT3 complexes have been identified in breast and prostate cancer cells [34, 35] and members of the STAT family are now considered as putative target in cancer studies [36, 37].

Clinically, Hsp27 level increases in a wide number of cancers such as breast [38], ovarian [39], glial [40], prostate [41, 42] and other tumors [43]. Hsp27 mediates tumorigenesis through inhibition of cell apoptosis, an essential phenomenon in cancer progression [15, 44], and is associated with metastasis and poor prognosis [8, 45]. Hsp27 expression is induced by hormone or chemotherapy and inhibits treatment-induced apoptosis through multiple mechanisms [8, 15, 35, 46-49].

In breast cancer for example, Hsp27 is highly expressed in biopsies as well as circulating in the serum of breast cancer patients [50-52]. Hsp27 is associated with estrogen-responsive malignancies and has been implicated in metastasis [53]. Overexpression of Hsp27 in human normal breast epithelial cells protects from senescence induced doxorubicin via suppression of p53 activation [54]. In murine breast carcinoma 4T1 cells, Hsp27 overexpression enhances tumor metastasis [55]. In prostate cancer, Hsp27 is an independent predictor for clinical outcome, its increase correlates with castrate resistant prostate cancer (CRPC) [35]. Hsp27 overexpression in human prostate LNCaP cells confers hormone resistance post castration via the activation of STAT3 dependent pathway [35] and also via the ability of the increase metalloproteinase type 2 activity [56]. Akt is another important factor modulated by Hsp27 [25], and is constitutively activated by loss of the PTEN tumor suppressor gene, one of the most frequently mutated genes in cancer including prostate cancer. Hsp27 is necessary for TGFβ-mediated increases in MMP-2 and cell invasion in human prostate cancer [57]. Interestingly, Hsp27 displaces Hsp90 as the predominant androgen receptor (AR) chaperone after androgen treatment, and shuttles the AR to the nucleus, and facilitating AR binding to androgen response element in AR-regulated genes [58].

This abbreviated review illustrates that Hsp27 is functions as a "HUB" in multiple signaling and transcriptional pathways, identifying it as an attractive therapeutic target by simultaneously affecting many pathways implicated in cancer progression and resistance.

4. Targeting Hsp27 by siRNA or Antisense Oligonucleotides

Hsp27 is a very appealing cancer target, however, acts through an ATP-independent mechanism making this target not easy amenable to inhibition by small molecules or antibodies, and so strategies to inhibit Hsp27 at the gene-expression level is attractive. Indeed, known nucleotide sequences of cancer-relevant genes offer the possibility to rapidly design short interfering RNA (siRNA) duplexes or an antisense (ASO) holds considerable promise [59].

4.1. Hsp27 siRNA: a Potent Technology with Drug Delivery Limitations

RNAi refers to the sequence specific degradation of mRNA as a consequence of formation of double stranded RNA (dsRNA) molecules with complementary RNA [60, 61]. In the cytoplasm, the enzyme Dicer cleaves the dsRNA into fragments of 21-23 in length called siRNAs, which can then associate with proteins to form an RNA-induced silencing complex, which in turn directs the siRNA to the target mRNA sequence for subsequent degradation [62]. Of consideration, there is usually no interferon response if the size of the dsRNA formed is < 30 nt [62, 63]. Cells can be treated directly with synthetic siRNAs or transfected with plasmids that express short-hairpin RNAs (shRNAs) that form dsRNA, which are subsequently cleaved by Dicer. One advantage of the shRNA approach is that when RNA polymerase (II and III) driven plasmid cassettes are used, one can essentially get constitutive expression of the siRNA, ensuring prolonged levels of dsRNA intracellularly [64, 65]. However, using shRNAs rather than chemically synthesized siRNAs imposes considerable vector restrictions with the necessity that promoter-based shRNAs must be delivered as genetic units, usually packaged in viruses.

Table 1. Targeting Hsp27 by different siRNA sequences in cancer

Cancer	siRNA sequence	Effect	Reference
Liver	CAUCCUUGGUCUUGACCGUdTdT'	Inhibition of migration and invasion	[88]
Breast	AATTCTCCGAACGTGTCACGT	Chemo-sensitization to herceptin	[82]
Colon	UUGGCGGCAGU, CUCAUCGGUU, UAGCCAUGCUCG, UCCUGCCUU, ACUGCGACCACU, CCUCCGGUU, ACAGGGAGGAGG, AAACUUGUU, AAACUUGUU, UUCUCUAUG	Inhibition of migration and motility	[89]
Melanoma	UUAUGGAUGUGAGUCAGCCtg	Enhances paclitaxel-inducinging caspase -3 activation	[86]
Cervical	AAGCTGCAAAATCCGATGAGA	Enhances apoptosis induced by cisplatin	[90]
Bladder	AAGCUGCAAAAUCCGAUGAGAC	Chemo-sensitization to paclitaxel	[84]
Prostate	AAGCUGCAAAAUCCGAUGAGAC, AAGGAUGGCGUGGUGGAGAUC, AACGAGAUCACCAUCCCAGUC, AAGAUCACCAUCCCAGUCACC	Inhibits cell growth and induces apoptosis	[85]
Prostate	TGAGACTGCCGCCAAGTAA	Inhibits proliferation assay, reduces GSH and chemo-sensitizes to 17-AAG	[91]
Breast	GGAATTCGTGGGCATCCGGGCTAAGG	Inhibits cell migration	[92]

Moreover, with delivery of plasmid DNA for shRNA by non-viral systems, nuclear translocation of the DNA is a prerequisite and is often inadequately achieved. Consequently, the cytoplasmic site of activity of chemically synthesized siRNA provides an important advantage [66]. Also, in practical terms, it is easier to titrate levels of synthetic siRNAs than levels of shRNAs. While the use of antisense oligonucleotides [67, 68] has shown considerable promise in clinical trials with patients with advanced prostate cancers, the siRNA approach has many intrinsic properties that are attractive [69]. Even sub-nanomolar concentrations of siRNAs can be surprisingly potent, achieving >90% reduction in target mRNA, whereas antisense oligonucleotides usually require nanomolar amounts to achieve a therapeutic effect [70]. Furthermore, the cellular handling of siRNAs is fundamentally hard-wired into the cell's natural process for sequence-specific gene silencing and is therefore easily accommodated and correctly processed by mammalian target cells [71]. While not rigorously demonstrated, the half life of siRNAs seems to be longer than that of antisense DNA oligonucleotides [72]. Although both RNA knockdown approaches are effective, RNAi may have some advantages with respect to sensitivity, specificity, and duration of effect. A major obstacle to successfully developing an RNAi-based therapy for clinical trials is the apparent absence of efficient and targeted systemic *in vivo* delivery systems for these molecules. The RNAi must be protected from nucleases in the serum and extracellular environment, be transported across the cell membrane and into the cell, and then be processed to interact with specific target mRNA. While naked siRNA can enter cells, it does so very inefficiently and is not a viable therapeutic delivery system since >99% of the injected dose is rapidly excreted by the kidneys and taken up by the liver [66, 73, 74]. While hydrodynamic iv injections into mice allow reasonably efficient access to liver tissue [75], they cause transient heart congestion and because of this, are unlikely to be used for therapeutic applications in humans [76]. While *in vivo* delivery has proven to be a major obstacle for therapeutic development of RNAi technologies, siRNAs are likely to be the most used in preclinical in vitro studies.

Many siRNAs have been developed to target different regions against Hsp27 (Table 1). Silencing Hsp27 using siRNA suppresses proliferation, migration and downregulates MMP-9 concomitantly with TMIP-1 up-regulation in murine breast cancer cells [77], and induces apoptosis inTF-1 erythroleukemic cells [78] via Bax activation in PI-3 kinase dependent mechanism in renal epithelial cells [18]. Moreover, knockdown of Hsp27 by siRNA decreases clonogenic survival and induces cell senescence in HCT116 human colon cancer [54] and markedly inhibited VEGF and PDGF -induced migration and tubulogenesis [79, 80]. Hsp27 knockdown inhibits migration and invasion induced by Secreted protein acidic and rich in cysteine in human glioma cells [81]. Hsp27 knockdown using siRNA chemo-sensitizes SK-BR-3 HR breast cancer cells to Herceptin [82], A549 lung cancer cells to 17-AAG [83], MUC-3 bladder cancer cells and PC-3 prostate cancer cells to paclitaxel [84, 85]. In Rac1N17 cells, Hsp27 downregulation significantly increased the paclitaxel-elicited caspase-3 activation and apoptosis [86] and radio-sensitizes SQ20B head-and-neck squamous carcinoma [87].

4.2. Hsp27 ASO: A Promising Cancer Therapy

ASO is a single-stranded, chemically modified DNA-like molecule that is 17–22 nucleotides in length of single-strand DNA complementary to mRNA regions of a target gene that inhibit translation by forming RNA/DNA duplexes, thereby reducing mRNA and protein levels of the target gene [93]. Therefore, the level of the target protein is reduced by blocking translation, and by altering the subsequent cascades regulating cellular proliferation, differentiation, homeostasis and apoptosis. These agents promise increased specificity and a more favorable side-effect profile owing to well-defined modes of action. The specificity of the antisense approach is based on the fidelity of Watson–Crick hybridization and on estimates that a particular sequence of 17 bases in DNA occurs only once within the human genome. Moreover, improved chemical modifications of the phosphothiodate backbone increase resistance to nuclease digestion; prolong tissue half-lives and improve scheduling [94]. Several mechanisms that explain how ASO inhibit translation have been proposed. The most accepted involves the formation of an mRNA–ASO duplex formed through Watson–Crick binding, leading to RNase-H-mediated cleavage of the target mRNA [95-97]. Other proposed mechanisms include prevention of mRNA transport, modulation or inhibition of splicing, translational arrest, and formation of a triple helix through ASO binding to duplex genomic DNA resulting in inhibition of transcription. Additionally, some ASOs, especially those containing a CpG motifs or strings of G might also possess immunostimulatory activity or other off target toxicities [98].

Clinically, phosphorothioate ASOs are water soluble, stable agents resistant to nuclease digestion through substitution of a non-bridging phosphoryl oxygen of DNA with sulfur [99]. In clinical trials, continuous or frequent intravenous infusions were required to administer the first generation phosphorothioate ASOs because of their short tissue half-life, a major technical limitation. Therefore, effort has been made to improve the stability and efficacy of ASO by modifications of the phophodiester-linkage, the heterocycle or the sugar. One such alteration is the 2'-*O*-(2-methoxy) ethyl (2'-MOE) modification to the 2'-position of the carbohydrate moiety. 2'MOE ASOs form duplexes with RNA with a significantly higher affinity relative to unmodified phosphorothioate ASOs. This increased affinity has been shown to result in improved antisense potency *in vitro* and *in vivo*. In addition, 2'MOE ASOs display significantly improved resistance against nuclease-mediated metabolism relative to first generation phosphorothioate ASOs resulting in an improved tissue half-life *in vivo*, which produces a longer duration of action and allows for more relaxed dosing regimen [100]. Finally, these second generation phosphorothioate ASOs have the potential for a more attractive safety profile relative to unmodified phosphorothioate ASOs [101].

Both phosphorothioate and the second-generation MOE gapmer Hsp27 ASO (OGX-427) have demonstrated anti-cancer activity *in vitro* and *vivo* [35, 45, 58, 102]. In prostate cancer, Hsp27 ASO induces AR degradation via the proteasome and decreases PSA activation in vivo [58] (Figure 2). Hsp27 ASO enhances apoptosis and delay prostate tumor progression [45]. Intravesical OGX-427 instillation therapy showed promising antitumour activity and minimal toxicity in an orthotopic mouse model of high-grade bladder cancer [102]. Hsp27 ASO chemo-sensitizes bladder, prostate, ovarian and uterine cancer to paclitaxel [45, 84, 103], colorectal cancer to irinotecan [104] and sgc7901/vcr to vincristine [105] and restores the apoptotic response to Dex in Dex-resistant MM cells by triggering the release of mitochondrial protein Smac, followed by activation of caspase-9 and caspase-3 [17]. Antisense strategies have also demonstrated that lymphomas and multiple myelomas can be sensitized to chemotherapeutic drugs like dexamethasone and the inhibitor of proteasome Velcade (PS-341). Hsp27 participates in proteins' ubiquitination/proteasomal degradation and that this effect contributes to its protective functions by enhancing the activity of proteins like NF-κB [19]. Velcade, currently tested in clinical trials involving multiple myelomas, has been shown in vitro to induce apoptosis in several cancer cell lines. HSP27 confers resistance to bortezomib, and an Hsp27 antisense sensitizes cells to bortezomib -induced apoptosis [17], suggesting that combinational therapy using bortezomib plus with an Hsp27 ASO like OGX-

427 will produce synergistic anti-cancer activity. Collectively, the biology and pre-clinical anticancer activity supports further development of novel therapeutic drugs targeting Hsp27.

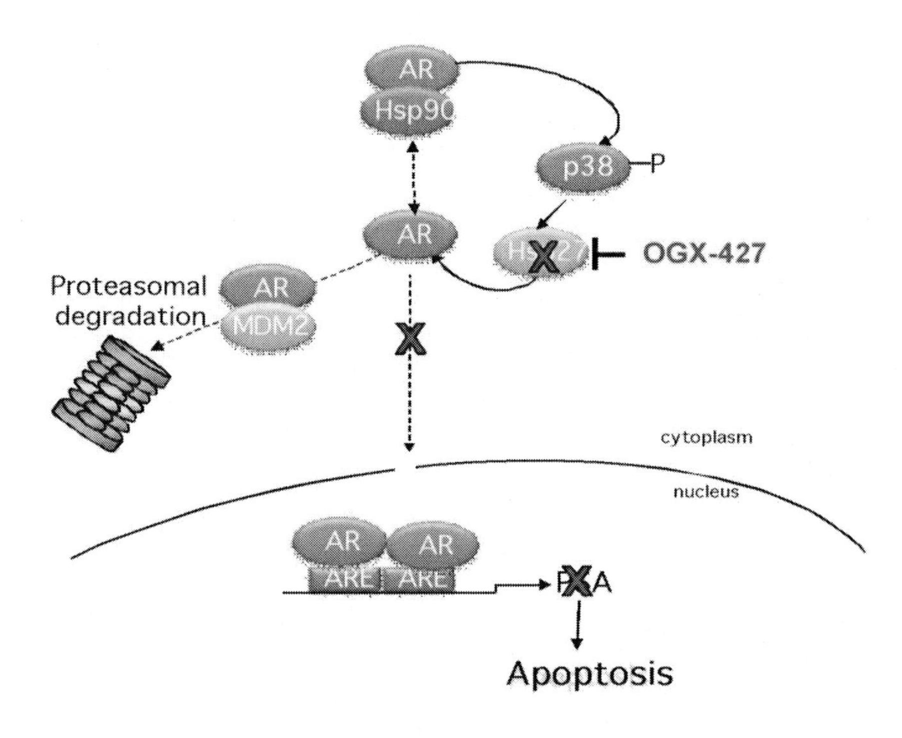

Figure 2: OGX-427 induces cell apoptosis via its ability to induce AR ubiquitination and degradation via the proteasome and hence inhibits its nuclear translocation and PSA activation leading to cell apoptosis.

4.3. Phase I clinical trial of Hsp27 Antisense OGX-427: Dose-Escalation Trial of OGX-427 in solid cancers expressing Hsp27

A second-generation MOE gapmer ASO targeting Hsp27 [OGX-427, OncoGeneX Technologies) is currently completing a single agent dose escalation Phase I trial in prostate, bladder, breast, and lung cancer. The primary objectives of this study are to define the pharmacokinetic and toxicity profile of OGX-427 given as a 2-hour IV infusion every week, to determine the recommended phase II dose (RP2D). Secondary endpoints are evaluating post-treatment PSA declines, measurable disease response and time to disease progression. Six patients per cohort were recruited at a starting dose of 200 mg IV weekly after an initial 3 loading doses given in the first week. Forty patients were enrolled after completion of the 1000 mg dose level OGX-427 monotherapy and MTD had not yet been reached. OGX-427 was also well tolerated at 800 mg dose in combination with docetaxel. Reduction in tumor markers were observed in patients with prostate (PSA) and ovarian (CA-125) cancer. In addition to pharmacokinetic and safety data, correlative studies using serial samples of circulating tumour cells (CTC) enumerated as an indicator of anti-tumour activity, as well as Hsp27 expression using immunofluorescence, were assessed. Decline of 50% or greater in both total and Hsp27+ CTCs were observed in over half the patients in this Phase I trial. Phase II trials of OGX-427 are planned to begin in prostate cancer and multiple myeloma in 2010.

Conclusion

Hsp27 is a multifunctional molecular chaperone that regulates activity of many important survival signaling and transcriptional pathways known to be relevant in many advanced cancers, including CRPC. Over-expression of Hsp27 confers broad spectrum treatment resistance and increased activity of AR, STAT3, Akt and NF-κB pathways. Hsp27 knockdown using ASO or siRNA induces apoptosis and chemo-sensitizes cancer cells to a wide variety of drugs. Pre-clinical proof-of-principle exists using OGX-427 in models of prostate, breast, lung, urothelial, MM, pancreas cancer. OGX-427, a 2nd generation MOE gapmer ASO targeting Hsp27, is now in clinical trials in solid cancers.

References

[1] Weir HK, Thun MJ, Hankey BF, Ries LA, Howe HL, Wingo PA, et al. Annual report to the nation on the status of cancer, 1975-2000, featuring the uses of surveillance data for cancer prevention and control. *J Natl Cancer Inst.* 2003 Sep 3;95(17):1276-99.

[2] Mosser DD, Morimoto RI. Molecular chaperones and the stress of oncogenesis. *Oncogene.* 2004 Apr 12;23(16):2907-18.

[3] Lindquist S, Craig EA. The heat-shock proteins. *Annu Rev Genet.* 1988;22:631-77.

[4] Takayama S, Reed JC, Homma S. Heat-shock proteins as regulators of apoptosis. *Oncogene.* 2003 Dec 8;22(56):9041-7.

[5] Soldes OS, Kuick RD, Thompson IA, 2nd, Hughes SJ, Orringer MB, Iannettoni MD, et al. Differential expression of Hsp27 in normal oesophagus, Barrett's metaplasia and oesophageal adenocarcinomas. Br J Cancer. 1999 Feb;79(3-4):595-603.

[6] Bruey JM, Ducasse C, Bonniaud P, Ravagnan L, Susin SA, Diaz-Latoud C, et al. Hsp27 negatively regulates cell death by interacting with cytochrome c. *Nat Cell Biol.* 2000 Sep;2(9):645-52.

[7] Garrido C. Size matters: of the small HSP27 and its large oligomers. *Cell Death Differ.* 2002 May;9(5):483-5.

[8] Garrido C, Fromentin A, Bonnotte B, Favre N, Moutet M, Arrigo AP, et al. Heat shock protein 27 enhances the tumorigenicity of immunogenic rat colon carcinoma cell clones. *Cancer Res.* 1998 Dec 1;58(23):5495-9.

[9] Ehrnsperger M, Graber S, Gaestel M, Buchner J. Binding of non-native protein to Hsp25 during heat shock creates a reservoir of folding intermediates for reactivation. *EMBO J.* 1997 Jan 15;16(2):221-9.

[10] Egeblad M, Werb Z. New functions for the matrix metalloproteinases in cancer progression. Nat Rev Cancer. 2002 Mar;2(3):161-74.

[11] Ricci JE, Maulon L, Battaglione-Hofman V, Bertolotto C, Luciano F, Mari B, et al. A Jurkat T cell variant resistant to death receptor-induced apoptosis. Correlation with heat shock protein (Hsp) 27 and 70 levels. Eur Cytokine Netw. 2001 Mar;12(1):126-34.

[12] Manero F, Ljubic-Thibal V, Moulin M, Goutagny N, Yvin JC, Arrigo AP. Stimulation of Fas agonistic antibody-mediated apoptosis by heparin-like agents suppresses Hsp27 but not Bcl-2 protective activity. Cell Stress Chaperones. 2004 Summer;9(2):150-66.

[13] Pandey P, Farber R, Nakazawa A, Kumar S, Bharti A, Nalin C, et al. Hsp27 functions as a negative regulator of cytochrome c-dependent activation of procaspase-3. *Oncogene*. 2000 Apr 13;19(16):1975-81.

[14] Concannon CG, Orrenius S, Samali A. Hsp27 inhibits cytochrome c-mediated caspase activation by sequestering both pro-caspase-3 and cytochrome c. *Gene Expr*. 2001;9(4-5):195-201.

[15] Paul C, Manero F, Gonin S, Kretz-Remy C, Virot S, Arrigo AP. Hsp27 as a negative regulator of cytochrome C release. Mol Cell Biol. 2002 Feb;22(3):816-34.

[16] Lavoie JN, Hickey E, Weber LA, Landry J. Modulation of actin microfilament dynamics and fluid phase pinocytosis by phosphorylation of heat shock protein 27. *J Biol Chem*. 1993 Nov 15;268(32):24210-4.

[17] Chauhan D, Li G, Hideshima T, Podar K, Mitsiades C, Mitsiades N, et al. Hsp27 inhibits release of mitochondrial protein Smac in multiple myeloma cells and confers dexamethasone resistance. Blood. 2003 Nov 1;102(9):3379-86.

[18] Havasi A, Li Z, Wang Z, Martin JL, Botla V, Ruchalski K, et al. Hsp27 inhibits Bax activation and apoptosis via a phosphatidylinositol 3-kinase-dependent mechanism. *J Biol Chem*. 2008 May 2;283(18):12305-13.

[19] Parcellier A, Schmitt E, Gurbuxani S, Seigneurin-Berny D, Pance A, Chantome A, et al. HSP27 is a ubiquitin-binding protein involved in I-kappaBalpha proteasomal degradation. *Mol Cell Biol*. 2003 Aug;23(16):5790-802.

[20] Neumark E, Sagi-Assif O, Shalmon B, Ben-Baruch A, Witz IP. Progression of mouse mammary tumors: MCP-1-TNFalpha cross-regulatory pathway and clonal expression of promalignancy and antimalignancy factors. Int J Cancer. 2003 Oct 10;106(6):879-86.

[21] Lirdprapamongkol K, Sakurai H, Kawasaki N, Choo MK, Saitoh Y, Aozuka Y, et al. Vanillin suppresses in vitro invasion and in vivo metastasis of mouse breast cancer cells. *Eur J Pharm Sci*. 2005 May;25(1):57-65.

[22] Nakagomi S, Suzuki Y, Namikawa K, Kiryu-Seo S, Kiyama H. Expression of the activating transcription factor 3 prevents c-Jun N-terminal kinase-induced neuronal death by promoting heat shock protein 27 expression and Akt activation. *J Neurosci*. 2003 Jun 15;23(12):5187-96.

[23] Arya R, Mallik M, Lakhotia SC. Heat shock genes - integrating cell survival and death. *J Biosci*. 2007 Apr;32(3):595-610.

[24] Fauconneau B, Petegnief V, Sanfeliu C, Piriou A, Planas AM. Induction of heat shock proteins (HSPs) by sodium arsenite in cultured astrocytes and reduction of hydrogen peroxide-induced cell death. *J Neurochem*. 2002 Dec;83(6):1338-48.

[25] Rane MJ, Pan Y, Singh S, Powell DW, Wu R, Cummins T, et al. Heat shock protein 27 controls apoptosis by regulating Akt activation. J Biol Chem. 2003 Jul 25;278(30):27828-35.

[26] Charette SJ, Lavoie JN, Lambert H, Landry J. Inhibition of Daxx-mediated apoptosis by heat shock protein 27. *Mol Cell Biol*. 2000 Oct;20(20):7602-12.

[27] Charette SJ, Landry J. The interaction of HSP27 with Daxx identifies a potential regulatory role of HSP27 in Fas-induced apoptosis. Ann N Y Acad Sci. 2000;926:126-31.

[28] Calderwood SK, Khaleque MA, Sawyer DB, Ciocca DR. Heat shock proteins in cancer: chaperones of tumorigenesis. *Trends Biochem Sci.* 2006 Mar;31(3):164-72.

[29] Taira T, Sawai M, Ikeda M, Tamai K, Iguchi-Ariga SM, Ariga H. Cell cycle-dependent switch of up-and down-regulation of human hsp70 gene expression by interaction between c-Myc and CBF/NF-Y. J Biol Chem. 1999 Aug 20;274(34):24270-9.

[30] Gao C, Zou Z, Xu L, Moul J, Seth P, Srivastava S. p53-dependent induction of heat shock protein 27 (HSP27) expression. *Int J Cancer*, 2000 Oct 15;88(2):191-4.

[31] Ceballos E, Munoz-Alonso MJ, Berwanger B, Acosta JC, Hernandez R, Krause M, et al. Inhibitory effect of c-Myc on p53-induced apoptosis in leukemia cells. Microarray analysis reveals defective induction of p53 target genes and upregulation of chaperone genes. *Oncogene.* 2005 Jun 30;24(28):4559-71.

[32] Brar BK, Stephanou A, Wagstaff MJ, Coffin RS, Marber MS, Engelmann G, et al. Heat shock proteins delivered with a virus vector can protect cardiac cells against apoptosis as well as against thermal or hypoxic stress. *J Mol Cell Cardiol.* 1999 Jan;31(1):135-46.

[33] Latchman DS. HSP27 and cell survival in neurones. Int J Hyperthermia. 2005 Aug;21(5):393-402.

[34] Song H, Ethier SP, Dziubinski ML, Lin J. Stat3 modulates heat shock 27kDa protein expression in breast epithelial cells. Biochem Biophys Res Commun. 2004 Jan 30;314(1):143-50.

[35] Rocchi P, Beraldi E, Ettinger S, Fazli L, Vessella RL, Nelson C, et al. Increased Hsp27 after androgen ablation facilitates androgen-independent progression in prostate cancer via signal transducers and activators of transcription 3-mediated suppression of apoptosis. *Cancer Res.* 2005 Dec 1;65(23):11083-93.

[36] Diaz N, Minton S, Cox C, Bowman T, Gritsko T, Garcia R, et al. Activation of stat3 in primary tumors from high-risk breast cancer patients is associated with elevated levels of activated SRC and survivin expression. Clin Cancer Res. 2006 Jan 1;12(1):20-8.

[37] Libermann TA, Zerbini LF. Targeting transcription factors for cancer gene therapy. *Curr Gene Ther.* 2006 Feb;6(1):17-33.

[38] Conroy SE, Sasieni PD, Amin V, Wang DY, Smith P, Fentiman IS, et al. Antibodies to heat-shock protein 27 are associated with improved survival in patients with breast cancer. *Br J Cancer,* 1998 Jun;77(11):1875-9.

[39] Arts HJ, Hollema H, Lemstra W, Willemse PH, De Vries EG, Kampinga HH, et al. Heat-shock-protein-27 (hsp27) expression in ovarian carcinoma: relation in response to chemotherapy and prognosis. *Int J Cancer*, 1999 Jun 21;84(3):234-8.

[40] Zhang R, Tremblay TL, McDermid A, Thibault P, Stanimirovic D. Identification of differentially expressed proteins in human glioblastoma cell lines and tumors. Glia. 2003 Apr 15;42(2):194-208.

[41] Bubendorf L, Kolmer M, Kononen J, Koivisto P, Mousses S, Chen Y, et al. Hormone therapy failure in human prostate cancer: analysis by complementary DNA and tissue microarrays. *J Natl Cancer Inst.* 1999 Oct 20;91(20):1758-64.

[42] Cornford PA, Dodson AR, Parsons KF, Desmond AD, Woolfenden A, Fordham M, et al. Heat shock protein expression independently predicts clinical outcome in prostate cancer. *Cancer Res.* 2000 Dec 15;60(24):7099-105.

[43] Bruey JM, Paul C, Fromentin A, Hilpert S, Arrigo AP, Solary E, et al. Differential regulation of HSP27 oligomerization in tumor cells grown in vitro and in vivo. *Oncogene*. 2000 Oct 5;19(42):4855-63.

[44] Tenniswood MP, Guenette RS, Lakins J, Mooibroek M, Wong P, Welsh JE. Active cell death in hormone-dependent tissues. Cancer Metastasis Rev. 1992 Sep;11(2):197-220.

[45] Rocchi P, So A, Kojima S, Signaevsky M, Beraldi E, Fazli L, et al. Heat shock protein 27 increases after androgen ablation and plays a cytoprotective role in hormone-refractory prostate cancer. Cancer Res. 2004 Sep 15;64(18):6595-602.

[46] Garrido C, Ottavi P, Fromentin A, Hammann A, Arrigo AP, Chauffert B, et al. HSP27 as a mediator of confluence-dependent resistance to cell death induced by anticancer drugs. *Cancer Res*. 1997 Jul 1;57(13):2661-7.

[47] Vargas-Roig LM, Gago FE, Tello O, Aznar JC, Ciocca DR. Heat shock protein expression and drug resistance in breast cancer patients treated with induction chemotherapy. *Int J Cancer,* 1998 Oct 23;79(5):468-75.

[48] Garrido C, Bruey JM, Fromentin A, Hammann A, Arrigo AP, Solary E. HSP27 inhibits cytochrome c-dependent activation of procaspase-9. FASEB J. 1999 Nov;13(14):2061-70.

[49] Parcellier A, Schmitt E, Brunet M, Hammann A, Solary E, Garrido C. Small heat shock proteins HSP27 and alphaB-crystallin: cytoprotective and oncogenic functions. *Antioxid Redox Signal,* 2005 Mar-Apr;7(3-4):404-13.

[50] Jantschitsch C, Trautinger F. Heat shock and UV-B-induced DNA damage and mutagenesis in skin. Photochem Photobiol Sci. 2003 Sep;2(9):899-903.

[51] Wieder R. Insurgent micrometastases: sleeper cells and harboring the enemy. *J Surg Oncol*. 2005 Mar 15;89(4):207-10.

[52] Oesterreich S, Hilsenbeck SG, Ciocca DR, Allred DC, Clark GM, Chamness GC, et al. The small heat shock protein HSP27 is not an independent prognostic marker in axillary lymph node-negative breast cancer patients. Clin Cancer Res. 1996 Jul;2(7):1199-206.

[53] Ciocca DR, Rozados VR, Cuello Carrion FD, Gervasoni SI, Matar P, Scharovsky OG. Hsp25 and Hsp70 in rodent tumors treated with doxorubicin and lovastatin. *Cell Stress Chaperones,* 2003 Spring;8(1):26-36.

[54] O'Callaghan-Sunol C, Gabai VL, Sherman MY. Hsp27 modulates p53 signaling and suppresses cellular senescence. Cancer Res. 2007 Dec 15;67(24):11779-88.

[55] Ciocca DR, Vargas-Roig LM. Hsp27 as a prognostic and predictive factor in cancer. *Prog Mol Subcell Biol*. 2002;28:205-18.

[56] Xu L, Bergan RC. Genistein inhibits matrix metalloproteinase type 2 activation and prostate cancer cell invasion by blocking the transforming growth factor beta-mediated activation of mitogen-activated protein kinase-activated protein kinase 2-27-kDa heat shock protein pathway. Mol Pharmacol. 2006 Sep;70(3):869-77.

[57] Di K, Wong YC, Wang X. Id-1 promotes TGF-beta1-induced cell motility through HSP27 activation and disassembly of adherens junction in prostate epithelial cells. *Exp Cell Res*. 2007 Nov 15;313(19):3983-99.

[58] Zoubeidi A, Zardan A, Beraldi E, Fazli L, Sowery R, Rennie P, et al. Cooperative interactions between androgen receptor (AR) and heat-shock protein 27 facilitate AR transcriptional activity. Cancer Res. 2007 Nov 1;67(21):10455-65.

[59] Zimmermann TS, Lee AC, Akinc A, Bramlage B, Bumcrot D, Fedoruk MN, et al. RNAi-mediated gene silencing in non-human primates. Nature. 2006 May 4;441(7089):111-4.

[60] Fire A, Xu S, Montgomery MK, Kostas SA, Driver SE, Mello CC. Potent and specific genetic interference by double-stranded RNA in Caenorhabditis elegans. *Nature,* 1998 Feb 19;391(6669):806-11.

[61] Medema RH. Optimizing RNA interference for application in mammalian cells. Biochem J. 2004 Jun 15;380(Pt 3):593-603.

[62] Elbashir SM, Harborth J, Lendeckel W, Yalcin A, Weber K, Tuschl T. Duplexes of 21-nucleotide RNAs mediate RNA interference in cultured mammalian cells. *Nature*, 2001 May 24;411(6836):494-8.

[63] Stark GR, Kerr IM, Williams BR, Silverman RH, Schreiber RD. How cells respond to interferons. Annu Rev Biochem. 1998;67:227-64.

[64] Sumimoto H, Yamagata S, Shimizu A, Miyoshi H, Mizuguchi H, Hayakawa T, et al. Gene therapy for human small-cell lung carcinoma by inactivation of Skp-2 with virally mediated RNA interference. Gene Ther. 2005 Jan;12(1):95-100.

[65] Uchida H, Tanaka T, Sasaki K, Kato K, Dehari H, Ito Y, et al. Adenovirus-mediated transfer of siRNA against survivin induced apoptosis and attenuated tumor cell growth in vitro and in vivo. Mol Ther. 2004 Jul;10(1):162-71.

[66] Oliveira S, Storm G, Schiffelers RM. Targeted Delivery of siRNA. *J Biomed Biotechnol.* 2006;2006(4):63675.

[67] Chi KN, Eisenhauer E, Fazli L, Jones EC, Goldenberg SL, Powers J, et al. A phase I pharmacokinetic and pharmacodynamic study of OGX-011, a 2'-methoxyethyl antisense oligonucleotide to clusterin, in patients with localized prostate cancer. *J Natl Cancer Inst*. 2005 Sep 7;97(17):1287-96.

[68] Syed S, Tolcher A. Innovative therapies for prostate cancer treatment. *Rev Urol*. 2003;5 Suppl 3:S78-84.

[69] Trougakos IP, So A, Jansen B, Gleave ME, Gonos ES. Silencing expression of the clusterin/apolipoprotein j gene in human cancer cells using small interfering RNA induces spontaneous apoptosis, reduced growth ability, and cell sensitization to genotoxic and oxidative stress. Cancer Res. 2004 Mar 1;64(5):1834-42.

[70] Behlke MA. Progress towards in vivo use of siRNAs. Mol Ther. 2006 Apr;13(4):644-70.

[71] Aigner A. Applications of RNA interference: current state and prospects for siRNA-based strategies in vivo. Appl Microbiol Biotechnol. 2007 Apr 25.

[72] Vickers TA, Koo S, Bennett CF, Crooke ST, Dean NM, Baker BF. Efficient reduction of target RNAs by small interfering RNA and RNase H-dependent antisense agents. A comparative analysis. J Biol Chem. 2003 Feb 28;278(9):7108-18.

[73] Lingor P, Michel U, Scholl U, Bahr M, Kugler S. Transfection of "naked" siRNA results in endosomal uptake and metabolic impairment in cultured neurons. *Biochem Biophys Res Commun*. 2004 Mar 19;315(4):1126-33.

[74] Overhoff M, Wunsche W, Sczakiel G. Quantitative detection of siRNA and single-stranded oligonucleotides: relationship between uptake and biological activity of siRNA. *Nucleic Acids Res*. 2004;32(21):e170.

[75] Lewis DL, Wolff JA. Delivery of siRNA and siRNA expression constructs to adult mammals by hydrodynamic intravascular injection. Methods Enzymol. 2005;392:336-50.

[76] Snove O, Jr., Rossi JJ. Expressing short hairpin RNAs in vivo. *Nat Methods*, 2006 Sep;3(9):689-95.

[77] Bausero MA, Page DT, Osinaga E, Asea A. Surface expression of Hsp25 and Hsp72 differentially regulates tumor growth and metastasis. *Tumour Biol.* 2004 Sep-Dec;25(5-6):243-51.

[78] Schepers H, Geugien M, van der Toorn M, Bryantsev AL, Kampinga HH, Eggen BJ, et al. HSP27 protects AML cells against VP-16-induced apoptosis through modulation of p38 and c-Jun. Exp Hematol. 2005 Jun;33(6):660-70.

[79] Evans IM, Britton G, Zachary IC. Vascular endothelial growth factor induces heat shock protein (HSP) 27 serine 82 phosphorylation and endothelial tubulogenesis via protein kinase D and independent of p38 kinase. Cell Signal. 2008 Jul;20(7):1375-84.

[80] Lee CK, Lee HM, Kim HJ, Park HJ, Won KJ, Roh HY, et al. Syk contributes to PDGF-BB-mediated migration of rat aortic smooth muscle cells via MAPK pathways. *Cardiovasc Res.* 2007 Apr 1;74(1):159-68.

[81] Golembieski WA, Thomas SL, Schultz CR, Yunker CK, McClung HM, Lemke N, et al. HSP27 mediates SPARC-induced changes in glioma morphology, migration, and invasion. *Glia.* 2008 Aug 1;56(10):1061-75.

[82] Kang SH, Kang KW, Kim KH, Kwon B, Kim SK, Lee HY, et al. Upregulated HSP27 in human breast cancer cells reduces Herceptin susceptibility by increasing Her2 protein stability. *BMC Cancer*, 2008;8:286.

[83] McCollum AK, TenEyck CJ, Stensgard B, Morlan BW, Ballman KV, Jenkins RB, et al. P-Glycoprotein-mediated resistance to Hsp90-directed therapy is eclipsed by the heat shock response. *Cancer Res.* 2008 Sep 15;68(18):7419-27.

[84] Kamada M, So A, Muramaki M, Rocchi P, Beraldi E, Gleave M. Hsp27 knockdown using nucleotide-based therapies inhibit tumor growth and enhance chemotherapy in human bladder cancer cells. *Mol Cancer Ther.* 2007 Jan;6(1):299-308.

[85] Rocchi P, Jugpal P, So A, Sinneman S, Ettinger S, Fazli L, et al. Small interference RNA targeting heat-shock protein 27 inhibits the growth of prostatic cell lines and induces apoptosis via caspase-3 activation in vitro. *BJU Int.* 2006 Nov;98(5):1082-9.

[86] Lee SC, Sim N, Clement MV, Yadav SK, Pervaiz S. Dominant negative Rac1 attenuates paclitaxel-induced apoptosis in human melanoma cells through upregulation of heat shock protein 27: a functional proteomic analysis. *Proteomics,* 2007 Nov;7(22):4112-22.

[87] Aloy MT, Hadchity E, Bionda C, Diaz-Latoud C, Claude L, Rousson R, et al. Protective role of Hsp27 protein against gamma radiation-induced apoptosis and radiosensitization effects of Hsp27 gene silencing in different human tumor cells. *Int J Radiat Oncol Biol Phys.* 2008 Feb 1;70(2):543-53.

[88] Guo K, Kang NX, Li Y, Sun L, Gan L, Cui FJ, et al. Regulation of HSP27 on NF-kappaB pathway activation may be involved in metastatic hepatocellular carcinoma cells apoptosis. BMC *Cancer*, 2009 Mar 31;9(1):100.

[89] Doshi BM, Hightower LE, Lee J. The role of Hsp27 and actin in the regulation of movement in human cancer cells responding to heat shock. *Cell Stress Chaperones,* 2009 Feb 18.

[90] Zhang Y, Shen X. Heat shock protein 27 protects L929 cells from cisplatin-induced apoptosis by enhancing Akt activation and abating suppression of thioredoxin reductase activity. *Clin Cancer Res*. 2007 May 15;13(10):2855-64.

[91] McCollum AK, Teneyck CJ, Sauer BM, Toft DO, Erlichman C. Up-regulation of heat shock protein 27 induces resistance to 17-allylamino-demethoxygeldanamycin through a glutathione-mediated mechanism. Cancer Res. 2006 Nov 15;66(22):10967-75.

[92] Shin KD, Lee MY, Shin DS, Lee S, Son KH, Koh S, et al. Blocking tumor cell migration and invasion with biphenyl isoxazole derivative KRIBB3, a synthetic molecule that inhibits Hsp27 phosphorylation. *J Biol Chem*. 2005 Dec 16;280(50):41439-48.

[93] Crooke ST. Therapeutic applications of oligonucleotides. *Annu Rev Pharmacol Toxicol*. 1992;32:329-76.

[94] Gleave ME, Monia BP. Antisense therapy for cancer. *Nat Rev Cancer*, 2005 Jun;5(6):468-79.

[95] Crooke ST. Molecular mechanisms of antisense drugs: RNase H. *Antisense Nucleic Acid Drug Dev*. 1998 Apr;8(2):133-4.

[96] Wu H, Lima WF, Zhang H, Fan A, Sun H, Crooke ST. Determination of the role of the human RNase H1 in the pharmacology of DNA-like antisense drugs. *J Biol Chem*. 2004 Apr 23;279(17):17181-9.

[97] Galarneau A, Min KL, Mangos MM, Damha MJ. Assay for evaluating ribonuclease H-mediated degradation of RNA-antisense oligonucleotide duplexes. *Methods Mol Biol*. 2005;288:65-80.

[98] Carpentier AF, Chen L, Maltonti F, Delattre JY. Oligodeoxynucleotides containing CpG motifs can induce rejection of a neuroblastoma in mice. Cancer Res. 1999 Nov 1;59(21):5429-32.

[99] Saijo Y, Perlaky L, Wang H, Busch H. Pharmacokinetics, tissue distribution, and stability of antisense oligodeoxynucleotide phosphorothioate ISIS 3466 in mice. Oncol Res. 1994;6(6):243-9.

[100] Zellweger T, Miyake H, Cooper S, Chi K, Conklin BS, Monia BP, et al. Antitumor activity of antisense clusterin oligonucleotides is improved in vitro and in vivo by incorporation of 2'-O-(2-methoxy)ethyl chemistry. *J Pharmacol Exp Ther*. 2001 Sep;298(3):934-40.

[101] Henry S, Stecker K, Brooks D, Monteith D, Conklin B, Bennett CF. Chemically modified oligonucleotides exhibit decreased immune stimulation in mice. *The Journal of pharmacology and experimental therapeutics,* 2000 Feb;292(2):468-79.

[102] Hadaschik BA, Jackson J, Fazli L, Zoubeidi A, Burt HM, Gleave ME, et al. Intravesically administered antisense oligonucleotides targeting heat-shock protein-27 inhibit the growth of non-muscle-invasive bladder cancer. *BJU Int*. 2008 Aug 5;102(5):610-6.

[103] Tanaka Y, Fujiwara K, Tanaka H, Maehata K, Kohno I. Paclitaxel inhibits expression of heat shock protein 27 in ovarian and uterine cancer cells. *Int J Gynecol Cancer,* 2004 Jul-Aug;14(4):616-20.

[104] Choi DH, Ha JS, Lee WH, Song JK, Kim GY, Park JH, et al. Heat shock protein 27 is associated with irinotecan resistance in human colorectal cancer cells. *FEBS Lett*. 2007 Apr 17;581(8):1649-56.

[105] Yang YX, Xiao ZQ, Chen ZC, Zhang GY, Yi H, Zhang PF, et al. Proteome analysis of multidrug resistance in vincristine-resistant human gastric cancer cell line SGC7901/VCR. *Proteomics,* 2006 Mar;6(6):2009-21.

In: Small Stress Proteins and Human Diseases
Editors: Stéphanie Simon et al.

ISBN: 978-1-61470-636-6

Chapter 4.3

BENEFICIAL AND DELETERIOUS, THE DUAL ROLE OF SMALL STRESS PROTEINS IN HUMAN DISEASES: IMPLICATIONS FOR THERAPEUTIC STRATEGIES

***Stéphanie Simon*[1] *and André-Patrick Arrigo*[1]**
Claude Bernard University Lyon 1, CGMC CNRS UMR 5534, 'Stress, Chaperones and Cell Death' Laboratory, 69 622 Villeurbanne Cedex, France

ABSTRACT

Several sHsps (HspB1, HspB4, HspB5 and HspB8) are molecular chaperones that act as holdase avoiding aggregation of misfolded proteins. They are also involved in mechanisms aimed at eliminating stress- or mutation-induced misfolded and/or aggregated polypeptides. By doing so, these sHsps provide the cell with some protection against the deleterious effects mediated by damaged proteins. These sHsps are also constitutively expressed in several mammalian cells, particularly in pathological conditions where their holdase activity provides cells with resistance to different types of stress-mediated injuries. Some of them (HspB1, HspB4, HspB5 and HspB8) also enhance cellular resistance to apoptotic conditions. Indeed, these sHsps are characterized by intriguing anti-apoptotic activities that have deleterious tumorigenic effect when they are expressed to high levels in cancer cells. Their ability to target specific proteins and regulate their activity or half-life, as for example caspase-3, is probably at the origin of their anti-apoptotic effect. This chapter summarizes the current knowledge about the holdase sHsps and their implications, either positive or deleterious, in pathologies such as neurodegenerative diseases, myopathies, asthma, cataracts and cancers. Current

[1] Stéphanie Simon and André-Patrick Arrigo
Claude Bernard University, LYON 1,
CGMC CNRS UMR 5534,
Stress, Chaperones and Cell death Laboratory,
16 rue Dubois, Bat. Gregor Mendel,
69622 Villeurbanne Cedex, France.
Phone: +33 (0) 472448595,
Fax : +33 (0) 472432385.
stef.labo@gmail.com
parrigo@me.com; arrigo@univ-lyon1.fr

therapeutic strategies aimed at modulating their expression and/or holdase activity are discussed in view of their ability to form complex, cell specific and dynamic mozaic oligomeric structures.

1. Small Stress Proteins Are Multi-Programmed Weapons to Fight Cell Death

1.1. Holdase Activity of sHsps

Several human sHsps (HspB1, HspB4, HspB5 and HspB8) plus the less conserved Hsp16.2 polypeptide share an ATP-independent chaperone « holdase » activity. This activity favors sHsps interaction with denatured polypeptides and their subsequent storage in a refolding competent state that attenuates and/or suppresses irreversible protein aggregation [1-6]. sHsps holdase activity is modulated by the dynamic ability of these proteins to change their oligomerization profile in order to trap denatured polypeptides. This activity differs from the well characterized ATP-dependent « foldase » chaperone machines (Hsp70, Hsp90 and Hsp60) aimed at refolding stress-induced misfolded polypeptides [7,8,9]. The denatured polypeptides, trapped within holdase large oligomeric structures [10,11], are then refolded through cooperation with the foldase machinery [12]. Hence, *in vivo,* the holdase and foldase chaperone systems are part of a coordinated cellular protein refolding network. The large oligomeric complexes (up to 700 kDa and more) which form between misfolded substrates and sHsps are very stable and act as dynamic storage reservoirs that can further display an increased size if more non-native proteins accumulate. Moreover, during drastic heat shock conditions, a fraction of HspB1 cellular content is recovered in the nucleus at the level of granules [13] containing denatured proteins [14]; a phenomenon which represents a situation of failed refolding where the nuclear substrates are stored for subsequent degradation [15]. Nuclear sHsps can also be recovered at the level of intranuclear lamin [16] or exert, as in the case of HspB1, a feedback inhibition of HSF1 (Heat Shock Factor 1) transactivation, through the binding and modification of this transcription factor [17]. Of interest, in unstressed cells, a small fraction of several sHsps has been identified as nuclear speckle components suggesting a link with the transcriptional status of the cells [18,19]. In some stress conditions, such as those encountered during heat shock, sHsps holdase activity may also indirectly be involved in the modulation of mRNA translation through the trapping of the initiation translation factor eIF4G in insoluble heat shock granules [20].

In unstressed cells, sHsps holdase activity is required to stabilize specific protein targets, as for example the cytoskeleton [21,22,23] or stimulate ubiquitination and/or sumoylation of specific substrates [17,24]. Moreover, the chaperone/client protein concept which is well characterized for Hsp90 [25,26], may also appear to be a fundamental property of sHsps. For example, caspase 3, which interacts with HspB1, shows a proteasome-dependent proteolytic degradation in HspB1 immunodepleted cells [27]. Hence, alteration in sHsp holdase activity can deregulate cellular homeostasis, by misfolding client proteins, prior to their ubiquitination and proteasome degradation. However, it is not yet known whether these particular activities of sHsps depend or not on the classical formation of large oligomeric structures trapping the targeted substrate: a property of HspB1 holdase activity. The problem is rather complex since the relationship between oligomerization, phosphorylation and holdase activity appears to be sHsp specific [28,29]. This has been extensively been studied in the case of HspB1 [30] and HspB4 and HspB5 (αA,B-crystallins) [31-33].

1.2. Direct and Indirect Anti-Apoptotic Functions of sHsps

In apoptotic cells, HspB1 and HspB5 protective effects rely on their interaction with specific protein targets located both upstream and downstream of mitochondria [34-40] or along the signal transduction pathways activated by death receptors [27,41-44]. Large oligomers of HspB1, that at least in heat shock and oxidative stress treated cells, bear holdase activity [30] appear to play a role in the downstream events that decrease cytochrome c mediated caspases activation [45]; a phenomenon that may result of the interaction of HspB1 with caspase-3. Large oligomers of HspB1 are also observed during the transient, but essential, expression of this protein during the differentiation of several cell types [46-55]. In that regard, a putative sHsp holdase activity could be involved at protecting the differentiating cells from massive apoptosis. It may act either directly towards the apoptosis machinery or indirectly against the cellular causes that induce apoptosis, such as the differentiation-induced drastic changes in the cellular protein content, structural organization and/or localization [49,50,54]. It could also prevent differentiating cells from the toxicity of misfolded protein aggregation or inaccurate protein interactions that may otherwise lead to accumulation of junk protein structures that could result in aberrant differentiation processes and apoptosis. Some proteins of the differentiating cells may also need to be protected from a transient hostile environment. In the opposite, sHsps holdase activity could also promote degradation of polypeptides that become undesirable for differentiated cells [54]. For example, during keratinocyte differentiation, the murine equivalent to HspB1 drives the disassembly of the keratin network through the sequestration and partial unfolding of keratin 5 and 14 subunits. This prevents undesirable interactions and aggregation and allows a switch from the keratin MK5/MK14 network to keratin MK1/MK10 network [53]. HspB1/MK5/MK14 structures may therefore represent a first step toward degradation or sequestration of keratins that can be reused once they have readopted their native state. Another example concerns myogenic precursor cells where the expression of HspB5, but not HspB1, renders them resistant to differentiation-induced apoptosis [56,57]. HspB5 negatively regulates apoptosis during myogenesis by directly inhibiting the proteolytic activation of caspase-3, whereas the corresponding R120G and pseudophosphorylation mutants are defective in this function. Indirect inhibition of apoptosis may also result, as described above in the case of HspB1, of activities towards differentiation induced toxicity of misfolded protein aggregation. Taken together, those findings indicate that several sHsps share the ability to protect differentiating cells from apoptosis.

Concerning the small oligomers of HspB1, they have been described to protect F-actin breakdown [37] and modulate apoptotic regulators, such as Daxx [41]. The interaction of HspB1 and/or HspB5 with other regulators such as Akt, Bax, Bcl-xs, cytochrome c as well as those involving involving PKCα, Raf/MEK/ERK and Akt signaling pathways [38,42,58] (see chapter 1-3 and part 3) may require specific, but unfortunately still unknown, structural organization of these proteins.

Whereas HspB1, HspB4 and HspB5 anti-apoptotic properties have been clearly demonstrated, HspB8 displays both pro- and anti-apoptotic behavior depending on the cell type where it is expressed. This particular aspect of HspB8 is discussed in chapters 2-5 and 3-4.

1.3. Household Activities of sHsps through Proteasome and Autophagic Degradation Machineries

sHsps are also active towards misfolded oxidized polypeptides. However, only limited repair mechanisms exist to refold drastically oxidized or heat-denatured polypeptides. In these conditions the degradation of these proteins is triggered through different proteolytic pathways [59,60,61,62]. One relates to the CHIP polypeptide that binds the Hsp70 foldase machine and targets irreversibly misfolded polypeptides to the ubiquin-26S proteasome [63]. sHsps are also involved in the process of protein degradation and triage. For example, HspB1 can interact with poly-ubiquitin chains [60] and, by promoting proteasome degradation of p27Kip1, it helps S-phase re-entry during stress recovery [24]. HspB5 promotes Fbx4-dependent ubiquitination of cyclin D1 by the SCF(Fbx4/HspB5) ligase machinery [61,62] and interacts with proteasome [59]. Another pathway is related to autophagy since this is an efficient mechanism to eliminate oxidized [64,65] and heat shock induced [66] aggregated polypeptides. Autophagy depends on the stress-inducible HspB8-Bag3 complex where HspB8 is responsible for recognizing the misfolded protein whereas Bag3 recruits and activates the macroautophagy machinery in close proximity to the chaperone-loaded substrates [67,68].

2. sHsps Protective Activities in Human Diseases: Are They Deleterious or Beneficial for Pathological Cells?

2.1. sHsps Can Be Friendly

As described in several chapters of this book, sHsps bearing ATP-independent holdase activity are considered as potent protective factors in cells where the disease-causing proteins have the tendency to aggregate and form deleterious and pathological large inclusions. Indeed, sHsps expression enhances the resistance of cells to the deleterious effects induced by agents or conditions that alter protein folding such as heat shock [69,70] and oxidative stress [30,71-73]. Mutations can also efficiently alter protein folding. In that regard, it has recently been reported that sHsps stimulate the cellular resistance to aggregation prone proteins, *i.e* α-synuclein, Alzheimer β-amyloid peptide, Huntingtin (polyQ mutants), intermediate filaments as well as HspB5 which, once mutated or oxidized, can lead to neurodegenerative, myopathic, cardiomyopathic or cataract diseases [74-79]. Another example concerns the holdase activity associated to HspB4 and HspB5 which is essential in maintaining lens transparency and preventing cataract and retina pathologies [80]. Hence, it is not surprising that high loads of sHsps have been detected in the muscles, neurones and lens fibers cells that are targeted by these pathologies. For example, HspB1 and/or HspB5 accumulate in Rosenthal fibers of

Alexander disease, cortical Lewy bodies, Alzheimer disease plaques, neurofibrillary tangles, Creutzfeldt-Jakob altered neurones as well as in synuclein deposit associated to Parkinson disease or myopathy-associated inclusion body [81-84] (see also chapter 1-4). The importance of the protective activity of sHsps was further confirmed by the discovery of mutations in their genes that result in diseases such as neuropathy, myofibrillar myopathy, cardiomyopathy and cataract [85,86]. Oxidative stress is a common feature that characterizes cells bearing aggregated polypeptides [87,88]. Indeed, the expression of several of the above described diseases related to aggregating proteins leads to the production of abnormally high levels of deleterious intracellular reactive oxygen species [88-92,93,94,95]. Hence, oxidative damages may aggravate the accumulation of aggregated proteins that occur in these diseases. This may in turn cause more oxidative damages by interfering for example with the function of the proteasome. Oxidative stress may result of altered mitochondrial function close to aggregates/inclusion bodies. These hypothesis seem realistic since in myofibrillar myopathies, several observations support a link between oxidative stress, specific protein aggregation, abnormal activities of the degradation machineries (both proteasomal and autophagic) and mitochondrial malfunctions [74,75,94,96,97]. Such a model is presented in chapter 2-3. In that regard it is interesting to note that Huntingtin, β-amyloid and α-synuclein are well known iron/copper binding or metal homeostasis modulating polypeptides [98,99]. These metals are well-known catalyzers of the hydroxyl radical generating Fenton reaction [100] which is probably highly deregulated in cells bearing aggregated polypeptides [101,102]. In regards to that issue, the high level of expression of HspB1 and HspB5 in pathological cells may partially counteract oxidative stress via their particular ability to modulate intracellular redox status via reduced glutathione and iron level regulation [74,75,94,103-106]. Other pathologies related to oxidative stress where sHsps expression is beneficial are of inflammatory origin [71,107-109]. One can cite for example, airway inflammation associated with asthma [110], alcoholic liver diseases characterized by the presence of Mallory bodies and ischemic related stroke injuries [111,112]. An important feature of sHsps concerns their ability to promote cardioprotection [113] and to enhance nerve survival [114] (see chapters 1-4 and 1-5). Other observations linked to the protective activity of sHsps deal with the presence of autoantibodies that probably neutralize their beneficial cellular effects. For example, autoantibody directed against HspB5 detected in multiple sclerosis [115] and retinal pathologies [116] probably contributes to the pathogenicity of those diseases.

2.2. sHsps Can Be Your Enemies

In contrast to the beneficial effect induced by the protective activity of sHsps in diseases characterized by pathological cell degeneration and death, the other side of the coin concerns their ability to protect pathological cells that evade cell death and proliferate, such as cancer cells (see chapters of part 3) [117,118]. Indeed, the number of reports dealing with the deleterious effects of sHsps in cancer pathologies has recently grown exponentially. These sHsps, particularly HspB1 and HspB5, that have been, until recently, the most studied ones, are often constitutively expressed to high levels in some cancer cells, particularly carcinoma. These sHsps, which are essential for the growth of these cells, protect them against apoptotic or other types of death triggered by the immune system in the aim of their elimination. In rodents, sHsps are tumorigenic, stimulate metastasis formation and dissemination and provide cancer cells with resistance to numerous anti-cancer drugs, which in turn can also stimulate sHsps expression [119,120-123]. Collectively, these phenomena decrease the effectiveness of chemotherapeutic agents. HspB1 large oligomers that bear chaperone holdase-like activity appear responsible of the tumorigenic activity [45]. Whether this reflects an effect towards client proteins that regulate tumorigenic and metastatic processes is not known and should be tested. Differential constitutive expression of sHsps is often observed between cancer cells, as for example the different profile of expression and putative involvement of HspB1 and HspB5 in anaplastic thyroid carcinomas and several types of brain cancer cells [124,125]. Taken together, these observations are clearly in favor of a deleterious role of sHsps in cancer cells. However, not all sHsps have yet been analyzed and it cannot be excluded that some of them could have a still unknown beneficial role. For example, can we exclude that sHsps, which are often described as membrane associated proteins [126-128], could, similarly to Hsp70 [129,130], elicit an immune response aimed at killing cancer cells through their association with immunogenic peptides?

2.3. Elements that Can Influence sHsps Behavior

In vivo, each type of cell within a specific tissue can have a specific pattern of basically expressed sHsps. Of interest, these proteins can play non redundant roles. For example, in the heart, HspB5 is essential to the structural remodeling associated with contractile performance whereas HspB2 is necessary to maintain the energetic cellular balance [131]. Nevertheless, one very important property of sHps is their ability to interact and form complex mosaic oligomeric structures. This feature, first described by Zantema et al. [132] in the case of HspB1 and HspB5, is illustrated in figure 1. For example, in lens fiber cells, HspB4 and HspB5 polypeptides are recovered in a 3 to 1 ratio in a unique large mozaic oligomer [133,134]. Hence, in cells expressing several sHsps, complex and multiple combinatorial oligomeric structures can be formed that may bear functions that are different from the original polypeptide. The phenomenon can be observed in muscle cells, where HspB1 interacts with HspB6 and by doing so these proteins mutually alter their own structure and consequently their ability to be phosphorylated [135]. Phosphorylation of sHsps may also result in conformational changes that modulate their interaction with other sHsps [136-138]. Moreover, sHsps that are present in a single cell do not interact with similar kinetics. In contrast, they form specific complexes that can contain different sHsp partners. Such situation is observed in myoblasts where HspB1 and HspB5, but not HspB2, form a complex localized on actin bundles specific to myotubes, underlying the fact that, in a single cell, structurally independent dynamic sHsps complexes may have specific interactions with distinct molecular targets. *In vivo,* only few of these structures have been characterized yet, thus future studies will have to characterize the multiple, specific and dynamic sHsps interactions and test whether they represent a highly adjustable system able to face rapidly with changes in cellular physiology. Functional importance of the mosaic oligomers formed between sHsps is reinforced by the fact that several sHsp mutations linked to human pathologies (in HspB1, HspB4, HspB5 and HspB8) induce specific alterations in their mutual interactions [137,139,140].

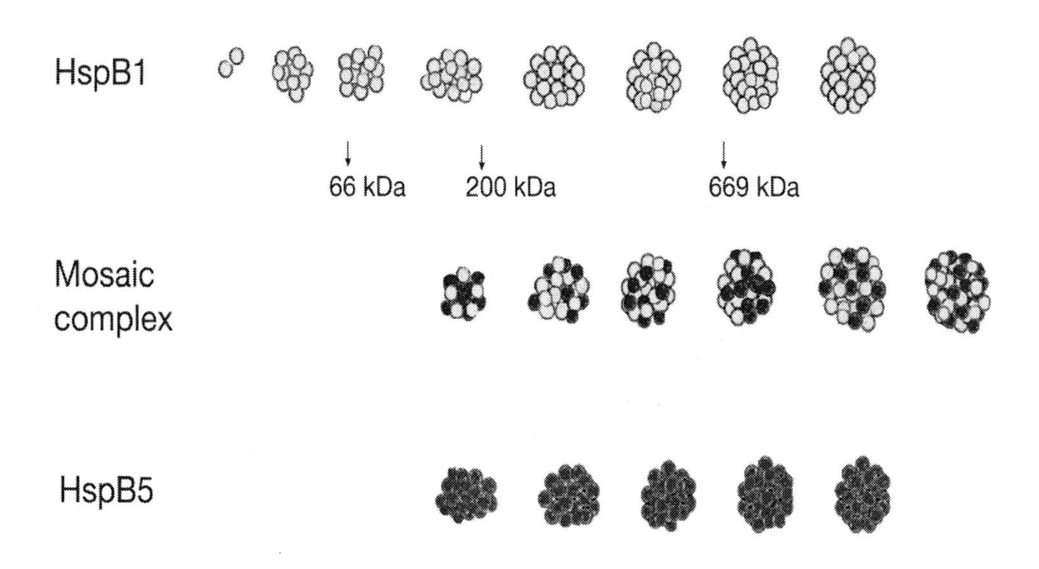

Figure 1. Mozaic oligomeric structures formed by HspB1 and HspB5. HeLa cells are characterized by the constitutive expression of HspB1. These cells were stably transfected with either an empty DNA vector or with a DNA vector encoding HspB5 gene under the control of the constitutive cytomegalovirus promoter. Stable HeLa clones expressing similar levels of HspB1 and HspB5 were analyzed. The cytosolic fraction of these cells which contains more than 90% of the cellular content of HspB1 and HspB5 was then analyzed on a Sepharose 6B sizing column. The presence of these proteins was analyzed in immunoblots probed with specific antibodies. Immunoblots were scanned and the size distribution of HspB1 and HspB5 was quantified. Immunoprecipitation was performed to confirm the formation of a 1:1 complex between HspB1 and HspB5. Schematic representation of the homo- and hetero-holigomeric size distribution of these proteins is presented. The subunits in the different structures symbolize monomers of HspB1 (light grey) or HspB5 (dark grey). HspB1 analysis was made in cells transfected with the empty vector. Analysis of HspB1-HspB5 mosaic complex was made using the stably transfected cell lines expressing similar levels of HspB1 and HspB5. Already published data were used to draw the cartoon describing the oligomeric profile of HspB5. Note that the mozaic structure has a different size distribution than HspB1 and HspB5.

3. Implications for Therapeutic Approaches

sHsps protective role can either be beneficial as it helps cells to better cope with pathologies that end up with cell loss or have deleterious effects by sustaining and even stimulating pathological cells that should be eliminated (Figure 2). The phenomenon is even more complex if the newly discovered property of sHsps, that is their ability to act extracellularly, is taken into account. For example, HspB1 has been described as a circulating extracellularly located protein associated with tumor progression and increased post-injury infection through monocyte-dependent immunoregulatory activities that contribute to immunopathology [141]. However, circulating HspB1 has also been reported to be artheroprotective [142]. Hence, similarly to intracellular sHsps, the few examples that are yet available concerning circulating sHsps again reflect their pathology-dependent dual, positive or negative, roles.

3.1. Modulation of sHsps Expression Levels

After the discovery of the sHsps chaperone holdase activity, the idea to use sHsps to treat conformational diseases has rapidly emerged. Since then, a large number of publications clearly demonstrated the beneficial effect sHsps over-expression on stressed cells. Nowadays, this concept is well accepted; an up-date of the more recent data is presented in the part 1 of this book. Nevertheless, efficient methods to promote sHsp expression *in vivo* using genetically based approaches do not yet exist. To short-cut this technical problem, efforts to find non toxic molecules that can specifically stimulate sHsps expression are currently made. For example, in transgenic mice suffering of desmin-related cardiomyopathy linked to the HspB5 R120G mutation, the protective effect of orally administered geranylgeranylacetone has been attributed to reduced amyloid oligomer levels and aggregates in the heart that correlated with an enhancement of HspB1 and HspB8 expression in this tissue [143]. Glucocorticoïd treatment may also be an interesting option [144]. Voluntary exercise has also been suggested to be beneficial to counteract the aggregation prone HspB5-R120G mutation that induces cardiomyopathy in transgenic mice [145]. This intriguing observation could have resulted of the stimulated expression of Hsps in the heart of exercising mice. Controversy nevertheless exists towards such an explanation since in the transgenic mice used by the authors, the gene encoding HspB5 mutant is under the control of a constitutive promoter. Therefore, in contrast to what should be normally expected in the heart of non transgenic animals suffering of the same pathology, the level of mutant HspB5 in the heart of exercising or treated transgenic mice will not increase. This could artificially increase the ratio between total Hsps to mutant HspB5. To generalize, in the case of inherited diseases caused by sHsps mutations, the strategy of using wild-type sHsps over-expression could be envisaged only if the level of mutated sHsps are not be stimulated by the treatment. This leads to an important point that refers to the cellular homeostasis protein folding machinery; a potent system that exists in every cells but which is limited and can be overwhelmed [146]. Hsps are important members of the folding cellular machinery and consequently an increase in their level should be well evaluated and controlled as it could favor the masking of refoldable mutations.

Approaches aimed at inhibiting HspB1 expression using anti-sense DNA vectors [147] or second generation of RNAi molecules, such as OGX 437 (Oncogenex Inc), have been reported to sensitize cancer cells to apoptotic inducers, anticancer drugs, radiations, and to reduce the tumorigenic potential of cancer cells [148-150] (see the chapters 3-4 and 4-2). These molecules destabilize the intracellular network formed by the different sHsps interacting with HspB1 and abolish the cellular protection generated by its presence. They could also induce the degradation of tumorigenic and/or metastatic client proteins that may bind HspB1 [27]. Ulterior studies will have to determine if decreasing sHsps level of expression or inhibiting their chaperone activity could unmask pre-existing mutations.

3.2. Modulation or Mimicking the sHsps Activities

For treating diseases caused by pathological inclusions, it is reasonable to propose that drugs that up-regulate sHsps holdase activity in a definite tissue will be interesting. An approach to mimic the holdase activity of HspB5 towards amyloid targets by peptides derived from crucial domains of HspB5 is presented in the chapter 4.1. This is a promising approach that may lead to peptidomimetic drugs that could be beneficial to cells accumulating aggregated polypeptides. Alternative studies have revealed that inhibitors of aldose reductase and nonenzymatic glycation delay cataract progression in diabetic condition through the positive modulation of HspB4-HspB5 holdase activity [151]. This drug acts without inducing deleterious side effects due to the modulation of these polypeptides in other tissues, in contrast to the effects induced by cyclosporine A on HspB1, HspB5 and myofibrillar cytoskeleton in rat heart [152]. It has also recently been shown that the chemical chaperones carnosine and its acetyl derivative are anticataract drugs when added to lubricant eye drops (Can-C™); an activity which probably results of their activity which resembles that of HspB4-HspB5 holdase [153,154].

Concerning the recently described mutations in HspB5, HspB1 and HspB8 that are responsible of myopathies, cardiomyopathies, neuropathies and cataracts, drugs that could rescue the chaperone holdase activity of the different mutant sHsps will be appropriated. It is reasonable to assume that specific peptide/RNA aptamers or small molecules acting as minichaperones that could refold and rescue the activity of point-mutated sHps, but have no effects towards the wild-type counterpart of the sHsps, may be promising approaches to delay the deleterious outcome of these pathologies.

For cancer diseases, the approach, not interfering with sHsps expression, will have to target and inactivate, in a definite cell type, the complex formed by the targeted sHsp with either itself or with other sHps. This procedure should not interfere with the activity of the targeted sHsp when it is expressed in other tissues. This is a very difficult task which will require i) a better knowledge of the different structural sHsp holdases activity that exist in human cells and ii) the search of agents or conditions that act as inhibitors of the holdase activity of the different sHsps. RNA/peptide aptamers or drugs that bind specific structural organizations of sHsps could be an approach to reduce the tumorigenic and metastatic activities of these proteins. Similarly, drugs that disrupt the anti-oxidant power of these proteins may prove useful to block their ability to counteract the killing efficiency of redox state dependent anti-cancer therapeutic drugs or conditions, such as 17AAG or X-rays irradiation [147,155].

CONCLUSION

The literature is filled with reports describing the expression and involvement of sHsps bearing holdase activity in human pathologies, suggesting that these proteins could be therapeutic targets that could interfere with pathologies as diverse as neurodegeneration, myopathies, cardiomyopathies, asthma, cataracts and cancers. Consequently, drugs that specifically up- or down-modulate their protective activities are urgently needed. This task is unfortunately hampered by several problems that need to be solved. First, a complex problem concerns sHsps property to oligomerize and form homo- and/or hetero-oligomeric structures that display, depending on cell homeostasis, hetero-dispersed, and rather unpredictable, native sizes and phosphorylation patterns [156]. Second, the dynamic exchanges between some, but not all, sHsps partners that, in a single cell, can occur. Hence, future works will have to unravel the precise role of the multiple combinatorial oligomeric structures formed by sHsps. Third, the tri-dimensional structure of human sHsps has long remained mysterious. In spite of these recurrent problems, strategy proposals aimed at either stimulating the beneficial properties of these sHsps or altering their deleterious roles are published in a regular basis. Fortunately, the dimeric structure of HspB5 has been recently deciphered [157] hence offering the opportunity to define new roads to define strategies aimed at designing active molecules that could interfere and modulate the activities of human sHsps. Hence, it is probable that in the near future new drugs will be testable in clinical trials for their effectiveness against different pathologies with the hope that they will not induce pathological side effects consequently of sHsps still uncontrolled dual roles.

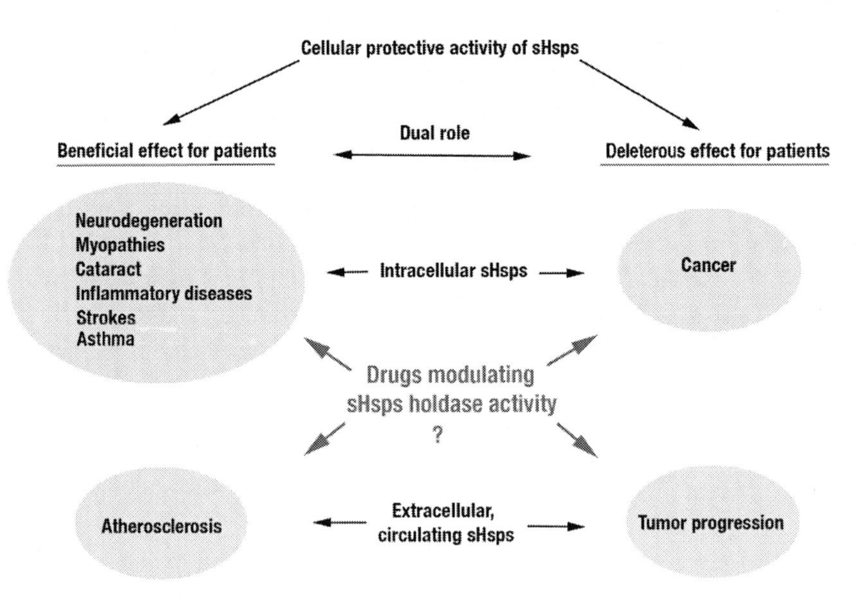

Figure 2. Dual, beneficial and deleterous, roles of sHsps in diseases. This figure summarizes the dual role that intracellular and extracellular sHsps can play in different pathologies. The impact of putative therapeutic drugs is enlightened.

ACKNOWLEDGMENTS

We wish to thank the Association Française pour les Myopathies (AFM) for supporting S.S post-doctoral fellowship. Research in the authors laboratory was supported by the Association pour la Recherche sur le Cancer, the Ligue contre le Cancer, AFM and the Région Rhône-Alpes.

REFERENCES

[1] Horwitz, J., Huang, Q.-L. and Ding, L.-L. (1992). Alpha-crystallin can function as a molecular chaperone. *Proc. Natl. Acad. Sci. USA.* 89, 10449-10453.

[2] Jakob, U., Gaestel, M., Engels, K. and Buchner, J. (1993). Small heat shock proteins are molecular chaperones. *J. Biol. Chem.* 268, 1517-1520.

[3] Ganea, E. (2001). Chaperone-like activity of alpha-crystallin and other small heat shock proteins. *Curr Protein Pept Sci.* 2, 205-225.

[4] Carra, S., Sivilotti, M., Chavez Zobel, A.T., Lambert, H. and Landry, J. (2005). HspB8, a small heat shock protein mutated in human neuromuscular disorders, has in vivo chaperone activity in cultured cells. *Hum Mol Genet.* 14, 1659-1669.

[5] Bellyei, S. et al. (2007). Preventing apoptotic cell death by a novel small heat shock protein. *Eur J Cell Biol.* 86, 161-171.

[6] Markossian, K.A., Yudin, I.K. and Kurganov, B.I. (2009). Mechanism of Suppression of Protein Aggregation by alpha-Crystallin. *Int J Mol Sci.* 10, 1314-45.

[7] Freeman, B.C. and Morimoto, R.I. (1996). The human cytosolic molecular chaperones hsp90, hsp70 (hsc70) and hdj-1 have distinct roles in recognition of a non-native protein and protein refolding. *Embo J.* 15, 2969-2979.

[8] Bukau, B. and Horwich, A.L. (1998). The Hsp70 and Hsp60 chaperone machines. *Cell,* 92, 351-366.

[9] Buchner, J. (1999). Hsp90 & Co. - a holding for folding. *Trends Biochem Sci* 24, 136-141.

[10] Lee, G.J., Roseman, A.M., Saibil, H.R. and Vierling, E. (1997). A small heat shock protein stably binds heat-denatured model substrates and can maintain a substrate in a folding-competent state. *EMBO J.* 16, 659-671.

[11] Ehrnsperger, M., Graber, S., Gaestel, M. and Buchner, J. (1997). Binding of non-native protein to Hsp25 during heat shock creates a reservoir of folding intermediates for reactivation. *EMBO J.* 16, 221-229.

[12] Lee, G.J. and Vierling, E. (2000). A small heat shock protein cooperates with heat shock protein 70 systems to reactivate a heat-denatured protein. *Plant Physiol.* 122, 189-198.

[13] Arrigo, A.-P., Suhan, J.P. and Welch, W.J. (1988). Dynamic changes in the structure and intracellular locale of the mammalian low-molecular-weight heat shock protein. *Mol. Cell. Biol.* 8, 5059-5071.

[14] Bryantsev, A.L., Loktionova, S.A., Ilyinskaya, O.P., Tararak, E.M., Kampinga, H.H. and Kabakov, A.E. (2002). Distribution, phosphorylation, and activities of Hsp25 in heat-stressed H9c2 myoblasts: a functional link to cytoprotection. *Cell Stress Chaperones* 7, 146-155.

[15] Bryantsev, A.L. et al. (2007). Regulation of stress-induced intracellular sorting and chaperone function of Hsp27 (HspB1) in mammalian cells. *Biochem J.* 407, 407-417.

[16] Adhikari, A.S., Sridhar Rao, K., Rangaraj, N., Parnaik, V.K. and Mohan Rao, C. (2004). Heat stress-induced localization of small heat shock proteins in mouse myoblasts: intranuclear lamin A/C speckles as target for alphaB-crystallin and Hsp25. *Exp Cell Res* 299, 393-403.

[17] Brunet Simioni, M. et al. (2009). Heat shock protein 27 is involved in SUMO-2/3 modification of heat shock factor 1 and thereby modulates the transcription factor activity. *Oncogene* 28, 3332-3344.

[18] van den, I.P., Wheelock, R., Prescott, A., Russell, P. and Quinlan, R.A. (2003). Nuclear speckle localisation of the small heat shock protein alpha B-crystallin and its inhibition by the R120G cardiomyopathy-linked mutation. *Exp Cell Res.* 287, 249-261.

[19] Vos, M.J., Kanon, B. and Kampinga, H.H. (2009). HSPB7 is a SC35 speckle resident small heat shock protein. *Biochim Biophys Acta,* 1793, 1343-1353.

[20] Cuesta, R., Laroia, G. and Schneider, R.J. (2000). Chaperone Hsp27 inhibits translation during heat shock by binding eIF4G and facilitating dissociation of cap-initiation complexes. *Genes Dev* 14, 1460-1470.

[21] Nicholl, I.D. and Quinlan, R.A. (1994). Chaperone Activity of alpha-Crystallins Modulates Intermediate Filament Assembly. *EMBO J.* 13, 945-953.

[22] Lavoie, J.N., Hickey, E., Weber, L.A. and Landry, J. (1993). Modulation of actin microfilament dynamics and fluid phase pinocytosis by phosphorylation of Heat Shock Protein 27. *J. Biol. Chem.* 268, 24210-24214.

[23] Perng, M.D., Cairns, L., van den, I.P., Prescott, A., Hutcheson, A.M. and Quinlan, R.A. (1999). Intermediate filament interactions can be altered by HSP27 and alphaB-crystallin. *J Cell Sci.* 112, 2099-2112.

[24] Parcellier, A. et al. (2006). HSP27 favors ubiquitination and proteasomal degradation of p27Kip1 and helps S-phase re-entry in stressed cells. *Faseb J* 20, 1179-1181.

[25] Neckers, L., Mimnaugh, E. and Schulte, T.W. (1999). Hsp90 as an anti-cancer target. *Drug Resist Updat.* 2, 165-172.

[26] Georgakis, G.V. and Younes, A. (2005). Heat-shock protein 90 inhibitors in cancer therapy: 17AAG and beyond. *Future Oncol.* 1, 273-281.

[27] Pandey, P. et al. (2000). Hsp27 functions as a negative regulator of cytochrome c-dependent activation of procaspase-3. *Oncogene* 19, 1975-1981.

[28] Koteiche, H.A. and McHaourab, H.S. (2003). Mechanism of chaperone function in small heat-shock proteins. Phosphorylation-induced activation of two-mode binding in alphaB-crystallin. *J Biol Chem.* 278, 10361-10367.

[29] Aquilina, J.A., Benesch, J.L., Ding, L.L., Yaron, O., Horwitz, J. and Robinson, C.V. (2004). Phosphorylation of alphaB-crystallin alters chaperone function through loss of dimeric substructure. *J Biol Chem.* 279, 28675-28680.

[30] Rogalla, T. et al. (1999). Regulation of Hsp27 oligomerization, chaperone function, and protective activity against oxidative stress/tumor necrosis factor alpha by phosphorylation. *J Biol Chem.* 274, 18947-18956.

[31] Liang, J.J. and Akhtar, N.J. (2000). Human lens high-molecular-weight alpha-crystallin aggregates. *Biochem Biophys Res Commun.* 275, 354-359.

[32] Horwitz, J., Huang, Q. and Ding, L. (2004). The native oligomeric organization of alpha-crystallin, is it necessary for its chaperone function? *Exp Eye Res* 79, 817-821.

[33] Ahmad, M.F., Raman, B., Ramakrishna, T. and Rao Ch, M. (2008). Effect of phosphorylation on alpha B-crystallin: differences in stability, subunit exchange and chaperone activity of homo and mixed oligomers of alpha B-crystallin and its phosphorylation-mimicking mutant. *J Mol Biol.* 375, 1040-1051.

[34] Garrido, C., Bruey, J.M., Fromentin, A., Hammann, A., Arrigo, A.P. and Solary, E. (1999). HSP27 inhibits cytochrome c-dependent activation of procaspase-9. *Faseb J* 13, 2061-2070.

[35] Bruey, J.M. et al. (2000). Hsp27 negatively regulates cell death by interacting with cytochrome c. *Nat Cell Biol.* 2, 645-652.

[36] Samali, A. et al. (2001). Hsp27 protects mitochondria of thermotolerant cells against apoptotic stimuli. *Cell Stress Chaperones* 6, 49-58.

[37] Paul, C., Manero, F., Gonin, S., Kretz-Remy, C., Virot, S. and Arrigo, A.P. (2002). Hsp27 as a negative regulator of cytochrome C release. *Mol Cell Biol.* 22, 816-834.

[38] Mao, Y.W., Liu, J.P., Xiang, H. and Li, D.W. (2004). Human alphaA- and alphaB-crystallins bind to Bax and Bcl-X(S) to sequester their translocation during staurosporine-induced apoptosis. *Cell Death Differ.* 11, 512-526.

[39] Li, D.W. et al. (2005). Calcium-activated RAF/MEK/ERK signaling pathway mediates p53-dependent apoptosis and is abrogated by alpha B-crystallin through inhibition of RAS activation. *Mol Biol Cell.* 16, 4437-4453.

[40] Havasi, A., Li, Z., Wang, Z., Martin, J.L., Botla, V., Ruchalski, K., Schwartz, J.H. and Borkan, S.C. (2008). Hsp27 inhibits Bax activation and apoptosis via a phosphatidylinositol 3-kinase-dependent mechanism. *J Biol Chem.* 283, 12305-12313.

[41] Charette, S.J., Lavoie, J.N., Lambert, H. and Landry, J. (2000). Inhibition of daxx-mediated apoptosis by heat shock protein 27. *Mol Cell Biol.* 20, 7602-7612.

[42] Rane, M.J. et al. (2003). Heat shock protein 27 controls apoptosis by regulating akt activation. J Biol Chem 278, 27828-27835.

[43] Kamradt, M.C. et al. (2005). The small heat shock protein alpha B-crystallin is a novel inhibitor of TRAIL-induced apoptosis that suppresses the activation of caspase-3. *J Biol Chem.* 280, 11059-11066.

[44] Wu, R., Kausar, H., Johnson, P., Montoya-Durango, D.E., Merchant, M. and Rane, M.J. (2007). Hsp27 regulates Akt activation and polymorphonuclear leukocyte apoptosis by scaffolding MK2 to Akt signal complex. *J Biol Chem.* 282, 21598-21608.

[45] Bruey, J.M., Paul, C., Fromentin, A., Hilpert, S., Arrigo, A.P., Solary, E. and Garrido, C. (2000). Differential regulation of HSP27 oligomerization in tumor cells grown in vitro and in vivo. *Oncogene* 19, 4855-4863.

[46] Spector, N.L. et al. (1994). Regulation of the 28kDa heat shock protein by retinoic acid during diferentiation of human leukemic HL-60 cells. *FEBS Lett.* 337, 184-188.

[47] Spector, N.L., Ryan, C., Samson, W., Levine, H., Nadler, L.M. and Arrigo, A.-P. (1993). Heat shock protein is a unique marker of growth arrest during macrophage differenciation of HL-60 cells. *J Cell Physiol* 156, 619-625.

[48] Chaufour, S., Mehlen, P. and Arrigo, A.P. (1996). Transient accumulation, phosphorylation and changes In the oligomerization of Hsp27 during retinoic acid-induced differentiation of HL-60 cells: Possible role in the control of cellular growth and differentiation. *Cell Stress Chaperones,* 1, 225-235.

[49] Mehlen, P., Mehlen, A., Godet, J. and Arrigo, A.-P. (1997). hsp27 as a Switch between Differentiation and Apoptosis in Murine Embryonic Stem Cells. *J. Biol Chem* 272, 31657-31665.

[50] Mehlen, P., Coronas, V., Ljubic-Thibal, V., Ducasse, C., Granger, L., Jourdan, F. and Arrigo, A.P. (1999). Small stress protein Hsp27 accumulation during dopamine-

mediated differentiation of rat olfactory neurons counteracts apoptosis. *Cell Death Differ.* 6, 227-233.

[51] Davidson, S.M. and Morange, M. (2000). Hsp25 and the p38 MAPK pathway are involved in differentiation of cardiomyocytes. *Dev Biol* 218, 146-160.

[52] Arrigo, A.P. and Ducasse, C. (2002). Expression of the anti-apoptotic protein Hsp27 during both the keratinocyte differentiation and dedifferentiation of HaCat cells: expression linked to changes in intracellular protein organization? *Exp Gerontol.* 37, 1247-1255.

[53] Duverger, O., Paslaru, L. and Morange, M. (2004). HSP25 is involved in two steps of the differentiation of PAM212 keratinocytes. *J Biol Chem.* 279, 10252-10260.

[54] Arrigo, A.P. (2005). In search of the molecular mechanism by which small stress proteins counteract apoptosis during cellular differentiation. *J Cell Biochem.* 94, 241-246.

[55] Brown, D.D., Christine, K.S., Showell, C. and Conlon, F.L. (2007). Small heat shock protein Hsp27 is required for proper heart tube formation. *Genesis,* 45, 667-678.

[56] Kamradt, M.C., Chen, F., Sam, S. and Cryns, V.L. (2002). The small heat shock protein alpha B-crystallin negatively regulates apoptosis during myogenic differentiation by inhibiting caspase-3 activation. *J Biol Chem* 277, 38731-38736.

[57] Ikeda, R. et al. (2006). The small heat shock protein alphaB-crystallin inhibits differentiation-induced caspase 3 activation and myogenic differentiation. *Biol Pharm Bull.* 29, 1815-1819.

[58] Liu, J.P. et al. (2004). Human alphaA- and alphaB-crystallins prevent UVA-induced apoptosis through regulation of PKCalpha, RAF/MEK/ERK and AKT signaling pathways. *Exp Eye Res.* 79, 393-403.

[59] Boelens, W.C., Croes, Y. and de Jong, W.W. (2001). Interaction between alphaB-crystallin and the human 20S proteasomal subunit C8/alpha7. *Biochim Biophys Acta,* 1544, 311-319.

[60] Parcellier, A. et al. (2003). HSP27 is a ubiquitin-binding protein involved in I-kappaBalpha proteasomal degradation. *Mol Cell Biol* 23, 5790-5802.

[61] den Engelsman, J., Keijsers, V., de Jong, W.W. and Boelens, W.C. (2003). The small heat-shock protein alpha B-crystallin promotes FBX4-dependent ubiquitination. *J Biol Chem.* 278, 4699-4704.

[62] Barbash, O., Lin, D.I. and Diehl, J.A. (2007). SCF Fbx4/alphaB-crystallin cyclin D1 ubiquitin ligase: a license to destroy. *Cell Div.* 2, 2.

[63] McDonough, H. and Patterson, C. (2003). CHIP: a link between the chaperone and proteasome systems. *Cell Stress Chaperones* 8, 303-308.

[64] Keller, J.N., Dimayuga, E., Chen, Q., Thorpe, J., Gee, J. and Ding, Q. (2004). Autophagy, proteasomes, lipofuscin, and oxidative stress in the aging brain. *Int J Biochem Cell Biol.* 36, 2376-2391.

[65] Kiffin, R., Bandyopadhyay, U. and Cuervo, A.M. (2006). Oxidative stress and autophagy. *Antioxid Redox Signal.* 8, 152-162.

[66] Nivon, M., Richet, E., Codogno, P., Arrigo, A.P. and Kretz-Remy, C. (2009). Autophagy activation by NFkappaB is essential for cell survival after heat shock. *Autophagy* 5, 766-783.

[67] Carra, S., Seguin, S.J. and Landry, J. (2008). HspB8 and Bag3: a new chaperone complex targeting misfolded proteins to macroautophagy. *Autophagy* 4, 237-239.

[68] Carra, S. (2009). The stress-inducible HspB8-Bag3 complex induces the eIF2alpha kinase pathway: implications for protein quality control and viral factory degradation? *Autophagy.* 5, 428-429.

[69] Landry, J., Chretien, P., Lambert, H., Hickey, E. and Weber, L.A. (1989). Heat shock resistance confered by expression of the human HSP 27 gene in rodent cells. *J. Cell Biol.* 109, 7-15.

[70] Aoyama, A., Frohli, E., Schafer, R. and Klemenz, R. (1993). Alpha B-crystallin expression in mouse NIH 3T3 fibroblasts: glucocorticoid responsiveness and involvement in thermal protection. *Mol Cell Biol.* 13, 1824-1835.

[71] Mehlen, P., Préville, X., Chareyron, P., Briolay, J., Klemenz, R. and Arrigo, A.-P. (1995). Constitutive expression of human hsp27, Drosophila hsp27, or human alpha B-crystallin confers resistance to TNF- and oxidative stress-induced cytotoxicity in stably transfected murine L929 fibroblasts. *J Immunol* 154, 363-374.

[72] Preville, X., Salvemini, F., Giraud, S., Chaufour, S., Paul, C., Stepien, G., Ursini, M.V. and Arrigo, A.P. (1999). Mammalian small stress proteins protect against oxidative stress through their ability to increase glucose-6-phosphate dehydrogenase activity and by maintaining optimal cellular detoxifying machinery. *Exp Cell Res.* 247, 61-78.

[73] Arrigo, A.P. (2001). Hsp27: novel regulator of intracellular redox state. *IUBMB Life* 52, 303-307.

[74] Wyttenbach, A., Sauvageot, O., Carmichael, J., Diaz-Latoud, C., Arrigo, A.P. and Rubinsztein, D.C. (2002). Heat shock protein 27 prevents cellular polyglutamine toxicity and suppresses the increase of reactive oxygen species caused by huntingtin. *Hum Mol Genet* 11, 1137-1151.

[75] Firdaus, W.J., Wyttenbach, A., Diaz-Latoud, C., Currie, R.W. and Arrigo, A.P. (2006). Analysis of oxidative events induced by expanded polyglutamine huntingtin exon 1 that are differentially restored by expression of heat shock proteins or treatment with an antioxidant. *Febs J.* 273, 3076-3093.

[76] Outeiro, T.F., Klucken, J., Strathearn, K.E., Liu, F., Nguyen, P., Rochet, J.C., Hyman, B.T. and McLean, P.J. (2006). Small heat shock proteins protect against alpha-synuclein-induced toxicity and aggregation. *Biochem Biophys Res Commun.* 351, 631-638.

[77] Lee, S., Carson, K., Rice-Ficht, A. and Good, T. (2006). Small heat shock proteins differentially affect Abeta aggregation and toxicity. *Biochem Biophys Res Commun.* 347, 527-533.

[78] Perrin, V., Regulier, E., Abbas-Terki, T., Hassig, R., Brouillet, E., Aebischer, P., Luthi-Carter, R. and Deglon, N. (2007). Neuroprotection by Hsp104 and Hsp27 in lentiviral-based rat models of Huntington's disease. *Mol Ther* 15, 903-911.

[79] Wilhelmus, M.M., Boelens, W.C., Otte-Holler, I., Kamps, B., Kusters, B., Maat-Schieman, M.L., de Waal, R.M. and Verbeek, M.M. (2006). Small heat shock protein HspB8: its distribution in Alzheimer's disease brains and its inhibition of amyloid-beta protein aggregation and cerebrovascular amyloid-beta toxicity. *Acta Neuropathol.* 111, 139-149.

[80] Andley, U.P. (2007). Crystallins in the eye: Function and pathology. *Prog Retin Eye Res.* 26, 78-98.

[81] Renkawek, K., Bosman, G.J. and de Jong, W.W. (1994). Expression of small heat-shock protein hsp 27 in reactive gliosis in Alzheimer disease and other types of dementia. *Acta Neuropathol.* (Berl) 87, 511-519.
[82] Muchowski, P.J. (2002). Protein misfolding, amyloid formation, and neurodegeneration: a critical role for molecular chaperones? *Neuron*, 35, 9-12.
[83] Wyttenbach, A. (2004). Role of heat shock proteins during polyglutamine neurodegeneration: mechanisms and hypothesis. *J Mol Neurosci.* 23, 69-96.
[84] Muchowski, P.J. and Wacker, J.L. (2005). Modulation of neurodegeneration by molecular chaperones. *Nat Rev Neurosci.* 6, 11-22.
[85] Vicart, P. et al. (1998). A missense mutation in the alphaB-crystallin chaperone gene causes a desmin-related myopathy. *Nat Genet* 20, 92-95.
[86] Elicker, K.S. and Hutson, L.D. (2007). Genome-wide analysis and expression profiling of the small heat shock proteins in zebrafish. *Gene,* 403, 60-69.
[87] Halliwell, B. (2001). Role of free radicals in the neurodegenerative diseases: therapeutic implications for antioxidant treatment. *Drugs Aging*, 18, 685-716.
[88] Bharath, S., Hsu, M., Kaur, D., Rajagopalan, S. and Andersen, J.K. (2002). Glutathione, iron and Parkinson's disease. *Biochem Pharmacol.* 64, 1037-1048.
[89] Jenner, P. and Olanow, C.W. (1996). Oxidative stress and the pathogenesis of Parkinson's disease. *Neurology,* 47, S161-S170.
[90] Browne, S.E., Ferrante, R.J. and Beal, M.F. (1999). Oxidative stress in Huntington's disease. *Brain Pathol* 9, 147-163.
[91] Tabner, B.J., Turnbull, S., El-Agnaf, O. and Allsop, D. (2001). Production of reactive oxygen species from aggregating proteins implicated in Alzheimer's disease, Parkinson's disease and other neurodegenerative diseases. *Curr Top Med Chem.* 1, 507-517.
[92] Turnbull, S., Tabner, B.J., Brown, D.R. and Allsop, D. (2003). Copper-dependent generation of hydrogen peroxide from the toxic prion protein fragment PrP106-126. *Neurosci Lett.* 336, 159-162.
[93] Choi, J., Rees, H.D., Weintraub, S.T., Levey, A.I., Chin, L.S. and Li, L. (2005). Oxidative modifications and aggregation of Cu,Zn-superoxide dismutase associated with Alzheimer and Parkinson diseases. *J Biol Chem.* 280, 11648-11655.
[94] Firdaus, W.J., Wyttenbach, A., Giuliano, P., Kretz-Remy, C., Currie, R.W. and Arrigo, A.P. (2006). Huntingtin inclusion bodies are iron-dependent centers of oxidative events. *Febs J.* 273, 5428-5441.
[95] Fox, J.H. et al. (2007). Mechanisms of copper ion mediated Huntington's disease progression. *PLoS ONE*, 2, e334.
[96] Janue, A., Olive, M. and Ferrer, I. (2007). Oxidative stress in desminopathies and myotilinopathies: a link between oxidative damage and abnormal protein aggregation. *Brain Pathol.* 17, 377-388.
[97] Liu, J., Chen, Q., Huang, W., Horak, K.M., Zheng, H., Mestril, R. and Wang, X. (2006). Impairment of the ubiquitin-proteasome system in desminopathy mouse hearts. *Faseb J.* 20, 362-364.
[98] Hilditch-Maguire, P., Trettel, F., Passani, L.A., Auerbach, A., Persichetti, F. and MacDonald, M.E. (2000). Huntingtin: an iron-regulated protein essential for normal nuclear and perinuclear organelles. *Hum Mol Genet.* 9, 2789-2797.

[99] Huang, X., Moir, R.D., Tanzi, R.E., Bush, A.I. and Rogers, J.T. (2004). Redox-active metals, oxidative stress, and Alzheimer's disease pathology. *Ann N Y Acad Sci.* 1012, 153-163.

[100] Halliwell, B. and Gutteridge, J. (1984). Role of iron in oxygen radical reactions. *Methods Enzymol.* 105, 47-56.

[101] Sayre, L.M., Perry, G., Atwood, C.S. and Smith, M.A. (2000). The role of metals in neurodegenerative diseases. *Cell Mol Biol* (Noisy-le-grand) 46, 731-741.

[102] Shoham, S. and Youdim, M.B. (2000). Iron involvement in neural damage and microgliosis in models of neurodegenerative diseases. *Cell Mol Biol* (Noisy-le-grand) 46, 743-760.

[103] Mehlen, P., Préville, X., Kretz-Remy, C. and Arrigo, A.-P. (1996). Human hsp27, Drosophila hsp27 and human αB-crystallin expression-mediated increase in glutathione is essential for the protective activity of these protein against TNFα-induced cell death. *EMBO J* 15, 2695-2706.

[104] Arrigo, A.P. (1998). Small stress proteins: chaperones that act as regulators of intracellular redox state and programmed cell death. Biol Chem 379, 19-26.

[105] Arrigo, A.P., Virot, S., Chaufour, S., Firdaus, W., Kretz-Remy, C. and Diaz-Latoud, C. (2005). Hsp27 consolidates intracellular redox homeostasis by upholding glutathione in its reduced form and by decreasing iron intracellular levels. *Antioxid Redox Signal,* 7, 414-422.

[106] Chen, H., Zheng, C., Zhang, Y., Chang, Y.Z., Qian, Z.M. and Shen, X. (2006). Heat shock protein 27 downregulates the transferrin receptor 1-mediated iron uptake. *Int J Biochem Cell Biol.* 38, 1402-1416.

[107] Kammanadiminti, S.J. and Chadee, K. (2006). Suppression of NF-kappaB activation by Entamoeba histolytica in intestinal epithelial cells is mediated by heat shock protein 27. *J Biol Chem.* 281, 26112-26120.

[108] Alford, K.A., Glennie, S., Turrell, B.R., Rawlinson, L., Saklatvala, J. and Dean, J.L. (2007). HSP27 functions in inflammatory gene expression and TAK1-mediated signalling. *J Biol Chem.* 282, 6232-6241.

[109] Dodd, S.L., Hain, B., Senf, S.M. and Judge, A.R. (2009). Hsp27 inhibits IKK{beta}-induced NF-{kappa}B activity and skeletal muscle atrophy. *Faseb J.* 23, 3415-3425.

[110] Merendino, A.M. et al. (2002). Heat shock protein-27 protects human bronchial epithelial cells against oxidative stress-mediated apoptosis: possible implication in asthma. *Cell Stress Chaperones*, 7, 269-280.

[111] Dillmann, W.H. (1999). Heat shock proteins and protection against ischemic injury. *Infect Dis Obstet Gynecol.* 7, 55-57.

[112] Efthymiou, C.A., Mocanu, M.M., de Belleroche, J., Wells, D.J., Latchmann, D.S. and Yellon, D.M. (2004). Heat shock protein 27 protects the heart against myocardial infarction. *Basic Res Cardiol.* 99, 392-394.

[113] Eaton, P., Awad, W.I., Miller, J.I., Hearse, D.J. and Shattock, M.J. (2000). Ischemic preconditioning: a potential role for constitutive low molecular weight stress protein translocation and phosphorylation? *J Mol Cell Cardiol.* 32, 961-971.

[114] Lewis, S.E. et al. (1999). A role for HSP27 in sensory neuron survival. J Neurosci 19, 8945-8953.

[115] Agius, M.A., Kirvan, C.A., Schafer, A.L., Gudipati, E. and Zhu, S. (1999). High prevalence of anti-alpha-crystallin antibodies in multiple sclerosis: correlation with severity and activity of disease. *Acta Neurol Scand.* 100, 139-147.

[116] Tezel, G. and Wax, M.B. (2000). The mechanisms of hsp27 antibody-mediated apoptosis in retinal neuronal cells. *J Neurosci.* 20, 3552-3562.

[117] Ciocca, D.R. and Calderwood, S.K. (2005). Heat shock proteins in cancer: diagnostic, prognostic, predictive, and treatment implications. *Cell Stress Chaperones,* 10, 86-103.

[118] Calderwood, S.K., Khaleque, M.A., Sawyer, D.B. and Ciocca, D.R. (2006). Heat shock proteins in cancer: chaperones of tumorigenesis. *Trends Biochem Sci.* 31, 164-172.

[119] Mehlen, P., Schulze-Osthoff, K. and Arrigo, A.P. (1996). Small stress proteins as novel regulators of apoptosis. Heat shock protein 27 blocks Fas/APO-1- and staurosporine-induced cell death. *The Journal of Biological Chemistry*, 271, 16510-16514.

[120] Richards, E.H., Hickey, E., Weber, L.A. and Master, J.R. (1996). Effect of overexpression of the small heat shock protein HSP27 on the heat and drug sensitivities of human testis tumor cells. *Cancer Res.* 56, 2446-2451.

[121] Arrigo, A.P. (2000). sHsp as novel regulators of programmed cell death and tumorigenicity. *Pathol Biol.* (Paris) 48, 280-288.

[122] Garrido, C., Gurbuxani, S., Ravagnan, L. and Kroemer, G. (2001). Heat shock proteins: endogenous modulators of apoptotic cell death. *Biochem Biophys Res Commun.* 286, 433-442.

[123] Kase, S., Parikh, J.G. and Rao, N.A. (2009). Expression of heat shock protein 27 and alpha-crystallins in human retinoblastoma after chemoreduction. *Br J Ophthalmol* 93, 541-544.

[124] Hitotsumatsu, T., Iwaki, T., Fukui, M. and Tateishi, J. (1996). Distinctive immunohistochemical profiles of small heat shock proteins (heat shock protein 27 and alpha B-crystallin) in human brain tumors. *Cancer*, 77, 352-361.

[125] Mineva, I. et al. (2005). Differential expression of alphaB-crystallin and Hsp27-1 in anaplastic thyroid carcinomas because of tumor-specific alphaB-crystallin gene (CRYAB) silencing. *Cell Stress Chaperones,* 10, 171-184.

[126] Mehlen, P. and Arrigo, A.-P. (1994). The serum-induced phosphorylation of mammalian hsp27 correlates with changes in its intracellular localization and levels of oligomerization. *Eur J Biochem* 221, 327-334.

[127] Tsvetkova, N.M. et al. (2002). Small heat-shock proteins regulate membrane lipid polymorphism. *Proc Natl Acad Sci U S A.* 99, 13504-13509.

[128] Chowdary, T.K., Bakthisaran, R., Tangirala, R. and Rao, M.C. (2006). Interaction of mammalian Hsp22 with lipid membranes. *Biochem J.* 401, 437-445.

[129] Basu, S. and Srivastava, P.K. (2000). Heat shock proteins: the fountainhead of innate and adaptive immune responses. *Cell Stress Chaperones*, 5, 443-451.

[130] Wells, A.D. and Malkovsky, M. (2000). Heat shock proteins, tumor immunogenicity and antigen presentation: an integrated view. *Immunol Today*. 21, 129-132.

[131] Pinz, I., Robbins, J., Rajasekaran, N.S., Benjamin, I.J. and Ingwall, J.S. (2008). Unmasking different mechanical and energetic roles for the small heat shock proteins CryAB and HSPB2 using genetically modified mouse hearts. *Faseb J.* 22, 84-92.

[132] Zantema, A., Vries, M.V.-D., Maasdam, D., Bol, S. and Eb, A.v.d. (1992). Heat shock protein 27 and αB-cristallin can form a complex, which dissociates by heat shock. *J. Biol. Chem.* 267, 12936-12941.

[133] Sun, T.X. and Liang, J.J. (1998). Intermolecular exchange and stabilization of recombinant human alphaA- and alphaB-crystallin. *J Biol Chem* 273, 286-290.

[134] Saha, S. and Das, K.P. (2004). Relationship between chaperone activity and oligomeric size of recombinant human alphaA- and alphaB-crystallin: a tryptic digestion study. *Proteins,* 57, 610-7.

[135] Bukach, O.V., Glukhova, A.E., Seit-Nebi, A.S. and Gusev, N.B. (2009). Heterooligomeric complexes formed by human small heat shock proteins HspB1 (Hsp27) and HspB6 (Hsp20). *Biochim Biophys Acta,* 1794, 486-495.

[136] Sun, X., Welsh, M.J. and Benndorf, R. (2006). Conformational changes resulting from pseudophosphorylation of mammalian small heat shock proteins--a two-hybrid study. *Cell Stress Chaperones,* 11, 61-70.

[137] Simon, S., Fontaine, J.M., Martin, J.L., Sun, X., Hoppe, A.D., Welsh, M.J., Benndorf, R. and Vicart, P. (2007). Myopathy-associated alpha B-crystallin mutants: Abnormal phosphorylation, intracellular location, and interactions with other small heat shock proteins. *J Biol Chem.* 82, 34276-34287.

[138] Michiel, M., Skouri-Panet, F., Duprat, E., Simon, S., Ferard, C., Tardieu, A. and Finet, S. (2009). Abnormal assemblies and subunit exchange of alphaB-crystallin R120 mutants could be associated with destabilization of the dimeric substructure. *Biochemistry* 48, 442-453.

[139] Fu, L. and Liang, J.J. (2003). Alteration of protein-protein interactions of congenital cataract crystallin mutants. *Invest Ophthalmol Vis Sci.* 44, 1155-1159.

[140] Fontaine, J.M., Sun, X., Hoppe, A.D., Simon, S., Vicart, P., Welsh, M.J. and Benndorf, R. (2006). Abnormal small heat shock protein interactions involving neuropathy-associated HSP22 (HSPB8) mutants. *Faseb J* 25, 25.

[141] Laudanski, K., De, A. and Miller-Graziano, C. (2007). Exogenous heat shock protein 27 uniquely blocks differentiation of monocytes to dendritic cells. *Eur J Immunol* 37, 2812-2824.

[142] Rayner, K., Chen, Y.X., McNulty, M., Simard, T., Zhao, X., Wells, D.J., de Belleroche, J. and O'Brien, E.R. (2008). Extracellular release of the atheroprotective heat shock protein 27 is mediated by estrogen and competitively inhibits acLDL binding to scavenger receptor-A. *Circ Res.* 103, 133-141.

[143] Sanbe, A. et al. (2009). Protective effect of geranylgeranylacetone via enhancement of HSPB8 induction in desmin-related cardiomyopathy. *PLoS One* 4, e5351.

[144] Nedellec, P., Edling, Y., Perret, E., Fardeau, M. and Vicart, P. (2002). Glucocorticoid treatment induces expression of small heat shock proteins in human satellite cell populations: consequences for a desmin-related myopathy involving the R120G alpha B-crystallin mutation. *Neuromuscul Disord.* 12, 457-465.

[145] Maloyan, A., Gulick, J., Glabe, C.G., Kayed, R. and Robbins, J. (2007). Exercise reverses preamyloid oligomer and prolongs survival in alphaB-crystallin-based desmin-related cardiomyopathy. *Proc Natl Acad Sci U S A.* 104, 5995-6000.

[146] Gidalevitz, T., Ben-Zvi, A., Ho, K.H., Brignull, H.R. and Morimoto, R.I. (2006). Progressive disruption of cellular protein folding in models of polyglutamine diseases. Science 311, 1471-1474.

[147] Aloy, M.T., Hadchity, E., Bionda, C., Diaz-Latoud, C., Claude, L., Rousson, R., Arrigo, A.P. and Rodriguez-Lafrasse, C. (2007). Protective Role of Hsp27 Protein Against

Gamma Radiation-Induced Apoptosis and Radiosensitization Effects of Hsp27 Gene Silencing in Different Human Tumor Cells. *Int J Radiat Oncol Biol Phys.* 70, 543-553.

[148] Rocchi, P., Jugpal, P., So, A., Sinneman, S., Ettinger, S., Fazli, L., Nelson, C. and Gleave, M. (2006). Small interference RNA targeting heat-shock protein 27 inhibits the growth of prostatic cell lines and induces apoptosis via caspase-3 activation in vitro. *BJU Int.* 28, 28.

[149] Bausero, M.A. et al. (2006). Silencing the hsp25 gene eliminates migration capability of the highly metastatic murine 4T1 breast adenocarcinoma cell. *Tumour Biol.* 27, 17-26.

[150] Kamada, M., So, A., Muramaki, M., Rocchi, P., Beraldi, E. and Gleave, M. (2007). Hsp27 knockdown using nucleotide-based therapies inhibit tumor growth and enhance chemotherapy in human bladder cancer cells. *Mol Cancer Ther.* 6, 299-308.

[151] Kumar, P.A. and Reddy, G.B. (2009). Modulation of alpha-crystallin chaperone activity: a target to prevent or delay cataract? *IUBMB Life*, 61, 485-95.

[152] Stacchiotti, A., Bonomini, F., Lavazza, A., Rodella, L.F. and Rezzani, R. (2009). Adverse effects of Cyclosporine A on HSP25, alpha B-crystallin and myofibrillar cytoskeleton in rat heart. *Toxicology.* 262, 192-198.

[153] Babizhayev, M.A. (2009). Current Ocular Drug Delivery Challenges for N-acetylcarnosine : Novel Patented Routes and Modes of Delivery, Design for Enhancement of Therapeutic Activity and Drug Delivery Relationships. Recent Pat Drug Deliv Formul 3, 229-265.

[154] Babizhayev, M.A., Burke, L., Micans, P. and Richer, S.P. (2009). N-Acetylcarnosine sustained drug delivery eye drops to control the signs of ageless vision: Glare sensitivity, cataract amelioration and quality of vision currently available treatment for the challenging 50,000-patient population. *Clin Interv Aging*, 4, 31-50.

[155] McCollum, A.K., Teneyck, C.J., Sauer, B.M., Toft, D.O. and Erlichman, C. (2006). Up-regulation of Heat Shock Protein 27 Induces Resistance to 17-Allylamino-Demethoxygeldanamycin through a Glutathione-Mediated Mechanism. *Cancer Res.* 66, 10967-10975.

[156] Paul, C., Simon, S., Gibert, B., Virot, S., Manero, F. and Arrigo, A.-P. (2010). Dynamic processes that reflect anti-apoptotic strategies set up by Hsp27. Exp. Cell Res. Epub. Ahead of print. March 15.

[157] Bagneris, C., Bateman, O.A., Naylor, C.E., Cronin, N., Boelens, W.C., Keep, N.H. and Slingsby, C. (2009). Crystal Structures of alpha-Crystallin Domain Dimers of alphaB-Crystallin and Hsp20. *J Mol Biol.* 392, 1242-1252; *Erratum: J. Mol. Biol.* 2009, 394, 588.

INDEX

A

B

D

E

F

G

H

I

K

N

O

P

T

U

V

W

X

Y

Z